THE INVERTEBRATES

THE INVERTEBRATES

R. McNEILL ALEXANDER

Department of Pure and Applied Zoology
University of Leeds

CAMBRIDGE UNIVERSITY PRESS

CAMBRIDGE

LONDON · NEW YORK · MELBOURNE

Published by the Syndics of the Cambridge University Press
The Pitt Building, Trumpington Street, Cambridge CB2 1RP
Bentley House, 200 Euston Road, London NW1 2DB
32 East 57th Street, New York, NY 10022, USA
296 Beaconsfield Parade, Middle Park, Melbourne 3206, Australia

First published 1979

Printed in Great Britain
at the University Press, Cambridge

Library of Congress Cataloging in Publication Data
Alexander, R. McNeill.
The invertebrates.
Includes bibliographies and index.
1. Invertebrates. I. Title.
QL362.A43 592 78-6275
ISBN 0 521 22120 X hard covers
ISBN 0 521 29361 8 paperback

Contents

Preface

This book is designed for undergraduates but I hope it will also interest other people. It is about the major groups of invertebrate animals, about their structure, physiology and ways of life. It is intended to complement my book on *The chordates* but the two books overlap slightly in subject matter, as their titles imply. Sea squirts and amphioxus are both invertebrate and chordate and appear in both books so that they can be discussed in relation both to the vertebrates and to other invertebrates.

Each chapter except the first and last deals with a different taxonomic group, usually a phylum or class. It consists of brief descriptions of a few examples of the group, followed by more detailed discussion of selected topics. Some of these topics are peculiarities of the groups (for instance, the shells of molluscs and the flight of insects). Others are more widespread features or properties of animals which can be illustrated particularly well by reference to the group (for instance, I have found it convenient to discuss reflexes in the chapter on crustaceans). Most chapters include descriptions of many experiments because I think it as important and interesting to know how zoological information is obtained, as to know the information itself. The first chapter explains some techniques which are referred to repeatedly in later chapters, and also explains how animals are classified. The final chapter is a brief discussion of the evolution of the invertebrates.

The diversity of the invertebrates is a major problem in writing about them. Any attempt at encyclopaedic coverage results in an enormous quantity of indigestible morphological and taxonomic information. This sort of information is available in existing textbooks and indeed dominates some of them so that they fail to reflect current trends in zoology. I have tried to overcome this problem by being selective. I have described only a few examples of each major group and I have limited description to points I consider interesting and important. I have hesitated to include any phylum which has fewer than a thousand known species, but I have described some small groups because knowledge of them improves our understanding of larger groups. I would rather give students a good understanding of a few familiar animals than tell them about obscure groups they may seldom or never see.

Though I have described rather few species, I am keen that students should appreciate the extraordinary diversity of animals. They will need to study many more animals than I have described, and it will be far better if they see them for themselves, than if they merely read about them in books. The selection of species they are able to study, in the laboratory and in the field, will depend on local opportunities.

One of the main aims of zoology is to explain the structure and physiology of animals in terms of physical science. For instance, explanations of nerve conduction, swimming and the composition of skeletons depend on physical chemistry, hydrodynamics and materials science, respectively. I have used many branches of physics and physical chemistry, but I have assumed that many readers will have little prior knowledge of them and have tried to explain them in simple terms. I have frequently used simple calculations to demonstrate the plausibility of explanations. There is little value in suggesting, for instance, that flatworms are less than a millimetre thick because oxygen could not diffuse into them fast enough if they were thicker, unless it can be shown by calculation that a flatworm a centimetre thick could not survive.

Professor M. Sleigh read the first draft of this book and made a very large number of helpful suggestions. Other experts read one or two chapters each, and saved me from numerous errors. I am extremely grateful to them all. I hope that readers who find further errors will tell me about them, so that I can correct them if demand justifies a second edition.

January 1978 R. McNeill Alexander

1
Introduction

Subsequent chapters describe a great many observations on animals, some of them anatomical and some physiological. This chapter explains some of the techniques used by zoologists to make these observations. It also explains how zoologists classify animals.

Many people seem still to think of a zoologist as a man with a microscope, a scalpel and a butterfly net. These are still important tools but a modern zoological laboratory requires a far more varied (and far more expensive) range of equipment. Some of the most important tools which will be mentioned repeatedly in later chapters are described in this one.

MICROSCOPY

It is convenient to start with the conventional light microscope (Fig. 1.1*a*). Light from a lamp passes through a condenser lens which brings it to a focus on the specimen S which is to be examined. The light travels on through the objective lens which forms an enlarged image of the specimen at I_1. This is viewed through an eyepiece used as a magnifying glass so that a greatly enlarged virtual image is seen at I_2. Each of the lenses shown in this simple diagram is multiple in real microscopes, especially the objective which may consist of as many as 14 lenses. This complexity is necessary in good microscopes to reduce to an acceptable level the distortions and other image faults which are known as aberrations.

Microscope technology has long been at the stage at which the capacity of the best microscopes to reveal fine detail is limited by the properties of light rather than by any imperfections of design. It would be easy to build microscopes with greater magnification than is generally used but this would not make finer detail visible any more than magnification will reveal finer detail in a newspaper photograph. If light of wavelength λ and glass lenses of refractive index 1.5 are used, objects less than $\lambda/3$ apart cannot be seen separately, however great the magnification and however perfect the lenses. Since visible light has wavelengths around 0.5 μm, details less than about 0.2 μm apart cannot be distinguished. This is expressed by saying that light microscopes are incapable of resolutions better than about 0.2 μm.

Even this resolution is only possible with an oil-immersion lens of high

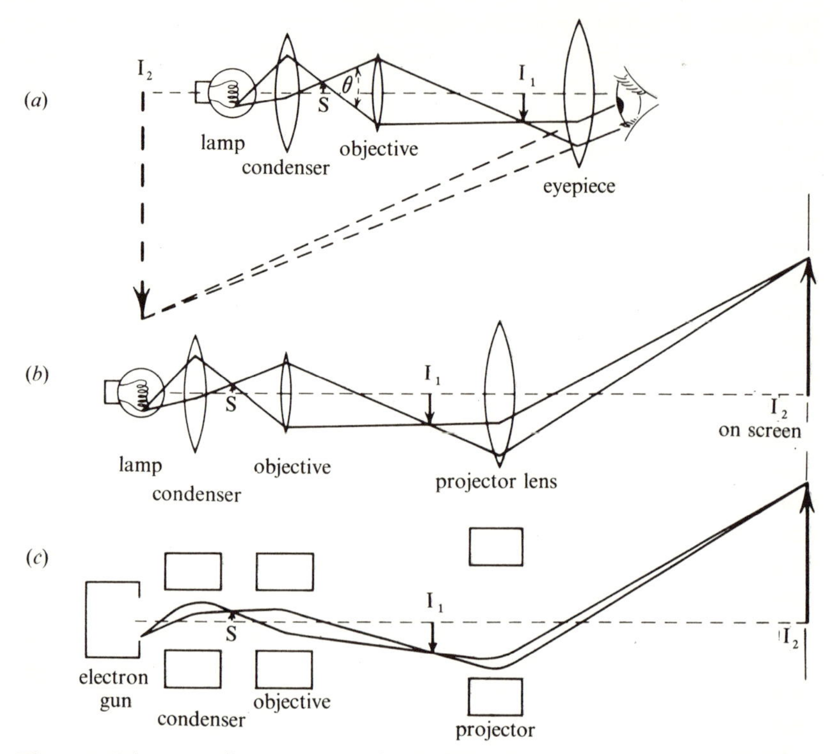

Fig. 1.1. Diagrams of (*a*) a conventional light microscope, (*b*) a projecting light microscope, and (*c*) a transmission electron microscope. The paths of a few rays (of light or of electrons) are indicated.

numerical aperture. An oil-immersion lens is one designed to have the space between it and the specimen filled by a drop of oil of high refractive index. High numerical aperture implies that light from a single point on the specimen may enter the objective at a wide range of angles, i.e. that the angle θ (Fig. 1.1*a*) is large.

Fig. 1.1(*b*) shows a projection microscope. The image I_1 is just outside the focal length of the projector lens whereas in a conventional microscope it is just inside the focal length of the eyepiece. Consequently the final image I_2 is real instead of virtual and can be projected onto a screen. The projection microscope has little use in zoology except in teaching, and is illustrated solely for comparison with the transmission electron microscope (Fig. 1.1*c*). This uses a beam of electrons instead of a beam of light. Magnetic fields set up by electric currents in coils of wire serve as lenses, refracting the rays of electrons in the same way as convex glass lenses refract rays of light. Electrons from the electron gun are focussed on the specimen by the condenser, and other magnetic lenses produce a greatly magnified final image in similar fashion to the lenses of a light projection microscope. The image is thrown onto a fluorescent screen which can be viewed directly, or onto photographic film.

The electron beam has a wavelength which depends on the potential difference used to accelerate it. Electron microscopes use large potential differences which make the wavelength exceedingly small, so that extremely fine resolution is possible in principle. The resolution actually achieved is much less good because even the best magnetic lenses are far less free from aberrations than the lenses used in light microscopes. Very small numerical apertures have to be used to reduce the aberrations to an acceptable level, making the resolution less good than is theoretically possible. It is still far better than for the light microscope. Resolutions around 1 nm (0.001 μm) are achieved in biological work.

The difference in resolution between the light and the electron microscope is illustrated by Fig. 14.12, which shows sections of the digestive gland of a cockle. Fig. 14.12(*a*) shows as much detail as can be seen by light microscopy, while (*b*), (*c*) and (*d*) have been drawn from electron micrographs which show much more detail than has been drawn. The flagella *f* have diameter 0.25 μm, so though they are visible by light microscopy, no detail can be seen within them. Fig. 14.12(*b*) includes the base of a flagellum and shows many strands within it and Fig. 2.10(*a*) shows some of the finer detail which can be resolved in flagella. Some large molecules such as the haemocyanin of snail blood (a protein: diameter about 30 nm) are large enough to be studied individually by electron microscopy.

To reveal fine internal detail, tissues must be cut into very thin slices. They cannot be cut thin enough without prior treatment. Frozen blocks of tissue can be cut thin enough for some purposes but most specimens for light microscopy are embedded in molten wax which is allowed to solidify and then sliced in a machine called a microtome. Sections only 2 μm thick can be cut. Thinner sections for electron microscopy are cut from specimens embedded in plastics such as Araldite. They can be cut 50 nm thick.

Preparation of specimens for sectioning is quite complex. Each specimen must first be treated with a fixative such as formaldehyde to give it the structural stability it needs to withstand further treatment. Formaldehyde seems to act by forming bridges between protein molecules (this is a process like vulcanization which converts liquid latex to solid rubber by inserting sulphur between the molecules). Next the water in the specimen must be removed and replaced by a liquid miscible with molten wax, so that the wax can permeate the specimen. This is usually done by transferring the specimen from water, through a series of water–ethyl alcohol mixtures, to pure alcohol and then to xylene (which is miscible both with alcohol and with wax, but not with water). This unfortunately tends to shrink the specimen. For instance, sea urchin eggs in xylene have been found to have only 48% of their initial volume. Specimens for electron microscopy have to be treated in rather similar fashion. They are

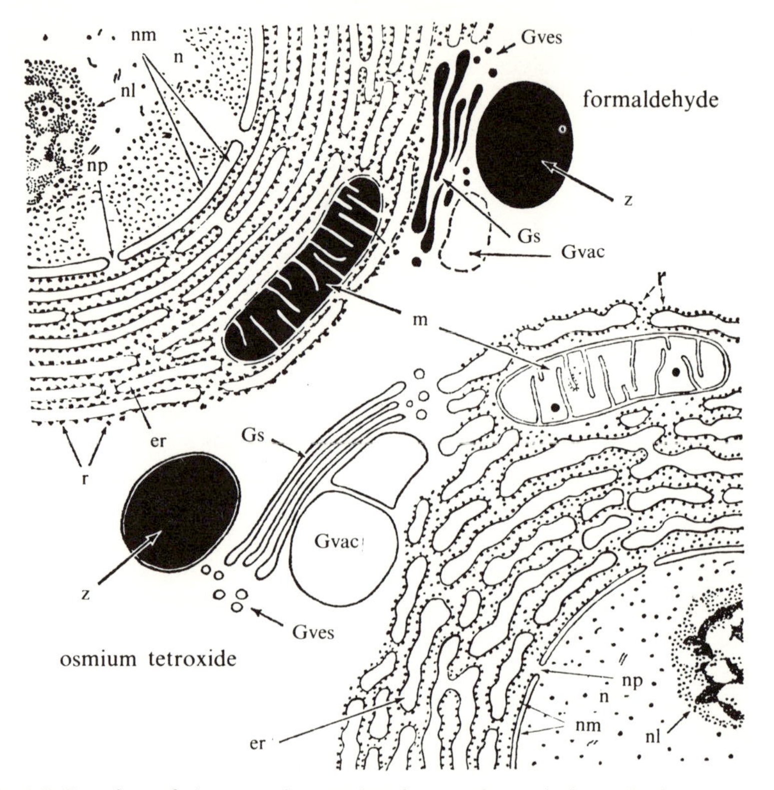

Fig. 1.2. Drawings of electron micrographs of parts of two similar cells (from a mouse pancreas). The one shown above and to the left was fixed by formaldehyde and the other (below and to the right) by osmium tetroxide. er, endoplasmic reticulum; Gs, Golgi saccule; Gvac, Golgi vacuole; Gves, Golgi vesicle; m, mitochondrion; n, nucleus; nl, nucleolus; nm, nuclear membranes; np, nuclear pore; r, ribosome; z, zymogen granule. From J. R. Baker (1966). *Cytological technique*, 5th edn. Methuen, London.

more often fixed with osmium tetroxide than with any other fixative, but they are passed through a series of concentrations of alcohol before being embedded in plastic.

Even all this processing produces sections in which little detail can be seen because most tissue constituents are transparent, especially in thin sections. Dyes are usually used in light microscopy, to colour different constituents different colours. The ways in which most of them work are not fully understood, but some of them are marvellously effective. For instance the Mallory technique colours muscle red, collagen blue and nerves lilac. Contrast in electron microscopy depends on different parts of the specimen scattering electrons to different extents. The greater the mass per unit area of section, the more electrons are scattered and the darker that part of the section appears in the image. Dyeing would be ineffective but contrast can be enhanced by treatment with compounds of heavy metals, such as phosphotungstic acid and

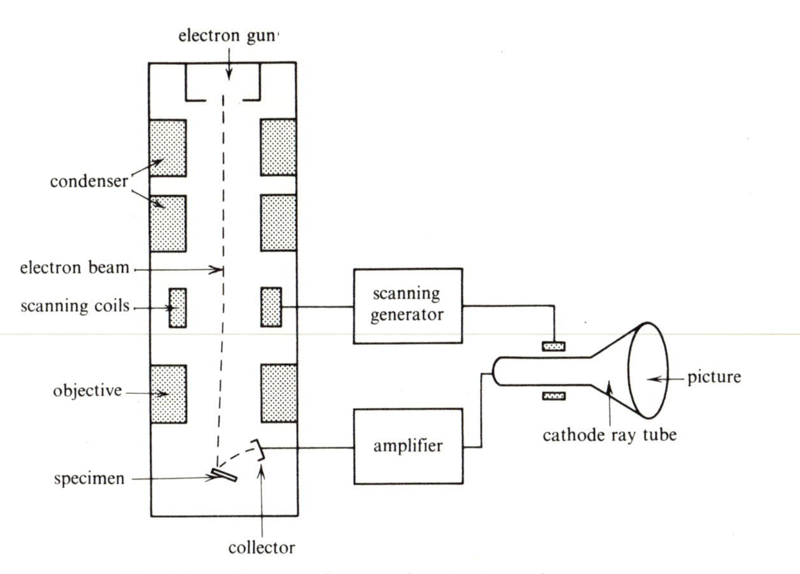

Fig. 1.3. A diagram of a scanning electron microscope.

uranyl acetate. These 'electron stains' attach preferentially to certain cell constituents.

How does all this treatment alter the structure of the specimens? Is the structure seen through the microscope more or less as in the living animal, or is it largely new structure produced by chemical treatment? Fig. 1.2 shows that the same material prepared by different but accepted techniques may look a little different. However, fairly similar appearances are obtained by grossly different techniques and it seems more likely that the structure which is seen was there initially than that it is formed independently by several different treatments.

The technique of serial sectioning is often used. A specimen is cut into sections all of which are examined, to discover the three-dimensional structure of the original specimen.

Protozoans and isolated cells can be examined alive and intact at high magnification, but little detail can be seen in them by ordinary light microscopy. Most of their parts are more or less equally colourless and transparent so details in them are as hard to see as glass beads in a tumbler of water. Fortunately there are differences of refractive index, and these are used to reveal more detail in the techniques of phase contrast and interference microscopy. The swallowing movements of the protozoan illustrated in Fig. 3.13 were revealed by phase contrast microscopy.

The polarizing microscope uses polarized light to show which parts of a specimen have the property of birefringence, due to alignment of molecules or larger structures parallel to each other. It also indicates the direction of alignment. It shows for instance that the molecules are more accurately aligned in the radial threads of spiders' webs than in the threads which form the sticky spiral (chapter 20).

Ordinary (transmission) electron microscopes are used for examining thin sections. Scanning electron microscopes are used for examining solid objects. They have been found particularly convenient for studying the hard parts of animals and have been used, for instance, to examine broken pieces of mollusc shell and find out how the crystals in them are arranged (Fig. 13.5). Fig. 1.3 shows how they work. The beam of electrons is focussed to a tiny spot on the specimen. One of its effects on the specimen is to release other electrons which are drawn to the positively-charged collector, where they are detected. The beam is deflected by the scanning coils which are used to move the spot systematically backwards and forwards over the specimen. The signal from the collector is used to control the brightness of a spot on a cathode ray tube (like the tube of a television set). This spot is moved backwards and forwards over the screen in precisely the same way as the spot is moving over the specimen, so it builds up a picture of the specimen in the same way as a television picture is built up by a bright, rapidly scanning spot. The pictures give an excellent three-dimensional effect: hollows look dark and projections throw shadows.

The scanning electron microscope cannot resolve detail finer than the diameter of the spot, which is typically 10 nm. Its resolution is therefore much less good than that of the transmission electron microscope, though very much better than that of the light microscope. It is sometimes used at very low magnifications to observe details which are easily visible by light microscopy, because it has much greater depth of field than the light microscope. When a thick specimen is examined by light microscopy it is necessary to focus up and down to see the detail at different levels. When a scanning electron microscope is used to view the same specimen, often everything can be seen simultaneously, all crisply in focus.

Though specimens do not need sectioning for scanning electron microscopy, they need some processing. Since a beam of electrons in air is scattered before it has travelled far, there has to be a vacuum in the microscope. In the vacuum, water would evaporate rapidly from a specimen which had not been previously dried. The specimen must be dried out without shrivelling it. This presents no problem when the specimen is a rigid one, like a piece of mollusc shell, but soft tissues need the special technique of critical point drying.

CHEMICAL ANALYSIS

Zoologists often want to know the chemical composition of a structure or fluid. If large enough samples are available ordinary analytical techniques can be used. Two instruments deserve special mention. One of them is the automatic amino acid analyser, which separates automatically the constituent amino acids of proteins and measures how much of each is present. The other is the X-ray diffraction spectrometer which measures the spacings of the repeating patterns in crystals. These two techniques have been used, for instance, to show that the horny skeletons of sea fans consist largely of a protein extremely like the collagen of which the tendons of vertebrates are made (chapter 6). Further analysis was needed to show why they are so much stiffer than tendons.

Often zoologists have only tiny samples available. This was the case, for instance, in an investigation of the contractile vacuole of an amoeba which is described in chapter 2. The contractile vacuole is a drop of fluid of diameter about 50 μm. The contents of contractile vacuoles were analysed individually.

The first problem in an investigation like this is to obtain the sample. This is often done by sucking it into a micropipette, a glass capillary drawn out to a very fine point. The tips of the micropipettes used in the contractile vacuole investigation had diameters less than 5 μm. They were stuck into the amoeba under a microscope. This would have needed a phenomenally steady hand if the micropipette had not been held in a micromanipulator, a device in which very fine movements in three dimensions are produced by quite coarse movements of knobs.

The zoologist often wants to know the osmotic concentration of his sample. It is usually most convenient to discover this by measuring the freezing point. If the osmotic concentration of the sample is X Osmol l^{-1} its freezing point is $-1.86X$ °C. A tiny sample is sucked into a fine glass capillary and frozen. It is watched through a microscope while it is warmed again in a specially designed bath, until the last of the ice disappears. The temperature at which this happens is measured by a Beckmann thermometer, graduated in hundredths of a degree.

The most plentiful cations in animals are sodium and potassium. Their concentrations in small samples are usually measured by flame photometry. If you heat a little of a sodium salt in a hot flame it emits yellow light of wavelength 589 nm. This is because electrons in the vaporized sodium atoms are temporarily excited to a higher energy level. When one falls back to its initial level it releases just enough energy to produce a photon of light of wavelength 589 nm. Potassium is similarly affected but the energy change is smaller so the wavelength of the light is longer, 766 nm. The intensity of the emitted light is used in flame photometry to measure the concentrations of the elements.

Fig. 1.4 shows a simple flame photometer. Air and fuel (often acetylene) are blown into the spray chamber. The air, entering through the nebulizer, draws the sample in through the long tube on the left and makes it into a spray which mixes with the fuel and so gets heated in the flame. Light from the flame is made to pass through a filter (to cut out wavelengths emitted by elements other than the one being measured) before falling on a photocell. The reading of the microammeter indicates the intensity of the light reaching the cell and the concentration of the sample can be calculated from it.

The most plentiful anion in animals is chloride. It is measured by titration with silver nitrate, which produces a precipitate of silver chloride. If the sample

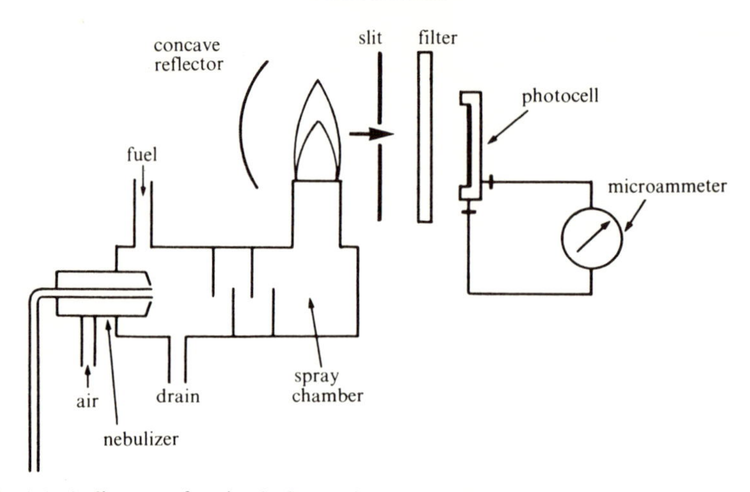

Fig. 1.4. A diagram of a simple flame photometer. From R. Ralph (1975). *Methods in experimental biology.* Blackie, Glasgow.

is large the titration can be done with a burette in the conventional way. If it is small the technique must be modified, for instance in the way shown in Fig. 1.5. The sample is a single drop which is stirred by a jet of air. Silver nitrate is added to it from a syringe operated by a micrometer. The endpoint is sensed electrically: the potential of the silver electrode changes rapidly at the endpoint as the concentration of silver ions in the drop rises.

Histochemistry is used to find out which cells in a microscope section contain a particular chemical compound. Substances are added which react with the compound to produce a coloured product which can be seen under the microscope. There is an account in chapter 8 of an investigation in which histochemical techniques were used to discover which cells produce which digestive enzymes in flatworms. Histochemical techniques have been devised for electron microscopy, as well as for light microscopy.

Another analytical technique which uses the electron microscope is electron probe microanalysis. It measures the concentrations of elements in selected parts of sections. An electron beam is focussed, in a modified electron microscope, onto a small spot on the specimen. The atoms bombarded with electrons emit X-rays of characteristic wavelengths, just as atoms in the flame photometer emit light of characteristic wavelengths (X-rays have shorter wavelengths than light so there is more energy in each quantum, and the electrons of an element have to be raised to even higher energy levels to make them emit X-rays when they fall back, than to make them emit light). X-rays of different wavelengths, characteristic of different elements, are separated by diffraction from crystals. The intensity of each characteristic wavelength is measured and used to calculate the concentration of the element in the part of the section on which the electron beam is focussed. Volumes of about 1 μm^3 can be analysed in this way. The technique is most easily applied if the

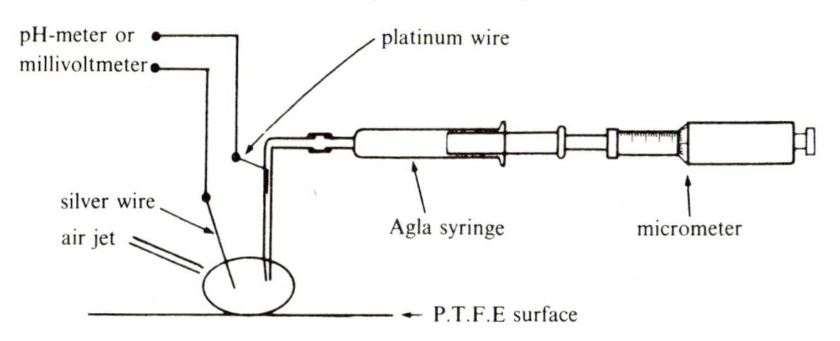

Fig. 1.5. Apparatus for titration of chloride in small samples. From R. Ralph (1975). *Methods in experimental biology*. Blackie, Glasgow.

elements under investigation are present as insoluble compounds. If they are soluble, special precautions have to be taken to make sure that they do not move from one part of a cell to another in the course of preparation. The tissue is frozen rapidly, sectioned while frozen, and examined while still frozen in a special chamber within the electron microscope. This has been done in an investigation of the concentrations of sodium and potassium in different parts of the cells of the excretory organs (Malpighian tubules) of insects (chapter 18).

There are two techniques of electron probe microanalysis, energy dispersive and wavelength dispersive. It is the latter which has been described.

USES OF RADIOACTIVITY

Many zoological experiments exploit the properties of radioactive isotopes. The atoms of an element are not all identical. For instance, carbon consists mainly of atoms containing six protons and six neutrons, giving a mass number of $6+6=12$. However, it includes about 1% of atoms with six protons and seven neutrons (mass number 13) and a tiny proportion with six protons and eight neutrons (mass number 14). These three types of atom are described as isotopes of carbon and are represented by the symbols ^{12}C, ^{13}C and ^{14}C. Their chemical properties are identical, apart from small differences in rates of reaction.

^{12}C and ^{13}C are stable but ^{14}C is not. One of the neutrons in its nucleus disintegrates, becoming a proton and an electron. This leaves the nucleus with seven protons and seven neutrons so that it is no longer carbon but the common isotope of nitrogen.

^{14}C	$\rightarrow$ ^{14}N	+ β^-
6 protons	7 protons	1 electron
8 neutrons	7 neutrons	

The electron leaves at high velocity because the change releases energy. Fast-moving electrons emitted like this by radioactive materials are known as β–rays.

Breakdown is a random process so the number of ^{14}C atoms in a sample falls exponentially. Half would vanish in the course of 5700 years but only a tiny proportion vanish during an ordinary experiment. The presence of the radioactive isotope is nevertheless easily detected because the β-rays can be detected individually, and counted.

^{14}C is present in normal carbon because ^{14}N atoms in the upper atmosphere, bombarded by neutrons in cosmic radiation, are converted to ^{14}C.

^{14}N	+neutron →	^{14}C	+proton
7 protons		6 protons	
7 neutrons		8 neutrons	

^{14}C is manufactured by bombarding nitrogen compounds with neutrons, in atomic reactors.

The special value of radioactive isotopes in research is due to their being detectable in tiny quantities and distinguishable from stable isotopes. For instance in experiments described in chapter 6 radioactive sodium bicarbonate ($NaH^{14}CO_3$) was put in the water with corals, and radioactive glycerol was later detected. The glycerol must have been made from the bicarbonate.

Radioactive isotopes of many other elements as well as carbon are available. A particularly useful one is the hydrogen isotope tritium, ^{3}H, which has a proton and two neutrons in its nucleus instead of the lone proton of the common isotope.

Some isotopes emit radiation which is easily detected by means of a Geiger counter. They include ^{24}Na, which was used in an experiment with squid axons described in chapter 15. Others, notably ^{14}C and ^{3}H, emit low-energy β-radiation which is best measured by the instrument called a liquid scintillation counter. The sample is mixed in solution with an organic compound which emits flashes of light when bombarded by β-radiation. Each β-ray causes a flash, and the instrument counts the flashes. The intensity of each flash corresponds to the energy of the β-ray which caused it. This makes it possible to count separately the dim flashes caused by the very low-energy β-rays from ^{3}H and the brighter flashes caused by β-rays from ^{14}C.

Radioactive isotopes can be detected in microscope sections by autoradiography. A photographic emulsion is laid over the section and left for a while in darkness before being developed and fixed. The β-rays from radioactive isotopes in the section affect the emulsion in the same way as light would. Clusters of silver crystals in the developed emulsion show where the radioactive isotopes are in the section. The emulsion is left in place on the section making it possible to see under a microscope which tissues contain the radioactive material. The technique has been used, for instance, to show that foodstuffs produced by photosynthesis by algal cells in corals pass into the cells of the coral itself.

The radiation emitted by radioactive isotopes is harmful to health, so appropriate precautions have to be taken in experiments. The stringency of the precautions which are needed depends on the isotope. ^{14}C and ^{3}H are the most

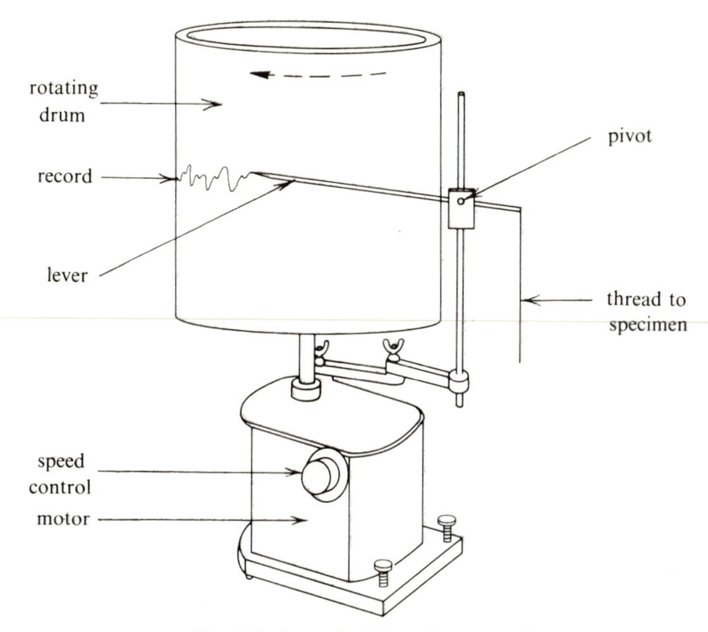

Fig. 1.6. A smoked drum kymograph.

useful isotopes in biology and are also, conveniently, among the least hazardous.

RECORDING EVENTS

Zoologists often want a permanent record of the movements of an animal, of the forces it exerts or of changes of temperature or pressure within it.

The most versatile instrument for recording movements is the cinematograph camera. Films are normally taken and shown at 18 or 24 frames (pictures) per second, which is just fast enough to avoid a flickering effect. The wings of flying insects would be blurred in films taken at this rate and no details of their movements could be made out, even if the frames were examined one by one. Films have been taken at rates up to at least 7000 frames per second, to show insect wing movements clearly (the outlines shown in Fig. 18.7(*b*) are traced from selected frames of a film taken at about 3500 frames per second). When films taken at high framing rates are projected at normal rates, the motion is seen slowed down. Conversely, films are sometimes taken at exceptionally low framing rates so that when they are projected, slow movements are seen speeded up. This is called time-lapse photography. It was used in an investigation of the development of sea urchins so that events which take 48 hours could be seen happening in a few minutes (Fig. 24.13).

Another important instrument for recording movement is the smoked drum kymograph (Fig. 1.6). It was invented in the nineteenth century and has been used in a great many significant physiological experiments, for instance on the

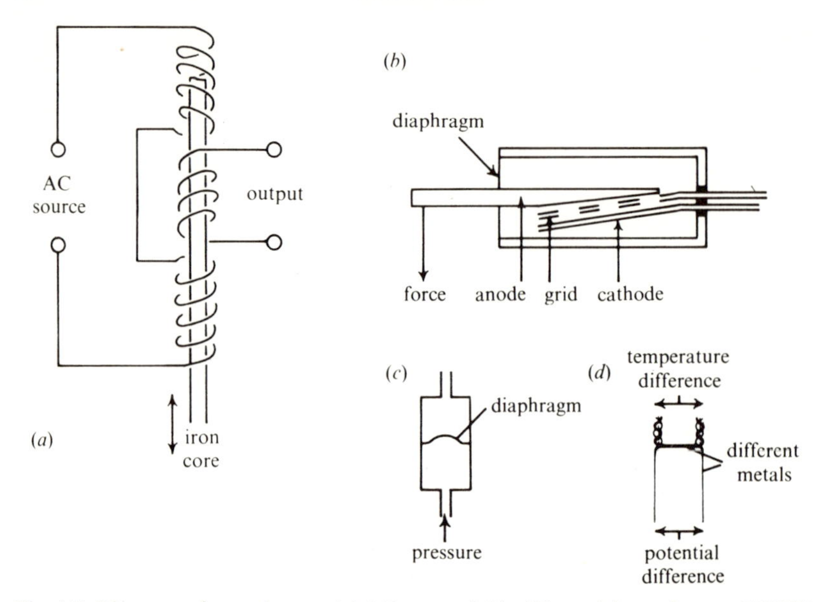

Fig. 1.7. Diagrams of transducers. (*a*) A linear variable differential transformer (LVDT). (*b*) A vacuum tube transducer (the RCA 5734). (*c*) A pressure transducer. (*d*) A thermocouple. (*c*) and (*d*) are from R. McN. Alexander (1975). *The chordates*. Cambridge University Press.

movements of sea anemones (Fig. 6.9). It involves a metal drum with a strip of glazed paper wrapped round it. A smoky flame is used to cover the paper with soot. A pivoted lever is attached at one end to the animal and at the other end bears a pointer which rests lightly against the smoked paper on the drum. A motor makes the drum revolve so that the pointer draws a line on it, which rises and falls as the animal moves. The rate of revolution is adjusted to suit the experiment: a single turn may take a few seconds or many hours. Afterwards the paper is removed from the drum and dipped in varnish to make the record permanent.

The smoked drum kymograph is generally thought of now as very old-fashioned. Displacement transducers are often used instead and have been used, for instance, in some of the more recent research on the movements of sea anemones. However, there are many experiments for which the kymograph still seems the best possible instrument.

Transducers are devices which translate one type of signal into another. For instance a microphone is a transducer which translates sound into electrical signals and a loudspeaker is one which does the reverse. Experimental zoologists make a lot of use of transducers which produce electrical signals which are displayed on an oscilloscope screen or recorded by a pen recorder (or an equivalent instrument). Displacement transducers produce an electrical output in response to movement. One type is the linear variable differential transformer (LVDT) which has been used, for instance, in experiments with spider silk (Fig. 20.7). Fig. 1.7(*a*) shows how it works. Two coils A and B, wound in opposite directions, are connected in series and set in line with each other. Alternating current is passed through them. A third coil C, between them, is connected to the output. An iron core runs through the coils and they work as a transformer. However if the core is symmetrically placed no current flows in C because the oppositely wound coils A and B have opposite effects. If the core is moved up or down, the device becomes asymmetrical, one of the coils A or B has more effect than the other, and current flows in C. Hence the output indicates the position of the core. When an LVDT is used the coils are kept stationary and the core is attached to the moving object which is being investigated.

Another system for recording movement uses a moving vane which interrupts a beam of light directed at a photoelectric cell (Fig. 14.7).

Force transducers give an electrical output in response to force. One type contains a piezoelectric crystal which develops a potential difference between its faces when it is distorted. Most gramophone pick-ups work this way, and a piezoelectric transducer was also used in an investigation of the properties of insect wing muscle (Fig. 18.11). Another type which has often been used by biologists is a vacuum tube transducer, the RCA 5734 (Fig. 1.7*b*). It was used, for instance, in an investigation of the properties of the muscles which oysters use to close their shells (Fig. 14.6*a*). It is an electronic valve with the anode mounted in a flexible diaphragm. Forces acting on the anode move it towards or away from the cathode, altering the electrical properties of the valve. An electrical output can be obtained which is proportional to the applied force.

There are also transducers designed to sense pressure. Steady pressures, or pressures which change only slowly, can be measured with the U-tube manometers, but such manometers do not respond very quickly to a change of pressure. They take an appreciable time to settle at the new equilibrium position, and cannot follow rapid fluctuations of pressure. Often, the pressures in which zoologists are interested fluctuate rather rapidly: the pressures in blood vessels, which fluctuate as the heart beats, are one example. Such pressures can generally be recorded satisfactorily by means of a pressure transducer incorporating a stiff metal diaphragm (Fig. 1.7*c*). The diaphragm is forced into a domed shape by the pressure which is to be measured. The higher the pressure, the more the diaphragm is distorted. The diaphragm has to be stiff to give a really fast response, so some sensitive device is needed to register its distortion. Various electrical devices have been used including strain gauges attached to the diaphragm. Pressure transducers have been used in several of the experiments described in this book, to record, among other things, the blood pressure of a snail (chapter 13).

It is often useful to have small devices to measure temperature changes, because small devices are easiest to fit into animals and because they heat and cool quickly. One such device is the thermocouple. If a conductor runs through a gradient of temperature, a potential difference is set up between its ends. This cannot be measured in a circuit made entirely of one metal, because the effect on one wire running down the gradient is cancelled out by the effect on the other wire running up the gradient to complete the circuit. This difficulty can be overcome by using two metals which are affected to different extents, as shown in Fig. 1.7(*d*). A potential difference is developed proportional to the difference in temperature betwen the two junctions where one metal joins the other. Thermocouples have been used, for instance to record the temperatures of different parts of the bodies of snails in the Negev Desert (Fig. 13.23).

Another small device for measuring temperature is the thermistor, which is simply a small bead of a metal oxide or oxide mixture with electrical leads attached. The electrical resistance of a thermistor decreases as its temperature rises. (Metals also change their resistance as the temperature changes, but not nearly as much and, as it happens, in the opposite direction.) Thermistors have been used to measure the temperatures in the thoraxes of insects flying in the apparatus shown in Fig. 18.8. Electrically-heated thermistors are sometimes used to measure velocities of moving air or water. The moving fluid tends to cool the thermistor, so the faster the flow, the lower the temperature, and the electrical output indicates the rate of flow.

ELECTRODES

Messages are transmitted around the bodies of animals as electrical signals in nerve cells, and zoologists often want to record these signals.

The most important type of signal is the action potential, a brief reversal of the potential difference across the cell membrane. While an action potential is travelling along a nerve cell, there are potential differences in the surrounding extracellular fluid (Fig. 15.13*a*). These are often detected by means of fine tungsten or platinum wires, connected through an amplifier to a cathode ray oscilloscope. In some experiments a fine nerve or a strand dissected from a nerve is picked up on a pair of tungsten or platinum hooks, which serve as electrodes (Fig. 1.8*a*). In others an electrode is pushed into a brain or ganglion to record the electrical activity within. Electrodes to be used in this way are insulated (often by varnish) with only the tip left bare. In yet other experiments electrodes are attached to the specimen by suction (Fig. 1.8*b*): this was done, for instance, in some experiments on sea anemones which are described in chapter 6. Recording electrodes are usually held in micromanipulators.

The potential differences recorded in nerves by extracellular electrodes are small, usually of the order of 1 mV. The changes of electrical potential which occur in nerve cells while an action potential is passing are much larger, of the order of 0.1 V, but they cannot be recorded unless an electrode is pushed into the cell. This electrode must be fine enough not to damage the cell too much, for the cell must remain alive and active with the electrode tip inside it. The electrodes most often used are glass micropipettes, drawn out to a diameter

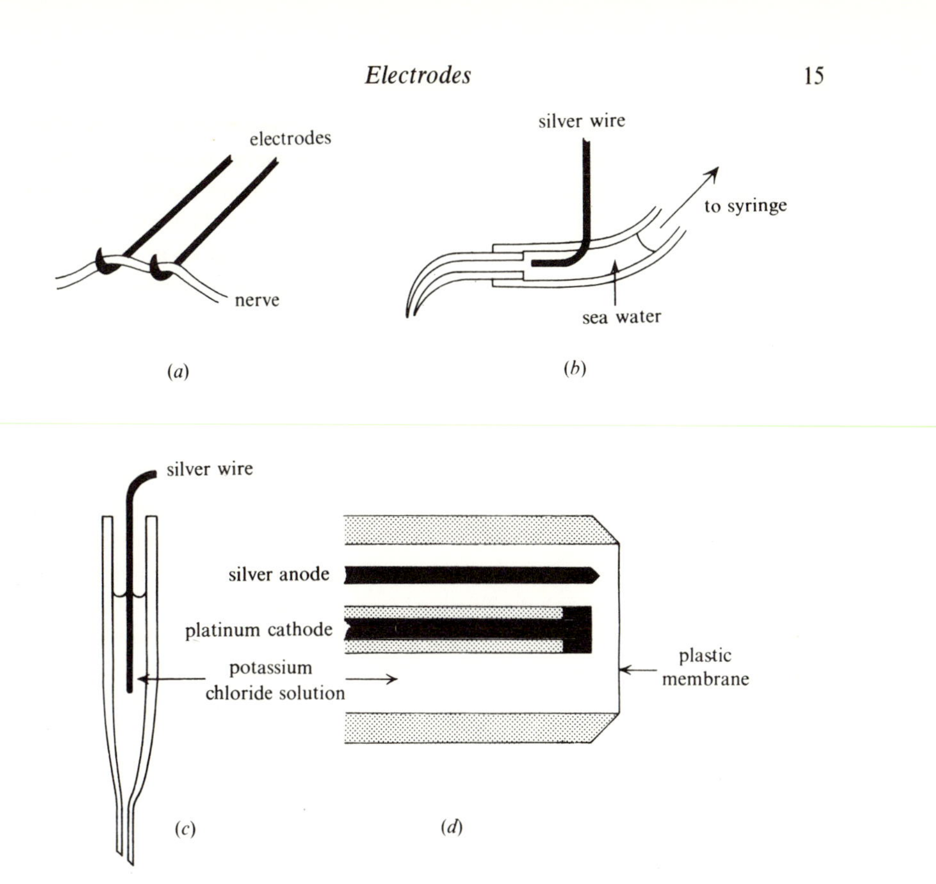

Fig. 1.8. Diagrams of: (*a*) metal hooks being used as electrodes to record from a nerve; (*b*) a suction electrode. The syringe is used to apply suction; (*c*) a micropipette electrode; and (*d*) a Clark oxygen electrode.

of 0.5 μm or less at the tip (Fig. 1.8*c*). They are filled with a strong aqueous solution of potassium chloride and connected to the recording equipment through a silver wire. The tip of the wire which dips into the potassium chloride is lightly coated (by electrolysis) with silver chloride.

This electrode may seem unduly complicated but all the details are important if the resting potential in the cell is to be measured, as well as the changes of potential which occur when the cell is excited. The potential recorded by a simple metal electrode would be affected by the (generally unknown) concentrations of salts in the fluid in the cell, so direct contact of metal with the cell contents is avoided. A potassium chloride solution is used because the most plentiful ions in cells are potassium and chloride, and also because the potential difference at a junction between two potassium chloride solutions of different concentrations is extremely small. The solution in the electrode is made strong, so that it has reasonably high electrical conductivity. The silver wire is coated with silver chloride to prevent it from becoming polarized.

A zoologist who has stuck an electrode into a brain or ganglion and recorded from a cell may want to identify the cell and investigate its anatomy. This can be done if the electrode is filled with a substance which can be injected and later seen in the cell. The recording is made, the substance is injected, the animal is killed and the ganglion is prepared for microscopy. The injected substance can be seen in the cell from which the records were made. A technique of this sort was used to show that certain action potentials recorded from crabs' brains were in the large cell shown in Fig. 17.21(*a*).

In many experiments, muscles or nerves are given small electric shocks to stimulate activity. Metal electrodes are often used.

Fig. 1.8(*d*) shows a device which is called an oxygen electrode, though it actually contains two electrodes. It is an instrument for measuring the partial pressure of oxygen, either in a gas mixture or in solution. It is often more convenient to use an oxygen electrode than to measure oxygen concentration by a chemical method. An oxygen electrode was used, for instance, to measure the partial pressure of oxygen in the part of a duck's gut where certain protozoan parasites live (chapter 4).

The Clark oxygen electrode (Fig. 1.8*d*) contains a platinum cathode and a chloride-coated silver anode. The cathode is kept around 0.5 V negative to the anode. Cathode and anode are in a potassium chloride solution, behind a plastic membrane. The device senses the partial pressure of oxygen in the gas or solution immediately *outside* the membrane. Oxygen diffuses through the membrane to the cathode where it is immediately reduced:

$$O_2+2H_2O+4e=4OH^-$$

or

$$O_2+2H_2O+2e=2OH^-+H_2O_2$$

where e represents an electron.

The current depends on the rate at which the oxygen diffuses in, which in turn depends on its partial pressure outside the membrane. Thus the current indicates the partial pressure.

FOSSILS

Most of the animals described in this book are modern but a few are extinct, known only as fossils. The next few paragraphs explain briefly how fossils are formed, and how their relative ages can be discovered.

The surface of the earth is continually being crinkled by the processes which produce mountains, and levelled again by processes of erosion and sedimentation. A variety of processes break down rocks, particularly on land. Heating of their surfaces by day and cooling by night sets up stresses due to thermal expansion and breaks fragments off. Water expands when it freezes, so water freezing in cracks in rocks is apt to split them. Streams carrying abrasive particles such as sand scour and erode the rocks over which they run. Water containing dissolved carbon dioxide removes calcium and other elements from rocks, carrying them away as a solution of carbonates and bicarbonates.

Materials removed from rocks in these ways are deposited as sediments in other places. Particles which are carried along by fast-flowing water settle out where flow is slower, for instance on the flood plains of rivers and around their mouths. The smaller the particles, the slower the flow must be before they will settle, so relatively large particles settle as gravel or sand in different places from the small particles which form mud. Dry sand may be blown by the wind, and accumulate as dunes. Dissolved calcium carbonate is apt to be precipitated out of water in areas where algae are removing carbon dioxide from the water by photosynthesis. There are also sediments which are not formed from products of erosion, but from animal or plant remains. Shell gravel and the ooze formed on the ocean floor by accumulation of the shells of planktonic Foraminifera are two examples of calcium carbonate sediments of animal origin. Peat is a deposit of incompletely decomposed plant material (in modern times mainly mosses).

Sediments are generally soft when they are formed but if they are not disturbed they tend in time to become rocks. Mud becomes shale, sand becomes sandstone and deposits of calcium carbonate become limestone. The change is partly due to the particles becoming more tightly packed and partly to processes which cement them together. Mud is a mixture of fine mineral particles and water. Initially up to 90% of its volume may be water, the particles are not in contact and it is sloppy. As more mud accumulates on top of it, it is subjected to pressure and water is squeezed out. Adjacent particles make contact when the water content is about 45% by volume and further compaction involves rearrangement and crushing of particles. Note that complete compaction of a layer of mud which initially contained 90% water involves reduction to one-tenth of its initial thickness. Settled sand contains only about 37% water by volume, so compaction cannot reduce its thickness much. The grains in sandstones do not generally seem to have been crushed, but to have dissolved at the points where they touch other grains, so that the grains fit more closely together. Generally silica (perhaps from the dissolved corners of the grains) or a deposit of calcium carbonate cements sandstone together.

An animal which dies where a sediment is being deposited, or is carried there by currents after death, is liable to become embedded in the sediment. If it does not decay completely, it becomes a fossil. Usually only hard parts such as the shells of molluscs survive. Occasionally traces of soft tissues are also found, as stains or impressions in the rock. The shells of molluscs and the hard parts of some other invertebrates are common fossils, but fossils of soft-bodied invertebrates such as worms and jellyfish are seldom found.

It is always interesting to know the age of the rock in which fossils are found. Sediments are formed in layers. Layers of the same material may be distinct because sedimentation was interrupted. Layers of different sediments may be formed on top of one another because local conditions changed. The layers mark time intervals. Successive layers sometimes followed one another immediately, but sometimes there were extremely long intervals of time while no sediment

TABLE 1.1. *The main divisions of time since the beginning of the Palaeozoic era*

The Present is at the top of the table. Age is the approximate time since the *beginning* of the period, estimated mainly from the decay of radioactive elements.

Era	Period	Age (million years)
Cenozoic	Quaternary	2
	Tertiary	70
Mesozoic	Cretaceous	140
	Jurassic	190
	Triassic	230
Palaeozoic	Permian	280
	Carboniferous	350
	Devonian	400
	Silurian	440
	Ordovician	500
	Cambrian	570

was formed, and erosion may even have occurred, at the locality in question. The order in which the sediments were formed at any particular locality is generally obvious since later sediments are on top of earlier ones, though later earth movements may fold sediments so that part of the sequence is upside-down. Sediments formed simultaneously at different places can often be matched, particularly if they contain similar fossils. Thus the relative ages of fossils which are being studied can generally be established.

Fossils are very rare in the oldest sedimentary rocks. The time spanned by rocks in which fossils are reasonably plentiful is divided into three eras and eleven periods, as shown in Table 1.1. It is usually possible to determine the period in which a particular sedimentary rock was formed, and the approximate position within the period.

CLASSIFICATION

This book is full of generalizations, some about large groups of animals (such as the molluscs) and some about small ones (such as the members of a single species). Names are needed for all these groups. A system of classification, if it is well designed, helps zoologists to marshal their knowledge of animals and provides them with names for most of the groups about which they wish to make generalizations.

The smallest unit of classification with which we are concerned is the species. This unit is notoriously difficult to define, and most readers will have

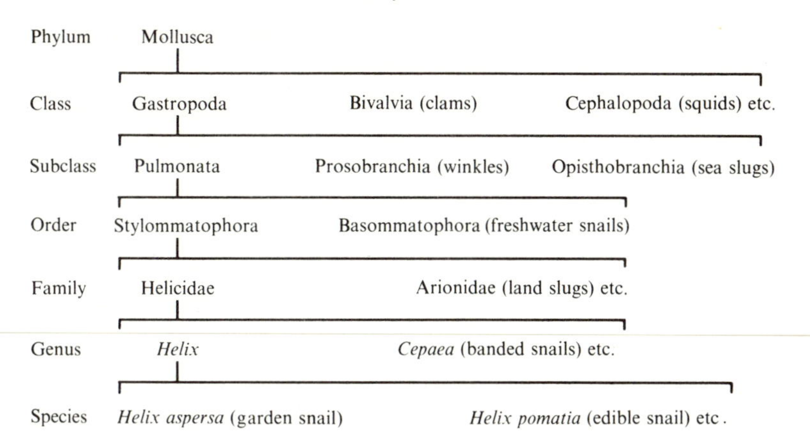

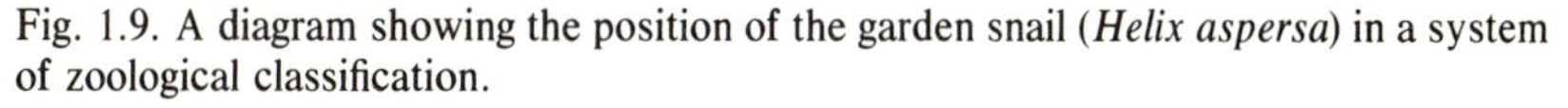

Fig. 1.9. A diagram showing the position of the garden snail (*Helix aspersa*) in a system of zoological classification.

at least an intuitive understanding of its scope. The garden snail is an example of a species. It is a snail with a brown and black shell, common in Britain and also in North America (where it has been introduced). Like other species it is assigned to a genus and has a Latin name (*Helix aspersa*) which consists of two words, of which the first is the name of the genus (*Helix*). Several other very similar snails are assigned to the same genus, including the edible snail *Helix pomatia*. This is the snail which is sold as food. It is larger than *Helix aspersa* and has a cream-coloured shell. Though the differences are quite small they are distinct, and *Helix aspersa* does not breed with *Helix pomatia*. The banded snail (*Cepaea nemoralis*, another common European species) is more different. Its shell is much flatter in shape as well as being smaller and differently coloured. It is universally judged too different from the species of *Helix* to be included in that genus and is therefore put in a separate genus *Cepaea*.

Precise rules are needed to ensure (as far as possible) that all zoologists call the garden snail *Helix aspersa* and that none use this name for other animals. These rules are incorporated in the *International Code of Zoological Nomenclature* and are administered by an international commission. If there is no other compelling reason to prefer one of two rival names, the one which was introduced first has priority.

Genera are grouped in families, families in orders, orders in classes and classes in phyla. Thus *Helix* and *Cepaea* belong (with various other similar snails) to the family Helicidae but the terrestrial slugs are put in a separate family Arionidae (Fig. 1.9). These two families and several other families of terrestrial snails are grouped together in an order Stylommatophora while most freshwater snails are put in a separate order Basommatophora. The Stylommatophora have four tentacles with eyes at the tips of the second pair (see *Helix* in Fig. 13.3 *a*) but the Basommatophora have only two tentacles with eyes at their bases (see

Lymnaea and *Planorbis* in Fig. 13.4*e, f*). The Stylommatophora and Basommatophora form the subclass Pulmonata while winkles and sea slugs are put in separate subclasses. All these subclasses are grouped together in the class Gastropoda. Finally the Gastropoda are grouped with other classes (which include the clams and squids) in the phylum Mollusca. A snail is not very obviously like a clam and still less like a squid but there are basic similarities of body plan which justify putting this strange assemblage of animals in a single phylum. This should be apparent from chapters 13 to 15.

Every animal is assigned to a species, genus, family, order, class and phylum, and additonal groupings are used when convenient. For instance, it has been found convenient to divide the large class Gastropoda into three subclasses, as shown in Fig. 1.9. Putting two species in the same genus implies close similarity between them, putting them in the same family implies rather less similarity and putting them in the same order, class or phylum imply successively lower degrees of similarity. There are no formal definitions of the degrees of similarity required, but experienced zoologists get a feel for them.

Diligent students will quickly find discrepancies between the classifications used in different books. Some of the discrepancies are mere differences of name: for instance, the class of molluscs which includes the clams is called the Bivalvia by some zoologists and the Pelecypoda by others. Such inconsistencies are more or less inevitable because the names of groups above the level of family are not subject to the *International Code*. Other discrepancies reflect differences or changes of opinion. For instances the Polyplacophora (the chitons, or coat-of-mail shells) and the Aplacophora (an obscure group of worm-like molluscs) are listed by some zoologists as separate classes of mollusc but by others, who are more impressed by the similarities between them, as subclasses of a united class Amphineura.

A classification is not right or wrong, but it may be good or bad. The best classification is generally the most useful one. It helps zoologists to marshal their thoughts in the most profitable way, it provides the names they need when they wish to make generalizations, and it does not unnecessarily change old names which they have grown accustomed to use.

It is generally agreed that a classification should not be inconsistent with existing knowledge of the course of evolution. In practice this means that if group *a* is believed to have evolved directly from class (or order or family) *A* and group *b* from class (or order or family) *B*, *a* and *b* will not be put together in the same class (or order or family) *C*. However, if *a* is believed to have evolved from order *x* of class *A* and *b* from order *y* of the same class, it would be permissible to include *a* and *b* in a single class *C*.

Fig. 1.10 will be used to illustrate the way in which these rules are applied. It represents four possible hypotheses about the evolution of the gastropods. Hypothesis (*a*) is the one immediately suggested by the classification of Fig. 1.9: it shows the ancestral gastropods (which are assumed to be members of the Prosobranchia) giving rise at about the same time to the Opisthobranchia

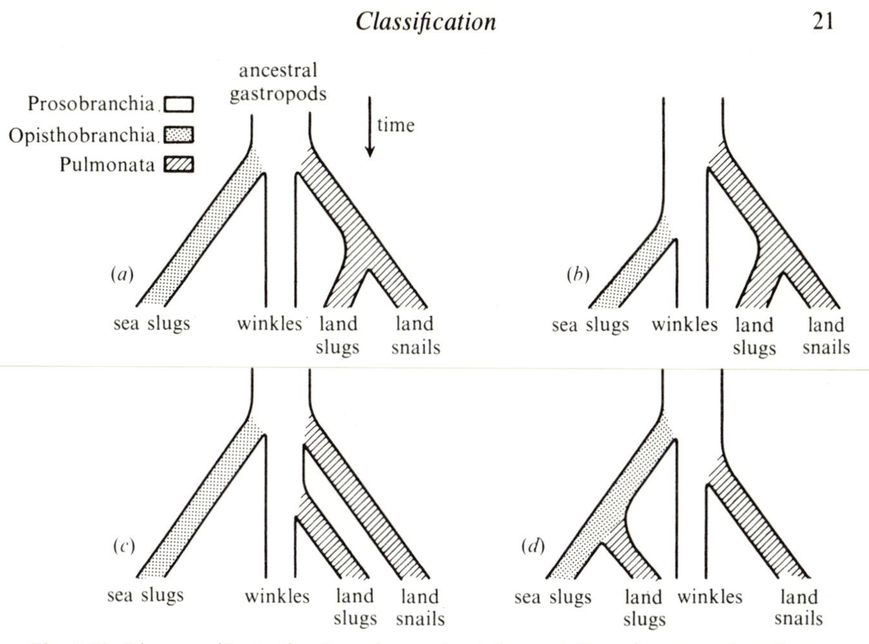

Fig. 1.10. Diagrams illustrating hypotheses about the evolution of gastropod molluscs.

and Pulmonata. However the classification is also consistent with (*b*), which shows the Pulmonata arising before the Opisthobranchia. Further, it is consistent with (*c*) which shows the terrestrial slugs and snails arising separately from different groups of Prosobranchia. However, it is not consistent with (*d*), which shows the snails evolving from Prosobranchia and the terrestrial slugs from the Opisthobranchia. A zoologist who believed that the gastropods had evolved as shown in Fig. 1.10(*d*) ought not to use the classification shown in Fig. 1.9, because this puts together in one subclass the snails (which he believes to have evolved from members of the subclass Prosobranchia) and the terrestrial slugs (which he believes to have evolved from members of the subclass Opisthobranchia).

Our fictitious zoologist who believes that the gastropods evolved as shown in Fig. 1.10(*d*) might argue that the terrestrial slugs and sea slugs (with inconspicuous shells or no shell at all) are so similar to each other that they must be more closely related to each other than either is to the other gastropods (with well developed shells). Most zoologists would disagree. They would argue that there are a great many close similarities between terrestrial slugs and snails, for instance in the arrangement of eyes and tentacles and in the possession of a lung instead of gills. There are so many features in which terrestrial slugs and snails resemble each other and differ from other gastropods that they must be closely related. Reduction of the shell in slugs and sea slugs does not necessarily indicate relationship; such a simple change could have occurred independently in two lines of evolution, as implied by Figs. 1.10(*a*)–(*c*).

In this book the animals are classified as follows. Minor phyla which are not mentioned elsewhere in the book are shown in parentheses.

PHYLUM PROTOZOA	
SUBPHYLUM SARCOMASTIGOPHORA	chapter 2
Superclass Mastigophora	
Class Phytomastigophorea plant flagellates	
Class Zoomastigophorea animal flagellates	
Superclass Opalinata	
Superclass Sarcodina	
Class Rhizopodea amoebas and foraminiferans	
Class Actinopodea radiolarians and heliozoans	
SUBPHYLUM CILIOPHORA	chapter 3
Class Ciliatea	
Subclass Holotrichia	
Subclass Peritrichia	
Subclass Suctoria	
Subclass Spirotrichia	
SUBPHYLUM SPOROZOA	chapter 4
Class Telosporea	
Subclass Gregarinia	
Subclass Coccidia	
Class Piroplasmea	
SUBPHYLUM CNIDOSPORA	chapter 4
Class Myxosporidea	
Class Microsporidea	
PHYLUM PORIFERA sponges	chapter 5
Class Calcarea	
Class Hexactinellida	
Class Demospongia	
(PHYLUM MESOZOA)	
PHYLUM CNIDARIA	
Class Anthozoa	chapter 6
Subclass Alcyonaria soft corals, sea fans, etc.	
Subclass Zoantharia stony corals, sea anemones, etc.	
Class Hydrozoa hydroids	chapter 7
Class Scyphozoa jellyfishes	chapter 7
(PHYLUM CTENOPHORA sea gooseberries)	
PHYLUM PLATYHELMINTHES	
Class Turbellaria flatworms	chapter 8
Order Acoela	
Order Rhabdocoela	
Order Tricladida	
Order Polycladida	
and other orders	
Class Monogenea	chapter 9
Order Monopisthocotylea skin flukes	
Order Polyopisthocotylea gill flukes	
Class Trematoda	chapter 9
Order Aspidobothrea	
Order Digenea gut, liver and blood flukes	
Class Cestoda tapeworms	chapter 9

PHYLUM NEMERTEA	chapter 10
(PHYLUM ACANTHOCEPHALA)	
PHYLUM ROTIFERA rotifers	chapter 11
Class Seisonacea	
Class Bdelloidea	
Class Monogononta	
(PHYLUM GASTROTRICHA)	
(PHYLUM KINORHYNCHA)	
(PHYLUM PRIAPULIDA)	
PHYLUM NEMATODA roundworms	chapter 12
(PHYLUM NEMATOMORPHA threadworms)	
(PHYLUM ENTOPROCTA)	
(PHYLUM CONODONTA extinct)	
PHYLUM MOLLUSCA	
Class Monoplacophora Neopilina	chapter 13
Class Aplacophora	chapter 13
Class Polyplacophora chitons	chapter 13
Class Gastropoda	chapter 13
Subclass Prosobranchia winkles etc.	
Subclass Opisthobranchia sea slugs etc.	
Subclass Pulmonata snails and slugs	
Class Scaphopoda tusk shells	chapter 13
Class Bivalvia clams etc.	chapter 14
Subclass Protobranchia *Nucula*	
Subclass Lamellibranchia most other genera	
Class Cephalopoda	chapter 15
Subclass Nautiloidea pearly nautilus etc.	
Subclass Ammonoidea ammonites etc., extinct	
Subclass Coleoidea octopus, squid, cuttlefish	
PHYLUM ANNELIDA	chapter 16
Class Polychaeta ragworms, lugworms, etc.	
Class Myzostomaria	
Class Oligochaeta earthworms etc.	
Class Hirudinea leeches	
(PHYLUM TARDIGRADA water bears)	
PHYLUM ARTHROPODA	
Class Trilobita extinct	chapter 21
Class Onychophora	chapter 21
Class Merostomata horseshoe crabs	chapter 20
Class Arachnida	chapter 20
Order Acari ticks and mites	
Order Scorpiones scorpions	
Order Araneida spiders	
and other orders	
Class Crustacea	chapter 17
Subclass Cephalocarida	
Subclass Branchiopoda water fleas etc.	

Subclass Mystacocarida
Subclass Ostracoda
Subclass Copepoda
Subclass Branchiura
Subclass Cirripedia barnacles
Subclass Malacostraca crabs, shrimps, woodlice, etc.
Class Insecta insects — chapters 18 and 19
numerous orders, of which the most important are listed in chapter 19
Class Chilopoda centipedes — chapter 21
Class Diplopoda millipedes
and other classes

PHYLUM PHORONIDA — chapter 22

PHYLUM BRYOZOA — chapter 22
Class Phylactolaemata
Class Stenolaemata
Class Gymnolaemata

PHYLUM BRACHIOPODA lamp shells — chapter 22
Class Inarticulata
Class Articulata
(PHYLUM SIPUNCULOIDEA)

(PHYLUM CHAETOGNATHA arrow worms)

(PHYLUM POGONOPHORA)

(PHYLUM GRAPTOLITA extinct)

PHYLUM HEMICHORDATA — chapter 23
Class Enteropneusta acorn worms
Class Pterobranchia

PHYLUM ECHINODERMATA — chapter 24
Class Crinoidea sea lilies and feather stars
Class Asteroidea starfish
Class Ophiuroidea brittle stars
Class Echinoidea sea urchins
Class Holothuroidea sea cucumbers
and other extinct classes

PHYLUM CHORDATA
SUBPHYLUM UROCHORDATA — chapter 25
Class Ascidiacea sea squirts
Class Larvacea
Class Thaliacea salps
SUBPHYLUM CEPHALOCHORDATA amphioxus — chapter 25
SUBPHYLUM VERTEBRATA — not included in this book

FURTHER READING

GENERAL BOOKS ON INVERTEBRATES

Barrington, E. J. W. (1967). *Invertebrates structure and function.* Nelson, London.
Cox, F. E. G., Dales, R. P., Green, J., Morton, J. E., Nichols, D. & Wakelin, D. (1969). *Practical invertebrate zoology.* Sidgwick & Jackson, London.

Grassé, P. P. (ed.) (1948–). *Traité de zoologie*, Many vols. Masson, Paris.
Hyman, L. H. (1940–67). *The invertebrates*, 6 vols. McGraw-Hill, New York.
Kaestner, A. (1967). *Invertebrate zoology*, 3 vols. Interscience, New York.
Meglitsch, P. A. (1972). *Invertebrate zoology*, 2nd edn. Oxford University Press.

OTHER BOOKS FOR GENERAL REFERENCE

Alexander, R. McN. (1968). *Animal mechanics*. Sidgwick & Jackson, London.
Ambrose, E. J. & Easty, D. M. (1977). *Cell biology*, 2nd edn. Nelson, London.
Holwill, M. E. & Silvester, N. R. (1973). *Introduction to biological physics*. Wiley, New York.
Welsch, U. & Storch, V. (1976). *Comparative animal cytology and histology*. Sidgwick & Jackson, London.

MICROSCOPY

Baker, J. R. (1966). *Cytological technique*, 5th edn. Methuen, London.
Grimstone, A. V. (1976). *The electron microscope in biology*, 2nd edn. Arnold, London.
Holwill, M. E. & Silvester, N. R. (1973). *Introduction to biological physics*. Wiley, New York.
Southworth, H. N. (1975). *Introduction to modern microscopy*. Wykeham, London.

CHEMICAL ANALYSIS

Ralph, R. (1975). *Methods in experimental biology*. Blackie, Glasgow.

USES OF RADIOACTIVITY

Wolf, G. (1964). *Isotopes in biology*. Academic Press, New York & London.

RECORDING EVENTS, ELECTRODES

Dewhurst, D. J. (1966). *Physical instrumentation in medicine and biology*. Pergamon, Oxford.
Geddes, L. A. (1972). *Electrodes and the measurement of bioelectric events*. Wiley-Interscience, New York.
Giles, A. F. (1966). *Electronic sensing devices*. Newnes, London.
Kay, R. H. (1964). *Experimental biology. Measurement and analysis*. Chapman & Hall, London.
Neubert, H. K. P. (1963). *Instrument transducers: an introduction to their performance and design*. Clarendon, Oxford.

FOSSILS

Raup, D. M. & Stanley, S. M. (1971). *Principles of palaeontology*. Freeman, San Francisco.

CLASSIFICATION

Blackwelder, R. E. (1967). *Taxonomy. A text and reference book*. Wiley, New York.
Savory, T. (1970). Animal taxonomy. Heinemann, London.

2

Flagellates and amoebas

Phylum Protozoa, Subphylum Sarcomastigophora
  Superclass Mastigophora
    Class Phytomastigophorea (plant flagellates)
    Class Zoomastigophorea (animal flagellates)
  Superclass Opalinata
  Superclass Sarcodina
    Class Rhizopodea (amoebas, foraminiferans, etc.)
    Class Actinopodea (radiolarians and heliozoans)
See also chapters 3 and 4

The phylum Protozoa consists of organisms which have only one cell in the body (some zoologists prefer to describe them as not divided into cells, which means much the same thing). All of them are small, but some are much smaller than others. One of the largest is the giant amoeba *Pelomyxa palustris* which is about 2 mm long. One of the smallest is a marine flagellate *Micromonas pusilla* which is only 1–1.5 μm long, about the size of a typical bacterium. This is of course a very wide range of sizes. *Pelomyxa* is over 1000 times as long as *Micromonas* and must be at least $(1000)^3 = 10^9$ times as heavy. A 100 tonne whale is only 5×10^7 times as heavy as a 2 g shrew.

Micromonas seems to be the smallest known eukaryote organism; that is, the smallest known organism to have a nucleus and other organelles enclosed by membranes within the cell. It is not by any means the smallest known organism. Viruses are much smaller but are incapable of independent life: they can only grow and multiply inside a living cell and they depend on the biochemical processes of that cell. However, mycoplasms only 0.3 μm in diameter have been grown on non-living media, and so shown to be capable of independent life.

The phylum Protozoa is divided into four subphyla. The subphylum Sarcomastigophora is the subject of this chapter, and includes the flagellates and amoebas. The subphylum Ciliophora consists of the ciliate Protozoa and is the subject of chapter 3. The subphyla Sporozoa and Cnidospora consist entirely of parasitic Protozoa and are considered (with parasitic members of the other subphyla) in chapter 4.

Nearly all Sarcomastigophora belong either to the superclass Mastigophora (the flagellates) or to the superclass Sarcodina (amoebas etc.). The third superclass, Opalinata, is needed for some peculiar protozoans found in the guts of frogs (Fig. 26.2*c*).

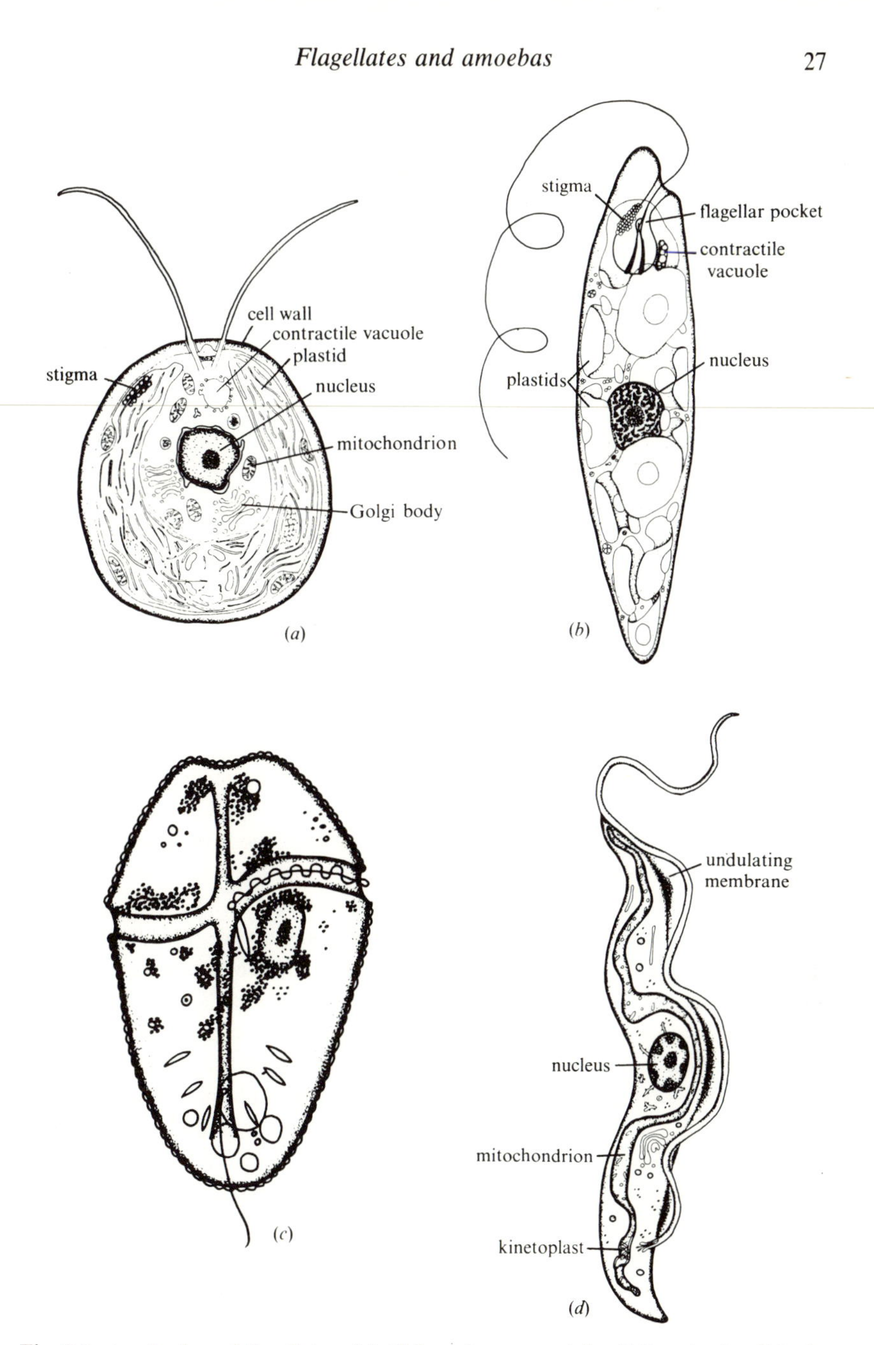

Fig. 2.1. A selection of flagellates. (*a*) *Chlamydomonas reinhardi* (length of cell body 10 μm); (*b*) *Euglena gracilis* (50 μm); (*c*) *Gymnodinium amphora* (30 μm); and (*d*) *Trypanosoma brucei* (20 μm). From M. A. Sleigh (1973). *The biology of Protozoa*. Edward Arnold, London.

A selection of flagellates is shown in Fig. 2.1. Fig. 2.1(*a*) to (*c*) are Phytomastigophorea and are capable of phytosynthesis. Look first at *Chlamydomonas* (Fig. 2.1 *a*) which is found in ponds and ditches. The drawing is based on electron microscope sections. The two long projections are the flagella, which are undulated to propel the organism through the water. A later section of this chapter is devoted to flagella. The rounded body of the cell has a rigid polysaccharide cell wall outside the cell membrane. The largest organelle in the cell is the plastid. This is green because it contains chlorophyll, which is used in photosynthesis. The stigma is a cluster of lipid globules which contrasts with the rest of the plastid because dissolved pigments give it a red colour. A food store of starch is laid down in the plastid. The protoplasm in the hollow of the cup-shaped plastid contains the nucleus, mitochondria, Golgi bodies and contractile vacuole. The nucleus of course contains the chromosomes, with the coded genetic information which enables the cell to synthesize its enzymes and other proteins. The mitochondria contain the enzymes required for the Krebs cycle which completes the oxidation of glucose to carbon dioxide and water and provides energy for vital processes in the cell. The Golgi bodies probably secrete the cell wall, just as the Golgi bodies of the cells of higher plants secrete their cell walls. The contractile vacuole is an organelle which pumps excess water (drawn in osmotically) out of the cell. Its working is discussed in a later section of the chapter.

Euglena (Fig. 2.1 *b*) is another green flagellate. It flourishes in pools visited by cattle, or polluted with organic matter in other ways. It has numerous plastids, each with a store of polysaccharide (paramylon, not starch). There are two flagella which have their bases in a pocket at one end of the body but one is very short and does not protrude from the pocket. There is no polysaccharide cell wall but a pellicle which is flexible enough to allow limited changes of shape of the body. It consists of long, narrow strips of a material which is mainly protein. Each strip is linked to its neighbours by a tongue-and-groove arrangement, like floor boards. The strips are not external to the cell membrane like the cell wall of *Chlamydomonas*, but internal to it.

Gymnodinium (Fig. 2.1 *c*) is one of the dinoflagellates, a group of flagellates which is abundant in the plankton of seas and lakes. The body is protected by plates of cellulose which are internal to the cell membrane, enclosed in membrane-lined cavities. There are two flagella of which one lies in a groove running round the equator of the body while the other points posteriorly. There are numerous brown plastids. *Gymnodinium amphora* is a free-living member of the plankton but another species of *Gymnodinium* lives symbiotically within the cells of corals (see chapter 6).

Trypanosoma brucei (Fig. 2.1 *d*) is one of the Zoomastigophorea. It causes a dangerous disease, African sleeping sickness. It lives in the blood of mammals (including man) and also in the gut of the tsetse fly (*Glossina*). There is an account of its life cycle in chapter 4. It has no plastids, which is not surprising since it lives in darkness. It has a single very long mitochondrion

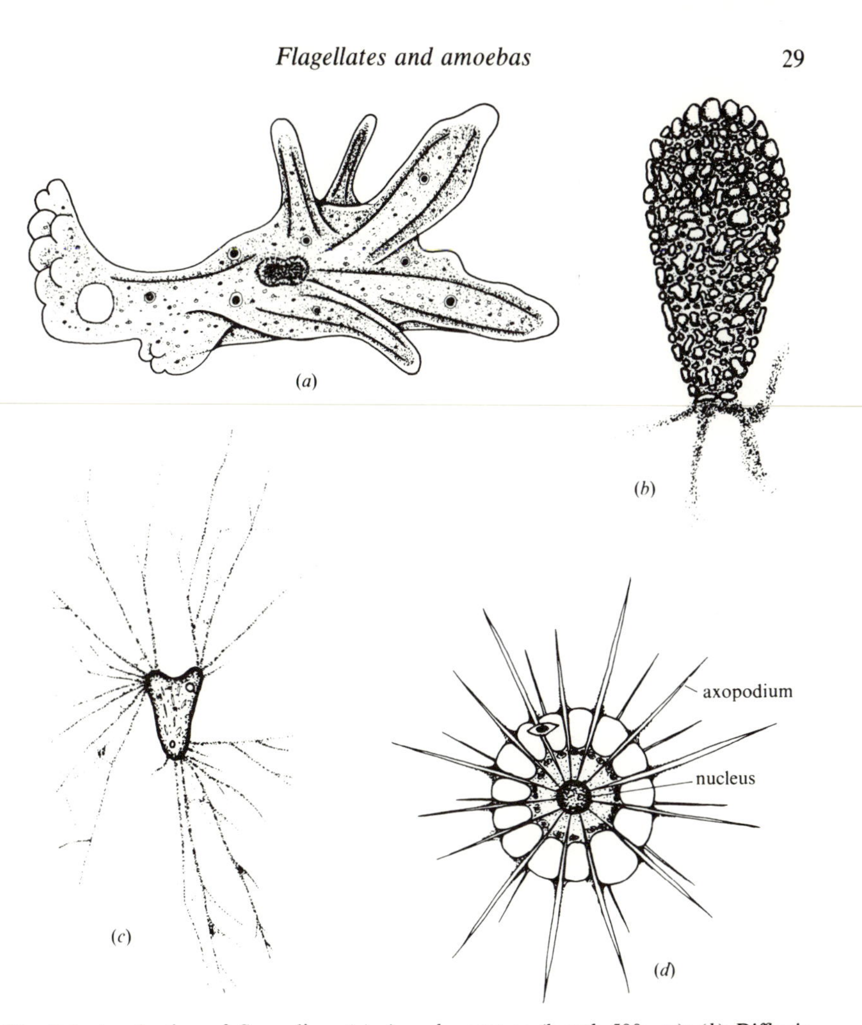

Fig. 2.2. A selection of Sarcodina. (*a*) *Amoeba proteus* (length 500 μm); (*b*) *Difflugia oblonga* (250 μm); (*c*) *Allogromia laticollaris* (600 μm); and (*d*) *Actinophrys sol* (diameter 40 μm). From M. A. Sleigh (1973). *The biology of Protozoa*. Edward Arnold, London.

with a DNA-rich region known as the kinetoplast. The surface of the body is flexible and there is a layer of microtubules, set side by side immediately under the cell membrane. They may help to cause the wriggling movements which trypanosomes make. Microtubules are found in flagella and many other protozoan structures, and some of their functions are discussed in later sections of this and the next chapter. The cell membrane adheres to the flagellum, forming an undulating membrane between the flagellum and the cell body.

Fig. 2.2 shows representatives of the Sarcodina. No member of this superclass has plastids. *Amoeba proteus* (Fig. 2.2 *a*) can be found on the muddy bottoms of ponds. It is not very common. It has an irregular shape which

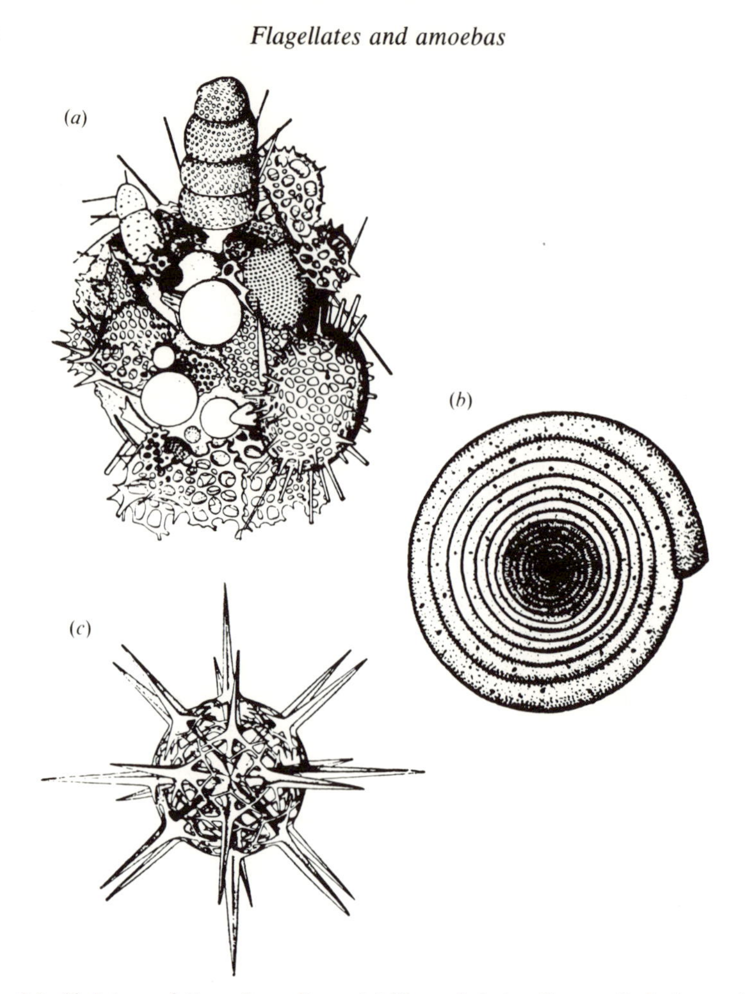

Fig. 2.3. Skeletons of three Sarcodinea. (*a*) The radiolarian *Cementella loricata* which builds a composite shell from the skeletons of other radiolarians, etc. (*b*) The foraminiferan *Ammodiscus incertus* (diameter 3 mm). (*c*) The radiolarian *Dictyacantha tetragonopa* (diameter to tips of spines 0.7 mm). From P. P. Grassé (ed.) (1953). *Traité de zoologie*, vol. 1, part 2. Masson, Paris.

changes constantly as it moves. The projections from its body (pseudopodia) are not permanent but are formed and disappear as it crawls. The specimen which is illustrated is crawling from left to right. The mechanism of crawling is discussed (but not explained, for it is not fully understood) in a later section of this chapter. There is a nucleus (the large organelle shown at the centre of the body) and a contractile vacuole (on the left). Also shown in the illustration are food vacuoles containing particles of food such as smaller Protozoa which have been engulfed by the animal and are being digested. *Difflugia* (Fig. 2.2 *b*) is found among decaying vegetation on the bottoms of ponds and ditches. It is rather like *Amoeba* but makes a shell out of sand grains cemented together with organic matter. *Amoeba* and *Difflugia* are both members of the subclass Rhizopodea. Both are large, but there are small amoebas as well, some of them as little as 20 μm long.

Allogromia (Fig. 2.2 *c*) is one of the foraminiferans. It can be found among the holdfasts of laminarian algae. Its body is enclosed in a casing or test of a material known as tectin, composed of protein and polysaccharide. From openings in the test project pseudopodia which are not stubby lobes like the pseudopodia of *Amoeba* but slender strands which branch and rejoin forming a network. Protozoa and other small animals, and bacteria, get caught on this network and are transported to the main body to be digested. *Actinophrys* Fig. 2.2 *d*) is a heliozoan. It is found in fresh water among decaying leaves. The outer layer of its cytoplasm has so many fluid-filled vacuoles in it that it looks frothy. Structures which look like spines, called axopodia, project from the body in all directions. They have a function in locomotion which is described later in this chapter.

The scales and shells of Sarcomastigophora are plentiful in some rocks. Coccoliths are scales, stiffened by incorporation of calcium carbonate, which cover the bodies of some of the small flagellates found in marine plankton. They are plentiful in the chalk which was laid down in Europe in the Cretaceous period. The test of *Allogromia* is made of an organic material, but many other foraminiferans have their tests reinforced by sand grains, sponge spicules or fragments of other materials, or by incorporation of calcium carbonate (Fig. 2.3 *b*). About one-third of the area of the floor of the oceans is covered by a deposit of the tests of planktonic foraminiferans. The calcareous tests of *Globigerina* are particularly plentiful and deposits in which they predominate are known as *Globigerina* ooze. Foraminiferan fossils are important constituents of some rocks (including chalk) and actually constitute the bulk of some limestones. Different species occur in different strata and can be used for identifying strata. They have been particularly useful to oil prospectors, who can identify the stratum they are drilling through by finding particular species of foraminiferan in the fragments of rock which are cut by the drill. Radiolarians are Actinopodea, with axopodia, which have internal skeletons of silica (Fig. 2.3 *c*). They live in the sea, most of them as plankton. Much of the floor of the ocean at depths greater than 4000 m is covered by radiolarian ooze, composed of the shells of dead radiolarians. Foraminiferans also live in the plankton over these areas, but it seems that their calcareous tests dissolve when they sink to great depths. High pressure increases the solubility of calcium carbonate in water.

Members of the superclass Mastigophora mostly have one or more flagella, and no pseudopodia. Members of the Sarcodina have pseudopodia or axopodia and no flagella. There is little similarity between typical members of the two superclasses, and readers may reasonably wonder why they are grouped together in the subphylum Sarcomastigophora. The reason is that there are Protozoa such as *Pedinella* which have both flagella and pseudopodia, and others

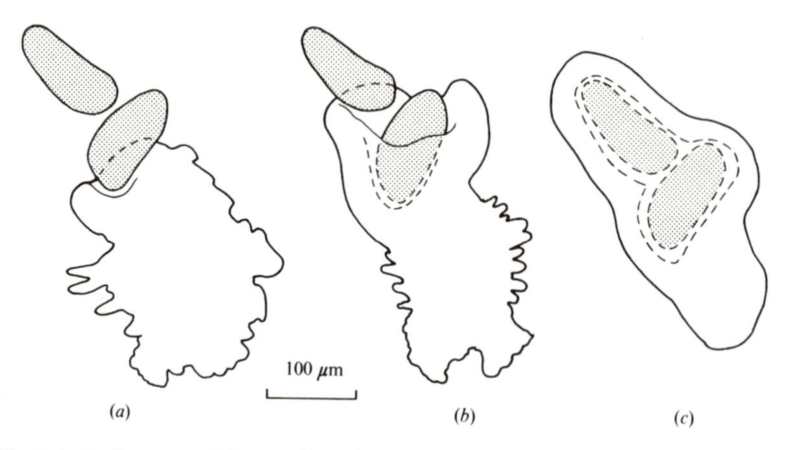

Fig. 2.4. Outlines traced from a film of *Amoeba proteus* ingesting two ciliate Protozoa (*Paramecium*). After K. G. Grell (1973). *Protozoology*. Springer, Berlin.

such as *Naegleria* which may change from an amoeboid to a flagellated form. *Naegleria* is an amoeba which lives in soil, and normally crawls about in amoeboid fashion (Fig. 2.12*a*). However, it may develop flagella and swim if flooding occurs. Animals like these could be placed quite plausibly either in the Mastigophora or in the Sarcodina. They indicate that the groups are closely related, and indeed make it difficult to draw a sharp dividing line between them.

In very dry conditions soil ameobae including *Naegleria* and *Acanthamoeba* round up and enclose themselves in protective cysts of carbohydrate (perhaps cellulose) and protein. While in the cyst they respire very slowly, and cysts can survive storage for years in dry bottles. The amoebas emerge from their cysts again when conditions are favourable.

PHOTOSYNTHESIS AND FEEDING

The subphylum Sarcomastigophora is the region of overlap between the animal and plant kingdoms. Many flagellates have plastids and produce foodstuffs by photosynthesis. They thus resemble plants, and indeed the higher plants are believed to have evolved from ancestors resembling *Chlamydomonas*. Amoebas feed in animal fashion, by engulfing and digesting small organisms including bacteria, diatoms and smaller Protozoa (Fig. 2.4.). A hollow forms in the surface of the amoeba, next to the food, and its edges close in so that the food is enclosed in a vacuole. This simple method of catching food is astonishingly effective. Add some fast-swimming ciliates to a dish containing *Amoeba* and within 10 minutes or so there are ciliates in vacuoles inside the amoebas. Granules called lysosomes coalesce with the food vacuole, releasing digestive enzymes into it. The prey is digested and eventually its indigestible remains are discarded by bringing the vacuole to the cell surface and reversing the process which formed it. Amoebas may form food vacuoles and discard remains at any point on the cell surface but many Protozoa have particular places for these processes.

The distinction between animals and plants among the flagellates is a blurred one. The dinoflagellate *Ceratium* has chlorophyll and practises photosynthesis, but vacuoles can be found in its cytoplasm containing the remains of bacteria, diatoms and blue-green algae which have apparently been engulfed. *Peranema* is a flagellate which is very similar to *Euglena*, but has no plastids. *Euglena* photosynthesizes but *Peranema* feeds by engulfing other organisms (including *Euglena*).

There is a third manner of feeding, by absorbing organic molecules dissolved in the surrounding water. This is the habit of *Polytoma*, which lives in polluted waters and can also be grown in artificial solutions which contain acetate as a source of energy and carbon. *Polytoma* has no chlorophyll but is otherwise very like *Chlamydomonas* which has no known needs for organic molecules. *Chlamydomonas* can apparently synthesize all the materials it needs from carbon dioxide, water and inorganic salts. *Euglena gracilis* grows in the light, needing no organic compounds except vitamins B1 and B12 which it cannot synthesize for itself. However, some strains can survive and grow in the dark in solutions of suitable organic compounds.

The ultimate source of energy which maintains all life on earth is the sun. Solar energy is captured by photosynthesis, by flagellates and higher plants. These are eaten by animals which may in turn be eaten by other animals, and if they die without being eaten their dead bodies will nourish the decomposers (such as bacteria and fungi) which break them down. On land, most of the photosynthesis is done by grasses and trees. In the sea and in lakes, it is done by planktonic organisms. The most important of them are the diatoms (which are always thought of as plants rather than as animals, and so are not described in this book), the dinoflagellates and very small flagellates broadly similar to *Chlamydomonas*. These photosynthesizing members of the plankton are known collectively as the phytoplankton, to distinguish them from the zooplankton which feed in animal fashion.

Some of the energy captured by photosynthesis is used by the photosynthesizing organisms for their own metabolism, to supply the energy they need for the maintenance of life. The remainder is used for growth or to build up food stores. The chemical energy accumulated in this way is referred to as the net primary production, and is energy potentially available to herbivores. It can be expressed in joules per square metre per year. Marine plankton and freshwater plankton in unpolluted waters achieve in favourable conditions a net primary production of about 4 MJ m^{-2} yr^{-2} (representing about 200 g dry organic matter m^{-2} yr^{-1}). However polluted waters produce more and as much as 90 MJ m^{-2} yr^{-1} has been achieved by cultivating algae in sewage ponds in California. This is comparable to the highest net primary production which can be obtained on land. Good agricultural grassland in New Zealand can produce 60 MJ m^{-2} yr^{-1} and tropical rain forest and sugar cane plantations can each produce about 120 MJ m^{-2} yr^{-1}.

If the energy which reaches the earth's atmosphere passed through it undepleted, around 10 GJ (10000 MJ) would fall on each square metre of level ground or water in the course of a year (rather more near the equator, and less near the poles). Much of this energy is absorbed or scattered by the atmosphere but far more reaches the ground than can be captured by photosynthesis. Even in the most favourable circumstances crops and plankton seem unable to capture more than about 3% of the radiation energy which falls on them.

The net primary production by plankton in unpolluted waters is much less than is possible in sewage ponds, with the same supply of light. It seems generally to be limited in unpolluted waters by the supplies of inorganic salts. Nitrate or an alternative nitrogen source is required to supply the nitrogen for protein synthesis. Phosphate is required for the synthesis of phospholipids for cell membranes and for many other essential constituents of cells. Diatoms need silicate, to build their silica tests. In temperate and polar lakes and seas nitrate, phosphate and silicate concentrations may fall from relatively high winter values to very low summer values.

The annual cycle of plankton population changes is closely linked to the phenomenon of thermal stratification. In spring, the temperature of surface waters is rising. They become warmer than the deeper water below them, and less dense than it. The difference in density tends to prevent mixing. Winds cause mixing down to a certain depth but at that depth there is a rather sharp change of temperature (the thermocline). The depth of the thermocline depends on the place and the season. It is often about 20 m from the surface in the English Channel, but may be as deep as 100 m in the North Atlantic. The thermocline persists through the summer but in the autumn the surface waters become cooler and denser, and are mixed with the deep water by convection.

The photosynthetic members of the plankton (the phytoplankton) can only grow and multiply where the light intensity is high enough for them to photosynthesize faster than they use energy in respiration. Thus they cannot survive for long in the perpetual darkness of the depths, and need to be reasonably near the surface. They can live deeper in clear oceanic water than in turbid coastal water, but it seems to be a general rule that little phytoplankton is found anywhere below the thermocline.

Consider an area where the water at the beginning of spring contains 0.1 mg nitrogen (as nitrate) per litre and where a thermocline forms at a depth of 50 m. These data are reasonably typical for many areas in the sea. There would be 100 mg available nitrogen in each cubic metre of water, and the total mass of available nitrogen between the surface and the thermocline would be 5 g m^{-2}. This is enough to make about 30 g protein m^{-2} or perhaps about 100 g (dry weight of plankton) m^{-2}. If no mixing occurred across the thermocline there could never be more than about this amount of plankton. The limit would be different if the initial concentration of available nitrogen, or the depth of

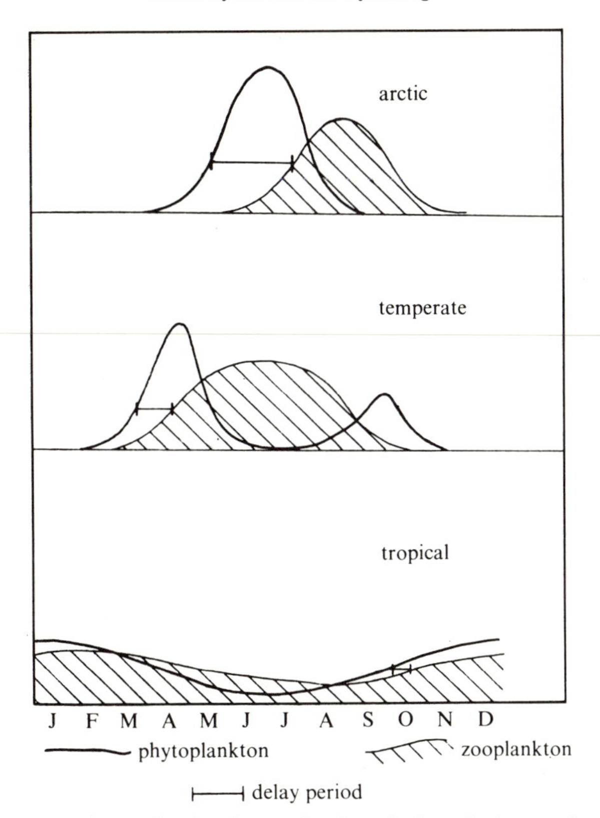

Fig. 2.5. Diagrammatic graphs showing production of phytoplankton and zooplankton in the course of a year at different latitudes in the northern hemisphere. From D. H. Cushing (1959). *J. Cons. perm. int. Explor. Mer.* **24**, 455–64.

the thermocline, were different. This is a limit to the amount of plankton which could be present at any one time. Total primary production during the season may be more than this because some mixing of water occurs across the thermocline and because decay of dead plankton makes its nitrogen available again. Mixing and decay are rather slow. This calculation helps us to understand why net annual primary production in the sea is seldom more than about 200 g dry weight m^{-2}.

The thermocline limits the amount of nitrate and other nutrients available to the plankton, but it is not until the thermocline forms in the spring that the plankton population starts rising. While there is no thermocline wind keeps the water circulating, and water from the surface is continually being carried down into the depths where photosynthesis cannot support life, taking plankton with it. The phytoplankton population cannot rise while the phytoplankton is being depleted in this way faster than it can reproduce.

Fig. 2.5 shows how marine plankton production generally fluctuates in the course of a year, at different latitudes. In arctic and temperate seas plankton is sparse in winter and plentiful in spring and summer. Phytoplankton production rises very rapidly in spring when the thermocline forms (which occurs earlier in temperate waters than in the arctic) and the phytoplankton population rises. This enables zooplankton which feed on the phytoplankton to multiply. The production of these herbivores (i.e. the rate at which they accumulate energy by growth) is shown in Fig. 2.5. There is a delay between the rise in phytoplankton production and the rise in herbivore production because the herbivores cannot reproduce until the phytoplankton population has reached a certain minimum density, and because the young herbivores are initially small. The delay is longer in the cold waters of the arctic than in temperate waters. In due course phytoplankton production falls because free nitrate and phosphate are running out, and because the herbivores have eaten a lot of the phytoplankton. Later still herbivore production falls because there is less phytoplankton for the herbivores to feed on. In temperate seas phytoplankton production often rises again for a while in the autumn, when there are fewer herbivores. It declines again as the thermocline breaks down and there is little plankton during the winter. The mixing of water which occurs in the absence of a thermocline brings fresh nitrate and phosphate up from the depths.

In tropical seas the thermocline persists all year and production fluctuates much less than in cooler seas. The concentrations of nitrate and phosphate are more constant and are generally low.

The factors considered in this explanation of seasonal fluctuation of plankton have been incorporated in mathematical models which provide more quantitative explanations.

AMOEBOID MOVEMENT

Amoebas crawl slowly, at speeds up to about 5 $\mu m\ s^{-1}$ (2 $cm\ h^{-1}$). Fig. 2.6(*a*) shows what seems to happen when an amoeba crawls. The outer layer of protoplasm (ectoplasm) appears to be relatively stiff and jelly-like. Granules which can be seen in it under the microscope remain stationary relative to each other and to the ground. The inner core (endoplasm) can be seen from the movement of granules in it to be fluid, flowing forwards. Ectoplasm must be converted to fluid endoplasm at A and endoplasm must be converted to jelly-like ectoplasm at B. A particular particle remains stationary while it is in the ectoplasm but in time it finds itself at the rear end of the animal, becomes endoplasm and flows forward.

Sections of amoebas examined by electron microscopy are found to contain two types of filament which are suspected of playing a part in locomotion. There are relatively short, thick filaments about 16 nm in diameter and long, thin filaments about 7 nm in diameter, scattered in the cytoplasm. These two types of filament are similar in diameter to the thick myosin filaments and the thin

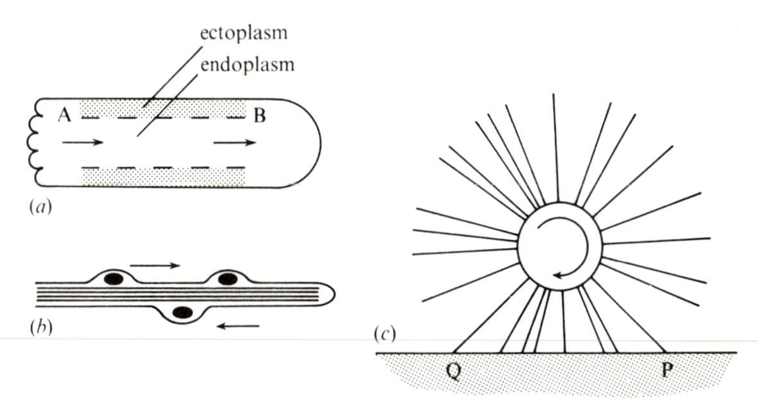

Fig. 2.6. (*a*) A diagrammatic section of an amoeba, crawling from left to right. (*b*) A diagram of a foraminiferan pseudopod, showing the movements of cytoplasm within it. (*c*) A diagram of *Actinophrys* crawling from left to right. Further explanation is given in the text.

actin filaments of vertebrate striated muscle, and there is good evidence that the thin microfilaments are actin. Treatment with heavy meromyosin gives the thin microfilaments a characteristic fringed appearance in electron micrographs, just as it does with actin filaments in muscle. (Heavy meromyosin is the part of the myosin molecule which forms the cross-bridges in muscle).

The cytoplasm of amoebas can be separated from the nuclei and cell membranes by high-speed centrifugation. Cytoplasm obtained in this way and kept at a low temperature (0.4 °C) was fixed, sectioned and examined by electron microscopy. It contained thick microfilaments, but very few thin ones. When the cytoplasm was warmed to 22 °C and ATP was added, it started moving in a most dramatic way. Streams of particles could be seen flowing in various directions through it. Apparently ATP can provide energy for amoeboid movement, as it does for muscle contraction. Cytoplasm fixed while in motion and examined by electron microscopy contained dense networks of thin microfilaments. It is suspected that the ectoplasm of living amoebas may be stiffened by a dense network of thin microfilaments which disappear when it is converted to endoplasm. If so, most of the thin microfilaments also disappear when the amoeba is prepared for electron microscopy, for no difference between ectoplasm and endoplasm is apparent in sections of intact amoebas. The isolated cytoplasm, with no cell membrane, may be less sensitive than the amoeba to the chemicals used for fixation, and this may be why the networks of thin microfilaments survive in sections of it.

The thick and thin microfilaments and the effect of ATP suggest that amoeboid movement may have a good deal in common with muscle contraction. Muscle works by the sliding of filaments along each other. Could amoeboid movement work in the same way? It seems likely but the details of the mechanism are still unknown, despite a great deal of research.

Difflugia with its heavy shell (Fig. 2.2*b*) crawls in a different way from *Amoeba*. A pseudopod is extended and attached to the surface on which it is crawling and then shortens, dragging the shell forward. It can be seen under a polarizing microscope that the pseudopodia are positively birefringent when contracting but not when extending. This suggests that they have microfilaments running lengthwise along them when they are contracting but not when they are extending. It has not so far been possible to obtain electron microscope sections of extending pseudopodia; the pseudopodia respond very rapidly to the fixative and always shorten a little before movement ceases. All sections which have been obtained are thus of contracting pseudopodia. They show both thick and thin microfilaments running lengthwise along the pseudopodia.

The pseudopodia of foraminiferans such as *Allogromia* (Fig. 2.2*c*) are quite stiff in spite of their slenderness. They show positive birefringence which indicates some sort of fibrous or filamentous material running along their length. Granules in the cytoplasm can be seen moving along them, often outwards along one side of the pseudopod and inwards along the other (Fig. 2.6*b*). Bacteria and other small food particles which collide with the pseudopodia and stick to them are carried to the main body of the cell in this way.

The fine structure of foraminiferan pseudopodia is hard to study because fixatives distort them, and even make them break up into separate droplets of cytoplasm. One cannot be confident that structures seen in sections of a distorted or fragmented pseudopod resemble structures present in life. However, it has been found that sea water with a high concentration of added magnesium chloride makes *Allogromia* cease movement almost instantaneously, without distortion. *Allogromia* immobilized in this way can be fixed, embedded and sectioned for electron microscopy, and preservation of gross structures is so perfect that the investigators who devised the technique found they could not distinguish photographs of treated *Allogromia* embedded in the plastic block from photographs of living animals. Sections of undistorted pseudopodia obtained in this way have bundles of microtubules running along them. These are hollow tubes, not solid filaments, of diameter about 20 nm. The movements of the cytoplasm along the pseudopodia probably involves active sliding along the microtubules. It will be shown in the next section of this chapter that flagella seem to work by active sliding of microtubules over each other.

There is also a core of microtubules in the axopodia of *Actinophrys* (Fig. 2.2*d*) and its relatives. These animals crawl in a peculiar way, rolling over and over (Fig. 2.6*c*). This movement has been filmed in side view, and it has been shown to involve shortening of axopodia at P and lengthening of axopodia at Q. Progress is slow: the top speed of *Actinosphaerium* (the genus which has been studied thoroughly) seems to be about 1.5 $\mu m\ s^{-1}$. This is not surprising since the shortening and elongation of the axopodia presumably involves breakdown and re-formation of microtubules.

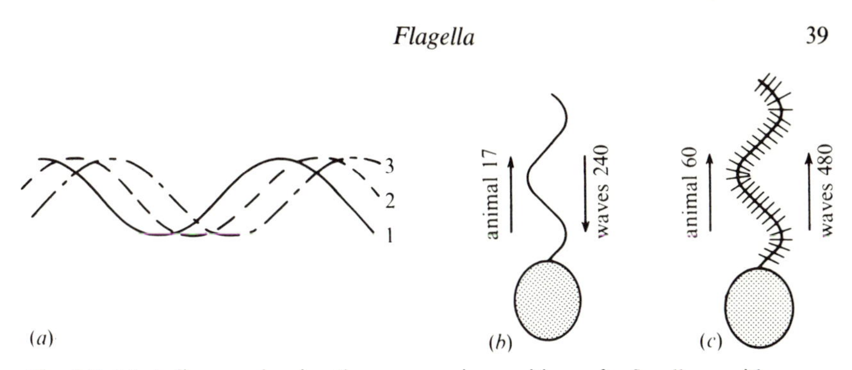

Fig. 2.7. (*a*) A diagram showing three successive positions of a flagellum, with waves travelling along it from left to right. (*b*), (*c*) Diagrams of *Strigomonas* and *Ochromonas*, respectively, swimming. The direction of swimming, and the direction in which waves travel along the flagellum, are shown. Typical speeds of swimming, and of movement of waves along the flagellum, are given in $\mu m\ s^{-1}$. These speeds were obtained from high-speed cinematograph films.

FLAGELLA

Most flagellates swim with their flagella, at speeds which are generally between 20 and 200 $\mu m\ s^{-1}$. A few are sessile, however, attached to plants or stones, and their flagella do not move them but set up currents of water which bring bacteria and other particles of potential food to be caught by other structures. One group of these flagellates (the choanoflagellates) is described in chapter 5, because they resemble one of the types of cells in sponges.

Flagella are about 0.25 μm thick but some are fringed by fine hairs (flimmer filaments) while others are not. They beat at up to at least 70 Hz (cycles per second), far too fast to be observed by the naked eye or recorded by conventional cinematography. High-speed cinematography is needed. Even to record flagella beating at 20 Hz it is desirable to film at at least 400 frames per second. Many flagellates swim with their flagella pulling them from in front, as in Fig. 2.7(*b*) and (*c*), though some swim with their flagella pushing them from behind. Some move their flagella in a single plane like the tail of an eel but some throw their flagella into helical waves. Waves of bending travel along the flagellum (Fig. 2.7 *a*). Figs. 2.7 (*b*) and (*c*) are diagrams of two flagellates which have been studied by filming. Each has only one flagellum, which is held in front and beaten with planar waves. *Strigomonas* has a smooth flagellum and waves travelling down it from tip to base pull it forwards; the animal travels in the opposite direction to the waves. *Ochromonas* has flimmer filaments and waves travelling from base to tip of the flagellum pull it forwards; the animal travels in the same direction as the waves. The flimmer filaments explain the difference, as will be seen.

The movements of flagella are very like the movements made by the tails of swimming fish. It is tempting to conclude that forces of the same nature propel flagellates and fish, but such a conclusion would be false. Water flows around a flagellum in quite a different way from its flow around a fish's tail, because

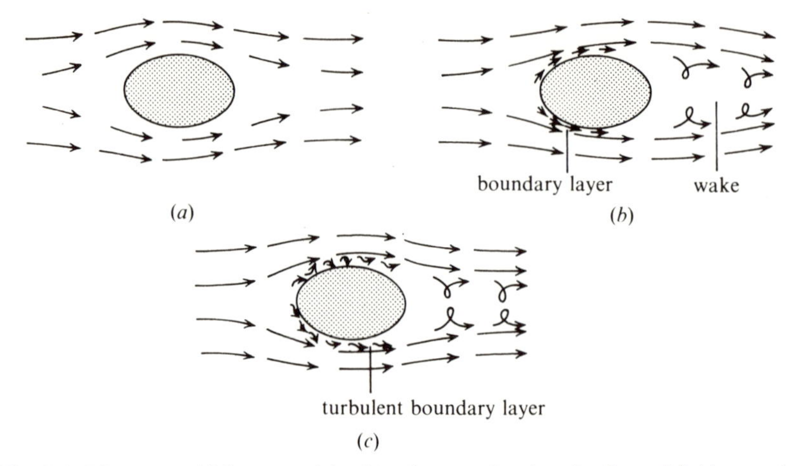

Fig. 2.8. Diagrams which are explained in the text showing the flow of fluid around a stationary object at three different Reynolds numbers. From R. McN. Alexander (1971). *Size and shape.* Arnold, London.

the flagellum is so much smaller and moves so much more slowly. This requires some explanation.

Fig. 2.8 shows three possible patterns of flow of fluid around a stationary object. The pattern of flow of fluid *relative to the object* would be the same in each case if the object were moving and the bulk of the fluid were stationary. In (*a*) the fluid is flowing smoothly around the body, parting in front of it and closing up behind it. The fluid in contact with the stationary body is itself stationary and the velocity increases gradually with distance from the body. Thus there are gradients of fluid velocity, and forces must act to overcome the viscosity of the fluid. The fluid exerts a force on the body in the direction of flow. This force is called drag. In the case shown in Fig. 2.8(*a*) nearly all the drag is due to viscosity.

Fig. 2.8(*b*) shows a different pattern of flow. Fluid in contact with the body is stationary but fluid quite close to it is moving almost as fast as the bulk of the fluid so steep gradients of velocity are confined to a thin boundary layer. The fluid does not close up smoothly behind the body but forms a wake of swirling eddies. Drag acts, partly due to the viscosity of the fluid in the boundary layer and partly due to the changed momentum of the fluid in the swirling wake (a force is needed to change momentum, according to Newton's Second Law of Motion). The component of drag due to viscosity would be less if the boundary layer were thicker because the velocity gradient would be less steep, but more fluid would be slowed down and more force would be needed to change its momentum.

Fig. 2.8(*c*) shows flow as in (*b*), except that the flow in the boundary layer is irregular (turbulent flow), not smooth and parallel to the surface (laminar flow).

The pattern of flow which will occur in a given situation can be predicted by calculating the Reynolds number which is:

$$\frac{(\text{density of fluid}) \times (\text{length of object}) \times (\text{velocity})}{(\text{viscosity of fluid})}$$

The density of water is 1000 kg m^{-3} and its viscosity is $10^{-3} \text{ kg m}^{-1} \text{ s}^{-1}$ so (density/viscosity) is 10^6 s m^{-2} and the Reynolds number in water is

$$10^6 \text{ (length in m)} \times (\text{velocity in m s}^{-1})$$

Note that the Reynolds number is dimensionless; it is not a quantity which has to be expressed in specified units. The length which is usually used in calculating the Reynolds number is the length of the object measured along the direction of movement. The pattern of flow shown in Fig. 2.8(*a*) occurs at Reynolds numbers less than 1. As the Reynolds number increases above 1 the pattern changes gradually to (*b*). The change to (*c*) occurs quite suddenly at a critical Reynolds number which lies somewhere between 2×10^5 and 2×10^6, depending on the shape of the body.

Strigomonas (Fig. 2.7*b*) has a body about 8 μm (8×10^{-6} m) long and swims at speeds around 17 $\mu\text{m s}^{-1}$ ($1.7 \times 10^{-5} \text{ m s}^{-1}$). The Reynolds number of its body is thus $10^6 \times 8 \times 10^{-6} \times 1.7 \times 10^{-5} = 1.4 \times 10^{-4}$. Its flagellum moves from side to side much faster than it moves forward so the length to be used in calculating the Reynolds number of the flagellum is its diameter, 0.25 μm (2.5×10^{-7} m). It moves from side to side at about 170 $\mu\text{m s}^{-1}$ ($1.7 \times 10^{-4} \text{ m s}^{-1}$) so the Reynolds number for the flagellum is $10^6 \times 2.5 \times 10^{-7} \times 1.7 \times 10^{-4} = 4 \times 10^{-5}$. Both these Reynolds numbers are much less than 1: the fluid must flow around both the body and the flagellum in the manner shown in Fig. 2.8(*a*). Larger, faster animals have higher Reynolds numbers. For instance, a water beetle 2 cm long swimming at 0.5 m s^{-1} has a Reynolds number of 10^4, and the water must flow around it as in Fig. 2.8(*b*). This pattern of flow also occurs around fish, except the largest and fastest (such as tunnies) which may reach high enough Reynolds numbers for the boundary layer to become turbulent (Fig. 2.8*c*).

In this chapter we are only concerned with Reynolds numbers less than 1 and flow of the type shown in Fig. 2.8(*a*). In such situations the drag on a body of length l moving at velocity u through a fluid of viscosity η is given by

$$\text{Drag} = \eta\, klu \tag{2.1}$$

where k is a constant which depends on the shape and orientation of the body. It has a value k_A for a cylinder moving lengthwise along its own axis and a different value k_N for the same cylinder moving broadside on, normal to its axis. It is found that $k_N \simeq 2\, k_A$.

Fig. 2.9(*a*) shows a flagellate passing waves along its flagellum from tip to base, that is towards the left of the diagram. Waves like this would propel *Strigomonas* (Fig. 2.6*b*) towards the right. Can we explain why?

We will suppose that the body of the flagellate is initially stationary, and

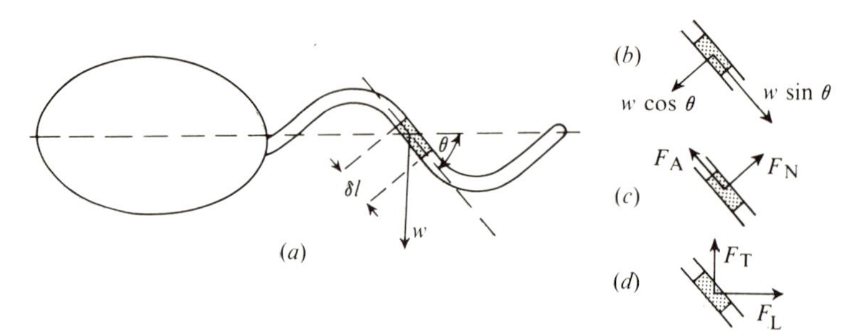

Fig. 2.9. Diagrams of a flagellum, which are explained in the text.

discover the direction of the force which the flagellum exerts on it. Consider the short segment of flagellum $\delta 1$ which at the instant illustrated is inclined at an angle θ to the longitudinal axis of the flagellate and is moving transversely with velocity w. This velocity can be resolved into a component $w \sin \theta$ along the axis of the segment and a component $w \cos \theta$ at right angles to the axis (Fig. 2.9*b*). Hence the force exerted by the water on the segment has components F_A, F_N (Fig. 2.9*c*). where, from equation (2.1)

$$F_A = \eta k_A w \sin \theta \, \delta l$$
$$F_N = \eta k_N w \cos \theta \, \delta l$$

This force can be resolved along different axes, into a transverse component F_T and a longitudinal component F_L (Fig. 2.9*d*). The transverse components cancel out over a cycle of beating but the longitudinal ones do not, and they propel the flagellate. At the instant we are considering the longitudinal component is

$$F_L = F_N \sin \theta - F_A \cos \theta = \eta w \sin \theta \cos \theta \, (k_N - k_A) \, \delta l \qquad (2.2)$$

Since, as we have seen, $k_N \simeq 2k_A$, F_L is positive: it acts towards the right in Fig. 2.9(*a*) and the flagellate is propelled towards the right.

Ochromonas (Fig. 2.7 *c*) has very fine, very numerous flimmer filaments. The total length of the flimmer filaments is about 20 times the length of the flagellum. They stand at right angles to the flagellum in the plane of beating, and because they are so long and numerous the forces on them must predominate over the forces on the main strand of the flagellum. For a naked flagellum, $k_N \simeq 2\,k_A$. For one with very long, numerous flimmer filaments $k_A \simeq 2k_N$ and F_L must (by equation 2.2) be negative. The flagellate will be propelled in the same direction as the waves on the flagellum, as is in fact observed.

Fig. 2.10(*a*) shows the structure of a flagellum, as seen in transverse section by electron microscopy. Note the microtubules which have about the same diameter as the microtubules found in the pseudopodia of foraminiferans, and in axopodia. Here, however, most of the microtubules are double. There is a

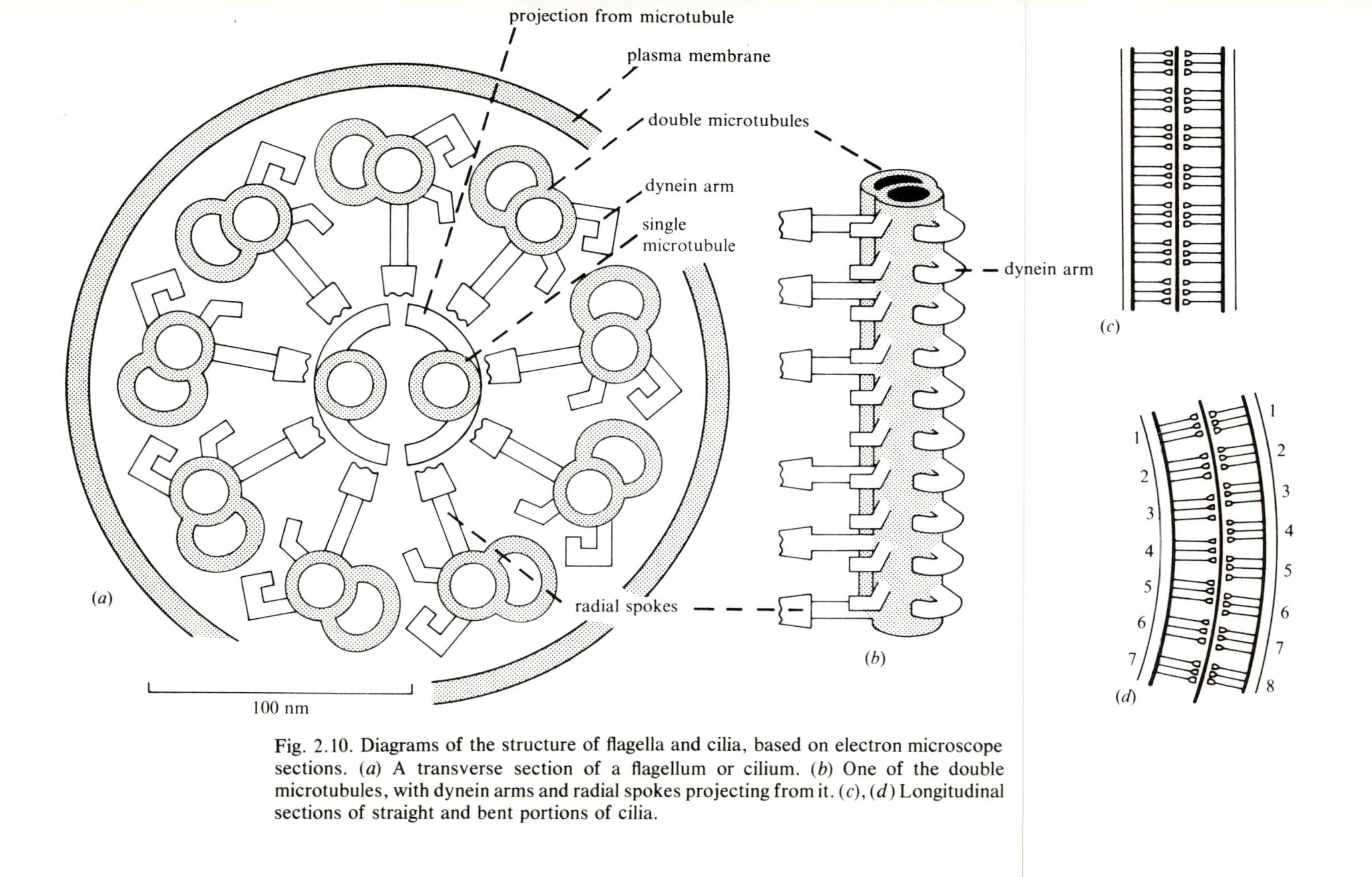

Fig. 2.10. Diagrams of the structure of flagella and cilia, based on electron microscope sections. (*a*) A transverse section of a flagellum or cilium. (*b*) One of the double microtubules, with dynein arms and radial spokes projecting from it. (*c*), (*d*) Longitudinal sections of straight and bent portions of cilia.

ring of nine double microtubules, with two single microtubules in the centre. Pairs of projections, dynein arms, project from one side of each double microtubule towards the next. Radial spokes run from the double microtubules towards the central ones. The whole flagellum is enclosed in an outer membrane. All flagella have these structures but some, including those of dinoflagellates, have an additional rod running parallel with the bundle of microtubules.

There are two ways in which a flagellum could bend. The microtubules on the inside of each bend could shorten and those on the outside could elongate. Alternatively, the microtubules could remain constant in length and slide relative to each other. The latter seems to be what happens. Fig. 2.10(*c*) and (*d*) are based on sections of cilia from freshwater mussels, examined at very high magnification under the electron microscope (cilia have the same structure as flagella and presumably work in the same way). These diagrams show that the radial spokes are arranged in regularly repeating groups of three, and that the spacing is not altered on either side of a bend. Bending involves sliding of microtubules and radial spokes on the outside of a bend, relative to those on the inside of the bend. Note the relative positions of the groups of spokes labelled 7 in Fig. 2.10(*d*).

How fast does sliding have to happen? Each double microtubule is about 70 nm from the next one so the radii of bending of adjacent double microtubules may differ by up to 70 nm. Bending through an angle θ radians must slide some of the double microtubules 70θ nm relative to their neighbours. Typical flagella bend through about 2 radians (114°), so a cycle of beating involves about 140 nm sliding, first in one direction and then in the other, making a total of 280 nm. A typical flagellum might beat at about 30 Hz (some beat faster, and some more slowly), which would require sliding rates of 30×280 nm $s^{-1} = 8$ $\mu m\ s^{-1}$. This is similar to the rates of sliding which occur between thick and thin filaments in moderately fast muscles.

The dynein arms are believed to make the microtubules slide, in the same way as the cross-bridges cause sliding in muscle. As in muscle, the energy is supplied by ATP. This has been demonstrated by experiments with flagella treated with glycerol, which disrupts the outer membrane. Any ATP that is in the flagella is thus allowed to escape, but when the treated flagella are put in suitable solutions containing ATP, this ATP can reach the microtubules, and the flagella beat. The experiment has been done, for instance, with flagella of *Polytoma* treated with glycerol and then broken off the body of the parent cell by centrifugation. In suitable solutions containing ATP these flagella beat more or less normally, with waves of bending starting at the base of the flagellum as in the intact flagellate. They swam along, base leading. Magnesium ions are needed in the medium, as well as ATP, and abnormal or irregular beating occur at ATP concentrations below about 10^{-5} mol l^{-1}.

OSMOTIC REGULATION

A protozoan must contain organic molecules which are not present in its environment. Its cell membrane must prevent these molecules from escaping, but it must also allow small molecules such as carbon dioxide and oxygen to pass through, to satisfy the needs of the cell. It seems inevitable that the cell membrane will be semipermeable, permeable to water and other small molecules but not to large organic molecules. The organism will thus tend to take up water from the environment and swell, due to the osmotic pressure of the organic molecules, unless there is an adaptation to prevent this.

The situation is rather more complicated than this because many of the organic molecules are ionized and because there are inorganic ions both in the cell and in the environment. Amino acids have molecules of the form $NH_2CHRCOOH$, where the part R of the molecule is different in different amino acids. At low (acid) pH they form positively charged ions (cations) while at high (alkaline) pH they form negatively charged ions (anions). Amino acids (and proteins and peptides) are mostly negatively charged at the values of pH (around 7) which are usual in cells. There must be an interaction between these anions which are trapped in the cell and any inorganic ions which pass freely through the cell membrane.

An uncharged molecule which passed freely through the membrane would reach equilibrium when its activity was the same on both sides of the membrane. (Activity is effective concentration. It may be less than the concentration determined by chemical analysis because of interaction between molecules.) In the case of an ion, however, the activities may not be the same at equilibrium; a difference in activity may be maintained by an electrical potential difference. Consider a cation X which has a activity X_o outside a cell and X_i inside it. If the cell membrane is freely permeable to X the work required to transfer a small quantity x ions of X against the activity gradient into the cell is $RTx \ln(X_i/X_o)$, where R is the universal gas constant and T the absolute temperature. If the electrical potential is E_o outside the cell and E_i inside, the electrical work required to transfer x ions of X into the cell is nFx $(E_i - E_o)$, where n is the valency of the ion and F is Faraday's constant. At equilibrium, no work is required to transfer small quantities of X across the membrane so

$$RT \ln(X_i/X_o) + nF(E_i - E_o) = 0$$

$$E_i - E_o = (RT/nF) \ln(X_o/X_i) \tag{2.3}$$

This is the Nernst equation. By inserting values for R, T and F and converting to logarithms to base 10, we find that in the range of temperatures at which animals live the membrane potential $(E_i - E_o)$ is about $(58/n) \log_{10} (X_o/X_i)$ mV for a cation. By a similar argument, it is $(58/n) \log_{10} (X_i/X_o)$ mV for an anion. A membrane potential of 58 mV can maintain a ten to one ratio of activity of a univalent ion, or a hundred to one ratio for a divalent ion.

0.5 Equiv l^{-1} Na^+	$[Na^+]_i$
0.5 Equiv l^{-1} Cl^-	$[Cl^-]_i$
	0.1 Equiv l^{-1} A^-

Fig. 2.11. A diagram of a cell in sea water, which is explained in the text.

Fig. 2.11 is a simplified representation of a cell in sea water. Sodium and chloride are much the most common ions in sea water and each has a concentration of about 0.5 Equiv l^{-1}. We will ignore the other inorganic ions. The organic particles trapped in the cell will vary in nature and in charge but we will suppose they are all univalent anions A^-, and that their concentration is 0.1 Equiv l^{-1}. This concentration has been chosen arbitrarily but lies within the range of concentrations of organic solutes which occur in living cells. What will the concentrations of sodium, $[Na^+]$, and chloride, $[Cl^-]_i$, in the cell be, when equilibrium has been reached? Assume that the activities of the ions are about equal to their concentrations. (This assumption is probably fairly accurate but could be badly wrong if ions were bound to organic compounds in the cell.) The membrane potential is the same for both sodium and chloride so

$$E_i - E_o = 58 \log_{10} (0.5/[Na^+]_i)$$
$$= 58 \log_{10} ([Cl^-]_i/0.5)$$

Also, since the concentrations of positive and negative charges in the cell must be (very nearly) equal

$$[Na^+]_i = [Cl^-]_i + 0.1$$

By solving these simultaneous equations we find $[Na^+]_i = 0.55$ Equiv l^{-1}, $[Cl^-]_i = 0.45$ Equiv l^{-1} and $(E_i - E_o) = -2.4$ mV. A very small negative membrane potential would develop, which would keep the sodium ions a little more concentrated inside the cell than outside and the chloride ions a little less concentrated. An equilibrium of this sort, involving some ions which pass freely through the membrane and others which are trapped on one side of it, is called a Donnan equilibrium.

This particular equilibrium would involve a total concentration of 1.1 Equiv l^{-1} ions inside the cell and 1.0 Equiv l^{-1} outside. There would be a difference in osmotic concentration 0.1 Osmol l^{-1} and the cell would tend to take up water. The osmotic pressure Π of a solution containing c moles or gram ions per unit volume is given by the equation

$$\Pi = RTc \tag{2.4}$$

If c is expressed in Osmol l^{-1}, Π works out as $2.4c$ MN m^{-2} ($24c$ atm) at temperatures around 15 °C. Thus a difference of 0.1 Osmol l^{-1} in osmotic concentration between cell and environment involves a pressure difference of 2.4 atm at equilibrium. If the cell were to be prevented from taking up water, its cell membrane would have to be tightly inflated, exerting substantial tension.

TABLE 2.1. *Concentrations of certain ions and of amino acids in a marine and a freshwater protozoan, and in the culture media in which they were kept*

	Concentrations (mEquiv kg^{-1} or mmol kg^{-1})			
	Na^+	K^+	Cl^-	Amino acids
Miamiensis avidus	88	74	61	317
Seawater medium	372	12	443	21
Amoeba proteus	1	25	10	—
Freshwater medium	0	0.08	0.08	—

Data from E. S. Kaneshiro, P. B. Dunham & G. G. Holz (1969). *Biol. Bull.* **136**, 63–75; E. S. Kaneshiro, G. G. Holz & P. B. Dunham (1969). *Biol. Bull.* **137**, 161–9; and R. D. Prusch & P. B. Dunham (1972). *J. exp. Biol.* **56**, 551–63.

This necessity could be avoided by making the cell membrane relatively impermeable to one of the inorganic ions and pumping that ion out of the cell to keep its concentration below the equilibrium concentration. Energy would be needed to pump the ion out, but the more impermeable the cell membrane was made to the ion, the less would leak in and the less energy would be needed. It is probably not too difficult to make a membrane less permeable to sodium ions than to potassium and chloride ions, as hydrated sodium ions have the largest radius of the three.

All that was theory. What are the concentrations of ions in real marine Protozoa? I have not been able to find data for marine Sarcomastigophora, but there is quite detailed information about *Miamiensis*, a marine ciliate. It was grown in culture, in sea water with added nutrients. Samples of the culture were centrifuged to obtain pellets of ciliates for analysis. The ciliates did not pack tightly enough to squeeze all the culture medium out from between them, but the amount of medium left in the pellets was measured and a correction was applied. Radioactive inulin (a polysaccharide) was added to some samples before centrifugation, and the radioactivity of the pellets was measured. Since the cell membrane is impermeable to inulin no inulin entered the cells, and the radioactivity of the pellet indicated the fraction of culture medium in it. The pellets were analysed by flame photometry (for cations), by titration (for chlorides) and in an automatic amino acid analyser (for amino acids). Some of the results are shown in Table 2.1. This table does not show a complete analysis, and there was presumably a substantial concentration of other organic molecules inside the ciliates. Notice that the concentration of potassium ions was about six times as high in the ciliates as in the medium, and the concentration of chloride ions was about seven times as high in the medium as in the ciliates. As $58 \log_{10} (1/7)$ is -49, then if the membrane

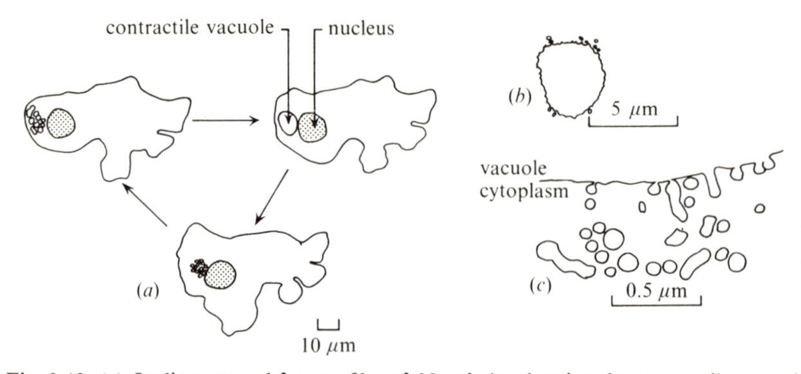

Fig. 2.12. (*a*) Outlines traced from a film of *Naegleria*, showing the contractile vacuole at three stages in its cycle. After K. G. Grell (1973). *Protozoology*. Springer, Berlin. (*b*), (*c*) A section through the contractile vacuole of *Acanthamoeba*, and part of the edge of the vacuole at a higher magnification. After B. Bowers & E. D. Korn (1968). *J. cell Biol.* **39**, 95–111.

potential was −49 mV the potassium and chloride ions were both more or less in equilibrium. If so, the sodium ions were plainly not in equilibrium and must have been being pumped out of the ciliates. The membrane potential would have to be measured to confirm this interpretation, and it is certainly not the whole story. Calcium and magnesium ions would have to be excluded as well as sodium, for otherwise they would have reached very high concentrations in the cells, which they did not do.

Table 2.1 also gives data for *Amoeba proteus*, which lives in fresh water. The concentration of potassium ions in the cell is much higher than in the medium, but so is the concentration of chloride ions. The membrane potential was measured by inserting a microelectrode into the amoeba and found to be −90 mV, enough to keep potassium 35 times as concentrated in the amoeba as outside. The concentration in the amoeba was much higher than this so potassium as well as chloride must have been being pumped into the cell.

Marine Protozoa may be able to prevent their osmotic concentration rising above that of the surrounding water, but freshwater Protozoa cannot do so. The osmotic concentration inside them is inevitably greater than the very low osmotic concentration of the water. Water will diffuse in, and must be pumped out if the animal is not to swell. It is pumped out by one or more contractile vacuoles.

Fig. 2.12(*a*) shows the action of the contractile vacuole of *Naegleria*. The vacuole attains a maximum diameter and then discharges its contents through the cell membrane, leaving only a cluster of much smaller vacuoles. These enlarge and coalesce and the cycle is repeated. Electron microscope sections of another amoeba (*Acanthamoeba*, Fig. 2.12*b*, *c*) show fine convoluted tubules around the vacuole and it is believed that the water which is excreted first enters the tubules and then drains from them into the vacuole. Microtubules

TABLE 2.2. *The composition of the cytoplasm, and of fluid from the contractile vacuole, of* Pelomyxa *in a medium of osmotic concentration less than 2 mOsmol l^{-1}*

	Cytoplasm	Vacuole fluid
Osmotic concentration (mOsmol l^{-1})	117	51
Sodium concentration (mEquiv l^{-1})	6	20
Potassium concentration (mEquiv l^{-1})	30	5

Data from D. H. Riddick (1968). *Am. J. Physiol.* **215**, 736–40.

have been found around the contractile vacuole of another amoeba (though not in *Acanthamoeba*) and may cause the contraction which drives the water out.

The action of the contractile vacuole of *Acanthamoeba* has been watched under the microscope. Measurements with a micrometer eyepiece showed that it reached an average diameter of 6 μm before discharging. It discharged on average once every 50 s in fresh water, and it is easy to calculate (since the vacuole was roughly spherical) that 8000 μm^3 water were being pumped out every hour. Most of this must have entered the amoeba by diffusion. Food vacuoles totalling 500 μm^3 were formed every hour. Water must also have been produced as a by-product of metabolism (which converts foodstuffs and oxygen to carbon dioxide and water) but the rate at which it was produced can be calculated, from the oxygen consumption, to be very small. Hence about 7500 μm^3 of water must have diffused into the amoeba every hour. The volume of the amoebas was determined by sucking them in and out of a pipette until they drew in their pseudopodia, measuring their diameter and calculating the volume. It was found to be 3000 μm^3, so the water diffusing into the amoebas every hour was 2½ times the volume of the body.

This may seem remarkable, but it must be remembered that *Acanthamoeba* is very small and so has a large ratio of surface area to volume. A real *Acanthamoeba* is about 50 μm long, but imagine one the size of a small fish, 5 cm long. It would be 10^3 times as long so it would have 10^6 times the surface area and so would presumably take up water 10^6 times as fast. However, it would be 10^9 times as heavy so the rate of uptake of water expressed in body volumes per hour would be only 10^{-3} times the value for the small amoeba. The fish-sized amoeba would take up 2.5×10^{-3} body volumes per hour or 2.5 g water (kg body weight)$^{-1}$ h^{-1}. Small freshwater fish produce urine as fast as this, or faster.

Samples of fluid from the contractile vacuoles of the giant amoeba, *Pelomyxa*, have been analysed. Micropipettes, 2–5 μm in diameter, were thrust into contractile vacuoles, and the fluid was sucked out. The freezing points of the tiny samples were determined, and used to calculate the osmotic concentration. The sodium and potassium concentrations were determined by flame photometry. Samples of cytoplasm were analysed in the same way. The results are shown in Table 2.2. The fluid in the vacuole has a lower osmotic concentration than the cytoplasm, and contains less potassium but more sodium.

Nearly all freshwater Protozoa have contractile vacuoles. Some marine forms and many which live as internal parasites do not; a contractile vacuole is unnecessary for a protozoan living in a medium of reasonably high osmotic concentration. However, some marine and parasitic Protozoa have contractile vacuoles, which generally beat rather slowly. Presumably their cytoplasm has a higher osmotic concentration than the medium they live in.

REPRODUCTION

Amoebas reproduce asexually, by binary fission. Duplicate copies of the chromosomes are made and the nucleus divides by the process of mitosis, producing two nuclei of the same genetic constitution as the original nucleus. The cytoplasm divides into two more or less equal halves, each containing one of the nuclei. This process produces two amoebas, each half the size of the parent. Each of them grows, in favourable conditions, doubling its size before dividing again. Sexual reproduction may also occur.

Most Protozoa reproduce by binary fission but also reproduce from time to time by sexual processes. Sexual reproduction involves two gamete nuclei each with a single (haploid) set of chromosomes uniting to form a zygote with a double (diploid) set. At some stage in the life cycle a diploid nucleus must divide in such a way as to produce haploid nuclei again. This is achieved by the process of meiosis in which (typically) a diploid nucleus divides twice while its chromosomes are duplicated only once. The four nuclei which are produced are thus haploid. Higher animals are diploid and meiosis occurs in the gonads, producing haploid ova and spermatozoa. Many flagellates are haploid while reproducing asexually, and the zygote divides by meiosis. This is the case with *Chlamydomonas*.

Chlamydomonas reinhardi can be cultured either in water or on agar jelly in petri dishes. They reproduce asexually so long as there is an adequate supply of nitrogen (as nitrate or ammonium ions) for protein synthesis. They become ready for sexual reproduction when transferred to a medium in which the concentrations of these ions are low.

A culture of genetically identical *Chlamydomonas* (a clone) can be produced by allowing a single individual to reproduce asexually. Members of the same clone will not join in sexual reproduction, and they will only join with members of 50% of other clones. It appears that there are two sexes and that all members of a clone have the same sex (as they should, being genetically identical). In some species of *Chlamydomonas* the sexes are similar in size but in others they are very different: in such cases the small sex may be regarded as male (by analogy with spermatozoa) and the large one as female (by analogy with ova).

The genetics of *Chlamydomonas* has been studied in detail. Members of two clones of opposite sex are mixed. Zygotes are taken from the mixture and set individually on agar jelly. If they are kept in suitable conditions meiosis occurs and the diploid zygote produces four haploid individuals in the course of a week. These four are genetically different. They can be separated by a fine glass loop under a dissecting microscope (a practiced experimenter can do this in less than a minute) and clones can be grown from them. Many mutations of *Chlamydomonas* are known, involving peculiarities of colour, requirements for nutrients, drug resistance, paralysis of flagella, etc. Clones showing these mutations have been bred together and it has been possible to demonstrate crossing over and to work out the relative positions of the mutant genes on the chromosomes.

Sexual reproduction may confer a strong selective advantage in a variable environment. Imagine a mixed population of sexually and asexually reproducing members of the same species, living in an environment in which the temperature changes at intervals of a few generations. Selection will occur in a cold period so that only cold-tolerant members of the species survive. When the environment becomes warmer the asexual reproducers will still produce only cold-tolerant offspring (except in the rare event of an appropriate mutation), and these offspring will probably be ill-adapted to the warm conditions. However, recombination of genes will probably produce some heat-tolerant individuals among the offspring of the sexual reproducers. In a varying environment the genes of sexual reproducers are more likely to be transmitted to subsequent generations than the genes of asexual reproducers.

Genes on separate chromosomes recombine at random. Genes on the same chromosome recombine less freely, by crossing over. This feature of sexual reproduction is illustrated by experiments which were undertaken to find out whether sexual reproduction occurs in dinoflagellates (it had not been observed). *Cryptothecodinium cohnii* is a dinoflagellate which is normally yellow, but several colourless clones were obtained and it was suspected that their lack of colour might be due to different genes. Colourless clones were mixed in a medium of low nitrate and phosphate content. Fifteen days later samples were taken from the mixture. It was found that some mixtures of two colourless clones produced up to 14% yellow offspring. Sexual reproduction with genetic recombination was plainly occurring between the clones.

Inheritance does not depend solely on the nucleus. Plastids and mitochondria also have DNA, and are believed to be involved in inheritance. Indeed, it is widely believed that plastids and mitochondria evolved from separate prokaryote (bacterium-like) organisms which lived within the cells of the ancestors of the eukaryotes. Particularly clear evidence of non-nuclear inheritance has come from experiments with *Amoeba*. It is possible to remove the nucleus from an *Amoeba* and replace it with a nucleus taken from another *Amoeba*, by an operation with appropriately shaped needles held in micromanipulators. For instance, *Amoeba proteus* has a large nucleus and *A. discoides* has a small one. When a nucleus is transferred from one species to the other, it changes its size: clones of amoebas with *proteus* cytoplasm and *discoides* nuclei have large nuclei, while the reverse combination have small nuclei. Nuclear size apparently depends on the cytoplasm, not on the nucleus. The phenomenon is not fully understood. Quite small injections of cytoplasm of either species into intact amoebas of the other alters the size of the nucleus, in clones grown from the injected specimens.

FURTHER READING

GENERAL

Grell, K. G. (1973). *Protozoology*. Springer, Berlin.
Honigberg, B. M. *et al.* (1964). A revised classification of the phylum Protozoa. *J. Protozool.* **11**, 7–20.
Jeon, K. W. (ed.) (1973). *The biology of Amoeba*. Academic Press, New York & London.
Sleigh, M. A. (1973). *The biology of Protozoa*. Edward Arnold, London.
Westphal, A. (1977). *Protozoa*. Blackie, Glasgow.

PHOTOSYNTHESIS AND FEEDING

Cushing, D. H. (1959). The seasonal variation in oceanic production as a problem in population dynamics. *J. Cons. perm. int. Explor. Mer* **24**, 455–64.
Fogg, G. E. (1965). *Algal cultures and phytoplankton ecology*. Athlone Press, London.
Jeon, K. W. & Jeon, M. S. (1976). Scanning electron microscope observations of *Amoeba proteus* during phagocytosis. *J. Protozool.* **23**, 83–6.
Westlake, D. F. (1963). Comparisons of plant productivity. *Biol. Rev.* **38**, 385–425.

AMOEBOID MOVEMENT

Eckert, B. S. & McGee-Russell, S. M. (1973). The patterned organization of thick and thin microfilaments in the contracting pseudopod of *Difflugia*. *J. Cell Sci.* **13**, 727–39.
Holberton, D. V. (1977). Locomotion of Protozoa and single cells. In *Mechanics and energetics of animal locomotion*, ed. R. McN. Alexander & G. Goldspink, pp. 279–332. Chapman & Hall, London.
McGee-Russell, S. M. & Allen, R. D. (1971). Reversible stabilization of labile microtubules in the reticulopodial network of *Allogromia*. *Adv. Cell mol. Biol.* **1**, 153–84.
Pollard, T. D. & Ito, S. (1970). Cytoplasmic filaments of *Amoeba proteus*. I. The role of filaments in consistency changes and movement. *J. Cell Biol.* **46**, 267–89.
Watters, C. (1968). Studies on the motility of the Heliozoa. I. The locomotion of *Actinosphaerium eichhorni* and *Actinophrys* sp. *J. Cell Sci.* **3**, 231–44.

FLAGELLA

Sleigh, M. A. (1973). *Cilia and flagella*. Academic Press, New York & London.
Warner, F. D. & Satir, P. (1974). The structural basis of ciliary bend formation. Radial spoke positional changes accompanying microtubule sliding. *J. Cell Biol.* **63**, 35–63.

OSMOTIC REGULATION

Pal, R. A. (1972). The osmoregulatory system of the amoeba, *Acanthamoeba castellanii*. *J. exp. Biol.* **57**, 55–76.
Prusch, R. D. & Dunham. P. B. (1972). Ionic distribution in *Amoeba proteus*. *J. exp. Biol.* **56**, 551–63.
Riddick, D. H. (1968). Contractile vacuole in the amoeba, *Pelomyxa carolinensis*. *Am. J. Physiol.* **215**, 736–40.

REPRODUCTION

Preer, J. R. (1969). Genetics of the Protozoa. In *Research in protozoology*, vol. 3, ed. T.-T. Chen, pp. 129–278. Pergamon, Oxford.
Treisman, M. (1976). The evolution of sexual reproduction: a model which assumes individual selection. *J. theoret. Biol.* **60**, 421–31.

3
Ciliates

Phylum Protozoa, Subphylum Ciliophora, Class Ciliatea
 Subclass Holotrichia
 Subclass Peritrichia
 Subclass Suctoria
 Subclass Spirotrichia

The ciliates are the most complex of the Protozoa. Some examples are shown in Figs. 3.1. and 3.2. Typical ciliates have their bodies covered by rows of cilia, which have the same internal structure as flagella but are generally short and numerous (Fig. 3.1 *a*) and beat asymmetrically. One section of this chapter is about cilia. Ciliates have two types of nucleus, large macronuclei and small micronuclei. Often, both types are more or less spherical (Fig. 3.1 *b*), but the macronucleus is sometimes long and slender (Fig. 3.1 *c*, *d*). The functions of the two types of nucleus are described in a later section of the chapter. A process of sexual reproduction called conjugation, also discussed later, is peculiar to the ciliates. Most ciliates have contractile vacuoles.

Colpoda and *Chilodonella* (Fig. 3.1 *a*, *b*) are relatively simple, primitive ciliates. *Colpoda* is common in soil and *Chilodonella* in various habitats including sewage works. *Colpoda* has a uniform covering of cilia with no grouping of cilia together in tight clumps. A depression of the surface of the body leads to the cytostome, the point at which particles of food are ingested. *Chilodonella* is similar but has very few cilia on the dorsal surface. It has a prominent cytopharyngeal basket: this is an arrangement of rods (made of bundles of microtubules) at the cytostome. The manner in which it is used in feeding will be described.

Many of the more advanced ciliates have some of their cilia grouped together to form compound structures. *Stentor* (Fig. 3.1 *c*) is found attached to weeds in ponds and slow streams. It is trumpet shaped, with its cytostome in the bell of the trumpet. There are ordinary cilia on the general body surface and in addition groups of about 70 long (25 μm) cilia form membranelles around the rim of the trumpet. *Stentor* is very large and individual cilia are not shown in the figure: each line in the fringe around the rim represents a membranelle. Each membranelle beats as a unit but it is not known what prevents the cilia from beating separately or fraying apart. Electron micrographs of sections of membranelles show that the cilia are packed closely, almost touching each

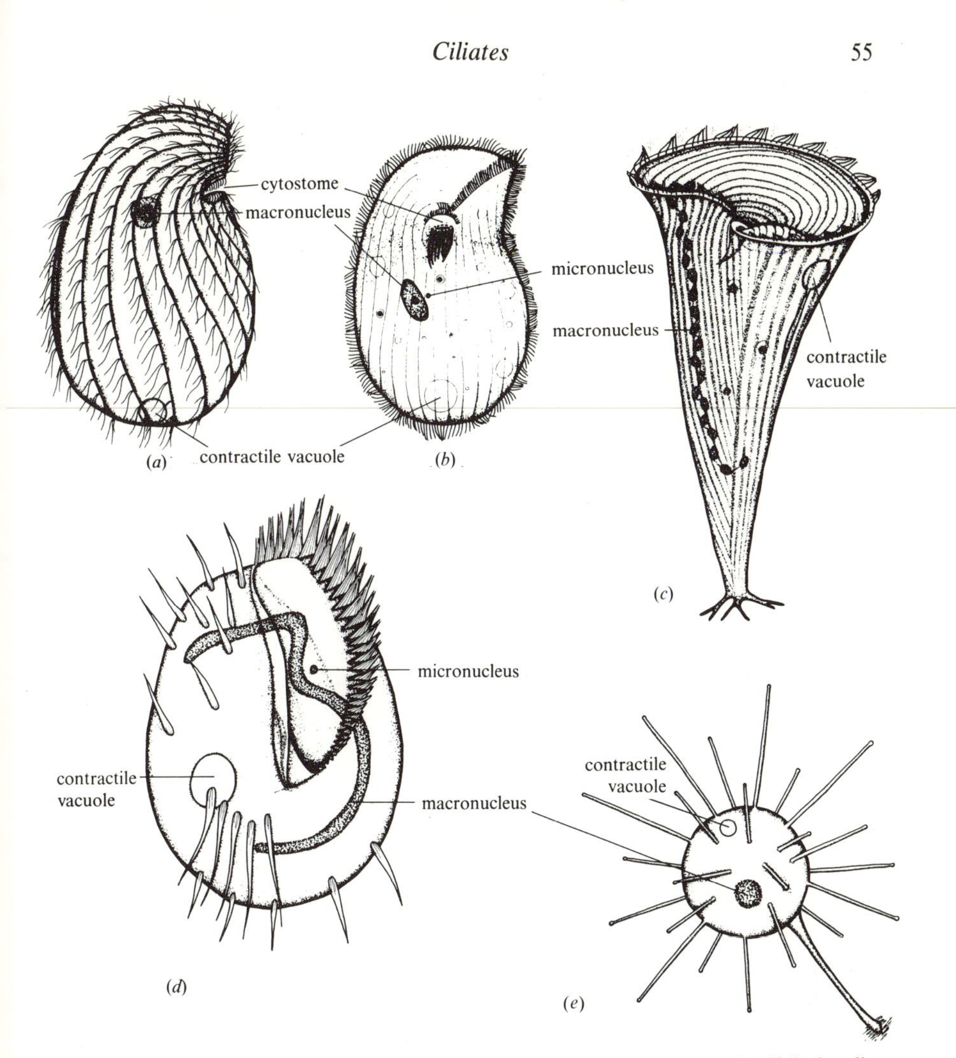

Fig. 3.1. A selection of ciliates. (*a*) *Colpoda cucullus* (length 75 μm); (*b*) *Chilodonella cucullulus* (140 μm); (*c*) *Stentor coeruleus* (1.5 mm); (*d*) *Euplotes patella* (100 μm); (*e*) *Podophrya collini* (50 μm). From M. A. Sleigh (1973). *The biology of Protozoa*. Arnold, London.

other, but do not show anything holding them together. The role of the membranelles in feeding will be described. *Stentor* and some other ciliates can change their shape rapidly: the mechanism of this will be discussed.

Euplotes (Fig. 3.1*d*), which is common in various marine and freshwater habitats, has no ordinary cilia on its ventral surface. All of its ventral cilia are grouped together in membranelles or similar structures. Each membranelle consists of two or three closely packed rows of cilia. Much of the surface of the body is bare. *Podophrya* (Fig. 3.1*e*) and other members of the subclass Suctoria have no cilia at all when adult. They generally live attached by a stalk

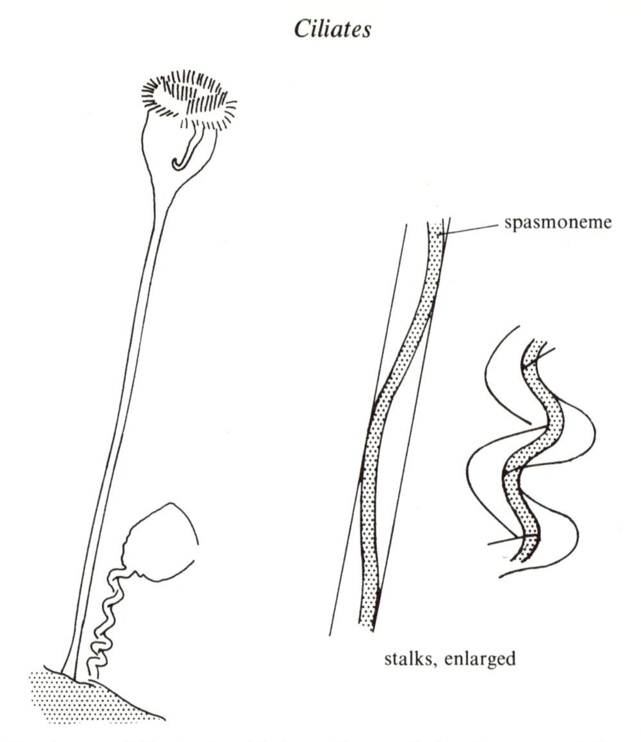

Fig. 3.2. Specimens of *Vorticella* with the stalk extended and contracted. Extended stalk about 400 μm long.

to some surface, and feed by sucking cytoplasm out of other cells through their long slender arms. They look very different from other ciliates but when they reproduce asexually they bud off ciliated swimming larvae. They have macro- and micronuclei and they practise conjugation, so it is clear that they should be classified as ciliates.

Vorticella (Fig. 3.2) is another stalked ciliate. It is common in sewage plants, and in many other habitats. The ecology of one type of sewage plant is discussed later in this chapter. *Vorticella* can shorten its stalk very rapidly by coiling it into a tight helix. Rows of cilia spiral outwards from the cytosome, which is set at the bottom of a deep hollow.

All members of the subphylum Ciliophora are included in a single class, which is divided into four subclasses. The subclass Suctoria, which includes *Podophrya* and its relatives, is clearly distinct from the others. The boundaries between the other subclasses are difficult to define precisely and will not be defined here. The more primitive ciliates are put in the subclass Holotrichia. They include *Colpoda* and *Chilodonella*, which have no membranelles, and also more advanced ciliates such as *Paramecium* and *Tetrahymena* which have some of the cilia around the cytosome grouped into membranelles and similar structures. The Peritrichia includes *Vorticella* and its relatives, which have rows of cilia spiralling outwards from the cytosome. The spiral is clockwise as seen from above, and generally makes several complete turns. The Spiro-

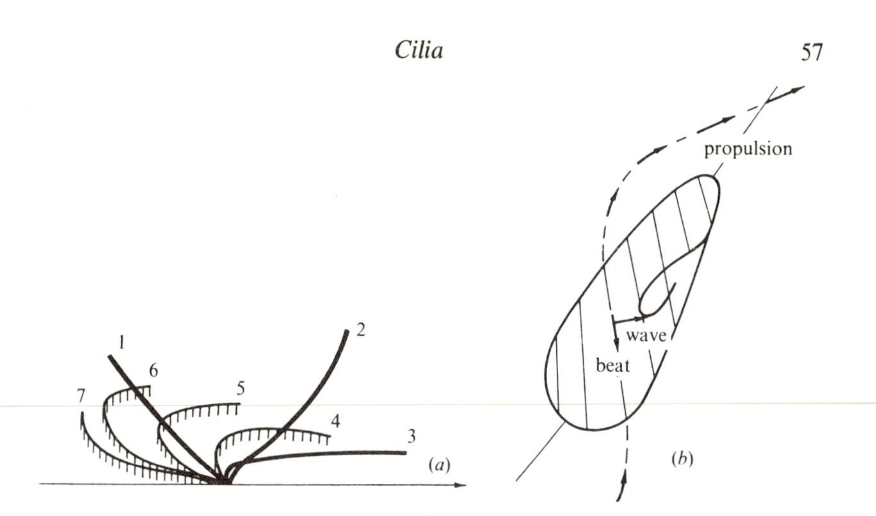

Fig. 3.3. Diagrams showing how the cilia of *Paramecium* beat. (*a*) Successive positions of a cilium. (*b*) The directions of the effective stroke of the ciliary beat, of metachronal waves and of propulsion. From J. R. Blake & M. A. Sleigh (1974). *Biol. Rev.* **49**, 85–125.

trichia (which includes *Stentor* and *Euplotes*) have an anticlockwise spiral of membranelles around the cytosome, making one turn or less.

CILIA

Flagella generally pass symmetrical planar waves, or helical waves, along their length. Cilia beat asymmetrically, as Fig. 3.3 (*a*) shows. The cilium is held fairly straight as it beats to the right (positions 1 to 3). It is bent as it returns more slowly to the left (positions 4 to 7). The effective stroke (1 to 3) drives water to the right and exerts relatively large forces because the movement is fast and because the cilium is moving at right angles to its long axis ($k_N \simeq 2k_A$, see p. 41). The recovery stroke (4 to 7) exerts forces to the left but they are relatively small because the cilium is moving more slowly and because much of it is moving more or less parallel to its length. The net effect of the cycle is to drive water to the right, parallel to the surface of the body. It will tend to drive the animal to the left.

In the effective stroke most of the cilium is held straight and only the base bends. In the recovery stroke a bend formed at the base travels out to the tip of the cilium, rather like a bending wave travelling along a flagellum. Many cilia do not bend in a single plane. In Fig. 3.3 (*a*) the effective stroke is in the plane of the paper but in the recovery stroke the cilium bends out of this plane, away from the reader.

Fig. 3.3 (*b*) shows the direction in which the cilia of *Paramecium* beat. This ciliate has a long groove (leading to the cytosome) along one side of the body. The cilia in the groove must be less effective than the others in propulsion. The grooved side thus travels more slowly than the other and the animal swims along a helix, with the groove always facing the axis of the helix.

Ciliates swim faster than flagellates. Speeds of 0.4 to 2 mm s^{-1} are usual, while flagellates can only achieve 20–200 μm s^{-1}. The reason seems clear. Compare *Euglena* (Fig. 2.1 *b*) with *Tetrahymena*, a ciliate of about the same size. *Euglena* has a single flagellum about 100 μm long. *Tetrahymena* has over 500 cilia. Each of them is only 5 μm long but their total length is several millimetres. Since the flagellum of *Euglena* and the cilia of *Tetrahymena* have the same diameter and internal structure, the ciliate has many times as much propulsive machinery as the flagellate.

The individual cilia do not beat at random, but in a highly organized way. In *Paramecium*, for instance, cilia along a line parallel to the direction of beating beat in phase with each other. To either side of this line, the cilia are more and more out of phase with the cilia on the original line (Fig. 3.4 *a*). The gradual change of phase in a direction at right angles to the beat makes the ciliated surface look as though waves were passing over it in this direction (Fig. 3.3 *b*). The pattern of beating is often described by speaking of metachronal waves travelling over the surface. There are usually about ten complete waves on a *Paramecium*, at a given instant.

The pattern is difficult to elucidate in detail, because the waves look different from different angles. One investigator found it useful to make models from wires stuck into a plastic base (Fig. 3.4). These were made to match electronic flash photographs of living animals, and could be viewed from different angles to check the interpretation of the pattern on parts of the animal which were seen obliquely. Fig. 3.4(*a*) shows the pattern normally observed all over the body of *Paramecium*. Note how the cilia bend to the side in their recovery stroke. Figs. 3.4(*b*), (*c*) and (*d*) show what would happen if the waves travelled in other directions. In each case there are places (indicated by arrows) where the cilia get in each other's way. The pattern which occurs seems to be the one which involves least interference between cilia, when the cilia bend to their left side in the recovery stroke. In some other animals cilia beat in a single plane, or bend to the right in the recovery stroke, and the metachronal waves move in other directions. The mechanism which controls metachrony has not been adequately explained.

Fig. 2.8(*a*) shows how fluid flows around an object at Reynolds numbers less than 1. The fluid in contact with the body is stationary relative to the body, and the velocity of the fluid changes gradually over long distances around the body. This is how fluid must flow around the bodies of flagellates when they are propelled by their flagella. Reynolds numbers for the bodies of swimming ciliates are also very low and water would flow around them similarly if they were inert bodies pulled along by external forces. The manner in which the velocity of the water would change with distance from the body is indicated by the line marked 'inert body' in Fig. 3.5. However, ciliates are not inert bodies moved by external forces but are propelled by their cilia. The maximum fluid velocity *relative to the body* must occur at the tips of the cilia and it can be shown mathematically that the velocity must change with distance from the body as indicated by the line marked 'self-propelling body' in Fig. 3.5.

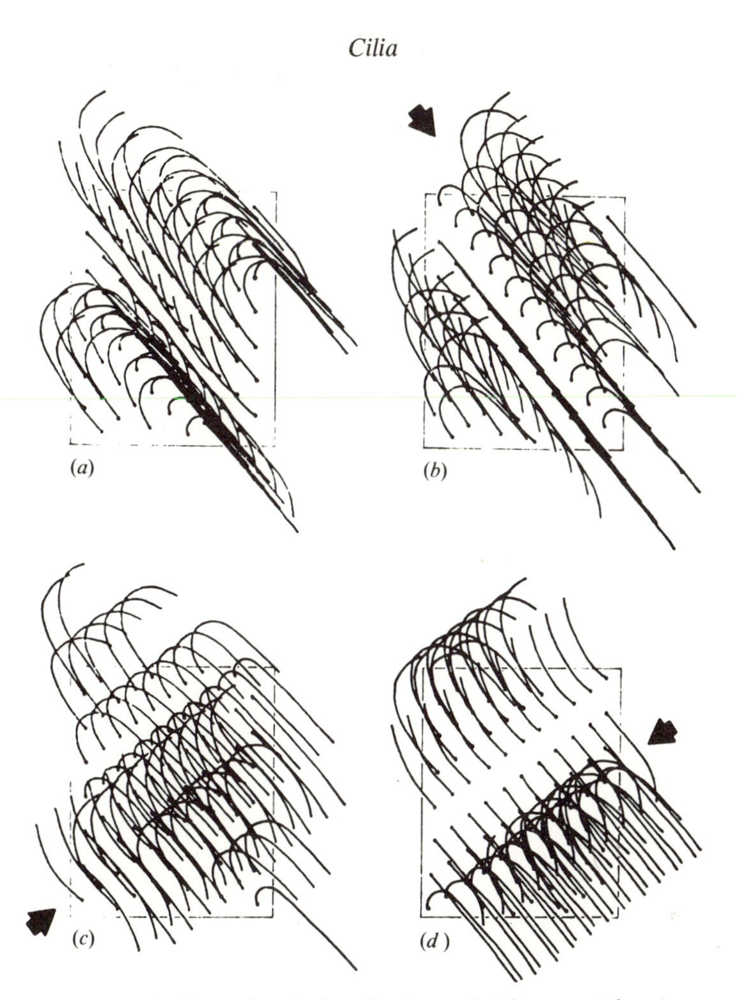

Fig. 3.4. Models of cilia beating (in the effective stroke) from top left to bottom right, showing different patterns of metachrony. (*a*) shows how the cilia of *Paramecium* beat. Metachronal waves are travelling from bottom left to top right. (*b*), (*c*) and (*d*) are hypothetical: they represent cilia beating in the same way, but with different patterns of metachrony. In (*b*) the waves are travelling towards the bottom left, in (*c*) towards bottom right and in (*d*) towards top left. The arrows point to places where cilia get in each other's way. From H. Machemer (1972). *J. exp. Biol.* **57**, 239–59.

Notice that the water at the tips of the cilia must move backwards relative to the ground. To see why this must be, consider a ciliate starting from rest. It exerts forces on the water and the water exerts forces on it but no external force acts on them so the combined momentum of the ciliate and the water must remain constant. If the ciliate is given forward momentum some of the water must be given backward momentum.

Fig. 3.5. shows that the velocity gradient between the body wall and the tips of the cilia must be much sharper than near an inert body moving at the same

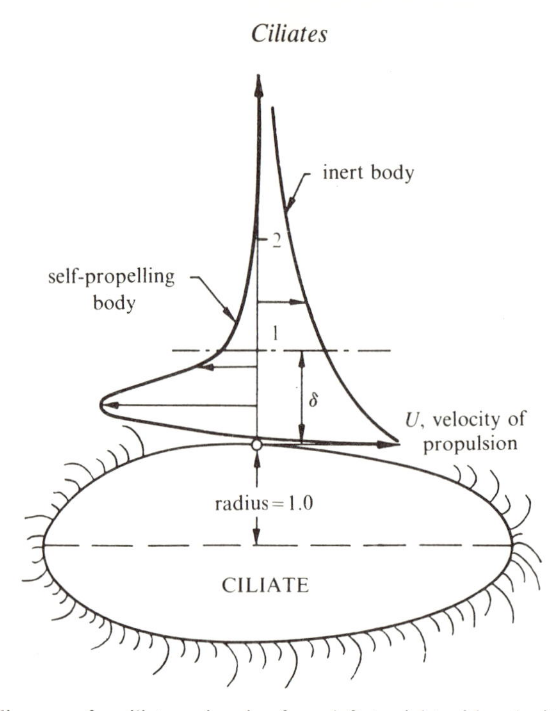

Fig. 3.5. A diagram of a ciliate swimming from left to right with velocity *U*. The lines above it show how water velocity must change with distance from the ciliate (i) if it is an inert body dragged passively through the water and (ii) if it propels itself by its own cilia. From J. R. Blake & M. A. Sleigh (1974). *Biol. Rev.* **49**, 85–125.

velocity. Since the force required to overcome viscosity is proportional to velocity gradient the total force required from all the cilia to propel a ciliate is several times the force needed to propel an inert body. The longer the cilia (within limits) the less sharp the velocity gradient and the less force and power are needed for swimming at a given speed. However, the force must act further from the base of the cilium so the bending moment at the base of the cilium may well be no less. Since there must be limits to the bending moments which can be exerted by cilia and to the number of cilia which can be fitted on a unit area of cell surface, there is probably a clear upper limit to the speed attainable by ciliated organisms, however long their cilia.

When *Paramecium* collides with an obstacle it reverses its cilia and swims backwards a little, before swimming forward again (Fig. 3.6*a*). This manoeuvre often gets it clear of the obstacle. How is it done, by an animal which has no nervous system? Poking the rear end of the animal does not cause reversal but (appropriately) faster forward swimming.

Paramecium has a negative membrane potential, like *Amoeba* (p. 48). This can be destroyed by sticking a microelectrode into *Paramecium* and applying a positive potential through it. When this is done, the beat of the cilia reverses. The negative membrane potential can also be destroyed by putting *Paramecium*

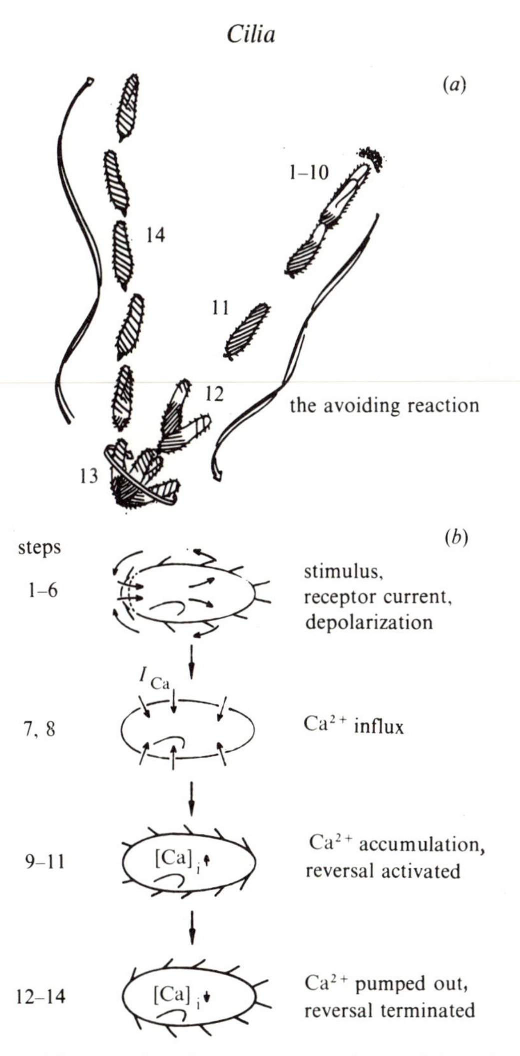

Fig. 3.6. (*a*) The avoiding reaction of *Paramecium*. After colliding with an obstacle the animal retreats (steps 1 to 13) and then swims forward again. (*b*) The sequence of events in the avoiding reaction, acording to a theory which is explained in the text. From R. Eckert (1972). *Science* **176**, 473–81.

in a solution containing a high concentration of potassium: this also reverses the beat. However, neither electrodes nor a changed solution will reverse the beat if there is no calcium in the water.

The role of calcium is shown more precisely by experiments on *Paramecium* treated with a detergent, Triton X. This has the same effect as the glycerol used in experiments with flagella (p. 44). It disrupts the membranes of the cell, making them freely permeable to inorganic ions and other small molecules. Treated *Paramecium* still swim, if they are kept in a suitable solution containing ATP.

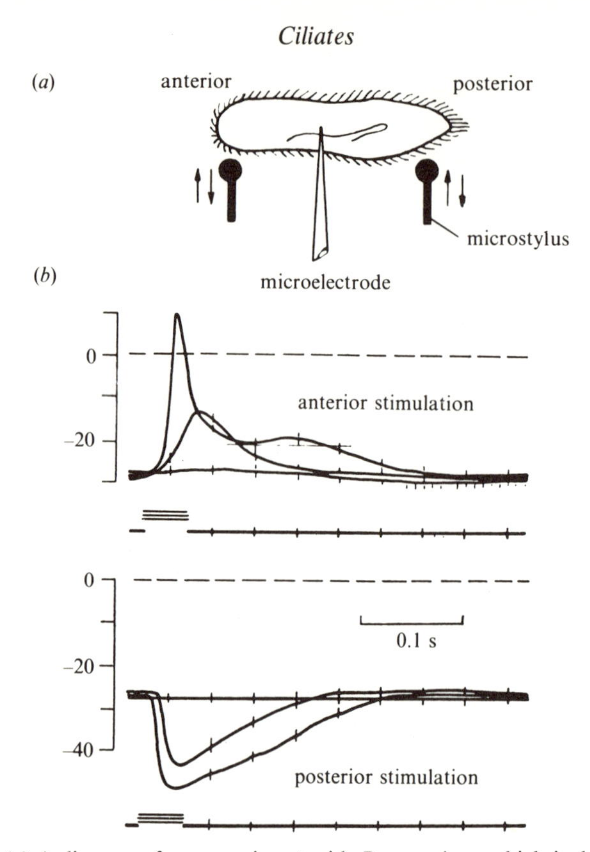

Fig. 3.7. (*a*) A diagram of an experiment with *Paramecium*, which is described in the text. (*b*) Records of membrane potential (mV) obtained in the experiment. The anterior and the posterior end were each stimulated by prodding with a microstylus, at the time indicated by the lines below the records. Records of three experiments, involving prodding at three different intensities, are superimposed in each case. From R. Eckert (1972). *Science* **176**, 473–81.

If the solution contains very little calcium, for instance 10^{-8} Equiv l^{-1}, the ciliates swim forwards. If the calcium concentration is increased above about 10^{-6} Equiv l^{-1} the ciliates swim backwards. These calcium concentrations are low: calcium concentrations in fresh water are generally 10^{-4} Equiv l^{-1} or higher. The experiment suggests that the concentration of calcium inside *Paramecium* is normally kept very low, but that calcium is admitted at appropriate times to reverse the cilia.

Another experiment is illustrated in Fig. 3.7. A *Paramecium* is impaled on a microelectrode, which is used to record its membrane potential. When the animal is undisturbed the membrane potential is −30 mV. The animal is prodded with an electrically driven microstylus which can be made to prod harder or less hard. When it is prodded very gently at the anterior end there is no effect. A moderate prod makes the membrane potential less negative and a harder one

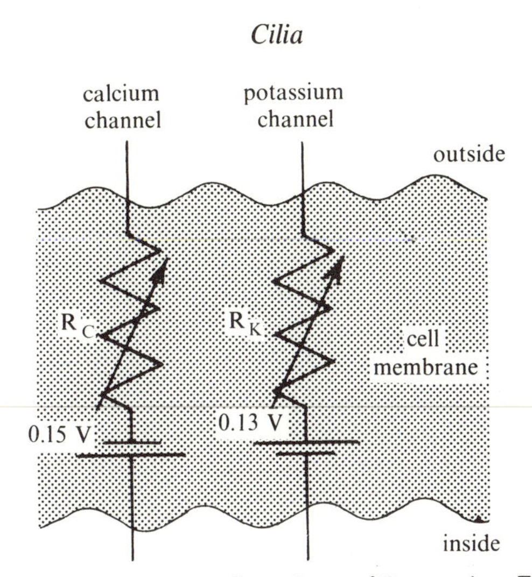

Fig. 3.8. A diagram representing the cell membrane of *Paramecium*. Further explanation is given in the text.

causes a bigger change and may even make the membrane potential positive. Conversely, prodding at the posterior end makes the membrane potential more negative. The harder the prod, the bigger the effect.

These phenomena are particularly interesting, because they have a lot in common with important processes in nerve cells. They have been partially explained by Dr Roger Eckert, but the explanation requires a little more physical chemistry than has been used so far. The Nernst equation (p. 45) refers to a situation where equilibrium has been reached. The ions which can penetrate the membrane have diffused through and achieved equilibrium, while the other ions cannot penetrate it at all. Now we have to consider a situation where ions penetrate the membrane, but not freely, and are not in equilibrium.

We have obviously got to consider calcium. It must have a very low concentration inside *Paramecium* (since *Paramecium* does not normally swim backwards!). It is probably usually about 10^5 times as concentrated outside the animal as inside it, and if so a membrane potential $V_C \simeq (58/2) \log_{10} (10^5) \simeq 150$ mV would be required to halt diffusion of calcium into the cell completely (see p. 45). If the membrane potential is less than this, a current of calcium ions will flow into the cell. We can represent this by an equivalent electrical circuit, a resistance and battery in series penetrating the cell membrane (Fig. 3.8.). The battery produces an electro-motive force V_C ($\simeq 150$ mV) which will make current flow into the animal unless the membrane potential exceeds 150 mV. The resistance R_C represents the resistance of the membrane to passage of calcium ions through it. It is shown as a variable resistance, for reasons which will be explained.

Potassium is likely to be important as well as calcium: it is almost certainly the most plentiful ion in *Paramecium*. There is evidence that the potassium concentration in *Paramecium* is about 20 mEquiv l^{-1}, a little less than in *Amoeba* (Table 2.1.). The potassium concentration of fresh water varies, but we will assume a value of 0.1 mEquiv l^{-1}, so that the concentration is 200 times as high in *Paramecium* as outside it. A membrane potential of $V_K = 58 \log_{10} (200) \simeq -130$ mV would be required to halt diffusion of potassium out of the cell. A potassium channel is shown in Fig. 3.8. with a 130 mV battery and a variable resistance R_K in it. Other channels could be drawn for other ions but these two are enough to explain the observed phenomena.

When the membrane potential is V_m the current of calcium ions out of the *Paramedium* will be $(V_m - V_C)/R_C$. The outward current of potassium ions will be $(V_m - V_K)/R_K$. (Inward currents are regarded as negative outward currents.) The net current out of the cell must be zero so

$$(V_m - V_C)/R_C + (V_m - V_K)/R_K = 0$$

and since $V_C = 150$ mV, $V_K = -130$ mV, the value of V_m is

$$\frac{150\ R_K - 130\ R_C}{R_K + R_C} \text{ mV}$$

If the calcium resistance R_C is much greater than the potassium resistance R_K, V_m will be approximately equal to the Nernst potential for potassium, -130 mV, If R_K is much greater than R_C, the membrane potential will be approximately equal to the Nernst potential for calcium, $+150$ mV. If neither R_K nor R_C is overwhelmingly greater than the other, the membrane potential will lie somewhere between these extremes. This seems to be the normal situation. In the experiment shown in Fig. 3.7, the membrane potential of the unstimulated *Paramecium* was -30 mV. Poking at the anterior end made the membrane potential less negative and poking at the posterior end made it more negative. Dr Eckert suggested that distortion of the cell membrane at the anterior end reduced R_C temporarily, while distortion at the posterior end reduced R_K.

The events which are believed to follow anterior stimulation are shown in more detail in Fig. 3.6(*b*). The initial effect is presumably local, reducing R_C and allowing calcium ions to flow in faster at the anterior end only. This inward current must be balanced by a current flowing outward through other parts of the cell membrane (Fig. 3.6*b*, stages 1 to 6). The latter current, flowing outwards through the resistance of the cell membrane, must raise V_m (make it less negative) and tend to stimulate a fall in R_C over the whole cell surface: we have already seen that when electrodes are used to raise V_m the beat of the cilia is reversed, presumably by an influx of calcium. Thus a fall in R_C tends to increase V_m, and an increase in V_m tends to reduce R_C. This is a positive feedback situation which could explain the effect of anterior stimulation spreading over the whole cell surface. A similar phenomenon (with sodium playing the part of calcium) is responsible for nerve conduction.

Calcium flows in (Fig. 3.6*b*, stages 7 and 8) until the calcium concentration in the *Paramecium* is high enough to reverse the cilia and make the animal swim backwards (stages 9–11). Presumably the extra calcium is then pumped out of the animal, so that forward swimming starts again (stages 12–14).

How much calcium gets into the *Paramecium* each time it is stimulated? The amount of charge required for a given change of membrane potential depends on the capacitance of the membrane. It has been found by experiment that cell membranes have capacitances around 10^{-2} F m^{-2}, which is about what one would expect for a layer of lipid 5 nm thick. A *Paramecium* 250 μm long and 50 μm in diameter would have a surface area of about 4×10^{-8} m^2, and a capacitance of 4×10^{-10} F. An increase in membrane potential of 0.04 V (a typical response to a strong prod at the anterior end) would require movement of a charge of $0.04\times4\times10^{-10} = 1.6\times10^{-11}$ coulombs across the membrane. One gram equivalent of an ion carried 10^5 coulombs so $1.6\times10^{-11}/10^5 = 1.6\times10^{-16}$ Equiv would have to cross the membrane. The volume of the *Paramecium* would be about 5×10^{-10} l so the concentration of calcium in the cell would be increased by about $1.6\times10^{-16}/5\times10^{-10} = 3\times10^{-7}$ Equiv l^{-1}. The experiments with *Paramecium* treated with detergent indicate that this change would be too little to reverse the cilia. What is wrong with the calculations?

The fault seems to be that we have considered only the general body surface, and ignored the cilia. The ratio of surface area to volume is much larger for the cilia than for the cell body, because they are so slender. Dr Eckert has shown that if calcium enters through the membranes of the cilia as well as through the cell membrane, the calcium concentration *in the cilia* should increase by almost 10^{-4} Equiv l^{-1} when the membrane potential rises by 0.04 V. If the direction of beating depends on the calcium concentration in the cilium itself, this is ample to explain the reversal.

The use which *Stentor* makes of its cilia in feeding is described in a later section of this chapter.

MYONEMES

Some ciliates can make large fast contractions, by means of organelles called myonemes. They include *Vorticella* (Fig. 3.2) which lives in fresh water attached to plants by a long stalk. When it is stimulated by vibration, or by poking, its stalk shortens suddenly and dramatically by coiling into a tight helix. It is made to shorten by a myoneme (known as the spasmoneme) which runs helically along the stalk (see Fig. 3.2). The sudden shortening is presumably an escape mechanism.

The mechanism of shortening has been investigated by experiments of *Vorticella* treated with glycerol or with detergents so that the cell membrane was no longer a barrier to small molecules. It was found that such preparations contracted in solutions containing more than about 10^{-7} Equiv l^{-1} calcium ions, and extended again in solutions containing less calcium than this. Surprisingly, no ATP or other energy-supplying compound was needed. Similar preparations of muscle will only contract in the presence of ATP and similar preparations of cilia will only beat in the presence of ATP.

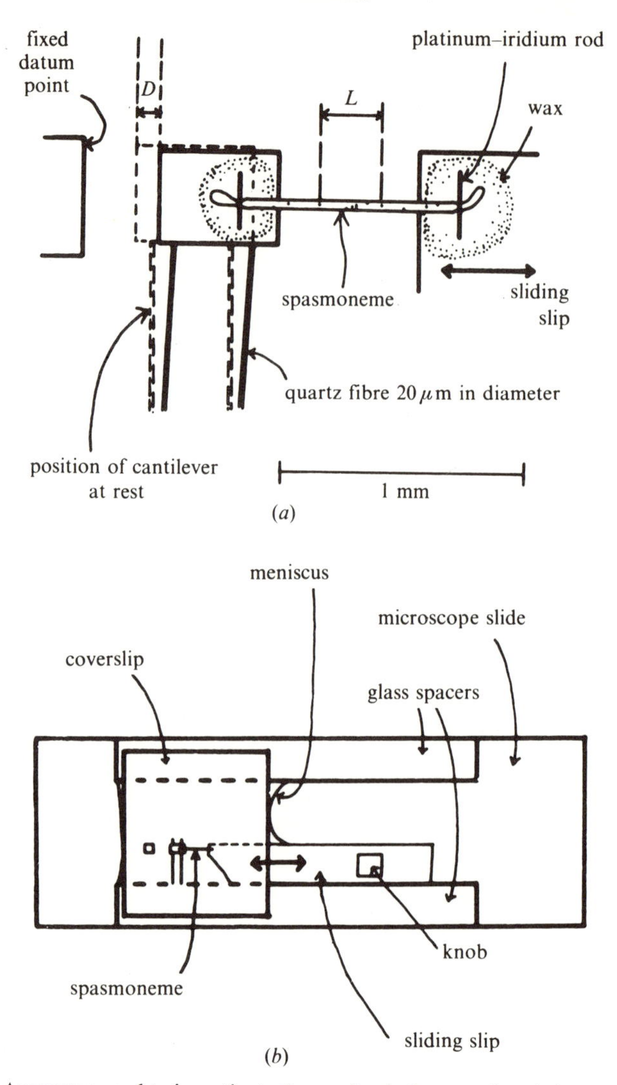

Fig. 3.9. Apparatus used to investigate the mechanical properties of the spasmoneme of *Zoothamnium*. The length L was measured between particles of graphite on the spasmoneme and the displacement D of the cantilevers was used to calculate the force. From T. Weis-Fogh & W. B. Amos (1972). *Nature, Lond.* **236**, 301–4.

Further experiments have been done by Professor Torkel Weis-Fogh and Dr Brad Amos. Their experiments would not have been feasible on the small spasmonemes of *Vorticella* so that they used its giant relative, *Zoothamnium*. This is a colonial ciliate: several thousand individuals, each similar in size to *Vorticella*, share a common stalk which is 2 mm long. Within this stalk is a spasmoneme about 1 mm long and 30 μm in diameter: this is too small to handle

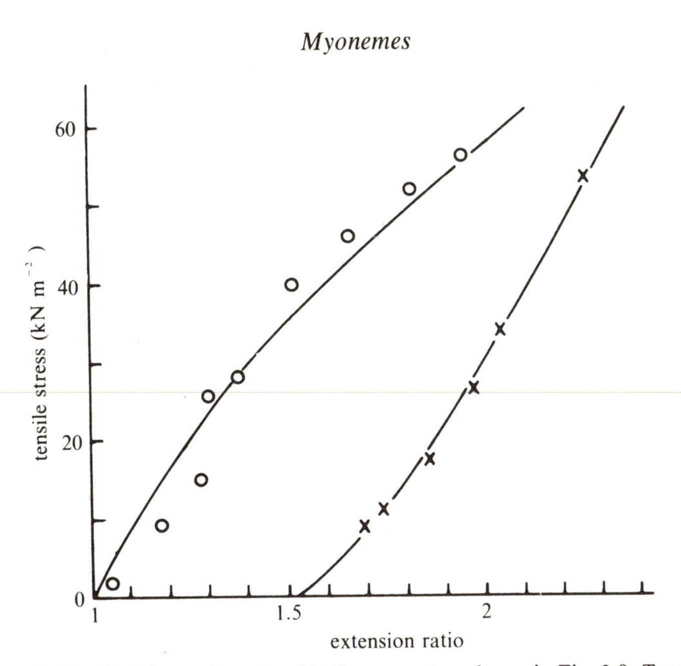

Fig. 3.10. Results of experiments with the apparatus shown in Fig. 3.9. Tensile stress is plotted against extension ratio for high (○) and low (×) calcium concentrations. The tensile stress is expressed as force per unit unstrained cross-sectional area at high calcium concentration. The extension ratio is the length divided by the unstrained length at high calcium concentration. From T. Weis-Fogh & W. B. Amos (1972). *Nature, Lond.* **236**, 301–4.

easily, but large enough for measurements of mechanical properties to be possible. Specimens of *Zoothamnium* were treated with glycerol. The spasmoneme was dissected out and fixed in the tiny apparatus shown in Fig. 3.9, which is built on an ordinary microscope slide. Its ends were fastened by trapping them under metal rods pushed down into blobs of wax. One end was attached to a larger slip of glass which could be slid along the slide. The spasmoneme could be stretched by moving the sliding slip. The force applied bent the quartz fibres and could be measured by observing the displacement *D*. Hence graphs of force against length, or stress against length, could be obtained (Fig. 3.10). They showed that the spasmoneme could be stretched greatly, and would recoil elastically, both at high and at low calcium concentrations. It was shorter, at any given stress, at high concentrations than at low ones. When the calcium was increased from the lower to the higher concentration (Fig. 3.10) the spasmoneme shortened by about a third if it was free to do so, or developed tension if its ends were held.

No ATP was required, but the spasmoneme did mechanical work. Where did the energy come from? Presumably from movements of calcium ions from the higher concentration in the surrounding solution to the lower concentration in the spasmoneme. How does the spasmoneme work? Electron microscope sections show that the spasmoneme contains fine filaments of protein, each like a chain of tiny beads. Among these are tubular structures which may store calcium in the living spasmoneme. Contracted Spasmoneme is not birefringent (except when stressed) and it seems probable that each bead on the filaments is a randomly-coiled protein molecule. The elastic properties of materials built of long, randomly coiled molecules linked together will be discussed later (chapter 6, in an account of mesogloea). Rubber is one such material, and the others tend to have similar properties. Extended spasmoneme is birefringent even when unstressed and it seems likely that reduction of the calcium concentration allows cross-links to form which hold the molecules in an extended configuration. When the cross-links are removed again by calcium ions, the molecules spring back to their shorter, random configurations.

Natural waters contain higher concentrations of calcium than are needed in the spasmoneme to make it contract. The calcium concentration in the spasmonemes of intact animals must normally be kept much lower than the concentration in the surrounding water. Much higher concentrations may be maintained in the tubules. Contraction is presumably initiated in much the same way as backwards swimming of *Paramecium*, by a change in the properties of the cell membrane or of the tubule membranes which allows calcium to diffuse into the contractile protein. If so, no metabolic process is required during the actual contraction. The animal does work pumping calcium out of its spasmoneme, and so stores energy which is released in the contraction.

The contraction is remarkably fast. High-speed cinematograph films of *Zoothamnium* show the spasmoneme shortening by 30% in 3 ms, at a rate equivalent to 100 lengths s^{-1}. No known muscle can shorten so fast: even an exceptionally fast toe muscle found in mice can only achieve 25 lengths s^{-1}. However, the spasmoneme takes several seconds to extend again.

Stentor can also contract, shortening its body to as little as a quarter of its extended length. As with *Vorticella* and *Zoothamnium* the contraction is very fast, so high-speed cinematograph films are needed to show the details. Films have been taken at rates up to 3000 frames s^{-1}. *Stentor* can be stimulated to contract by poking, or by an electric shock. It is not just a stalk that contracts, but the whole body. The films show that electrical stimulation makes the whole body contract simultaneously, but poking with a glass needle does not: contraction starts where the animal was poked and spreads to other parts of the body with a velocity of 10–20 cm s^{-1}. Action potentials travel at similar rates in the nerve nets of Cnidaria (chapter 6).

Stentor has myonemes, running lengthwise along the body close under the cell membrane. They apparently make it contract, and they are believed to work in the same way as the spasmonemes of *Vorticella* etc. After contraction the myonemes elongate, throwing themselves into folds but failing to extend the animal. Extension is a separate process and is believed to be caused by other structures, the km fibres, which remain straight throughout. Each km fibre is a ribbon of (usually) 21 parallel microtubules, arranged side by side. One of

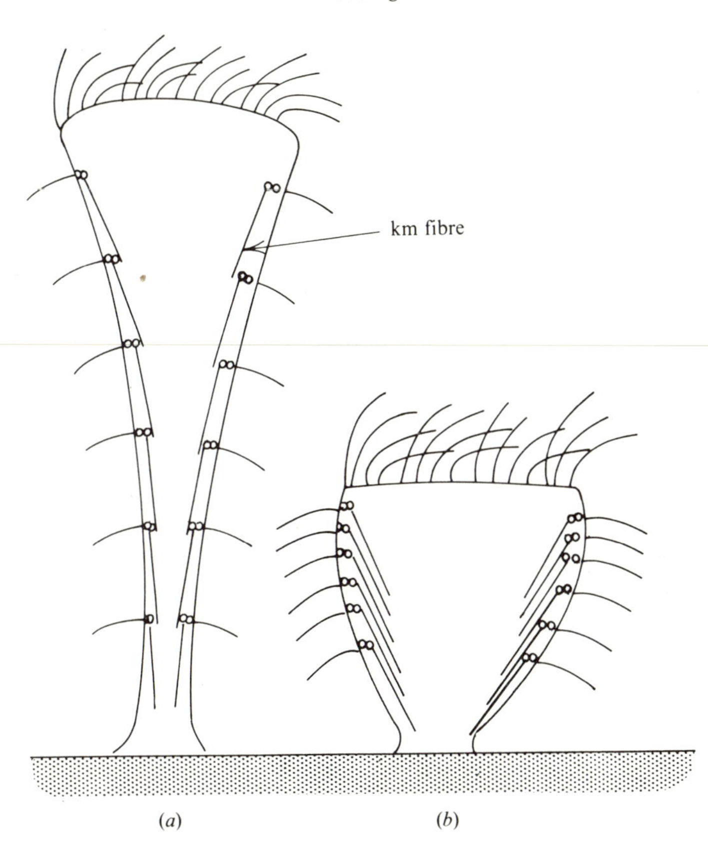

Fig. 3.11. Diagrams showing how the km fibres slide over each other as *Stentor* contracts.

these fibres starts near the base of each of the cilia on the side of the body (Fig. 3.11). The cilia are arranged in longitudinal rows, and the km fibres associated with successive cilia in a row overlap each other. Electron microscope sections of contracted and extended *Stentor* show that the animal shortens like a telescope: the km fibres remain constant in length, but slide over each other. It is believed that the animal is extended by active sliding of the microtubules of km fibres over each other. We have already seen how cilia and flagella are bent by active sliding of microtubules over each other (p. 44).

FEEDING

Some ciliates have quite elaborate structures which are used for taking in food. The cytopharyngeal basket of *Chilodonella* has already been mentioned (Fig. 3.1 *b*). *Nassula* has a similar structure which has been studied in more detail (Fig. 3.12). It contains about 30 rods, each of them a bundle of parallel

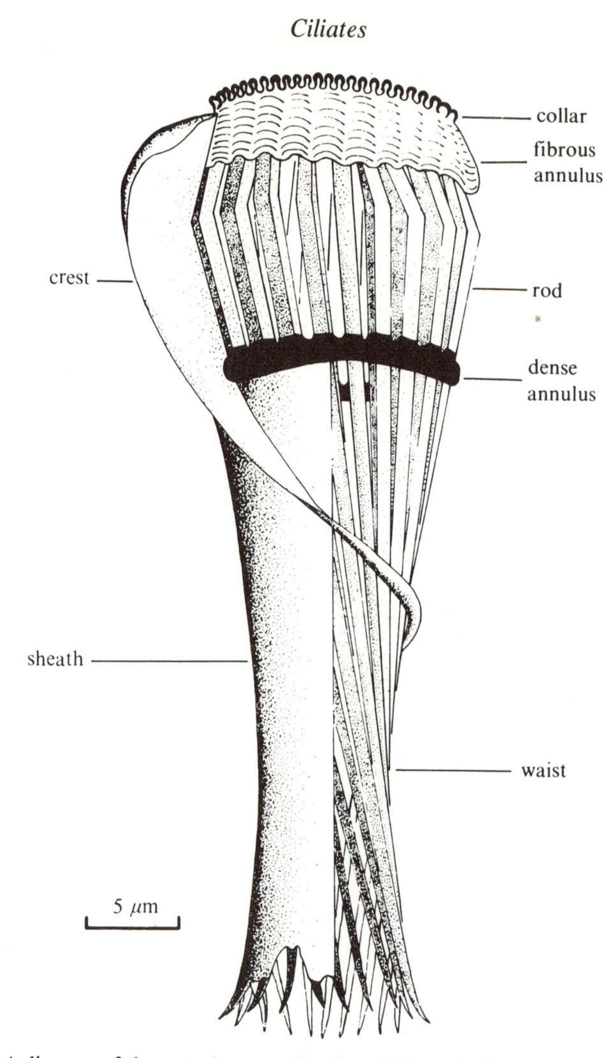

Fig. 3.12. A diagram of the cytopharyngeal basket of *Nassula*. From J. B. Tucker (1968). *J. Cell Sci.* **3**, 493–514.

microtubules. An equal number of crests formed by microtubules run helically around the rods. Two rings of material are also involved in holding the rods together: a fibrous annulus made of material which looks very similar in electron microscope sections to the material of myonemes, and another ring of different material known as the dense annulus. The whole of this remarkably complicated structure is embedded in the cytoplasm.

Nassula feeds on filamentous blue-green algae. Its feeding movements have been watched and photographed using phase contrast microscopy. Fig. 3.13(*a*) shows the basket in its resting position. The first movement in feeding (*b*) draws out the mouth of the basket from its initial circular shape into an ellipse, with

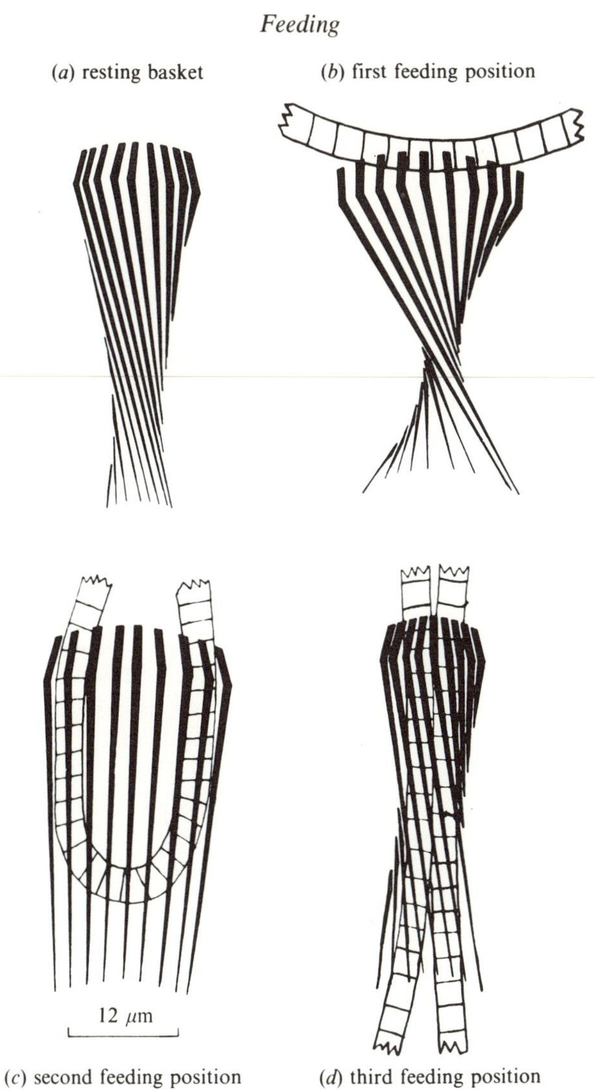

Fig. 3.13. Diagrams showing successive positions of the cytopharyngeal basket of *Nassula* during feeding on an algal filament. From J. B. Tucker (1968). *J. Cell Sci.* **3**, 493–514.

the long axis of the ellipse parallel to the algal filament. The cytoplasm bulges out and engulfs the filament, drawing it down into the basket. Movements (*c*) and (*d*) follow quite rapidly and a loop of filament is drawn down into the basket. The basket is elliptical in section in position (*c*) but circular again in (*d*). All this occurs without any bending of the rods or any changes in their length. It is assumed that it is achieved by sliding of microtubules relative to each other and by contraction, at the appropriate time, of the myoneme-like fibrous

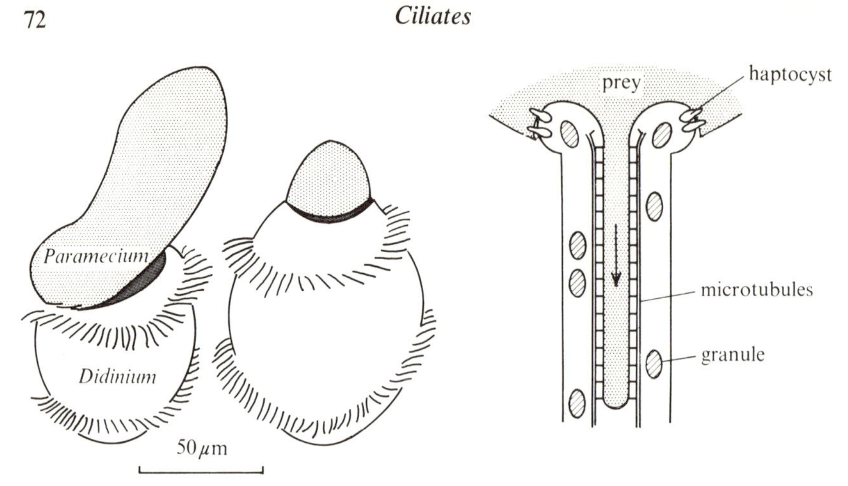

Fig. 3.14. (*a*) Sketches of *Didinium* ingesting *Paramecium*. Based on scanning electron micrographs by H. Wessenberg & G. Antipa (1970). *J. Protozool.* **17**, 250–70. (*b*) A diagrammatic section of a tentacle of a feeding suctorian. This diagram is based on electron microscope sections.

annulus. However, the mechanism of the movements has not been explained in detail.

Once the end of the loop has been pulled down into the basket, ingestion continues without movements of the basket. A possible mechanism for this will be discussed shortly. As much as 1.6 mm of filament may be ingested by a 0.2 mm *Nassula* in less than 15 min. Long filaments are coiled in the body: as many as three double coils may be formed, deforming the ciliate from a prolate spheroid (the shape of a Rugby or American football) to a disc. Excessively long filaments may be broken off rather than swallowed completely. The ingested filament breaks up and becomes enclosed in several separate food vacuoles.

Didinium is an even more voracious feeder. It is a predator which attacks and engulfs *Paramecium* larger than itself. It collides with a *Paramecium*, bumping into it with its pointed snout. During the moment of contact its snout discharges trichocysts. These are threads which penetrate the *Paramecium* but remain firmly fixed at the other end to the *Didinium*: they thus attach the animals to each other. Once this has happened the *Paramecium* stops swimming, and it dies in a few minutes even if it is not ingested. Ingestion may be completed in a minute but two hours is needed for digestion. The process of ingestion has been studied by scanning electron microscopy of *Didinium* fixed while they were feeding (Fig. 3.14*a*). A cavity opens around the point of attachment of the *Paramecium*, which is drawn down into it. There are microtubules around the snout which may be responsible for opening the cavity. Though the victim is initially larger than the predator it is believed that its volume may be reduced during ingestion by removing water, which the predator gets rid of by means of its contractile vacuole.

There are also suctorians which eat ciliates of about their own size, but their manner of feeding is rather different (Fig. 3.14 *b*). A passing ciliate happens to collide with the knob on the end of a tentacle and sticks fast, held by discharging haptocysts which are rather like the trichocysts of *Didinium*. Cilia around the point of contact stop moving and if the prey struggles before it is totally immobilized, its struggles only bring it into contact with more tentacles. The cell membranes of predator and prey merge where they are in contact so that only a single membrane separates the cytoplasm of the two animals. This membrane is drawn down into the tentacle, carrying prey cytoplasm with it into a food vacuole in the cell body.

This sequence of events is inferred from electron microscope sections of suctorians fixed during feeding. The sections also show that each tentacle has a core of microtubules, which presumably give it its stiffness. During feeding the microtubules form themselves into a tube, down which the prey flows. Bridges have been seen in electron micrographs between the microtubules and the membrane covering the prey cytoplasm, and it is believed that they propel the prey down the tentacles, working like the dynein arms of the microtubules in cilia.

New membrane has to be formed at the tip of the tentacle as feeding proceeds, because the diameter of the stream of protoplasm which flows down the tentacle (about 0.7 μm) is only a tiny fraction of the diameter of the prey. A long slender rod has a much larger surface area than a sphere of the same volume. Electron microscope sections of the tentacles of feeding suctorians show granules which are believed to be materials for constructing cell membrane travelling outwards to the place where they are needed (Fig. 3.14 *b*). These granules have a striped appearance which suggests they contain alternate layers of protein and lipid, the materials of cell membranes.

More typical ciliates such as *Nassula* may also depend on bridges formed by microtubules to draw food into the body. Projections which look like dynein arms have been found on some of the microtubules of the cytopharyngeal basket of at least one species.

Many different sorts of trichocyst are found on ciliates. Some (such as those *Didinium*) are used to capture prey while others seem to be purely defensive in function. *Paramecium* discharges trichocysts when it is attacked by *Didinium*, when it is poked and when acetic acid (among other substances) is put in the water. There is evidence that the process which discharges the trichocysts of *Paramecium*, like the process which reverses the cilia (p. 60), involves entry of calcium through the cell membrane.

Nassula, *Didinium* and the suctorians are all examples of ciliates which eat relatively large food. Many other ciliates feed on bacteria which are captured from a stream of water kept flowing by ciliary action. *Stentor* (Fig. 3.1 *c*) and *Vorticella* (Fig. 3.2) both feed in this way, though *Stentor* also captures larger prey. The membranelles around the rim of *Stentor* beat outwards. They beat in turn so that at any instant some are vertical while others quite close to them

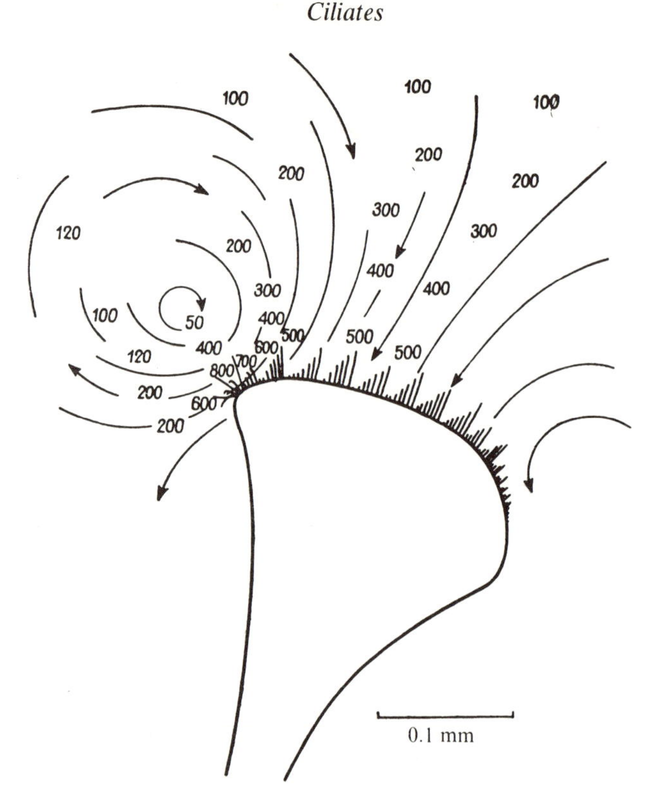

Fig. 3.15. Water movements around *Stentor*, due to the beating of the membranelles. The numbers indicate average speeds of particle movement in μm s^{-1}. From M. A. Sleigh & E. Aiello (1972). *Acta Protozool.* **11**(31), 265–77.

are horizontal. The movements they set up in the water have been investigated by taking film through a microscope of *Stentor* in a dilute suspension of tiny latex particles. These particles were about 3 μm in diameter, only a little bigger than typical bacteria. The films showed that the membranelles beat at about 30 Hz, their tips reaching velocities around 4 mm s^{-1}. The movements of the particles showed that the water was flowing in the pattern shown in Fig. 3.15, reaching maximum velocities around 1 mm s^{-1} near the membranelles. Food particles are apparently caught on the end of the animal, within the ring of membranelles: this is where the cytostome is.

NUCLEI AND REPRODUCTION

All known ciliates except the primitive *Stephanopogon* have two kinds of nuclei: macronuclei and micronuclei. Typically, there is one nucleus of each type. The micronucleus is diploid. It contains DNA but cytochemical tests show no RNA in it. The macronucleus is generally much larger and contains both DNA and RNA. Only the micronucleus takes part in sexual reproduction (the process of conjugation): the old macronucleus disappears and is replaced by a new one formed by division of the zygote micronucleus. Thus the micronucleus alone is responsible for transmission of genetic information at sexual reproduction. However it is presumably not involved in the ordinary processes of running the cell, in providing templates for protein synthesis, etc., since it produces no RNA. The macronucleus, which does produce RNA, presumably has this responsibility.

Since the macronucleus is formed from the micronucleus it cannot carry more genetic information than the micronucleus. However, it contains far more DNA. This has been shown by staining the DNA by the Feulgen method and measuring photometrically the amount of stain taken up by each nucleus. Various species have been investigated. In most of them the macronucleus contains at least 8 times as much DNA as the micronucleus, and in some over 6000 times as much. Either the macronucleus contains a lot of random DNA which carries no genetic information or else it contains many duplicate sets of chromosomes (i.e. it is polyploid). The latter alternative seems the more probable, and there is further evidence for it. It is possible to dissect out parts of the macronucleus of *Stentor*. Even when 90–95% of the nucleus is removed the remaining fragment can regenerate an apparently normal macronucleus and the animal regains a normal appearance. It seems that any reasonably-sized piece of macronucleus contains a complete set of genes.

Ciliates generally reproduce asexually. The micronucleus undergoes a normal mitosis. The macronucleus also divides, not necessarily into equal halves, but apparently in such a way that each part receives complete sets of genes. There is evidence which suggests that the chromosomes of each set are joined together end to end to form a composite chromosome which is passed intact to one daughter cell or the other. The cytoplasm divides after complicated processes which provide each half with the proper complement of organelles. In *Tetrahymena*, for instance, the daughter formed by the anterior half of the parent retains the original cytostome and associated membranelles but develops a new contractile vacuole. The posterior daughter retains the original contractile vacuole but develops a new cytosome.

From time to time sexual reproduction occurs. In wild populations it often occurs regularly at a particular season. In laboratory cultures it generally only occurs when several asexual generations have occurred since the previous conjugation. Conjugation does not seem to be essential for survival of a clone: clones of *Tetrahymena* which have no micronucleus have been kept successfully for over 30 years and show no sign of abnormality apart from this lack. However conjugation, like other forms of sexual reproduction, gives a selective advantage in a changing environment (p. 51).

The details of conjugation vary between species, but the conjugation of *Paramecium caudatum*, shown in Fig. 3.16, is reasonably typical. The figure was drawn from preparations stained by the Feulgen method, so the DNA in the nuclei appears dark. Two individuals come together and become attached by their ventral surfaces (i.e. by the surfaces in which the cytosome lies). In the first stage which is illustrated (*a*) the diploid micronuclei have extended into

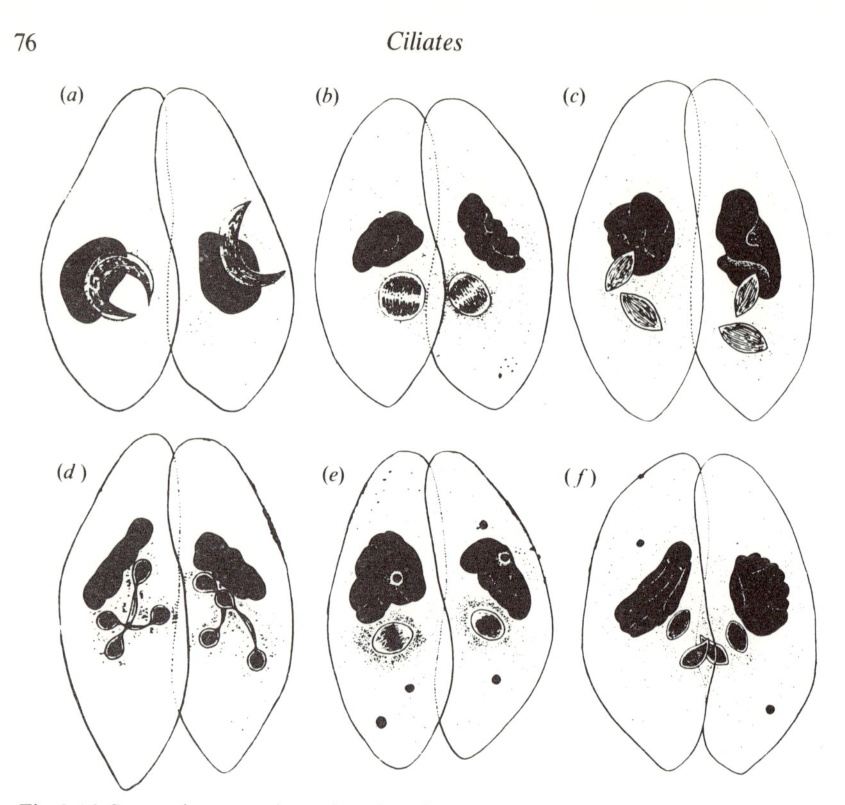

Fig. 3.16. Successive stages in conjugation of *Paramecium caudatum*. From I. B. Raikov (1972). In *Research in protozoology*, ed. T.-T. Chen, vol. 4, pp. 147–289. Pergamon, Oxford.

crescents before dividing. The large black objects are the macronuclei. The micronuclei divide twice (*b* to *d*) undergoing a typical meiosis to produce four haploid nuclei in each *Paramecium*. Three of these disappear and the remaining one divides again (*f*) producing two haploid nuclei of which one stays where it is while the other moves across into the other animal and fuses with the stationary nucleus there. This produces a diploid nucleus which divides mitotically (*i*) as the conjugating animals separate. Two more mitotic divisions occur so that each *Paramecium* contains eight nuclei (*k*). At this stage the macronucleus, which has remained intact through stages (*a*) to (*j*), breaks up. One of the eight nuclei becomes the micronucleus, three disappear and four (the ones at the anterior end in (*k*)) become macronuclei. Two of these four are passed to each of the offspring of the next division (*n*) and one to each of the offspring of the division after that.

This is plainly a sexual process. The haploid nuclei of stage (*f*) are comparable to gamete nuclei of other animals: the stationary one may be thought of as the female gamete and the migrating one as the male. The diploid nucleus of stage (*h*) which is formed by the fusion of gamete nuclei from the

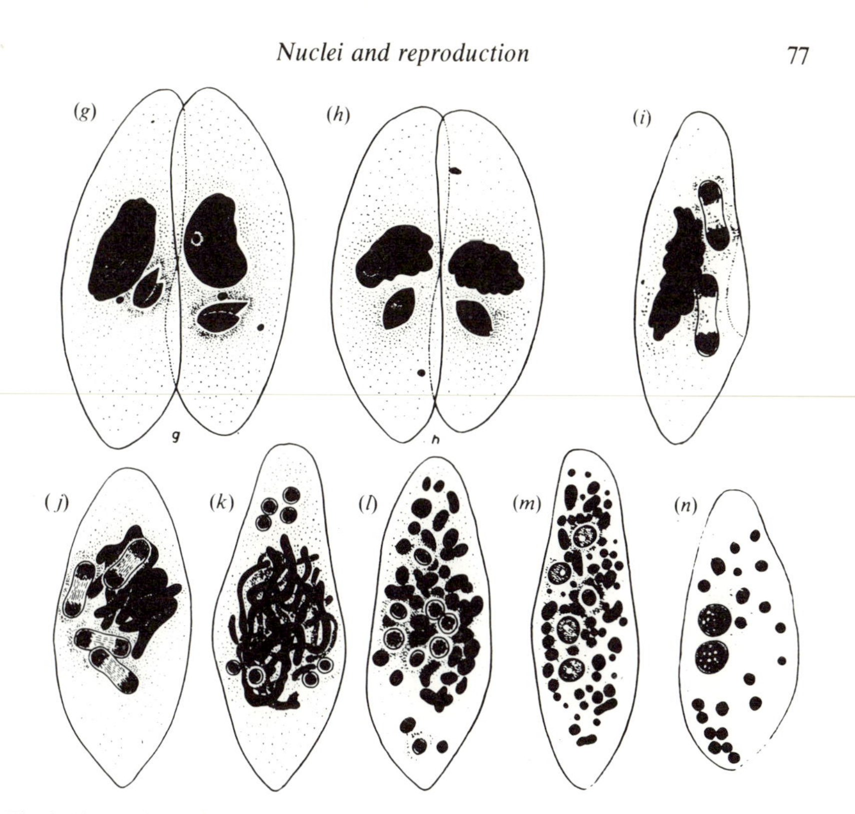

Fig. 3.16 (*cont.*).

two animals is a zygote nucleus. Half of its set of chromosomes comes from each parent.

The exchange of the migrating haploid nuclei is an essential part of the process and it seems important to be absolutely sure that it occurs. It can be observed directly in living *Paramecium* and *Tetrahymena*. Also, there are strains of *Paramecium* in which the micronucleus is polyploid so that it and the gamete nuclei derived from it take up more Feulgen stain than the corresponding nuclei of normal *Paramecium*. When a polyploid strain conjugates with a normal one it can be seen at stage (*g*) that the nuclei which fuse together are dissimilar.

A great deal of research has been done on the genetics of *Paramecium*, which can be kept and bred quite easily. Culture media can be made, for instance, by boiling chopped hay in water. This produces a sterile medium with nutrients in it. A little of a culture of suitable bacteria is added, and these bacteria use the nutrients as food and multiply. *Paramecium* are then added. They feed on the bacteria and multiply in turn. Clones of genetically identical *Paramecium* can be grown from a single parent which reproduces asexually.

Members of a clone will not conjugate with each other but they will conjugate with members of certain other clones. Animals from different clones can be picked up in pipettes and put together in a drop of water on a microscope slide. In this way clones can be sorted out, finding which will conjugate with which. Such experiments have shown that some supposed species are actually groups of species. For instance, *Paramecium bursaria* can be divided into six syngens, and no member of one syngen will conjugate with any member of any other. The syngens should be regarded as separate species since they cannot interbreed, although they look identical. Within each syngen, clones may be classified as belonging to one of several mating types. In *P. bursaria* syngen 1, for instance, there are four mating types, A, B, C, and D. Conjugation occurs only between members of different mating types. For instance, type A will conjugate with B, C and D but not with A. The mating types are not sexes: individual *Paramecium* are hermaphrodites which produce both male and female gamete nuclei.

Mating type is determined genetically, and the manner of its inheritance has been demonstrated by breeding experiments. In a pioneer investigation, one clone of each of the four mating types of *P. bursaria* syngen 1 was chosen. All possible combinations of these four clones were crossed and the offspring were tested for mating type. The ratios of mating type found in the offspring were consistent with the following hypothesis.

Mating type in this syngen is determined by pairs of alleles X,x and Y,y. X is dominant to x and Y to y. Mating type A includes all genotypes in which X and Y are both expressed (XXYY, XxYY, XXYy and XxYy; XXYY does not normally occur because it can only be produced by an abnormal conjugation between two parents of type A). Type B includes the genotypes in which x and Y are expressed (xxYY and xxYy). Type C is the double recessive (xxyy) and type D includes the genotypes in which X and y are expressed (XXyy and Xxyy). The clones used in the original tests happened to be XxYy(A), xxYy (B), xxyy (C) and Xxyy (D). Thus crosses between the B clone and the C clone produced roughly equal numbers of B and C, and crosses between A clone and C clone produced roughly equal numbers of all four types. The other crosses confirmed the hypothesis. There is no evidence of linkage between the X and Y loci, so they are presumably on separate chromosomes.

There is apparently an X substance produced by *Paramecium* which have the dominant X gene and a complementary x substance produced only by ones which lack it. Similarly there is a complementary pair of substances Y and y. X sticks to x and Y sticks to y, and a pair of animals must have opposite members of at least one of these pairs of substances if they are to be able to attach to each other for conjugation. The substances are present on the cilia of the ventral surface. This has been inferred from experiments with cilia isolated from *Paramecium*. When these are added to a culture of *Paramecium* of different mating type they attach to the ventral surfaces of the intact

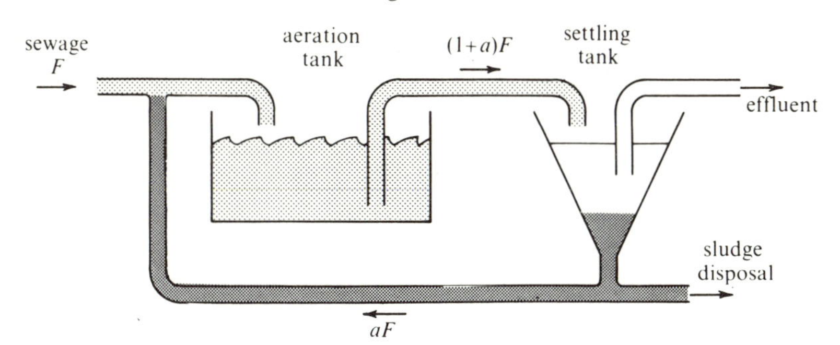

Fig. 3.17. A diagram illustrating the activated sludge process of sewage treatment.

Paramecium (provided both lots of *Paramecium* are in a state of readiness for sexual reproduction).

Mating type substances have apparently evolved as a means of sticking ciliates together for conjugation. A consequence of this is that a ciliate cannot conjugate with all members of the same species. The proportion with which it can conjugate is increased if there are several complementary pairs of substances of which only one need be different in the conjugating partners. *P. bursaria* syngen 1 has two pairs, and some ciliates have more than two.

After conjugation a new macronucleus is formed from division products of the same nucleus as the micronucleus, so the macro- and micronuclei presumably carry the same genes. *Paramecium* in which the macro- and micronuclei are genetically different have however been produced by brief heating to 38 °C after conjugation. This treatment retards development of the new macronuclei so that some of the offspring are left without one. These animals still have fragments of the degenerating parental macronucleus, which re-form gradually over several asexual generations until a clone is obtained with a superficially normal macronucleus, genetically different from the micronucleus. Such clones show features dictated by the macronucleus rather than those dictated by the micronucleus. This is as might be expected, since only the macronucleus produces RNA.

SEWAGE TREATMENT

Ciliate Protozoa, particularly *Vorticella* and other peritrichs, are abundant in sewage works. They probably play a useful part in the purification process. One of the two widely-used processes is the activated sludge process, and it has been observed that the effluent from plants which lack ciliates is apt to be turbid and to contain more bacteria than normal.

The activated sludge process is illustrated in Fig. 3.17. The sewage enters a large tank where it is kept stirred and well aerated by bubbling air through it. From this aeration tank it flows to a settling tank where the sludge (solid matter) settles out. The relatively clear fluid from the top of the tank is drained off and sludge is also drained from the bottom. Some of the sludge is returned to the aeration tank and some is removed for drying and disposal. Bacteria thrive in the tanks, breaking down the organic waste, and the ciliates feed on them.

Purified effluent and sludge are constantly leaving the system and must carry ciliates away with them. If the ciliates do not reproduce fast enough they will be washed out of the system. How fast must they reproduce?

Let sewage flow into the aeration tank at a rate F and let sludge from the settling tank be returned to the aeration tank at a rate aF. The rate of flow from the aeration tank must be $(1+a)F$ if the level of fluid in the tank is kept constant. Let the aeration tank have volume V and contain C ciliates per unit volume. Let the ciliates reproduce at a rate r so that if there were no dilution the population density would increase by rC and the population by rCV in unit time.

Consider first a free-swimming species which has the same population density C in aeration tank, effluent and sludge. In unit time the aeration tank gains rCV ciliates by reproduction and aFC in returned sludge, but loses $(1+a)FC$ in its outflow. At equilibrium the number of ciliates in the tank is constant and

$$rCV + aFC - (1+a)FC = 0$$

$$r = F/V \tag{3.1}$$

The ciliates must be able to reproduce at a rate F/V if they are not to be washed out of the system.

Now suppose that the ciliates are attached to the sludge particles (as *Vorticella* etc. generally are) and that settlement increases the concentration of solid matter in the sludge by a factor b. Then the population density of the ciliates in the settled sludge will be bC and the rate of return of ciliates from the settlement tank $aFbC$. The equation for equilibrium is now

$$rCV + aFbC - (1+a)FC = 0$$

$$r = (1+a-ab)F/V \tag{3.2}$$

In typical operating conditions F/V would be about $0.1\ \text{h}^{-1}$, a about 1 and b about 1.9. Hence the reproduction rate r required for equilibrium is $0.1\ \text{h}^{-1}$ for free-swimming ciliates (equation 3.1) but only $0.01\ \text{h}^{-1}$ for attached ciliates (equation 3.2).

Reproduction rates of ciliates have been measured in laboratory experiments. Small numbers of ciliates were counted out from cultures, kept for 24 hours at 25 °C in the presence of ample bacteria and counted again. Experiments were done with a selection of the species of ciliate which are common in activated sludge plants, fed on a variety of species of bacteria. Certain ciliates did not grow well on certain species of bacteria but when the most favourable bacteria were used there were generally about four times as many ciliates at the end of the 24 hour period as at the beginning. The reproduction rate r must have been about $0.06\ \text{h}^{-1}$ ($1.06^{24} = 4$). If this is the maximum rate which can be achieved in a sewage plant attached ciliates should be able to survive but free-swimming ones should not, when F/V, a and b have the typical values which were assumed. The commonest ciliates in activated sludge plants are attached ones such as *Vorticella* and others such as *Aspidisca* (rather similar to *Euplotes*, Fig. 3.1 *d*) which crawl over the sludge particles rather than swim freely in the fluid.

The actual reproduction rate in a sewage plant is likely to be less than the maximum. It must depend on the population density of bacteria which must in turn depend on dilution effects and on the concentration of nutrients in the incoming sewage. Account was taken of this in a computer simulation which reached broadly similar conclusions to those of this simple discussion.

RUMEN CILIATES

The ruminant mammals (cattle, deer, etc.) feed almost entirely on plants which contain a lot of cellulose and hemicellulose in the cell walls. For instance, these materials constitute about 65% of the organic content of mature ryegrass. The ruminants have no enzymes which will digest them, but they have in their stomachs bacteria and ciliates which do. Their stomachs are huge, with a particularly large chamber called the rumen where the micro-organisms flourish. No hydrochloric acid or gastric enzymes are secreted into this chamber so the pH is close to neutrality and the micro-organisms are not digested there. Ciliates (especially *Entodinium*) are numerous in the rumen, but not all of them can digest cellulose.

Like the sewage works described in the previous section of this chapter, the rumen is a habitat through which fluid flows. Bacteria and ciliates are carried out of the rumen with the food into more posterior parts of the gut, where they are digested. If they are to maintain their numbers in the rumen they must reproduce fast enough to compensate for the losses. The division rates of various rumen ciliates have been measured in cultures kept in appropriate conditions. Different rates have been found for different species but rates around 0.04 h^{-1} are typical. Ciliates unable to divide faster than this could only maintain their numbers in the rumen while the dilution rate (F/V, equation 3.1) was less than 0.04 h^{-1}. The dilution rate of the fluid in the rumen of cattle has been measured in various ways, and values around 0.07 h^{-1} have been obtained for cattle on normal rations. However, most of the ciliates are probably closely associated with the solid matter in the rumen which passes through much more slowly: it is changed at rates around 0.02 h^{-1} when the cattle are eating hay. Ground and pelleted food passes through the rumen faster than this, and cattle which are fed on it are liable to lose their ciliates.

Foodstuffs which are used by the rumen bacteria and ciliates for growth become available to their host in due course, when the micro-organisms are carried through to more posterior parts of the gut and digested. Energy which they use for metabolism is lost to the host, but they can only use a small fraction of the energy of the food passing through the rumen because the partial pressure of oxygen in the rumen is extremely low. They convert cellulose and other carbohydrates to fatty acids such as acetic acid by reactions such as

$$C_6H_{12}O_6 = 2CH_3COOH + CO_2 + CH_4$$

Note that this requires no oxygen, but produces carbon dioxide and methane. The heat of combustion of one gram molecule of glucose is 2.9 MJ. That of the two gram molecules of acetic acid which would be formed from it in this reaction is 1.8 MJ. This acetic acid is available to the ruminant. Thus the ruminant loses only a fraction of the energy content of its food to the micro-organisms. It even gets most of the energy from the cellulose and hemicellulose which it could not itself digest. Both the micro-organisms and the ruminant benefit from their association.

However, the ciliates are not essential to the ruminant, and ruminants which have been artificially deprived of ciliates survive with the bacteria alone digesting their cellulose. The bacteria become more plentiful in the absence of ciliates, presumably because there are no ciliates competing with them for the food and because ciliates feed on the bacteria as well as on plant material. In an experiment, lambs with both ciliates and bacteria grew 30% faster than ones with bacteria alone.

Ruminants have no bacteria or ciliates in the rumen when they are born, but they quickly acquire them. The habit of chewing the cud brings rumen contents (including the micro-organisms) into the mother's mouth. When she licks and grooms her young it is likely to swallow a little of her saliva and so acquire enough bacteria and ciliates to start a population in the rumen. Also, it may swallow small quantities of another animal's saliva and micro-organisms accidentally by grazing beside it. Rumen ciliates have been found on the grass in a sheep pasture.

FURTHER READING

GENERAL

See the list for chapter 2

CILIA

Blake, J. R. & Sleigh, M. A. (1974). Mechanics of ciliary locomotion. *Biol. Rev.* **49**, 85–125.

Eckert, R. (1972). Bioelectric control of ciliary activity. *Science* **176**, 473–81.

Machemer, H. (1972). Ciliary activity and the origin of metachrony in *Paramecium*: effects of increased viscosity. *J. exp. Biol.* **57**, 239–59.

MYONEMES

Huang, B. & Pitelka, D. R. (1973). The contractile process in the ciliate *Stentor coeruleus*. I. The role of microtubules and filaments. *J. Cell Biol.* **57**, 704–28.

Newman, E. (1972). Contraction in *Stentor coeruleus*: a cinematic analysis. *Science* **177**, 447–9.

Weis-Fogh, T. & Amos, W. B. (1972). Evidence for a new mechanism of cell motility. *Nature, Lond.* **236**, 301–4.

FEEDING

Bardele, C.F. (1974). Transport of materials in the suctorian tentacle. *Symp. Soc. exp. Biol.* **28**, 191–208.
Sleigh, M. A. & Aiello, E. (1972). The movement of water by cilia. *Acta Protozool.* **11**(31), 265–77.
Tucker, J. B. (1968). Fine structure and function of the cytopharyngeal basket in the ciliate *Nassula. J. Cell. Sci.* **3**, 493–514.
Tucker, J. B. (1972). Microtubule-arms and propulsion of food particles inside a large feeding organelle in the ciliate *Phascolodon vorticella. J. Cell Sci.* **10**, 883–903.
Wessenberg, H. & Antipa, G. (1970). Capture and ingestion of *Paramecium* by *Didinium nasutum. J. Protozool.* **17**, 250–70.

NUCLEI AND REPRODUCTION

Preer, J. R. (1969). Genetics of the Protozoa. In *Research in protozoology*, ed. T.-T. Chen, vol. 3, pp. 129–278. Pergamon. Oxford.
Raikov, I. B. (1969). The macronucleus of ciliates. In *Research in protozoology*. ed. T.-T. Chen, vol. 3, pp. 1–128. Pergamon. Oxford.
Raikov, I. B. (1972). Nuclear phenomena during conjugation and autogamy in ciliates. In *Research in protozoology*, ed. T.-T. Chen, vol. 4, pp. 147–290. Pergamon, Oxford.

SEWAGE TREATMENT

Curds, C. R. (1971). Computer simulations of microbial population dynamics in the activated-sludge process. *Water Res.* **5**, 1049–66.
Curds, C. R. & Vandyke, J. M. (1966). The feeding habits and growth rates of some fresh-water ciliates found in activated-sludge plants. *J. appl. Ecol.* **3**, 127–37.

RUMEN CILIATES

Hungate, R. E. (1966). *The rumen and its microbes.* Academic Press, New York & London.

4
Parasitic protozoa

Phylum Protozoa
  Subphylum Sporozoa
    Class Telosporea
      Subclass Gregarinea
      Subclass Coccidia
    Class Piroplasmea
  Subphylum Cnidospora
    Class Myxosporidea
    Class Microsporidea
(Parasitic members of other subphyla are also discussed in this chapter)

Two subphyla of Protozoa, the Sporozoa and the Cnidospora, consist exclusively of parasites. However, they include little more than half of the known parasitic species of Protozoa. There are plenty of parasites among the flagellates and ciliates, and a few among the amoebas.

A few species of Protozoa cause serious diseases of man and domestic animals. This chapter is largely about them because I expect readers will be particularly interested in them and because research has been concentrated on them. Some of them are Sporozoa. These include *Plasmodium* which causes malaria, *Eimeria* spp. which cause coccidiosis of cattle, sheep and poultry, and *Babesia* which causes red water fever of cattle. Others belong to the Sarcomastigophora. *Trypanosoma* is a genus of flagellates with species which cause sleeping sickness in Africa and Chagas' disease in S. America. *Leishmania* is another genus of flagellates which causes a variety of extremely unpleasant tropical diseases. *Entamoeba histolytica* is an amoeba and causes amoebic dysentery. The next section of this chapter describes the Sporozoa and (much more briefly) the Cnidospora. Succeeding sections deal more generally with the habitats and life cycles of parasitic Protozoa, and with interaction between them and their hosts.

SPOROZOA AND CNIDOSPORA

Fig. 4.1(*a*) shows the structure of *Eimeria*, a typical sporozoan, as seen in electron microscope sections. The anterior part of the body (at the top of the figure) contains two long organelles called rhoptries which look like glands, with small bodies called micronemes scattered around them. At the extreme

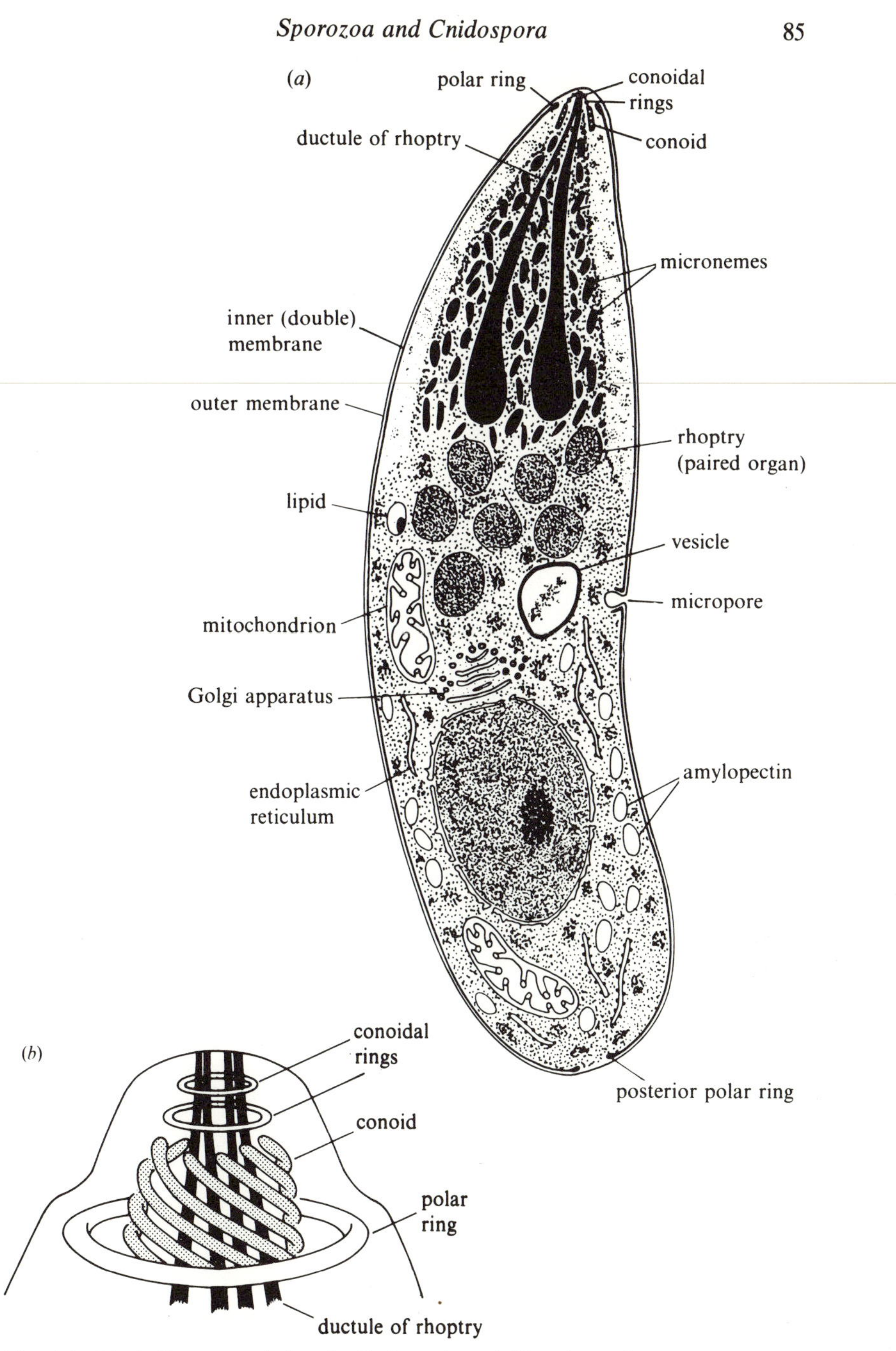

Fig. 4.1. (*a*) A diagrammatic longitudinal section of *Eimeria perforans*, a sporozoan parasite of rabbits. This is the stage in the life cycle known as the merozoite (see Fig. 4.2). From E. Scholtyseck (1973). In *The Coccidia* ed. D. M. Hammond & P. L. Long. University Park Press, Baltimore. (*b*) A diagram of a conoid in the protruded position.

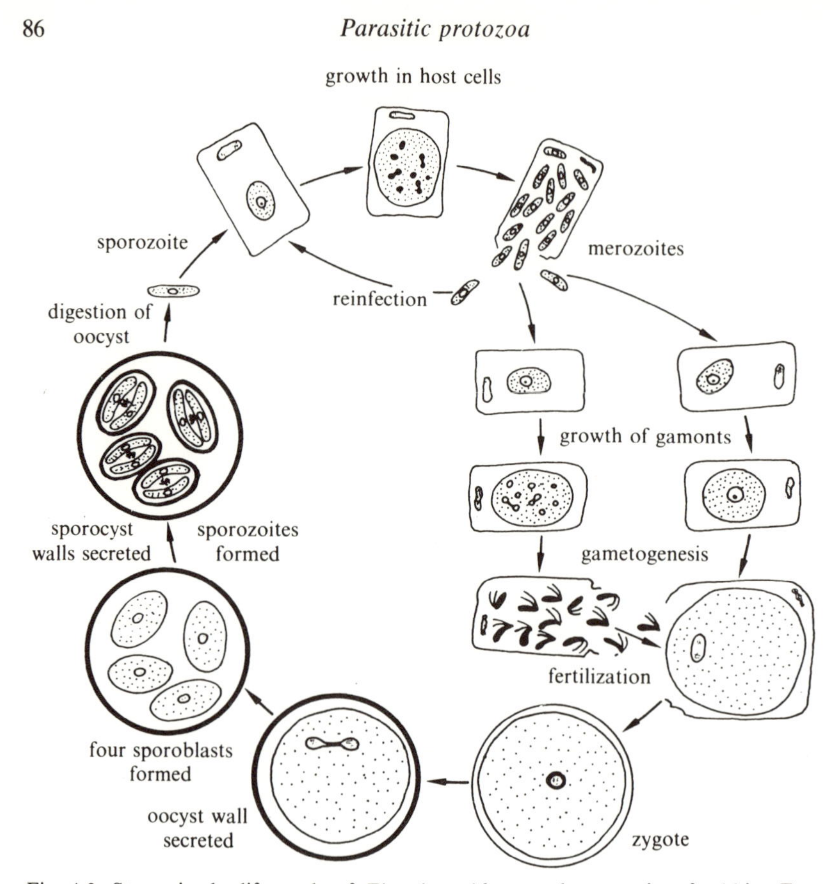

Fig. 4.2. Stages in the life cycle of *Eimeria steidae*, another parasite of rabbits. From M. A. Sleigh (1973). *The biology of Protozoa*. Edward Arnold, London.

anterior end is a small structure known as the conoid which is shown in more detail in Fig. 4.1(*b*). There is a depression in the side of the body called the micropore. There is an outer cell membrane enclosing the cell in the usual way but within it is a double layer of similar membrane. This inner membrane is perforated at the conoid, the micropore and (usually) the posterior end of the body. These structures are characteristic of the Sporozoa but remarkably little is known about their functions, as we shall see.

The form shown in Fig. 4.1 is only one stage in the life cycle which is shown (for another species of *Eimeria*) in Fig. 4.2. The host (in this case a rabbit) ingests the stage known as the oocyst accidentally with its food. The oocyst has an extraordinarily impermeable wall which seems not to have been analysed satisfactorily, and contains sporocysts each containing two haploid sporozoites. The sporozoites are very well protected while in the oocyst: they are unharmed when the oocyst is put in strong disinfectants or acids. However, when the oocyst reaches the small intestine of the host openings are formed

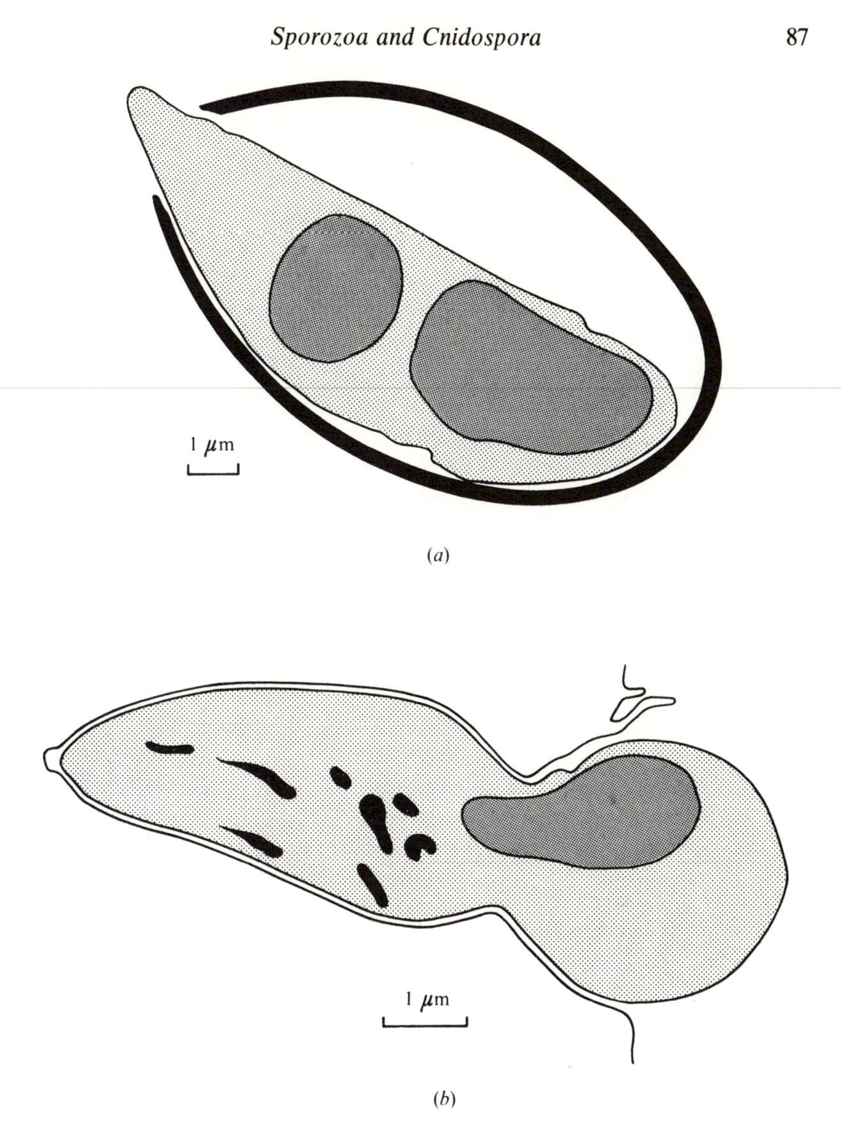

Fig. 4.3. Sporozoites of *Eimeria larimerensis* (*a*) escaping from a sporocyst and (*b*) entering a host cell. Based on electron microscope sections by W. L. Roberts, C. A. Speer & D. M. Hammond: (1970) *J. Parasitol*, **56**, 918–26; and (1971) *J. Parasitol*, **57**, 615–25, respectively.

in its wall and in the walls of the sporocysts, and the sporozoites escape (Fig. 4.3 *a*). Oocysts can be made to release their sporozoites outside a host by exposing them first for several hours to a high partial pressure of carbon dioxide (such as would be found in all parts of the gut of a host) and then to a solution containing trypsin and bile (which are found in the small intestine). Exposure to carbon dioxide weakens or splits the end of the oocyst, probably by stimulating production of an enzyme or by activating an enzyme. Trypsin digests a plug in one end of each sporocyst, so that the sporozoites are free to escape.

There are a very large number of species of *Eimeria*, which are parasites of numerous species of mammal and bird. The sporozoites are released from the oocyst in the gut of any mammal or bird (or presumably any other animal where the conditions are appropriate). However, the life cycle can only be completed in one species, or a small number of species. Oocysts obtained from the faeces of one species can generally infect other species of the same genus and eventually produce a new generation of oocysts, but transmission from one host genus to another generally fails.

The sporozoites enter cells in the body of the host. Each species of *Eimeria* tends to enter a particular type of cell, in a particular part of the gut or liver. *Eimeria steidae* (Fig. 4.2.) enter the epithelial cells of tributaries of the bile duct. They apparently get there by making their way into the blood vessels of the intestine and getting carried in the hepatic portal circulation to the liver.

Various species of *Eimeria* have been watched and filmed entering cells in tissue culture. The sporozoites were generally moving along immediately before entering cells. Only a few seconds were needed to get into the cell, and their motion was scarcely interrupted. Sometimes a slender protuberance from the anterior end of the body is thrust repeatedly into the cell before penetration. The protuberance apparently contains the conoid. Some electron microscope sections of *Eimeria* show the conoid retracted (as in Fig. 4.1 *a*) while others show it protruded through the polar ring (as in Fig. 4.1 *b*). The conoid presumably plays a part in the process of penetration. However, *Plasmodium* has no conoid and enters red blood corpuscles about as quickly as *Eimeria* enters other cells. Both *Eimeria* and *Plasmodium* have rhoptries, which may secrete a substance which aids penetration.

Details of penetration which cannot be seen by light microscopy and in films, are revealed by electron microscope sections (Fig. 4.3 *b*). The parasite does not simply pass through the host cell membrane. The membrane may be broken in places but it caves in so that the parasite is eventually enclosed in a membrane-lined vacuole within the cell. While entering the cell, the parasite is constricted at the point of entry. (This can also be seen by light microscopy.)

Inside the cell, the parasite grows and becomes rounded or irregular in shape, instead of spindle-shaped. The conoid, inner membrane, micronemes and rhoptries gradually disappear. The parasite feeds on the host cell, using the micropore as a cytosome.

As the parasite grows its nucleus divides repeatedly. Eventually the whole parasite divides into a large number of small spindle-shaped merozoites, each with a conoid, inner membrane, micronemes and rhoptries (Fig. 4.1 *a*). These escape from the host cell. They enter other cells and the process of growth and division is repeated. Division of a relatively large intracellular parasite into many small merozoites is a characteristic feature of most sporozoan life cycles. It is called schizogony.

After a small number of these asexual generations the new merozoites become gamonts. They grow within the host cell, but do not divide to form merozoites again. Instead they form gametes, either a single large (female) gamete or numerous small (male) ones. The male gametes consist of a nucleus, a mitochondrion, two or three flagella and very little else. Apart from having more than one flagellum they are similar to vertebrate spermatozoa (and to many invertebrate ones). The male gametes escape from the host cells and fertilize any female gametes which may be available, producing diploid zygotes. Each zygote secretes an oocyst wall around itself, and divides to form eight sporozoites. This division involves a meiosis, so that the sporozoites are haploid. The oocysts are discharged from the host cells, and eventually leave the host in the faeces. If they are ingested with food by another animal of the host species, the life cycle begins again.

The likelihood that a particular oocyst will be eaten by the host species may be very low, but if it is eaten it may produce an enormous number of oocysts of the next generation. *Eimeria bovis*, which causes a troublesome disease of cattle, provides a striking example. Each oocyst contains eight sporozoites. Each of these may produce, in the first asexual generation, 100000 merozoites. There is a second asexual generation in which each of these merozoites may produce about 30 merozoites, which produce gamonts. If no parasite cells died, one oocyst could produce $8 \times 100000 \times 30 = 24$ million gamonts. Only 50% of gamonts can be expected to be female so n oocysts could, in principle, give rise to 12 million n oocysts of the next generation. The number actually produced is likely to be a lot less than this but it seems clear that each oocyst needs only a small probability of entering an appropriate host, for the species to survive. *E. bovis* grows exceptionally large in the host cells and produces exceptionally large numbers of merozoites, but many other species of *Eimeria* have more than two asexual generations and may have as high a potential for multiplication within the host.

The structure and life cycle of *Eimeria* has been described in some detail, as an example of the Sporozoa. The next few paragraphs are about major groups within the Sporozoa.

The subclass Gregarinia consists of parasites of the guts and body cavities of invertebrates. Sporozoites enter the cells of the host and may produce one or more generations of merozoites which infect other cells, but eventually the parasite leaves the host cells and lives outside them for some time (a striking difference from the Coccidia). It grows to a large size before becoming a gamont. Species of *Selenidium* which live in the guts of polychaete worms grow over 200 μm long, and some gregarines grow much larger. The extracellular stages of gregarines are often attached to the tissues of the host by the conoid end of the body. Electron microscope sections of the attachment points (Fig. 4.4 *a*) show no material connection between parasite and host cells: the cell

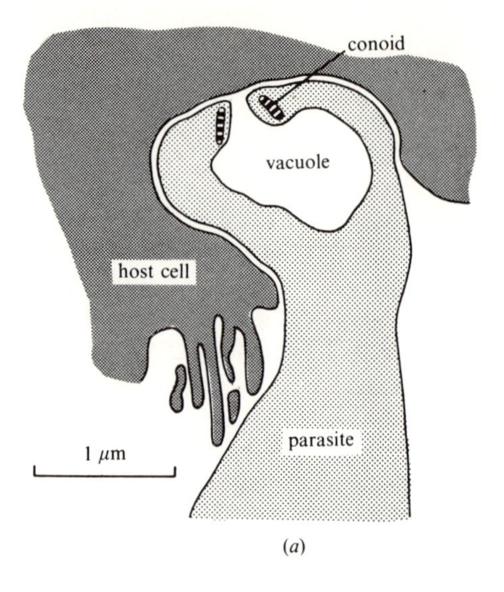

(*a*)

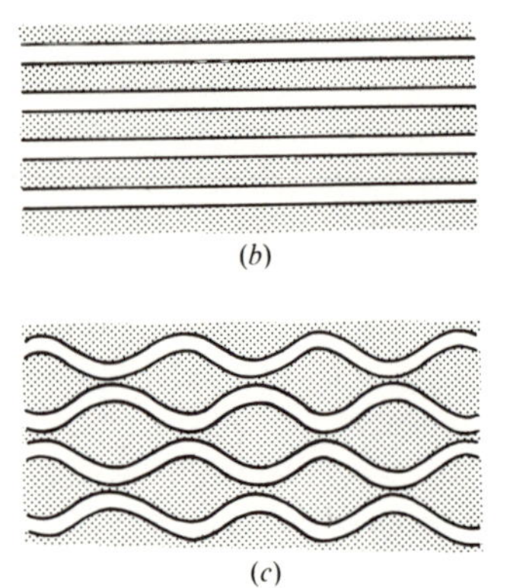

(*b*)

(*c*)

Fig. 4.4. (*a*) A longitudinal section of the anterior end of a gregarine (*Selenidium*) attached to a cell in the gut epithelium of the worm *Sabellaria*. Based on an electron micrograph in J. Schrevel (1968). *J. Microsc.* **7**, 391–410. (*b*), (*c*) Diagrams of ridges seen on the surfaces of gregarines by scanning electron microscopy.

membranes run parallel to each other, about 15 nm apart. The mechanisms of this and other types of attachment between cells are discussed in chapter 5. Fig. 4.4(*a*) also seems to show a food vacuole forming, taking in fragments of host cells through the conoid. Feeding through the conoid rather than through the micropores may be a common habit of the extracellular stages of gregarines. The large extracellular stages of gregarines become gamonts which come together in pairs, presumably of opposite sex. Each pair encloses itself in a protective cyst. The gamonts divide to form large numbers of gametes. Fertilization occurs, producing zygotes which produce the sporozoites of the

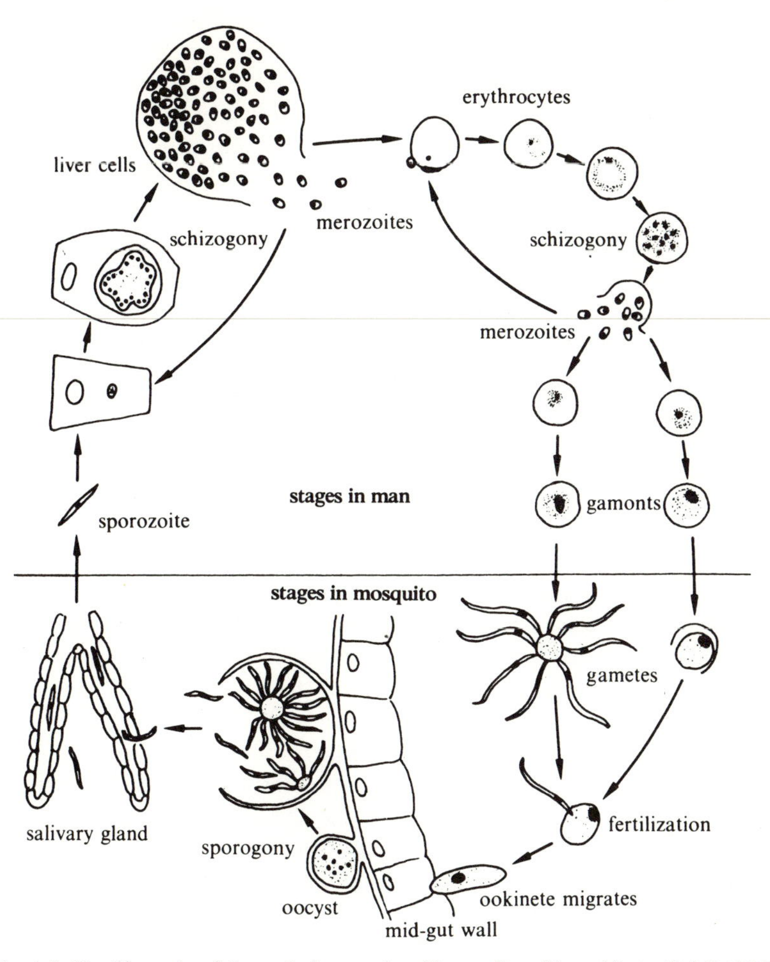

Fig. 4.5. The life cycle of the malaria parasite, *Plasmodium*. From M. A. Sleigh (1973). *The biology of Protozoa*. Edward Arnold, London.

next generation. New hosts are presumably infected by cysts voided in the faeces of the original host.

Gregarines glide around in a most mysterious-looking way. No wriggling or other propulsive movements can be seen by light microscopy but the animals move along at 5 $\mu m\ s^{-1}$ or more. The body has parallel ridges running along it, about 0.3 μm apart. These ridges are scarcely visible by light microscopy, but they are clearly visible by scanning electron microscopy. In some of the specimens which have been examined the ridges are straight (Fig. 4.4 *b*) but in others they undulate in the manner shown in Fig. 4.4(*c*). It is supposed that the waves travel backwards along the ridges, pushing the surrounding fluid backwards and the animal forwards. However, the supposed movement of the waves has not been seen because the scanning electron microscope can only be used to examine dead specimens.

The subclass Coccidia includes parasites both of vertebrates and of invertebrates. *Eimeria* belongs to this subclass and so does *Plasmodium*, which causes malaria. *Plasmodium* species spend part of their life cycle in a vertebrate and part in a mosquito (Fig. 4.5). The mosquito becomes infected by feeding on the blood of an infected vertebrate. It infects other vertebrates by injecting infected saliva into them when it feeds. Asexual reproduction occurs in the vertebrate, and sexual reproduction in the mosquito. The parasite is injected into the vertebrate as a sporozoite. The species which cause malaria in man enter liver cells where they grow and then undergo schizogony, dividing into very large numbers of merozoites. Some of these may enter other liver cells but more enter red blood corpuscles. It takes them only a few seconds to get into a corpuscle. Once inside the corpuscles they feed on haemoglobin, taking it in through the micropore, but they also use glucose and other foodstuffs which diffuse into the corpuscle from the plasma. The merozoites cannot survive long outside the corpuscles, and there is evidence that they cannot synthesize their own coenzyme A. Once in a corpuscle, they can obtain this vital material from the cytoplasm of the corpuscle.

Infected corpuscles are less dense than uninfected ones and so can be separated almost completely from them by repeated centrifugation of blood from an infected animal. When they are separated in this way it can be shown that infected corpuscles use oxygen and glucose many times faster than uninfected ones, presumably owing to the metabolism of the parasites.

The parasites apparently increase the permeability of the red cell membrane, making it easier for foodstuffs to diffuse in. This has been demonstrated by experiments with infected and uninfected cells from the same animal, separated by centrifugation. The cells were suspended for a short time in a solution of L-glucose labelled with ^{14}C. They were centrifuged out again, and the radioactivity of the pellet was measured. The pellet retained a little of the solution between the cells, but the amount was measured by an experiment with radioactive inulin (see p. 47) and corrected for. It was found that much more L-glucose had entered the infected cells than the uninfected ones. The parasites had apparently made the cell membrane much more permeable to L-glucose and presumably also to D-glucose and other small molecules (D-glucose is the form which occurs naturally in blood, and is metabolized).

Thus the parasite feeds on materials from the contents of the corpuscle and from the plasma. It grows (but not very large, since the diameter of the corpuscle is only 7.5 μm) and divides into 6–24 second-generation merozoites which escape and infect other corpuscles. Some go through more asexual generations but some become gamonts which remain dormant in the corpuscles for as long as the corpuscles survive, unless a mosquito swallows them with a meal of blood. Fertilization occurs in the gut of the mosquito, forming a zygote which becomes a worm-like ookinete, burrows through the wall of the gut and

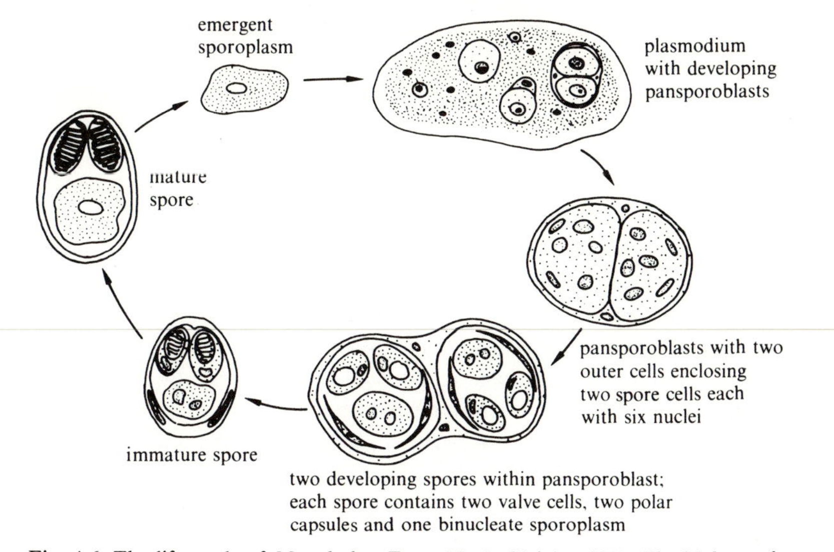

Fig. 4.6. The life cycle of *Myxobolus*. From M. A. Sleigh (1973). *The biology of Protozoa*. Edward Arnold, London.

forms an oocyst. This divides into large numbers of sporozoites which are released into the blood of the insect. Some of them get into the salivary glands and may be injected into another vertebrate when the mosquito feeds. The saliva has no digestive function, but contains a substance which prevents the vertebrate's blood from clotting at the bite.

It should be clear from comparisons of Figs. 4.2 and 4.5. that there is close similarity between the life cycles of *Eimeria* and *Plasmodium*, though one is completed in a single host and the other requires two. In each case there is repeated schizogony in the vertebrate host, but a new vertebrate host can only be infected by sporozoites produced by sexual reproduction.

The class Piroplasmea is a group of parasites of the red blood corpuscles of vertebrates. Species of *Babesia* cause serious diseases of domestic animals, which are transmitted by blood-sucking ticks. The intracellular stages of other Sporozoa are normally contained in vacuoles lined by host cell membrane (Fig. 4.3 *b*), but the cell membrane of Piroplasmea is directly in contact with the cytoplasm of the host cell.

So far, this section has been about Sporozoa. This final paragraph is about the Cnidospora, which are generally treated as a subphylum of the Protozoa though some of them (the Myxosporea) include multicellular stages in their life cycle. *Myxobolus* (Fig. 4.6) is one of the Myxosporea. It infects cyprinid fish such as roach, causing boils in which spores develop as multicellular structures. Spores are released when the boil bursts and infect other fish which swallow them. Both classes of Cnidospora have polar filaments which are remarkably like the nematocysts of Cnidaria (chapter 6). They are long tubular structures which are initially helically coiled within the spore but turn inside-out to become long projections from it. There do not seem to be any species of Cnidospora which normally infect man but there are species which cause troublesome diseases of honey bees, silkworms and fish.

HABITATS FOR PARASITES

'Parasite' is a term used both in a broad sense and in a narrow one. In the broad sense, it means any organism which lives in or on another. The bacteria and ciliates in the rumens of cattle are parasites in this sense, though the cattle benefit from their presence. So are the green flagellates which live in corals, which also confer an advantage (chapter 6). Very often, however, the term 'parasite' is restricted to organisms which harm their host by feeding on its tissues or by robbing it of food without giving a compensating advantage. *Eimeria* is plainly a parasite in this sense, but rumen ciliates are not. This section is about some of the habitats available to parasites (in the broadest sense) and about the problems they present to the Protozoa which inhabit them.

Ectoparasites live on the external surfaces of their hosts. *Vorticella* living attached to plants could conceivably be regarded as ectoparasites, but this would be rather an extreme view since they might just as advantageously be attached to some inanimate object. The gills of aquatic animals provide a particularly favourable habitat for some Protozoa because the host keeps water flowing over them, bearing bacteria and other particles of food. Firm attachment is often necessary, so that the parasite is not washed off the gills by the water currents. Various Protozoa such as *Epistylis* (a peritrich ciliate) can often be found attached to the gill plates of freshwater species of the crustacean *Gammarus*. Many species of Protozoa live on the gills of molluscs and of fish. Three examples from fish will illustrate their variety. *Brooklynella* is a ciliate which crawls over the gills but has a cavity in its body opening by a pore, from which it secretes a substance which is suspected of being an adhesive. Perhaps it sticks itself temporarily to the gills and feeds there for a while, before releasing itself and moving on. It feeds on the epithelial cells of the gills, and may cause inflammation, bleeding and the death of the host. It has been a very troublesome pest in at least one public aquarium where it infested a variety of marine fish. *Trichophrya* is a suctorian which attaches itself (presumably permanently) to fish gills, by means of a cement. It seems to cause no irritation, and it feeds on other ciliates. It seems to be harmless to its host. *Cryptobia* is a flagellate which is attached to the gills by one of its flagella. The attachment seems to depend on close apposition of the flagellar membrane to the epithelial cell membrane, just as the attachment of gregarines to gut cells depends on close apposition of membranes (p. 89).

Ectoparasites which live attached to their hosts must of course have in their life history a detached stage which can infect new hosts.

Endoparasites live inside their hosts in a very wide variety of habitats. Many live in the guts of vertebrates and have to live more or less without oxygen. The ciliates which live in the rumen have already been discussed (p. 81). They are digested when they are passed to more posterior parts of the gut so they must leave the host by its mouth if they are to infect other hosts. They are brought to the mouth by the process of chewing the cud and are particularly likely to infect a calf when it is licked by its mother.

Many other parasites which live in the gut or in the cells of the gut wall are transmitted via faeces which contaminate the food of new hosts. *Eimeria* is an example (p. 84). Another is *Entamoeba histolytica* which causes amoebic dysentry in man. It lives in the large intestine, which is also inhabited by bacteria. The bacteria feed on such remnants of food as reach the large intestine without having been digested, and the *Entomoeba* feed on them. *Entomoeba* can be kept and will multiply in cultures of suitable bacteria but it has been found very difficult to devise media in which it can grow without bacteria. While it lives in the lumen of the intestine it does no harm but it sometimes invades the gut wall and other tissues, feeding on them and causing disease. The response of the large intestine to damage is diarrhoea. This may seem inconvenient to the sufferer, but is a very appropriate response because it increases the dilution rate of the contents of the intestine. If the dilution rate exceeds the rate of reproduction of the amoeba, the population of amoebae in the lumen of the intestine must fall.

When the host is healthy he forms normal faeces, removing water from them in the large intestine. *Entamoeba* in the faeces respond to dehydration in the same way as *Acanthamoeba* (p. 32), by enclosing themselves in protective cysts. These survive in the faeces and may enter another man when he eats contaminated food or drinks contaminated water. Food may be contaminated by handlers who take insufficient care over the cleanliness of their hands, or by flies which land on food after visiting faeces. The *Entamoeba* emerge from their cysts in the small intestine of their new host.

Parasites which live free in the gut lumen can only maintain their population if their reproduction rate balances the dilution rate. Those that attach themselves to the gut wall need not reproduce so fast. *Giardia* is a common parasite of the human intestine, which attaches itself to the intestine wall and causes inflammation. It is a flagellate, and attaches itself by a suction device worked by its flagella.

Though the partial pressure of dissolved oxygen is very low indeed in most of the fluid in the gut, it is higher very close to the gut wall where *Giardia* (for instance) lives. This is because oxygen diffuses in from the blood vessels of the gut wall. The partial pressure of oxygen in the intestine of ducks has been measured by inserting a small oxygen electrode. It was found to be less than 0.001 atm in the centre of the lumen but about 0.03 atm close to the wall. It is about 0.21 atm in air. Parasites in the centre must depend on anaerobic

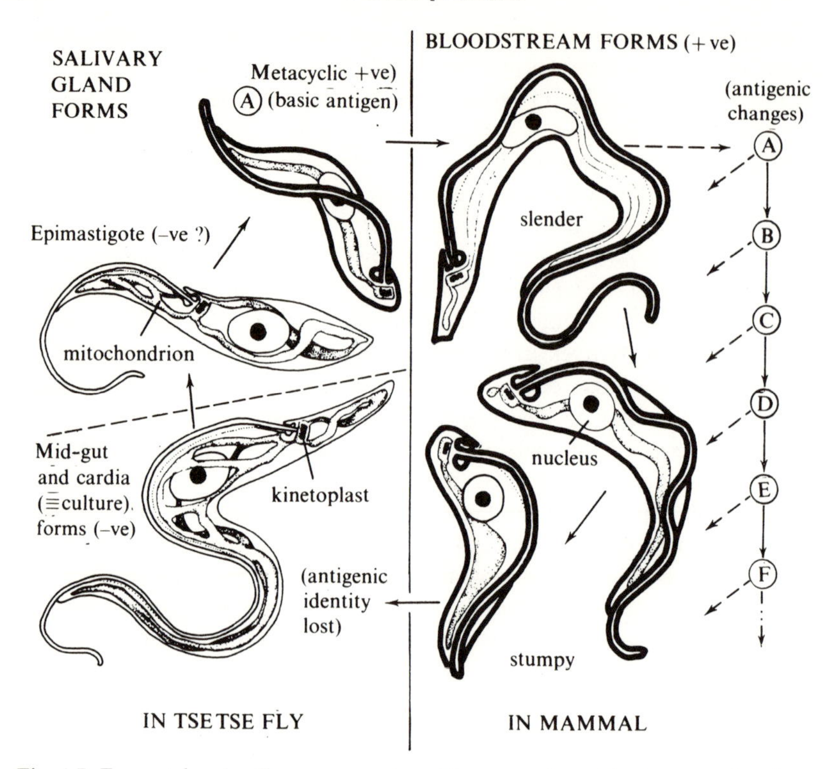

Fig. 4.7. Forms taken by *Trypanosoma brucei* in mammals and in the tsetse fly. From K. Vickerman (1969). *J. Cell Sci.* **5**, 163–93.

respiration like the micro-organisms of the rumen in cattle, but those which attach themselves closely to the wall may be able to oxidize some or all of their food.

A gut parasite cannot survive unless it can withstand the digestive enzymes of its part of the gut. Rumen ciliates are not harmed by the saliva in the rumen but are digested in more posterior parts of the gut.

Many parasites spend at least part of their life cycle in the blood of vertebrates. Blood-sucking insects carry many of them from one host to another. The malaria parasites (*Plasmodium*) are transmitted by mosquitoes. African sleeping sickness is transmitted by the tsetse flies (*Glossina*) in essentially the same way. It is caused by *Trypanosoma brucei* (Fig. 2.1 *d*) which lives in the blood plasma of men and antelopes. The trypanosomes are swallowed by tsetse flies when they take meals of blood and multiply in the flies' guts. They infect the salivary glands and so are likely to be injected with saliva into a new host. Another trypanosome, *Trypanosoma cruzi*, causes Chagas' disease which is an important disease in S. America. It parasitizes dogs, cats, opossums and monkeys as well as man and is transmitted by blood-sucking bugs including *Rhodnius prolixus*. It establishes itself in the gut of the bug and is passed in the faeces. If the bug defecates while feeding on a man the faeces may get into the bite or other abrasions and infect him with the disease. The bugs flourish in mud huts: they spend the day in cracks in the walls and come out at night to feed on sleeping people. Improved housing has done a lot to reduce the prevalence of the disease. *Leishmania* is transmitted by biting sandflies (*Phlebotomus*) and *Babesia* by ticks.

Trypanosoma brucei occupies very different habitats, in the blood of mammals and in the gut of the tsetse fly. It has to adapt itself to each, as it is transferred from one to the other. Fig. 4.7 shows anatomical changes which occur in the two hosts. In mammals, the trypanosomes have long but unbranched mitochondria, and the cell membrane is covered by an external coat about 15 nm thick. In the tsetse fly the mitochondria are larger and more elaborate in shape, and the surface coat is lost. The trypanosomes change back to the bloodstream form in the insect's salivary glands. Trypanosomes can be kept in culture as well as in host animals, and when they are kept this way they take the form found in the gut of the tsetse fly. Trypanosomes swim free in the blood, but attach themselves by their flagella to the wall of the tsetse fly's gut. The attachment is by close apposition, like the attachment of *Cryptobia* to fish gills. Asexual reproduction occurs both in the mammal and in the tsetse fly; trypanosomes have not been observed to reproduce sexually.

There is a difference in metabolism between the bloodstream and culture (or tsetse fly) forms of trypanosomes. This has been demonstrated by experiments with suspensions of trypanosomes from the blood of rats and from cultures. The bloodstream forms were separated from blood corpuscles by centrifugation. The respiratory rates of both forms were measured in Warburg manometers, in solutions of various foodstuffs, and the solutions were analysed afterwards to discover how much of each foodstuff was used and what substances were produced.

Both forms ceased moving and stopped using oxygen in solutions containing no foodstuffs; they seem to have no food reserves in their bodies. They could use glycerol, glucose and certain other sugars. The culture form could also use succinate and other substances involved in the Krebs cycle, but the bloodstream form could not. The culture form oxidizes most of the glucose it uses completely, to carbon dioxide and water

$$C_6H_{12}O_6+6O_2=6CO_2+6H_2O+38(\sim)$$

where $(\sim)$ represents an energy-rich phosphate bond. However, the bloodstream form cannot oxidize glucose completely, but even in the presence of ample oxygen converts it mainly to pyruvic acid

$$C_6H_{12}O_6+O_2=2CH_3COCOOH+2H_2O+8(\sim)$$

It apparently lacks the enzymes of the Krebs cycle, which oxidize pyruvic acid to carbon dioxide and are normally located in mitochondria. This may explain why the mitochondria of the bloodstream form are relatively small.

When glucose is oxidized to carbon dioxide, 6 molecules of oxygen yield 38 energy-rich bonds, 6.3 bonds per molecule. When it is oxidized to pyruvic acid each molecule of oxygen yields 8 energy-rich bonds so slightly less oxygen is needed to obtain a given amount of energy (though very much more glucose is used). The partial pressure of oxygen is high in arteries but may be quite low in veins, so there may be some advantage to the bloodstream form in using oxygen economically. In any case, glucose is very plentiful in mammal blood so there is no need to use it economically. Conversion to pyruvic acid leaves much of the energy of the glucose unused (and available to the host) but this does not matter to the parasite if plenty of glucose is available. In the gut of the tsetse fly glucose and alternative foods are probably not as constantly plentiful, so there is an advantage to the parasite in using them economically. The partial pressure of oxygen in the guts of tsetse flies is not known, but is probably much higher than in the guts of vertebrates. Otherwise the culture (and tsetse fly) form of the trypanosome would presumably not have the ability to oxidize glucose completely.

The external coat which is present in the bloodstream form but is shed in the tsetse fly probably plays an important role in the defence of the parasite against the immune system of the mammal host. This is discussed in the next section of this chapter.

This section has dealt with some ectoparasites, gut parasites and blood parasites, and the special problems of life in their habitats. Sporozoa are intracellular parasites (at least for part of their life cycle) and were discussed in the previous section, where some of the problems of life inside the cells of a host were considered. They will not be considered further here although there are other intracellular protozoan parasites, which do not belong to the Sporozoa. They include *Trypanosoma cruzi* and *Leishmania*. *T. cruzi*, the cause of Chagas' disease, invades muscle cells (including heart cells) and reticulo-endothelial cells in the liver and spleen. Symptoms include swelling of the heart, liver and spleen. *Leishmania* invades reticulo-endothelial cells, and in kala-azar (one of the human diseases it causes) the spleen may swell from its normal weight of 150 g to 1 kg and even, in an extreme case, 3.5 kg.

IMMUNITY TO PROTOZOAN DISEASES

Parasites of vertebrates have to withstand the immune responses of the host, which tend to destroy them and any other foreign cells which enter the body. The vertebrate host produces proteins known as antibodies which react specifically with particular molecules on the surfaces of the foreign cells, but not with the surfaces of host cells. The foreign molecules against which antibodies are produced are known as antigens. Antibodies attack foreign cells in various ways. Some agglutinate them (stick them together in clumps). Others make them swell and burst. These antibodies attach complement (a group of proteins) to the cell membrane, and the complement apparently makes small

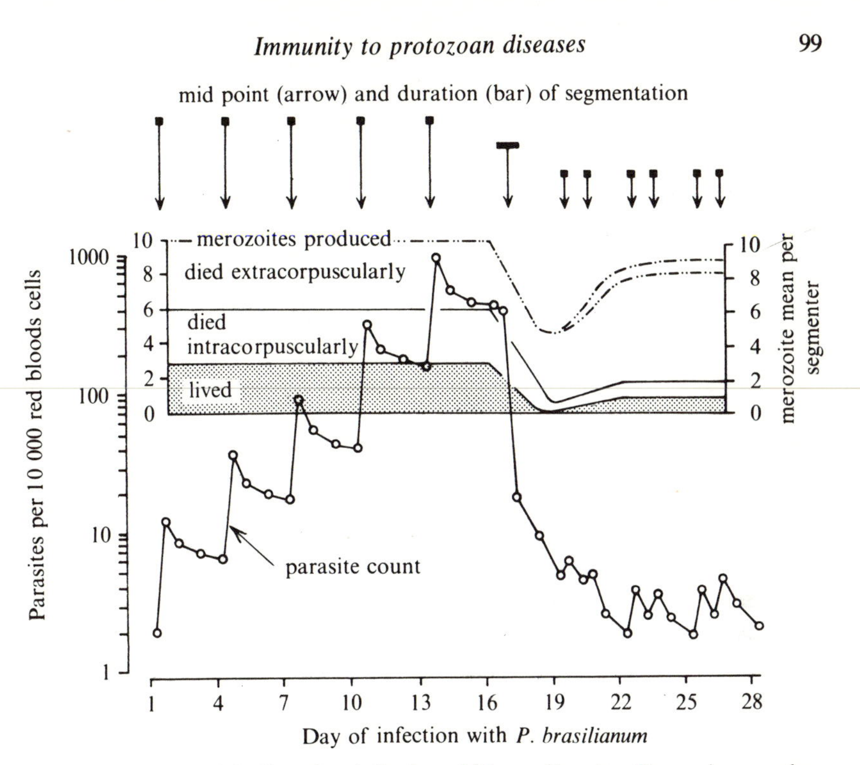

Fig. 4.8. Growth and decline of an infection of *Plasmodium brasilianum* in a monkey. From W. H. Taliaferro & L. Stauber (1969). In *Research in protozoology*, ed. T.-T. Chen, vol. 3, pp. 505–64. Pergamon, Oxford.

holes in the membrane by enzyme action. Ions can diffuse through these holes but the large organic molecules of the cell cannot, so the cell approaches Donnan equilibrium. Since equilibrium would involve a considerably higher pressure in the cell than outside it, the cell swells up and bursts (see p. 46). Foreign cells may also be engulfed and digested by macrophages, much as *Amoeba* engulfs food. The macrophages seem to 'recognize' the foreign cells by antibody attached to them.

All these processes involve antibodies synthesized by the host to attack the particular foreign cells which have invaded it. They are produced by the reticulo-endothelial cells in the spleen, lymph nodes and bone marrow and can be produced quite quickly, within a few days. The host also responds to foreign cells by producing more macrophages in the spleen and lymph nodes, and releasing them into the blood.

The immune responses of the host can have a dramatic effect on a protozoan infection, as Fig. 4.8 shows. A monkey was infected with a malarial parasite. Blood samples were taken daily and the number of parasites in them was counted. The number of parasites rose very rapidly. On the seventeenth day of the infection one red corpuscle in 25 had a parasite in it. This rise was achieved in a series of steps, because all the parasites in red corpuscles undergo schizogony more or less simultaneously. Each schizogony multiplied the number of parasites by about 10, but 4 out of every 10 merozoites apparently died without entering a blood corpuscle and a further 3 died in corpuscles leaving only 3 to undergo schizogony. The net effect of each cycle of growth and schizogony was multiplication of the number of parasites by about three, so the number was increased by a factor of nearly 3^5 in five cycles. A sixth schizogony occurred on the seventeenth day and the number of parasites would have risen again, had the immune response of the host not become effective at this stage. However, the immune response did become effective and the parasite population fell very rapidly indeed.

Fever in malaria patients recurs every 48 or 72 hours (according to the species of parasite), at the time when the red corpuscles burst and release merozoites into the blood. The symptoms of African sleeping sickness are also apt to fluctuate at intervals of a few days, for a quite different reason. Partial recoveries, during which the number of trypanosomes in the blood is quite low, alternative with relapses during which the number is much higher. It seems that the immune responses of the host are relatively effective during the recovery phases and ineffective during the relapses. The host's antibodies presumably act against the macromolecules of the external coat of the trypanosomes. It is believed that the trypanosomes alter the chemical constitution of this coat periodically, so that the antibodies produced by the host against previous coats are no longer effective. A relapse occurs in the interval between adoption of a new coat and production of new antibody to attack it. Each change of coat gives the parasite temporary respite from the attacks of host antibodies.

Evidence for this comes from experiments with goats and rabbits. The animals (which were initially uninfected) were infected by a single bite of the tsetse fly. Blood samples were taken at intervals of a few days and used to prepare samples of serum and trypanosomes. The trypanosomes were rather sparse in the blood of the goats and rabbits, so larger numbers were obtained by injecting samples of the blood into mice. The trypanosomes multiplied in the mice. Blood from the first mouse was used to infect a second one after 2–3 days, a third mouse was infected from the second 2–3 days after that, and so on. In this way blood containing very large numbers of trypanosomes was obtained, without allowing enough time in any one mouse for that mouse to produce antibodies or for the trypanosome to change its coat. Trypanosomes were separated from the mouse blood by centrifugation and suspended in a saline solution. The goat and rabbit serum and the suspensions of trypanosomes were stored deep frozen until they were required for agglutination tests.

In each test, a drop of trypanosome suspension and a drop of diluted serum were mixed on a microscope slide. They were examined half an hour later to see whether the trypanosomes had formed clumps. If they had, the serum must have contained antibodies effective against the surface coats of those particular trypanosomes. The tests were repeated with more and more dilute sera, to assess the concentration of the antibodies, which was expressed as the

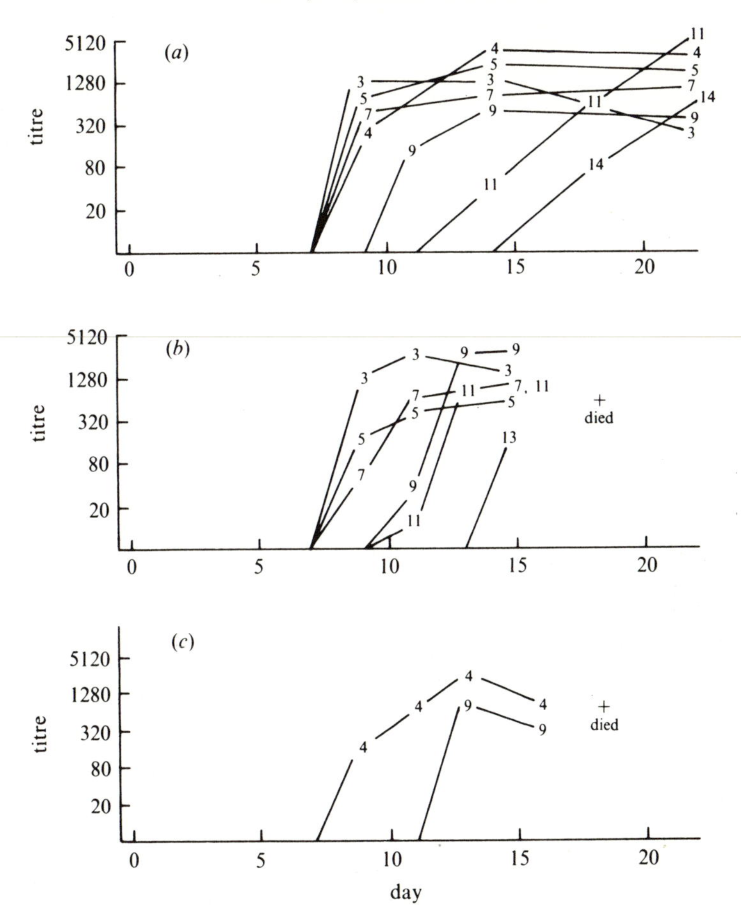

Fig. 4.9. Graphs of titre of antibodies against *Trypanosoma brucei*, against number of days since initial infection. (*a*) Titres of antibodies in the serum of a rabbit to trypanosomes from the same rabbit; (*b*) titres in a goat to trypanosomes from the goat; and (*c*) titres in the goat to trypanosomes from the rabbit. In each case separate lines show titres to trypanosomes taken from the host on the days indicated by numbers on the lines; for instance, the line 3–3 refers to trypanosomes taken on day 3. Re-drawn from A. R. Gray (1965). *J. gen. Microbiol.* **41**, 195–214.

titre. If the trypanosomes were clumped by serum diluted to 10 times its initial volume, but not by serum diluted 20 times, the titre was 10. If another sample of serum agglutinated the trypanosomes when diluted 1280 times but not 2560 times, its titre was 1280. A high titre indicates a high concentration of antibodies.

Fig. 4.9 shows some of the results of these experiments. In (*a*) are the results of an experiment with a rabbit. Samples of blood were taken every few days and serum from each sample was tested for antibodies against samples of trypanosomes taken from the same animal on the same or different days.

Up to 7 days after infection, no evidence of antibodies against any of the trypanosomes could be found. Serum from day 9 had high titres of antibody effective against trypanosomes from days 3, 4, 5 and 7 but failed to agglutinate trypanosomes from day 9 or later days. By day 11, antibodies against day 9 trypanosomes had appeared. Antibodies against day 11 and day 14 trypanosomes first appeared on days 14 and 18, respectively.

These results can be explained as follows. The trypanosomes injected by the tsetse fly had antigen A in their surface coats. No antibodies were produced by the rabbit until between days 7 and 9 when antibody A (effective against antigen A) was produced. By day 9, however, the trypanosomes had changed their antigen to B so that antibody A was no longer effective against them. By day 11 the blood had antibody B as well as antibody A, but the trypanosomes had changed to antigen C. On day 14 the blood had antibodies A, B and C but the trypanosome had changed again to antigen D.

Fig. 4.9(*d*) shows very similar results from an experiment with a goat. Successive antibodies were detected on days 8, 11 and 15. Fig. 4.9(*c*) shows the results of testing serum from the goat for antibodies against trypanosomes from the rabbit. The antibody which was first detected on day 8 was effective against day 4 trypanosomes from the rabbit and the one detected on day 11 against day 9 trypanosomes from the rabbit. It seems that the trypanosomes injected into the rabbit and goat (by different tsetse flies) both had the same antigen A, and that the first change of antigen in each case was the same antigen B. Further experiments with rabbits and goats confirmed what this one suggests, that *Trypanosoma brucei* generally goes through the same sequence of antigen changes (A, B, C, etc.) in mammal hosts but reverts to A when ingested by a tsetse fly. This is indicated in Fig. 4.7.

FURTHER READING

GENERAL

Marcial-Rojas, R. A. (ed.) (1971). *Pathology of protozoal and helmintic diseases.* Baltimore, Williams & Wilkins.

von Brand, T. (1973). *Biochemistry of parasites*, 2nd edn. Academic Press, New York & London.

Vickerman, K. (1972). The host–parasite interface of parasitic Protozoa. *Symp. Br. Soc. Parasitol.* **10**, 71–91.

SPOROZOA AND CNIDOSPORA

Frenkel, J. K. (1974). Advances in the biology of Sporozoa. *Z. Parasitenk.* **45**, 125–62.

Hammond, D. M. & Long, P. L. (1973). *The Coccidia.* University Park Press, Baltimore.

Homewood, C. A. & Neame, K. D. (1974). Malaria and the permeability of the host erythrocyte. *Nature, Lond.* **252**, 718–19.

Ladda, R., Masamichi, A. & Sprinz, H. (1969). Penetration of erythrocytes by merozoites of mammalian and avian malarial parasites. *J. Parasitol.* **55**, 633–44.

Vavra, J. & Small, E. B. (1969). Scanning electron microscopy of gregarines (Protozoa, Sporozoa) and its contribution to the theory of gregarine movement. *J. Protozool.* **16**, 745–57.

IMMUNITY TO PROTOZOAN DISEASES

Gray, A. R. (1965). Antigenic variation in a strain of *Tryapnosoma brucei* transmitted by *Glossinia morsitans* and *G. palpalis. J. gen. Microbiol.* **41**, 195–214.

Taliaferro, W. H. & Stauber, L. A. (1969). Immunology of protozoan infections. In *Research in protozoology*, ed. T.-T. Chen, vol. 3, pp. 505–64. Pergamon, Oxford.

5

Sponges

Phylum Porifera
  Class Calcarea
  Class Hexactinellida
  Class Demospongia

The sponges are very simple multicellular animals. They are more plentiful in the sea than many people realize, and there are a few freshwater species. Some form thin layers of living matter encrusting rocks. Others stand up from their points of attachment, as vase-shaped structures of various sizes or as irregularly shaped lumps. All live attached to rocks or other submerged objects. Many of them are brightly coloured. They make no very obvious movements but some of them contract the openings in their bodies when they are touched. They are common on shores, and they cover the rock surfaces of some submerged caves almost to the exclusion of other attached animals. They occur at great depths in the oceans, as well as in shallow water.

The structure of *Leucosolenia* is shown in Fig. 5.1. This is a small sponge, about 2 cm high, and an unusually simple one. It is white, and is common in pools on British shores. It has a central cavity connected to the surrounding water by many small pores (ostia) and one large one (the osculum). It is easy to show by releasing dye into the water near a sponge that water flows continuously in through the ostia and out through the osculum. The outer surface of *Leucosolenia* is covered by an epithelium of flat cells (like paving stones). The inner surface is covered by choanocytes which are cells with flagella and which will be described in the next paragraph. Between these two layers of cells is the mesogloea, a jelly-like material which has other cells embedded in it. Spicules of calcium carbonate, which are described in a later section of this chapter, are embedded in the mesogloea. Some are completely embedded in it, and stiffen the sponge. Others have only their bases embedded in it and project from the animal, probably helping to protect it. Any animal which tries to eat the sponge is likely to get an unpleasantly spiky mouthful. Cells called porocytes, shaped like napkin rings, form walls for the ostia.

The structure of choanocytes is shown in more detail in Fig. 5.2(*a*) and (*b*). Each has a single flagellum, edged by a delicate vane. Around the base of the flagellum is a collar which can be seen by electron microscopy to consist of a ring of microvilli about 0.2 μm in diameter spaced 0.2 μm apart. Very fine

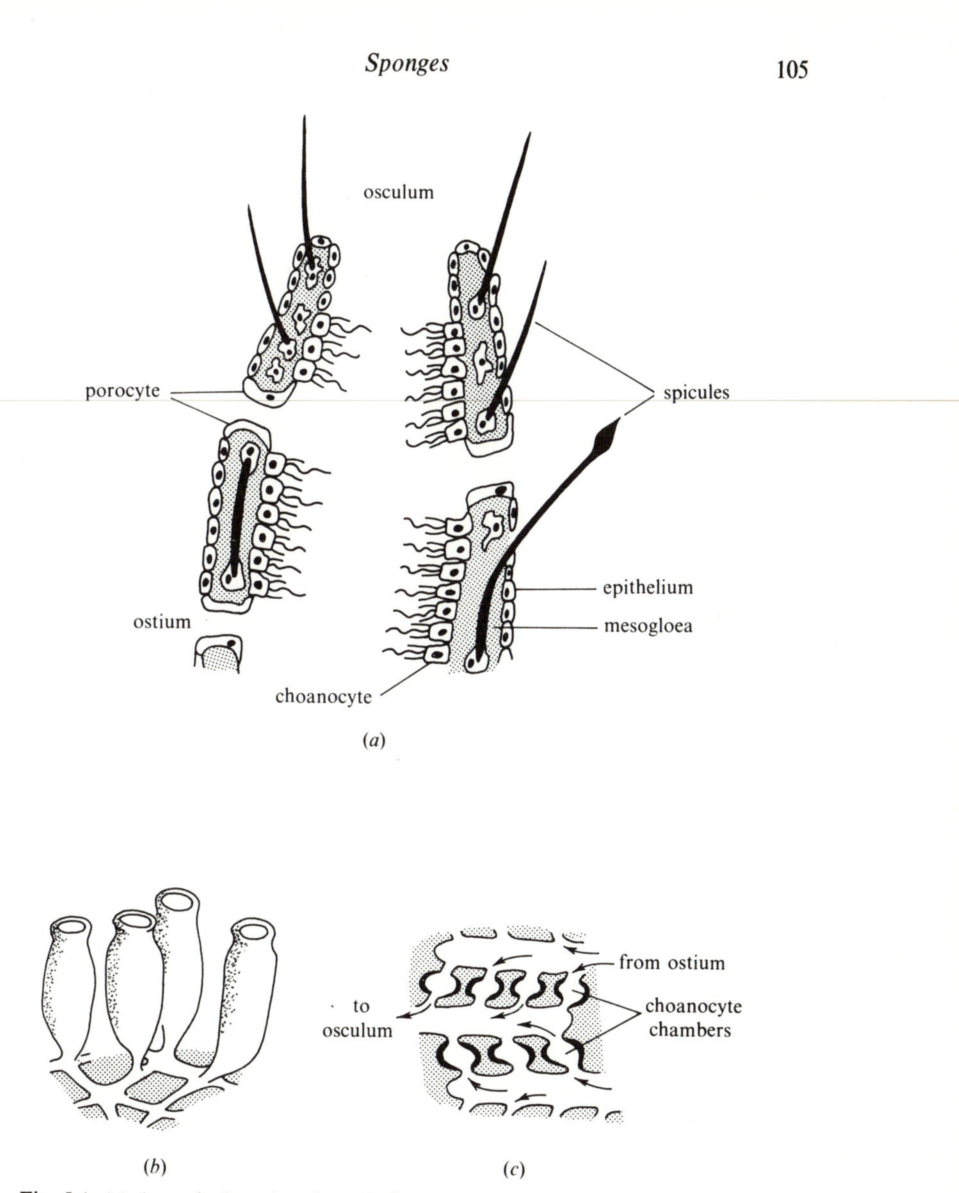

Fig. 5.1. (*a*) A vertical section through the sponge *Leucosolenia*. (*b*) A colony of *Leucosolenia*. (*c*) A diagrammatic section through a more complex sponge.

strands connect adjacent microvilli. Waves of bending travel along the flagellum from base to tip. Water is drawn through the meshes of the collar and propelled towards the tip of the flagellum. Particles of food which are too large to pass through the mesh stick to the collar. They are carried to the base of the collar, perhaps by processes like those which carry food particles along the pseudopodia of foraminiferans (p. 38). At the base of the collar they are enclosed in food vacuoles. Large particles of food are sometimes ingested by

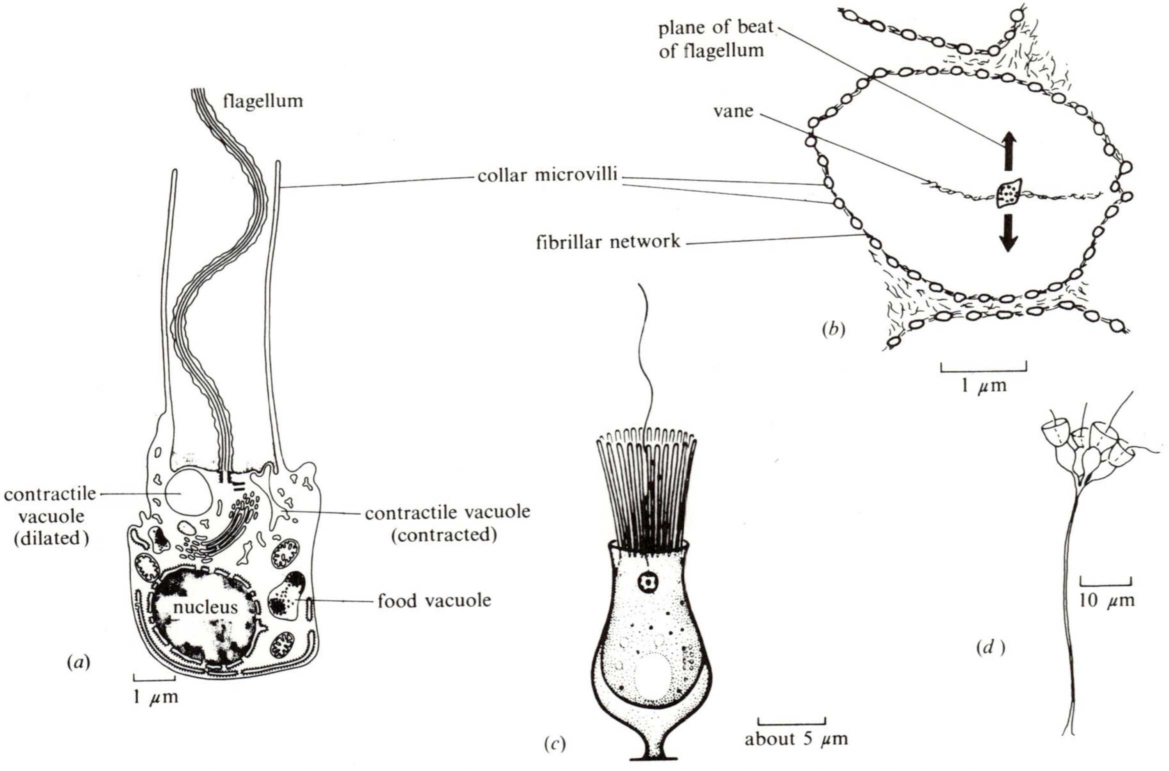

Fig. 5.2. (*a*) A diagrammatic section of a choanocyte of the sponge *Ephydatia*, based on electron micrographs. (*b*) A section through the collar of a choanocyte of *Ephydatia* with parts of two neighbouring collars. (*c*) A solitary choanoflagellate, *Salpingoeca amphoroideum*. (*d*) A colonial choanoflagellate, *Codosiga pyriformis*. (*a*) and (*b*) after B. Brill (1973). *Z. Zellforsch.* **144**, 231–45. (*c*) from K. G. Grell (1973). *Protozoology*. Springer, Berlin. (*d*) after W. S. Kent (1880–2). *A manual of the Infusoria*. Bogue,

other cells. Also, food taken in by a choanocyte can apparently be passed to other cells. Some electron microscope sections of sponges seem to show material being passed between adjacent cells.

The choanocytes are very like the entire bodies of members of a group of flagellate Protozoa, the choanoflagellates (Fig. 5.2 *c*, *d*). They have collars, and flagella with vanes, very like choanocytes and they obtain fine particles of food from the water in the same way. Fig. 5.2(*c*) shows a solitary choanoflagellate and Fig. 5.2(*d*) a colonial one consisting of several individuals on a single stalk. It is possible that sponges may have evolved from choanoflagellates.

Leucosolenia is a simple sac with a wall consisting of two layers of cells separated by mesogloea. This structure is suitable enough for a small sponge, 1 or 2 cm high. It would not be suitable for large sponges which grow up to 1 m in diameter. Large sponges (and some small ones) are more complicated than *Leucosolenia.* They have large numbers of small chambers lined with choanocytes, with branching tubes carrying water to them from the ostia and from them to one or several oscula (Fig. 5.1 *c*). Even in large sponges the chambers which contain the choanocytes are generally much smaller than the central cavity of *Leucosolenia.* They often have diameters less than 100 μm, so that the tips of the flagella are near the centres of the chambers.

Both sexual and asexual reproduction occur. Sexual reproduction produces rounded larvae with flagella on their outside cells, which swim about for a while before settling. Similar larvae are produced by asexual processes, but asexual reproduction by budding also occurs. This is how colonies of *Leucosolenia,* for instance, are formed (Fig. 5.1 *b*). New individuals develop from outgrowths of existing ones, until there is a colony of individuals attached side by side to the same rock, linked together by strands of tissue.

The sponges are put in a phylum by themselves, and are divided into three classes. The first class, the Calcarea, consists of the sponges which have spicules of calcium carbonate. *Leucosolenia* belongs to this class and so do many other small shallow-water sponges such as *Sycon* and *Grantia.* The Hexactinellida or glass sponges have spicules of silica, which consist of three bars joined together mutually at right angles (Fig. 5.3 *a*). The spicules are often fused together to form a three-dimensional network (Fig. 5.3 *b*). In such cases the skeleton of spicules may remain intact when the soft tissues have decayed, and can be strikingly beautiful. Venus's flower basket (*Euplectella*) is a member of this class and its skeleton is one of the loveliest things in many museums of zoology. Hexactinellida live on the sea bottom at substantial depths (from about 100 m to over 5000 m) and are collected by dredging. The third class, the Demospongia, includes some sponges which have silica spicules and some which have no spicules. If there are spicules, they do not consist of three bars meeting at right angles. The large spicules are either simple needles, or consist of four bars which do not meet at right angles but point towards the apices of an (imaginary) tetrahedron like the bonds radiating from a carbon atom (Fig. 5.3 *c*, *d*). There are generally small spicules as well as large ones, and the small

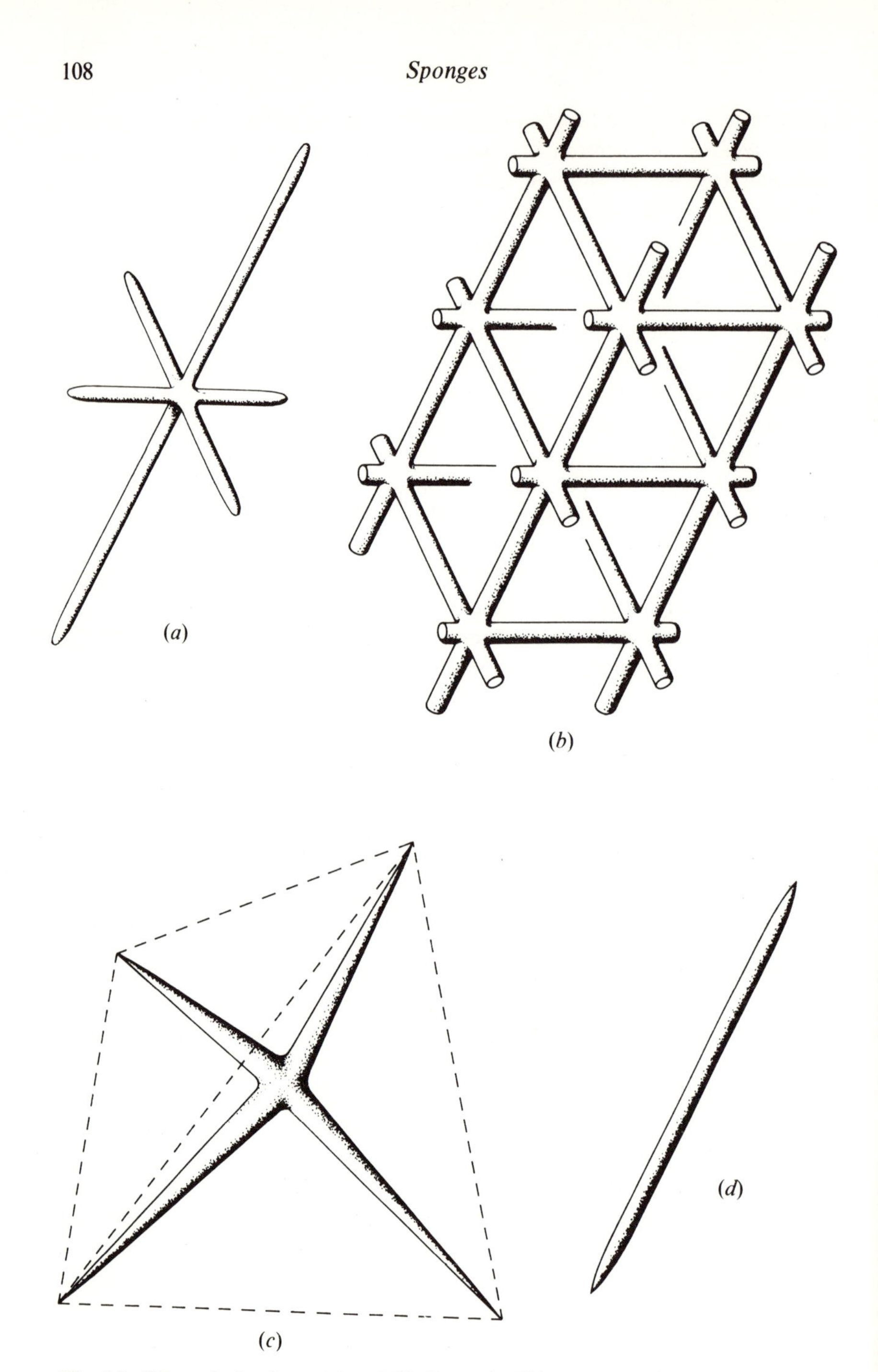

Fig. 5.3. Silica spicules from, (*a*) and (*b*), Hexactinellida; and (*c*) and (*d*), Demospongia.

ones have a great variety of shapes. If there are no spicules (and sometimes if there are) the sponge is strengthened by a tangled mass of fibres of the protein spongin. The Demospongia range from shores to depths of 5000 m or more. They include *Halichondria* (which is common on British shores) and the bath sponge, *Spongia.* Bath sponges are found in the Mediterranean, mainly at depths between 5 and 50 m. They are collected by divers and allowed to decompose so that the sponge which is eventually sold and used in the bath consists solely of spongin fibres.

In cases of doubt, Demospongia which have spicules can be distinguished from Calcarea by putting pieces into hydrochloric acid. The calcium carbonate spicules of Calcarea are dissolved by acid but the silica ones of Demospongia are not.

FEEDING CURRENTS

Sponges in still water (for instance in bowls or aquaria) keep water moving constantly through them, in at the ostia and out at the oscula. It is presumably propelled by the flagella of the choanocytes. If small particles are suspended in the water, they are filtered out by the collars of the choanocytes. Suspensions of graphite seem to be filtered as assiduously as suspensions of particles which can serve as food, and it has been shown that graphite particles of diameter 1 μm are filtered out effectively.

This happens even in still water. However, sponges do not seem to flourish in still water: they are found on shores where they are exposed to waves and tidal currents, and in deeper waters only where currents flow. One zoologist judged from his observations of sponges in their natural habitats that current speeds of 0.5–1 m s^{-1} were most favourable to them. In such currents, water will flow through the sponges whether the flagella beat or not, as has been demonstrated by experiments. The flagella may still be needed to make water flow through the collars; since the collars do not completely block its passage, water can flow from ostia to osculum without being filtered.

Experiments on the effect of external currents on flow through sponges have been performed in a flow table. This is an apparatus designed to produce smooth, even currents of water across a shallow tank. Specimens of the sponge *Halichondria* were used: they were finger-shaped, about 3 cm tall with an osculum of diameter 5 mm at the top. A fine rod with a heated bead thermistor on its end (see p. 14) was slipped into the osculum, so that the velocity of flow of water out of the osculum could be measured. It was measured while the velocity of flow of water across the flow table was varied.

Fig. 5.4(*a*) shows the results of these experiments. In still water, the velocity of flow through the osculum was less than 3 cm s^{-1}. It increased when there was a current across the flow table and was nearly 7 cm s^{-1} when the current was 10 cm s^{-1}. Faster currents would presumably have caused faster flow through the sponge but could not be obtained with this particular flow table. The choanocyte flagella could be stopped by putting the sponge into fresh water

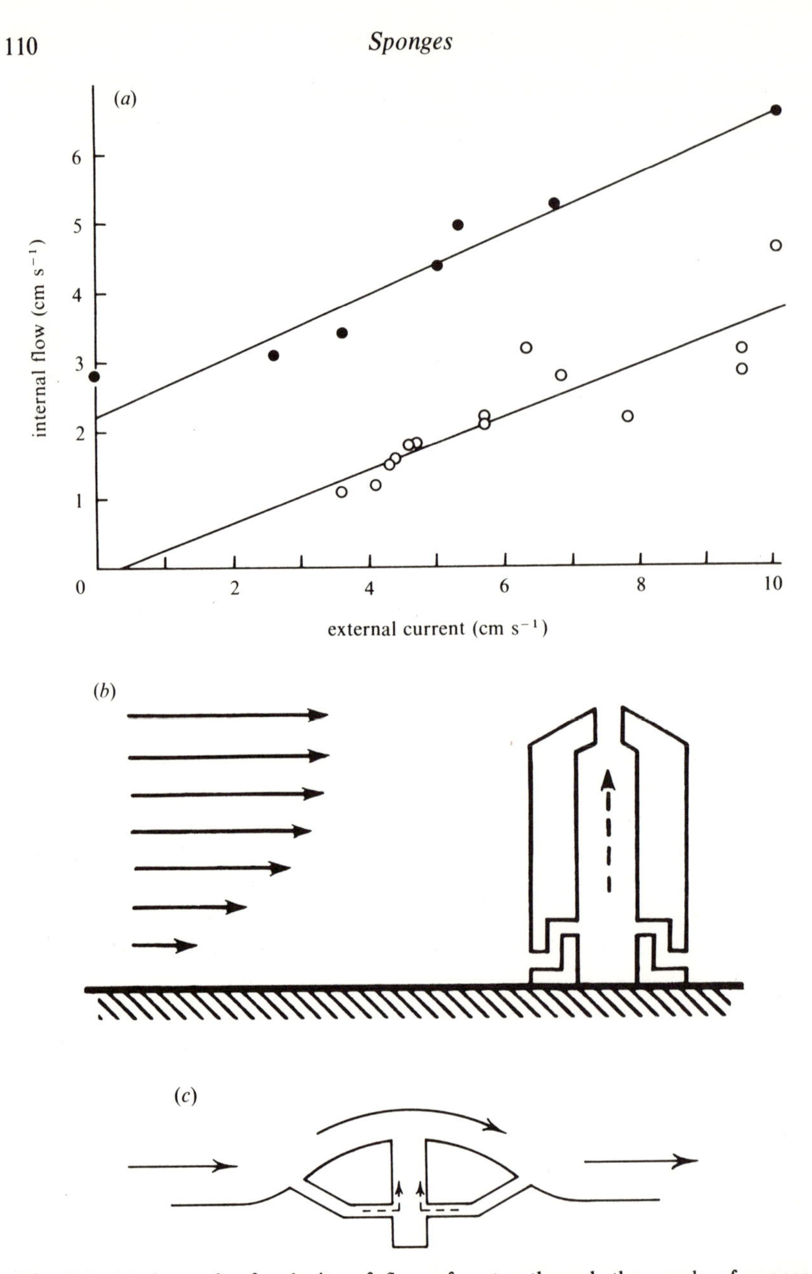

Fig. 5.4. (*a*) A graph of velocity of flow of water through the oscula of sponges (*Halichondria*) against the velocity of the current flowing over them in a flow table. Separate graphs are shown for living sponges (●) and for ones treated with fresh water to stop the flagella (○). From S. Vogel (1974). *Biol. Bull.* **147**, 443–56. (*b*) A vertical section through a plastic model, about 2 cm high, which was used to demonstrate the principle of viscous entrainment. At the base of the model, inner and outer rings of radial holes connect with an annular slot. From S. Vogel & W. L. Bretz (1972). *Science* **175**, 210–11. (*c*) A diagram showing how water could be made to flow through a sponge by the Bernoulli effect.

for a while. Sponges treated in this way were returned to the flow table. It was found that no water flowed through their oscula in still water but it flowed through at 4 cm s^{-1} in a current of 10 cm s^{-1}.

Further experiments were done with plastic models, including the one shown in Fig. 5.4(*b*). It is cylindrical, with a hole representing the osculum at the top and radial holes representing ostia round the base. When it was tested in the flow table it was found that currents across the table made water flow in through the 'ostia' and out through the 'osculum' as if it were a real sponge.

This flow is probably largely due to the phenomenon of viscous entrainment. Fluid flowing over a surface tends to draw water out of any hole in the surface; it tends to make the fluid in the hole move with it because of its viscosity. The effect is greatest where the fluid is flowing fastest. The water in contact with the bottom of a flow table is stationary and water near the bottom is slowed down by its viscosity, so there is a gradient of viscosity in the bottom layer of water. This is the boundary layer. Above it all the water flows at the same velocity. The lengths of the horizontal arrows in Fig. 5.4(*b*) represent water velocities. The thickness of the boundary layer depends on various factors including the velocity of the water and is represented in this diagram as a little less than the height of the model.

The phenomenon of viscous entrainment must have tended to draw water out of both the 'ostia' and the 'osculum' of the model. However, the tendency must have been stronger at the osculum because the flow past it was faster. Therefore water entered the ostia and flowed out through the osculum. The same thing happens for the same reason in real sponges.

Fig. 5.4(*c*) shows another way in which a similar effect could be produced. Fluid is flowing over a mound and since the mound projects into the channel, reducing its cross-section, flow is faster over the mound that it is either in front of it or behind it. I am not referring now to velocities at different heights in the boundary layer, but to a change in velocity of the diverted fluid. By Bernoulli's theorem the pressure is lower where the velocity is higher so fluid is drawn through the pores in the mound, as indicated by the broken arrows.

The effects represented in Fig. 5.4(*b*) and (*c*) probably combine to cause flow through sponges and sponge models. Viscous entrainment is probably the more important for tall chimney-like sponges but the Bernoulli effect may be more important for mound-like ones. The two effects are quite distinct. The viscous entrainment effect depends on differences of velocity at different levels in a boundary layer but Bernoulli's principle does not apply to such differences: it applies only to differences in velocity along a streamline (line of flow).

Sponges living in surf generally form thin layers, encrusting the rocks. A tall sponge would be easily damaged by surf and since the water flows fast around the sponge, very little height may be needed to get adequate flow through the sponge. In calmer places the same species may be tall and chimney-like. Height may be desirable in a place where the water moves slowly, to raise the osculum as far up the boundary layer as possible and get the greatest possible viscous entrainment effect.

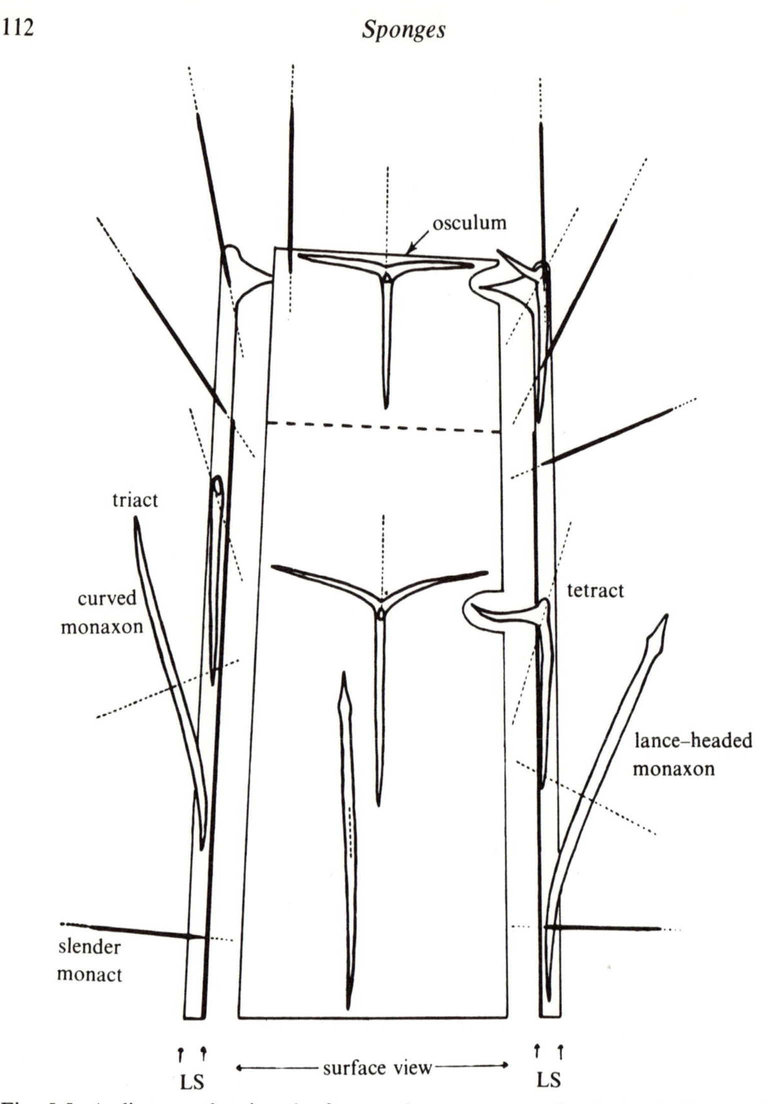

Fig. 5.5. A diagram showing the form and arrangement of spicules in *Leucosolenia complicata*. The optic axes of the spicules are indicated by broken lines. The diagram shows a surface view and, on either side, longitudinal sections. From W. C. Jones (1954). *Q. J. microsc. Sci.* **95**, 33–48.

SPICULES

Such variations in the external form of sponges often make identification difficult. However, the spicules are often characteristic of the species so identification involves boiling a piece of sponge in potassium hydroxide solution to get a sample of spicules. Partly because of their importance in taxonomy, spicules have been studied intensively.

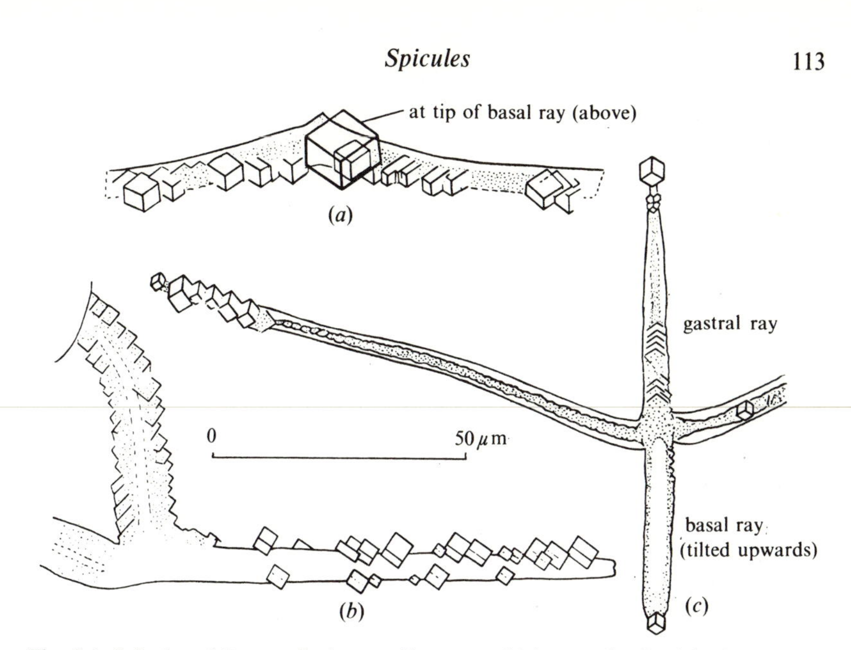

Fig. 5.6. Spicules of *Leucosolenia complicata* on which crystals of calcite have been allowed to form. From W. C. Jones (1955). *Q. J. microsc. Sci.* **96**, 129–49.

The spicules of Calcarea are made of calcite (one of the crystalline forms of calcium carbonate) with a small proportion of magnesium ions. Spicules of many different shapes may occur in the same sponge, as Fig. 5.5 shows. The triacts are Y-shaped and the tetracts have an additional ray projecting inwards. Both serve to stiffen the mesogloea. Several types of unbranched, needle-like spicules project from the surface of the sponge. The largest spicules of this particular species are about 0.2 mm long.

If isolated spicules of Calcarea are put in a saturated solution of calcium bicarbonate, calcite crystallizes out onto their surfaces (Fig. 5.6). All the crystals which form on a given spicule are parallel to each other. Since they presumably form as extensions of the crystal lattice of the spicule, this implies that the lattice is oriented in the same way throughout: it is as though the spicule had been carved out of a single large crystal. The direction of the crystal lattice relative to the rays is constant for each type of spicule. Broken lines through the spicules in Fig. 5.5 indicate a particular direction within the crystal lattice (the optic axis).

Silica spicules, unlike calcite ones, have strands of organic matter running along the centres of their rays. Apart from this they consist of very pure silica. Though they have been studied by X-ray diffraction and electron microscopy their structure has not been elucidated as fully as the structure of calcite spicules.

MULTICELLULAR STRUCTURE

The nucleus synthesizes the messenger RNA which moves out into the cytoplasm and serves as a template for the synthesis of enzymes and other proteins. It seems reasonable to suppose that there must a limit to the size of cell which can be served by an ordinary diploid (or haploid) nucleus. If the cell is too large the nucleus may be unable to produce RNA fast enough, and parts of the cell may be a long way from the nucleus so that RNA has to travel long distances. Study of Protozoa seems to confirm that there is indeed a practical limit for most purposes. The giant amoeba *Pelomyxa* has a length of about 2 mm but has many nuclei. The dinoflagellate *Noctiluca* is about the same size and has only one nucleus but much of its volume is occupied by fluid-filled vacuoles. Ciliates such as *Stentor* and *Spirostomum* have lengths of 1 mm or more but their RNA is produced by macronuclei which are polyploid and so have many copies of each gene which can produce RNA simultaneously. Also, the macronucleus tends to be long and to wind through the body so that none of the cytoplasm is very far from it. Other Protozoa grow large by forming colonies of small individuals, each with its own nucleus. A *Zoothamnium* colony may be several millimetres in diameter though each of its component individuals is only the size of a small *Vorticella*, with a body of diameter about 30 μm. *Volvox* is a flagellate which forms hollow spherical colonies 1 mm or more in diameter, but the individual cells (which resemble *Chlamydomonas*) are only 6–9 μm long.

Thus Protozoa with a single haploid or diploid nucleus are seldom more than about 100 μm long, and ciliates with a single compact macronucleus are seldom more than about 200 μm long. The smallest known Protozoa are 1–1.5 μm long which is probably about as small as is practicable for a eukaryote cell. Multicellular animals tend to consist of cells between about 20 and 30 μm in diameter, though there are of course exceptional much larger cells. Human striated muscle fibres can be as much as 12 cm long but their diameter is only 10–100 μm and they have nuclei spaced along them at short intervals. Human nerve cells with only a single nucleus have axons up to 1 m long. The yolk of an ostrich egg is a single cell of diameter 5 cm but the part where the embryo forms divides into a large number of cells, each with its own nucleus, in the early stages of development. Sponges range in diameter from a few millimetres to a metre but are apparently all constructed of cells within the normal range of sizes.

CELL ADHESION

The cells of a multicellular organism must obviously be attached to each other, but it is not so obvious how they are attached. Fig. 5.7 shows three types of attachment which are found. Cell membranes are shown in this diagram as double black lines, which is how they appear in electron microscope sections of material treated with osmic acid. (The osmic acid attaches to the polar groups

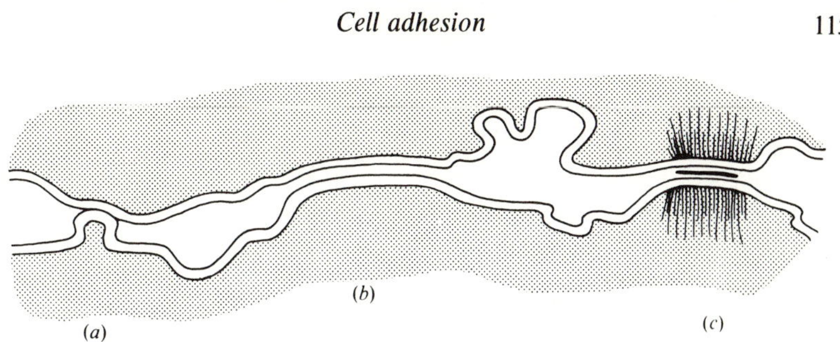

Fig. 5.7. A diagrammatic section through two adjacent cells, showing three types of attachment between cells which are described in the text. This diagram is based on electron microscope sections.

which are concentrated near the two faces of the membrane.) They are about 8 nm thick. At junction (*a*) the cell membranes of the two cells seem to be in close contact but at (*b*) and (*c*) they run parallel to each other, with a gap of 15 nm between them. This seems to be a genuine gap, not just a band of unstained material: it has been shown that molecules of proteins such as haemoglobin percolate quite freely into the gaps. The type of attachment shown in (*c*) is known as a desmosome. There are filaments in each cell, ending at the desmosome, and there seems to be some material in the gap betwen the cell membranes. Nevertheless the gap is not blocked, for protein molecules can enter it. Cells with attachments of type (*a*) between them are damaged if they are pulled apart but attachments of types (*b*) and (*c*) can be broken quite easily without damaging the cells. Most attachments of sponge cells to each other are of types (*b*) and (*c*), and sponges can be broken up into their component cells by squeezing them through cloth or by putting them in solutions of EDTA (ethylene diamine tetraacetic acid, which reduces the concentration of calcium and magnesium ions to a very low level). However, some attachments of type (*a*) are found in sponges. Most of the attachments of parasitic Protozoa to their hosts which were described in chapter 4 are of types (*b*) and (*c*), but the suctorian *Trichophrya* attaches itself to fish gills by a layer of some cementing substance about 100 nm thick.

Attachments of types (*b*) and (*c*) seem rather puzzling. It is easy to envisage mechanisms which could hold two membranes in contact, but how can membranes be held 15 nm apart? The answer seems to involve interaction between two opposing forces, one of mutual attraction between the cells and one of mutual repulsion. The force of attraction is the London–van der Waals force, which should be familiar to readers who have studied chemistry. The force of repulsion is an electrostatic one, due to the outer surfaces of the cell membranes being negatively charged. The charge can be demonstrated by applying a potential across a suspension of cells, which move towards the anode. It is due to neuraminic acids (which are acid sugars) being attached to the membrane and being partially ionized. This negative charge on the outside of the cell should not be confused with the potential difference (inside negative)

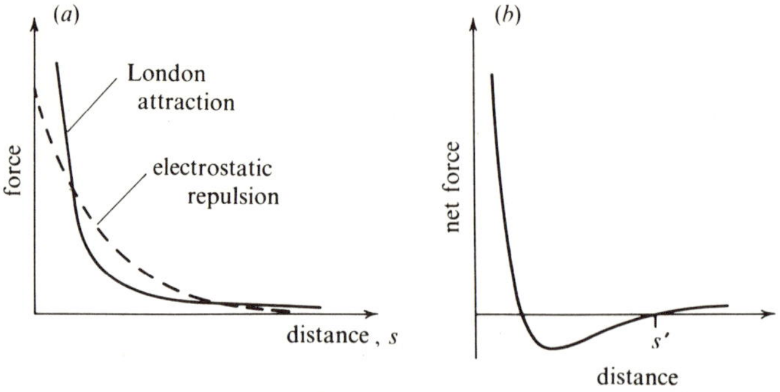

Fig. 5.8. Diagrammatic graphs of forces acting between two cells, against their distance apart. (*a*) The London–van der Waals attraction and the electrostatic repulsion shown separately; (*b*) the net attraction.

which occurs across the cell membrane because of the distribution of ions.

Both the London–van der Waals attraction and the electrostatic repulsion diminish as the distance s between the cell membranes increases (Fig. 5.8*a*). The London attraction is roughly proportional to $1/s^3$ while the electrostatic repulsion is proportional to e^{-ks} (where k is a constant). The attraction is always greater than the repulsion when s is very small ($1/s^3 = \infty$ when $s = 0$) and also when s is large, but if there is enough charge per unit area on the cell membranes there is an intermediate range of values of s for which the repulsion is greater than the attraction. Thus the net force between two cells is likely to depend on their distance apart in the manner shown in Fig. 5.8(*b*). The cells are weakly attracted to each other when well separated but there is a distance s' at which they neither attract nor repel each other. At slightly shorter distances they repel each other but at very short distances they attract each other strongly. If cells are brought together they will tend to stop and stick at a distance s' apart. If they are somehow brought really close together they will adhere very strongly in close contact. Cell junctions of type (*a*) (Fig. 5.7) involve close contact while those of types (*b*) and (*c*) involve separation by, presumably, a distance s'.

The cells of sponges and of many other tissues separate in solutions from which the calcium and magnesium ions have been removed by EDTA. It has been suggested that calcium and magnesium, being divalent, link cells by forming bridges between univalent ions in the two cell membranes. However, the effect of divalent ions can also be explained in terms of their effect on the electrostatic forces.

Sponge cells which have been separated by passing through cloth or by EDTA will form aggregates in ordinary sea water (i.e. in sea water with a normal calcium content). If a suspension of sponge cells in sea water is allowed to settle the cells move about, clump together and eventually form a new sponge. Large

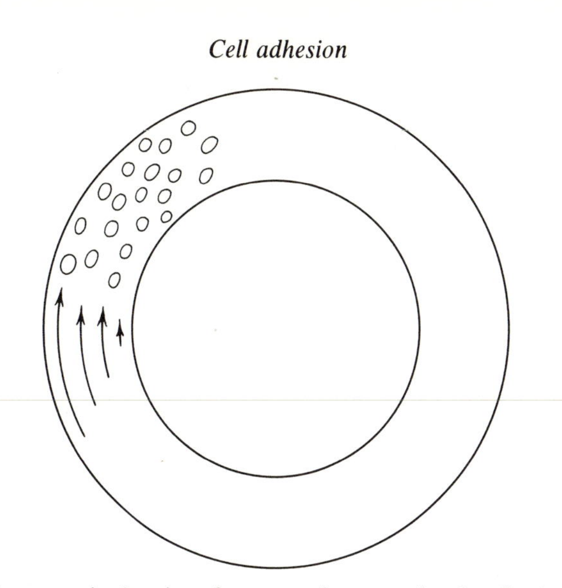

Fig. 5.9. A diagrammatic plan view of apparatus for measuring the adhesiveness of cells. For explanation see text.

clumps of cells also form in suspensions which are prevented from settling by shaking.

An experiment which has often been tried is to mix suspensions of sponge cells of two species. Clumps form and in some cases it can be shown that there is a strong tendency for cells of the same species to clump together. For instance, *Microciona prolifera* has red cells and *Haliclona oculata* has brown ones. The clumps which form in a mixed suspension of cells of these two species consist either of red cells alone or of brown ones alone. Some other pairs of species form mixed clumps initially but sort themselves out later.

A possible explanation for the formation of single-species clumps is that the cells show specific adhesion, that is that they stick more readily to cells of the same species than to cells of other species. So that this hypothesis could be tested a method was devised for measuring the adhesiveness of cells. A suspension of cells is put in a narrow gap between two concentric stainless steel cylinders (Fig. 5.9). The outer cylinder rotates at a known rate while the inner one remains stationary. Fluid near the outer wall moves faster than fluid near the inner one so some cells move faster than others. Collisions occur as cells overtake others which have their centres slightly nearer the inner wall, and the frequency of collisions can be calculated. As clumps form the suspension comes to contain fewer but larger particles and counts of the number of particles (cells or clumps) at any instant can be used to calculate the percentage of collisions that have resulted in adhesion. This fraction is called the collision efficiency. For instance, about 80% of collisions between cells in a pure suspension of *Microciona fallax* resulted in adhesion while 20% of collisions in a pure suspension of *Halichondria panicea* resulted in adhesion. The collision

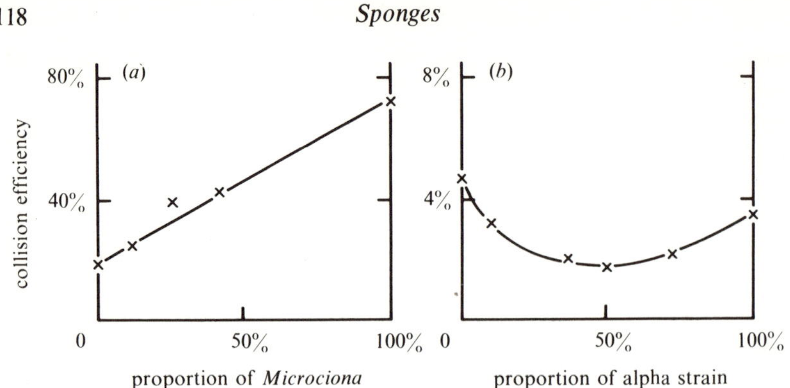

Fig. 5.10. Results of experiments with sponge cells in the apparatus illustrated in Fig. 5.9. (*a*) A graph of collision efficiency against the proportion of *Microciona*, in mixed suspensions of cells of *Microciona fallax* and *Halichondria panicea*. (*b*) A graph of collision efficiency against the proportion of the alpha strain in a mixed suspension of two strains of *Ephydatia fluviatilis*. Redrawn from (*a*) A. S. G. Curtis (1970). *Symp. zool. Soc. Lond.* **25**, 335–52; and (*b*) A. S. G. Curtis & G. van de Vyver (1971). *J. Embryol. exp. Morph.* **26**, 295–312.

efficiencies for pure suspensions were thus 80% and 20%, respectively. Mixed suspensions gave intermediate collision efficiencies and a graph of collision efficiency against the proportion of *Microciona* in the suspension was found to be a straight line (Fig. 5.10). This is what one would expect if adhesion were non-specific. What would happen if adhesion were strictly specific, occurring only between cells of the same species? Consider a mixture of equal numbers of cells of the two species. Half the collisions would be between cells of different species and would not lead to adhesion, a quarter would be between *Microciona* cells and 80% of them would result in adhesion, and another quarter would be between *Halichondria* cells and 20% of them would result in adhesion. The overall collision efficiency would be $(\frac{1}{4}\times 80)+(\frac{1}{4}\times 20) = 25\%$, only half of the value of 50% which is actually observed. It seems clear that adhesion between cells of these two species is non-specific, at least in the conditions of the experiment. Similar results have been obtained with two other pairs of species. All these experiments were performed with suspensions of cells which had been centrifuged and re-suspended in fresh solutions several times to wash them clean of the solution in which the sponge was broken up.

Very different results were obtained when the experiment was repeated with unwashed cells from two strains of *Ephydatia fluviatilis* (a freshwater sponge). It had previously been discovered that when two sponges of this species are placed in contact they first adhere to each other and then either separate or coalesce. When coalescence occurs the canals of one individual join up with those of the other so that the two become, in effect, one animal. It seems that there are several strains of this species and that coalescence only occurs between members of the same strain. Members of two strains were broken up into separate cells with EDTA, and the apparatus of Fig. 5.9 was used to measure the collision efficiencies of separate and mixed suspensions of their cells. The results are shown in Fig. 5.10(*b*) which shows that mixtures had lower collision efficiencies than either strain alone. This seems at first sight to indicate specific adhesion, but further experiments led to a different explanation. In the experiments with unwashed cells the collision efficiencies were between 2 and 5% (Fig. 5.10*b*). In other experiments great care was taken to wash the cells clean of the solution in which the sponges had been broken up, and much lower collision efficiencies were found. This suggests that there is a soluble substance produced by sponge cells which favours adhesion. To test this, extracts of *Ephydatia* cells were prepared by disrupting suspended sponge cells with ultrasound and filtering the resulting suspension. When a solution prepared in this way from one strain was added to a suspension of cells of the same strain, the collision efficiency was greatly increased. When it was added to a suspension of cells of another strain the collision efficiency was decreased. Cells treated with the solution prepared from the other strain never recovered their normal adhesiveness, even when washed clean of the solution. In the experiment shown in Fig. 5.10(*b*) the suspensions of cells of each species were contaminated with the substance released by their own species. When the suspensions were mixed the adhesiveness of each species was reduced by the substance from the other species. This, rather than specific adhesion, explains the results of the experiment.

Ephydatia of different strains presumably fail to coalesce because at their point of contact the cells of each are exposed to the substance produced by the other. If different species of sponges produce different substances this may explain why cells of two species which at first clump indiscriminately later sort themselves out. A few cells of one species in a clump of the other would become less adhesive and tend to drop off. It is possible that similar substances may play an important part in the development of more advanced animals, preventing different tissues within a single animal from merging.

FURTHER READING

GENERAL

Fry, W. G. (ed.) (1970). The biology of the Porifera. *Symp. zool. Soc. Lond.* **25**, 1–512.

FEEDING CURRENTS

Vogel, S. (1974). Current-induced flow through the sponge *Halichondria*. *Biol. Bull.* **147**, 443–56.

MULTICELLULAR STRUCTURE AND CELL ADHESION

Curtis, A. S. G. (1973). Cell adhesion. *Prog. Biophys. mol. Biol.* **27**, 317–86.

Curtis, A. S. G. & van de Vyver, G. (1971). The control of cell adhesion in a morphogenetic system. *J. Embryol. exp. Morph.* **26**, 295–312.

6

Sea anemones and corals

Phylum Cnidaria, Class Anthozoa
  Subclass Alcyonaria (soft corals, sea fans, etc.)
  Subclass Zoantharia (stony corals, sea anemones, etc.)
(For other classes see chapter 7)

This chapter and the next are about the phylum Cnidaria, which includes the sea anemones, corals, hydroids and jellyfishes. The phylum is sometimes called Coelenterata but it is probably best to avoid that name since two phyla, the Cnidaria and the Ctenophora (sea gooseberries etc.), are sometimes referred to together as the coelenterates.

This chapter is about the sea anemones and corals, which are all included in the class Anthozoa. They are more complicated than some of the other Cnidaria but it seems convenient to describe them first because a great deal of the research which has thrown light on the phylum in general has been done on sea anemones.

Most Anthozoa live in the sea but *Diadumene*, a sea anemone, is also found in brackish water in estuaries.

Fig. 6.1 is a highly simplified diagram showing the structure of a sea anemone. It is a hollow cylindrical animal, normally attached by its base (pedal disc) to a rock or some other firm foundation. There is a ring of tentacles at the other end, surrounding a mouth which opens into a central gastric cavity. This cavity is incompletely partitioned by radial mesenteries. The body wall is turned in at the mouth to form a pharynx, and some or all of the mesenteries are attached to the pharynx as well as to the outer body wall. Two grooves, the siphonoglyphs, run longitudinally along the pharynx. Their cells bear cilia. The mesenteries have thickenings (mesenteric filaments) along their free edges, and some anemones have thread-like acontia which are continuations of these.

The body wall in Cnidaria is formed by only two layers of cells, an outer epidermis and an inner gastrodermis. Between them is a jelly-like mesogloea (Figs. 6.6*b*, 7.2). This arrangement is reminiscent of the sponges, which have an outer layer of epithelium and an inner layer of choanocytes, with mesogloea between them. However Cnidaria have very few cells in the mesogloea, while sponges have quite a lot. The mesogloea has an important role in sea anemones in stiffening the anemone without preventing changes of size and shape. This chapter has a section about it.

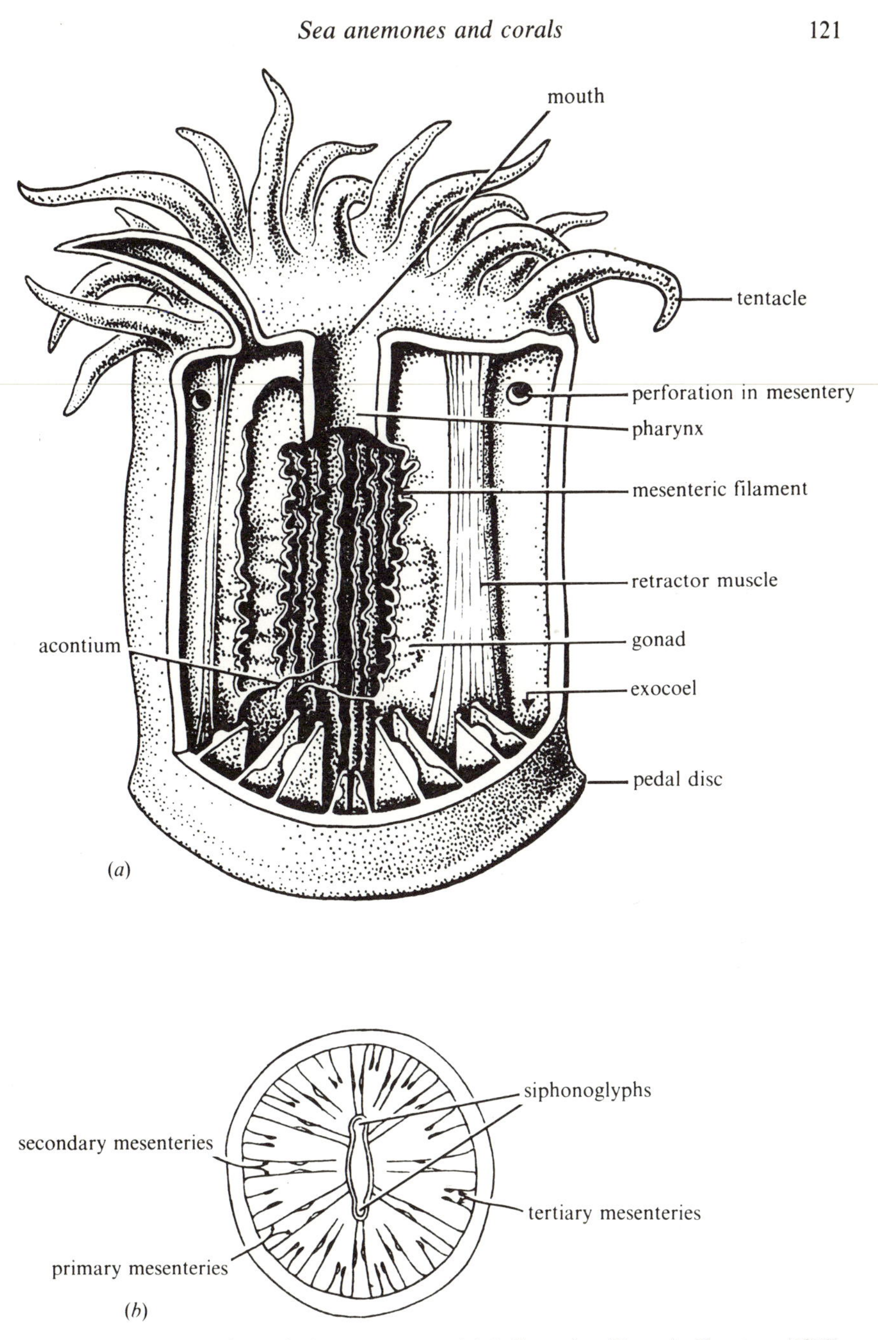

Fig. 6.1. Diagrams of a typical sea anemone. (*a*) A dissection. From A. Kaestner (1967). *Invertebrate zoology*. Interscience, New York. (*b*) A section through the pharyngeal region. From F. E. G. Cox, R. P. Dales, J. Green, J. E. Morton, D. Nichols & D. Wakelin (1967). *Practical invertebrate zoology*. Sidgwick & Jackson, London.

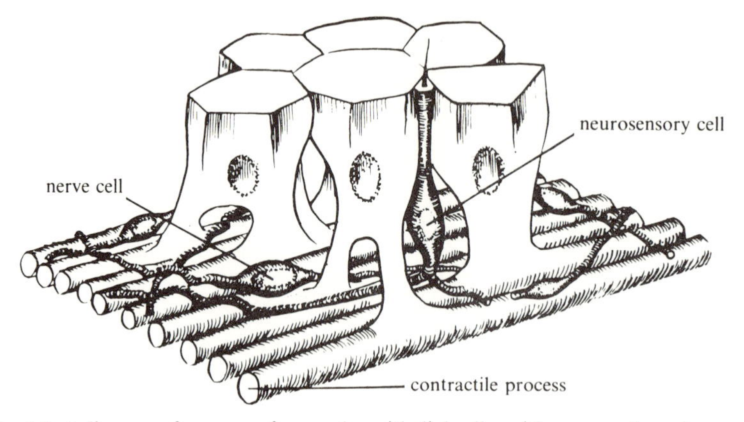

Fig. 6.2. A diagram of a group of musculo-epithelial cells, with nerve cells and a sensory cell. From G. O. Mackie & L. M. Passano (1968). *J. gen. Physiol.* **52**, 600–21.

Cnidaria have nerve and muscle cells, which sponges do not. These enable them to move the parts of their bodies in a co-ordinated way. Many of the muscle cells have the curious and characteristic form shown in Fig. 6.2. Where they occur, they are the principal cells of the gastrodermis. Their bodies are packed neatly together in a single layer but their bases (next to the mesogloea) are drawn out into long muscle fibres (contractile processes). The length of these fibres is not obvious in ordinary microscope sections but can be seen when the cells are separated by chemical treatment. In the sea anemone *Metridium* the fibres are up to 1 mm long though only about 1 μm in diameter. Networks of nerve cells run among the bases of the epithelial cells. The muscles and nerve cells of Anthozoa and the range of behaviour they make possible are described in the following section of this chapter.

Interspersed in places among the epithelial cells are cells called cnidoblasts, which contain nematocysts (Fig. 6.19). These are rather like the trichocysts of ciliates and the polar capsules of sporozoans, but are distinct from both and characteristic of the Cnidaria. They contain hollow coiled threads which can be extruded, entangling prey or enemies or penetrating them and injecting toxin into them. They are plentiful in the epidermis of the tentacles of sea anemones, in the mesenteric filaments and in the acontia. There is a section of this chapter about nematocysts.

Sea anemones reproduce sexually but some also reproduce asexually, by splitting in half or by breaking off pieces of the pedal disc which develop into complete anemones. Sexual reproduction involves gonads which develop on the mesenteries. The eggs or sperm are released into the gastric cavity whence they can escape through the mouth into the open sea, where fertilization generally occurs. The zygote forms a ciliated larva which settles on a rock or other suitable site and in due course attains the adult form.

Many Anthozoa are colonial. A colony consists of a large number of polyps, each like a little sea anemone, joined together by living tissue.

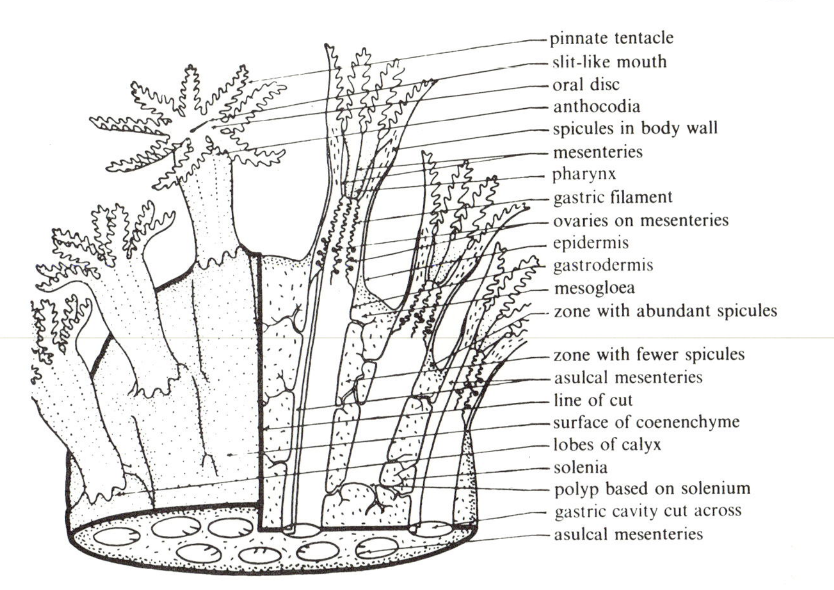

Fig. 6.3. Part of a colony of *Alcyonium*. The external surface is shown on the left but on the right the colony has ben cut to show internal features. The polyps project about 1 cm from the surface of the colony when extended. From W. S. Bullough (1950). *Practical invertebrate anatomy*. Macmillan, London.

There are two subclasses of Anthozoa, the Alcyonaria and the Zoantharia. Alcyonaria have eight pennate tentacles on each polyp, and eight mesenteries (Fig. 6.3). There is only one siphonoglyph. Zoantharia have tentacles which vary in number (but are often a multiple of six) and are not pennate. There are at least twelve mesenteries and often many more, and the muscles on the mesenteries are arranged differently from those of Alcyonaria. There are usually two siphonoglyphs.

The simplest Alcyonaria are colonies of polyps connected together by stolons in the same way as the sponge *Leucosolenia* (Fig. 5.1). Other Alcyonaria form more massive colonies. The only common British member of the subclass is *Alcyonium digitatum*, which is called dead men's fingers because its colonies can look unpleasantly like a swollen hand, drained of blood. Fig. 6.3 shows the tip of a single branch ('finger') of a colony. *Alcyonium* is found attached to rocks near and below the low tide mark. The bulk of the colony is mesogloea which has spicules of calcium carbonate embedded in it but still has a soft, fleshy texture. The gastric cavities of the polyps extend well down into this mesogloea, and are connected by branching tubes. The polyps normally project from the surface of the colony as shown in Fig. 6.3, but they can withdraw into their cavities where they are well protected.

The gorgonians or sea fans are also Alcyonaria, but form much more delicate colonies which are branched or form networks (Fig. 6.4*c*, *d*). The delicate structure is possible because of the stiffening effect of a material called

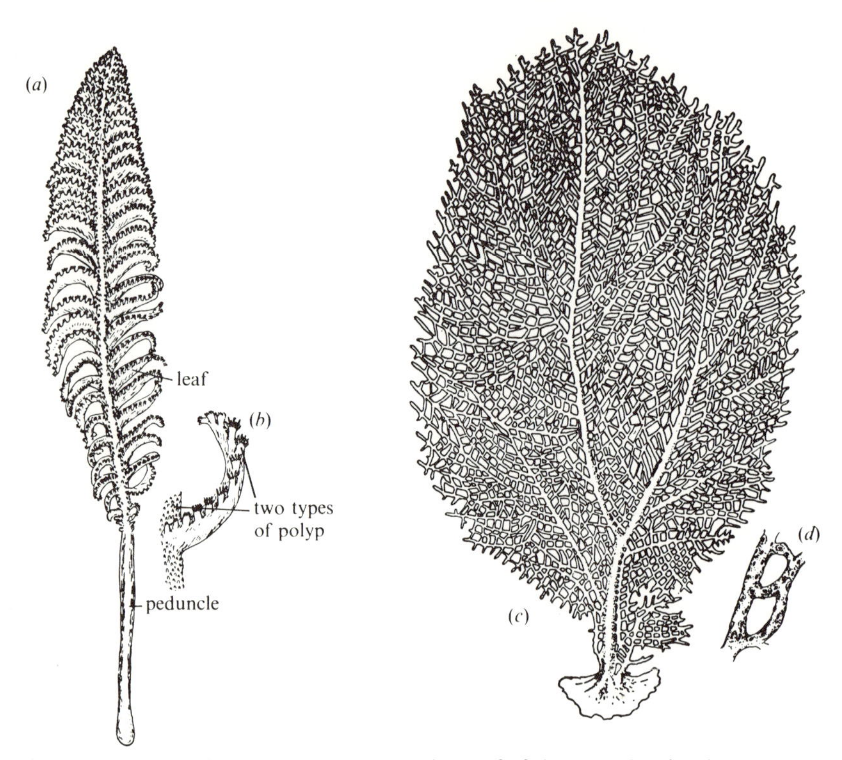

Fig. 6.4. (*a*) *Pennatula* (length up to 40 cm); (*b*) a leaf of the same showing the two types of polyp; (*c*) *Gorgonia*, a sea fan (height about 50 cm); (*d*) a portion of the same showing the holes for the polyps. From L. H. Hyman (1940). *The invertebrates*, vol. 1. McGraw-Hill, New York.

gorgonin in the mesogloea of the colony. Gorgonin is discussed later in this chapter. Gorgonians are common on Atlantic coral reefs, where groups of their fan-shaped colonies are found aligned more or less parallel to each other. The reason for the alignment will be discussed. Another group of Alcyonaria includes *Pennatula* and the other sea pens (Fig. 6.4 *a*, *b*). They live on sandy and muddy bottoms at depths of 40 m or more, with the peduncle thrust into the sand or mud. *Pennatula* has two types of polyps: normal ones on the leaves and others which are incapable of feeding but have enlarged siphonoglyphs and maintain a flow of water through the colony.

Among the Zoantharia, the typical sea anemones are solitary while most stony corals are colonial. Sea anemones (Fig. 6.7) are common on rocks, on the shore and in deeper waters. Some of them are brightly coloured, and are perhaps the most spectacular inhabitants of rock pools. As well as anemones which attach to rocks there are others which burrow. *Cerianthus*, for instance, lives with its body buried in sand and only the tentacles projecting into the water above.

Some stony corals form massive reefs, which are discussed in the final

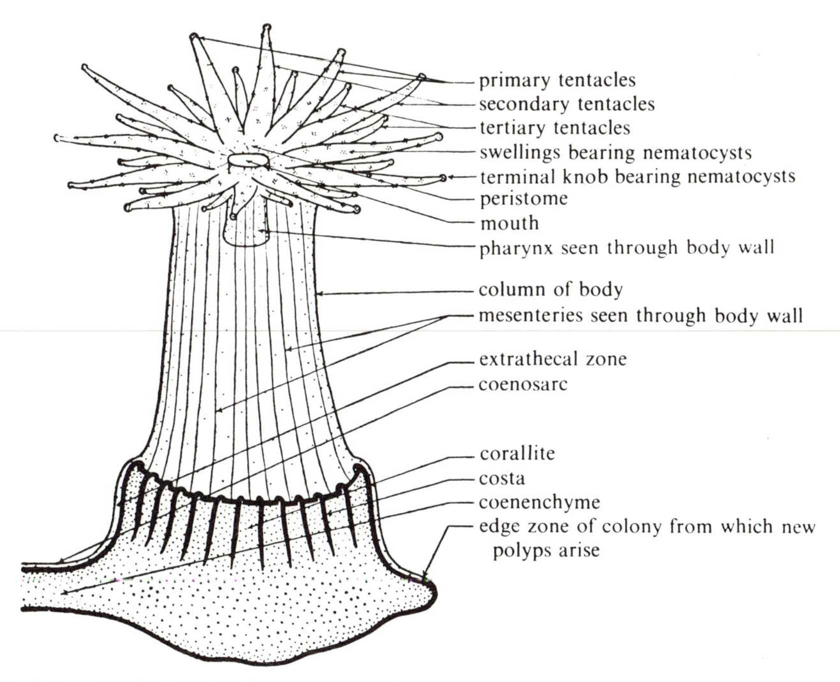

Fig. 6.5. A single polyp from the edge of a colony of the coral *Astrangia danae*. From W. S. Bullough (1950). *Practical invertebrate anatomy*. Macmillan, London.

section of this chapter. Most of them live in the tropics and few are found deeper than 45 m. The others which do not form reefs are found in all latitudes, mainly at depths of 180–550 m but sometimes in shallower water. *Astrangia danae* (Fig. 6.5) is one of the shallow-water species. It forms small encrusting colonies up to about 10 cm across, and is found on stones on the Atlantic coast of the United States. The illustration shows how the polyps grow on a base of calcium carbonate, secreted by the colony. Each polyp stands in a cup (corallite) which has radial ribs (costae) alternating with its mesenteries. In many corals (but not in *Astrangia*) the polyps can retract so as to be contained entirely within their cups, where they are well protected.

MUSCLES AND NERVES

The musculo-epithelial cells of cnidarians have already been described and illustrated (Fig. 6.2). We have a far less clear understanding of how they work than we have of the working of vertebrate striated muscle, but they apparently depend on interaction between actin and myosin filaments. Proteins resembling actin and myosin have been extracted from sea anemones. It is not obvious in electron microscope sections of anemone muscle that there are two kinds of filament, but sections of jellyfish muscle show filaments of diameter about

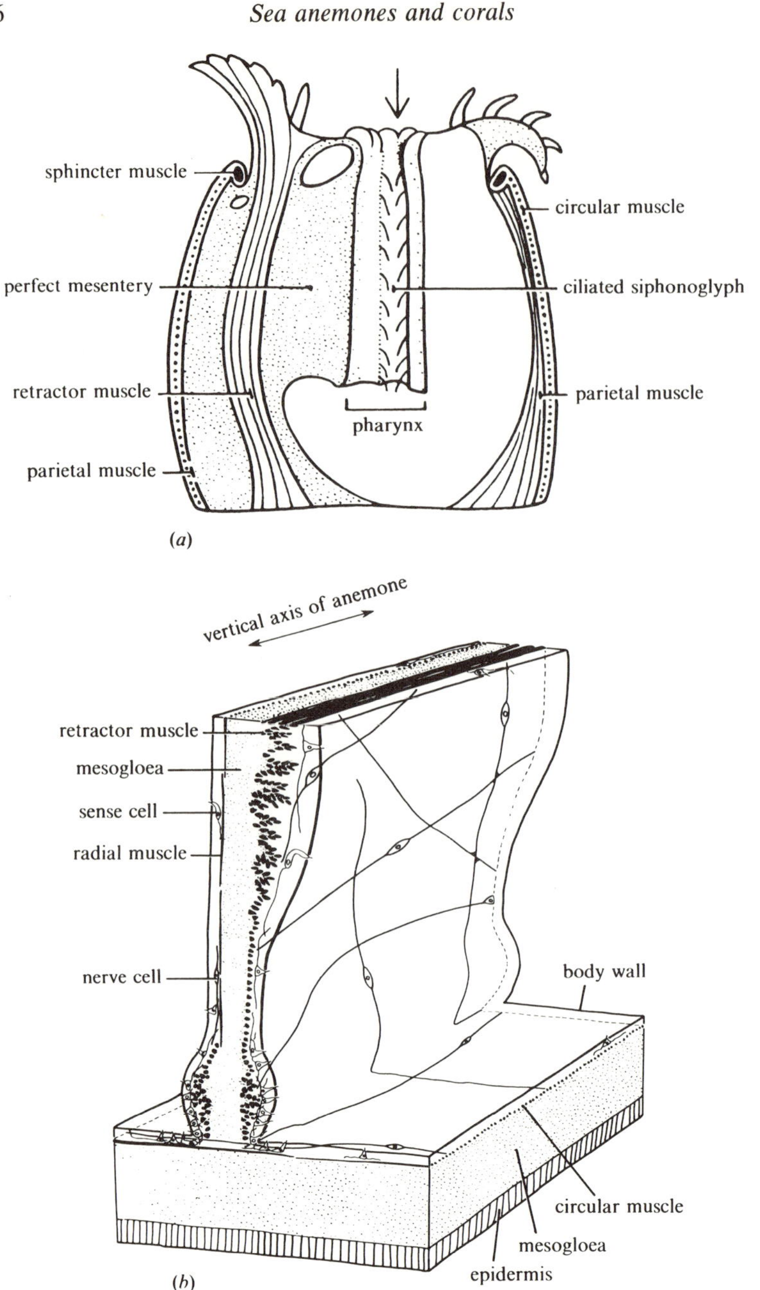

Fig. 6.6. (*a*) A diagram of a sea anemone cut in half to show the principal muscles. E. J. Batham & C. F. A. Pantin (1950). *J. exp. Biol.* **27**, 264–88. (*b*) A diagram of part of the outer body wall and a mesentery of a sea anemone, showing muscle fibres and nerve cells. From T. H. Bullock & G. A. Horridge (1965). *Structure and function of the nervous system of invertebrates.* Freeman, San Francisco.

5 nm (presumably actin) and about 15 nm (presumably myosin). Sea anemones such as *Metridium* can make dramatic changes of size (Fig. 6.7) and their muscle fibres must be capable of shortening to as little as 20% of their extended length.

Fig. 6.6 shows how the muscle cells are arranged in a typical sea anemone. The epidermis of the column is not muscular but the gastrodermis consists largely of musculo-epithelial cells. The gastrodermis of the body wall has muscle fibres which run circumferentially round the column, forming the circular muscle. The sphincter is a large bundle of these fibres. Each mesentery has longitudinal muscle fibres on both sides, close to the body wall. These form the parietal muscles. There is another strip of longitudinal fibres forming the retractor muscle on one face of each mesentery, and there are radial muscle fibres (fibres running at right angles to the body wall) on the other face. Under the retractor muscles, the interface between the gastrodermis and the mesogloea is deeply folded (Fig. 6.6*b*). This greatly increases the number of musculo-epithelial cells which can be fitted as a single layer onto a given area of mesentery. The oral disc has radial and circular muscle fibres and the tentacles have longitudinal and circular ones.

Most of the muscle fibres are outgrowths of typical musculo-epithelial cells but some have their cell bodies reduced to a small mass of cytoplasm containing the nucleus, and so resemble the muscle fibres found in most other phyla. The sphincter muscle is like this, and so are some of the muscle fibres of the oral disc and tentacles.

Muscles can shorten, but they cannot forcibly extend: they cannot make a contracted anemone expand again by pushing it up. The movements of anemones do not depend on the muscles alone but on interaction between them, the water in the gastric cavity, the siphonoglyphs and the mesogloea. The role of the mesogloea is discussed in the next section of this chapter. The siphonoglyphs drive water into the anemone and are apparently used to inflate it to large sizes. This would not work if the mouth were a gaping hole: water could leave by the mouth as fast as it entered by the siphonoglyphs. However, the flexible pharynx must act as a valve, which tends to be closed if the pressure in the gastric cavity rises above the pressure of the surrounding water.

The pressure in the gastric cavity has been measured, by means of a manometer or a pressure transducer connected to the cavity by a glass tube pushed through a hole in the body wall. In one investigation the anemone *Tealia* was found to maintain pressures up to 3.5 cm water (350 N m^{-2}) while inactive. Higher pressures up to 10 cm water occurred during natural movement and up to 15 cm water when the anemone was made to contract by poking it.

When an anemone is inflated, contraction of the circular muscle of the column must tend to make it tall and thin, while contraction of the parietal muscles must tend to make it squat and fat (Fig. 6.7*e*, *f*). These movements will occur with no change of volume if the mouth remains closed. The parietal and circular muscles thus have opposite effects (they are antagonistic to one another) so long as the volume of water in the gastric cavity remains constant. If both

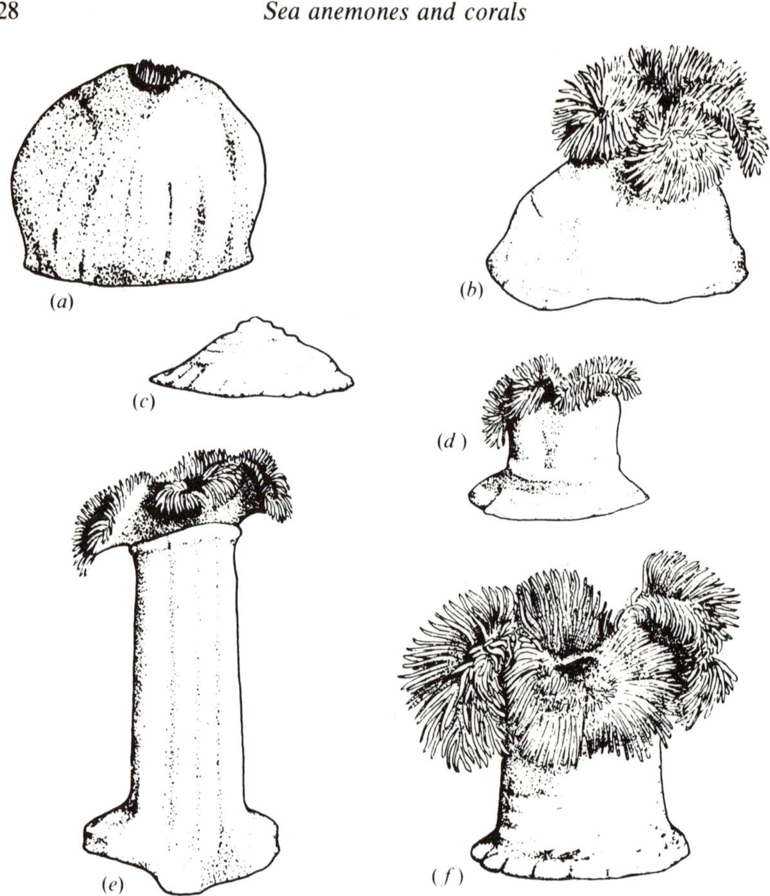

Fig. 6.7. Outlines traced from photographs to the same scale of the same individual of *Metridium* on different occasions. Diameter in (*a*) 4 cm. From R. B. Clark (1964). *Dynamics in metazoan evolution*. Oxford University Press (after E. J. Batham & C. F. A. Pantin).

contract together water must be driven out of the mouth and siphonoglyphs and the size of the anemone must be reduced (Fig. 6.7*d*). Anemones can also bend to one side (presumably by contracting the parietal muscles of that side only) and they sometimes contract only part of the circular muscle, producing a waisted effect. Contraction of the retractor muscles pulls the oral disc and tentacles down into the column, and if the sphincter then contracts it closes over the tentacles (Fig. 6.7*a*, *c*). An anemone contracted like this is much less vulnerable than when it has its tentacles spread. Extreme contraction of all the muscles buckles the body wall (Fig. 6.7*c*).

Fig. 6.2 shows neurones (nerve cells) and a neurosensory cell, as well as musculo-epithelial cells. Each neurone has a central body containing the nucleus and two (or more) long processes which are called neurites. (The more familiar term 'axon' is generally reserved for single long processes.) Particularly long neurones, many of them over 5 mm long in *Metridium*, occur on that face of each mesentery which bears the retractor muscles. The neurosensory cells are presumed to be sensory because they resemble known sensory cells in other animals but we do not know what senses they serve, whether different cells are sensitive to touch, chemical stimuli, etc., or whether each responds to a variety of stimuli. These cells bear cilia and also have neurites like the neurones.

The neurones form loose networks. The networks on mesenteries can be shown particularly clearly by stretching the mesentery on a microscope slide and staining it by a technique which impregnates the neurones with silver. The neurones then appear black, in an otherwise transparent tissue. Where the neurites of two neurones cross, their surfaces are brought into close contact. These junctions are presumably synapses, places where information can be passed from cell to cell. Electron microscope sections of similar junctions in jellyfish show granules in the cytoplasm of one or both neurones at the synapses. These presumably contain the transmitter substance, which is released by one cell and stimulates the other to activity. Synapses which have granules in both neurones can presumably transmit information in either direction.

The neurones apparently transmit information by means of the electrical changes called action potentials which also occur in other animals. A record believed to show action potentials in the neurones of a sea anemone will be presented shortly. Apart from such records we have little knowledge from direct experiment of how cnidarian neurones work. It is assumed that they work in the same general way as the giant neurones of squids, which have been studied in great detail (chapter 15).

So that the significance of some important experiments on sea anemones can be appreciated, it is necessary to say a little about the working of neurones as discovered by research on other animals. When no message is being transmitted, neurones have a negative membrane potential which is typically about -50 mV. This is due to differences in ion concentrations inside and outside the cell, and to the permeabilities of the membrane to the various ions; in other words it is due to the same causes as the negative membrane potentials of Protozoa. An action potential is a brief reversal of the membrane potential, like the reversal of the membrane potential which occurs when the anterior end of *Paramecium* is stimulated (p. 60). As in *Paramecium* it is due to brief changes in the permeability of the cell membrane to ions. Details are given in chapter 15. Once an action potential has been started in a neurone it travels along it to its end unless it 'collides' with an action potential travelling in the opposite direction.

An action potential is started if the membrane potential is increased (made less negative) in any part of the cell, provided the increase is large enough. There is a threshold value of membrane potential which must be passed, if an action

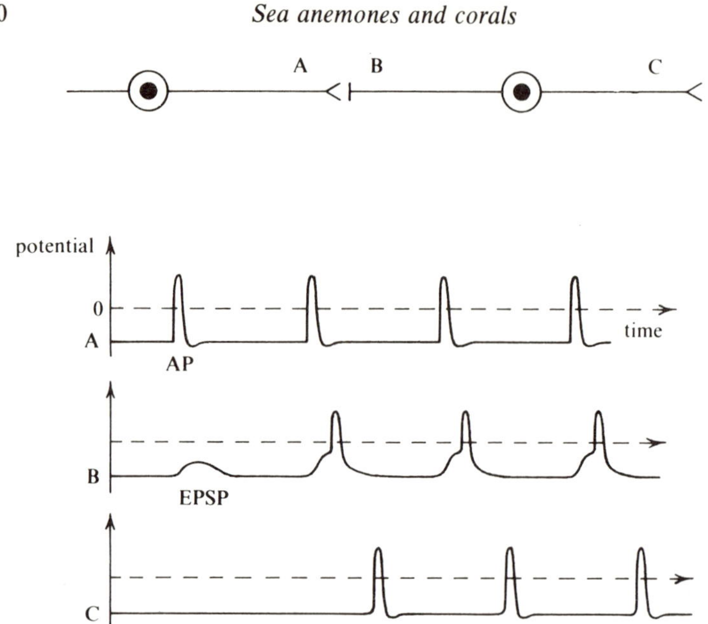

Fig. 6.8. A diagram showing two neurones which make a synapse, and graphs of membrane potential against time for points A, B and C. Action potentials (AP) arrive from the left of the diagram, reach the synapse and set up excitatory post-synaptic potentials (EPSP) in the second neurone. The second and subsequent EPSPs give rise to action potentials which travel along the second neurone.

potential is to occur. The increase can be brought about artificially, by electrical stimulation. It occurs naturally at synapses where action potentials in one neurone increase the membrane potential in the other, either by direct electrical effects or by stimulating release of a transmitter substance.

Fig. 6.8 is designed to show some features of the action of synapses. Records of electrical events at A, B and C are shown: each record is a graph of membrane potential against time. A series of action potentials arrive from the left. The first causes a temporary, localized increase in membrane potential (an excitatory post-synaptic potential or EPSP) in the second neurone, at B. This fails to reach the threshold so no action potential is caused and no electrical event is recorded at C. The second and subsequent action potentials cause larger EPSPs which pass the threshold and set up action potentials. These travel along the second neurone and so are recorded both at B and at C. They cause larger EPSPs because the size of an EPSP depends on the time which has elapsed since its predecessors: it is largest if it occurs soon after previous EPSPs. This phenomenon is called facilitation. There are many synapses at which a single action potential causes an EPSP which passes the threshold, but in the example shown facilitation has to occur before the threshold is reached.

Corresponding action potentials occur a little later at B than at A, and at C than at B. This is because action potentials travel at finite velocity and because delays occur at the synapse.

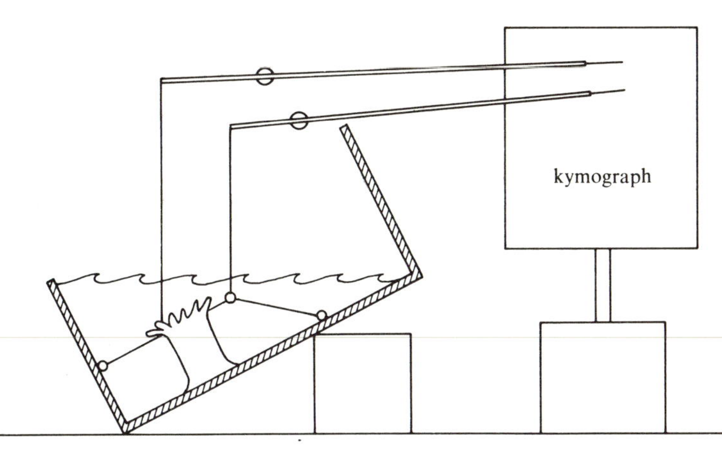

Fig. 6.9. Apparatus used for investigating the responses of sea anemones to electrical stimuli.

The neuromuscular junctions at which neurones join muscles and stimulate them to contract work in the same way as synapses. Action potentials in the neurone set up changes equivalent to EPSPs in the muscle and in some (but not all) muscles these give rise to action potentials which travel along the muscle fibre, stimulating the contractile material all along the fibre.

Professor Carl Pantin carried out a classic series of experiments in the 1930s on the sea anemone *Calliactis*. The apparatus he used for some of these experiments is shown in Fig. 6.9. Two threads attach the top of the column of the anemone to light levers. They are attached in such a way that one lever is moved if the column shortens (i.e. if the parietal muscles contract) and the other if the sphincter muscle contracts. Fine pointers on the ends of the levers rest lightly on the surface of a kymograph drum (see p. 11), so that they scribe lines on the smoked surface as the drum rotates. Deflections of the lines show changes in length of the parietal and sphincter muscles. The apparatus can be used to record the spontaneous behaviour of the anemone, or its responses to stimuli.

Calliactis withdraws its tentacles (by contracting the retractor muscles) and contracts its sphincter if it is given a series of electric shocks of sufficient intensity. One of Pantin's discoveries was that the first shock caused no contraction of the sphincter, the second a small contraction (provided the interval between shocks was neither too short nor too long) and succeeding shocks a series of stepwise contractions.

Pantin suggested that the first stimulus produced impulses in the neurones which reached the muscle fibres but failed to stimulate contraction. Nevertheless they had a facilitating effect at the neuromuscular junctions, making the muscle fibres more receptive to subsequent impulses. This explanation is still acceptable but the evidence for it has been strengthened by later experiments with more modern equipment.

An investigation which gave particularly clear evidence was performed on

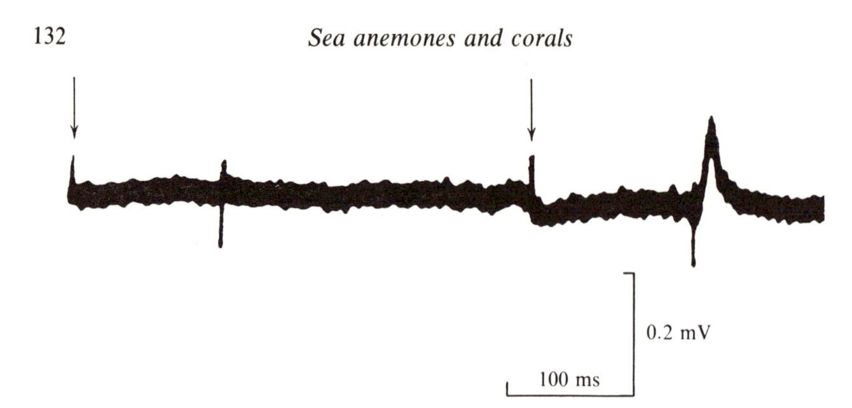

Fig. 6.10. An oscilloscope record of electrical activity in a mesentery of *Calamactis*, in response to two electric shocks given to a tentacle. Arrows show when the shocks were given. From P. E. Pickens (1969). *J. exp. Biol.* **51**, 513–28.

Calamactis. This is a burrowing anemone which lives in sand with only its oral disc and tentacles projecting into the water above. Whereas *Calliactis* protects itself by contracting its sphincter, *Calamactis* protects itself by contracting its retractor muscles and so pulling itself down into its burrow. Its retractor muscles respond to electric shock in the same way as sphincter muscle of *Calliactis*. In the experiments which will be described, a specimen of *Calamactis* was pinned down by the aboral end (i.e. the end furthest from the mouth) to the bottom of a dish of sea water. A thread sewn to the oral disc was connected to a displacement transducer, which thus made a record of any contractions of the retractor muscles. The animal was slit open to expose the mesenteries and a suction electrode (Fig. 1.8*b*) was placed on one of the retractor muscles.

Fig. 6.10 shows a record obtained from such an electrode, while two electric shocks were delivered to a tentacle. Two artefacts appear on the record, due to the direct effect of the shocks on the recording electrode. They show when the shocks occurred. One hundred and twenty milliseconds after each shock is a brief event which is believed to be a nerve action potential. After the second one only there is a distinct, larger event, which is believed to be a muscle potential. Muscle contraction does not begin until the second shock.

In an action potential, the membrane potential changes by (typically) about 0.1 V. This can be recorded by an electrode inside the cell, if an electrode can be inserted. Action potentials can also be detected by electrodes outside the cell, but the signal obtained in this way is only a small fraction of the change in membrane potential. This is why the action potentials in Fig. 6.10 seem so small. Also, an action potential recorded extracellularly may seem positive or negative, depending on the relative positions of cell and electrode. The nerve action potentials appear as downward (positive) deflections in Fig. 6.10 and the muscle potential as an upward (negative) one.

Fig. 6.11 shows records from the displacement transducer as well as from

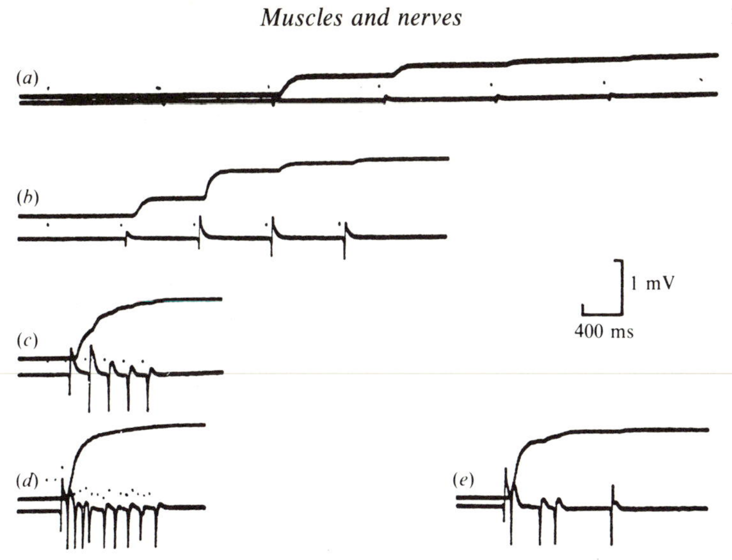

Fig. 6.11. Oscilloscope records of responses of *Calamactis* to electrical stimulation of a tentacle, (*a*) to (*d*), and mechanical stimulation, (*e*). In each record the upper trace shows the output from a displacement transducer and the lower one the potential of a recording electrode. From P. E. Pickens (1969). *J. exp. Biol.* **51**, 513–28.

the electrode. In this experiment (as in most) nerve action potentials were not recorded: the electrode must have been particularly favourably placed in the experiment of Fig. 6.10.

Look first at Fig. 6.11(*b*). Electric shocks were given to a tentacle, at the instants indicated by dots. The first produced no muscle potential and no contraction of the muscle. The second produced a small potential and a small contraction: facilitation has apparently occurred. The third produced a larger potential, showing that further facilitation had occurred, and a large contraction. Subsequent shocks produced large muscle potentials but only small further contractions, presumably because the animal had already contracted nearly as far as it could.

Action potentials in any nerve or muscle fibre are uniform in size but EPSPs vary in size, depending on the state of facilitation. The muscle potentials vary in size, and must thus be EPSPs. In contrast, the potentials identified in Fig. 6.10 as nerve action potentials are uniform in size. It seems that every shock of sufficient size sets off action potentials in the nerve net (unless it follows too soon after its predecessor) but that the effect on the muscle fibres depends on the state of facilitation of the nerve–muscle junction. It is not clear why the first of a group of nerve action potentials produces, in this case, no measurable EPSP.

Fig. 6.11(*b*) shows the effect of shocks delivered at intervals of 0.8 s. Less frequent shocks produce a slower, incomplete contraction (Fig. 6.11 *a*). More frequent ones produce faster contractions (Fig. 6.11 *c*, *d*). Notice also that larger muscle potentials are obtained when stimulation is more frequent. Fig. 6.11(*e*) shows the effect of prodding the anemone. No electrical stimulus was applied but a series of muscle potentials and contractions occurred. Presumably sensory cells sensitive to touch started action potentials in the nerve net, which in turn stimulated the muscle. A gentler touch produces fewer, more widely spaced, muscle potentials and a smaller contraction. Thus the size of the anemone's response depends, appropriately, on the intensity of the stimulus.

The velocity of conduction in the nerve net has been measured. A tentacle was stimulated by electric shocks, and muscle potentials were recorded simultaneously at two points on a retractor muscle. Corresponding potentials occurred at slightly different times at the two points and the interval was used to calculate the velocity of conduction. Near the oral ends of the mesenteries the velocity was 0.8 m s^{-1} but it was less near their aboral ends where the neurites are generally thinner.

The withdrawal responses of *Calliactis* and *Calamactis*, which have been described, are not the only possible responses to electrical or mechanical stimulation. They only occur in response to fairly strong mechanical stimulation or to electric shocks at fairly short intervals. A series of gentle touches at the base of a *Calliactis* tentacle, or a series of shocks at long intervals (for instance, 4 s), evoke a different response. First, the stimulated tentacle turns in towards the mouth. As the sequence of stimuli continues, neighbouring tentacles join in the movement. Often the edge of the oral disc bends inward at the point of stimulation and if stimulation continues this response (due, probably, to the radial muscles of the mesenteries) spreads round the disc. This is a feeding response: movements like this normally carry food caught by the tentacles to the mouth.

Fig. 6.12 shows a different spreading response. An electrode has been placed on a coral polyp at *S*, and stimuli are delivered through it at regular intervals. The first stimulus has no effect but the second stimulus makes some neighbouring polyps retract and with subsequent stimuli the effect spreads progressively further over the coral colony. Similar spreading responses have been demonstrated for many species of coral.

Though several stimuli are needed before a distant polyp retracts, action potentials apparently travel over wide areas of the colony from the first stimulus. This has been demonstrated by recording with suction electrodes. The electrical response to a single stimulus travels over the colony at about 0.2 m s^{-1} and is no smaller far from the site of stimulation than close to it. Why, then, do distant polyps not retract so soon as near ones, when a sequence of stimuli is given?

It has been shown that the electrical response to the second stimulus of a series travels more slowly over the colony than the response to the first, the response to the third travels more slowly still, and so on. This slowing is most marked close to the site of stimulation. It is not known why it happens, but it seems to explain why retraction spreads progressively. Fig. 6.13 illustrates

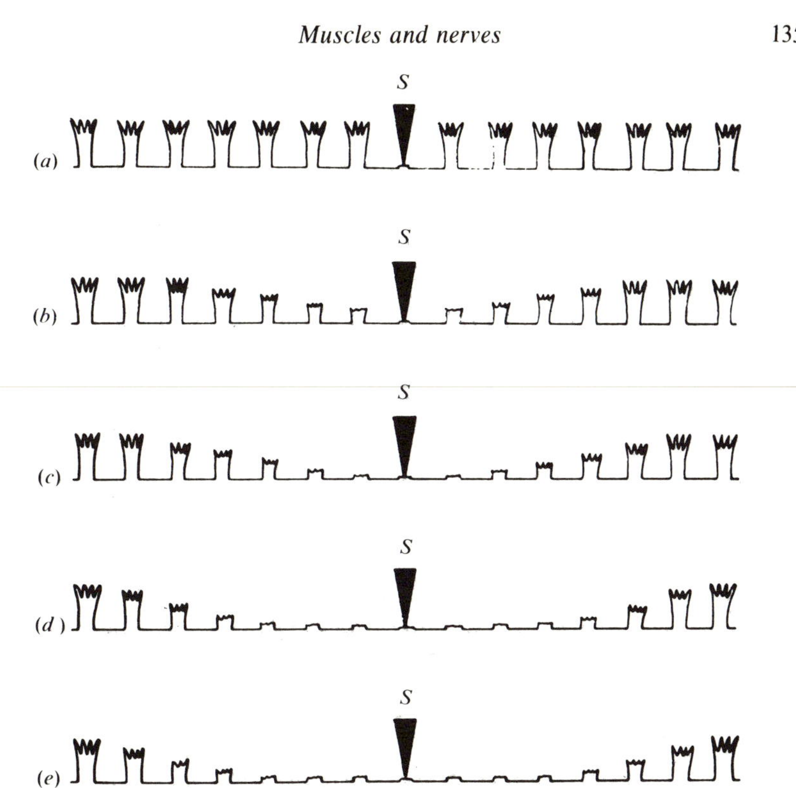

Fig. 6.12. A diagram of a stimulating electrode *S* on a colony of the coral *Porites*. (*a*) to (*e*) show the colony immediately after each of a series of five electrical stimuli, delivered at equal intervals. From G. A. B. Shelton (1975). *Proc. R. Soc. Lond.* **190B**, 239–56.

a hypothetical experiment involving more electrodes than it was feasible to use in the actual experiments on which it is based. It is supposed that a stimulating electrode and five recording electrodes are placed in line, equally spaced, on a coral colony. Eight stimuli are delivered through the stimulating electrode, at intervals of 1 s, making action potentials set out from the point of stimulation at times 1 s, 2 s, etc. The graph shows the times at which action potentials would be recorded at the five recording electrodes. The action potentials resulting from the first stimulus travel at a uniform velocity which is relatively high (so that the gradient of the graph is low). Those resulting from subsequent stimuli travel progressively more slowly (indicated by steeper gradients) for the first few centimetres. Consequently the time interval between successive action potentials increases with increasing distance from the point of stimulation. The action potentials started at intervals of 1 s but they arrive at recording electrode (5) at intervals of about 2.3 s.

Suppose that the facilitating effect of an action potential decays in 2 s, so that each polyp retracts as soon as it has received two action potentials at an

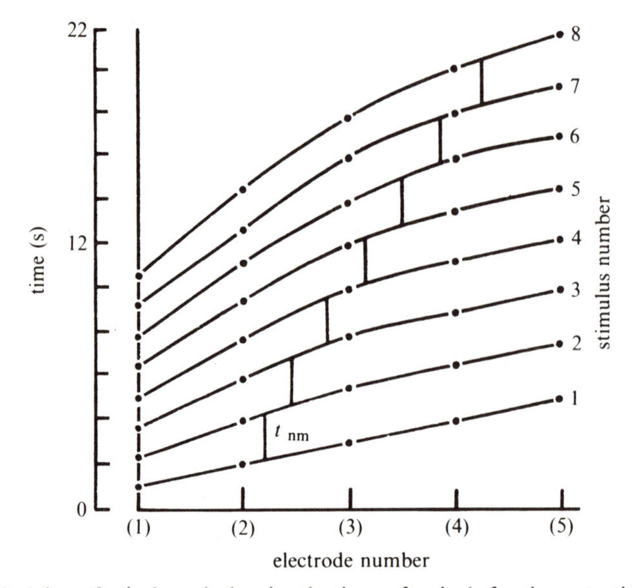

Fig. 6.13. A hypothetical graph showing the times of arrival of action potentials resulting from a series of stimuli, at equally spaced recording electrodes on a coral colony. Further explanation is given in the text. Modified from G. A. B. Shelton (1975). *Proc. R. Soc. Lond.* **190B**, 239–56.

interval of 2 s or less. A polyp near electrode (1) receives its second action potential about 1.5 s after the first and will retract then. One near electrode (5) receives all its action potentials at intervals of about 2.3 s and will not retract at all.

The vertical lines t_{nm} ('time for the decay of neuromuscular facilitation') are all of equal length, representing 2 s. They show, for each pair of successive action potentials, the distance from the point of stimulation at which they are exactly 2 s apart. To the left of them the intervals are shorter and polyps will retract. To the right the intervals are longer and polyps will not retract. Hence polyps at (1) and (2) will retract after the second stimulus, a polyp at (3) after the fourth, and one at (4) after the eighth. Retraction will spread gradually over the colony.

To explain fully the way in which retraction is actually observed to spread over colonies (Fig. 6.12) we have to make the additional assumption that an interval only slightly shorter than 2 s results in only partial retraction.

So far we have considered only some simple effects which are achieved with nerve nets. Sea anemones are capable of more complex behaviour and a particularly complex behaviour pattern is shown in Fig. 6.14. The hermit crab *Pagurus bernhardus* occupies old shells of the whelk *Buccinum. Calliactis* is sometimes found on stones but it is also very commonly found attached to *Buccinum* shells occupied by *Pagurus.* This is probably advantageous to both animals. The anemone gets particles of food scattered from the crab's meals

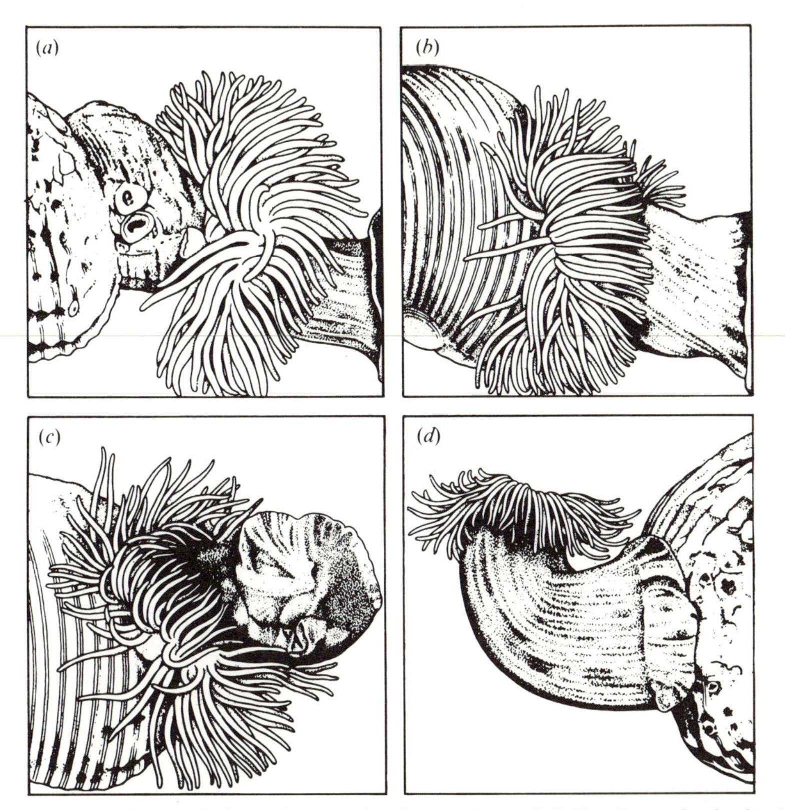

Fig. 6.14. Drawings made from photographs of a specimen of *Calliactis* transferring itself from the wall of an aquarium to a *Buccinum* shell occupied by *Pagurus bernhardus*. The photographs were taken at (*a*), 0 min; (*b*), 5 min; (*c*), 9 min; (*d*), 22 min. From E. J. W. Barrington (1967). *Invertebrate structure and function.* Nelson, London (after Ross).

and the crab is probably given some protection from predators by the nematocysts of the anemone. Fig. 6.14(*a*) shows a *Calliactis* attached to the wall of an aquarium. A *Buccinum* shell occupied by a hermit crab has come into contact with it. The *Calliactis* grips the shell with its tentacles, discharging nematocysts to get a firm grip (Fig. 6.14*b*). It releases its pedal disc from the aquarium wall and bends its body into a hoop (Fig. 6.14*c*). Then it attaches its pedal disc to the shell and releases its tentacles (Fig. 6.14*d*). It has transferred itself from the tank wall to the shell by a complex sequence of movements in (in this case) 22 min.

Calliactis will try to transfer itself from glass, plastic or stone to a *Buccinum* shell, whether the shell is occupied by a hermit crab or a whelk or is empty. However, if there is a whelk in the shell it generally shakes the anemone off by violent movements. *Calliactis* will not transfer themselves to *Buccinum* shells which have been boiled in alkali (to remove organic matter) or painted over with plastic. They apparently use a chemical sense to recognize the shells.

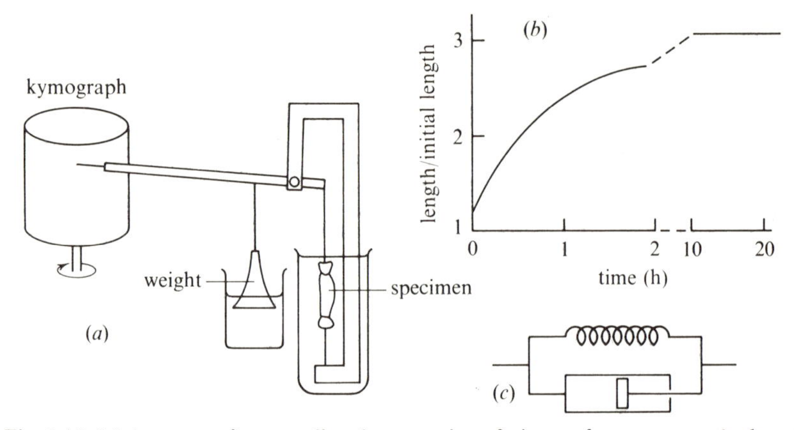

Fig. 6.15. (*a*) Apparatus for recording the extension of pieces of sea anemone body wall under constant stress. (*b*) A graph of the length of a strip of *Metridium* body wall against time, in an experiment using the apparatus. At the beginning of the experiment the strip was 2 cm wide and the load was 0.05 N. (*c*) This diagram is explained in the text.

Sea anemones often move spontaneously, without obvious stimulus. This is particularly clear in films made by time-lapse photography, which show the movements speeded up. Anemones expand, often taking an hour or so to complete the movement, and contract again. They bend from side to side and even crawl. Several species have been observed to crawl at rates of 0.01–0.1 mm s^{-1} by passing waves of muscular contraction over the pedal disc in the manner of snails and slugs (chapter 13).

There is evidence that some of the behaviour patterns of sea anemones are not controlled by action potentials in specialized nerve cells, but by similar electrical events which are transmitted through the epidermis in the ordinary epidermal cells.

MESOGLOEA

Some sea anemones such as *Metridium* can make remarkable changes of size (Fig. 6.7). Their mesogloea must be stretched greatly when the animal swells. How large are the forces required to stretch it? What are its mechanical properties, and what role does it play in the movements of the animal?

The mechanical properties of the mesogloea have been investigated by stretching strips of body wall. Cutting a sea anemone makes its muscles contract violently, and the mechanical properties of strips cut from untreated anemones owe as much to the muscle as to the mesogloea. However, the muscles can be made inactive by leaving the anemone for several hours in a mixture of equal volumes of sea water and magnesium chloride of the same ionic strength, or in a solution of menthol in sea water (many other marine invertebrates can be immobilized in the same ways). The anemone becomes almost entirely unresponsive to mechanical stimulation. Strips of body wall from anemones narcotized in these ways have been used in tests of mechanical properties.

The apparatus which was used for the tests is shown in Fig. 6.15(*a*). The piece of tissue is being stretched by a specially shaped (hyperboloid) weight, and the course of stretching over a period of many hours is recorded on a slowly-rotating kymograph drum. There is a reason for the oddly-shaped weight. The strips of body wall often stretched to as much as three times their initial length and when they did their cross-sectional area must have been reduced to about a third of its initial value. If the stretching force were constant the stress (i.e. force/cross-sectional area) would increase by a factor of three. However, as the strip stretches the weight sinks into water and the upthrust exerted on it by the water diminishes the force. The weight was shaped in such a way as to keep the stress constant, as the specimen stretched and the weight sank deeper into the water. The experiments were designed in this way because it was expected to simplify analysis.

The results of a typical experiment are shown in Fig. 6.15(*b*). The specimen stretched at a gradually diminishing rate. After half an hour it was nearly twice its initial length, after 10 hours it was still stretching perceptibly and it was eventually three times its initial length. In some experiments the weight was removed at this stage and it was found that the strip returned gradually, over many hours, to approximately its initial length.

Fig. 6.15(*c*) shows a simple mechanical device which would stretch and recoil in much the same way. It consists of a spring and a dashpot, connected in parallel. The dashpot is a cylinder with a leaky piston, filled with viscous fluid. The piston can be moved along the cylinder because fluid can leak round it, but its movements are resisted by the viscosity of the fluid. The spring by itself would extend immediately a force was applied and recoil as soon as it was released. The extension would be proportional to the force. The dashpot by itself would extend at constant rate under constant force, and would not recoil. The *rate* of extension would be proportional to the force. The two elements together, connected as shown, would extend under constant force at a gadually decreasing rate. At first the spring would be unstretched and all the applied force would act on the dashpot, which would extend relatively fast. As extension continued the spring would take up more of the force, leaving less and less to overcome the viscosity of the fluid in the dashpot, so extension would become gradually slower. Elastic recoil would occur when the force was removed: it would start quickly and slow down gradually.

The body wall of *Metridium*, like the combination of spring and dashpot, has a combination of elastic and viscous properties. These seem to be properties of the mesogloea, rather than of other tissues. The viscous component makes the animal rather resistant to brief forces, though it is easily deformed by forces lasting several hours. The animal resists waves, for instance, as though it were rather rigid, but a small pressure exerted for an hour or more by the siphonoglyph may inflate it enormously. (Unlike many other sea anemones,

Metridium has only one siphonoglyph.) *Metridium* seems unable to inflate itself quickly: inflation from the condition shown in Fig. 6.7(*d*) to that shown in Fig. 6.7(*e*) takes an hour or so. Contraction is effected by muscles, assisted by elastic forces in the stretched mesogloea, and can be much faster. A fully expanded anemone can be stimulated by vigorous poking to contract to the condition shown in Fig. 6.7(*c*) in a few seconds.

Since the mesogloea is elastic it will always return (slowly) to the same size, when it is unstressed. This is presumably the size eventually reached when the animal is narcotized in magnesium chloride or menthol solution. Narcotized *Metridium* generally look slightly more inflated than the specimen shown in Fig. 6.7(*d*). A fully inflated *Metridium* which halts the cilia of its siphonoglyph will tend to shrink to this size and a fully contracted one which relaxes its muscles will spring up to this size.

What pressure must the siphonoglyph be able to produce in order to inflate *Metridium*? Living anemones often inflate themselves to double their narcotized diameter. The experiments with strips of body wall showed that 4 mN (0.4 g wt) was enough to keep a strip 1 cm wide stretched to double its initial length. The tension (force per unit width) was then 4 mN cm^{-1} or 0.4 N m^{-1}. The pressure in a cylinder of radius r with a circumferential tension T in its wall is T/r. The radius of a typical inflated *Metridium* would be about 2 cm (0.02 m) so the pressure needed is $0.4/0.02 = 20$ N m^{-2} (only 2 mm water). Measurements with manometers have shown that *Metridium* often maintain pressures of this magnitude for long periods.

Now we will examine the structure of mesogloea and see how it relates to the mechanical properties. *Metridium* mesogloea contains about 8% protein and 1% polysaccharide. The rest is water and salts. The mesogloea gives X-ray diffraction patterns just like the pattern given by rat tendons, so it apparently contains a substantial proportion of collagen. Other evidence suggests that most of the protein in it is collagen. The mesogloea is not homogeneous, but has fibrils running through it in various directions. These are seen particularly clearly when a polarizing microscope is used. They are believed to consist of collagen. However, collagen fibres from vertebrate tendons are highly inextensible, compared to mesogloea. They can only be stretched by 10–20% before breaking. How can mesogloea with collagen fibres running through it stretch to three times its unstressed length?

At this stage we need to know more about the mechanical properties of proteins and high polymers in general. Proteins consist of long chains of amino acid residues. Polysaccharides consist of long chains of sugar groups. Rubber consists of long chains of isoprene units. Plastics are also high polymers. All these materials have molecules which are long flexible chains of more or less similar units. This structure makes certain properties possible.

Fig. 6.16 shows some of the possibilities. Fig. 6.16(*a*) represents a material consisting of long-chain molecules which are not attached to each other. Latex is like this. When the material is stretched, the molecules simply slide past each

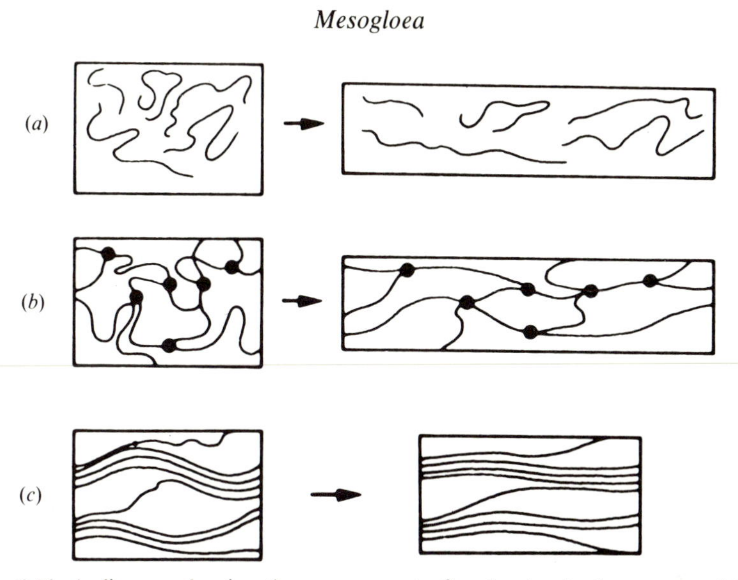

Fig. 6.16. A diagram showing the arrangement of molecules in three types of high polymer, and the effects of stretching. (*a*) An amorphous polymer which is not cross-linked, (*b*) an amorphous cross-linked polymer, and (*c*) a fibre. From R. McN. Alexander (1975). *Biomechanics*. Chapman & Hall, London.

other. The material may have a high viscosity but it is not elastic (except for transient elastic effects due to molecules getting tangled with each other). Fig. 6.16(*b*) shows a similar material, but with the molecules connected together in a three-dimensional network. Rubber is like this: the process of vulcanization inserts sulphur bridges between the latex molecules. The molecules cannot slide past each other but a great deal of stretching is possible, unravelling the molecules and drawing them out in the direction of stretch. When the material is released it recoils elastically, for the following reason. The long molecules are continually coiling and uncoiling, in Brownian motion. A given molecule will sometimes be rather extended and sometimes be rolled up into quite a tight ball. The distance between its ends fluctuates, but will generally not be too far from a certain most probable value. Stretching draws the molecules out in the direction of stretch but Brownian motion tends to restore them to their most probable lengths when the stretching force is removed, causing elastic recoil. Very large extensions are possible: for instance a rubber band can be stretched to about four times its initial length. Elastic moduli are fairly low (i.e. quite small stresses can produce large distortions). Young's modulus is commonly of the order of 1 MN m^{-2} but may be very much lower, particularly for dilute gels. The more cross-links there are, the higher the modulus. In (*b*) the material is amorphous but in (*c*) parts of the molecules are lined up parallel with each other in a crystalline array. This severely limits the amount of stretching that is possible without breaking molecules. Some fibres including collagen seem to have this sort of structure. They cannot generally be stretched by more than

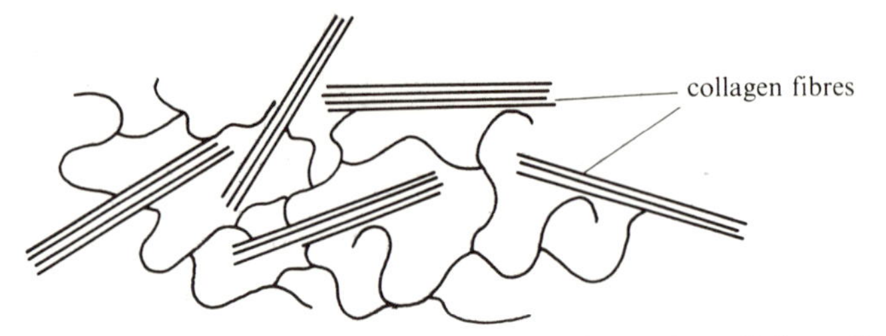

Fig. 6.17. A diagram illustrating the probable structure of mesogloea. Collagen fibres (here shown much too short relative to their width) are scattered in a continuous but dilute protein/polysaccharide gel. The lines represent molecules.

20% and Young's modulus is quite high, generally of the order of 1 or 10 GN m^{-2}. It has been assumed throughout this paragraph that the temperature is not too low. Rubber and other polymers become glass-like at low temperatures.

Mesogloea has a Young's modulus of the order of 1 kN m^{-2} (this is calculated from the extension after many hours at constant stress) and can be stretched to at least three times its initial length. These are likely properties for a gel containing a very tenuous network of cross-linked polymer molecules, but are certainly not the properties one would expect of a mesh of fairly straight collagen fibres.

The most plausible explanation seems to be that the collagen fibres are not continuous but are relatively short. Only the dilute protein/polysaccharide gel between them forms a continuous network of linked molecules extending throughout the mesogloea (Fig. 6.17). This has not been demonstrated directly. No method has been devised for measuring the length of the fibres and it is not at all obvious when they are examined under a microscope, that individual fibres do not run right through the mesogloea.

GORGONIN

Gorgonians have a delicate fan-like structure (Fig. 6.4*c*) but grow upright on coral reefs and have to withstand wave action. Their blades are stiffened only by mesogloea with calcareous spicules in it, and are quite flexible. Their slender stems have a core of material called gorgonin, which is not part of the mesogloea. This makes the stems stiff, and remarkably strong. To collect a large gorgonian from a coral reef it is far easier to hack away the reef limestone at its base than to break or cut the stem.

Gorgonin is fibrous, with fibres running predominantly parallel to the axis of the stem. X-ray diffraction patterns and amino acid analysis show that it consists largely of collagen, but its Young's modulus is several times higher than that of ordinary tendon collagen. Also, ordinary collagen can be dissolved by autoclaving (heating with water under pressure) but gorgonin cannot. The high modulus and the insolubility suggest that gorgonin has additional cross-links which are more resistant to heat than the cross-links of tendon collagen. The only known way of dissolving gorgonin (except by really drastic action with concentrated acids or alkalis) is by putting it in a 10% solution of sodium hypochlorite. It shares this property with some other proteins (including the protein of insect cuticle) which are known to be cross-linked by a process called quinone tanning. It is therefore presumed that gorgonin is a quinone-tanned collagen.

The process seems to work like this. It starts with an *ortho* diphenol

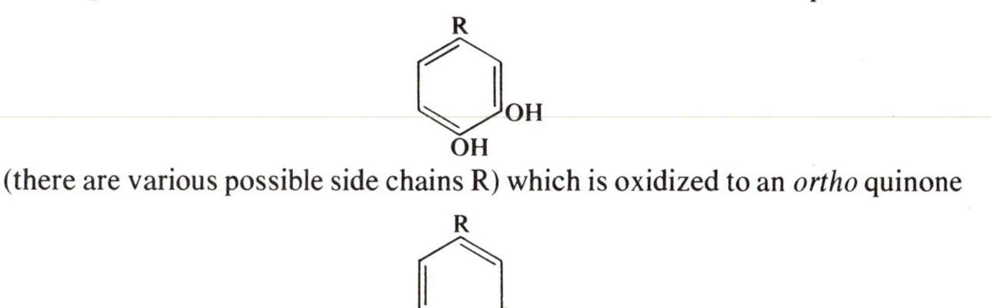

(there are various possible side chains R) which is oxidized to an *ortho* quinone

R
O
O

This reacts with two free NH_2 groups in adjacent peptide chains, linking the chains together. For instance, it may attach to the second NH_2 group of lysine, thus

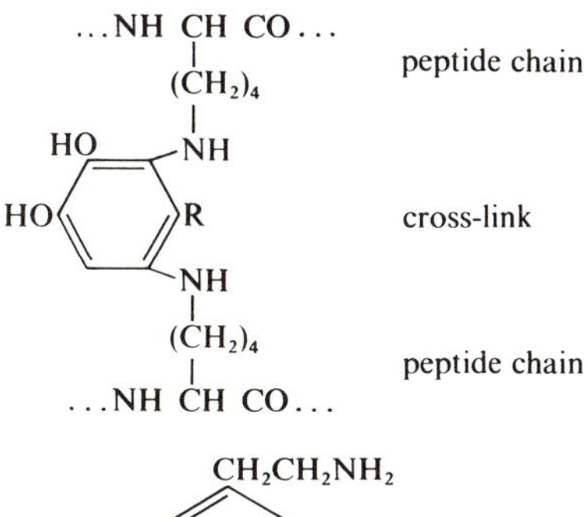

Dopamine

$CH_2CH_2NH_2$
OH
OH

is an *ortho* diphenol used in quinone tanning by various animals. It is synthesized from the amino acid tyrosine,

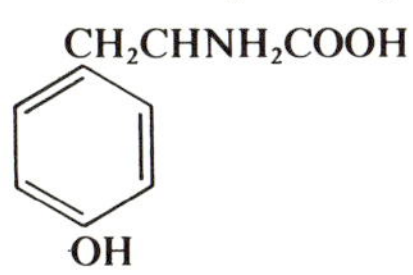

Clumps of sea fans are common on Atlantic reefs. They grow upright, and the fans in any clump tend to be parallel to each other. Divers on reefs off Florida measured the compass bearings of individual fans and found that large fans (0.5 m or more high) seldom deviated by more than 20° from the average direction of the clump. Smaller fans tended to deviate more, and sections of the stems of some large fans seemed to show that they had twisted as they grew.

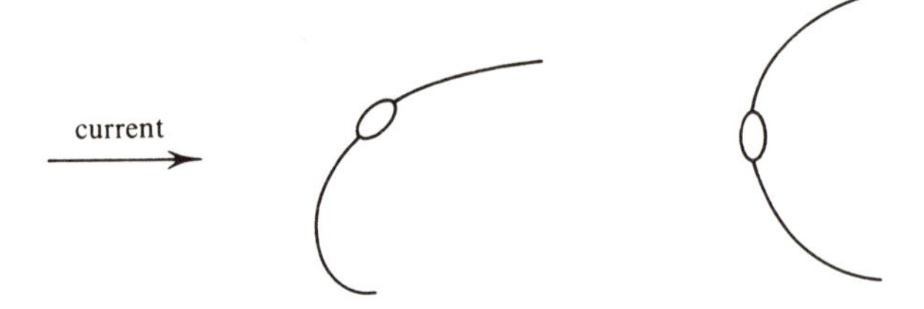

Fig. 6.18. Diagrammatic horizontal sections of a sea fan in a water current.

The large fans generally stand at right angles to the prevailing wave motion. Thus sea fans apparently start their life oriented at random and gradually twist at right angles to the direction of water movement.

Living sea fans are flexible, unlike the dried specimens kept in museums. Fig. 6.18 shows sections of a flexible blade in a current. When the blade is set obliquely in the current, it bends in such a way that its upstream edge is more nearly at right angles to the current than the downstream edge. Hence greater forces act on the upstream than on the downstream half of the blade, and the asymmetrical forces tend to twist the blade more nearly at right angles to the current. If the gorgonin is not fully cross-linked it will flow very slowly in viscous fashion and a constant current (or one which flows backwards and forwards along the same line, like wave motion) will gradually twist it. It is believed that this is what happens.

Gorgonians and stony corals both have to withstand wave action but gorgonians have slender, relatively flexible skeletons whereas corals have massive, rigid ones. The corals have only a tiny proportion of organic matter in their skeletons, which are consequently brittle like pottery. The physical basis of brittleness is discussed in chapter 13, in an account of mollusc shells. Stony corals cannot form delicate branching structures, because they are so brittle.

NEMATOCYSTS

Nematocysts have long hollow threads which coil inside them in the course of development. The thread turns inside-out as it is discharged (Fig. 6.19), and this process cannot be reversed. A nematocyst can only be used once and must then be discarded.

The process of discharge is difficult to study, because it is fast and because some important details of nematocyst structure are too small to be seen by optical microscopy. However, it has been possible to obtain large nematocysts which have stopped short in the process of discharge, by wiping the anemone *Corynactis* over a sheet of plastic. Electron microscope sections of these nematocysts have greatly improved our understanding of the process of discharge.

Fig. 6.20(*a*) shows part of one of these sections. It shows the tip of the incompletely discharged thread, where the thread was turning inside-out when the action stopped. The barbs are on the outside of the discharged part of the

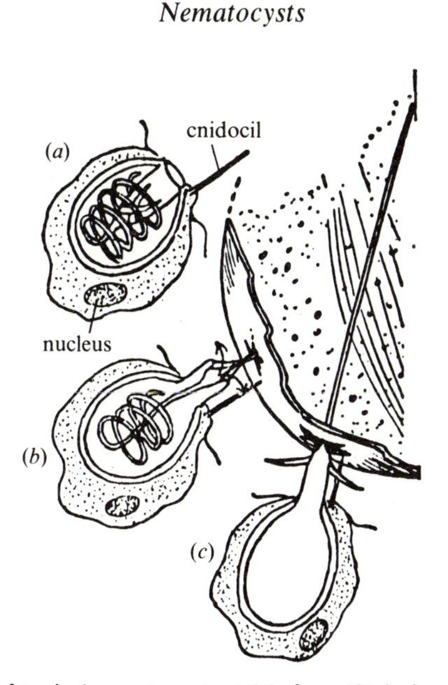

Fig. 6.19. Sketches of typical nematocysts, (*a*) before, (*b*) during and (*c*) after discharge. Note how the emerging spines in (*b*) cut the cuticle of the prey. From A. C. Hardy (1956). *The open sea, its natural history*, part 1, *The world of plankton.* Collins, London.

thread but inside the undischarged part. They are much closer together on the undischarged part because it is deeply pleated. The section shows that when discharge occurs and the pleats are eliminated, the thread becomes about three times as long as before.

The model shown in Fig. 6.20(*b*) shows that the pleats are helical: three deep pleats run round and round the undischarged thread. If you take a tube of paper, hold it by its ends and twist it, helical pleats appear in it and it becomes shorter and more slender. The undischarged thread is pleated in a deeper and more regular way than the paper tube. Careful measurement on the sections used to construct Fig. 6.20(*b*) shows that the area of the thread wall is scarcely changed when the nematocyst discharges. However, the total volume of the nematocyst must increase greatly, since the capsule does not collapse as the thread emerges from it. Fluid must enter the capsule during discharge. It may be drawn in by an osmotic mechanism.

To investigate this possibility, the freezing point of the fluid in undischarged nematocysts was determined. The nematocysts were cut open under liquid paraffin with a splinter of silica, and the contents were sucked into micropipettes. The freezing point of fluid from undischarged capsules was about −5 °C. This is well below the freezing point of sea water (about −1.9 °C) so the undischarged capsules have an osmotic concentration well above that of sea water. It would be feasible for them to draw water in osmotically as they discharge.

When a nematocyst has discharged, droplets of fluid emerge from the open

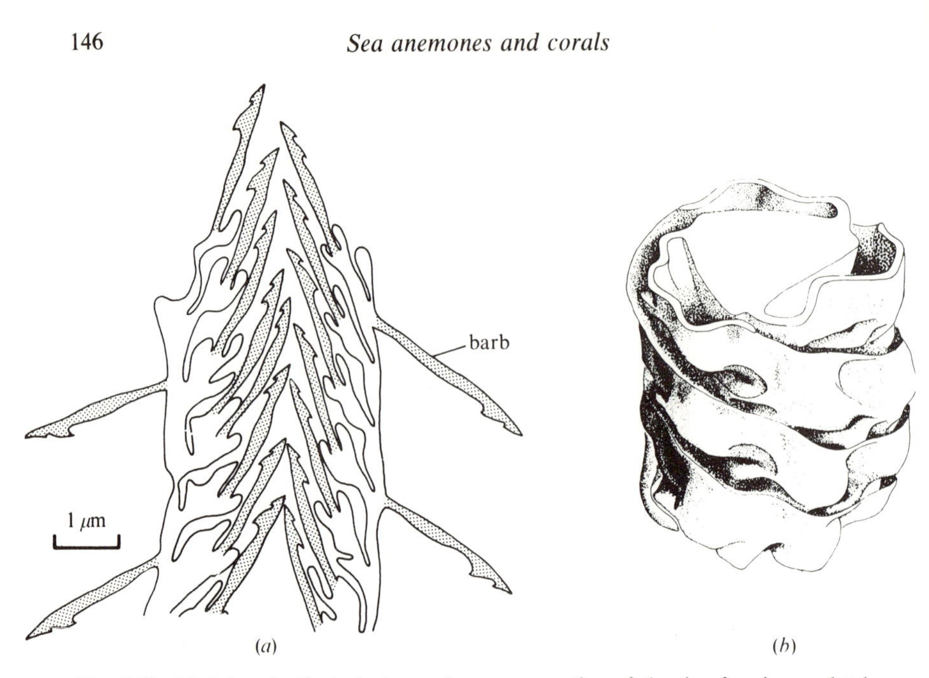

Fig. 6.20. (*a*) A longitudinal electron microscope section of the tip of an incompletely discharged *Corynactis* nematocyst. Drawn from a photograph in R. J. Skaer & L. E. R. Picken (1965). *Phil. Trans. R. Soc.* **250B**, 131–64. The external barbs had broken off the original specimen, but have been drawn in approximately their proper positions. (*b*) A model based on serial transverse electron microscope sections of part of the undischarged thread of a *Corynactis* nematocyst. From L. E. R. Picken & R. J. Skaer (1966). *Symp. zool. Soc. Lond.* **16**, 19–50.

tip of the thread. Such droplets were collected and found to have a freezing point of −3 °C. This confirms that the osmotic concentration of the nematocyst contents falls during discharge, as it would if water were being taken up osmotically.

Presumably discharge is started by some process which increases the permeability of the capsule wall, letting water leak in. Changes in cell membrane permeability can occur rapidly, for instance in the reversal mechanism of *Paramecium* (p. 60) and in nerve conduction. We have already seen that both these processes can be brought about by electrical stimulation. Similarly, nematocysts of *Corynactis* can be fired by an electric shock. When a tentacle is put between electrodes and a brief shock is given, only the nematocysts on the side towards the cathode discharge. This suggests that the stimulus affects the nematocysts directly, rather than through the nervous system. There is no evidence that the nervous system is involved in firing nematocysts (though it may modify their sensitivity to stimuli). There is a structure on the cnidoblast which is believed to have a sensory function. This is the cnidocil, a cilium surrounded by a tuft of microvilli (Fig. 6.19). The microvilli are similar in diameter to the cilium but do not contain microtubules.

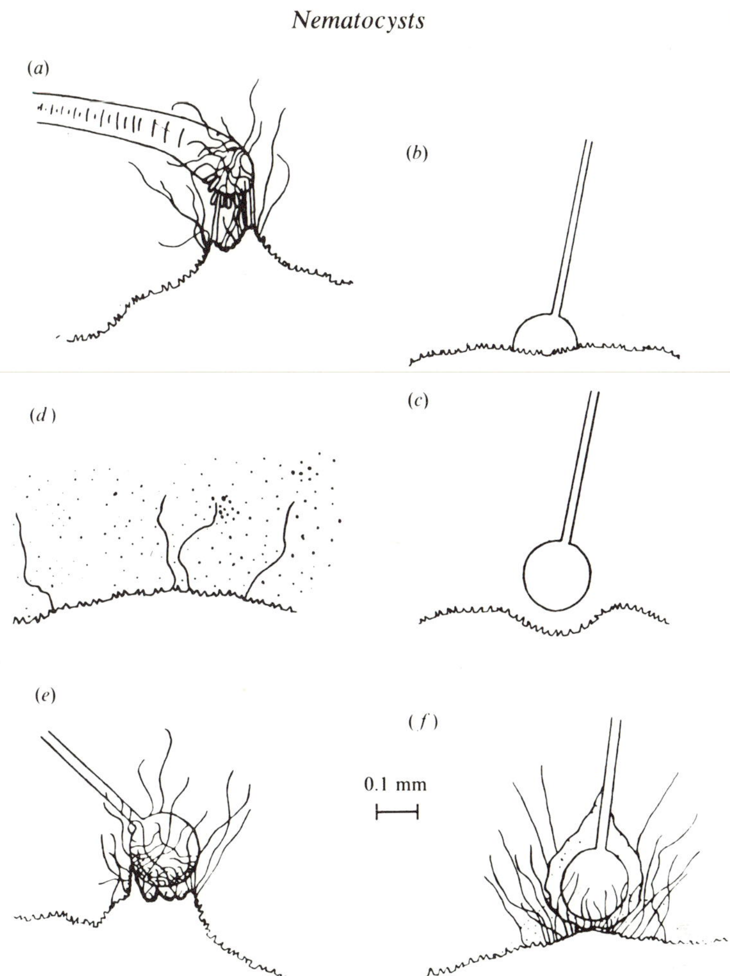

Fig. 6.21. Responses of nematocysts on the tentacles of the sea anemone *Anemonia* to various stimuli which are described in the text. From C. F. A. Pantin (1942). *J. exp. Biol.* **19**, 294–310.

Fig. 6.21 shows some experiments designed to discover the natural stimuli for nematocyst discharge. When a tentacle of *Anemonia* was touched with a human hair, nematocysts at the point of contact discharged (*a*). Touching with a clean glass rod did not cause discharge (*b*, *c*), and adding saliva to the water stimulated only a few nematocysts to discharge (*d*). However, large numbers were discharged by touching with the glass rod when there was saliva in the water (*e*) or when the rod had been smeared with a preparation of fatty material from the scallop, *Pecten* (*f*). These experiments seem to show that a combination of mechanical and chemical stimulation is normally required to cause discharge. The nematocysts (or at least the epidermis near them) must be touched, and some substance or substances must be present. The substances have not been identified but since they are present on hair, in saliva and in scallop tissues, they are probably widespread in and on animals. If the nematocysts discharged in response to touch or to chemical stimulation alone they would be likely to be wasted by being discharged at inappropriate times. As it is, they are likely to discharge and capture any potential food object which happens to touch them.

Fluid oozes from the tips of discharged nematocysts. This fluid is injected into prey which has been penetrated by nematocysts, and may cause paralysis or death. Thus cnidarians are enabled to master prey which would otherwise escape. Some cnidarians can inflict painful stings on man and an Australian jellyfish (*Chironex fleckeri*) can kill him.

Nematocyst venom can be obtained in quantities large enough for analysis from the acontia of the anemone *Aiptasia.* The acontia are put in a 1 M solution of sodium citrate. This makes the cnidoblasts extrude their nematocysts in an undischarged state. The remains of the acontia are strained off and the strained solution is centrifuged, yielding a pellet of nematocysts. This is put into a small quantity of distilled water which stimulates the nematocysts to discharge, releasing fluid from the tips of their threads. The discharged nematocysts are removed by centrifugation, leaving the water into which they have shed some of their venom.

This solution has been separated into components by column chromatography, yielding four proteins. These have been injected into fiddler crabs (*Uca*) to find out whether they are toxic. They have also been added to the solution bathing a crayfish (*Procambarus*) nerve cord, to test for effects on the transmission of action potentials. The first protein kills fiddler crabs, causing a tranquil death. The second (which has not been completely separated from the first) is the enzyme phospholipase A. It breaks down cell membranes and destroys the ability of crayfish nerve cord to transmit action potentials. The third does not kill fiddler crabs but stimulates autotomy (shedding of limbs). The fourth does kill them, causing violent convulsions which are apparently due to its effect on the nervous system. It turns single action potentials in crayfish nerve cord into bursts of action potentials following each other in rapid succession. Thus the nematocyst fluid which is injected into prey contains a mixture of toxins of different types.

FEEDING AND DIGESTION

Many sea anemones eat quite large crustaceans, such as shrimps, and even small fish. Others, and other members of the Anthozoa, eat smaller members of the zooplankton. When potential prey touches the tentacles nematocysts discharge, attaching the prey to the tentacles and injecting venom into it. Next, the tentacles and the edge of the oral disc bend inward, carrying the food towards the mouth which is opened to receive it. The mouth is presumably opened by contraction of the radial muscles of the oral disc.

Experiments designed to discover the stimulus for nematocyst discharge have already been described. The next paragraph describes experiments designed to discover the stimulus for the quite distinct process of passing food to the mouth. The experiments were performed on *Palythoa*, which looks like a colonial sea anemone but belongs to a different group of Anthozoa.

Pieces of filter paper were placed on the tentacles of *Palythoa* to see whether they were swallowed or not. Clean pieces were not swallowed, but pieces impregnated with an extract of brine shrimps (*Artemia*) were. The extract was prepared by shaking ground brine shimps with alcohol. It was of course a complex mixture but it was separated into components by paper chromatography. This process yielded a sheet of filter paper impregnated at different places with different compounds. Pieces were cut from the sheet and offered to *Palythoa*. It was found that the part of the sheet which was impregnated with the amino acid proline was always swallowed, the part containing alanine was sometimes swallowed and other parts were seldom swallowed. Further experiments showed that paper impregnated with pure proline from other sources was swallowed, as was paper impregnated with glutathione (a peptide consisting of one residue each of glutamine, cysteine and glycine). Glutathione is present in crustacean body fluids and so is likely to leak or diffuse out of potential prey.

The polyps of sea fans such as *Gorgonia* (Fig. 6.4*c*, *d*) are spaced in such a way that almost any particle passing through the meshes of the fan must pass within reach of a polyp. Hence a sea fan standing in a current should be very effective in catching plankton, particularly if the fans stand at right angles to currents as they seem generally to do. However, divers who have watched sea fans on reefs have seldom seen them feed and it is not clear whether food caught by the polyps is really important to them. Like reef-building corals (described in the next section of this chapter) they have symbiotic algae in them and the products of photosynthesis of the algae may provide a large proportion of their food requirements.

REEF-BUILDING CORALS

Three groups of Cnidaria contribute to coral reefs. The most important is a group of stony corals which are members of the Zoantharia. The blue coral (*Heliopora*) and the organ pipe coral (*Tubipora*) are Alcyonaria and can be recognized as such by the eight pennate tentacles on each polyp. Finally the milleporine corals are not Anthozoa at all, but members of the class Hydrozoa which is considered in the next chapter.

Reef-building corals have unicellular algae in their cells but other corals such as *Astrangia* (Fig. 6.5) do not. These algae are generally found only in the gastrodermis, but they may contain as much as 50% of the total protein content of the coral. They are oval cells about 10 μm long with chloroplasts and thick cell walls. They can be separated from the host tissues by grinding and centrifuging and grown in culture in suitable solutions. When this is done some of them grow flagella and take the shape of a typical dinoflagellate. The alga usually (and perhaps universally) found in reef-building corals is the dinoflagellate *Gymnodinium microadriaticum*, another species of the genus illustrated in Fig. 2.1(*c*).

The algae grow satisfactorily in solutions of inorganic salts, provided traces of vitamins B1 and B12 are added. They are thus just as capable of independent life as *Euglena gracilis* (p. 33). The coral can also live without the algae. If a coral is kept in darkness for long enough the algae, unable to photosynthesize, degenerate and are expelled by the host cells. In such experiments, the coral must be fed to keep it alive. However corals kept in the light have survived without food for over a year, which suggests that they can benefit from the photosynthesis of the algae. This has been confirmed in experiments in which corals were kept in the light in the presence of $^{14}CO_2$. Polyps were subsequently sectioned and examined by autoradiography to find the ^{14}C. Radioactivity was found in the tissues of the coral, as well as in the algae. It was found even in the epidermis where there were no algae. It could only have been captured by photosynthesis so it must have been captured by the algae and passed to the coral.

This could happen if the coral periodically digested some of the algae, but there is no good evidence that this happens. Another possibility is suggested by the observation that when the algae are kept in culture, glycerol and a few other organic compounds diffuse out into the culture medium. If these compounds also diffuse out when the alga is in a coral cell, they will become available to the coral. Evidence that they do has been obtained by rather a subtle experiment. Pieces were cut off corals (using wire cutters) and put in sea water to which $NaH^{14}CO_3$ had been added. They were kept in this solution in bright shade out of doors so that the algae took up ^{14}C by photosynthesis. After a while the water was analysed. Acid was added to it to drive off the $^{14}CO_2$ from the bicarbonate, so that any remaining radioactivity could only be due to organic compounds produced by photosynthesis. Very little radioactive organic matter was found in the water, presumably because any organic compounds which diffused out of the algae were taken up by the coral cells before they could escape to the water. The experiment was repeated, adding (non-radioactive) glycerol as well as $NaH^{14}CO_3$ to the initial solution. When this was done a radioactive organic compound escaped into the water. It was shown by chromatography that this compound was [^{14}C]glycerol. In this experiment, so much glycerol was diffusing into the coral cells that they could not use it all, and some of the radioactive glycerol diffused right through them into the water. Similarly in the presence of glucose and of the amino acid alanine, [^{14}C]glucose and [^{14}C]alanine escape into the medium. Fructose, glutamic acid and several other compounds which were tested had no effect: they did not cause release of radioactivity into the water. It is concluded that glycerol, glucose and alanine leak out of the algae into the coral cells. The quantities of these compounds are quite large. In typical experiments with glycerol, the glycerol in the water at the end of the experiment contained 10%

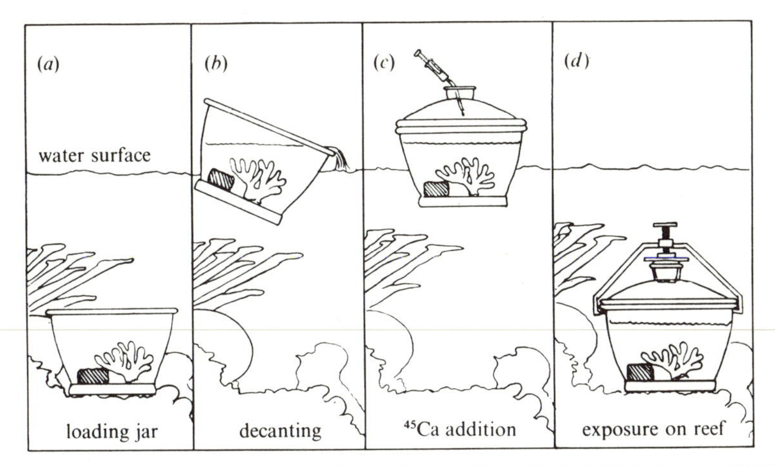

Fig. 6.22. Experiments on the growth of coral skeleton. From T. F. & N. I. Goreau (1959). *Biol. Bull.* **117**, 239–50. For explanation see text.

of the ^{14}C fixed by photosynthesis (the rest was in the coral). The glycerol and glucose which the coral gets from its algae presumably serve as energy sources. The alanine may be used in part to synthesize other amino acids and build proteins, providing for growth as well as metabolism.

A reef coral generally consists of a very thin layer of tissues on a thick block of calcium carbonate. One effect of the symbiotic algae is to speed the deposition of calcium carbonate. This has been demonstrated by the experiments illustrated in Fig. 6.22. Pieces were cut off corals and put in glass jars with sea water containing $^{45}CaCl_2$. The jars were weighted and put back on the reef for several hours, at the site from which the coral had been taken. Afterwards they were recovered, the pieces of coral were washed and their radioactive content was measured. This showed how much calcium carbonate had been laid down during the experiment. It was shown that in sunlight, the skeleton was thickened at rates of the order of 0.5 $\mu m\ h^{-1}$ in several species. The rates were lower in cloudy weather, and in darkness (in jars painted black) the rate was typically only 10% of the rate in sunlight. It seems clear that the photosynthesis of the algae aids skeleton formation. The algae remove carbon dioxide from the tissues around them, and removal of carbon dioxide from a bicarbonate solution produces carbonate, thus

$$2\,HCO_3^- = CO_3^{2-} + CO_2 + H_2O$$

If the concentrations of calcium and carbonate are high enough, calcium carbonate will be precipitated.

It is probably because of this effect that the same species of coral may form much more massive skeletons in shallow than in deeper water. In shallow water it may form great convex lumps but in deeper water where the light is dimmer it may form shelves like bracket fungi.

The cells of the coral may be a particularly favourable habitat for the algae, because the metabolism of the coral releases nitrogenous wastes and nitrates are apt to be scarce in the sea, especially in the tropics (p. 36). The association between coral and alga seems beneficial to both. Such associations are called symbioses (*singular*, symbiosis).

FURTHER READING

GENERAL

Muscatine, L. & Lenhoff, M. M. (1974). *Coelenterate biology. Reviews and new perspectives.* Academic Press, New York & London.

MUSCLES AND NERVES

Pantin, C. F. A. (1935). The nerve net of the Actinozoa [three papers]. *J. exp. Biol.* **12**, 119–64.

Pickens, P. E. (1969). Rapid contractions and associated potentials in a sand-dwelling anemone. *J. exp. Biol.* **51**, 513–28.

Ross, D. M. & Sutton, L. (1961). The response of the sea anemone *Caliactis parasitica* to shells of the hermit crab *Pagurus bernhardus*. *Proc. R. Soc.* **155B**, 266–81.

Shelton, G. A. B. (1975). Colonial behaviour and electrical activity in the Hexacorallia. *Proc. R. Soc.* **190B**, 239–56.

Trueman, E. R. (1966). Continuous recording of the hydrostatic pressure in a sea anemone. *Nature, Lond.* **209**, 830 only.

MESOGLOEA

Alexander, R. McN. (1962). Visco-elastic properties of the body wall of sea anemones. *J. exp. Biol.* **39**, 373–86.

Gosline, J. M. (1971). Connective tissue mechanics of *Metridium senile* [two papers]. *J. exp. Biol.* **55**, 763–95.

Koehl, M. A. R. (1977). Effects of sea anemones on the flow forces they encounter. *J. exp. Biol.* **69**, 87–105.

GORGONIN

Goldberg, W. M. (1974). Evidence of a sclerotized collagen from the skeleton of a gorgonian coral. *Comp. Biochem. Physiol.* **49B**, 525–9.

Wainwright, S. A. & Dillon, J. R. (1969). On the orientation of sea fans (genus *Gorgonia*). *Biol. Bull.* **136**, 130–9.

NEMATOCYSTS

Hessinger, D. A., Lenhoff, H. M. & Kahan, L. B. (1973). Haemolytic, phospholipase A and nerve-affecting activities of sea anemone nematocyst venom. *Nature New Biol.* **241**, 125–7.

Picken, L. E. R. & Skaer, R. J. (1966). A review of researches on nematocysts. *Symp. zool. Soc. Lond.* **16**, 19–50.

FEEDING AND DIGESTION

Leversee, G. J. (1976). Flow and feeding in fan-shaped colonies of the gorgonian coral *Leptogorgia*. *Biol. Bull.* **151**, 344–56.
Nicol, J. A. C. (1959). Digestion in sea anemones. *J. mar. Biol. Ass. UK* **38**, 469–76.
Reimer, A. A. (1971). Chemical control of feeding behaviour in *Palythoa* (Zoanthidea, Coelenterata). *Comp. Biochem. Physiol.* **40A**, 19–38.

REEF-BUILDING CORALS

Lewis, D. H. & Smith, D. C. (1971). The autotrophic nutrition of symbiotic marine coelenterates with special reference to hermatypic corals. I. Movement of photosynthetic products between symbionts. *Proc. R. Soc.* **178B**, 111–29.
Lewis, J. B. (1977). Processes of organic production on coral reefs. *Biol. Rev.* **52**, 305–47.
Taylor, D. L. (1973). The cellular interactions of algal–invertebrate symbiosis. *Adv. mar. Biol.* **11**, 1–56.
Yonge, C. M. (1968). Living corals. *Proc. R. Soc.* **169B**, 329–44.

7

Hydroids and jellyfishes

Phylum Cnidaria (cont.)
Class Hydrozoa (hydroids)
Class Scyphozoa (jellyfishes)

The Cnidaria described in chapter 6 all had the form of polyps: they were hollow cylindrical animals with a mouth at one end, surrounded by tentacles. Some of the polyps were squat and some slender, some were solitary and others grouped in colonies, but the basic pattern of the polyp was always apparent. The classes of Cnidaria described in this chapter have two body forms, the polyp and the medusa, which generally both occur at different stages in the life cycle.

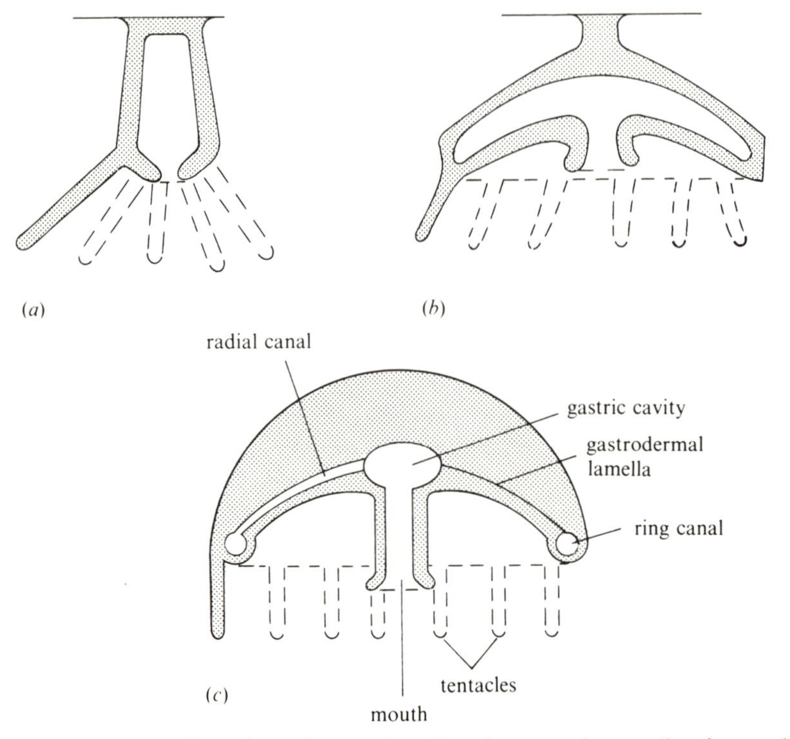

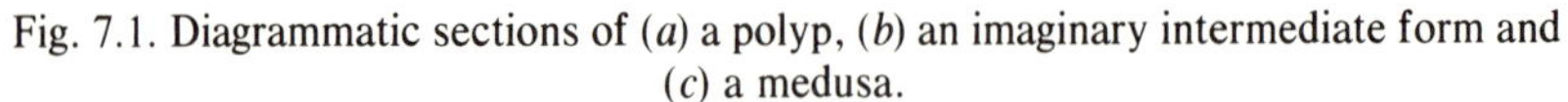
Fig. 7.1. Diagrammatic sections of (*a*) a polyp, (*b*) an imaginary intermediate form and (*c*) a medusa.

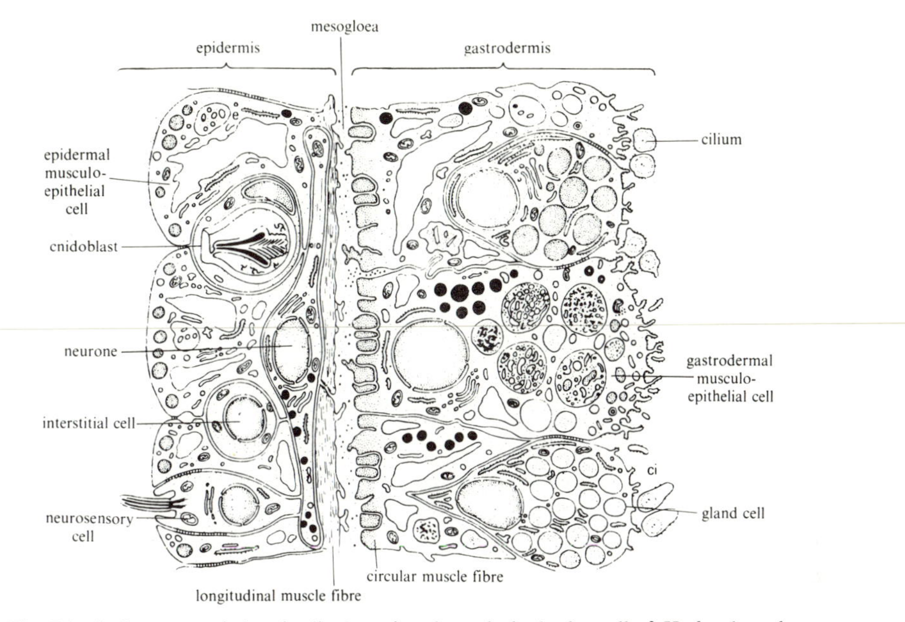

Fig. 7.2. A diagrammatic longitudinal section through the body wall of *Hydra*, based on electron micrographs. After T. L. Lentz (1966). *The cell biology of Hydra*. North-Holland, Amsterdam.

Fig. 7.1 shows a typical hydrozoan polyp and a medusa, and the relationship between them. Hydrozoan polyps differ from anthozoan ones (Fig. 6.1) in having no mesenteries and no inturned pharynx. The medusa is shaped like a bell, with its mouth at the end of a tube (manubrium) which hangs down like the clapper of the bell. There are tentacles round the edge of the bell. There is an epidermis covering the outer surface of the body and a gastrodermis lining the gastric cavity, each consisting of a single layer of cells. These two layers of cells are separated by mesogloea, which is much thicker than in polyps and represents most of the volume of the medusa. The gastric cavity is not a simple sac, but consists of a central cavity, a ring canal round the circumference of the medusa and radial canals connecting this to the central cavity. A sheet of cells known as the gastrodermal lamella fills the gaps between the radial canals. This sheet is only one cell thick.

Fig. 7.1(*b*) is wholly imaginary but helps to stress a basic similarity between the polyp and the medusa. A soft clay model of a polyp could be distorted to the medusa form in the manner indicated by the series of diagrams. The gastrodermis of opposite faces of the gastric cavity could be made to meet and merge in places, forming a gastrodermal lamella but leaving radial and ring canals.

Anthozoa have musculo-epithelial cells only in the gastrodermis. There are circular muscle fibres on the musculo-epithelial cells of the body wall and

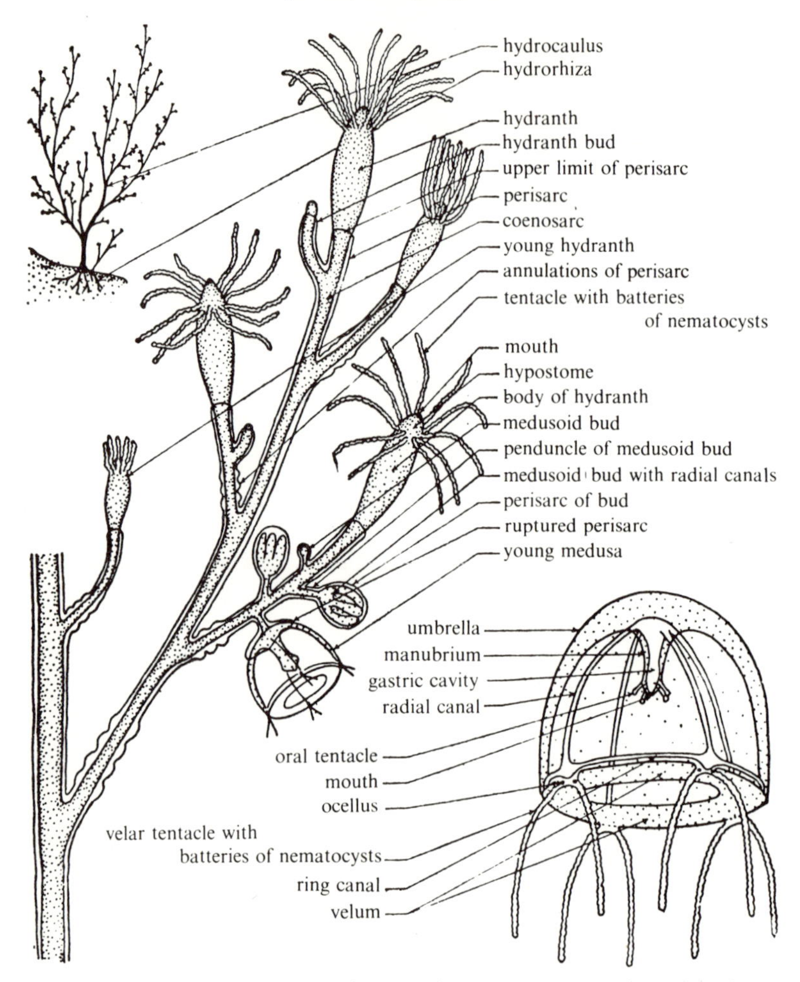

Fig. 7.3. *Bougainvillea*: a colony (top left), part of a colony (centre) and a medusa (bottom right). The colony was probably about 4 cm high and the diameter of the medusa about 3 mm. From W. S. Bullough (1951). *Practical invertebrate anatomy*. Macmillan, London.

longitudinal and radial ones on the mesenteries (Fig. 6.6). Hydrozoa have no mesenteries, but have musculo-epithelial cells bearing longitudinal muscle fibres in the epidermis (Fig. 7.2).

Bougainvillea (Fig. 7.3) is a typical marine member of the Hydrozoa. Its polyps grow in bushy colonies on stones and shells and on the surfaces of sponges, sea squirts, etc. Many colonies may grow alongside one another forming quite a thick carpet. Each colony consists of a branching root-like hydrorhiza and a branching stem-like hydrocaulus which bears the polyps. The hydrocaulus has a tubular outer casing, the perisarc, which seems to consist of chitin and quinone-tanned protein. The same combination of materials is

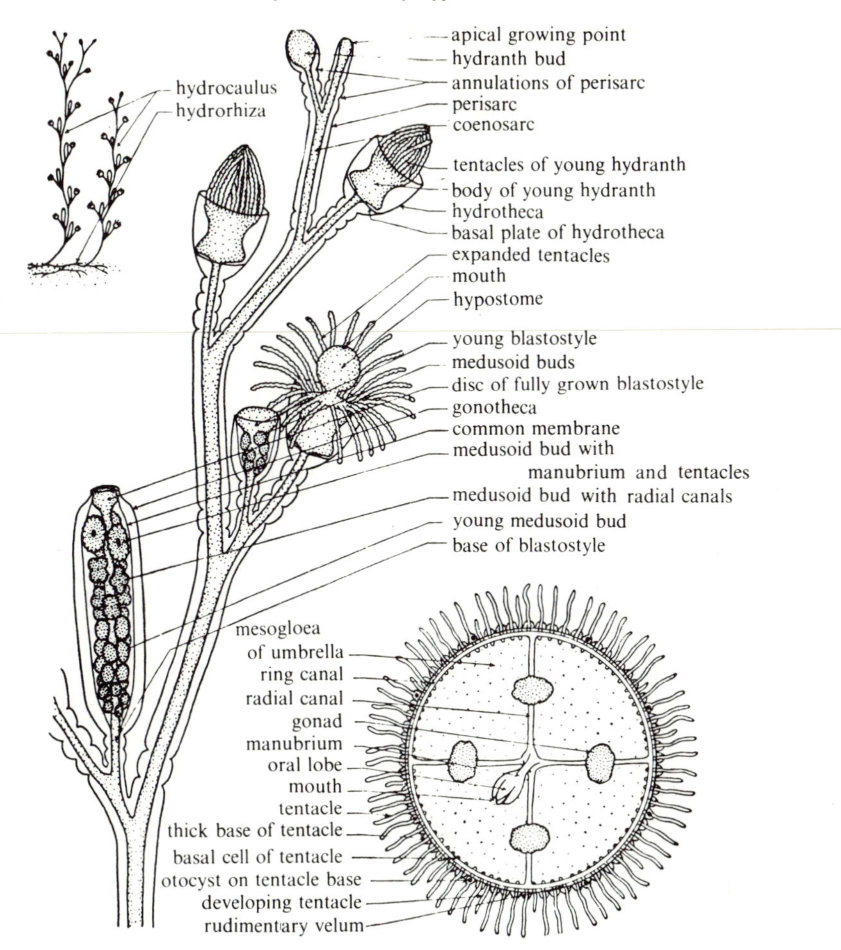

Fig. 7.4. *Obelia*: a colony (top left), part of a colony (centre) and a medusa (bottom right). The colony was probably about 2 cm high and the diameter of the medusa was probably about 5 mm. From W. S. Bullough (1951). *Practical invertebrate anatomy*. Macmillan, London.

found in insect cuticle. Chitin is a polysaccharide built up from acetylglucosamine units: that is, from glucose units in which one of the OH groups has been replaced by $NHCOCH_3$. Its presence can be demonstrated by histochemical tests. The evidence that the protein is quinone-tanned comes from observations of *Campanularia*, another colonial hydroid. Dopamine has been extracted from *Campanularia* colonies and it has been shown by histochemical methods that it and the enzyme phenol oxidase (which can oxidize it to a quinone) are present in some of the epidermal cells, under the perisarc. It seems probable that they are used to tan the perisarc. Within the perisarc, tubes of tissue connect the gastric cavities of the polyps.

The medusae of *Bougainvillea* develop as buds on the colony but detach from it and swim freely. Fig. 7.3 shows that they have the basic structure already described and in addition that they have ocelli and a velum. The ocelli are believed to function as very simple eyes, and are described in a later section of this chapter. The velum is a shelf around the inside edge of the bell. Medusae swim by alternate expansion and contraction of the bell. The mechanism of swimming is considered later in the chapter.

The medusa shown in Fig. 7.3 is immature. An adult medusa would have four gonads (all of the same sex) on the sides of the manubrium. Eggs released into the sea by female medusae are fertilized by spermatozoa released by male ones, giving rise to larvae which settle and develop into colonies of polyps. Thus polyps and medusae alternate in the life cycle: polyps give rise to medusae by the asexual process of budding and medusae give rise to polyps by sexual reproduction.

Fig. 7.4 shows *Obelia*, another member of the Hydrozoa. It is found growing on marine algae and on driftwood. In many ways it is like *Bougainvillea*, but there are some obvious differences. The perisarc forms a cup-like extension (a hydrotheca) round each polyp. The medusae bud from specialized structures called blastostyles, which have a protective sheath of perisarc (the gonotheca). The medusae are flatter than those of *Bougainvillea*, and their gonads form on the underside of the bell rather than on the manubrium. There are around the edge of the bell a few statocysts, which are gravity receptors. They are discussed further in a later section of this chapter.

Bougainvillea and *Obelia* are typical members of two large groups of Hydrozoa, the suborders Gymnoblastea and Calyptoblastea, respectively. The differences between them which have been mentioned are characteristic differences between these two groups.

Hydra is a very simple member of the Hydrozoa, often studied in schools. It is one of the few genera of Cnidaria which lives in fresh water. Its polyps are solitary, not colonial. It does not produce medusae: instead, gonads develop on the sides of the polyps which can reproduce both asexually (by budding) and sexually.

Unlike the Hydrozoa mentioned so far the siphonophores have no sessile polyp stage. Some, such as *Physalia*, the Portuguese man-o'-war, float at the surface of the sea. Others such as *Nanomia* (Fig. 7.5) swim below the surface. Siphonophores form polymorphic colonies: that is, colonies of which the individual members have different forms. Sea pens such as *Pennatula* (Fig. 6.4) have two types of polyp but there is far more variety in a siphonophore colony. At the top of a *Nanomia* colony there is a gas-filled float, which seems to be a modified, inverted medusa. The opening of the bell of the medusa is reduced to a small pore, closed by a sphincter muscle, at the top of the float. The gas has been analysed and found to be mainly carbon monoxide. Below the float there is a cluster of nectophores which are also modified medusae. They contract and expand like free-swimming medusae, propelling the colony

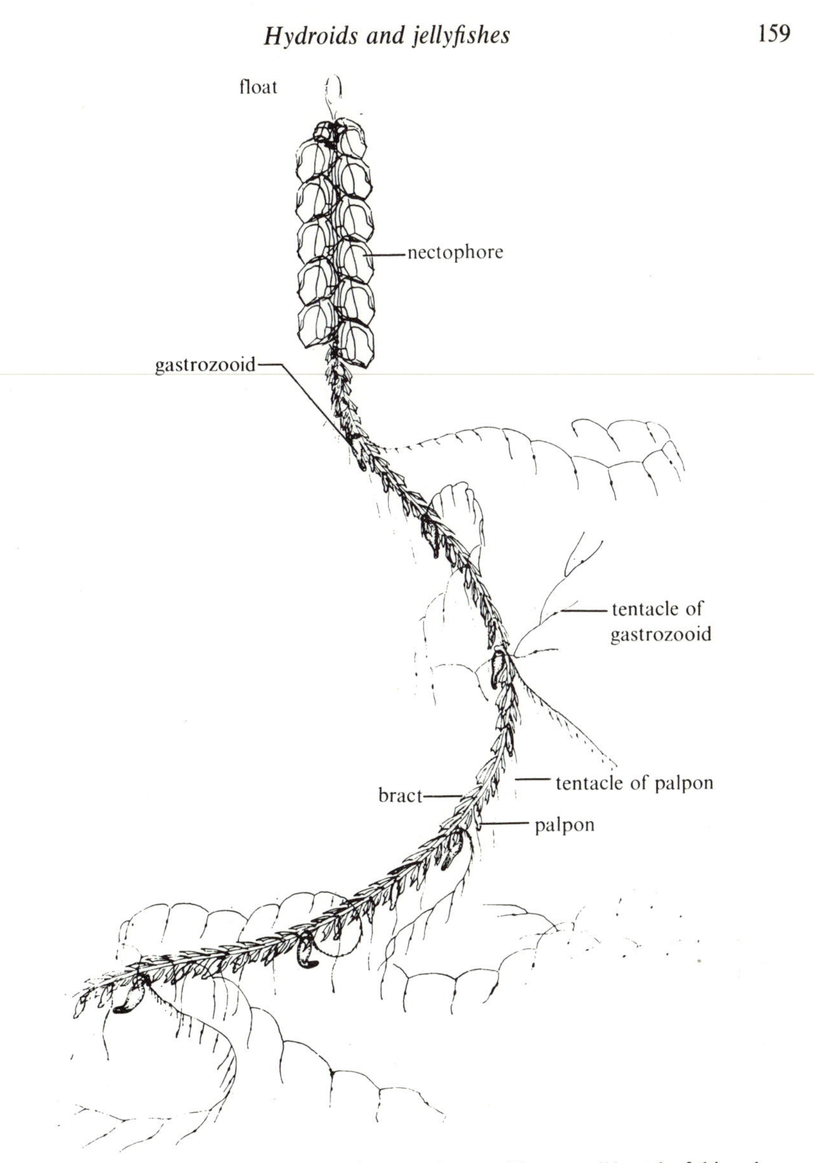

Fig. 7.5. The top two-thirds of a colony of *Nanomia cara*. The overall length of this colony was probably about 25 cm. From G. O. Mackie (1964). *Proc. R. Soc.* **159B**, 366–91.

through the water. Further down the colony the gastrozooids are polyps, each with a single long tentacle. They are the only members of the colony capable of swallowing food. The palpons are polyps with a tentacle but no mouth. The bracts are leaf-shaped, but seem to be modified medusae. Yet other members of the colony bear gonads. The colony is a highly integrated structure, whose different members have distinct functions and would be incapable of independent life. *Nanomia* and related siphonophores are very common at substantial

Fig. 7.6. Two members of the Scyphozoa. Left, *Aurelia aurita* (diameter up to 0.4 m); right, *Cyanea capillata* (diameter up to 1 m). Both species are common in the North Atlantic. From W. de Haas & F. Knorr (1966). *The young specialist looks at marine life.* Burke, London.

depths in the oceans. A later section of this chapter is concerned with their distribution and movements.

Aurelia and *Cyanea* (Fig. 7.6) are examples of the third class of Cnidaria, the Scyphozoa. Their polyps are small and inconspicuous, but their medusae are large: *Cyanea arctica* grow to diameters up to 2 m. Their radial canals are more numerous than those of hydrozoan medusae, and are branched. There is no velum. The mouth is surrounded by extensions of the manubrium called

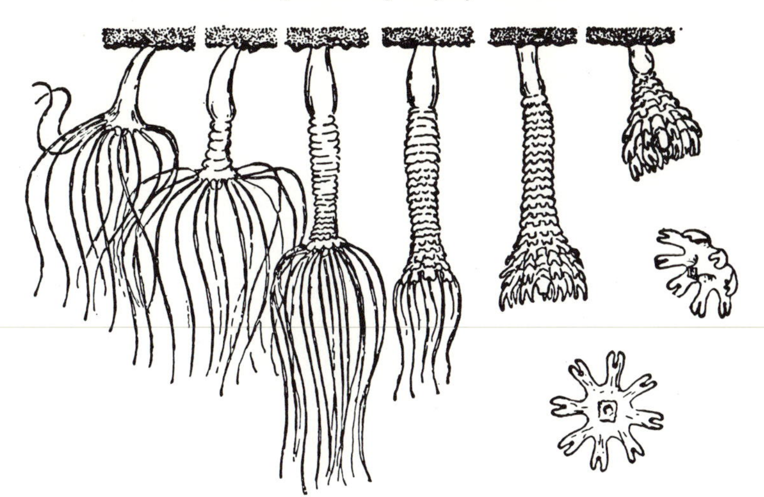

Fig. 7.7. Stages in the development of *Aurelia* from the polyp (scyphistoma) to free ephyrae. From A. C. Hardy (1956). *The open sea, its natural history*, part 1, *The world of plankton*. Collins, London.

oral arms. The gonads develop internally, on the floor of the gastric cavity. Fertilized eggs form ciliated, swimming larvae which settle on stones or algae and become polyps, which may form new polyps by budding (Fig. 7.7). These polyps seldom grow more than 1 cm high. They live for several years and then split up into ephyra larvae which are small medusae. Each cylindrical polyp forms a pile of saucer-shaped ephyrae which break free and grow to become adult medusae.

SWIMMING AND BUOYANCY OF MEDUSAE

The mesogloea of medusae contains protein and polysaccharide, but in much smaller concentrations than in sea anemones. The mesogloea of *Aurelia*, for instance, contains less than 1% organic matter so it is a much more dilute gel than table jelly, which must contain about 5% gelatin to be reasonably rigid. Even a very dilute gel made up in sea water would be denser than sea water, but the mesogloea of medusae is less dense than sea water. An explanation for this was found when the concentrations of ions in jellyfish mesogloea were measured. It was found that the total osmotic concentration was about the same as sea water, but that the mesogloea contained less sulphate and correspondingly more of other ions. (Sodium sulphate solutions are denser than isotonic solutions of sodium chloride.) Since sea water contains less than 3 g sulphate kg^{-1}, reduced sulphate content cannot reduce the density of mesogloea very much. In fact the specific gravity of jellyfish mesogloea is only about 0.001 less than sea water (1.025 compared to 1.026).

The mesogloea is less dense than sea water but the cells of marine medusae

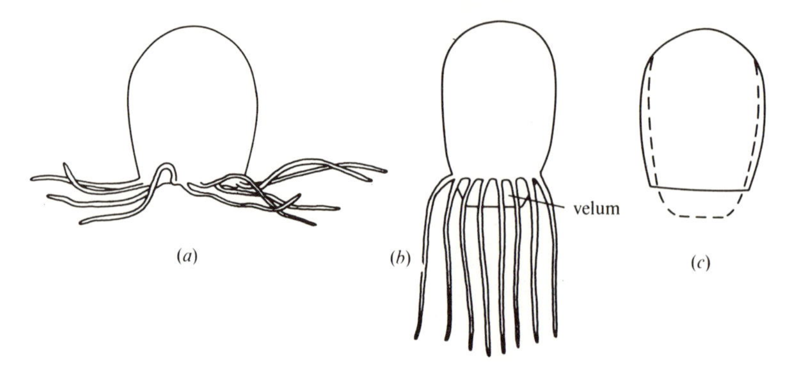

Fig. 7.8. A medusa of *Polyorchis montereyensis* (*a*) relaxed, sinking with its tentacles spread, and (*b*) at the end of a swimming contraction. (*c*) shows outlines of the relaxed and contracted bell superimposed. Based on photographs in W. B. Gladfelter (1972). *Helgolander wiss. Meeresunters.* **23**, 38–79.

are denser than sea water. Some medusae with very bulky mesogloea, such as *Pelagia* (Scyphozoa), have overall approximately the same density as sea water. Others such as *Polyorchis* (Hydrozoa) sink quite rapidly whenever they stop swimming.

Swimming involves alternate contraction and expansion of the bell (Fig. 7.8). Contraction drives water downwards out of the bell and propels the medusa upwards. Expansion draws water up into the bell and must tend to pull the medusa down again, but a swimming medusa can make progress because the contraction is more rapid than the expansion.

Polyorchis swims upwards by contracting a few times and then allows itself to sink before swimming up again. As it sinks its flexible tentacles tend to spread out around it (Fig. 7.8 *a*). Food objects in the water which passes between them can be caught. It may be an advantage to *Polyorchis* to be considerably denser than the water. If it sank more slowly it would not have to use so much energy for swimming, but it would have less opportunity to catch food. However, it would be disadvantageous to sink too fast.

The buoyant mesogloea of *Polyorchis* must tend to keep it the right way up as it sinks. If the edge of the bell (with the tentacles and velum) is cut off, the remainder floats. Thus the denser parts of the animal are concentrated around the edge of the bell.

The outer surface of the bell is covered by plain epithelial cells with no muscle fibres. The inner surface has musculo-epithelial cells. *Polyorchis* has a narrow band of radial muscle fibres under each radial canal, and sheets of circular fibres filling the quadrants between the canals (Fig. 7.9). The circular fibres probably provide most of the power for swimming. They are striated, while the radial muscles (like all the other muscles discussed so far) are unstriated. The structure of striated muscle and the changes which occur when it contracts are considered in chapter 12.

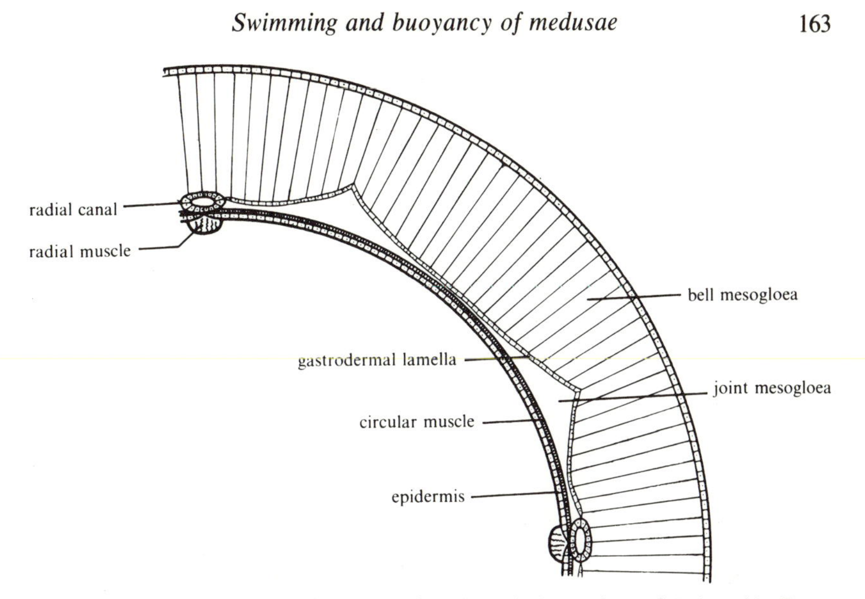

Fig. 7.9. One quarter of a horizontal section through the medusa of *Polyorchis*. From W. B. Gladfelter (1972). *Helgolander wiss. Meeresunters.* **23**, 38–79.

Some other medusae have their muscle differently arranged. *Obelia* has extraordinary cells bearing two sets of muscle fibres at right angles to each other. Scyphozoan medusae such as *Cyanea* have bands of radial and circular muscle which are thickened by folding of the epidermis, like the retractor muscles of sea anemones (Fig. 6.6*b*). The muscle fibres are about 0.3 μm in diameter, both in small hydrozoan medusae such as *Obelia* and in large scyphozoan medusae such as *Cyanea*. If large medusae had only a single unfolded layer of musculo-epithelial cells, the proportion of muscle in the body would be very small.

When the muscles contract in swimming, the mesogloea is deformed. When they relax again, the bell is expanded by elastic recoil of the mesogloea. *Polyorchis* has two types of mesogloea, separated by the gastrodermal lamella (Fig. 7.9). The bell mesogloea is relatively stiff. The joint mesogloea is much less stiff, but still gelatinous rather than fluid: dye injected into it does not spread through it, but remains at the place where it was injected, even when the medusa swims. The joint mesogloea is arranged in eight thick radial wedges, which are referred to as joints. When the muscles contract the bell bends at the joints into a more or less octagonal shape, because the joint mesogloea is less stiff than the rest (Fig. 7.10).

Polyorchis medusae usually remain upright, but can turn over and swim downwards. They are apt to do this when a bright light is fixed over their tank. They turn over by contracting the muscles on one side of the velum, so that the jet of water leaves the bell asymmetrically.

Scyphozoan medusae such as *Cyanea* are much flatter than *Polyorchis* (they

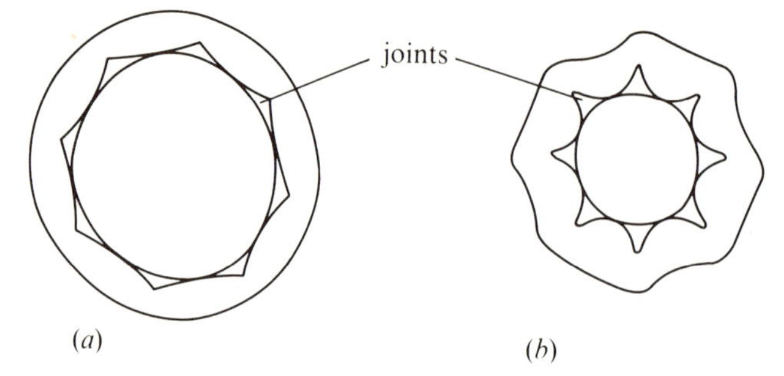

Fig. 7.10. A medusa of *Polyorchis*, seen from above, (*a*) relaxed and (*b*) contracted, showing bending at the joints. Traced from photographs in W. B. Gladfelter (1972). *Helgolander wiss. Meeresunters.* **23**, 38–79.

are shaped more like saucers than bowls) but the mechanism of swimming is similar. They have joints, but these are formed by extensions of the gastric cavity rather than by wedges of joint mesogloea. They have no velum, and turn by contracting the bell asymmetrically.

RHYTHMIC MOVEMENT OF MEDUSAE

Movements which are repeated at regular intervals over long periods are involved in the locomotion, respiration and feeding of many animals. In Metozoa they are often controlled by cells called pacemakers in the nervous system, which produce action potentials spontaneously at regular intervals. Pacemakers have been discovered in sea anemones and have a role in (for instance) the process of attachment of *Calliactis* to whelk shells, but discussion of pacemakers has been deferred to this chapter because the swimming of medusae is a conspicuous rhythmic activity which has been well studied.

As well as a nerve net which extends throughout the epidermis, hydrozoan medusae have two interconnected nerve rings which run round the edge of the bell. These are bundles of neurites attached to cell bodies in the epidermis. Most scyphozoan medusae do not have these rings but have four or eight ganglia (clumps of nerve cells) spaced out round the edge of the bell. The ganglia are easy to locate, because they are close to the ocelli and statocysts. The nerve rings of Hydrozoa and the ganglia of Scyphozoa apparently contain the pacemaker neurones which initiate swimming contractions. A hydrozoan medusa will continue to beat after most of both nerve rings have been cut away, but not if they are completely removed. A scyphozoan medusa will continue to beat when all but one of its ganglia have been cut away. If the last ganglion is removed spontaneous beating ceases, but it is still possible to evoke single beats by electrical stimulation.

The precise mechanism of the pacemakers is not known but it appears that

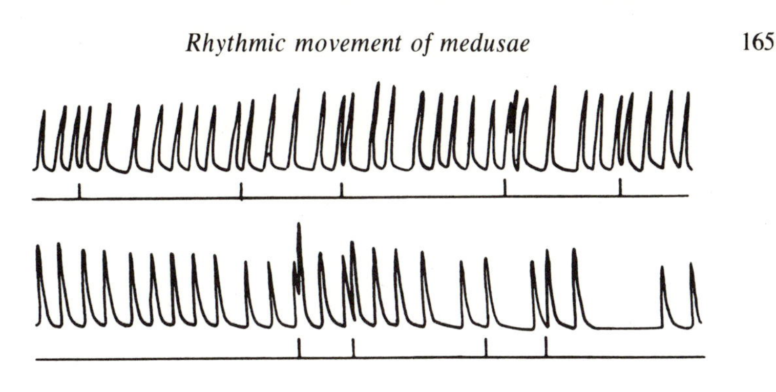

Fig. 7.11. Records of experiments with pieces of *Cassiopea* containing only one ganglion. In each case the upper trace is a kymograph record of contractions, and the lower trace shows when electrical stimuli were delivered. Time reads from left to right, with each division on the scale representing 2 s. From G. A. Horridge (1959). *J. exp. Biol.* **36**, 72–91.

after each action potential a state of excitation builds up gradually until a threshold is passed and another action potential is produced. If excitation builds up at a constant rate, and if each action potential returns it to the same starting point, action potentials will be produced at equal intervals. A medusa has many pacemakers and it is believed that a swimming contraction occurs when any one of them produces an action potential. That action potential spreads through the nerve net and reaches all the other pacemaker cells, re-setting them by taking their states of excitation back to the starting point.

Fig. 7.11 shows some results from an experiment which supports this interpretation. A piece containing only one ganglion was cut from a scyphozoan medusa, *Cassiopea.* It was attached to a kymograph so that its contractions were recorded. Normally it contracted at rather irregular intervals of about 4 s. It was stimulated electrically to contract prematurely, at the times indicated. The interval immediately following the electrically evoked contraction was about normal in length or even rather longer than usual. If the stimulus had not re-set the pacemaker, this interval would have been a short one. In some cases very long intervals occurred soon after a stimulus: it is not clear why.

There are far more pacemakers than seems necessary: one per animal would suffice. Does the redundancy confer some advantage? Plainly it ensures that no one ganglion is indispensable, and medusae with one or more damaged ganglia are sometimes found. However, eight ganglia seems an excessive provision of spare parts. Another suggestion which has been made is that multiple ganglia might make the beat more regular (which might or might not be advantageous). An isolated ganglion fires rather irregularly (Fig. 7.11) so if a single ganglion controlled the beat there would be some unusually long intervals between beats. Exceptionally long intervals would be less likely with eight ganglia, any one of which can initiate a beat. For instance, an isolated

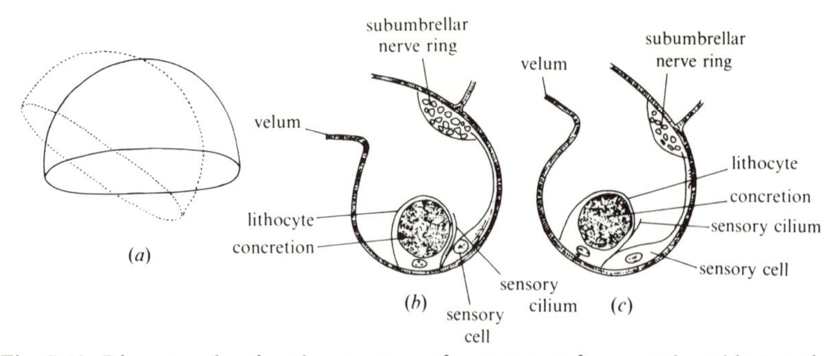

Fig. 7.12. Diagrams showing the structure of a statocyst from a calyptoblast medusa, and the effect of tilting on it. (*a*) shows the medusa horizontal and tilted. (*b*) and (*c*) are sections of the statocyst when the medusa is horizontal and tilted, respectively. From C. L. Singla (1975). *Cell Tissue Res.* **158**, 391–407.

ganglion was found to fire at intervals of 6.9±2.5 s (mean and standard deviation). It was calculated that a medusa which had eight similar ganglia would contract at intervals of 4.2±0.6 s: the mean interval would be less than for a single ganglion and the standard deviation would be greatly reduced. An attempt to confirm this effect experimentally gave inconclusive results. A whole jellyfish generally beats more regularly than a piece with only one ganglion but it is not clear to what extent this is due to redundancy of pacemakers.

SENSE ORGANS

Medusae have ocelli or statocysts or both and are the first animals encountered in this book to have multicellular sense organs.

Statocysts occur around the margin of the bell but vary in number and structure. A simple type found in medusae of calyptoblast hydrozoans (the group which includes *Obelia*) is shown in Fig. 7.12(*b*). It occupies a pocket at the base of the velum. The most conspicuous structure of it is a large cell (the lithocyte) containing a concretion which is stained by alizarin red and is therefore presumably an insoluble calcium salt. Next to the base of the lithocyte is a sensory cell with a cilium and with an axon leading to one of the nerve rings. The cilium has the same 9+2 arrangement of microtubules as ordinary motile cilia. There are other similar cells nearby but only this one adjacent to the lithocyte seems to be involved in gravity reception so it is the only one only shown in Fig. 7.12.

Evidence that the statocysts are gravity receptors has been obtained in experiments with *Aequorea* which is an unusually large member of this group of medusae, commonly attaining a diameter of 10 cm. Specimens were tilted and shown to be capable of righting themselves. They were anaesthetized with magnesium chloride, their statocysts were carefully removed and they were allowed to recover. They could still swim but no longer made correcting movements when tilted.

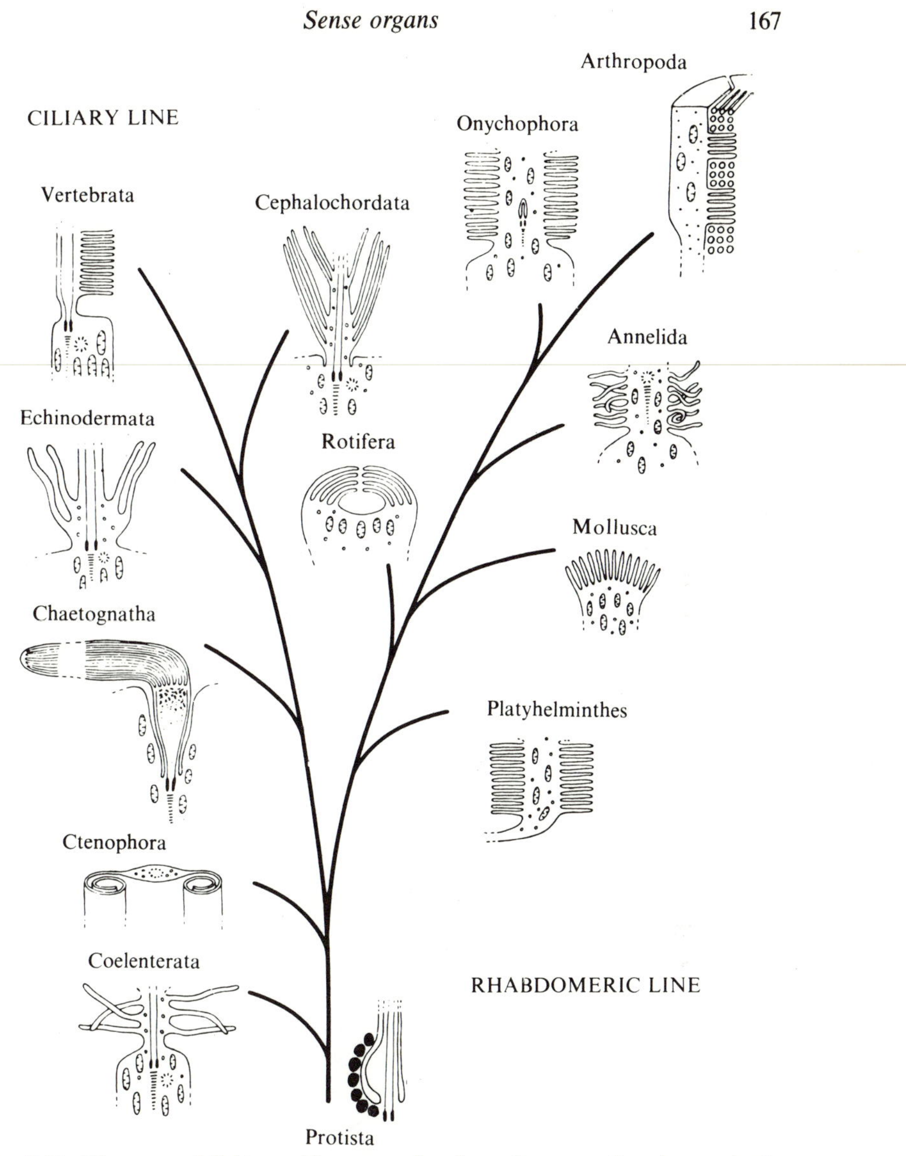

Fig. 7.13. Diagrams of light-sensitive organelles from the eyes of various animals, showing possible lines of evolution. The examples on the left are modified flagella but those on the right are not. From R. M. Eakin (1968). *Evolutionary Biol.* **2**, 194–242.

Fig. 7.12 shows how the statocyst is believed to work. When the medusa is erect the lithocyte stands clear of the sensory cilium. However, the lithocyte tends to bend when tilted because the concretion is denser than water. It presses against the cilium as shown in Fig. 7.12(*c*), distorting the cell membrane of the sensory cell. It seems probable that this alters the membrane potential and, if a threshold is passed, sets off action potentials in the sensory cell.

The ocelli of the medusa of *Polyorchis* have been studied by electron microscopy. They are cup-shaped and contain two types of cells mingled together, sensory cells and pigment cells. The latter are filled with dark pigment so that the sensory cells only receive light entering the mouth of the cup. The sensory cells have axons and also strangely modified cilia of which the base of one is shown in Fig. 7.13 (labelled Coelenterata). Note the long microvilli projecting at right angles from it. These cilia have a fairly normal 9+2 structure but seem to lack dynein arms.

Large numbers of ocelli have been collected from medusae of another hydrozoan (*Spirocodon*) and a light-sensitive pigment has been extracted from them. This pigment is believed to come from sensory cells, not pigment cells. After exposure to light it absorbs blue light less strongly but ultra-violet radiation more strongly. Pigments with rather similar properties which occur in the eyes of vertebrates and some higher invertebrates are known to play an essential role in vision.

Evidence that the ocelli function as simple eyes comes from experiments in which they have been removed. For instance, medusae of *Sarsia* (another member of the Hydrozoa) are attracted to light and will collect in a light beam. If their ocelli are removed this behaviour is abolished. However, some species of medusa lack ocelli and yet are responsive to light: they must have some other means of detecting it.

Many types of sensory cell bear modified cilia. The sensory cells of Cnidaria bear cilia (Figs. 6.2 and 7.2). The cnidocil of nematocysts is a cilium surrounded by a tuft of microvilli. The sensory cells of the statocysts of medusae have cilia and in one type of statocyst (not the type shown in Fig. 7.12) some of them have the cilium surrounded by a bunch of microvilli. The sensory cells of vertebrate ears similarly bear a cilium and a bunch of microvilli but have the cilium at the edge of the bunch. The ocelli of *Polyorchis* have sensory cells with modified cilia. Many other animals have light-sensitive cells bearing cilia with extensions of one sort or another on them, as the left side of Fig. 7.13 shows. Even the rods and cones of vertebrate eyes bear cilia, with stacks of discs attached to them. However, the light-sensitive cells in the eyes of many invertebrates have microvilli but no trace of cilia (Fig. 7.13, right side). Cilia and microvilli are both common in sensory cells and it may be significant that both have high ratios of surface area to volume.

Cilia and flagella may have subsidiary sensory functions even in Protozoa. It is tempting to suppose that the cilia of *Paramecium* play the same sort of role in the avoiding reaction, as cilia do in the statocysts of medusae. *Euglena* and related flagellates have a swelling at the base of one flagellum which is believed to serve as a light receptor (Fig. 7.13, labelled Protista).

DEEP-SEA SIPHONOPHORES

Siphonophores such as *Nanomia* (Fig. 7.5) have long been known from specimens caught by nets in the oceans, but it has only recently been realized how widespread and numerous they are. The realization has come by a curiously indirect route.

Most readers probably know the principle of echo-sounding. A ship emits a sound pulse and receives an echo from the bottom of the sea. The length of time from the emission of the pulse to the return of the echo is recorded, and indicates the depth of the bottom. Echo-sounding was originally developed to measure the depth of the sea, and found an additional use in the Second World War when it was used to detect submarines. However, it was found that some other less easily identified targets also returned echoes to ships. Among them were layers in mid water, continuous at the same depth over large areas of water. They did not remain at constant depth, but came up near the surface at dusk each evening and descended again at dawn to spend the day at depths of 300 m or more. These sources of echoes are known as deep scattering layers.

The echoes might come from planktonic animals, if the animals were numerous enough and occupied a restricted range of depths. Animals containing bubbles of gas would be particularly likely to return strong echoes. The acoustic properties of animal tissues are not much different from those of water so any echoes from them would be very weak, but submerged bubbles of gas scatter sound strongly. The sound is scattered in all directions (if the bubbles are small compared to the wavelength of the sound) and some of the scattered sound would return as an echo to an echo-sounding ship.

There are two groups of animals with gas-filled cavities which are commonly caught in nets at appropriate depths, and could be the sources of echoes. There are lantern fishes (Myctophidae) up to about 10 cm long, many of them with a gas-filled swimbladder. There are also physonectid siphonophores with gas-filled floats; *Nanomia* (Fig. 7.5) is the most common genus. Certain Crustacea are often caught in deep scattering layers but are unlikely to contribute much to echoes because they have no gas-filled cavities.

A submerged bubble has a resonant frequency which depends on its size and on the pressure. It scatters sound of this frequency particularly strongly. Most echo-sounders emit a pure tone but some experiments have been carried out using noises produced by small explosions, involving a wide range of frequencies. The frequency composition of the echoes was investigated as well as their time of arrival. It was found that particular frequencies were predominant in the echoes from deep scattering layers and it was concluded that these were the resonant frequencies of gas bubbles. Some deep scattering layers scattered high frequencies (5–25 kHz, depending on depth) and it was calculated that they could be due to bubbles of radius about 1 mm Others scattered lower frequencies, indicative of larger bubbles. The former layers could contain siphonophores and the latter ones lantern fishes.

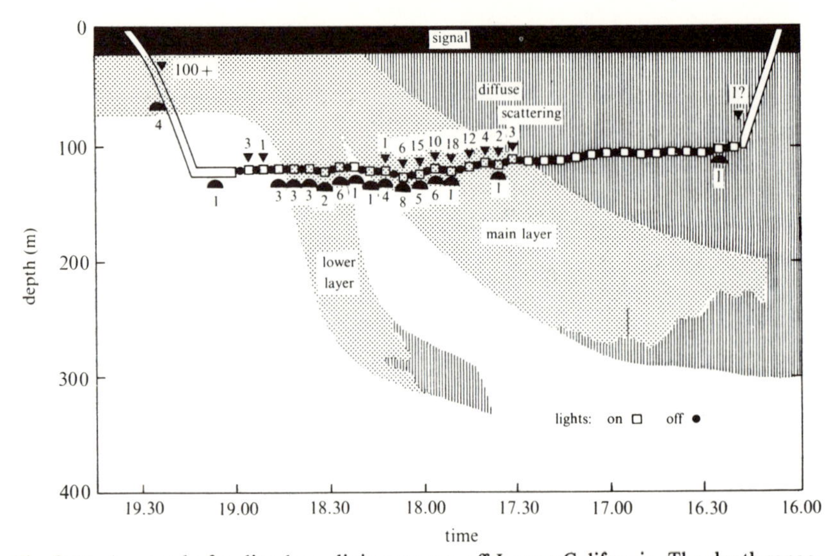

Fig. 7.14. A record of a dive by a diving saucer, off Lower California. The depth range of the deep scattering layer (determined by echo-sounding) is indicated by hatching. The semicircular and triangular symbols represent sightings of siphonophores and myctophid fishes respectively: the numbers are the numbers seen. Note that time reads from right to left. From E. G. Barham (1966). *Science* **151**, 1399–403.

More recently, deep scattering layers have been visited by observers in a bathyscaphe, and in a diving saucer. The latter is a craft which can take two men to a maximum depth of 300 m, and can hover while observations are made. Fig. 7.14 is a record of a dive off Lower California. At 16.00 h the diving saucer descended to 100 m and hovered. Echo soundings showed that the main deep scattering layer was then at 200–300 m but the diving saucer stayed at about 100 m while the layer rose past it to its night-time level in the top 80 m. The observers switched lights on for a while every few minutes, but they saw only one siphonophore and one possible lantern fish until 17.30 h, when the deep scattering layer reached them. They saw plenty of lantern fishes as it passed, and quite a lot of siphonophores near its bottom and in the lower deep scattering layer which followed it. They returned to the surface at about 19.30 h, seeing over 100 lantern fishes as they passed through the deep scattering layer. Similar observations were made as the deep scattering layer descended again in the morning. These observations confirm that lantern fishes and siphonophores are common in deep scattering layers. It was estimated in another series of dives that the population density of *Nanomia* in the deep scattering layer was 0.3 m^{-3}. The largest of these *Nanomia* were 75 cm long.

Siphonophores observed at night were usually motionless, but some were disturbed by the bathyscaphe. These swam away so fast that they could be mistaken for fish. Siphonophores seen ascending at dusk or descending at dawn were swimming actively.

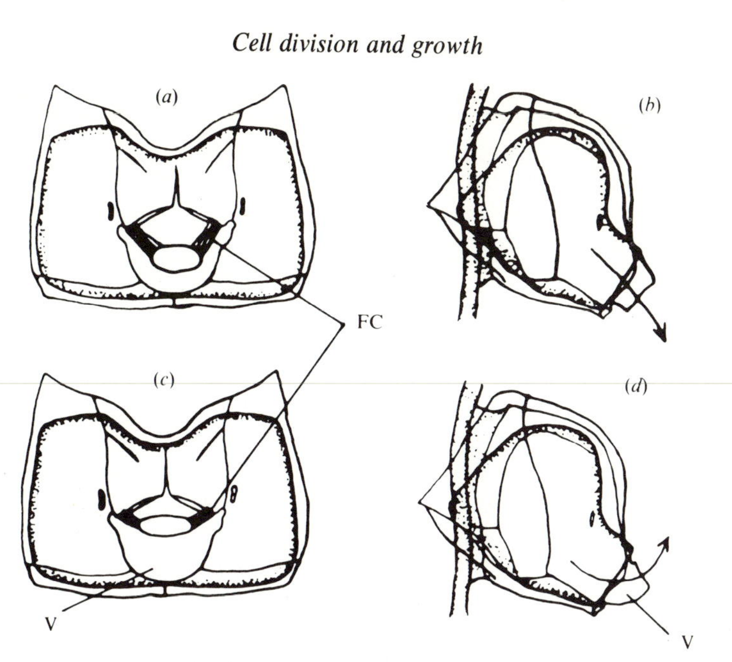

Fig. 7.15. Nectophores of *Nanomia* swimming normally (*a*, *b*) and in reverse (*c*, *d*). The arrows show the direction of the jet of water ejected at each contraction. FC, muscles of the velum; V, velum. From G. O. Mackie (1974). In *Coelenterate biology*, ed. L. Muscatine & M. M. Lenhoff. Academic Press, New York & London.

More detailed observations of swimming have been made on captive *Nanomia* in large tanks. They kept still for long periods in dim light. When they were touched they made sudden darts, but simultaneous contraction of all the nectophores. They also sometimes swam at fairly steady speeds of 8–10 cm s^{-1} (for 25 cm colonies) by asynchronous beating of the nectophores. On hitting the surface or some other obstacle they would reverse for a while, using the velum to change the direction of the jet from each nectophore (Fig. 7.15).

CELL DIVISION AND GROWTH

Nanomia is one of the most complex of the Hydrozoa but *Hydra* is the simplest. Its simple structure has made it a popular object of study among developmental biologists who hope to formulate general principles of metazoan growth and development. Some of their work is described in the remainder of this chapter.

A *Hydra* is an assembly of different types of cell (Fig. 7.2) which must grow and divide at appropriate rates as the animal grows and reproduces. The cells can be separated for counting by first soaking and then shaking *Hydra* in an aqueous solution of glycerol and acetic acid. Large *Hydra attenuata* are found to consist of about 130000 cells. The largest of these are the 25000 or so epidermal and gastrodermal cells which make up most of the total volume of cells. There are about 35000 mature cnidoblasts (most of them in the tentacles), about 5000 nerve cells and 5000 gland cells. The remainder are the small cells known as interstitial cells, some of them in clumps which are not broken up by pipetting. Cnidoblasts develop in clumps of 4, 8, 16 or 32 and interstitial cells in clumps of these numbers are presumably developing cnidoblasts.

A culture of well-fed *Hydra attentuata*, kept in uncrowded conditions at 20 °C, reproduces by the asexual process of budding and thereby doubles its numbers every 3½ days. The average size of the individuals remains constant so the numbers of each type of cell must double every 3½ days. Additional cells are needed to replace used nematocysts and to replace cells which are sloughed off the pedal disc and the ends of the tentacles.

Rates of cell division have been investigated by experiments with radioactive thymidine. Division of a cell into two daughter cells involves doubling the quantity of DNA in its nucleus. Thymidine is a constituent of DNA but not of RNA and is taken up by cells in the period prior to division when they are synthesizing DNA (the so-called S phase). A dilute solution of radioactive thymidine is injected through the mouth of a living *Hydra*, into the gastric cavity. Any cells that are in the S phase at the time or within a few minutes thereafter take up some of the radioactive thymidine and incorporate it into their DNA. The *Hydra* is broken up then or later, and samples of its cells are spread out on slides. The cells which have radioactive DNA in their nuclei can be identified by autoradiography.

If the cells are examined immediately, a proportion of the epidermal, gastrodermal, gland and interstitial cells are found to be radioactive. All these types of cell divide and some of each type are in the S phase at any instant. No cnidoblasts or nerve cells are found to be radioactive so it seems that these types of cell do not divide. However, if a long enough interval is left between injection and autoradiography, radioactive cnidoblasts and nerve cells are found. The parents of these cells must have been in the S phase at the time of injection.

Further experiments with radioactive thymidine showed how fast each type of cells was dividing. It was found that epidermal/gastrodermal cells divide (and so double their numbers) every 3½ days, or about once for every doubling of the population. Interstitial cells divide about once every 24 h, or even more frequently in the case of the clumps of developing cnidoblasts. They give rise to the new cnidoblasts and nerve cells in addition to doubling their own numbers every 3½ days: this is why they have to divide so frequently.

There seem to be among the interstitial cells of a large *Hydra* about 5000 'stem cells' which are not yet committed to becoming cnidoblasts or nerve cells. Each of these may divide to form two more stem cells or to form two incipient nerve cells or to form a cluster of cnidoblasts.

Table 7.1 shows the arithmetic of cell division. If the population is doubling every 3½ days the number of each type of cell must increase daily by a factor

TABLE 7.1. *Cell division in* Hydra

Cell type		Today	Tomorrow		Cell type
Epidermal/gastrodermal		25000	30500		Epidermal/gastrodermal
Gland		5000	6100		Gland
Nerve		5000	5000	6100 (5000 + 1100)	Nerve
	5000	550	1100		
Stem		3050	6100		Stem
		1400	2800		Cnidoblast precursors
Cnidoblast precursors[a]		7700	7700	42700 (7700 + 35000)	Mature cnidoblasts
Mature cnidoblasts		35000	35000		

The numbers in the 'Today' column are the approximate number of each type of cell in a large *Hydra attenuata.* Those in the 'Tomorrow' column are the numbers required 24 h later (1.22 times present numbers) if the population is doubling every 3½ days. No allowance is made for replacement of cells which are sloughed off or for used nematocysts which are discarded.

[a] There is in addition a large number of cnidoblast precursors which will not mature tomorrow.

of 1.22 ($1.22^{3.5} = 2$). Thus 25000 epidermal/gastrodermal cells must give rise to $25000 \times 1.22 = 30500$ cells. Similarly the 5000 stem cells must give rise in 24 h to 6100 stem cells. If the stem cells are dividing daily only 3050 of them need divide to form stem cells: the remaining 1950 are available to form nerve cells and cnidoblasts. There are 5000 nerve cells, which do not divide. 6100 nerve cells will be required after 24 h, so the 5000 existing cells must have 1100 added to them by division of 550 stem cells. This leaves 1400 stem cells to become cnidoblast precursors in the course of the 24 h. Each of these will give rise to a clump of (in most cases) 16 cnidoblasts which will mature 6 days from now, producing a total of 20000–25000 new cnidoblasts. This is far more than the 7700 new cnidoblasts required tomorrow but that is as it should be since there will be about three times as many cells in the growing population, 6 days from now.

This arithmetic shows why the interstitial cells have to divide so much more rapidly than the epidermal/gastrodermal cells.

PATTERN AND REGENERATION

Hydra is much simpler in structure than most metazoans but it has, nevertheless, a clear pattern, with a mouth and tentacles at one end and a pedal disc at the other. The manner in which this pattern is maintained has been

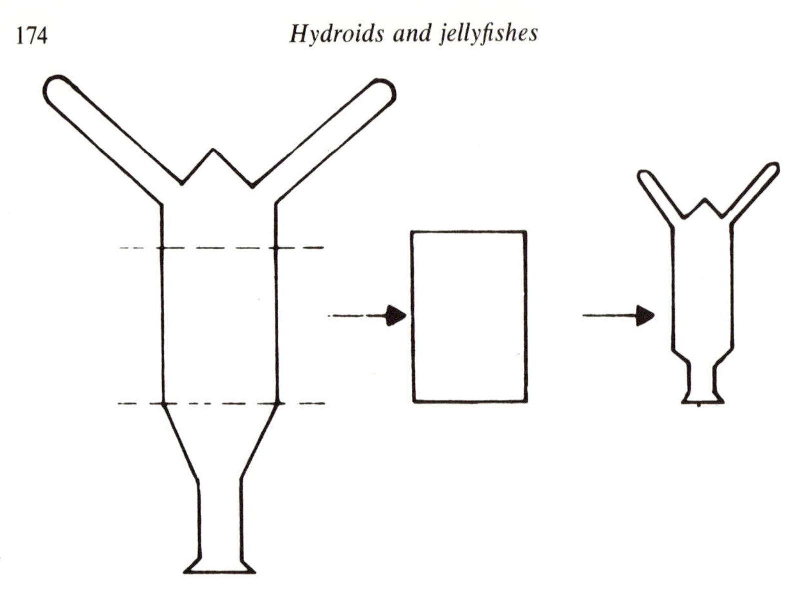

Fig. 7.16. Regeneration of a piece cut from *Hydra*. From D. R. Garrod (1973). *Cellular development*. Chapman & Hall, London.

investigated by experiments in which it has been disturbed by surgery. The simplest experiments consist simply of cutting pieces off *Hydra*, which have remarkable powers of regeneration. They do not simply re-grow lost parts as an earthworm, for instance, replaces an amputated posterior end, but re-form the remaining parts of the body to form a complete but smaller animal. Even when a piece is cut with neither an oral end nor a pedal disc, the end which was originally nearer the oral end becomes the new oral end (Fig. 7.16).

More subtle experiments have been performed by grafting. If a small piece cut from one *Hydra* is slipped into an incision in the side of another it generally becomes incorporated in it. Fig. 7.17 shows the results of some such grafts. A piece from the extreme oral end of one *Hydra* grafted into the side of another forms a new oral end with a mouth and tentacles (*a*). A piece from just below the tentacles is simply absorbed into the existing structure of the host animal (*b*) unless the oral end of the host is removed (*c*) or the graft is placed near the pedal disc (*e*). It appears that pieces from below the tentacles have the capacity to form a new oral end but are normally inhibited from doing so by the oral end nearby. In experiment (*d*) the oral end of the donor animal was cut off 4 h before the piece was taken for the graft. No new tentacles or mouth were formed in this time but the processes of regeneration must have started, for the cut end acquired the ability to overcome the inhibitory effect of the host's oral end.

Attempts are being made to explain the results of these and other grafting experiments but they have not so far been wholly successful. There is evidence that the oral end exerts its inhibiting influence by releasing a substance which diffuses rapidly along the animal and is destroyed. Thus a gradient of inhibitor

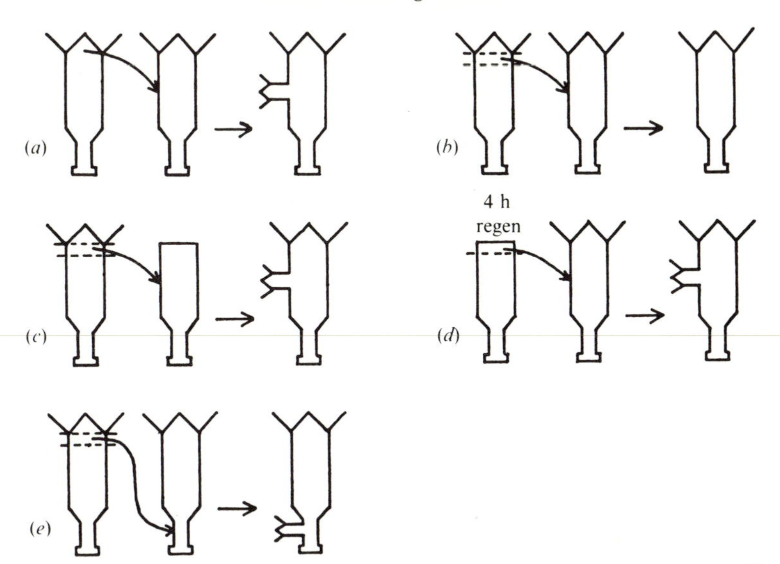

Fig. 7.17. Grafting experiments on *Hydra*. Further explanation is given in the text. From L. Wolpert, J. Hicklin & A. Hornbruch (1971). *Symp. Soc. exp. Biol.* **25**, 391–415.

concentration is set up, highest at the oral end. It also appears that individual cells must have some more stable property which has been called positional information, which determines their aptitude to form a new oral end if the inhibitor concentration is not too high.

Detailed suggestions have been made as to how the inhibitor and the positional information may interact, and their consequences have been investigated by computer modelling, but we do not know how realistic they are.

FURTHER READING

GENERAL

See the list at the end of chapter 6

SWIMMING AND BUOYANCY OF MEDUSAE

Gladfelter, W. B. (1972). Structure and function of the locomotory system of *Polyorchis montereyenis* (Cnidaria, Hydrozoa). *Helgoländer wiss. Meeresunters.* **23**, 38–79.

RHYTHMIC MOVEMENT OF MEDUSAE

Lerner, J., Mellen, S. A., Waldron, I. & Factor, R. M. (1971). Neural redundancy and regularity of swimming beats in scyphozoan medusae. *J. exp. Biol.* **55**, 177–84.

SENSE ORGANS

Eakin, R. M. (1968). Evolution of photoreceptors. *Evolutionary Biol.* **2**, 194–242.
Singla, C. L. (1975). Statocysts of hydromedusae. *Cell Tissue Res.* **158**, 391–407.

DEEP-SEA SIPHONOPHORES

Barham, E. G. (1966). Deep scattering layer migration and composition: observations from a diving saucer. *Science* **151**, 1399–403.

Mackie, G. O. (1964). Analysis of locomotion in a siphonophore colony. *Proc. R. Soc.* **159B**, 364–91.

CELL DIVISION AND GROWTH

David, C. N. & Gierer, A. (1974). Cell cycle kinetics and development of *Hydra attenuata*. III. Nerve and nematocyte differentiation. *J. Cell Sci.* **16**, 359–75.

Gierer, A. (1974). *Hydra* as a model for the development of biological form. *Sci. Am.* **231**(6), 44–54.

PATTERN AND REGENERATION

Hicklin, J., Hornbruch, A., Wolpert, L. & Clarke, M. (1973). Positional information and pattern regulation in *Hydra*: the formation of boundary regions following grafts. *J. Embryol. exp. Morph.* **30**, 701–25.

Wolpert, L., Hicklin, J. & Hornbruch, A. (1971). Positional information and pattern regulation in regeneration of *Hydra*. *Symp. Soc. exp. Biol.* **25**, 391–415.

8
Flatworms

Phylum Platyhelminthes, Class Turbellaria
  Order Acoela
  Order Rhabdocoela
  Order Tricladida
  Order Polycladida
  and other orders
(For other classes see chapter 9)

The phylum Platyhelminthes has four classes of which three consist of parasites, and are dealt with in chapter 9. This chapter is about the members of the class Turbellaria, which are nearly all free-living. They are small, primitive worms which crawl along on the ventral surface of the body or in some cases swim. One end of the body, the anterior end, normally takes the lead. This is a very different way of life from that of a sessile polyp or of a medusa which alternately swims up and sinks down but undertakes little directed horizontal movement. There is a related difference in the symmetry of the body.

A circular cake is radially symmetrical if it can be divided into slices (cut as sectors in the usual way) which are identical with each other. A vertical line through the centre of the cake is its axis of symmetry. A loaf is bilaterally symmetrical if the left hand half of every slice (cut transversely, as bread is sliced) is a mirror image of the right hand half. It would not matter if one end of the loaf were different from the other: the loaf would still be bilaterally symmetrical. A bilaterally symmetrical object has no axis of symmetry but only a plane of symmetry. This divides it into two equal halves which are mirror images of each other.

Fig. 8.1(*a*) shows that an individual polyp is more or less radially symmetrical. This seems appropriate to its way of life. If it has a mouth at one end (the oral end) and attaches the other (aboral) end to the bottom there is no obvious reason why it should not be more or less radially symmetrical about an oral–aboral axis. Anthozoan polyps are not perfectly radially symmetrical (their symmetry is spoilt by the siphonoglyphs, for instance) but the departures from radial symmetry are quite small. Medusae are similarly more or less radially symmetrical (Fig. 8.1 *b*).

A well-adapted flatworm requires differences of structure between the

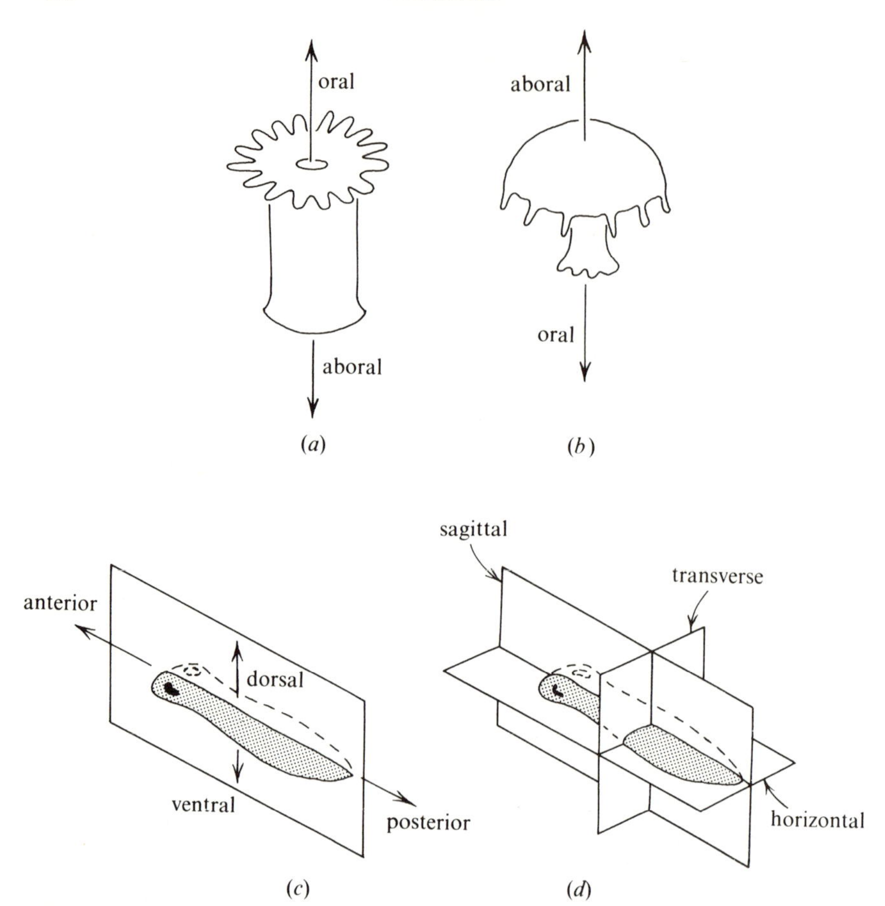

Fig. 8.1. (*a*), (*b*) Diagrams of a polyp and a medusa showing their axes of radial symmetry (arrows). (*c*) A diagram of a turbellarian showing its plane of bilateral symmetry. (*d*) A diagram showing terms used to describe sections through bilaterally symmetrical animals. Each term can be applied to any section parallel to the plane indicated.

anterior end which leads as it crawls and the posterior end which follows behind. (For instance, its principal sense organs will be most useful at the anterior end.) It also requires differences between the ventral surface which rests on the ground and the dorsal one which does not. These requirements cannot both be met by a radially symmetrical design, and flatworms are in fact bilaterally symmetrical (Fig. 8.1 *c*). The plane of symmetry is known as the median plane. Fig. 8.1(*d*) shows the meanings of terms used to describe sections through bilaterally symmetrical animals.

Most of the animals described in the remainder of this book are more or less bilaterally symmetrical. There are, however, animals which are far from symmetrical, either radially or bilaterally. *Euplotes* (Fig. 3.1 *d*) is one example, and a snail is another.

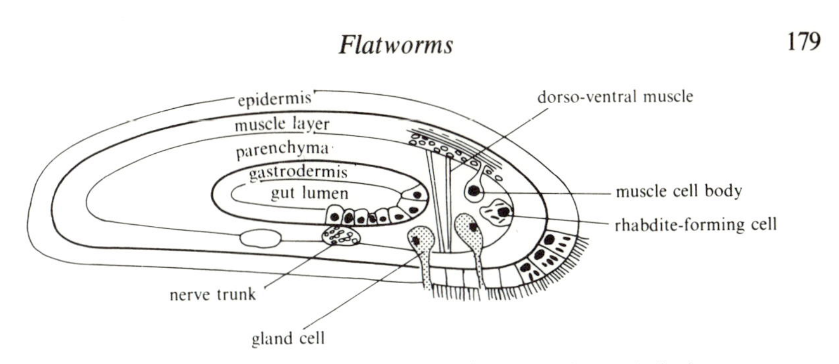

Fig. 8.2. A diagrammatic transverse section through a turbellarian.

The body walls of Cnidaria consist of an outer epidermis and an inner gastrodermis, each a single layer of cells, with a more or less cell-free mesogloea between (Fig. 7.2). Those of sponges are similar, but with more cells in the mesogloea (Fig. 5.1). The Platyhelminthes have no mesogloea. The epidermis is a single layer of cells on the outside of the animal and the gastrodermis is a single layer lining the gut, but the space between them is packed with cells (Fig. 8.2).

The epidermal cells of the ventral surface bear cilia but those of the dorsal surface often do not. Most of the epidermal cells contain rod-shaped bodies called rhabdites which are discharged when the worm is injured and swell up to form a gelatinous covering for the body. There is usually a thin layer of fibrous material (the basement membrane) underlying the epidermis and another underlying the gastrodermis.

Immediately under the epidermis is muscle, typically an outer layer of circular muscle fibres and an inner layer of longitudinal ones. Musculo-epithelial cells have been found in a few turbellarians but in most species the muscle cell bodies are under the epidermis. Between the muscle layers and the gastrodermis is a tissue called parenchyma, which contains various types of cell. There are gland cells with ducts running through the muscle layers and epidermis to openings in the body surface. There are cells with rhabdites forming in them which seem to be developing epidermal cells. There are dorso-ventral muscle fibres and cell bodies of muscle cells which have their fibres in the muscle layers. There are also cells of no obvious function.

Dugesia (Figs. 8.3, 8.4) will serve as a typical example of the Turbellaria. It is a member of the order Tricladida. It is a flattened worm which grows to lengths around 2 cm and is common in lakes, ponds and slow streams where it crawls over the surface of the mud. It crawls at speeds up to about 1.5 mm s^{-1} simply by beating its ventral cilia. It can travel a little faster by muscular action, using essentially the same technique as crawling snails (see chapter 13). It has a pair of eyes at the anterior end of the body. The light-sensitive cells have microvilli but no flagellum (Fig. 7.13) and those of each eye are enclosed in a cup of pigment cells so that the left eye receives light only from the left and the right eye only from the right. On either side of the head are ciliated

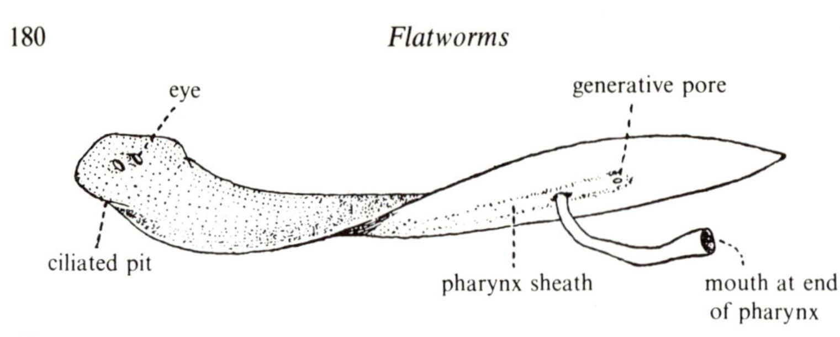

Fig. 8.3. *Dugesia*, a typical triclad 1–2 cm long. From A. E. Shipley & E. W. McBride (1901). *Zoology*. Cambridge University Press.

pits containing sensory cells which respond to chemical stimuli. The mouth is not at the anterior end of the body but at the end of a muscular pharynx which can be protruded from the ventral surface. The gut has three main branches which meet at the base of the pharynx. One branch runs anteriorly and two posteriorly from that point. They give rise to subsidiary branches, which are shown complete only for the right posterior main branch in Fig. 8.4(*a*). Feeding and digestion are discussed in a later section of the chapter. There is an extremely complicated hermaphrodite reproductive system, also described in later sections.

The nervous system (Fig. 8.4 *b*) includes a pair of connected ganglia at the anterior end of the body and a pair of main ventral nerve trunks which run posteriorly from them along the length of the body. There are branches from the main trunks and connections between them. The trunks and their main branches lie immediately internal to the muscle layers of the body wall. The protonephridial system (Fig. 8.4 *c*) consists of flame cells (described later in this chapter) and branching tubules which lead from them to pores at the surface of the body. The system probably functions as a kidney, getting rid of excess water and waste products.

The Acoela are tiny marine worms, most of them less than 2 mm long. Some live in the sand of shores, in the spaces between sand grains. They have no permanent gut cavity but the body has a core of digestive cells in which spaces appear to accommodate food taken in at the mouth. They have no protonephridia and often no rhabdites. They are the simplest of the Turbellaria and may be the most primitive. However, most of them have a statocyst though the more advanced flatworms (Tricladida and Polycladida) do not. The other orders of Turbellaria all have permanent gut cavities. Examples are shown in Fig. 8.5. Most Rhabdocoela are 0.3–3 mm long and more or less circular in cross-section, and are found in ponds and on marine shores. Their guts are simple unbranched tubes. Tricladida are generally larger, often 10–20 mm long, with bodies which are much wider than they are deep. They are found in streams, ponds and lakes and in the sea and there are even species which live on land in moist habitats. Their guts have three main branches, which in turn have side branches. The Polycladida are mostly broader and even more flattened, with more complex branching of the gut. They are nearly all marine

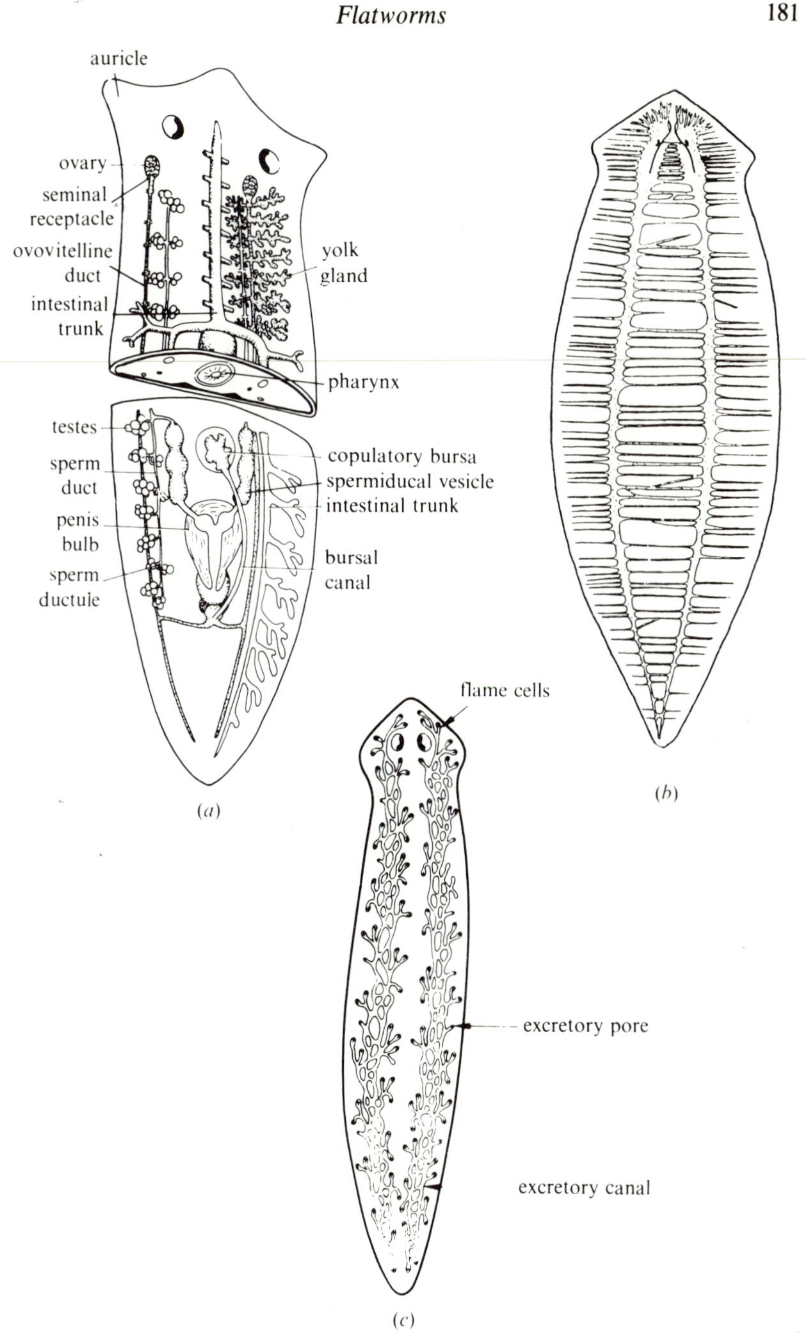

Fig. 8.4. Diagrams showing the internal structure of *Dugesia*. (*a*) The gut and reproductive system. From P. A. Meglitsch (1972). *Invertebrate zoology*, 2nd edn. Oxford University Press. (*b*) The nervous system. From J. Ude (1908). *Z. wiss. Zool.* **89**, 308–70. (*c*) The protonephridial system, with the bore of the tubules exaggerated. From K. Schmidt-Nielsen (1975). *Animal physiology: adaptation and environment*. Cambridge University Press.

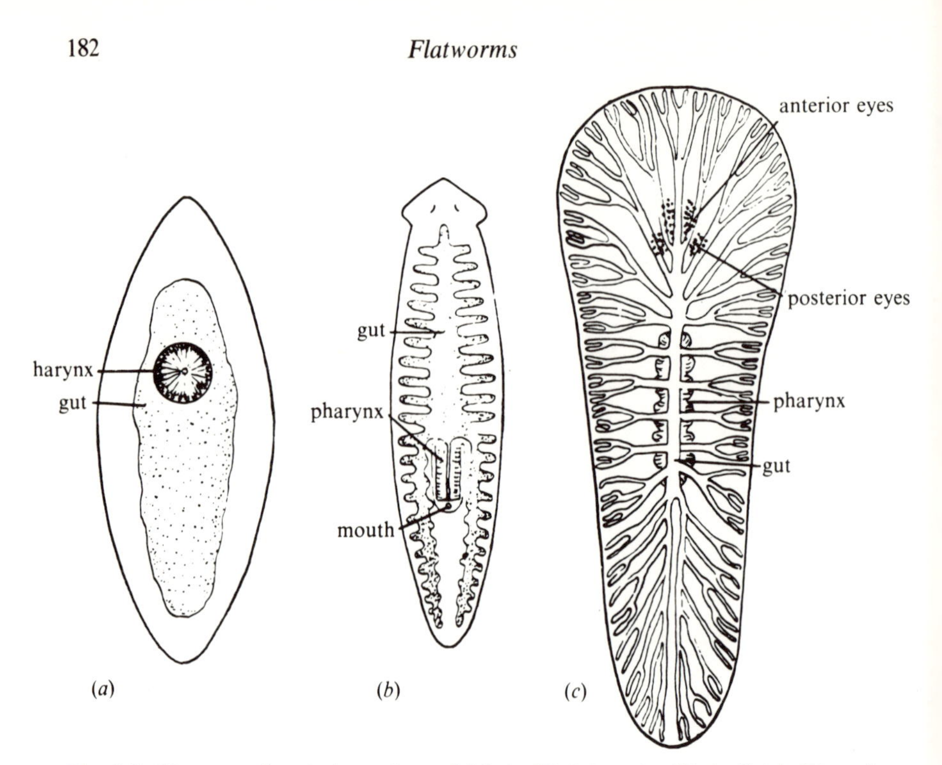

Fig. 8.5. Diagrams of typical members of (*a*) the Rhabdocoela, (*b*) the Tricladida and (*c*) the Polycladida. From L. H. Hyman (1951). *The invertebrates*, vol. 2. McGraw-Hill, New York.

and can be found on shores, under stones at low tide. Polyclads and some triclads have large numbers of eyes.

The Turbellaria cover a wide range of sizes, with the smallest no larger than many ciliate Protozoa and the largest many centimetres long. Large Turbellaria are generally more flattened than small ones and their guts branch more. Possible explanations will be offered later in the chapter.

The classification of the Turbellaria which is used in this chapter is traditional but not very satisfactory. Several alternative schemes involving large numbers of orders have been proposed.

RESPIRATION AND DIFFUSION OF GASES

The rates at which turbellarians use oxygen have been measured by putting specimens in a jar of water and analysing samples of the water from time to time to determine the concentration of dissolved oxygen in it. The results of such experiments are shown in Fig. 8.6 together with data obtained by other methods for other animals. Within groups of similar animals oxygen consumption increases with body mass, but not in direct proportion to body mass. It is generally about proportional to (body mass)$^{0.75}$, so that graphs on logarithmic

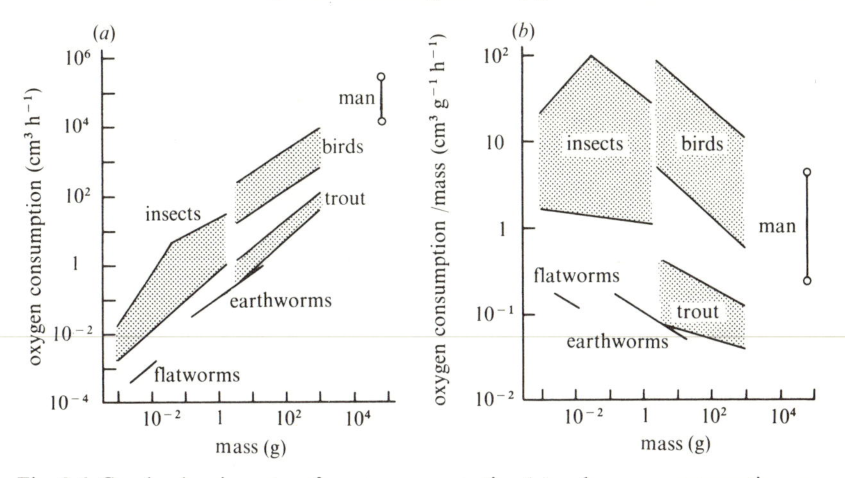

Fig. 8.6. Graphs showing rates of oxygen consumption (*a*) and oxygen consumption per unit body mass (*b*) plotted against body mass. Note that logarithmic co-ordinates have been used. Where a range of rates of oxygen consumption is indicated for animals of the same type and mass the bottom of the range is the rate at rest and the top is the maximum rate recorded in fast swimming, running or flight. From R. McN. Alexander (1971). *Size and shape.* Arnold, London.

co-ordinates of oxygen consumption against body mass tend to be straight lines of gradient 0.75 (Fig. 8.6*a*). Oxygen consumption per unit body mass therefore tends to be smaller for large animals than for small ones, but it lies between 0.1 and 0.2 $cm^3\ g^{-1}\ h^{-1}$ for all the flatworms represented in Fig. 8.6.

Flatworms have no gills or other special respiratory organs, and they have no blood system to distribute oxygen round the body. The oxygen they use must presumably reach the tissues by diffusion from the surface of the body. No part of the body can be too far from the surface, or oxygen would not diffuse to it fast enough. This is probably why large flatworms are flattened, as the following discussion will show.

Gases diffuse from regions where their partial pressure is high to regions where it is lower. The partial pressure of a gas in a mixture of gases is the pressure it would exert if it occupied the whole volume of the mixture. Air contains 21% oxygen by volume so the partial pressure of oxygen in air at atmospheric pressure is 0.21 atm. The partial pressure of a gas in solution is its partial pressure in the mixture of gases which would be in equilibrium with the solution. Hence the partial pressure of oxygen in well-aerated water is 0.21 atm.

The rate of diffusion of a gas is proportional to the gradient of partial pressure (that is to the gradient of a graph of partial pressure against distance). Consider a gas diffusing at a rate J (volume per unit time) across a surface of area A. Then if the gradient of partial pressure is dP/dx

$$J = -AD \cdot dP/dx \qquad (8.1)$$

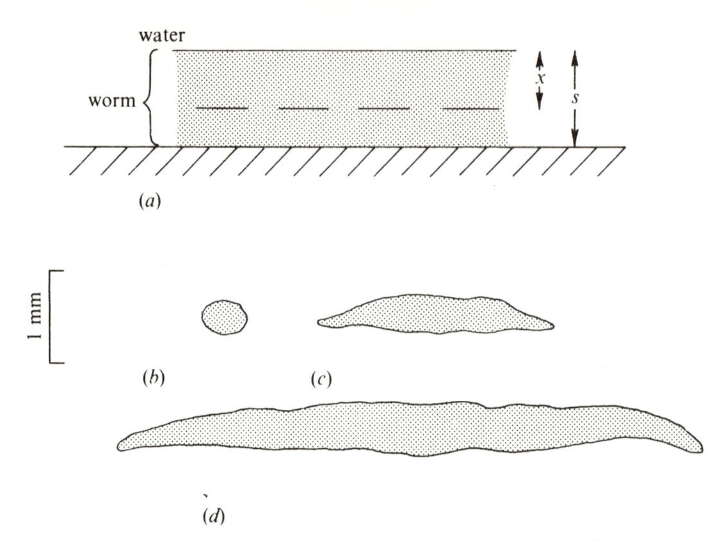

Fig. 8.7. (*a*) A diagram illustrating the account of diffusion through flatworms. (*b*), (*c*), (*d*) Transverse sections drawn to the same scale through a rhabdocoel, a triclad and a polyclad, respectively. From R. McN. Alexander (1971). *Size and shape.* Arnold, London.

The negative sign indicates that diffusion occurs down the gradient. D is a constant, for a given gas diffusing through a given medium, and is known as the diffusion constant. If J is expressed in $mm^3\, s^{-1}$, A in mm^2 and dP/dx in $atm\, mm^{-1}$, D has units $mm^2\, atm^{-1}\, s^{-1}$. The diffusion constant for oxygen diffusing through water is $6\times10^{-5}\, mm^2\, atm^{-1}\, s^{-1}$ and for oxygen diffusing through frog muscle and connective tissue about $2\times10^{-5}\, mm^2\, atm^{-1}\, s^{-1}$.

Consider a flatworm of thickness s and area A (Fig. 8.7). Oxygen diffuses in from the dorsal surface but not from the ventral surface because it is resting on a rock or some other impermeable base. It will be convenient to assume that the length and breadth of the flatworm are large compared to s so that diffusion through the vertical surfaces at the edges of the animal can be ignored. All parts of the worm use oxygen at a rate m per unit volume of tissue. Consider the horizontal plane at a distance x below the dorsal surface of the worm. The tissue below it has volume $A\,(s-x)$ and uses oxygen at a rate $Am\,(s-x)$ so oxygen must diffuse through the plane at this rate. Using this rate as the value for J in the diffusion equation we find

$$Am\,(s-x) = -AD\cdot dP/dx$$
$$m\,(s-x)\,dx = -D\cdot dP$$

If the partial pressure of dissolved oxygen is P_0 at the dorsal surface (where $x=0$) and P_s at the ventral surface (where $x=s$)

$$m\int_0^s (s-x)\,dx = -D(P_s-P_0)$$
$$\tfrac{1}{2}\,m\,s^2 = D(P_s-P_0)$$

and since P_0 cannot be negative

$$s \leqslant (2DP_s/m)^{\frac{1}{2}} \tag{8.2}$$

Fig. 8.6 shows that m for flatworms is 0.1 cm^3 oxygen $g^{-1}\,h^{-1}$, or a little more. Since the density of flatworm tissue is about 1 g cm^{-3} this is about 10^{-1} mm^{-3} oxygen $mm^{-3}\,h^{-1}$ or $3\times10^{-5}\,s^{-1}$. If the flatworm is in well aerated water $P_s = 0.21$ atm. Assume that the diffusion constant is about the same as for frog tissues. Then

$$\begin{aligned} s &\leqslant (2\times2\times10^{-5}\times0.21/3\times10^{-5})^{\frac{1}{2}} \\ &\leqslant 0.5 \text{ mm} \end{aligned}$$

This calculation indicates that the maximum possible thickness for a flatworm using oxygen at the observed rate is about 0.5 mm if oxygen diffuses in only from the dorsal surface, or 1.0 mm if it diffuses in equally from the ventral surface (as it well might if the worm lived on the surface of sand or mud). A similar calculation for a cylindrical turbellarian indicates that the maximum possible diameter would be 1.5 mm.

Fig. 8.7(*b*), (*c*), (*d*) are transverse sections through three turbellarians. Though very different in size all are about 0.5 mm thick. Oxygen would probably not diffuse into them fast enough for their requirements, if they were much thicker than this. This is probably why large flatworms are flat.

Some sea anemones and jellyfish are of course much larger than flatworms, but the diffusion distances for the oxygen they use in metabolism are small because their cells form single layers of epidermis and gastrodermis. A *Cyanea* medusa 30 cm in diameter may be 3 cm thick but most of its thickness is occupied by mesogloea and almost all the cells are very close to the outside surface of the animal or to the gastric cavity and canals. Water is circulated through the canals by the cilia of the gastrodermis. Even the folded swimming muscles of the *Cyanea* would be less than 5 μm thick.

FEEDING

Most Turbellaria are predators which attack and feed on other animals but there are several different methods of feeding corresponding to differences in the structure of the pharynx.

The simplest type of pharynx is found in rhabdocoels such as *Macrostomum* (Fig. 8.8*a*). It is a simple tube leading from the slit-shaped mouth to the gut. It is surrounded by layers of muscle which are not particularly thick and are continuous with the muscle layers of the general body surface. It can be enlarged greatly to take in food, which is swallowed intact. Small crustaceans such as *Daphnia*, small annelid worms and many other animals are eaten. Microscope sections of *Macrostomum* killed at various times after eating *Daphnia* showed digestion well advanced after 10 hours, and only an empty *Daphnia* skeleton after 24 hours. Even the empty skeleton preserved the general shape of the intact *Daphnia*, so food is not broken up prior to digestion.

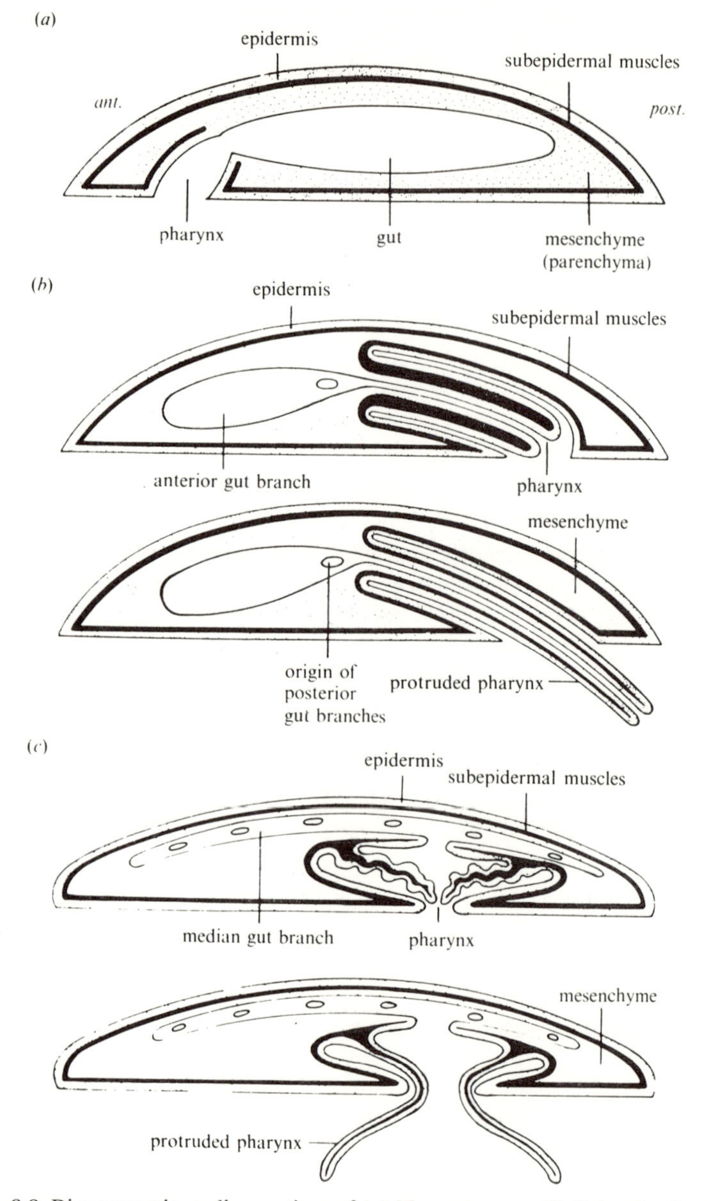

Fig. 8.8. Diagrammatic median sections of (*a*) *Macrostomum*, (*b*) *Polycelis*, showing the pharynx retracted and protruded, and (*c*) *Leptoplana* showing the pharynx retracted and protruded. *ant.*, anterior; *post.*, posterior. From J. B. Jennings (1957). *Biol. Bull.* **112**, 63–80.

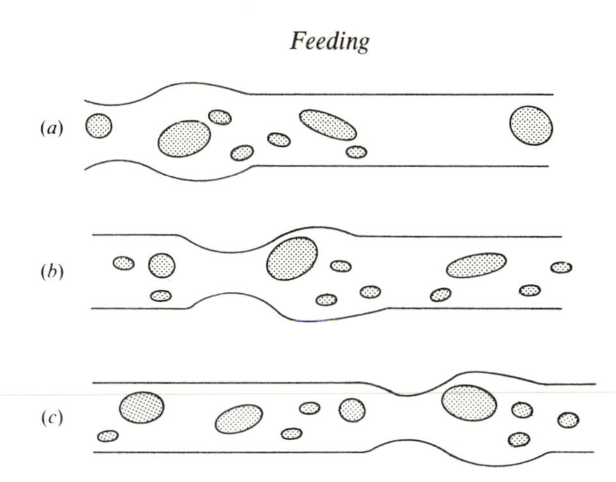

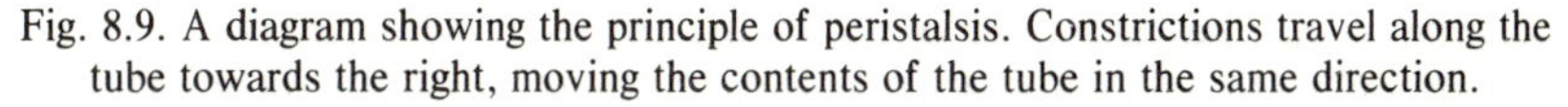

Fig. 8.9. A diagram showing the principle of peristalsis. Constrictions travel along the tube towards the right, moving the contents of the tube in the same direction.

Triclads have a different type of pharynx with a thick muscular wall (Fig. 8.8*b*). It is housed in a cavity in the body but can be protruded for feeding; presumably its circular muscles contract and make it elongate. Triclads feed on various invertebrates including crustaceans, annelids and snails. They leave trails of mucus behind them as they crawl, and potential prey sometimes get caught in this mucus; the success of triclads in capturing relatively large prey may depend largely on this. The pharynx is protruded and inserted into the prey. It enters crustaceans through the relatively soft cuticle between skeletal plates. It moves about inside the prey, sucking out the contents by peristalsis (Fig. 8.9). The food is already in a semifluid state when it enters the gut.

Some polyclads have pharynxes like triclads but others such as *Leptoplana* have the type shown in Fig. 8.8(*c*). This type can also be protruded but forms a broad funnel rather than a slender tube. *Leptoplana* seizes worms and crustaceans by wrapping its body round them. If they are small it swallows them whole but if large it spreads its pharynx over them and they gradually disintegrate over a period of an hour or more. The disintegration is presumably due to enzymes secreted by the polyclad. Fragments break off the prey and are taken into the gut where their digestion is completed.

It is usually possible to identify the natural foods of animals by capturing a sample and examining their gut contents. This method is little use for triclads because most of their food is already unrecognizable by normal methods, when it is swallowed. Triclads attack and eat a wide variety of prey when it is offered in the laboratory and seem to prefer certain types of prey, but this is not necessarily a reliable indication of the food actually taken in their natural habitat.

These difficulties were overcome in a study of some British lake-dwelling triclads, by using an immunological method. It seemed likely that these triclads might feed on certain crustaceans (*Gammarus* and *Asellus*), annelid worms and

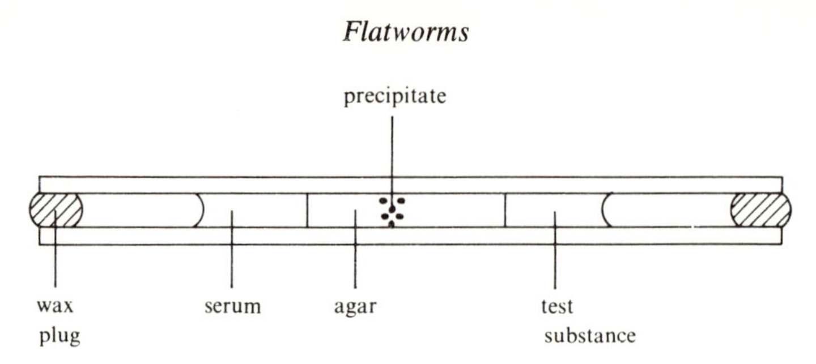

Fig. 8.10. The immunological test described in the text. A reaction between antibodies in the serum and antigens in the preparation under test is indicated by a white precipitate in the agar jelly.

water snails, which were common in the lakes. Preparations of proteins were made from each of these likely foods and injected into rabbits which produced antibodies against them. The reactions of the antibodies were tested in short lengths of capillary tubing, as shown in Fig. 8.10. Serum from the immunized rabbit was put on one side of the agar jelly, and a preparation of the test substance on the other. The antibody and the test substance diffuse from opposite ends into the jelly and if they react a band of precipitate appears in the jelly. It was found that the anti-*Asellus* serum reacted with *Asellus* proteins but not with proteins from *Gammarus*, earthworms or snails. Similarly the anti-*Gammarus* serum reacted only with *Gammarus* protein, anti-earthworm serum reacted with protein from various annelid worms but nothing else and anti-water-snail serum reacted with protein from various gastropod molluscs but nothing else. Further tests showed that each antiserum reacted with preparations made from triclads which had recently fed on the food in question. Hence the antibodies could be used to identify the gut contents of triclads.

Triclads were collected from lakes and an extract made from each individual was tested against all four antisera. It usually reacted with one of them, and occasionally with two. The results were used to calculate the relative frequency with which each species of triclad fed on each of the four types of prey. It had been found in the preliminary tests that *Gammarus* remained detectable for longer after a meal than the other prey, so the data were corrected to eliminate this source of error.

The investigation was concerned with four species of triclad which are often found in the same part of the same lake, and even on the same stone. They coexist, yet none of them eliminates the others by competition. It was found that all four species ate at least three of the foods but that each had a preferred food which was eaten much more commonly than the others. *Dendrocoelom lacteum* ate mainly *Asellus*, *Dugesia polychroa* ate mainly gastropods, and two species of *Polycelis* ate mainly oligochaete worms. Conveniently, oligochaete worms are among the few foods which can be recognized visually after being eaten because some of their chaetae (bristles) are swallowed. Different families of oligochaetes have chaetae of different shape. Specimens of the two species of *Polycelis* were squashed and the squashes were examined for chaetae under a microscope. It was found that the two species ate different proportions of the various families of oligochaetes.

The results illustrate the wide variety of prey attacked by triclads. They also show that each of the four coexisting species had a principal food which was relatively seldom eaten by the other three. This may be why they can coexist without one eliminating the others.

DIGESTION

Fragments of food can often be seen in microscope sections of flatworms, enclosed in vacuoles in gastrodermal cells. Sections of *Polycelis* which had been fed starch or clotted blood showed starch grains or blood corpuscles in vacuoles. Presumably digestion occurs in vacuoles, as in Protozoa. Digestion also occurs in the gut cavity: this is particularly obvious in rhabdocoels which swallow their prey whole. What are the roles of the two types of digestion? Histochemical tests have been made on triclads, to find out.

Amino acids have the general formula $NH_2CHRCOOH$, where R is one of many alternative groups. The large molecules of proteins and the smaller ones of peptides consist of amino acids linked together, the carboxyl (COOH) group of one joining the amino (NH_2) group of the next

$$NH_2CHRCOOH + NH_2CHR'COOH = NH_2CHRCO.NHCHR'COOH + H_2O$$

Digestion involves breaking proteins down to their constituent amino acids, and several types of enzyme are involved. Aminopeptidases remove the terminal amino acid from the end of the chain which has the free amino group, carboxypeptidases remove the terminal amino acid from the other end and endopeptidases break the chain further from its ends. Arrows in the diagram below show the points of attack of the enzymes.

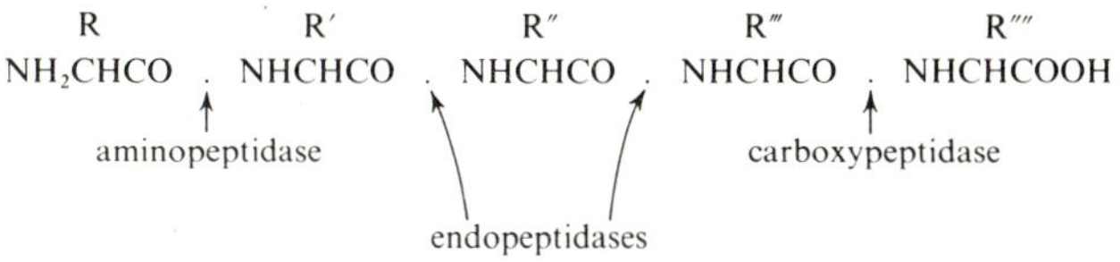

$$\overset{R}{NH_2CHCO} \, . \, \overset{R'}{NHCHCO} \, . \, \overset{R''}{NHCHCO} \, . \, \overset{R'''}{NHCHCO} \, . \, \overset{R''''}{NHCHCOOH}$$

aminopeptidase (first bond) — endopeptidases (second and third bonds) — carboxypeptidase (last bond)

As well as breaking amino acids off the ends of peptides, aminopeptidases can split compounds in which amino acids are combined with amines. For instance, they can split leucyl-β-napthylamide into the amino acid leucine and β-napthylamine

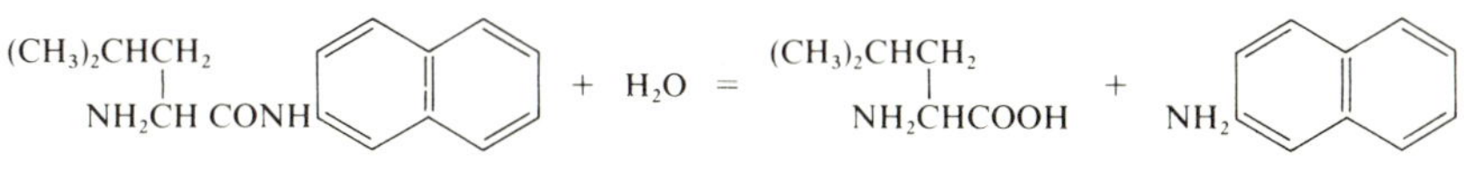

$$\begin{array}{l}(CH_3)_2CHCH_2 \\ \quad | \\ NH_2CH\,CONH\text{-}C_{10}H_7\end{array} + H_2O = \begin{array}{l}(CH_3)_2CHCH_2 \\ \quad | \\ NH_2CHCOOH\end{array} + NH_2\text{-}C_{10}H_7$$

This is the basis of a histochemical test for aminopeptidases which can be applied to microscope sections provided they have been prepared by a method which does not inactivate the enzymes. The sections are left for a few hours in a solution containing leucyl-β-napthylamide and Garnet GBC, a diazonium salt. They are then rinsed and examined. Wherever aminopeptidases are present they will have broken down the leucyl-β-napthylamide, releasing β-napthylamine which combines immediately with Garnet GBC to give a bright red precipitate (an azo dye). The precipitate shows precisely where in the section the aminopeptidase is. The test is not wholly reliable: there are probably aminopeptidases which will not attack leucyl-β-napthylamide and there may well be enzymes which break down leucyl-β-napthylamide but have no effect on peptides. Nevertheless the test has been found a very useful guide to the distribution of aminopeptidases. Other tests designed on similar principles are used to locate endopeptidases and many other enzymes.

These tests have been applied to sections of triclads (mainly *Polycelis*). The pharynx has gland cells which open to its outer surface, especially near its tip. The tests showed that many of these cells contained endopeptidase, and were shrunken after feeding. Presumably they secrete endopeptidases which attack proteins in the prey, enabling the pharynx to penetrate it and disrupting the tissues so that they can be sucked into the gut.

There are two types of cell in the gastrodermis, columnar cells which engulf food particles and gland cells which do not. The gland cells contain numerous small spheres which react strongly to the histochemical test for endopeptidases. There are traces of endopeptidase in the gut immediately after feeding (perhaps due to endopeptidase secreted by the pharynx and taken in with the food), but the concentration builds up to a maximum in the next 4 hours. During the same period the gland cells lose a lot of their spheres. It seems that the gland cells produce endopeptidase, and that they secrete it into the gut lumen after a meal. No other enzymes have been found in the gut lumen. Once the gland cells have released their spheres of enzyme they need 1 or 2 days to make a new supply.

Endopeptidase can also be detected in the cytoplasm of the columnar cells, and it is present in food vacuoles for the first 10 hours or so after a meal. Thereafter it disappears from the vacuoles, and aminopeptidase and lipase appear in them (lipases are enzymes which break down fats). Starch is also digested in the vacuoles at this stage. Digestion may not be complete until 2 days or more after a meal.

Thus digestion starts externally (around the tip of the pharynx) and in the gut lumen. It continues in vacuoles in the columnar cells. Initially the food is attacked only by endopeptidases but later other enzymes are involved. The endopeptidases reduce protein molecules to larger numbers of shorter peptide chains and so increase the number of chain ends which can be attacked by aminopeptidases (and carboxypeptidases if they are present).

The final products of digestion in the vacuoles are presumably amino acids

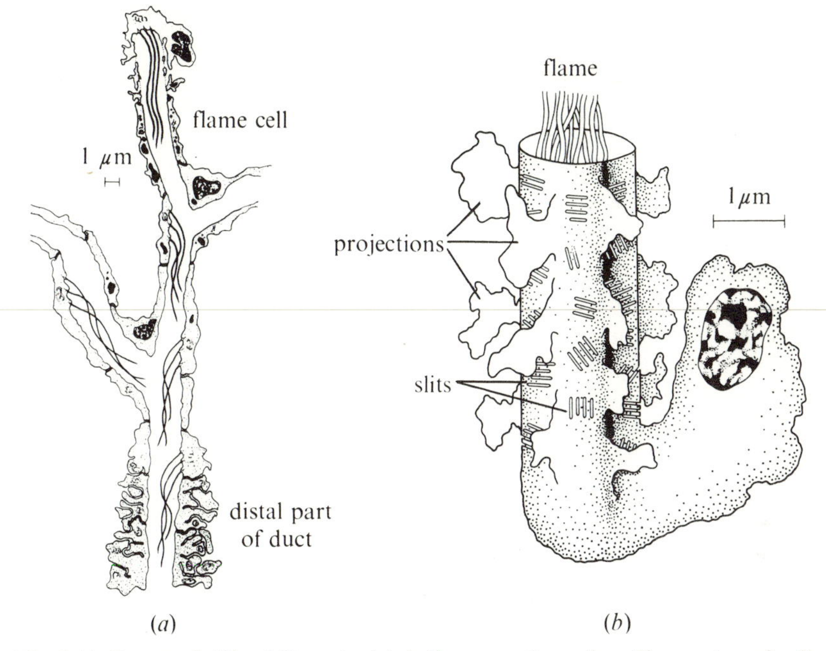

Fig. 8.11. Protonephridia of *Dugesia*. (*a*) A diagrammatic section. The number of cells has been reduced so that the duct is shown much shorter than it actually is, and the number of flagella has also been reduced. (*b*) A diagram of the flame cell. From J. A. McKanna (1968). *Z. Zellforsch.* **92**, 524–35 and 509–23, respectively.

from proteins, fatty acids and glycerol from fats, and sugars from polysaccharides. These substances must travel to the tissues which need them, throughout the body. There is no blood circulation to carry them and it is likely that they travel by diffusion (though active transport may occur). The larger flatworms (triclads and polyclads) have branching guts arranged in such a way that no part of the body is far from a branch (Fig. 8.5). Consequently the products of digestion need not diffuse far, provided the food is distributed reasonably uniformly among the branches of the gut before being taken into food vacuoles.

PROTONEPHRIDIAL SYSTEM

Fig. 8.11 (*a*) is a diagram of part of the protonephridial system of *Dugesia*, based on electron microscope sections. Flame cells close the blind ends of the branching ducts. Each contains a bundle of 35–90 flagella which are packed close together and beat in unison with a frequency of about 1.5 s^{-1}. The bundle is visible by light microscopy in living animals and is called the flame because of its flickering appearance.

Fig. 8.11(*b*) shows the flame cell in more detail. It is perforated by groups of slits which connect the intercellular space outside the cell to the lumen within.

The slits are about 35 nm wide, and seem to be crossed by fine, closely spaced filaments. Projections from the flame cell hold adjacent cells at a distance so that they do not block the slits.

Waves of bending travel along the flame from its base to its tip, tending to drive fluid down the tubule and reduce the pressure at the base of the flame. Fluid is probably drawn from the intercellular spaces of the parenchyma through the slits and so into the lumen. Ultrafiltration probably occurs: the fine filaments probably allow water and salts to pass through but stop larger molecules such as proteins. (Protein molecules would pass easily through the slits themselves, for even the huge molecules of snail haemocyanin have diameters of only about 7 nm.)

There are bundles of flagella at intervals along the protonephridial ducts as well as in the flame cells at the ends of the ducts. The cells which line the distal parts of the duct have many mitochondria and the cell membrane of their outer surface is deeply infolded. Cells which look like this in various other animals are known to secrete salts actively against a concentration gradient across the cell membrane (Figs. 15.9, 18.14, 18.15). The ducts presumably lead to external openings but a zoologist who examined several thousand sections by electron microscopy failed to find a single opening.

Proteins and other large molecules contribute to the osmotic pressure of fluids containing them: their contribution is known as the colloid osmotic pressure. Where ultrafiltration occurs the colloid osmotic pressure tends to draw fluid back, and a pressure difference is needed to overcome it. Blood pressure serves this purpose in kidneys, and the pressure difference set up by the flame may suffice in protonephridia.

There is little direct evidence about the functions of flatworm nephridia. More is known about the rather similar protonephridia of rotifers. The rest of this section describes experiments on rotifer protonephridia because for the present we can only assume that flatworm protonephridia work in the same way.

The rotifer *Asplanchna priodonta* is only 1 mm long but its protonephridia open into a single duct and it is possible to collect tiny samples of fluid from this duct by drawing the fluid into a micropipette. We will call this fluid urine, because the protonephridia seem to serve as kidneys. The osmotic concentration of the urine has been measured, by determining its freezing point.

Like other rotifers, *Asplanchna* has a fluid-filled space in its body. Fluid from this was found to have an osmotic concentration of 80 mOsmol l^{-1}. The osmotic concentration of the urine was 42 mOsmol l^{-1}, when the rotifer was kept in a medium of concentration 18 mOsmol l^{-1}. When the rotifer was put into distilled water the osmotic concentration of the body fluid was little changed but that of the urine fell to 15 mOsmol l^{-1}, and urine was produced a little faster.

These results seem to show that at least in the rotifer, protonephridia are involved in osmotic and ionic regulation. A rotifer living in fresh water must lose salts by diffusion and take up water by osmosis. To compensate for this it must have means of taking up salts and getting rid of water. *Asplanchna* eats Protozoa and other small animals. If it excretes a urine which has a lower concentration of salts than this food it can get rid of excess water without at the same time losing all the salts from the food. In distilled water it will tend to gain water and lose salts faster so it must excrete a larger volume of more dilute urine.

REPRODUCTION

Turbellaria have remarkably complicated reproductive organs. There are differences between and even within the orders, but the reproductive system of *Dugesia* (Fig. 8.4 *a*) will serve as an example. Like nearly all other Turbellaria *Dugesia* is hermaphrodite: that is to say, each individual has both male and female gonads. There is a single external genital opening.

There is a line of testes along each side of the body. The sperm they produce have two flagella which do not have the usual 9+2 arrangement of microtubules: the two central microtubules are replaced by a single rod. Ducts lead from the testes to spermiducal vesicles where ripe sperm are stored, and from there to the muscular penis. This normally lies in a cavity just within the genital pore but it can be protruded like the pharynx.

There is an ovary on each side of the body, near the anterior end. The ducts which lead from them to the genital pore are lined with yolk glands which produce yolk cells. In most animals the store of food required for development is incorporated as yolk in the ovum itself, but in triclads and many other platyhelminths it is kept in separate yolk cells. The copulatory bursa receives sperm from another animal in copulation.

When they copulate, two animals place their genital openings in contact with each other and remain thus for a few minutes (Fig. 8.12 *a*). The details of copulation cannot be seen by examining intact animals but are revealed by microscope sections of pairs killed in the act (Fig. 8.12 *b*). Each partner protrudes its penis and inserts it through the genital pore of the other, into the duct of the copulatory bursa.

The sperm of triclads are inactive in the spermiducal vesicle but become active on entering the copulatory bursa. They travel from it up the ovovitelline ducts where they are joined by ripe ova, which they fertilize. The fertilized ova travel down the ducts, mingling with yolk cells. Several ova and hundreds of yolk glands are enclosed in a capsule formed from droplets enclosed in the yolk cells. The capsules are given a sticky coating (perhaps by the cement glands, Fig. 8.12 *b*) and adhere after they have been laid onto surfaces such as the undersides of stones. The embryos ingest the yolk cells as they develop and minute worms hatch from the capsules after a few weeks.

The egg capsules of triclads are pale and soft initially but soon become brown and hard. The change may well be due to quinone-tanning and there is clear evidence from histochemical studies of various parasitic platyhelminths that their capsules are quinone-tanned.

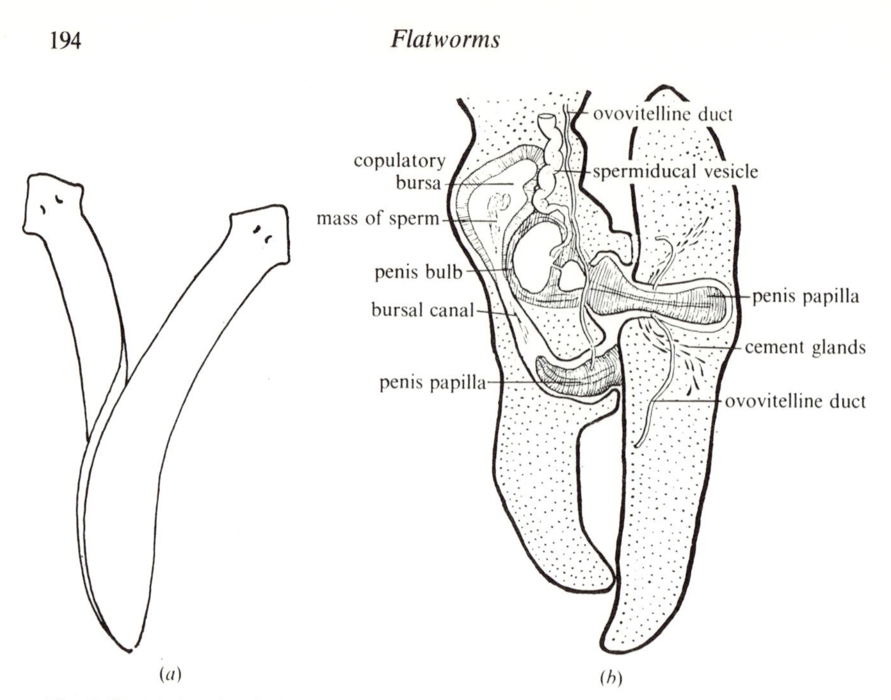

Fig. 8.12. (*a*) A pair of *Dugesia* copulating. (*b*) A section through a copulating pair of *Dugesia*. From L. H. Hyman (1951). *The invertebrates*, vol. 2. McGraw-Hill, New York.

Though turbellarians are hermaphrodite self-fertilization seems to be rare. This may be due to the mechanism which keeps the sperm inactive in the spermiducal vesicles. Triclads kept in isolation sometimes lay capsules but these do not hatch: presumably the ova are unfertilized.

Reproduction is not exclusively sexual. Some rhabdocoels elongate and form a chain of individuals which eventually separate (Fig. 8.13). Some triclads split in two transversely. The offspring may be provided with individual sets of organs before splitting occurs, or they may not: if not, each develops after the division whatever it lacks.

Distinct sexual and asexual forms occur in *Dugesia*. The sexual forms do not reproduce asexually though both halves regenerate to form complete animals if they are cut in two. The asexual forms have only rudimentary sex organs. Some wild populations consist exclusively of asexual specimens, others are mixed and some seem to be exclusively sexual. Offspring of asexual reproduction are asexual but the offspring of sexual reproduction of specimens from a mixed population were sexual and asexual in approximately equal numbers.

SPIRAL CLEAVAGE

Most turbellarians are unusual in having separate yolk cells instead of incorporating the yolk in their ova. In these turbellarians, development of the embryo proceeds in rather peculiar ways. Polyclads do not have separate yolk

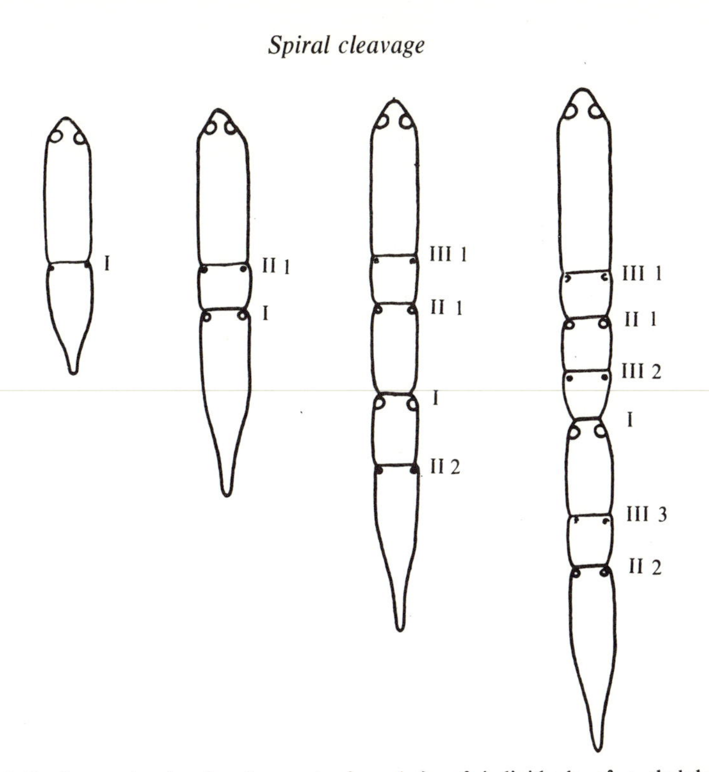

Fig. 8.13. Stages in the development of a chain of individuals of a rhabdocoel, *Stenostomum*. Numbers indicate the sequence of demarcation of boundaries. From C. M. Child (1941). *Patterns and problems of development*. Chicago University Press, Chicago.

cells and their early development proceeds in a strikingly orderly fashion, according to a programme which is very closely paralleled in molluscs and in annelid and nemertean worms. The polyclad zygote divides into cells in a manner described as spiral cleavage. It divides first into two cells each of which divides again, producing four cells arranged like the segments of an orange (Fig. 8.14*a*). Each of these cells divides obliquely so that the four upper cells are not directly over the lower ones but in the grooves between. This obliquity of division is described by the term spiral cleavage. Subsequent divisions are also oblique.

Zygotes which divide by spiral cleavage, in whatever phylum, also have determinate development. This means that the fate of each individual cell is already decided at the time of the division which gives it its identity. The first four cells are designated A, B, C and D. Fig. 8.14(*b*) shows how cell D of a polyclad divides and what happens to its progeny. It first divides into cells designated 1d and 1D. The progeny of 1d all go to form the nervous system and the anterior part of the epidermis. Cell 1D divides again, producing 2d and 2D, and 2D divides into 3d and 3D. The progeny of 2d and 3d form epidermis and the muscle and parenchyma of the pharynx. 3D divides to form 4D, which

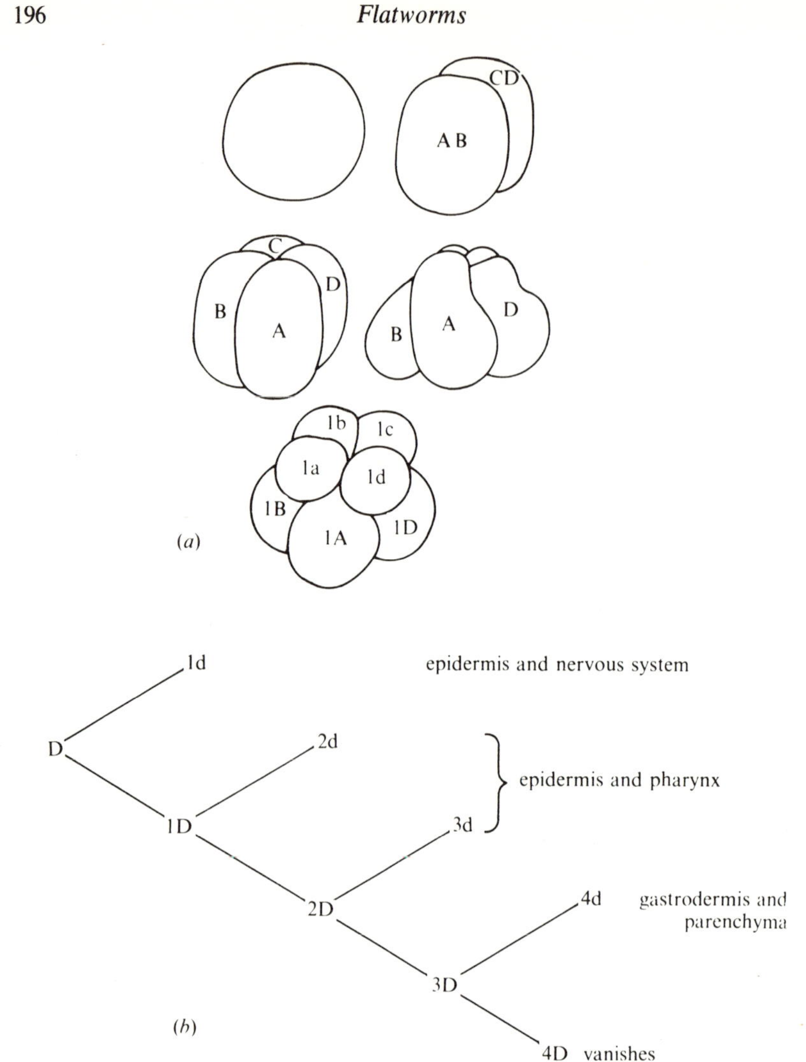

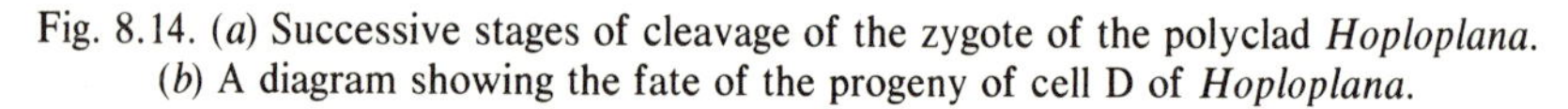

Fig. 8.14. (*a*) Successive stages of cleavage of the zygote of the polyclad *Hoploplana*. (*b*) A diagram showing the fate of the progeny of cell D of *Hoploplana*.

degenerates and disappears, and 4d, which forms the gastrodermis and parenchyma. Cells A, B and C divide in the same way except that 4a, 4b and 4c (the equivalents of 4d) vanish: all the gastrodermis and all the parenchyma except some in the pharynx is formed by the progeny of 4d.

FURTHER READING

GENERAL

Hyman, L. H. (1951). *The invertebrates*, vol. 2, *Platyhelminthes and Rhyncocoela, the acoelomate Bilateria.* McGraw-Hill, New York.

Riser, N. W. & Morse, M. P. (1974). *Biology of the Turbellaria.* McGraw-Hill, New York.

Skaer, R. J. (1961). Some aspects of the cytology of *Polycelis nigra. Q. J. microsc. Sci.* **102**, 295–317.

FEEDING

Jennings, J. B. (1957). Studies on feeding, digestion and food storage in free-living flatworms (Platyhelminthes: Turbellaria). *Biol. Bull.* **112**, 63–80.

Reynoldson, T. B. & Davies, R. W. (1970). Food niche and co-existence in lake-dwelling triclads. *J. anim. Ecol.* **39**, 599–617.

DIGESTION

Jennings, J. B. (1962). Further studies on feeding and digestion in triclad Turbellaria. *Biol. Bull.* **123**, 571–81.

PROTONEPHRIDIAL SYSTEM

Braun, G., Kummel, G. & Mangos, J. A. (1966). Studies on the ultrastructure and function of a primitive excretory organ, the protonephridia of the rotifer *Asplanchna priodonta. Pflügers Arch.* **289**, 141–54.

McKanna, J. A. (1968). Fine structure of the protonephridial system in planaria. I and II. *Z. Zellforsch.* **92**, 509–23 and 524–35.

Wilson, R. A. & Webster, L. A. (1974). Protonephridia. *Biol. Rev.* **49**, 127–60.

REPRODUCTION

Henley, C. (1974). Platyhelminthes (Turbellaria). In *Reproduction of marine invertebrates*, vol. 1, ed. A. C. Giese & J. S. Pearse, pp. 267–343. Academic Press, New York & London.

9
Flukes and tapeworms

Phylum Platyhelminthes (cont.)
  Class Turbellaria (see chapter 8)
  Class Monogenea
    Order Monopisthocotylea (skin flukes)
    Order Polyopisthocotylea (gill flukes)
  Class Trematoda
    Order Aspidobothrea
    Order Digenea (gut, liver and blood flukes)
  Class Cestoda (tapeworms)

There are only a few turbellarians which are parasitic in the strict sense. They include the rhabdocoels of the genus *Fecampia* which are found in the haemocoels (body cavities: see chapter 17) of crabs and other crustaceans. There are rather more turbellarians which live on the surfaces of other animals without feeding on their tissues. These include the members of another genus, *Temnocephala*, which live attached to the gills of shrimps and feed on protozoans and other small animals which they capture from the water flowing over the gills.

The great majority of parasitic platyhelminths belong to the classes Monogenea, Trematoda and Cestoda, which are the subjects of this chapter. Older classifications (and some modern ones) include the Monogenea in the Trematoda, but they seem different enough to be given a separate class.

Most Monogenea and Trematoda are leaf-shaped, like the larger Turbellaria. They are the flukes. The Cestoda have a characteristic ribbon-like (or tape-like) shape, and are called tapeworms. Most of them have constrictions which divide the body into segments called proglottids. All three parasitic classes have parenchyma and nervous, protonephridial and reproductive systems similar to those of Turbellaria. They also have attachment organs of one sort or another, and they have a peculiar epidermis which is known as a tegument (Fig. 9.1).

The tegument is syncytial: that is, it has no cell boundaries although it has many nuclei. Its outer zone is a thin layer of cytoplasm without nuclei, 20 μm thick in the trematode *Fasciola* on which Fig. 9.1(*a*) is based. This layer often includes spines which seem to consist of crystalline protein, and it is perforated by the endings of sensory cells. Immediately under the outer zone is a basement membrane (basal lamina) and under that again are layers of circular and longitudinal muscle fibres. The inner zone of the tegument is under

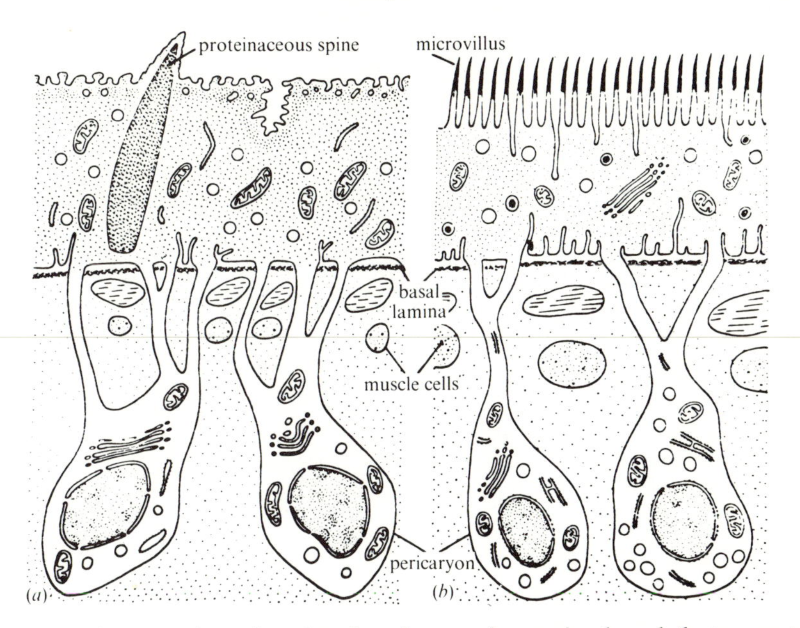

Fig. 9.1. Diagrammatic sections, based on electron micrographs, through the teguments of (*a*) the trematode *Fasciola* and (*b*) the cestode *Abothrium*. From U. Welsch & V. Storch (1976). *Comparative animal cytology and histology*. Sidgwick & Jackson, London.

all these. It consists of tegumentary cell bodies (pericarya), each containing a nucleus and each connected to the outer zone by several fine strands of cytoplasm. The cell bodies extend deep into the parenchyma.

SKIN FLUKES

The Monogenea consists mainly of ectoparasites which live on the external surfaces of fishes and amphibians. The first of its two orders, the Monopisthocotylea, consists largely of parasites which live on the host's skin but includes a few which attach themselves to some of the more accessible internal surfaces such as the walls of the mouth cavity and the cloaca.

One species which has been particularly thoroughly studied will serve as an example of the Monopisthocotylea. It is *Entobdella soleae* (Fig. 9.2) which lives on the skin of the common sole, *Solea solea*. It is a flattened hermaphrodite worm with a conspicuous attachment organ, the haptor, at the posterior end. Its internal structure is like that of *Dugesia* (Fig. 8.4) in many ways. There is a mouth with a short protrusible pharynx at the anterior end. The gut has two main branches which run posteriorly from the mouth, giving off subsidiary branches. The gut is shown only on the left side of Fig. 9.2 while the yolk glands (vitelline follicles) are shown only on the right. The yolk glands are scattered widely through the body, as in *Dugesia*. There is only one ovary. The sperm

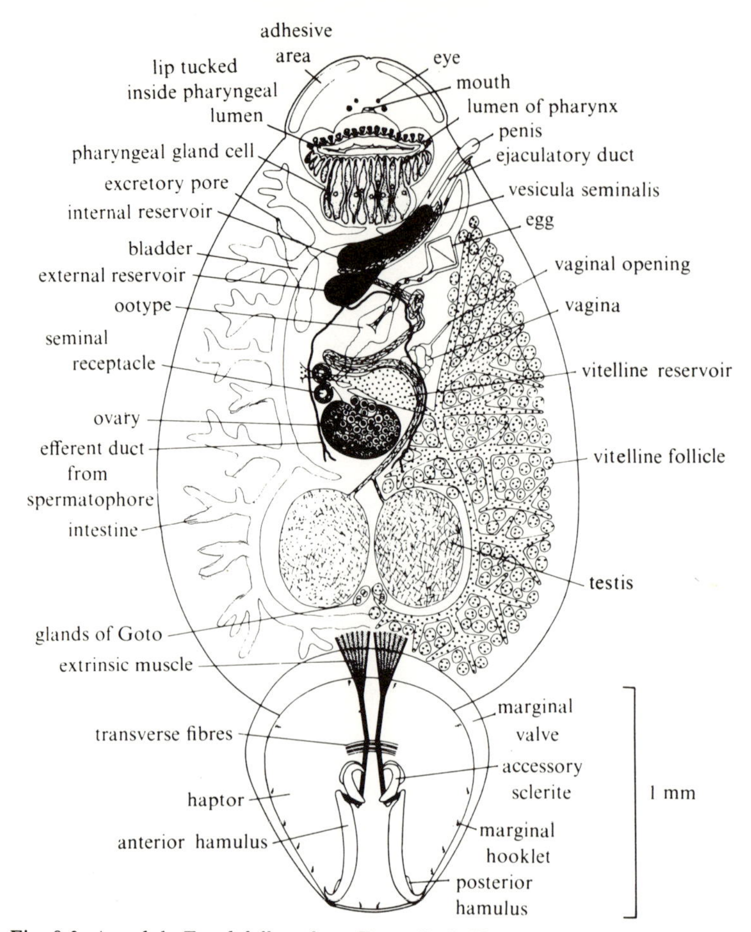

Fig. 9.2. An adult *Entobdella soleae*. From G. C. Kearn (1971). In *Ecology and physiology of parasites*, ed. A. M. Fallis, pp. 161–87. Hilger, London.

ducts from the testes lead to a protrusible penis. The female reproductive system has an external opening beside the penis and an additional vaginal opening.

The haptor has a number of sclerites (skeletal elements) in it, ranging in size from the large anterior hamuli to the tiny marginal hooklets (Fig. 9.2). Sclerites have been separated from the other tissues by allowing the latter to decay. The quantities obtained were small, but just large enough for analysis. They were found to consist of a protein incorporating a good deal of the amino acid cysteine $CH_2SH.CHNH_2.COOH$. They gave an X-ray diffraction pattern very like that of α keratin, the protein of hair and horn. Keratin owes its rigidity to cross-links between cysteine residues in adjacent peptide chains: the –SH groups of two cysteine residues are oxidized and linked together, thus

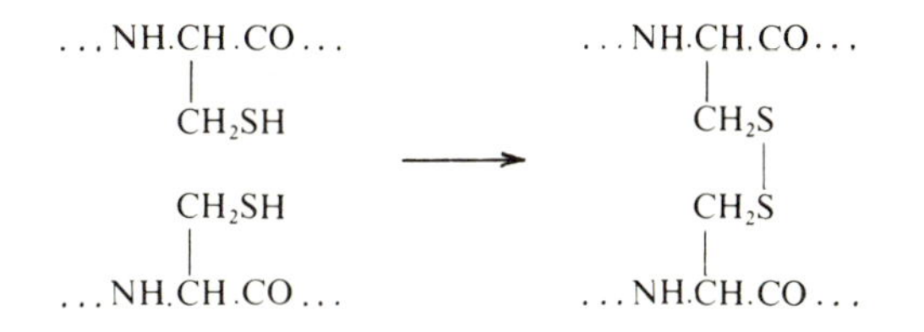

The protein of the sclerites is probably similar to keratin, and cross-linked in the same way.

Entobdella attaches itself to the scales of its host. The hooks of the anterior hamuli penetrate the host tissue but the haptor seems to grip mainly by suction. If its edge is lifted with a needle it releases immediately (except for the hooks) just as a rubber sucker can be released by raising its edge. Fig. 9.3 shows how it is believed to work. There is a pair of muscles with tendons which pass under loops of connective tissue fibres and through notches in the accessory sclerites to insert on the anterior hamuli. When these muscles contract they tend to lift the centre of the haptor in several places: they lift the points of attachment of the fibre loops and also the anterior ends of the anterior hamuli. The marginal valve is a flap of tissue round the edge of the haptor which presumably forms a good seal with the scale, so contraction of the muscle applies suction and attaches the haptor. When the muscle relaxes the haptor is released.

Entobdella can also attach itself for a while by the adhesive areas at the anterior end (Fig. 9.2). These seem not to attach by suction, but by the adhesive action of a secretion from their gland cells. They can attach quite firmly. *Entobdella* removed from soles will attach to the glass surface of a dish and cannot easily be dislodged by a jet of water from a pipette, even when attached only by the adhesive areas. They move about in the dish, attaching themselves by the adhesive areas while moving the haptor to a new site.

Large *Entobdella* are normally found only on the lower surface of the sole, so that they are not easily observed in an ordinary aquarium. Infected sole have been kept in glass-bottomed tanks and examined from below every few hours. It was found that the flukes moved about over their skin.

Feeding is most easily observed by removing flukes from a sole, starving them for a day and replacing them on the skin of a freshly killed sole. In these circumstances they generally feed soon. While attached by the haptor they protrude the pharynx and hold it for 5 minutes or so against a single spot on the skin of the host (Fig. 9.3 *b*). During this time, peristaltic movements of the pharynx draw fluid into the mouth.

Sole skin consists of a delicate superficial epidermis, typically about six cells thick, and a tougher underlying dermis which contains a feltwork of collagen fibres. The scales are embedded in the dermis. If the skin where an *Entobdella* has been feeding is examined afterwards a circular area is found to be damaged. Microscope sections through this area show that the epidermis has

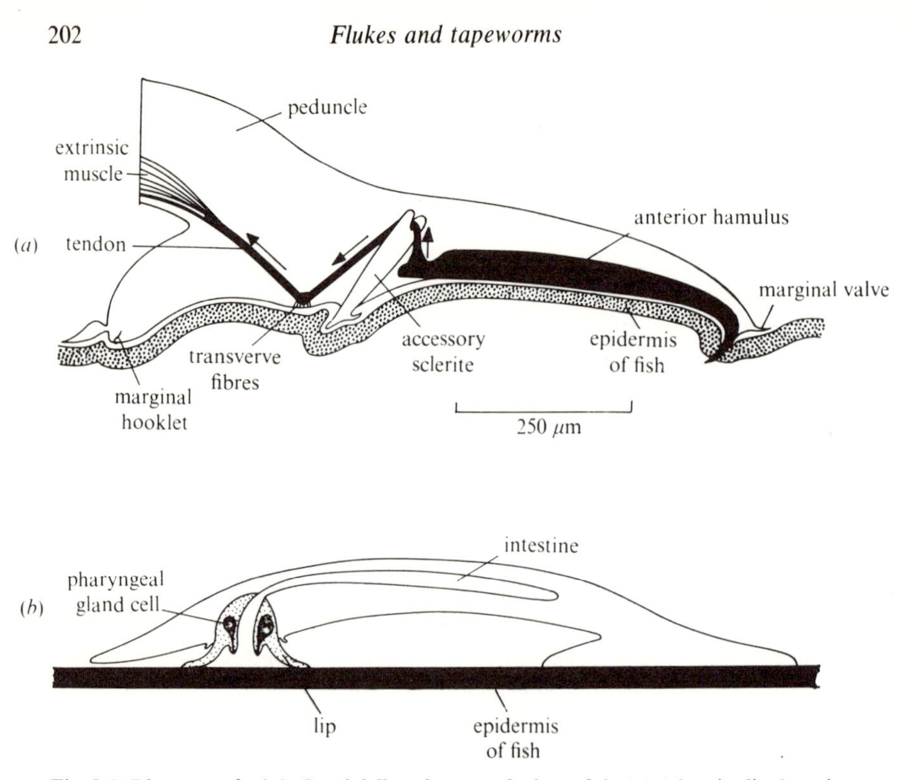

Fig. 9.3. Diagrams of adult *Entobdella soleae* attached to a fish. (*a*) A longitudinal section of the haptor, with arrows showing how the tendon moves when the extrinsic muscle contracts. (*b*) A longitudinal section through the whole worm. The pharyngeal lip is spread, not folded in as in Fig. 9.2. From G. C. Kearn (1971). In *Ecology and physiology of parasites*, ed. A. M. Fallis, pp. 161–87. Hilger, London.

vanished but that the dermis below it is intact. The epidermis has apparently been digested off. There are large gland cells in the pharynx and it is believed that they secrete a peptidase which breaks down the epidermal cells. Evidence that they contain a peptidase has been obtained by slicing through the head of an *Entobdella* and holding the cut surface against the gelatin layer of a piece of photographic film. The gelatin was dissolved where pharyngeal glands were cut through, but not elsewhere.

The wounds made by *Entobdella* apparently heal rapidly. Careful examination of a sole bearing over 100 *Entobdella* revealed only 21 of the tiny wounds, all but one of them near one of the parasites. This was an exceptionally heavily infected fish from an aquarium. Freshly-caught sole from the English Channel seldom have as many as 10 *Entobdella* on them.

Mating has been observed. When two *Entobdella* come in contact they touch heads repeatedly and then entwine with the penis of each near the vaginal opening of the other. The penis is not actually inserted but releases a spermatophore, a packet of sperm enclosed in a sheath of jelly. This attaches

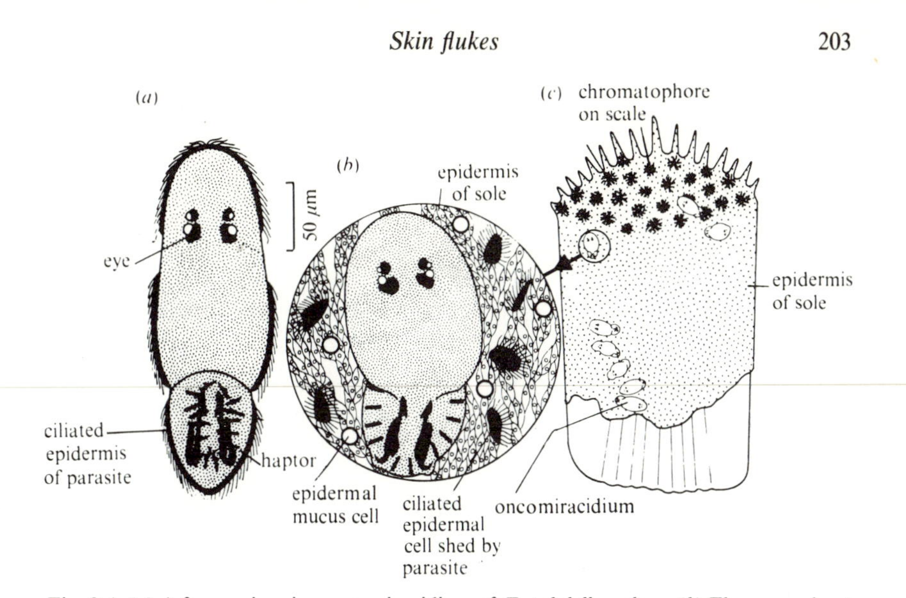

Fig. 9.4. (*a*) A free-swimming oncomiracidium of *Entobdella soleae*. (*b*) The same about 30 s after attachment to the skin of a sole. (*c*) A sole scale from the experiment described in the text, with eight *Entobdella larvae* attached to it. From G. C. Kearn (1967). *Parasitology*, **57**, 585–605.

to the partner and is drawn into the vaginal opening after the flukes have separated. There are other Monopisthocotylea which insert the penis into the vaginal opening when they copulate.

Active sperm have been observed in the ovaries of *Entobdella*, so fertilization presumably occurs there. Zygotes and yolk cells meet in the ootype, where each zygote is enclosed with a group of yolk cells in a quinone-tanned shell formed from materials released by the yolk cells. An egg is shown leaving the ootype in Fig. 9.2. It is tetrahedral with a long 'tail'. Mature *Entobdella* release about two eggs per hour. The eggs are denser than sea water, and their tails are sticky. When infected sole are kept in a bare glass tank the eggs stick to the glass bottom. Eggs can also be found stuck to sand grains, in sand from aquaria containing infected sole.

Ciliated larvae called oncomiracidia, about 250 μm long, hatch from the eggs (Fig. 9.4 *a*). They swim about at around 5 mm s^{-1} and can remain active, if they find no sole, for about a day.

Scales overlap each other in the skin of fish. Only the part of each scale which is not overlapped by another bears epidermis. Hatching *Entobdella* eggs have been put in dishes of sea water with sole scales, so that attachment could be observed. The oncomiracidia always settled on the part of the scale which was covered by epidermis. They attached first by the anterior end but then unfolded the haptor (which is initially folded) and attached by it. Within a minute the ciliated cells were discarded (Fig. 9.4 *b*) leaving the young fluke looking much more like a miniature adult.

Most Monogenea are strictly specific to a single host species, and *Entobdella* is no exception. Divers report that sole live in the sea in close proximity to other flatfish such as plaice (*Pleuronectes*) and dab (*Limanda*), but *Entobdella soleae* is found only on sole. Two specimens were once found on a ray (*Raia*, an elasmobranch fish) but this ray came from a tightly-packed trawl containing infected sole and it is suspected that the flukes moved onto the ray only in the net.

Do *Entobdella soleae* distinguish between sole and other fish and attach only to sole, or do specimens which settle on other species quickly die? Simple experiments have been performed to try to find out. Oncomiracidia were released in the centre of dishes of sea water, with scales of various species of fish arranged around the edge. Nearly all of them settled on sole scales. For instance, in an experiment in which they were given the choice between sole and dab scales, 49 settled on sole scales and only 2 on dab. In other experiments the larvae showed a clear preference for common sole (*Solea solea*) over thickback sole (*S. variegata*).

It was suspected that the larvae might be distinguishing between the species by chemical means, and a further experiment seemed to confirm this. Sheets of agar jelly were strapped to a sole and to a flounder (*Platichthys*) and left in place for half an hour so that any substance diffusing out of the skin would diffuse into the jelly. The jelly was then removed and put in dishes with *Entobdella* oncomiracidia, with the face which had not been against the fish uppermost. Oncomiracidia attached themselves readily to jelly from the sole but very few attached to jelly from the flounder or to clean jelly.

Adult *Entobdella* removed from sole will attach for a while to the skin of other species of flatfish but detach again in a day or so. Surprisingly, they remain attached to ray skin for longer.

Though *Entobdella* is in many ways typical of the Monopisthocotylea it should be remembered that it is only one species among many.

GILL FLUKES

The Polyopisthocotylea are easily distinguished from the Monopisthocotylea, because they have numerous small posterior attachment organs instead of a single large haptor (Fig. 9.5*a*). Most of them live on the gills of fish but *Polystoma* lives in the urinary bladder of frogs. Most species are found only on a single host species.

Diclidophora is typical of the order. Separate species live on the gills of whiting (*Merlangus merlangus*), coalfish (*Pollachius virens*) and pout (*Trisopterus luscus*). Each species tends to occupy a characteristic position on the gills of its host: *D merlangi* is usually found on the first gill of the whiting but *D. lusci* is usually on the second or third gill of the pout.

Fish gills have a complicated structure which gives them the very large surface area needed for respiration. Each gill consists of numerous more or

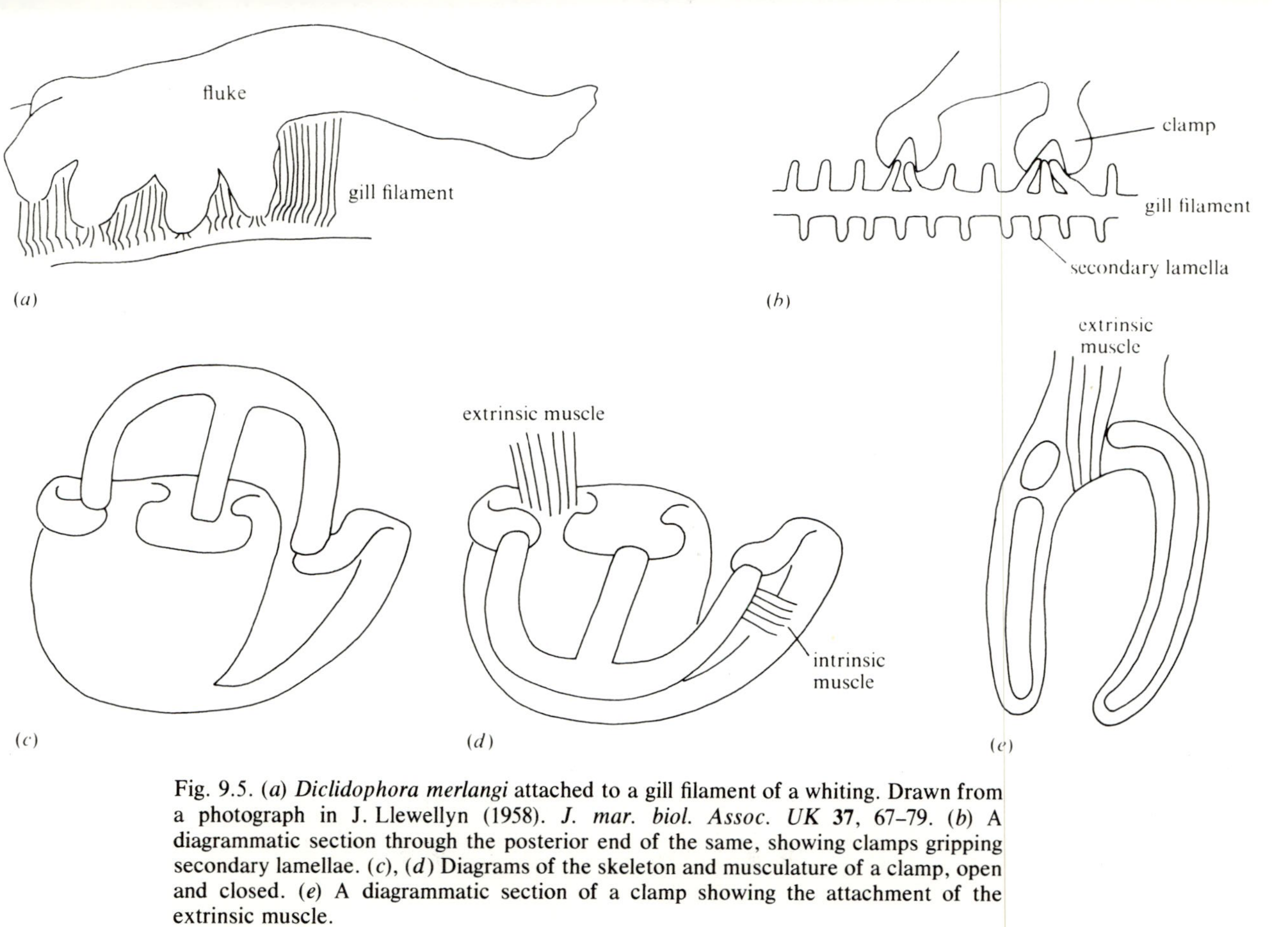

Fig. 9.5. (*a*) *Diclidophora merlangi* attached to a gill filament of a whiting. Drawn from a photograph in J. Llewellyn (1958). *J. mar. biol. Assoc. UK* **37**, 67–79. (*b*) A diagrammatic section through the posterior end of the same, showing clamps gripping secondary lamellae. (*c*), (*d*) Diagrams of the skeleton and musculature of a clamp, open and closed. (*e*) A diagrammatic section of a clamp showing the attachment of the extrinsic muscle.

less horizontal plates, the gill filaments, stacked one above the other. Smaller closely-spaced plates, the secondary lamellae, project up from the upper surfaces of the filaments and down from the lower surfaces (Fig. 9.5 *b*). A single large sucker could not obtain a firm grip on such a structure, but *Diclidophora* has eight small clamps which grip the secondary lamellae. Fig. 9.5 (*c*), (*d*) show a clamp open and closed. It has a relatively stiff skeleton of protein which presumably owes its stiffness to cross-links of some sort. However, histochemical tests show that it is not quinone-tanned, and it has not got disulphide linkages. This skeleton is formed as a pair of jaws which are hinged together and can be opened and closed by small intrinsic muscles. A *Diclidophora* detached from its host repeatedly opens the jaws and snaps them shut, apparently seeking a grip. A much larger extrinsic muscle enters the clamp through an aperture beside the hinge and attaches to the tissue covering the gripping surface (Fig. 9.5 *e*). By pulling on this tissue it reduces the pressure between the jaws and fixes the clamp more tightly shut.

Skin flukes feed on epidermis but gill flukes feed mainly on blood. In the secondary lamellae of most fish, less than 5 μm thickness of tissue separates the external water from the blood. This enables rapid diffusion of oxygen and carbon dioxide to occur between the water and the blood. It also makes the blood easily accessible to gill flukes. Microscope sections of *Diclidophora merlangi* taken from whiting showed mainly blood in the gut, with small numbers of cells from other tissues. The pharynx is not protrusible and has no gland cells, so digestion does not begin until the food is in the gut.

A large proportion of the potential food material in blood is made up by the protein haemoglobin. For instance, carp (*Cyprinus*) blood contains 6–7% by weight of haemoglobin and only 4% of plasma proteins. Haemoglobin contains 0.3% iron. *Diclidophora* digests haemoglobin, producing as a waste product the iron compound haematin (8% iron). Microscope sections show haematin granules, which can be identified by histochemical tests, in the cells of the gastrodermis. Some of the cells contain only a little haematin but others are tightly packed with the granules. There are gaps in the gastrodermis where there are no gastrodermal cells and the basement membrane is exposed. It seems that the gastrodermal cells gradually accumulate haematin. Old cells distintegrate leaving gaps in the gastrodermis which are later filled by new cells. Usable constituents of the disintegrated cells are probably digested and re-absorbed but the haematin is eventually voided through the mouth.

DIGENEAN FLUKES

Most of the Monogenea are ectoparasites of fish but the Digenea are, as adults, endoparasites of all classes of vertebrates including mammals and birds. They also have larval stages parasitic in molluscs (usually gastropods) and often have in addition other larval stages which are parasitic in other hosts. There are even species which pass through four host species in the course of their life cycle.

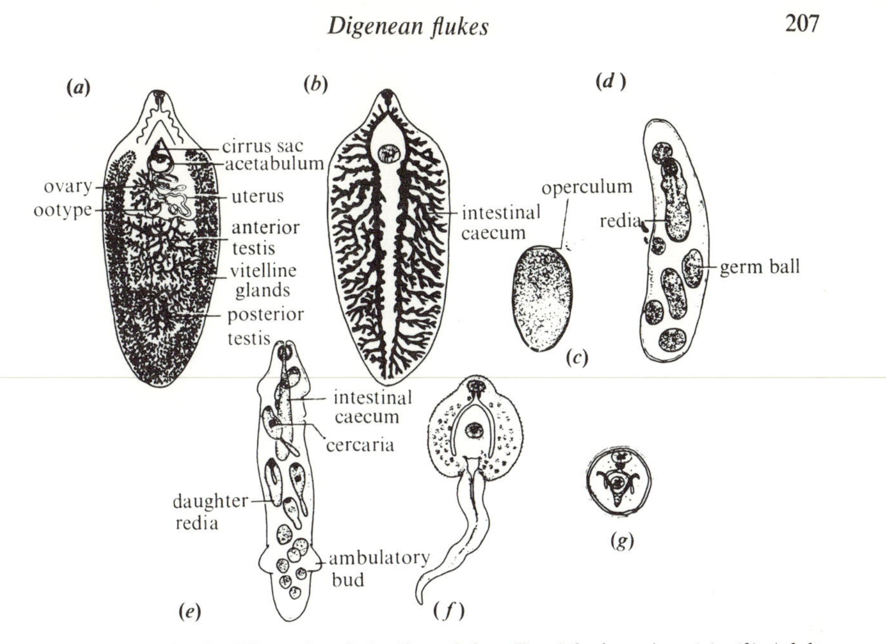

Fig. 9.6. Stages in the life cycle of the liver fluke, *Fasciola hepatica*. (*a*), (*b*) Adult specimens showing the reproductive system and the gut, respectively; (*c*) egg; (*d*) sporocyst; (*e*) redia; (*f*), cercaria; (*g*), metacercaria in cyst. From T. C. Cheng (1973). *General parasitology*. Academic Press, New York & London.

Several digenean flukes cause important diseases of man and his domestic animals. Three species of *Schistosoma* infect man and the disease they cause, schistosomiasis, is one of the most serious human diseases. It is widespread in the tropics and subtropics and it has been estimated recently that 200 million people suffer from it. Two species of *Fasciola*, the liver flukes, infect sheep and cattle all over the world. They also infect some wild mammals and occasionally man. They cause a disease called liver rot which kills many sheep in epidemics. They are also involved in another fatal disease of sheep, caused by bacterial infection of livers which they have damaged. In addition, very large numbers of animal livers in many countries are condemned as unfit for human consumption because they are infected with *Fasciola*. In 1962, 64% of cattle livers and 12% of sheep livers in N. Ireland abattoirs were condemned for this reason.

Fasciola hepatica is not a typical member of the Digenea for it is unusually large, but it will serve as an example. The adult is found in the bile ducts of its mammal hosts and grows to a maximum length of about 5 cm. Its structure is shown in Fig. 9.6(*a*) (which shows the reproductive system) and (*b*) (which shows the gut). There is no haptor at the posterior end of the body but there is an attachment organ called the acetabulum on the ventral surface and another round the mouth, which is at the anterior end of the body. The pharynx is not protrusible and has no gland cells. Otherwise the structure is much like that of *Entobdella* (Fig. 9.2). The body is leaf-shaped. The gut has a main branch down each side of the body, and side branches. There is a pair of diffuse yolk glands (vitelline glands), an ovary, two testes and a cirrus. (A cirrus is an organ with the function of a penis, but whereas a platyhelminth penis is simply lengthened by muscle action to make it protrude from the body a cirrus is protruded by being turned inside-out). *Fasciola* feeds mainly on blood but liver cells are also found in its gut.

The eggs are enclosed in quinone-tanned shells. They pass down the bile duct and eventually leave the host's gut in the faeces. While they remain in the faeces, little or no development occurs. If they are washed clean of faeces and remain moist, hatching occurs in a few weeks (depending on the temperature). The operculum (Fig. 9.6*c*) opens like a lid and a ciliated larva hatches out. This larva, the miracidium, can only survive a few hours unless it encounters and enters an appropriate snail of the genus *Lymnaea*. Different species of *Lymnaea* serve as hosts in different countries. In Britain the host is *L. truncatula* which flourishes in the muddy areas around gateways in damp fields, where there are puddles and deep hoof prints. Wet conditions like this particularly favour the spread of the disease.

If the miracidium encounters a suitable snail it bores its way in through the body wall. The damage to the snail tissues, which is shown by microscope sections of infected snails, indicates that the miracidia secrete enzymes and digest their way through. As the larva enters the snail it loses its cilia and enters the stage in the life cycle known as the sporocyst. This has no mouth and presumably absorbs soluble foodstuffs through its body wall. It is initially only about 70 μm long but it grows to a length of 500–700 μm and as it grows several larvae of the next stage develop inside it (Fig. 9.6*d*). These rediae have a mouth and a simple gut. They break through the wall of the sporocyst and escape from it. They travel to the snail's digestive gland where they feed on its tissues, fragments of which can be found in their guts. They grow and eventually reach a length of about 2 mm. Another generation of asexually produced larvae develop within the rediae: they may be the tailed larvae known as cercariae (Fig. 9.6*f*) or they may be a second generation of rediae. Cercariae and daughter rediae may even develop within the same redia as shown in Fig. 9.6(*e*). Laboratory experiments have shown that cercariae develop at normal room temperatures but the low temperatures promote development of rediae.

Cercariae have a two-branched gut, a sucker round the mouth and an acetabulum. They escape from the rediae in which they develop and congregate in a boil-like structure near the anus of the snail. They are squirted out from this when the snail is in water. They swim until they reach submerged vegetation where they attach, lose their tail to become a metacercaria and form a cyst protected by an outer layer of tanned protein (and several inner layers). The cyst can survive for several months. It need not remain submerged, provided the humidity is high. If it is swallowed by a grazing mammal it releases the metacercaria. Microscope sections of recently infected guinea pigs show the metacercariae boring through the wall of the gut. The metacercariae reach the liver and tunnel through it, eventually reaching the bile ducts. By this time they are adult flukes.

The life history of *Schistosoma mansoni* is rather different, and less typical of the class. The adults are 6–20 mm long and live in the veins of the hepatic portal system. They are not hermaphrodite but males and females are generally found together with the male clasping the more slender female in a groove in his body. The females lay eggs, each bearing a spine. These break through the walls of the veins and through other tissues, causing damage and inflammation which may be largely responsible for the ill-effects of the disease. Some of the eggs get into the gut and are voided with the faeces. They hatch if the faeces are diluted by water, releasing miracidia which infect freshwater snails. It has been shown that the snails release magnesium ions into the surrounding water and that these ions attract the miracidia. There are two generations of sporocysts in the snail, and eventually cercariae escape through the body wall (there is no redia stage). The cercariae do not encyst, but if they come in contact with the skin of a man they bore through it and infect him.

The cercariae have been watched entering human skin, and the skin of other animals. A cercaria placed on the skin in a drop of water crawls about exploring the skin, attaching itself alternately by the acetabulum and the mouth sucker. It usually effects an entry in a skin wrinkle or a hair follicle. In the process it wriggles and lashes its tail and discharges the contents of certain glands: these may contain enzymes which attack the skin. The average time required to enter human skin is about 7 minutes. Once in the skin, the cercaria sheds its tail. It makes its way to a blood vessel and is carried to the liver, whence it later moves into hepatic portal vessels.

EVASION OF THE IMMUNE RESPONSE

Endoparasites, particularly blood parasites, are exposed to attack by the immune response of the host. Trypanosomes evade attack by repeatedly changing the immunological character of their surface coat. Schistosomes have an entirely different technique.

Schistosoma mansoni infects rodents and monkeys as well as man. Man does not develop immunity to it but Rhesus monkeys (*Macaca mulatta*) infected with schistosomes are highly resistant to further infections. The worms already living in them survive and indeed flourish but any additional schistosome cercariae which are injected fail to establish themselves.

Experiments have been done in which worms grown in one animal have been transplanted surgically to the veins of another. Worms were transferred from one Rhesus monkey to another previously uninfected one. The faeces of the new host were examined and schistosome eggs were found in them within a few days, and regularly thereafter. Some of these monkeys were killed after a month and their veins were searched for schistosomes. Most of the worms which had been inserted were found, still alive. Worms transferred from mice to monkeys survived equally well but laid few eggs in the first month. Worms transferred from hamsters to monkeys laid very few eggs and most of them died within a fortnight. Plainly worms from other monkeys have an advantage not shared by worms from other species.

Evidence of the nature of the advantage was obtained by further experiments. Monkeys were immunized against mouse tissues by injections of mouse liver and spleen or of mouse red blood corpuscles. Schistosomes transplanted into these monkeys from other monkeys survived well, as would be expected. Ones transplanted from mice did not and if they were recovered even after only a few hours and examined by electron microscopy, they were found to have the tegument damaged.

This has been explained as follows. Schistosomes cover their bodies with a layer of host material so that their own antigens are not exposed to attack by host antibodies: in effect, the fluke disguises itself as a part of its host. Flukes from monkeys have a layer of monkey material and survive well, continuing to lay eggs, in other monkeys. Worms from mice have a layer of mouse material which offers no protection in monkeys, and is vigorously attacked by anti-mouse antibodies in monkeys immunized against mouse tissues. Further experiments indicate that the materials used for the immunological disguise may be glycolipids. (The A and B antigens of human blood belong to this group of compounds.)

When monkeys are infected by cercariae they are stimulated to develop anti-schistosome antibodies, before the flukes gather their coating of host material. By the time the antibodies are plentiful the worms have acquired their protective covering but any new invading cercariae will be attacked by the antibodies. The established parasite is safe from antibody attack and it is also protected from possible competition from later invasions of schistosomes.

TAPEWORMS

The tapeworms live as adults in the small intestine of vertebrates. *Taenia-rhynchus saginatus*, which infects man, is a slender worm up to 12 m long. This is considerably longer than the human small intestine, but since the worm is rather tangled the intestine can contain it. It has a larval stage which lives in cattle and the adult infects over 10% of the human population in parts of Africa where infected beef is eaten without adequate cooking. Though so large it usually does little harm, and may cause no symptoms other than pieces of worm in the faeces. In other cases it may cause pain, nausea, weakness and loss of weight. Most other species likewise have little obvious effect on the host.

Hymenolepis diminuta and *H. nana* have been studied more thoroughly than other tapeworms and will be used as examples of the group (Figs. 9.7, 9.9). Their adults live in rodents and are easily kept in the laboratory as parasites in rats or mice. *H. nana* also infects man. *H. nana* grows only to about 10 cm

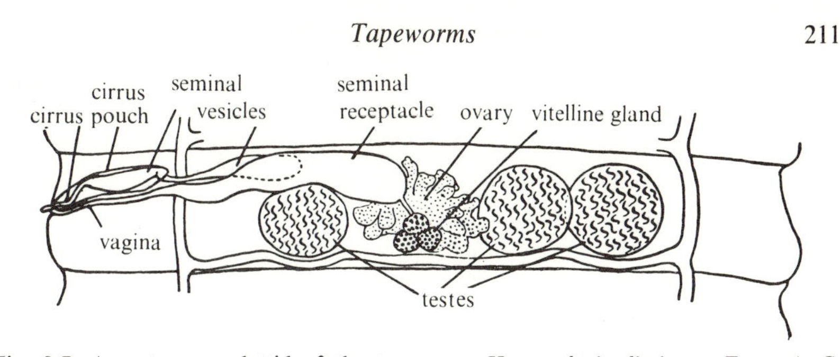

Fig. 9.7. A mature proglottid of the tapeworm *Hymenolepis diminuta*. From A. C. Chandler & C. P. Read (1961). *Introduction to parasitology*, 10th edn. Wiley, New York.

but *H. diminuta* grows to a maximum length of about 1 m, about the length of the small intestine of a 150 g rat. They consist of a scolex or attachment region and several hundred segments called proglottids. The proglottids are flattened so the worm is tape-like, but it is a tapering tape because proglottids near the scolex are narrower than the rest. The scolex has a ring of hooks in *H. nana* (but not in *H. diminuta*) and four suckers. The wall of the mammal intestine has finger-like projections (villi) with hollows (crypts of Lieberkühn) between them, and the scolex is lodged in one of the crypts (Fig. 9.8). The hooks penetrate the epithelium and serve as anchors. Also, host tissue is drawn into the suckers (tissue plug, Fig. 9.8). The proglottids lie in the lumen of the intestine, attached to the wall only through the scolex. The scolex is usually near the anterior end of the small intestine and the proglottids extend posteriorly down the intestine from it in rather tangled fashion.

There is no gut, but in other respects the internal structure of the proglottids is much like the internal structure of other platyhelminths. There is a tegument which is very like the tegument of monogeneans and trematodes (Fig. 9.1) except that it has microvilli on the external surface. There are circular and longitudinal muscle layers, dorso-ventral muscle fibres and a parenchyma. Several main nerve trunks run lengthwise along the body, from the scolex through all the proglottids. They are connected together by nerve rings in each proglottid. Two main excretory canals run lengthwise along the body from proglottid to proglottid and are connected together by a transverse canal in each proglottid. Ducts from flame cells open into them.

Though the nerve trunks and excretory canals run unbroken along the body each proglottid has a separate hermaphrodite reproductive system very much like the reproductive system of *Dugesia* (Fig. 8.4), *Entobdella* (Fig. 9.2) or *Fasciola* (Fig. 9.6).

The life history of *Hymenolepis nana* is illustrated in Fig. 9.9. Mature eggs are found in the proglottids furthest from the scolex. Proglottids containing mature eggs burst, releasing the eggs which are passed out in the host's faeces. Experiments have shown that the eggs will only hatch in the presence of carbon dioxide and peptidases, and that hatching is aided by prior breakage of the eggshell. In the natural habitat some of the eggs may get eaten by *Tribolium*

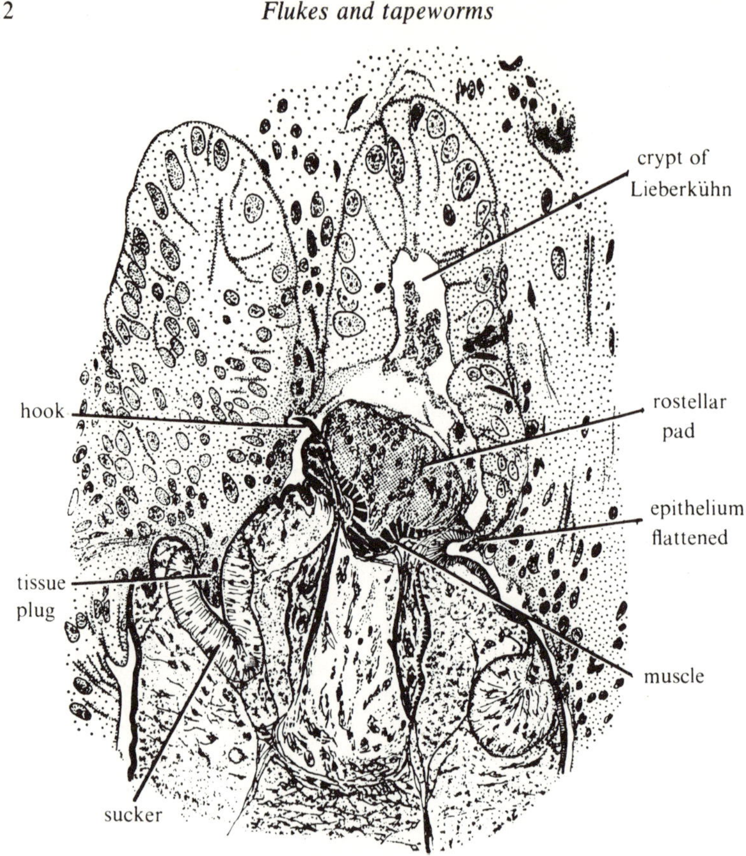

Fig. 9.8. A section through the scolex of *Hymenolepis nana* attached to the wall of the small intestine of a mouse. From J. D. Smyth (1969). *The physiology of cestodes*. Oliver & Boyd, Edinburgh.

(a grain beetle) or by certain other insects which can serve as alternative hosts. (Mice and grain beetles are of course found together in certain habitats.) The eggshell is likely to be broken by the insects' jaws and the enzymes and carbon dioxide in the gut stimulate hatching. The larva that hatches is an oncosphere, a more or less spherical larva with six projecting hooks. It passes through the gut wall (probably with the aid of enzymes) and enters the haemocoel where it develops to become a cysticercoid larva containing a scolex enclosed in a cavity. Cysticercoid larvae can be made to evaginate (i.e. the cavity can be made to open up, exposing the scolex) by putting them in a solution containing bile and the enzyme trypsin. If infected *Tribolium* are eaten by a mammal the cysticercoids are exposed to bile and trypsin in the small intestine and are stimulated to evaginate. The scolex attaches to the intestine wall and grows a long chain of proglottids, becoming an adult tapeworm. Though mice are primarily herbivorous they eat a wide range of foods as opportunity offers, and eat insects often enough for this life history to be feasible.

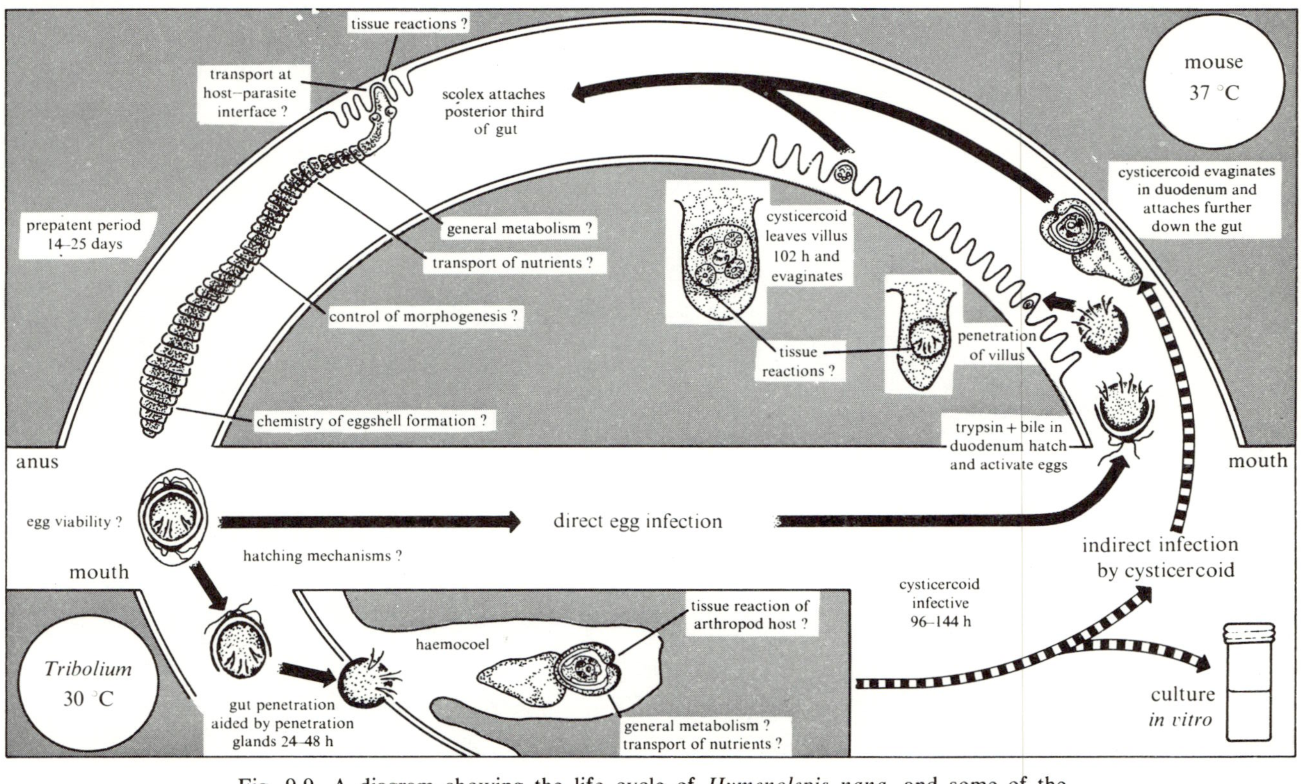

Fig. 9.9. A diagram showing the life cycle of *Hymenolepis nana*, and some of the physiological problems associated with it. From J. D. Smyth (1969). *The physiology of cestodes*, Oliver & Boyd, Edinburgh.

Nearly all tapeworms require at least two host species to complete their life cycle. The host for the adult is nearly always a vertebrate but the larvae may inhabit vertebrates or invertebrates: *Taeniarhynchus* has its larval stage in cattle, but *Hymenolepis* normally use insects. *Hymenolepis nana* is exceptional among tapeworms in that it can complete its life cycle in a single host species. If the eggs are swallowed by a mouse instead of an insect they nevertheless hatch, and the oncosphere invades the gut wall and develops to a cysticeroid there (Fig. 9.9). Mice are apt to eat mouse faeces when food is short, making themselves especially liable to infection.

As adult tapeworms grow, new proglottids form immediately behind the scolex, so the proglottid nearest the scolex is always the youngest one. The sequence of formation of proglottids is thus quite different from that of division of *Stenostomum* into a chain of individuals (Fig. 8.13) though the proglottids are in some ways comparable to the individuals of the chain.

NUTRITION AND METABOLISM OF TAPEWORMS

Carbohydrates seem to be particularly important for the nutrition of tapeworms. In one series of experiments rats infected with about 10 *Hymenolepis diminuta* each were fed equal rations (in terms of energy content) of various diets. After a week they were killed and the worms were dissected out and weighed. It was found that worms in rats fed on a protein-free, starch-rich diet had grown larger than those in rats fed a normal diet, or a protein-free fat-rich diet. Omission of proteins from the diet does not imply absence of peptides and amino acids from the intestine because the digestive enzymes of the host are proteins and because amino acids diffuse out of the gut wall into the lumen when their concentrations in the lumen are low.

Glycogen seems to be the principal food reserve of tapeworms and is often present in remarkably large quantities. For instance, it was found in one investigation that glycogen made up 46% of the dry weight of *Hymenolepis diminuta*.

Tapeworms have no gut and seem not to produce digestive enzymes. Nevertheless, host enzymes may work faster at the surface of a tapeworm. This possibility was demonstrated by experiments on the digestion of starch by the enzyme amylase, which breaks it down to sugars. A starch solution was made up and samples were kept at 37 °C with additions as follows:

(i) starch solution alone;
(ii) starch solution plus a tapeworm (*Hymenolepis* or *Moniezia*);
(iii) starch solution plus amylase;
(iv) starch solution plus amylase and a tapeworm.

After a period the solutions were analysed to find out how much starch remained in them. It was found that little or no starch had been destroyed in (i), and also in (ii): tapeworms by themselves have little or no ability to digest starch. A good deal of the starch had been destroyed in (iii) and even more

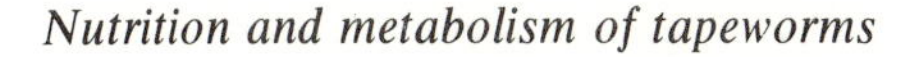

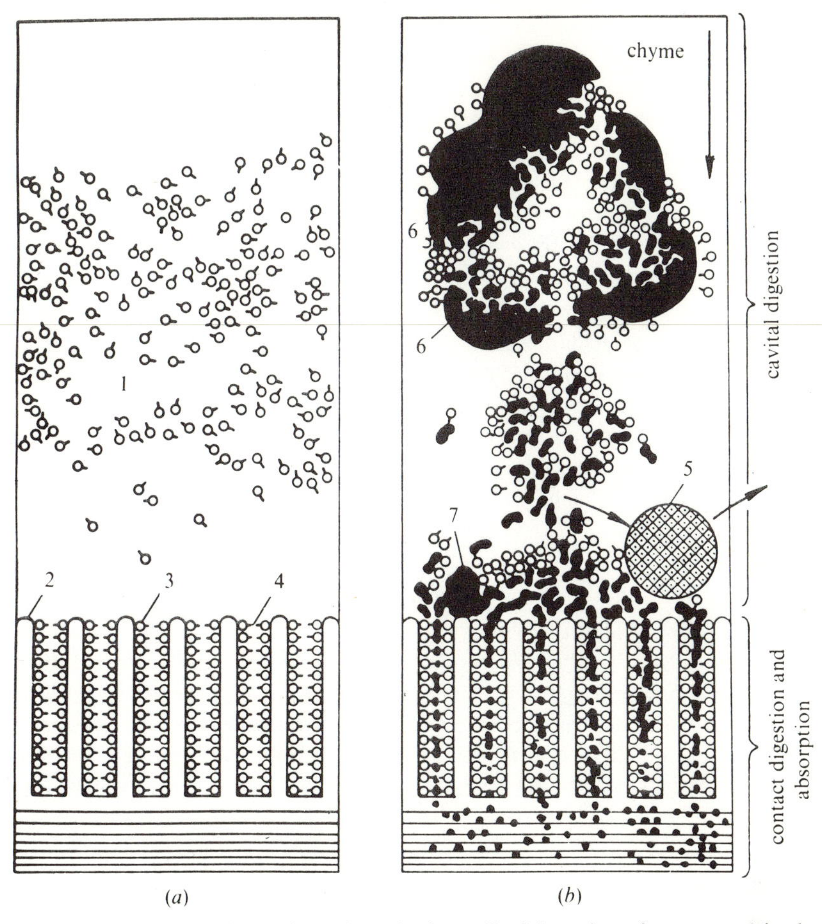

Fig. 9.10. Diagrammatic sections through the wall of intestine of a mammal in the absence (*a*) and presence (*b*) of food. *1*, enzyme molecules in lumen; *2*, microvilli; *3*, enzymes on the surface of microvilli; *4*, spaces between microvilli which cannot be entered by *5*, micro-organisms; *6*, *7*, food particles. From A. M. Ugolev (1965). *Physiol. Rev.* **45**, 555–95.

in (iv): presence of a worm as well as amylase increased the rate of digestion by 20–50%. Similar effects have been demonstrated with pieces of small intestine: various enzymes have been shown to work faster in the presence of pieces of small intestine, even if the pieces had been treated with fixatives to destroy their own enzymes.

Both the outer surface of tapeworms and the inner surface of the wall of the intestine bear microvilli. In each case the diameter of the microvilli is of the order of 100 nm and the length of the order of 1 μm. It is believed that digestive enzymes are adsorbed onto the microvilli and that this is responsible for the catalytic effect of the gut wall or of the tapeworm (Fig. 9.10). The phenomenon is known as membrane digestion. Certain inorganic catalysts such as spongy platinum are believed to work in similar fashion by adsorbing reagents on their surfaces.

If the microvilli of the gut wall do work in this way, much of the digestion may occur in the narrow spaces between them. If the products of digestion are taken up rapidly by the cells of the gut wall, little may diffuse out into the lumen of the gut and become available to tapeworms. The tapeworms seem to have overcome this difficulty by evolving their own microvilli which adsorb host enzymes, so that fast digestion also occurs at the surface of the tapeworm.

Each product of digestion will tend to diffuse into a tapeworm if its concentration C_l in the lumen is greater than its concentration C_w in the worm. The rate of diffusion will be proportional to $(C_l - C_w)$. Products of digestion might also be taken up by active transport, a process which uses energy and can operate against a concentration gradient. This would presumably involve an enzyme so its rate should depend on substrate concentration in the manner predicted by the Michaelis–Menten equation which applies to most enzyme-catalysed reactions.

$$V = V_{max}\, C_l / (K_m + C_l) \tag{9.1}$$

where C_l is the concentration of the substrate on which the enzyme is acting (in this case the digestion product in the gut lumen), V is the rate of the reaction (in this case the rate of uptake) and K_m is the Michaelis constant. If C_l is much smaller than K_m, V is more or less proportional to C_l. If, however, C_l is much larger than K_m, V is nearly equal to V_{max}, the maximum rate: there is plenty of substrate and the reaction proceeds as fast as the enzyme can operate. Further explanation of the equation can be found in textbooks of biochemistry. The equation can also be written

$$1/V = [(K_m/C_l) + 1]/V_{max} \tag{9.1 a}$$

so for reactions which proceed according to the equation, a graph of $1/V$ against $1/C_l$ is a straight line.

Fig. 9.11 shows the results of an experiment on glucose uptake by *Hymenolepis*. The tapeworms were removed from their rat hosts and put in a solution of radioactive glucose for 1 minute. They were removed and washed, and their radioactivity was measured. This indicated the amount of glucose which had been absorbed. The experiment was repeated with different concentrations of glucose and the graph of $1/V$ against $1/C_l$ is more or less a straight line. This result would not be obtained if glucose entered the worms mainly by diffusion, but it would if glucose were taken up mainly by active transport. Further experiments showed that the worms could take up glucose until the concentration in the body was greater than the concentration in the external solution. Plainly, active transport must occur. It has also been shown that amino acids are taken up by active transport.

All these experiments used tapeworms removed from their hosts. More elaborate experiments on tapeworms *in situ* in the intestines of rats showed

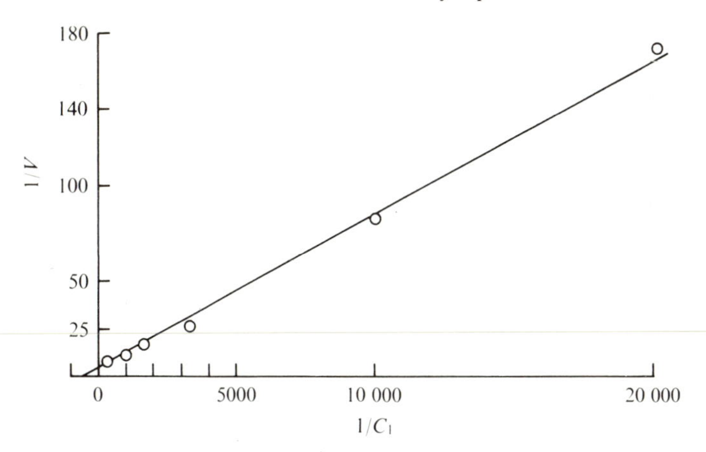

Fig. 9.11. A graph of the reciprocal of the rate of uptake of glucose (in arbitrary units) by *Hymenolepis*, against the reciprocal of the concentration of the glucose solution (in mol l^{-1}). From K. Phifer (1960). *J. Parasitol.* **46**, 51–62.

that glucose was taken up mainly by active transport but partly by the solvent drag effect: when the worm absorbs water, dissolved glucose tends to enter as well. Diffusion was relatively unimportant. Similarly the cells of the wall of the small intestine take up glucose mainly by active transport.

The small intestine is a habitat in which foodstuffs are often plentiful, the concentration of carbon dioxide is high and oxygen is often scarce. Dissolved oxygen is not completely absent for it diffuses in from the blood vessels of the wall of the intestine and is present in the secretions of the digestive glands. Tapeworms can use oxygen or survive without it, depending on its availability. Tapeworms of various species kept in well aerated solutions at 37 °C (the body temperature of their natural hosts) have been found to use oxygen at rates around 0.2 cm^3 (g body weight)$^{-1}$ h^{-1}. If there were free-living flatworms of similar size which lived at such high temperatures, they might be expected to use oxygen at about this rate. (See Fig. 8.6 which gives data for smaller flatworms at lower temperatures. Note that oxygen consumption per unit body weight tends to increase as temperature increases and to decrease as body size increases.) However, *Hymenolepis nana* has been grown from cysticercoid to maturity in solutions from which oxygen was excluded.

When oxygen is absent animals cannot get energy by oxidizing glucose, but they can get a smaller amount of energy from a given quantity of glucose by converting it to lactic acid. This does not require oxygen.

$$\underset{\text{glucose}}{C_6H_{12}O_6} = 2\,\underset{\text{lactic acid}}{CH_3.CHOH.COOH} + 2\sim$$

where ~ represents an energy-rich phosphate bond.

Some of the steps in the process, as it is usually performed by animals, are shown in Fig. 9.12(*a*). Oxidation of a molecule of glucose to two molecules of

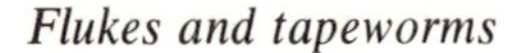

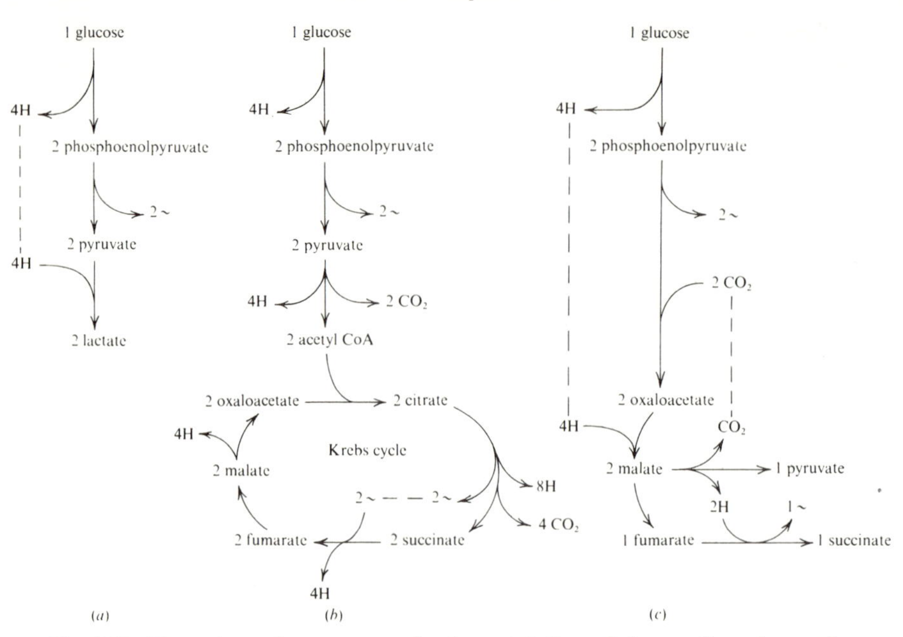

Fig. 9.12. Three alternative processes for the metabolism of glucose. In each case the products of metabolism of one molecule of glucose are shown. Some intermediate steps are omitted. (*a*) shows anaerobic conversion to lactate, (*b*) shows oxidation to carbon dioxide and water and (*c*) shows the anaerobic process involving fixation of carbon dioxide which occurs in tapeworms.

phosphoenolpyruvate releases four hydrogen atoms (attached to a coenzyme) which were used up again when two molecules of pyruvate are reduced to lactate. The intermediate step in which two molecules of phosphoenolpyruvate are converted to two molecules of pyruvate changes two molecules of ADP to ATP. Hence each molecule of glucose converted to lactate yields two energy-rich phosphate bonds which are available for use as required, as an energy supply.

Most animals oxidize glucose in the manner shown in Fig. 9.12(*b*). Glucose is oxidized to pyruvate as in Fig. 9.12(*a*) but the pyruvate is not reduced to lactate: instead it is oxidized to carbon dioxide and water by a circuit of the Krebs cycle. This rather complicated process is explained in textbooks of biochemistry and will not be explained in detail here. Conversion of two molecules of citrate to two molecules of succinate yields two energy-rich phosphate bonds which are destroyed again when the succinate is converted to fumarate. Apart from this, oxidation of a molecule of glucose yields two energy-rich phosphate bonds and 24 hydrogen atoms (attached to coenzyme molecules). Oxidation of each pair of hydrogen atoms to water yields three energy-rich phosphate bonds so the final yield is $2+(3\times 12)=38$ energy-rich phosphate bonds.

Tapeworms excrete lactate which seems to be produced by the process shown in Fig. 9.12(*a*). It is known that tapeworms use oxygen when it is available and there is some evidence that *Echinococcus* (and perhaps other tapeworms) oxidize glucose in the manner illustrated in Fig. 9.12(*b*). There is also another major process whereby tapeworms get energy from glucose. It uses carbon dioxide (which is present in high concentrations in the small intestine) and produces succinate. Tapeworms often excrete as much succinate as lactate.

This strange process is shown in Fig. 9.12(*c*). Its details have been worked out by measurements of enzyme activity, and by experiments with radioactive materials. *Hymenolepis* were kept in anaerobic conditions in solutions containing $NaH^{14}CO_3$ or glucose incorporating ^{14}C at one position only in the molecule. They excreted radioactive succinate which was analysed to discover which position in its molecule was occupied by ^{14}C. It was concluded that glucose is oxidized to phosphoenolpyruvate which is then converted to oxaloacetate by combination with carbon dioxide. The oxaloacetate is in turn converted to malate by the reverse of a reaction which occurs in the Krebs cycle. Some of the malate is converted by further reversed Krebs cycle reactions through fumarate to succinate, a process of reduction. Some is oxidized and loses carbon dioxide to become pyruvate. The oxidations which occur at some stages in this scheme are balanced by reductions at others, so no oxygen is required. The whole process yields three energy-rich phosphate bonds per molecule of glucose, one bond more than conversion to lactate. The whole process is

$$\underset{\text{glucose}}{C_6H_{12}O_6}+CO_2=\underset{\substack{\text{succinic}\\ \text{acid}}}{(CH_2.COOH)_2}+\underset{\substack{\text{pyruvic}\\ \text{acid}}}{CH_3.CO.COOH}+H_2O+3\sim$$

It does not seem to be clear what finally happens to the pyruvic acid.

The lactate and succinate which are produced by anaerobic metabolism are excreted by the protonephridial system. The excretory canals of *Hymenolepis* are up to 0.5 mm in diameter so it is possible to insert micropipettes into them and draw out samples of fluid for analysis. It has been discovered in this way that the main cation in the fluid is sodium and the main anions are chloride, lactate and succinate. In the conditions of the experiment the concentration of each of these anions was about 50 mEquiv l^{-1}.

Rats infected with *Hymenolepis* have been anaesthetized and opened so that the worms could be observed. Peristaltic waves travelled along the worms, squeezing the contents of the excretory canals towards the free end of the worm and so, eventually, out of the body. Colloidal graphite was injected into the canals and its subsequent movement observed, to find out how fast the fluid was being passed. It was found that the rate was sufficient to account for most or all of the succinate and lactate excretion. So far as is known, succinate and lactate excretion is the main function of the protonephridial system in tapeworms. It might be expected by analogy with rotifers (p. 192) that the system would be responsible for osmotic regulation, but tapeworms seems to

TABLE 9.1. *Food intake and growth during 24 h of 150 g rats and of tapeworms* (Hymenolepis diminuta) *in some of them*

	Uninfected	Infected
Food intake (g)	11.54	10.91
Increase in rat weight (g)	2.17	1.77
Increase in tapeworm weight (g)	—	0.10

Data from D. F. Mettrick (1973). *Can. J. Public Health* **64**, monograph supplement, 70–82.

be incapable of osmotic regulation. *Hymenolepis* swells in dilute solutions and shrinks in concentrated ones.

To survive in the intestine, tapeworms need some protection against digestive enzymes. Since their outer surfaces consist of protein and lipid, protection against peptidases and lipases is necessary. There is experimental evidence of such protection. It has been shown that solutions of trypsin (one of the peptidases of mammal intestines) digest protein less fast if a *Hymenolepis diminuta* is left in them for a while, and then removed, before the protein is added. The effect is probably not simply due to adsorption of the enzyme on the worm, as the same worm put successively into a series of trypsin solutions has the same effect on each. Similar results have been obtained in experiments with α- and β-chymotrypsin (which are other intestinal peptidases).

Most tapeworms producc no obvious symptoms of disease but they must deprive their host of nutrients. An experiment was carried out with rats and *Hymenolepis diminuta*, to find out how severe the deprivation was (Table 9.1). One group of rats was infected with *Hymenolepis* which were allowed to grow to maturity, and another group was kept free from infection. Both groups were allowed as much food as they wanted and the amount they took was recorded. It was about the same in both cases, and analysis of the faeces showed that both groups excreted the same fraction of the energy intake. The infected rats grew 0.4 g per day less than the uninfected ones but their worms grew only 0.1 g per day. Some of the lost growth must have been due to energy used in the metabolism of the tapeworms. Also, tapeworms alter the pH and chemical composition of the gut contents and may increase the energy the host has to use in absorbing products of digestion from the lumen.

FURTHER READING

GENERAL

Cheng, T. C. (1973). *General parasitology.* Academic Press, New York & London.

Erasmus, D. A. (1972). *The biology of trematodes.* Edward Arnold, London. (This book includes the monogeneans among the trematodes.)

SKIN FLUKES AND GILL FLUKES

Halton, D. W. & Jennings, J. B. (1965). Observations on the nutrition of monogenetic trematodes. *Biol. Bull.* **129**, 257–72.
Kearn, G. C. (1971). The physiology and behaviour of the monogenean skin parasite *Entobdella soleae* in relation to its host (*Solea solea*). In *Ecology and physiology of parasites*, ed. A. M. Fallis, pp. 161–87. Hilger, London.
Llewellyn, J. (1958). The adhesive mechanisms of monogenetic trematodes: the attachment of species of the Diclidophoridae to the gills of gadoid fishes. *J. mar. Biol. Assoc. UK* **37**, 67–79.

DIGENEAN FLUKES

Erasmus, D. A. (1977). The host–parasite interface of trematodes. *Adv. Parasitol.* **15**, 201–42.
Kendall, S. B. (1965). Relationships between the species of *Fasciola* and their molluscan hosts. *Adv. Parasitol.* **3**, 59–98.
Halton, D. W. (1967). Observations on the nutrition of digenetic trematodes. *Parasitology* **57**, 639–60.
Pantelouris, E. M. (1965). *The common liver fluke.* Pergamon, Oxford.

EVASION OF THE IMMUNE RESPONSE

Clegg, J. A. (1972). The schistosome surface in relation to parasitism. *Symp. Br. Soc. Parasitol.* **10**, 19–40.
Smithers, S. R., Terry, R. J. & Hockley, D. J. (1969). Host antigens in schistosomiasis. *Proc. R. Soc.* **171B**, 483–94.

TAPEWORMS

Arme, C. (1975). Tapeworm–host interactions. *Symp. Soc. exp. Biol.* **29**, 505–32.
Bryant, C. (1975). Carbon dioxide utilization and the regulation of respiratory pathways in parasitic helminths. *Adv. Parasitol.* **13**, 35–69.
Mettrick, D. F. (1973). Competition for ingested nutrients between the tapeworm *Hymenolepis diminuta* and the rat host. *Can. J. Public Health* **64**, monograph supplement, 70–82.
Mettrick, D. F. & Podesta, R. B. (1974). Ecological and physiological aspects of helminth–host interactions in the mammalian gastrointestinal canal. *Adv. Parasitol.* **12**, 183–278.
Pappas, P. W. & Read, C. P. (1972). Trypsin inactivation by intact *Hymenolepis diminuta. J. Parasitol.* **58**, 864–71.
Scheibel, L. W. & Saz, H. J. (1966). The pathway for anaerobic carbohydrate dissimilation in *Hymenolepis diminuta. Comp. Biochem. Physiol.* **18**, 151–62.
Smyth, J. D. (1969). *The physiology of cestodes.* Oliver & Boyd, Edinburgh.
Taylor, E. W. & Thomas, J. N. (1968). Membrane (contact) digestion in the three species of tapeworm *Hymenolepis diminuta, Hymenolepis microstoma* and *Moniezia expansa. Parasitology* **58**, 535–46.
Webster, L. A. (1971). The flow of fluid in the protonephridial canals of *Hymenolepis diminuta. Comp. Biochem. Physiol.* **39A**, 785–93.
Webster, L. A. (1972). Succinic and lactic acids present in the protonephridial canal fluid of *Hymenolepis diminuta. J. Parasitol.* **58**, 410–11.

10
Nemertean worms

Phylum Nemertea (Rhynchocoela)

Fewer than 1000 species of Nemertea have been described so the phylum is a small one. None of the species is particularly conspicuous and none seems to play a major role in any ecosystem. The phylum is interesting enough to deserve a chapter in this book, but the chapter is a short one.

Most nemertean worms are long and slender, dorso-ventrally flattened (Fig. 10.1). Some are astonishingly long. The longest known is the bootlace worm, *Lineus longissimus*, which is quite commonly found under stones exposed at low tide on W. European shores. It is only a few millimetres wide and it is generally found in a fairly compact tangled mass but when specimens are unravelled they are often found to be several metres long. One specimen was estimated to be 30 m long. This was presumably its fully extended length: nemerteans can make themselves longer and thinner, or shorter and fatter, to a remarkable degree. A section of this chapter is about their ability to change their length.

In many features of their internal construction, nemerteans are very like turbellarians. A parenchyma fills the space between epidermis and gastrodermis. There are layers of circular and longitudinal muscle fibres, and dorso-ventral muscles. There is a protonephridial system. The principal components of the nervous system are longitudinal nerves starting from a ring of ganglia at the anterior end of the body. Two of these nerves, one on each side of the body, are particularly large.

There are also some striking differences from the turbellarians. The reproductive system is quite simple and only a few species are hermaphrodite: most have separate sexes. There is an anus at the posterior end of the body as well as a mouth at the anterior end. There is a simple blood system. Finally, there is a characteristic proboscis which is normally housed in a cavity in the body but can be extended to attack prey (Fig. 10.1 *b*).

Details of proboscis structure (and also of the structure of other organs) differ from group to group. Fig. 10.2 is based largely on *Paranemertes*, the common Pacific nemertean which is also illustrated in Fig. 10.1 (*a*).

The proboscis cavity is about one quarter the length of the body, in *Paranemertes*. It contains the proboscis which is a tube closed at one end like the finger of a glove. When it is protruded, the proboscis is turned inside-out

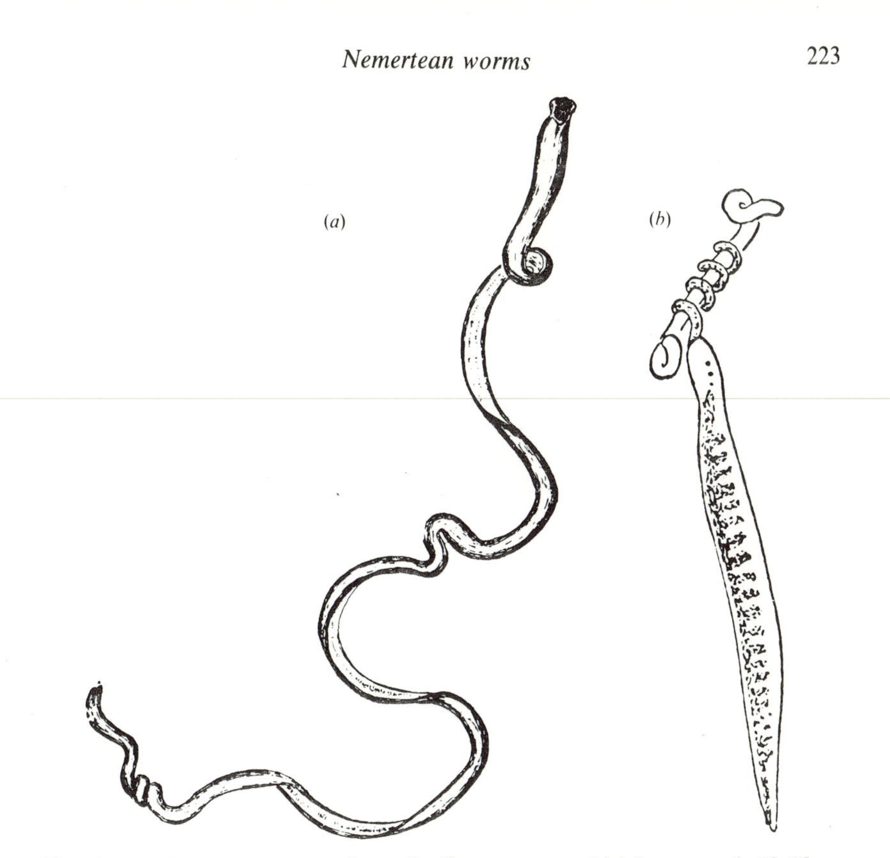

Fig. 10.1. (*a*) *Paranemertes peregrina*, a Pacific nemertean which is commonly 10–15 cm long. (*b*) *Prostoma*, another nemertean, capturing an annelid worm by means of its proboscis. From L. H. Hyman (1951). *The invertebrates*, vol. 2. McGraw-Hill, New York.

(Fig. 10.2 *b*). This is presumably caused by the muscles of the wall of the proboscis cavity which squeeze the cavity and force the proboscis out. The proboscis is drawn back into the cavity by the proboscis retractor muscle. *Paranemertes* and many other nemerteans have a stylet, a hard needle-like structure. Species which have a stylet do not turn the proboscis completely inside-out, but only as far as the stylet.

The proboscis has a muscular wall so it can be coiled round prey (Fig. 10.1 *b*), which are stabbed with the stylet. Polychaete worms attacked by *Paranemertes* are often paralysed in a few seconds. It is presumed that a toxin is secreted by gland cells in the posterior part of the proboscis (the part which is not turned inside-out) and gets squeezed into wounds made by the stylet.

Most nemerteans live near shores, crawling about on the bottom. Most do not allow themselves to be exposed at low tide but hide under stones or weed, or in crevices. *Paranemertes* is unusual. It crawls over exposed mud at low tide, blundering into potential prey. It seems to have no means of finding prey except by colliding with it. In habitats where polychaete worms of the family Nereidae

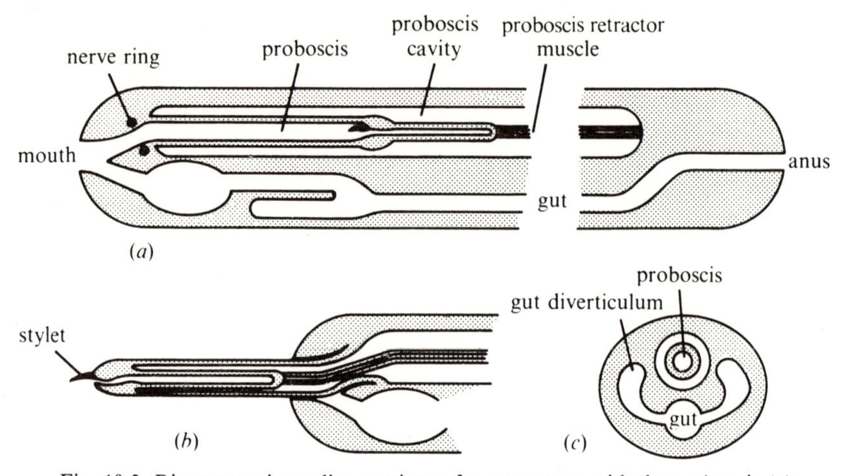

Fig. 10.2. Diagrammatic median sections of a nemertean with the proboscis (*a*) contracted and (*b*) extended, and (*c*) a transverse section.

are plentiful it feeds almost entirely on them. It can swallow surprisingly large worms. Digestion occurs very much as in turbellarians: extracellular digestion by endopeptidases is followed by intracellular digestion by other enzymes. Faeces are passed 12–36 hours after a meal. Freshly-caught *Paranemertes* have been put into sea water in separate, clean containers and any faeces they passed collected and examined. The abundance of chaetae (bristles) of nereid worms in these faeces showed the importance of nereids in the diet. Other nemerteans which have been studied eat more varied food.

The blood system is very simple. In most nemerteans it consists of two or three blood vessels running the length of the body, with connections between them. The major vessels have a layer of muscle in their walls. Contractions of this muscle and of the main body muscles move the blood along the vessels, but not in a constant direction: the blood may flow in either direction in each vessel. The blood is colourless in most species and has cells (blood corpuscles) suspended in it. Its functions are unknown.

LOCOMOTION

A nemertean cannot change the volume of fluid inside itself in the way a sea anemone can. It can, however, change its shape. It can make itself long and thin by contracting its circular muscles or short and fat by contracting its longitudinal ones. It can also make parts of the body thin and parts fat, by contracting circular muscles in some parts of the body and longitudinal ones in others. This happens in crawling. Fat regions form at the anterior end and pass posteriorly along the body. The peristaltic waves travelling backward drive the body forward. In the film sequence shown in Fig. 10.3 and analysed in Fig.

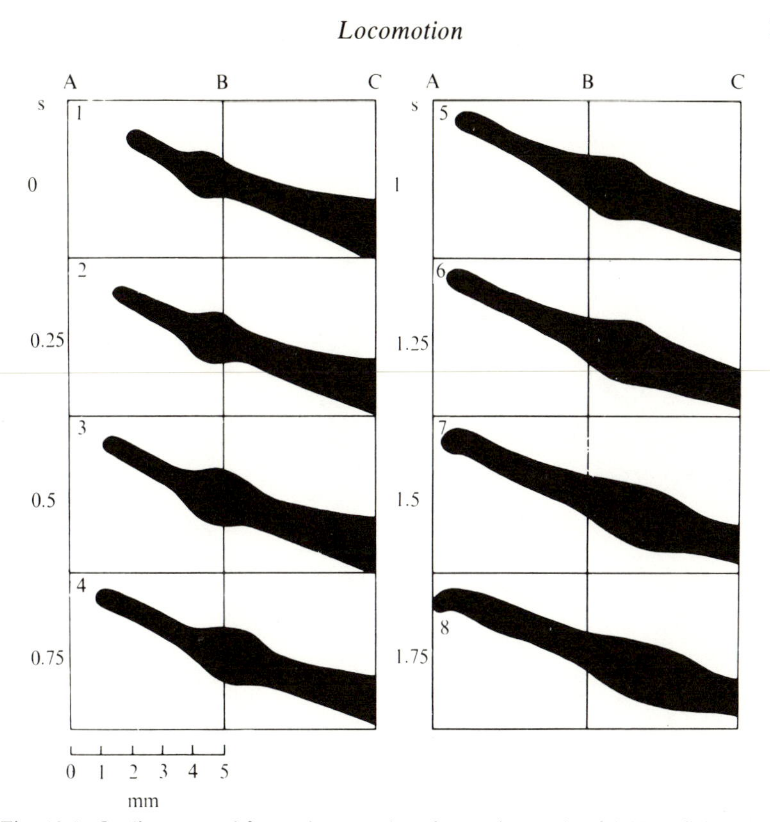

Fig. 10.3. Outlines traced from photographs taken at intervals of 0.25 s of the anterior end of a *Lineus* crawling. From J. Gray (1968). *Animal locomotion.* Weidenfeld & Nicolson, London.

10.4, the waves were travelling posteriorly at 4.7 mm s^{-1} relative to the ground while the head of the worm travelled forward at 1.7 mm s^{-1}. (The waves were therefore travelling at $4.7+1.7=6.4$ mm s^{-1} relative to the worm.) Fig. 10.4 shows the progress of marked points on the body. Note that their speed was not constant. They were stationary or even slid back a little whenever they were fat, and advanced faster than 1.7 mm s^{-1} whenever they were thin.

Fig. 10.5 is a grossly simplified diagram of a crawling worm. The body is shown divided into imaginary segments, numbered 1, 2, 3, etc. (the bodies of nemerteans are not segmented but those of earthworms, which crawl in the same way, are). Each segment is shown either long and thin, or short and fat. Between the instants shown in Fig. 10.5(*a*) and (*b*), the peristaltic waves have moved posteriorly one segment: segment 3 has elongated and segment 9 has shortened. Since segments 4 to 8 have not moved, segments 1 and 2 have been pushed forward by the elongation of 3 and segments 10, 11, etc. have been drawn forward by the shortening of 9. Segments 4 to 8 have been kept stationary by friction: the fat segments support most of the weight of the worm. This

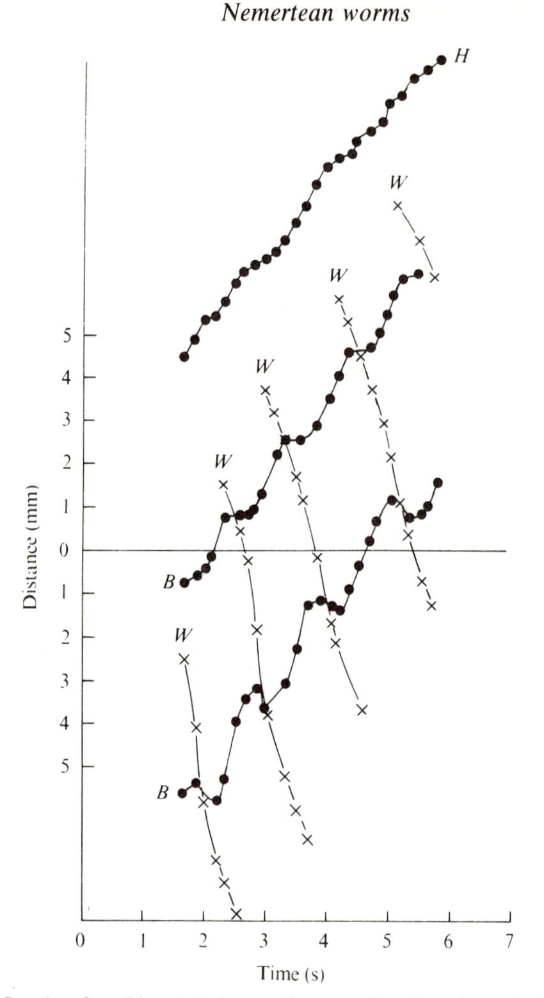

Fig. 10.4. Graphs of position (relative to the ground) against time, obtained from the same sequence of photographs as Fig. 10.3. The line H shows the position of the head, lines B show the positions of two marked points on the body and lines W show peristaltic thickenings of the body. From J. Gray (1968). *Animal locomotion.* Weidenfeld & Nicolson, London.

technique of locomotion is very effective for burrowing, since the fat segments are jammed against the wall of the burrow.

Let the worm crawl forward with velocity u, by means of peristaltic waves which travel posteriorly at velocity U relative to the ground. Let each segment have length l when contracted and length $(l+\Delta l)$ when elongated, and let a fraction q of the segments be elongated at any instant. Fig. 10.5(*b*) shows the worm at a time l/U after Fig. 10.5(*a*); the waves travelling with velocity U have moved a distance l relative to the ground. In this interval of time the head of the worm has advanced a distance Δl, due to elongation of segment 3, so the

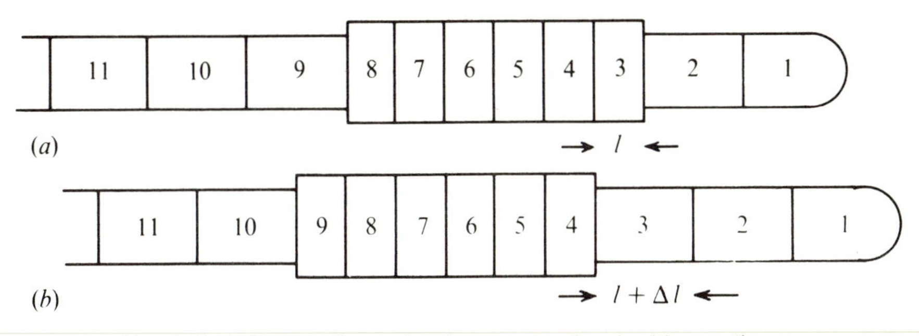

Fig. 10.5. A diagram illustrating the account of peristaltic crawling.

velocity of segment 1 is $U \cdot \Delta l/l$. Segment 1 has this velocity for a fraction q of the time, but for the remaining fraction $(1-q)$ it is shortened and stationary. Hence the mean velocity of segment 1, and of the whole worm, is given by the equation

$$u = qU \cdot \Delta l/l \tag{10.1}$$

A worm crawling in this manner can increase its speed by making the peristaltic waves travel faster (increasing U), by making each part of the body extend and shorten more (increasing $\Delta l/l$) or by increasing the fraction of the body which is extended at any instant (increasing q).

Nemerteans crawl over solid surfaces, whether they are dry or submerged. Some also swim by movements like those of a flagellum (Fig. 2.7) or an eel. However, swimming eels bend their bodies from side to side but nemerteans bend their bodies dorsally and ventrally. This seems the appropriate way for a dorso-ventrally flattened worm to swim: lateral bending would be relatively ineffective because the narrow edges of the worm would impinge on the water.

CHANGES OF LENGTH

As a nemertean crawls, each part of its body lengthens and shortens. The whole body can also change in length, and to a remarkable extent. For instance, the length of specimens of *Amphiporus* was measured after they had been poked repeatedly to make them shorten as much as possible. They were measured again after being anaesthetized in a magnesium chloride solution and stretched. The stretched length was found to be five times the contracted length. The limits of extension and contraction seem to be set by connective tissue fibres (possible collagen) which form a basement membrane immediately under the epidermis. These fibres cannot be stretched appreciably, even by pulling on them with a micromanipulator until they break. They are laid down in layers. All the fibres in each layer are parallel to each other and the fibres of successive layers run round the body in left-handed and right-handed geodesics which resemble helices (they are not strictly helices because the body is not circular in cross-section).

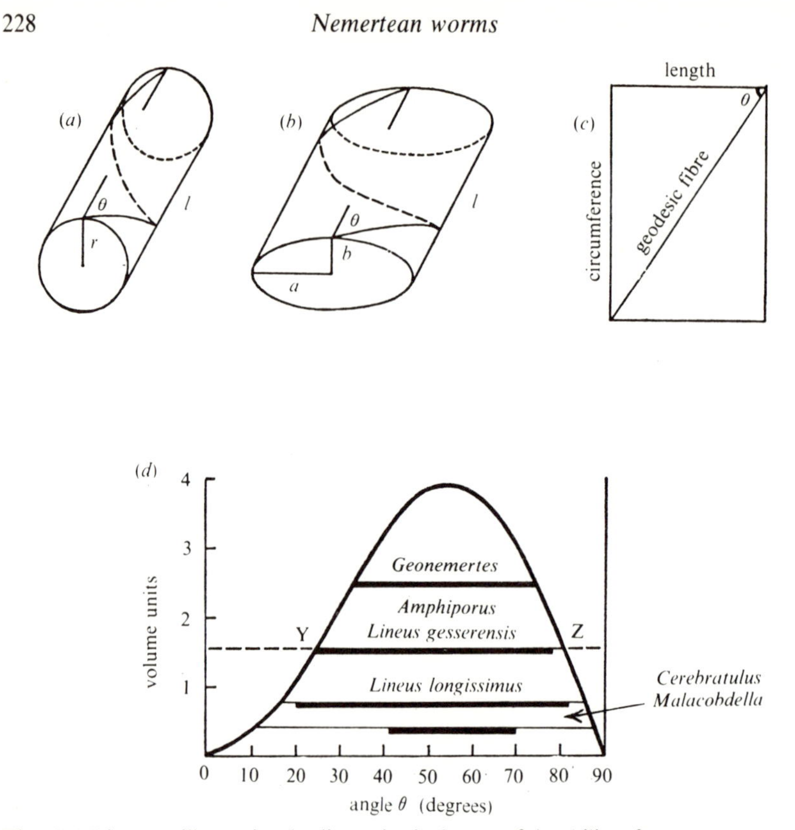

Fig. 10.6. Diagrams illustrating the discussion in the text of the ability of nemertean worms to change their length. From R. Gibson (1972). *Nemerteans*. Hutchinson, London.

If an *Amphiporus* is anaesthetized it adopts a length intermediate between its maximum and minimum lengths. If it is fixed at this length and sectioned for microscopy, it is found that the fibres of the basement membrane run at about 55° to the long axis of the body. If it is anaesthetized, stretched as far as possible and fixed while stretched, the angle is about 23°. If it is fixed in the fully contracted state the angle is 70–80°. The latticework of fibres allows the worm to lengthen and shorten by changes of fibre angle, just as a pair of lazy tongs is extended and shortened by changing the angle of its lattice.

What limits should the fibres set to changes of length? Consider a cylindrical tube of radius r, length l (Fig. 10.6). A fibre wound around it at an angle θ to its long axis makes exactly one turn of a helix. Let the length of this fibre be D. Now imagine the tube cut along its length and unrolled, making a rectangle of diagonal D. The sides of the rectangle are $2\pi r$ (the circumference of the original tube) and l, so

$$l = D\cos\theta \tag{10.2}$$

and

$$2\pi r = D\sin\theta$$

whence

$$r = (D/2\pi)\sin\theta \tag{10.3}$$

The volume V of the cylinder is given by

$$V = \pi r^2 l \tag{10.4}$$

Inserting the values of l and r from equations (10.2) and (10.3)

$$V = (D^3/4\pi) \sin^2\theta \cos\theta \tag{10.5}$$

The curve in Fig. 10.6(*d*) is a graph of V against θ for constant D, calculated from this equation. V is of course zero when $\theta = 0$ (which makes $r = 0$) and when $\theta = 90°$ (which makes $l = 0$). V has a maximum value when $\theta = 55°$.

Now apply this geometry to a nemertean worm. V is not the volume of the worm but the maximum volume which the lattice of fibres can contain, at given θ. When V is greater than the volume of the worm, the worm adopts a non-circular cross-section (Fig. 10.6*b*). Transverse sections of anaesthetized *Amphiporus* are approximately elliptical, of width five times their height. The area of an ellipse of these proportions is only 0.38 that of a circle of the same circumference. $\theta \simeq 55°$ in anaesthetized *Amphiporus*. Hence the volume of *Amphiporus* is 0.38 times the value which V has when $\theta = 55°$. The line YZ has been drawn at this level in Fig. 10.6(*d*). It intersects the graph of V when $\theta = 24°$ and when $\theta = 82°$. Hence when $\theta = 24°$ and when $\theta = 82°$ the lattice can only just contain the worm and the cross-section of the worm must be circular. These angles correspond to the maximum and minimum lengths of the worm. For all intermediate angles and lengths V is greater than the volume of the worm and the cross-section is flattened. These geometrical predictions agree quite closely with the observations on extended and contracted worms.

The thick horizontal lines in Fig. 10.6(*d*) show the volumes of various nemerteans and the ranges of fibre angle which occur between their fully extended and fully contracted states. For all except *Cerebratulus* and *Malacobdella* there is reasonably good agreement between the observed range of angles and the range indicated by the graph of V. The ranges of length of *Cerebratulus* and *Malacobdella* must be limited in some other way.

FURTHER READING

Clark, R. B. & Cowey, J. B. (1958). Factors controlling the change of shape of certain nemertean and turbellarian worms. *J. exp. Biol.* **35**, 731–48.

Cowey, J. B. (1952). The structure and function of the basement membrane muscle system in *Amphiporus lactifloreus* (Nemertea). *Q. J. microsc. Sci.* **93**, 1–15.

Gibson, R. (1972). *Nemerteans*. Hutchinson, London.

Roe, P. (1976). Life history and predator–prey interaction of the nemertean *Paranemertes peregrina* Coe. *Biol. Bull.* **150**, 80–106.

Roe, P. & Gibson, R. (1970). The nutrition of *Paranemertes peregrina* (Rhynchocoela; Hoplonemertea) [2 papers]. *Biol. Bull.* **139**, 80–106.

11
Rotifers

Phylum Rotifera
  Class Seisonacea
  Class Bdelloidea
  Class Monogononta

The rotifers are among the smallest of the multicellular animals. Most are between 100 and 500 μm long, similar in size to ciliate Protozoa. Most of them live in fresh water, and they are very numerous. Almost anyone who has looked at pond water under a microscope will have seen rotifers.

Epiphanes (Fig. 11.1, a member of the class Monogononta) is a fairly typical rotifer, which is particularly common in pools contaminated with manure. It feeds mainly on green flagellates which are abundant in the same pools. Like many other rotifers it is superficially rather like *Stentor* (Fig. 3.1*c*). It has a ciliated corona at its anterior end and tapers to a narrow foot at the posterior end. The cilia of the corona are arranged more or less in two rings, with the mouth in the gap between them. The outer ring (the circumapical band) consists of single cilia but the inner one (the pseudotroch) consists of membranelles which, like the membranelles of *Stentor*, are clumps of closely-packed cilia. The cilia and membranelles beat outwards setting up currents in the surrounding water, just as in *Stentor* (Fig. 3.15). They beat metachronally and the metachronal waves travel round the corona making it look as though it were rotating. Because of this rotifers used to be called 'wheel animalcules'. The rotifer can attach its foot to solid surfaces, probably by means of the secretion of the pedal glands. When it is attached the water currents driven by the corona serve simply as feeding currents bringing potential food near the mouth. When it is detached the currents propel it through the water.

Stentor has only one micronucleus and one macronucleus. *Epiphanes*, which is smaller, has almost 1000 nuclei. The discrepancy is less remarkable than it might at first appear, for the large macronucleus of *Stentor* contains a great many duplicate sets of chromosomes. Some of the tissues of *Epiphanes* are divided up into separate cells but others are syncytial. The number of nuclei in each tissue is constant and the position of each nucleus is more or less identical in different individuals. Thus there are 280 nuclei in the epidermis, 157 in the gut, 246 in the nervous system, 104 in the muscles and the rest elsewhere. This indicates a remarkably inflexible pattern of development. The

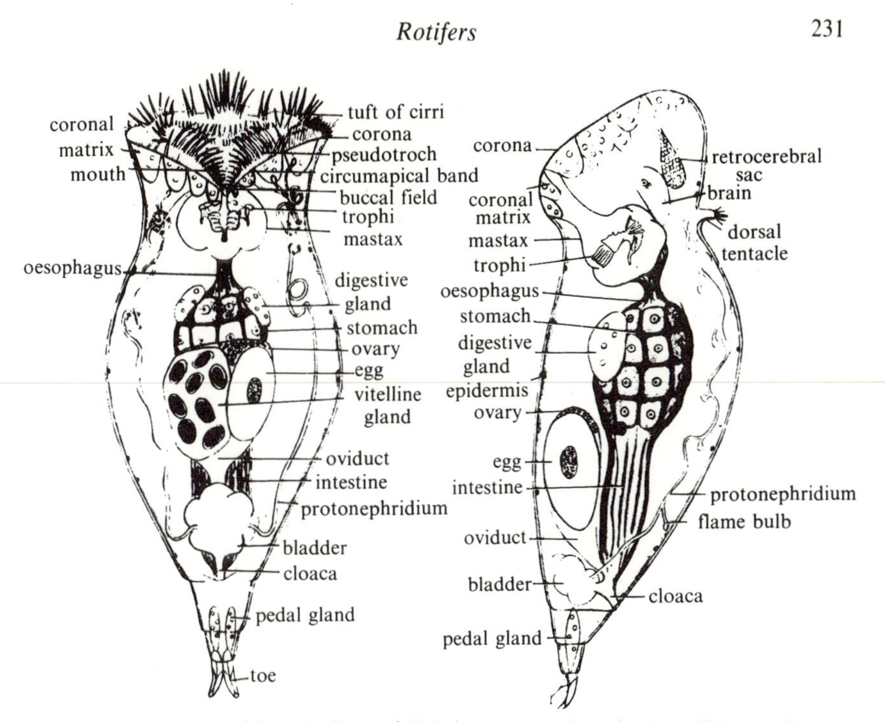

Fig. 11.1. Ventral and lateral views of *Epiphanes senta* (length about 500 μm). From P. A. Meglitsch (1972). *Invertebrate zoology*, 2nd edn. Oxford University Press.

nuclei do not divide after the rotifer has hatched from its egg and any that are lost if the animal is damaged cannot be replaced. This is a striking difference from *Hydra* and triclad flatworms which readily regenerate lost parts.

The epidermis is one of the syncytial tissues. It is thin over most of the body but thicker under the corona and is covered externally by a thin cuticle which seems to consist of protein. There is no parenchyma such as platyhelminths and nemerteans have. Instead there is a fluid-filled space known as the pseudocoel, between the epidermis and the internal organs (Fig. 11.2). There are bands of longitudinal and circular muscle attached at their ends to the epidermis (or in some cases to internal organs). Most of the longitudinal muscles run from the middle of the body to the epidermis of the corona, or of the foot. When they contract they pull the corona into the body and retract the foot like a telescope. A sphincter muscle draws the anterior end closed over the retracted corona, just as the sphincter of sea anemones closes over the retracted tentacles. The other bands of circular muscle do not encircle the animal completely (see Fig. 11.2). They are antagonistic to the longitudinal muscles and serve to lengthen the animal.

The most conspicuous parts of the gut are the pharynx or mastax and the stomach. The mastax contains complicated jaws (trophi) which are used to grind small food and can also be protruded from the mouth to seize large Protozoa.

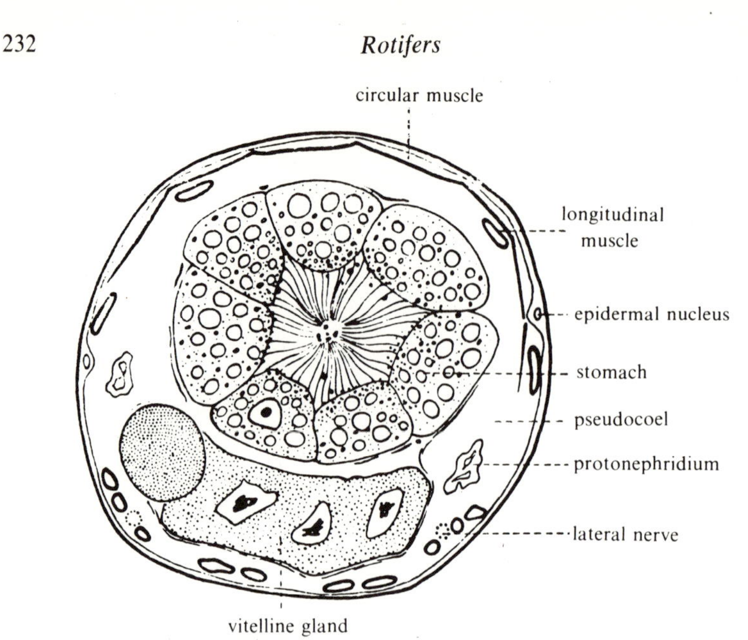

Fig. 11.2. Semi-diagrammatic transverse section through *Epiphanes senta*. After P.-P. Grassé (ed.) (1965). *Traité de zoologie*, vol. 4. Masson, Paris.

Digestion occurs in the stomach, which is enclosed in a network of muscles. These constrict it from time to time, stirring its contents. The posterior end of the gut opens into a cloaca, a combined external opening of the gut and of the excretory and reproductive systems.

The excretory system consists of a pair of syncytial protonephridia with flame bulbs very like the flame cells of platyhelminths. The protonephridia are connected to a bladder which discharges periodically through the cloaca. Four muscle cells around the bladder constrict it and expel its contents. Analyses of urine samples taken from the bladder of a rotifer have already been described (p. 192).

The nervous system consists of a principal ganglion or brain with nerves radiating to all parts of the body. There are a few ciliated sensory cells on the corona and on three short tentacles, one of which is shown in Fig. 11.1(*b*). The eyes of *Epiphanes* are rudimentary but other rotifers have better-developed eyes.

REPRODUCTION

The three classes of rotifer have different patterns of reproduction. The only members of the class Seisonacea are the species of *Seison* which are ectoparasites of marine Crustacea. They reproduce in normal sexual fashion. The other two classes are both large. The Bdelloidea consists solely of females whose eggs develop parthenogenetically (i.e. without having been fertilized). Males have never been found. The Monogononta have both males and

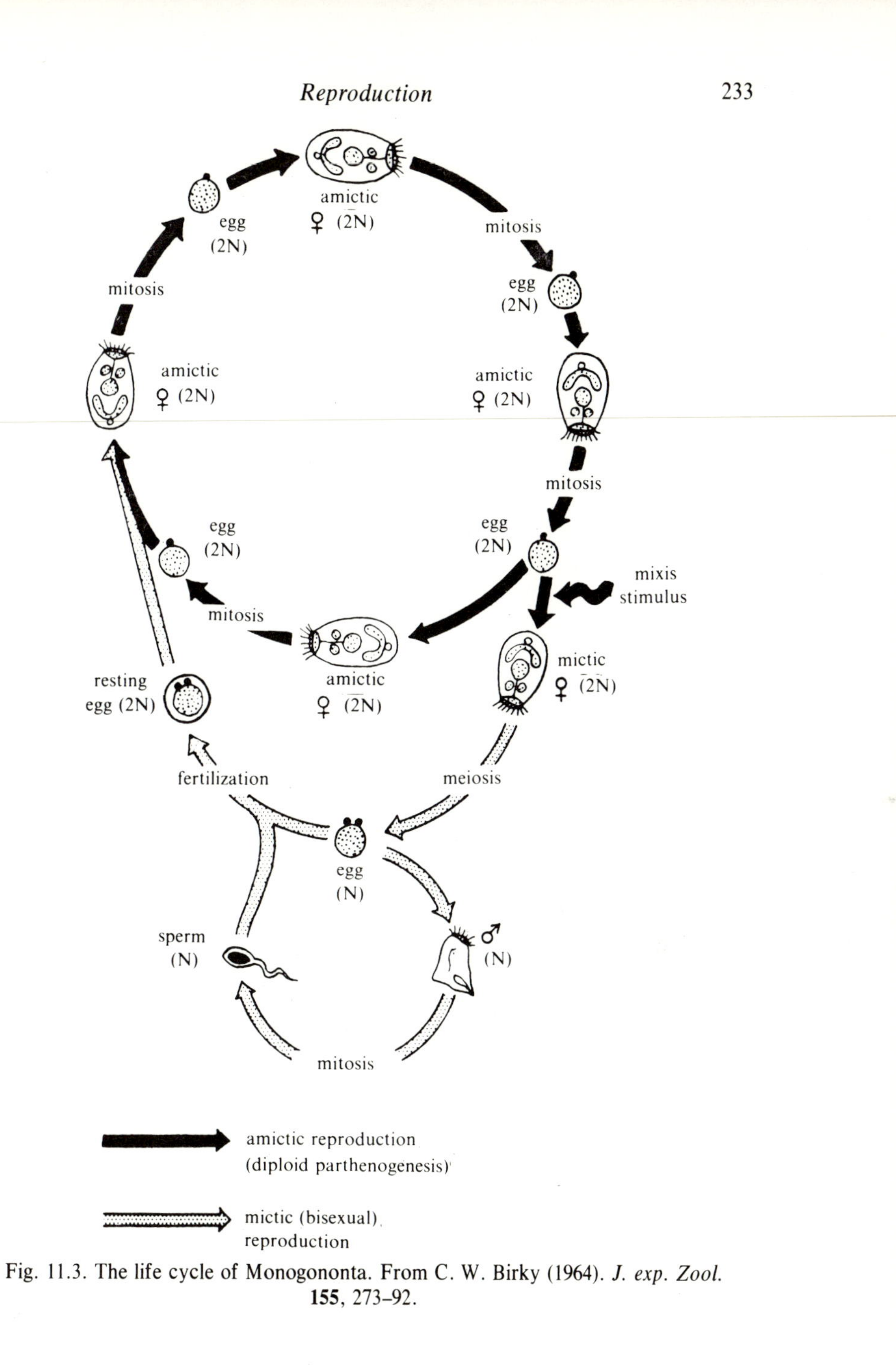

Fig. 11.3. The life cycle of Monogononta. From C. W. Birky (1964). *J. exp. Zool.* **155**, 273–92.

females, but males appear only sporadically. For much of the time a given population consists entirely of females capable only of parthenogenetic reproduction (amictic females). From time to time, especially when the population density is high, males appear in the population and also females capable of sexual reproduction (mictic females). The males are smaller than the females, live a brief life, and do not feed. Mictic females differ in shape from amictic ones in at least some species. The rest of this section is about Monogononta.

Egg production in multicellular animals which reproduce by normal sexual processes involves meiosis. The nucleus of a diploid oocyte divides twice. At each division one of the daughter nuclei is expelled from the oocyte as a polar body so each oocyte produces one ovum and two polar bodies, with all of the original cytoplasm in the ovum. Though the nucleus divides twice the chromosomes divide only once; the diploid oocyte produces a haploid ovum. Fig. 11.3 shows the life cycle typical of Monogononta. The females, whether mictic or amictic, are presumably diploid. Mictic females apparently produce ova by normal meiosis, forming two polar bodies. These haploid ova are fertilized in the usual way by haploid sperm, producing diploid zygotes. Meiosis seems not to occur in amictic females. Their oocytes divide once, not twice, and form only one polar body. The ova are presumably diploid and so can give rise to diploid embryos without being fertilized.

Until males appear in the population, the eggs of mictic females cannot be fertilized. Some of these haploid eggs develop directly without being fertilized, becoming males. There has been some dispute about the nature of the male nucleus but it seems likely that it is haploid (as shown in Fig. 11.3) and produces haploid sperm by mitosis. If so, sex determination works in the same way in these rotifers as in bees. Unfertilized bee eggs develop into males (drones) which are haploid. Fertilized ones develop into females (workers or queens) which are diploid.

Fertilization in rotifers is internal. In most of the cases which have been observed the male does not insert his penis into an aperture, but simply pierces the female's body wall and injects sperm into the pseudocoel. The female can only be fertilized during the first few hours after birth, while she is still growing. Later her cuticle hardens, and may be too hard for the male to pierce. Once fertilized a female generally produces only fertilized eggs, known as resting eggs. These do not hatch for at least a few days after being laid, and can remain dormant for many months. In temperate climates rotifers survive the winter either as amictic females or as resting eggs which hatch in the spring.

Fig. 11.3 shows a 'mixis stimulus' causing formation at certain times of mictic females. A great deal of research has been done, to find out what this stimulus is. Some of the results have been rather confusing. Separate experiments seemed to show that temperature, diet and population density all affect the proportion of mictic females produced. One discovery was that *Epiphanes* produces more mictic females if it is fed on *Chlamydomonas* (a green flagellate) than if it is fed on *Polytoma* (a colourless one). This started a search for a constituent of green plants which would stimulate production of mictic females. It was found that amictic females fed non-green food could not produce mictic daughters unless vitamin E was added to the water. Vitamin E is a constituent of green plants which is essential for male fertility in vertebrates and insects. If it is essential also in rotifers, there is nothing to be gained by producing mictic females while it is not available. Male rotifers do not feed but experiments with radioactive vitamin E have shown that they receive the vitamin from their mothers, in the egg.

Wild rotifers normally have green algae available and it seems unlikely that they are often prevented from producing mictic females by lack of vitamin E. On the other hand there is plenty of evidence that mictic females tend to appear in crowded conditions. Most natural rotifer populations fluctuate greatly in numbers. Numbers are low in the winter, and rise and decline again several times each year in the warmer months. Mictic females tend to appear only when the population density is high. In laboratory experiments the rotifer *Asplanchna* has been kept at different population densities, in solutions containing vitamin E. Larger proportions of mictic females appeared in crowded cultures than in less crowded ones. Further, rotifers kept at low density in solutions in which rotifers had previously been reared at high density, produced high proportions of mictic females. It seems that rotifers release some substance into the water and that if the concentration of this substance gets high enough mictic females are produced, provided vitamin E is available.

Rotifers can reproduce very rapidly: natural populations have been known to multiply tenfold in a week. Their ability to reproduce fast seems to be partly due to parthenogenesis. If each female produces $2n$ offspring by sexual reproduction and if half of these are male and half female, the population can multiply by a factor n in a generation. If, however, each female produces $2n$ offspring, all female, by parthenogenesis, the population can multiply by a factor $2n$ in each generation. The rapidity with which rotifers can multiply is also partly due to their very short generation times. Amictic female *Asplanchna* in laboratory cultures produce about 10 offspring each in a life of only 3–4 days. The first daughter may be born only 30 hours after the birth of the mother. This is not as fast as the reproduction of the ciliate Protozoa described on p. 80, which divided every 12 hours and so produced 4 offspring in a day or 16 in 2 days. It is, however, remarkably fast for an animal with about 1000 nuclei. Only one mitosis is needed when a protozoan divides but the production of 1000 nuclei in a developing rotifer requires at least 10 rounds of mitosis ($2^{10} = 1024$). These 10 rounds of mitosis occur in *Asplanchna* in only 6 hours, an average of only 36 minutes per mitosis.

DNA must be synthesized between nuclear divisions, to replicate the chromosomes. In addition in normal growing tissues RNA must be synthesized, so that proteins can in turn be made. DNA and RNA synthesis each occupy the chromosomes for substantial times. Division is followed by RNA synthesis which is followed in turn by DNA synthesis before division occurs again.

Synthesis of nucleic acids and protein in rotifers has been investigated by experiments with radioactive precursors. Thymidine is a constituent of DNA (but not of RNA) and if rotifers are kept in a solution containing [^{3}H]thymidine while they are synthesizing DNA they incorporate some of this thymidine in their new DNA and become radioactive. Their radioactivity can be detected afterwards by autoradiography. Similarly [^{3}H]uridine can be used to detect RNA synthesis and [^{3}H]leucine (an amino acid) to detect protein synthesis. It has been shown in this way that developing rotifers synthesize DNA and protein, but not RNA.

The cells of the vitelline gland supply the egg with a large stock of food materials and it seems that they also supply all the RNA needed for development. Though RNA is not synthesized by the developing oocyte it is synthesized rapidly in the nuclei of the vitelline gland. The cytoplasm which streams from the gland into the maturing oocyte contains polyribosomes, which presumably contain the RNA needed for development. Since this RNA is supplied ready-made there is no need for the embryo to make its own. The time which would otherwise be needed for RNA synthesis between divisions is saved.

Some rotifers live in lakes where their food supply fluctuates as the plankton blooms and dwindles. Others live in pools and puddles which grow and shrink as the weather changes and may even dry up altogether. Yet others live among damp moss which may dry out. All these environments are unstable. There are times when rotifers can flourish and others when they must dwindle to very small numbers. Their ability to reproduce rapidly enables them to take advantage of the favourable periods.

There are three main types of adaptation which help species to succeed in competition against others. First, there are adaptations which increase the rate at which they are capable of reproducing, by increasing fecundity or reducing generation time. Secondly, there are adaptations which enable them to make more economical use of food, space or other resources. Thirdly, there are adaptations which make them less susceptible to competition, for instance by enabling them to use foods not available to the competitor. It can be shown mathematically that the first type of adaptation is particularly effective in unstable environments in which disasters (such as a pool drying up) periodically destroy a large proportion of the population in an unselective way. This type of adaptation is relatively ineffective in stable environments, in which the second type of adaptation is particularly effective.

CRYPTOBIOSIS

Some rotifer habitats are liable to dry out completely and at least some species of rotifer can survive even this. When a pool dries out it often leaves a crust of dry algae. If this crust is put in water, rotifers will generally emerge from it within an hour. In the same way laboratory cultures of certain rotifers can be dried out and most of the dried animals subsequently revived by adding water. Dry rotifers do not form a protective cyst, but lose most of their water and shrivel up. Survival in this state is called cryptobiosis. Many seeds such as peas are capable of it, as well as rotifers and some other animals.

It is difficult to investigate the internal structure of dried rotifers because normal methods of preparation of specimens for sectioning involve treatment with aqueous solutions. These solutions tend to make the specimens swell up. However it has been possible to prepare electron microscope sections of rotifers very nearly in the fully shrunken state, by using a quick-acting fixative. The sections showed that the cell membranes were shrivelled, as well as the outer cuticle. The cilia of the gut were intact, but packed closely side-by-side. The mitochondria were shrivelled but their cristae could still be distinguished. It seems that a great deal of the structure of the cells survives in dried rotifers although the cells are severely distorted and greatly reduced in volume.

FURTHER READING

Birky, C. W. & Gilbert, J. J. (1971). Parthenogenesis in rotifers: the control of sexual and asexual reproduction. *Am. Zool.* **11**, 245–66.

Dickson, M. R. & Mercer, E. H. (1967). Fine structural changes accompanying desiccation in *Philodina roseola* (Rotifera). *J. Microsc.* **6**, 331–48.

Shorrocks, B. & Begon, M. (1975). A model of competition. *Oecologia* (*Berl.*) **20**, 363–7.

12
Roundworms

Phylum Nematoda

The nematodes or roundworms are generally slender worms, circular in cross-section. They are remarkably uniform in structure but vary a lot in size and in way of life. Some live as parasites in other animals, some are parasites of plants and some live free in soil, marine muds and decaying organic matter.

Some of the parasitic nematodes cause serious diseases, of man or of domestic animals. Elephantiasis is a particularly nasty human disease, caused by *Wuchereria bancrofti.* This nematode is no more than about 8 cm long and 0.3 mm in diameter but it blocks lymph ducts and causes swellings quite disproportionate to its size. Infected legs may swell to a circumference of 75 cm or the scrotum may swell till its mass is 25 kg. The hookworms *Ancylostoma* and *Necator* are more widespread, and more insidious. It has been estimated that 450 million people suffer from hookworm disease, most of them in underdeveloped countries in the tropics and subtropics. The worms live in the intestine, abrading it and feeding on blood. They cause anaemia, indigestion and debility. Patients are apt to be weak and apathetic, with slow mental processes. The 'poor white trash' of the southern United States owed their sad condition to hookworm disease. Eggs are passed in the faeces of the host and larvae enter new hosts through the skin, so a great deal can be done to combat the disease by improving sanitation and providing shoes. *Trichinella* is another nematode which causes human disease. Adults live in the intestines of various mammals including men, pigs and rats, and cause digestive disturbances. Their larvae bore through the wall of the gut and travel to muscles where they form cysts and cause pain, swelling and even death. The disease is transmitted when infected flesh is eaten. Pigs catch it by eating rats and men catch it by eating undercooked pork or raw sausage. It is common in eastern Europe but rare in Britain where raw sausage is less popular and pig swill is usually well cooked.

Nematode parasites of plants cause heavy losses in agriculture. *Globodera rostochiensis* is an important pest of potato crops, almost everywhere they are grown. It damages the roots and the plants become stunted. *Ditylenchus dipsaci* is a pest of oats, rye, onions, sugar beet and many other crops and also of tulips and narcissus. It attacks the stem, so that the plants become stunted and twisted with swollen stem bases. Strains of oats and of some other crops have been bred which are resistant to it. *Meloidogyne* causes serious damage to tobacco crops. As well as causing damage directly, nematodes transmit some virus diseases of plants.

Small nematodes are generally plentiful in the top few centimetres of soils and marine sediments. They belong to the interstitial fauna which consists of animals small enough to fit into the spaces between one sand grain and the next. When they crawl they do not have to force the grains apart, but wend their way through the network of existing spaces. They are particularly abundant in agricultural soils where there may be as many as 10 million individuals, with a total mass exceeding 10 g per square metre. Some of the nematodes found in soil are juveniles of species which in later life become parasites of animals or plants. Others spend their whole life in the soil.

Soil nematodes and nematode parasites of plants are generally small. The smallest species are only about 200 μm long and since they are slender they are much smaller in volume than many ciliate Protozoa. In contrast, some nematode parasites of animals are quite large. *Ascaris lumbricoides* which lives in the intestines of pigs and men grows to a length of 40 cm and a diameter of 6 mm. The Guinea worm *Dracunculus medinensis* grows even longer, up to 1.2 m, although its diameter is less than 2 mm. It lives as a parasite under the skin of man and causes severe pain.

Fig. 12.1 shows the structure of typical small nematodes, which have been drawn rather stout for clarity. Most nematodes are more slender, relative to their length, like the ones shown in Fig. 12.8. All are circular in cross-section, as shown in Fig. 12.2. The body is enclosed in a cuticle. Within this is a thin layer of cells, or a syncytium, called the hypodermis. Within this again is a layer of longitudinal muscle divided into quadrants by four ridges which project from the hypodermis. There are no circular muscles. Later sections of this chapter discuss the muscles and the cuticle and how they interact when the worm moves.

There is a mouth at the anterior end of the body and an anus near the posterior end. There is a muscular pharynx. The mechanisms of feeding and defecation are discussed later in the chapter.

There is a nerve ring around the pharynx, from which nerves run posteriorly along the body. The largest of these nerves are a median dorsal one and a median ventral one. There are no motor nerves to the muscles: instead the muscle cells have slender processes which run to the nerve ring or to the dorsal or ventral nerve and synapse with neurones there.

Most species have separate sexes. The females generally have two ovaries and the males only one testis. The ovaries and testes are unbranched, blind-ended tubes. In small nematodes they are relatively short, as shown in Fig. 12.1. In large nematodes they are not proportionately thicker but are relatively slender, long and coiled. For instance male *Parascaris equorum*, less than 30 cm long, have testes which, when unravelled, are around 1.8 m long. Females of the same species, up to 50 cm long, have ovaries around 2.3 m long.

Spermatozoa produced in the testis collect in a seminal vesicle (labelled

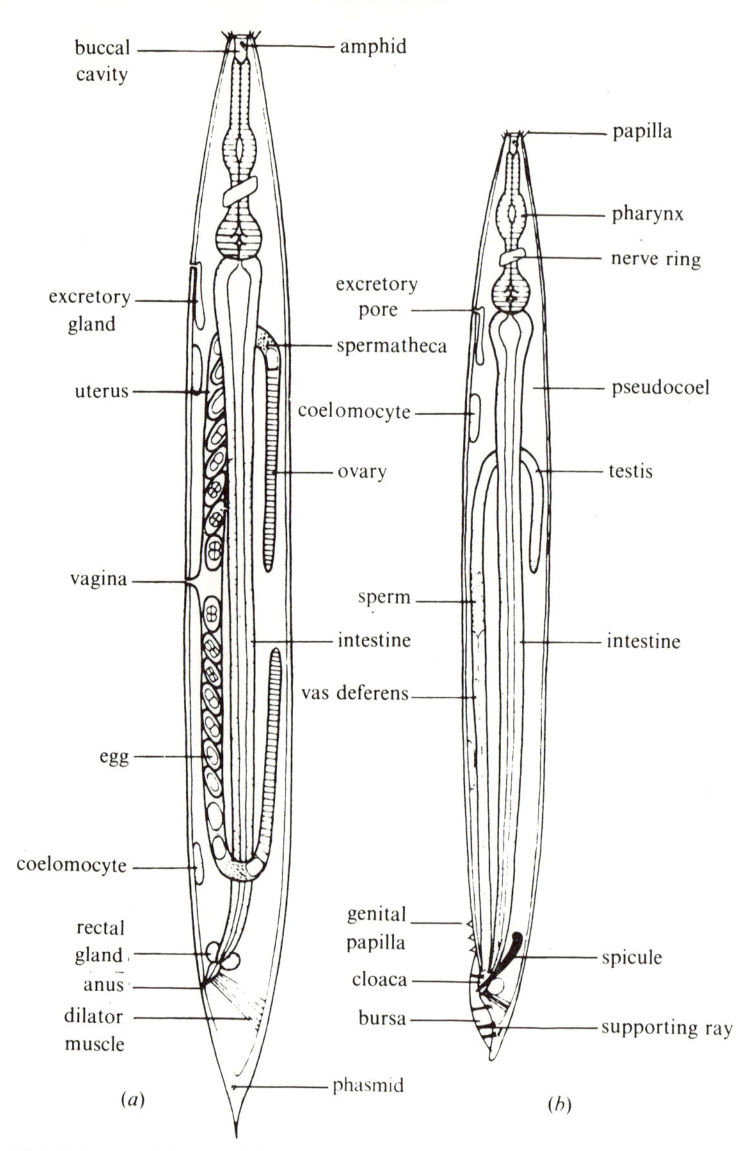

Fig. 12.1. Diagrams showing the structure of typical nematodes. (*a*) represents a female and (*b*) a male. From D. L. Lee & H. J. Atkinson (1976). *The physiology of nematodes*, 2nd edn. Macmillan, London.

'sperm' in Fig. 12.1 *b*) and travel from there down the vas deferens to the cloaca. The males of most species have two copulatory spicules, one on each side of the cloaca, which are inserted into the genital opening of the female in copulation. They do not serve as a penis but simply keep the female opening open and opposite the male one, so that sperm ejected by the male will enter

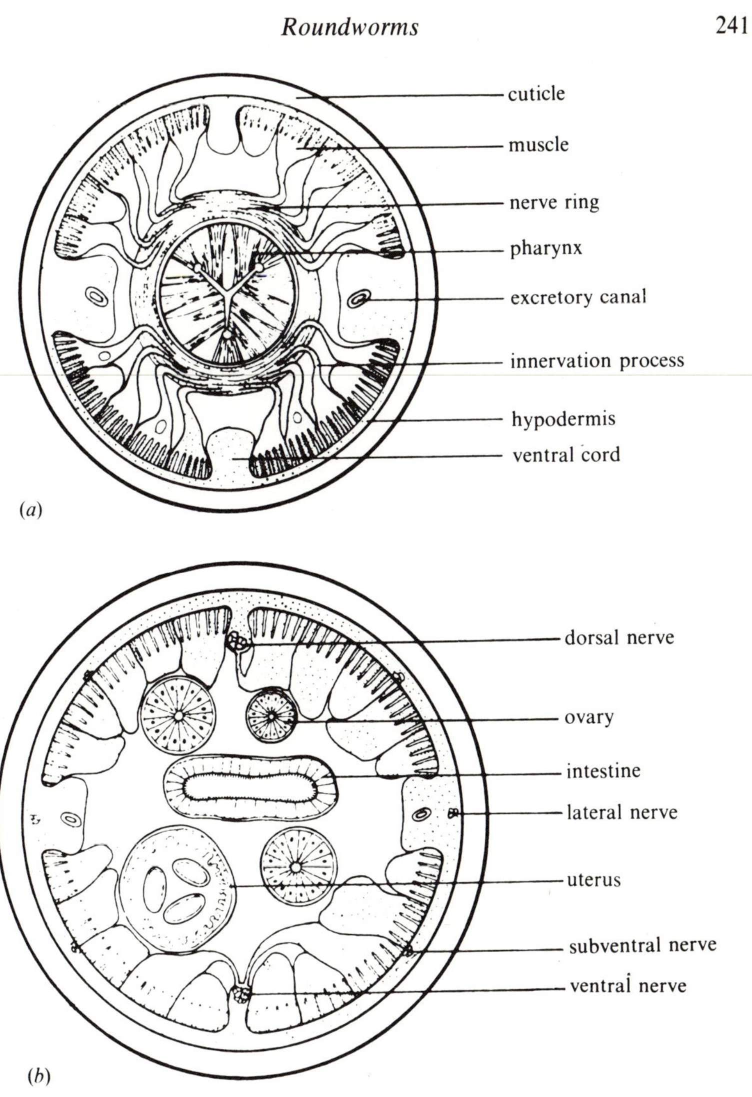

Fig. 12.2. Diagrammatic transverse sections through (*a*) the pharyngeal region and (*b*) the intestinal region of a nematode. From D. L. Lee & H. J. Atkinson (1976). *The physiology of nematodes*, 2nd edn. Macmillan, London.

the female. In many species the male also has a bowl-shaped bursa around the cloaca, which grasps the female during copulation. Once in the female, the sperm travel to her spermathecae where fertilization occurs.

The spermatozoa of nematodes are unusual in having no flagella. In other respects they vary a lot between species. The eggs are usually between 50 and 100 μm long and rather less in diameter, irrespective of the size of the nematode. They are enclosed in eggshells which have been shown by X-ray diffraction and by chemical tests to contain chitin. Large parasitic nematodes produce very large numbers of eggs. It has been estimated by counting eggs in samples of the faeces of infected pigs that a large *Ascaris* may lay as many as 1.6 million eggs per day. Juvenile nematodes look like miniature adults.

Many nematodes have one or two cells called excretory glands, with a duct leading to the exterior. Many have excretory tubes which run along the sides of the body and open through the same duct. There is very little evidence that these structures have in fact an excretory function.

The cavity which contains the gut and reproductive organ is a fluid-filled pseudocoel, as in rotifers. It often contains a few cells called coelomocytes which may be very large – about 5 mm long in *Ascaris*.

There is a pair of anterior sense organs called amphids and often a pair of posterior ones called phasmids (Fig. 12.1). The amphids are pits containing sensory cells with modified cilia and the phasmids are rather similar. It is thought likely that amphids and phasmids detect chemical stimuli, but there is no conclusive evidence.

In nematodes as in rotifers the number and arrangement of cells in the nervous system are constant, in normal members of the same species. This fact has been exploited in an investigation of the inheritance of structure which is described in the final section of this chapter.

MUSCLES

The muscles of nematodes are of the type called obliquely striated, which is also found in molluscs and annelid worms. It is most easily described by comparison with the better-known striated muscle of vertebrates. Fig. 12.3(*a*) shows part of a vertebrate striated muscle fibre. It can be seen under the microscope to be banded, with the bands at right angles to its length. This banding is in the contractile material. There are also mitochondria in the fibre, and nuclei. (Only one nucleus is shown but the fibres are syncytial.) Fig. 12.3(*d*) shows part of a typical nematode muscle cell. The contractile material has oblique bands, not transverse ones, and forms a trough with mitochondria in the hollow. The single nucleus is in a bag of glycogen-rich cytoplasm which gives rise to the process to the nervous system.

Electron microscope sections show that in both types of muscle the banding is due to a regular pattern of thick and thin filaments which run parallel to the length of the fibres. The thin filaments have been shown to consist largely of the protein actin and their diameter is about 6 nm. The thick filaments of vertebrate striated muscle consist of the protein myosin, and their diameter is about 11 nm. Those of nematode obliquely striated muscle are stouter (about 20 nm) and presumably have the same constitution as the thick filaments of mollusc obliquely striated muscle which have their myosin arranged around a core of another protein, paramyosin. The thick filaments have projections which can attach to the thin filaments, forming cross-bridges. In vertebrate striated muscle there are twice and in nematode muscle six times as many thin filaments as thick ones.

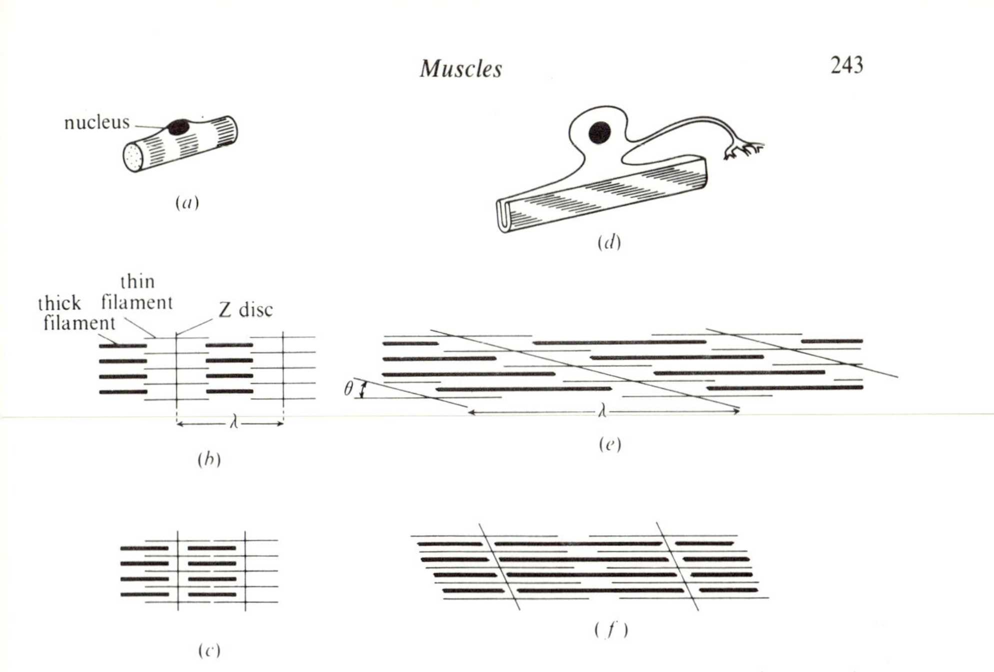

Fig. 12.3. Diagrams of (*a*), (*b*), (*c*) vertebrate striated muscle and (*d*), (*e*), (*f*) nematode obliquely-striated muscle. (*a*) and (*d*) each represent part of a muscle fibre. The other diagrams represent longitudinal sections at much higher magnification, and are based on electron microscope sections. (*b*) and (*e*) represent extended muscle and (*c*) and (*f*) contracted muscle.

Fig. 12.3(*b*) and (*e*) show the thick and thin filaments are arranged so as to produce a transverse or oblique banding pattern. Notice how the thin filaments interdigitate with the thick ones. Transverse partitions called Z discs join the thin filaments together in vertebrate striated muscle and there are similar structures in some nematodes. In other nematodes there are no continuous partitions (Fig. 12.4). The unit of structure from one Z disc to the next is known as a sarcomere. It corresponds to a single unit of the banding pattern visible under the light microscope.

When a muscle contracts its sarcomeres shorten and (in obliquely striated muscle) the angle of banding changes so that the bands run more transversely. The changes involved have been studied in more detail by examining electron microscope sections of muscle fixed while extended (Fig. 12.3*b*, *e*) or contracted (Fig. 12.3*c*, *f*). These sections show that contraction does not involve any change in the length of the myofilaments. The sarcomere length λ diminishes and in obliquely striated muscle the angle θ increases but the thick and thin filaments slide past each other with no change in length. They are apparently made to slide by the cross-bridges, which work like the dynein arms of flagella (p. 44). When the muscle becomes active the cross-bridges attach to the thin filaments and 'walk' along them.

The increase in θ is more difficult to demonstrate than the decrease in λ, because the angle seen in sections depends on the direction of the cut. A section cut in the XZ plane (Fig. 12.4) shows the angle θ but one cut in the

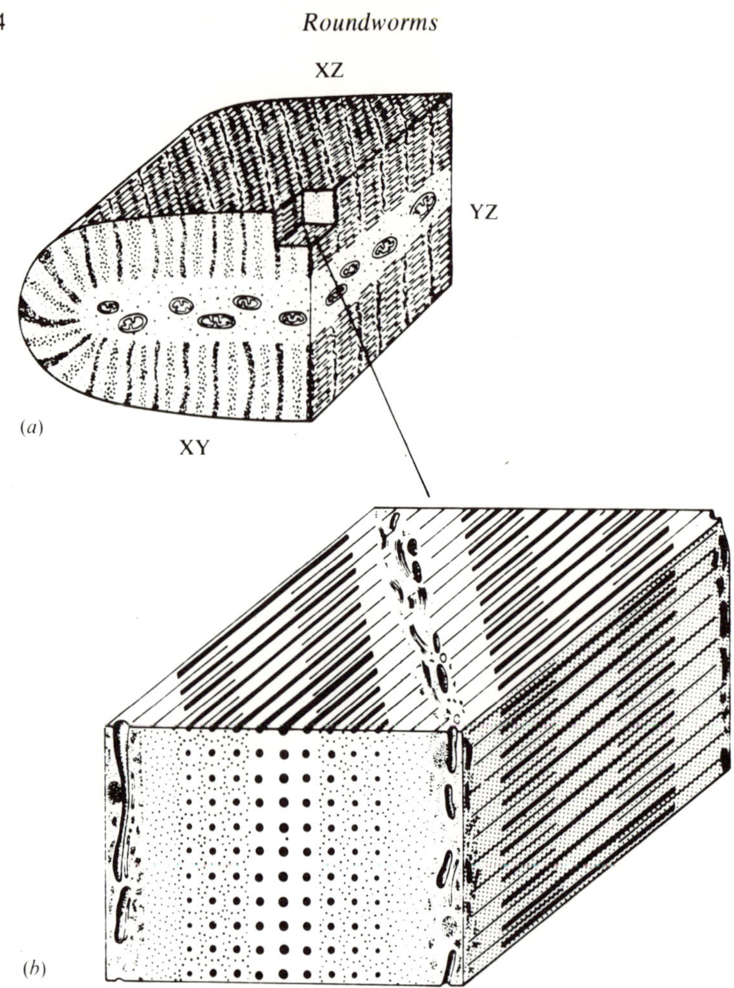

Fig. 12.4. A diagram showing the structure of *Ascaris* muscle, based on electron micrographs. The thick filaments are about 6 μm long. From J. Rosenbluth (1965). *J. Cell Biol.* **25**, 495–515.

YZ plane shows transverse striations and ones cut in the other longitudinal planes show other angles. The difficulty can be overcome by examining transverse sections (i.e. sections cut parallel to the XY plane). Such sections show a band of thick filaments in each sarcomere. In contracted muscle there are more rows of thick filaments in the band than in relaxed muscle. This shows that θ must be larger, making more rows of thick filaments overlap. The process which increases θ has not been explained.

Vertebrate striated muscle can work only over a limited range of lengths. Suppose for simplicity that the thick and thin filaments have equal length l. The muscle can exert a contractile force only when the thick and thin filaments overlap, that is when the sarcomere length is less than $2l$. It cannot shorten to sarcomere lengths much less than l because at this length the thick filaments abut on the Z discs. Thus the maximum working length is little more than double the minimum working length. Experiments with frog striated muscle fibres have shown that the maximum length at which a force can be exerted is about 2.8 times the minimum length.

The changes in θ which occur in obliquely striated muscle make larger changes of length possible.

CUTICLE

Whatever the size of the nematode the thickness of the cuticle is generally about 0.07 times the radius of the body. The structure of the cuticle varies a lot between species but two examples are shown in Fig. 12.5. Fig. 12.5(*a*) shows the structure typical of the juveniles of parasitic nematodes, living in the soil prior to becoming parasitic. Juveniles of many parasites both of animals and of plants have cuticle like this. Fig. 12.5(*b*) shows a type of cuticle found in adult *Ascaris* and various other nematode parasites of animals. In both cases the cuticle has external and internal cortical layers, a median layer and a basal layer. The external cortical layer is very thin. It stains strongly with osmic acid (which makes it appear black in electron microscope sections) so it presumably consists largely of lipid. The other layers of the cuticle are mainly protein. *Ascaris* cuticle, and presumably the cuticle of other nematodes, contains a protein similar to vertebrate collagen. It contains a similar mix of amino acids and gives similar X-ray diffraction patterns. Histochemical tests on several nematodes have indicated that the internal cortical layer is quinone-tanned but that the deeper layers are not.

The principal difference between the two types of cuticle illustrated is in the basal layer. In the larval cuticle it contains radially arranged rods about 20 nm apart. In *Ascaris*, however, it consists of three layers of stout fibres running helically at about 75° to the long axis of the body. In successive layers the fibres form left- and right-handed helices.

The mechanical properties of the cuticle probably depends largely on the rods or fibres in the basal layer. Consider *Ascaris* first. The fibres run at angles greater than 55° to the axis of the body so any shortening of the body must tend to diminish the volume the cuticle can contain (Fig. 10.6). The cross-section of the body is circular even in relaxed *Ascaris* so the body already fills the cuticle to capacity. Hence the body cannot shorten unless the fibres stretch. When the longitudinal muscles contract and shorten the body they must stretch the fibres and when they relax elastic recoil of the fibres must extend the body again. The elasticity of the cuticle is antagonistic to the longitudinal muscles so the animal can function without circular muscles. Further, if the dorsal longitudinal muscles contract they will tend not to shorten the body but to bend it, stretching the ventral muscles. Similarly contraction of the ventral muscles must tend to stretch the dorsal ones.

Now consider the type of cuticle shown in Fig. 12.5(*a*). To see how it works

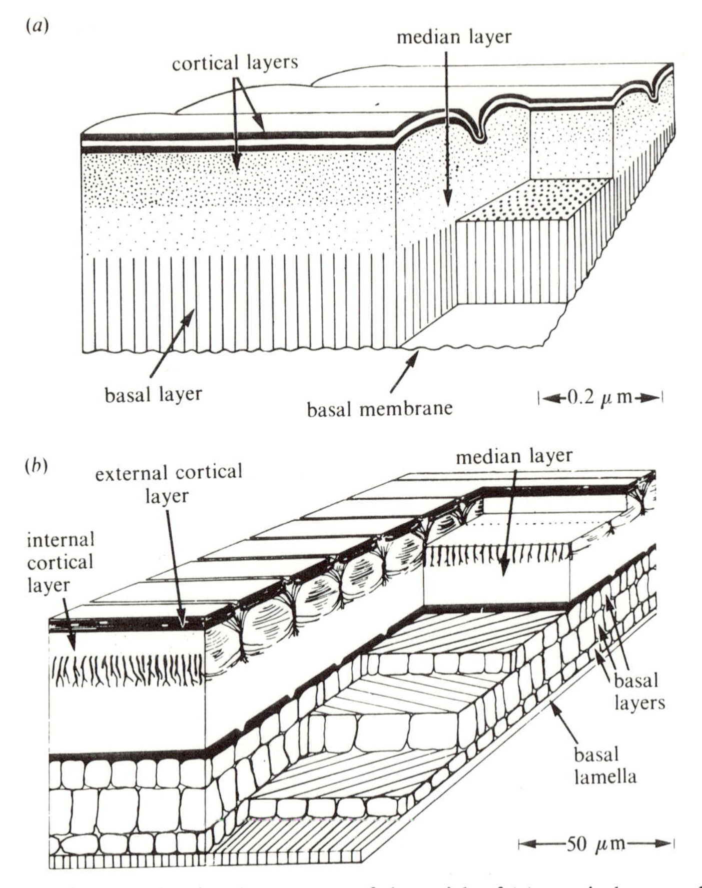

Fig. 12.5. Diagrams showing the structure of the cuticle of (*a*) a typical nematode larva and (*b*) adult *Ascaris*. These diagrams are based on electron microscope sections cut in various directions. From A. F. Bird (1971). *The structure of nematodes*. Academic Press, New York & London.

we have to consider the change of area which occurs when a nematode shortens its body. Consider a cylindrical nematode of length l and radius r. Suppose it shortens to a fraction a of its original length, to a length al. To keep its volume constant its cross-sectional area must increase by a factor $1/a$ so its radius must become $r/a^{\frac{1}{2}}$. The surface area of the cuticle was initially $2\pi rl$ but is now only $2\pi(r/a^{\frac{1}{2}})al = 2\pi rla^{\frac{1}{2}}$. Shortening decreases the surface area of the cuticle, making it thicker. The mechanical properties of the radial rods of the basal layer are not known but it seems likely that these rods are relatively inextensible fibres. Shortening of the body must tend to stretch them and will be opposed by their elastic stiffness. Thus the radial rods of larval cuticle and the lattice of adult *Ascaris* cuticle are probably equivalent in mechanical effect, though they are arranged so differently.

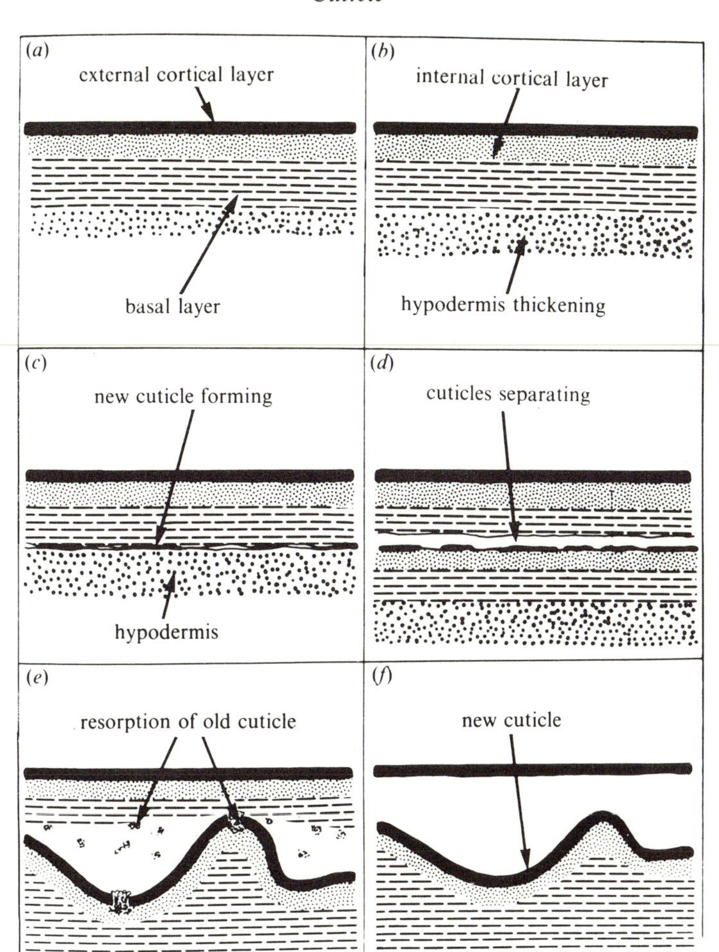

Fig. 12.6. Diagrammatic sections through the cuticle of *Meloidogyne*, showing the sequence of changes which occurs during moulting. From A. F. Bird (1971). *The structure of nematodes*. Academic Press, New York & London.

Nematodes grow a new cuticle and shed the old one several times in the course of their life history. This is called moulting, and occurs four times in most species. It is not clear why it occurs, since the cuticle can grow: nematodes grow between moults and after the final moult. In the extreme case of *Ascaris* the body may be as little as 6 mm long after the final moult but grows to 20 cm or more without further change of cuticle. Nematodes do not become sexually mature until after the final moult and all stages prior to this moult are regarded as juveniles.

Fig. 12.6 shows the sequence of changes which occurs when a nematode moults. The hypodermis thickens and separates from the old cuticle and then

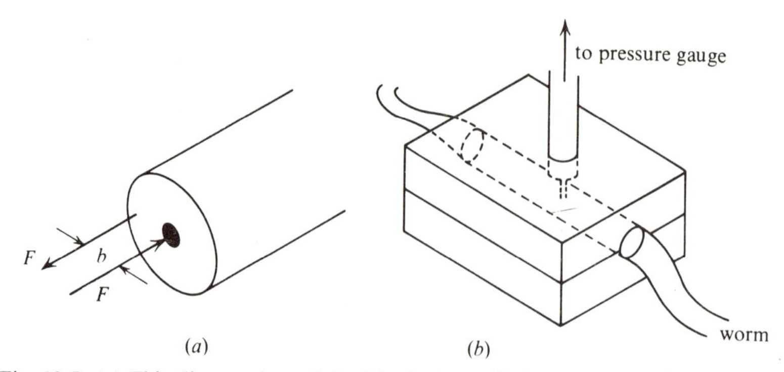

Fig. 12.7. (*a*) This diagram is explained in the text. (*b*) Apparatus used to measure the pressure in *Ascaris*. The worm is threaded through a hole in a block of plastic and the hypodermic needle is inserted from above, through a hole at right angles to the first one.

lays down a new cuticle, starting with the external cortical layer and working inwards. The outer layers of the new cuticle become wrinkled but flatten again as the worm grows after the moult. The inner layers of the old cuticle break down in some nematodes, forming fragments which pass into the hypodermis before the outer layers are shed. In many and perhaps most species, however, the old cuticle seems to be shed intact.

INTERNAL PRESSURE AND LOCOMOTION

Fig. 12.7(*a*) shows forces which act in a vertebrate when it bends its body. Longitudinal muscles exert a tensile force F at a distance b from the vertebral column. A corresponding compressive force F is developed in the vertebral column so that there is a couple Fb tending to bend the body. Nematodes also bend their bodies by means of longitudinal muscles but they have no vertebral column. Contraction of the longitudinal muscles tends to deform the cuticle as has already been shown, and is resisted by the elastic stiffness of the cuticle. A pressure must be developed in the body to balance the force F exerted by the muscles. If the cross-sectional area of the body is A, the pressure will be F/A. This pressure acts uniformly over the whole cross-section so the force F due to it can be considered to act at the centroid (geometrical centre) of the cross-section. The force in the muscle and the force due to the pressure form a couple Fb, tending to bend the body. Pressure will be developed without a couple if the longitudinal muscles contract symmetrically.

The pressure in the pseudocoel of *Ascaris* has been measured. This was first done about 1956, before pressure transducers were readily available, and a simple but effective pressure gauge made from a helix of fine glass tubing was used. It was attached to a hypodermic needle which had been cut short, which was pushed through the body wall so that its tip was in the pseudocoel (Fig. 12.7*b*). It was found that the contents of *Ascaris* were under pressure. In a

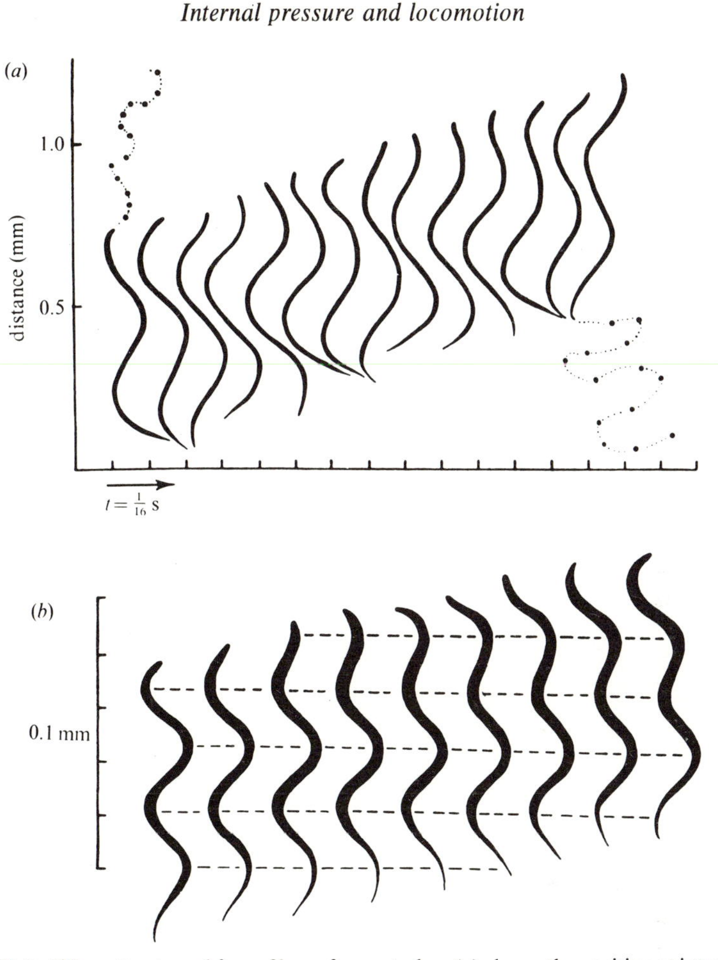

Fig. 12.8. Silhouettes traced from films of nematodes. (*a*) shows the position at intervals of 1/16 s of *Turbatrix* swimming, and (*b*) shows the position at intervals of about 1/3 s of *Haemonchus* crawling on the surface of an agar gel. In each case successive silhouettes have been displaced laterally so as not to overlap their predecessors. From J. Gray & H. W. Lissmann (1964). *J. exp. Biol.* **41**, 135–54.

typical record from a fresh worm the pressure fluctuated rhythmically, with a period of about 30 s, about an average value of about 95 cm water (9 kN m^{-2}). It was found that the length fluctuated with the pressure: whenever the worm shortened the pressure rose. From time to time the pressure rose far above the mean value and one worm maintained a pressure of 30 kN m^{-2} for a few seconds. The longitudinal muscle fibres fill only about 5% of the cross-sectional area of the worm so the stress in them must have been 600 kN m^{-2}, much higher than is possible in vertebrate striated muscle. This is only possible because the thick filaments are longer than in vertebrates, as will be explained in chapter 14.

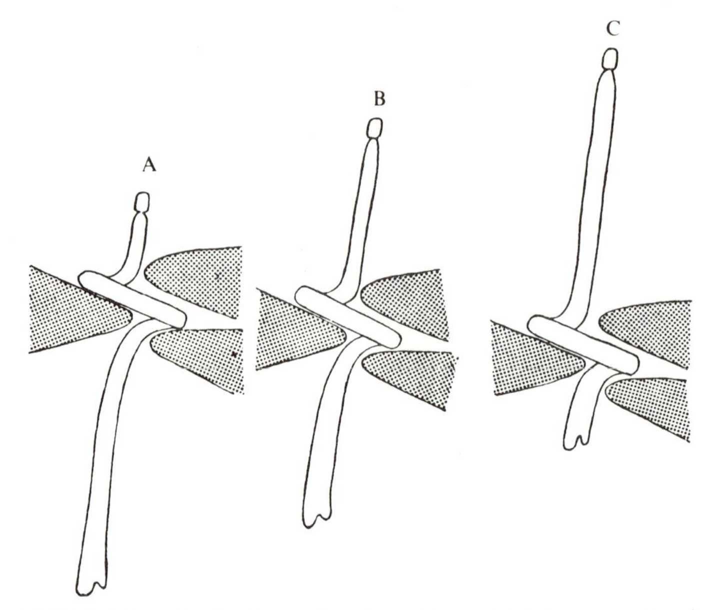

Fig. 12.9. Sketches showing how *Nippostrongylus* crawls between sand grains. From D. L. Lee & H. J. Atkinson (1976). *The physiology of nematodes*, 2nd edn. Macmillan, London.

Nematodes swim and crawl by passing waves of bending posteriorly along the body. Eels and snakes swim and crawl by bending from side to side but nematodes bend dorsally and ventrally (a nematode crawling on a surface is lying on its side). The nematode *Turbatrix* which lives in vinegar swims quite well at an average speed of about 0.7 mm s^{-1} (Fig. 12.8*a*). The waves of bending travel backwards relative to the liquid, as in a swimming flagellate. Most other nematodes are ineffectual swimmers which thrash about in water but cannot keep themselves off the bottom. However, they can crawl over solid surfaces (Fig. 12.8*b*). In crawling as in swimming waves of bending travel posteriorly along the body, but they are stationary relative to the ground (compare Fig. 12.8*b* with 12.8*a*). The same is true of snakes crawling. It can be shown by oblique lighting that a nematode crawling on agar jelly leaves behind it a shallow sinuous groove which marks its path. When waves of bending move posteriorly along the worm, the worm must either slide sideways out of the groove or it must slide forwards, extending the groove forwards. The latter requires less energy so it is what happens.

Fig. 12.9 shows an unusual method of crawling which is used by *Nippostrongylus* when put among sand grains. This nematode forms a loop near its anterior end, and then passes the loop posteriorly along its body. The loop is prevented from moving by the sand grains, so the worm moves forward. Adult *Nippostrongylus* are parasites in the intestines of rats so they do not normally have to crawl through sand, but they probably use the technique shown in Fig.

12.9 to crawl among the intestinal villi of their hosts. Most soil nematodes crawl by waves of bending which are less regular than in Fig. 12.8(*b*): the bends must be fitted to the tortuous paths between the soil particles.

Hookworm larvae live in the soil, and become infective when about 5 mm long. They may climb some distance up a damp human leg before penetrating the skin. Similarly some nematodes which feed on plants climb up damp stems and many nematodes can crawl on damp glass. Surface tension holds the worm against the glass and gives it a purchase.

FEEDING

Nematodes eat many types of food. Many soil nematodes feed on bacteria or decaying organic matter. Their mouths are small, simple openings as shown in Fig. 12.1. Others feed on fungi and roots and have a stylet (a structure like a hypodermic needle) which they thrust into plant cells. They inject enzymes through the stylet and then suck out the partly-digested cell contents through it. The process cannot easily be watched underground but can be watched more conveniently if the worms are allowed to feed on the roots of seedlings grown on agar jelly. A small piece of jelly containing both roots and nematodes can be cut out and put under a microscope for observation. Some other nematodes which live in the soil are carnivores, feeding on protozoans, rotifers, small oligochaete worms or other nematodes. Some suck out the contents of their prey through a stylet and others swallow it whole or have tooth-like structures which are used to break the prey open or to bite pieces off it.

Some parasitic nematodes which live in the guts of animals feed on the gut contents, and their mouths are simple openings like those of soil nematodes which feed on bacteria. *Ascaris*, which is found in the intestines of men and pigs, and *Ascaridia*, in the domestic fowl, are examples. Infected hosts have been given food containing charcoal or radioactive (^{32}P) phosphate. Charcoal, or radioactivity, was subsequently found in the guts of the worms. However, radioactive phosphate injected into the blood of fowl infected with *Ascaridia* did not appear in the worms: these worms do not feed on their hosts' blood. By contrast the hookworms *Ancylostoma* and *Necator* which also live in the intestine feed on host tissues and blood. They have large mouth cavities with sharp teeth.

Ancylostoma can drink blood very fast, as has been shown by the experiment illustrated in Fig. 12.10. A hypodermic needle was used to pierce the rubber membrane and thread the worm through. The head of the worm was immersed in blood and the tail in a saline solution. The worm drank the blood and incompletely digested blood could be seen being squirted out of its anus. The saline gradually became stained with blood. No staining occurred if the apparatus was set up the other way round with the worm's tail in the blood and its head in the saline, so all the blood was presumably passing through the worm and not leaking through some other way.

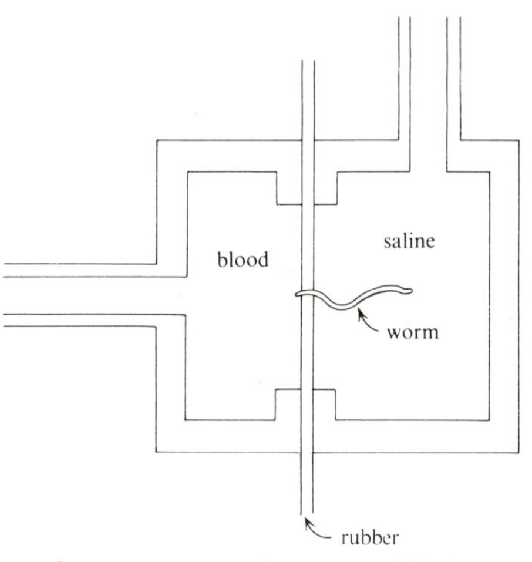

Fig. 12.10. An experiment to measure the rate at which the hookworm *Ancylostoma* drinks blood.

The average mass of the worms was 1.8 mg. The rate at which blood cells accumulated in the saline was measured and it was found that the cells from 30 mg blood passed through the worm in the course of a day. In addition, blood cells were digested by the worms. A radioactive chromium salt was added to the blood and the rate at which radioactivity appeared in the saline was measured. It was calculated that each worm was drinking a total of 60 mg blood each day. Thus in the course of a day a hookworm can drink over 30 times its own mass of blood, but may only digest half the blood corpuscles in it.

Many (and perhaps most) nematodes have a high pressure in the pseudocoel, so they have to pump food into their bodies against a pressure gradient. *Hexatylus* is different, for the pressure in it seems to be low. Evidence for this can be seen when the worm is bent by contact with some other subject: its body kinks, which it would not do if it had a high internal pressure. *Hexatylus* feeds on fungal hyphae and can be watched feeding on moulds growing on agar jelly. It seems to feed only on turgid hyphae, which are flaccid by the time it leaves them. It feeds through a stylet like other plant-feeding nematodes but probably relies on the pressure in the turgid cells to drive cytoplasm into its gut. If disturbed while feeding it withdraws its stylet and cytoplasm squirts from the hole thus left in the fungal cell wall.

Other nematodes use the muscular pharynx to pump food into the gut, against the pressure in the pseudocoel. The movements involved are not easily observed even on film but the pharynx probably dilates at the anterior end, taking in food, and then the dilation travels posteriorly taking the food to the intestine (Fig. 12.11). When *Ascaris* is feeding, its pharynx makes about four cycles of pumping movements per second.

Fig. 12.2 shows the structure of the pharynx. It is circular in cross-section

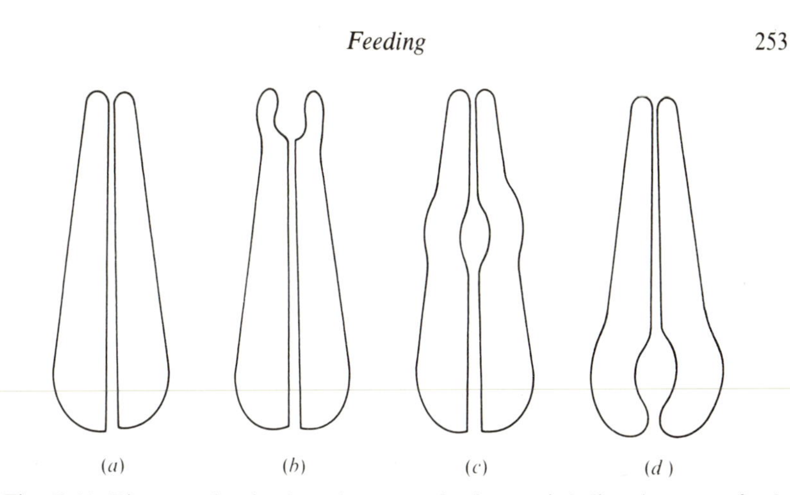

Fig. 12.11. Diagrams showing how the nematode pharynx is believed to pump food into the gut. Dilations form at the anterior end and travel posteriorly.

but its lumen (when empty) is Y-shaped. Its wall is highly muscular with all the muscle fibres running radially: there are no circular or longitudinal muscle fibres. It is lined by a cuticle which is, in *Ascaris*, about 3 μm thick.

The pharynx of *Ascaris* is about 1 cm long so it is big enough to dissect out and manipulate. It can be stretched to about 1.4 times its initial length and springs back elastically when released. The elasticity seems to reside in the cuticle, which can be cleaned of other tissues by dissolving them off in a concentrated solution of phenol. It is a protein with rubber-like properties and a Young's modulus of about 10 MN m^{-2}. It shows no birefringence when examined in a polarizing microscope and it shows no sign of fibrous structure when it is torn. These properties are typical of amorphous, cross-linked polymers (Fig. 6.16*b*). Other proteins with similar properties have important functions in bivalve molluscs (abductin, chapter 14) and in insects (resilin, chapter 18).

How does the pharynx dilate and contract? Dr Henry Bennet-Clark has suggested a possible mechanism which depends on the elasticity of the cuticle. The cuticle is stretched by the action of the muscles, storing elastic strain energy which is then used to dilate the pharynx. In the same way, energy can be stored in a stretched spring and subsequently used to do work. Fig. 12.12(*a*) shows a piece of the pharynx in its relaxed state. The radial muscles shorten, making the pharynx longer and more slender (Fig. 12.12*b*). This stretches its lining of elastic cuticle, storing elastic strain energy. Eventually the pharynx dilates (Fig. 12.12*c*), releasing most of the strain energy and doing work against the pressure in the pseudocoel. The radial muscle fibres have the same length in state (*c*) as in state (*b*) but the pharynx is shorter and fatter and its lumen has opened so that its shape in cross-section is a trefoil.

The muscular wall of the pharynx has a constant volume, V_0, but transformation to state (*c*) draws a volume $0.125\,V_0$ of fluid into the lumen,

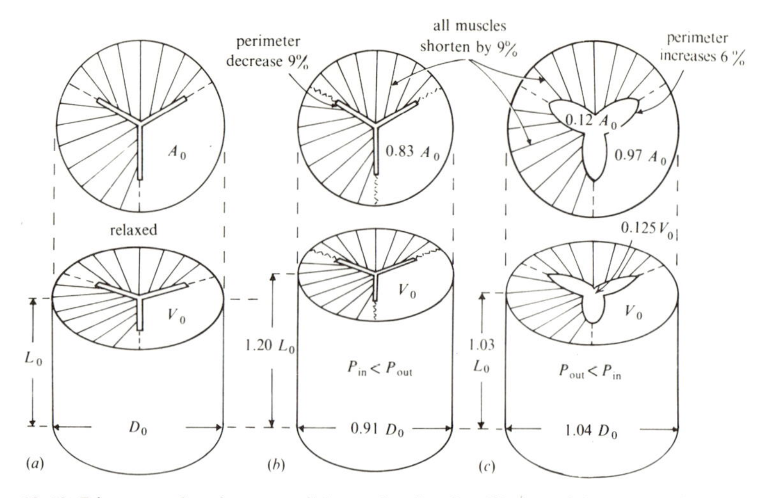

Fig. 12.12. Diagrams showing a possible mechanism for dilation of the nematode pharynx. When relaxed (*a*) the section of pharynx has length L_0, diameter D_0, cross-sectional area A_0 and volume V_0. (*b*) and (*c*) show two other positions which are discussed in the text. From H. C. Bennet-Clark (1976). In *Organisation of nematodes*, ed. N. A. Croll, Academic Press, New York & London.

increasing the total volume of the pharynx by that amount. This is resisted by the high pressure in the pseudocoel. If this pressure is ΔP above the pressure outside the worm the work required is $0.125\ V_0\Delta P$. The transformation from state (*b*) to state (*c*) will occur as soon as the cuticle is stretched enough for the transformation to release elastic strain energy greater than $0.125\ V_0\Delta P$. Thus the muscles do work stretching the cuticle, storing elastic strain energy which builds up until the critical stage is reached at which a lot of the energy is released and used to draw fluid into the pharynx.

The pharynx can be emptied again by allowing the muscles to relax, when the residual strain energy in the cuticle makes the lumen collapse.

Now consider defecation. In nematodes in which the pressure in the pseudocoel is high, it will tend to squirt the gut contents out of the anus unless there is either a valve or a sphincter muscle to prevent this. Fig. 12.13 shows the arrangement in one particular (rather unusual) species. There is a long tubular cloaca lined with cuticle stiff enough not to collapse under the pressure of the pseudocoel. At the anterior end of the cloaca there is a sphincter muscle which encircles the gut and prevents the gut contents from escaping except when it is relaxed. The anus at the posterior end of the cloaca does not close tightly but there is nevertheless a muscle capable of opening it widely (Fig. 12.1). Fig. 12.13 also shows that there is a network of muscle fibres all along the posterior part of the intestine.

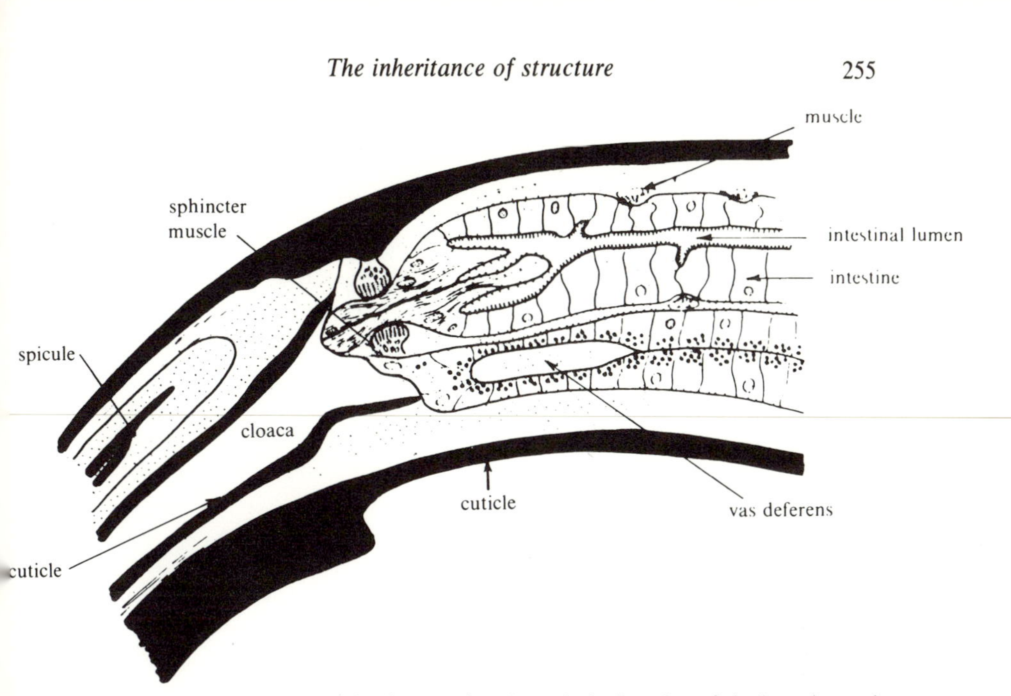

Fig. 12.13. Diagrammatic longitudinal section through the junction of the intestine and cloaca of a male *Heterakis gallinarum*. From D. L. Lee (1975). *Parasitology* **70**, 389–96.

THE INHERITANCE OF STRUCTURE

A great deal of progress has been made in learning how genes specify the structure of enzymes and other proteins. We know how different groups of three nucleotides in the DNA molecule specify different amino acids, and we know that the groups are arranged in appropriate order to specify the order of amino acids in the protein chain. Most of the information has been obtained by experiments on bacteriophages and bacteria but it is believed that the same code applies in higher organisms.

It is much less clear how details of gross structure (as opposed to molecular structure) are specified by genes. The problem is being investigated by Dr Sidney Brenner and his colleagues. They hope to discover principles which apply to all organisms but they have chosen to study a nematode (*Caenorhabditis*) which is found in soil. Several features make this animal particularly favourable for the study. It has only about 600 cells, 300 of them in the nervous system. As in other nematodes, the arrangement of the cells in the nervous system is more or less identical in all normal members of the species: identifiable neurones occupy similar positions in different individuals and make the same connections to other cells. Mutant worms with quite small defects in the nervous system can often be spotted easily, because they are paralysed or move oddly. *Caenorhabditis* is only about 1 mm long and completes its life cycle in only 3½ days. It feeds and grows satisfactorily on bacteria cultured on plates of agar jelly. Most specimens are hermaphrodite and fertilize themselves but males appear occasionally in cultures and are useful for certain genetic tests.

Caenorhabditis were put for a few hours into a solution of ethylmethane-sulphonate, which induces large numbers of mutations in the germ cells. It was most unlikely that the same mutation would occur in an ovum and in the spermatozoon which fertilized it, so the mutations in the first (F1) generation were heterozygous. Most of the mutations were recessive so they did not become apparent until the second (F2) generation.

Many mutants were obtained. Some were shorter and fatter or longer and more slender than normal worms. Some had abnormal (blistered) cuticle. Some were paralysed or moved oddly. In a few cases it was possible to show that defects of movement were associated with peculiarities of the central nervous system: cells were missing or did not make their usual connections. This is being investigated further.

The DNA has been extracted from *Caenorhabditis* and the total amount in an individual measured. It was found that there are about 10^{11} nucleotide pairs and as the individual has about 600 nuclei, each with a diploid set of chromosomes, there must be $10^{11}/(600\times 2) = 8\times 10^7$ nucleotide pairs in each haploid set. An average gene might contain about 1000 nucleotide pairs, enough to specify a polypeptide chain of about 330 amino acids. Hence there is enough DNA for about 80000 genes. It is of course possible for long sequences of DNA to be repeated but tests have shown that even when repetitions are discounted *Caenorhabditis* has enough DNA for 70000 genes. Are so many really necessary?

The experiments make a partial answer possible. A very large number of mutant clones were obtained but many of them were indistinguishable. These could be identical mutations which had occurred in different worms. Alternatively, they could be mutations in different genes which happen to have similar effects. The rare males were useful in a test to distinguish the alternatives. (The test is not applicable to genes on the sex chromosome.) Suppose m_1 and m_2 are recessive mutant genes with similar effects. The homozygotes m_1/m_1 and m_2/m_2 are indistinguishable. First m_1/m_1 is bred with a normal (wild type) male. The offspring have genotype $+/m_1$ and some of them are males. These are bred with m_2/m_2 females giving equal numbers of m_1/m_2 and $+/m_2$ offspring. If m_1 and m_2 are different genes, m_1/m_2 will not show the mutant character because it is heterozygous for both. If they are the same gene m_1/m_2 will be homozygous and have the mutant form. In this way it was shown that the mutants which could be identified simply by looking at the cultures involved only about 77 genes.

Three hundred and eighteen F1 offspring of parents treated with the mutagenic agent were put singly on agar plates. In 26 cases their F2 offspring showed one or other of the 77 identifiable mutations. This implies that the probability of a mutation occurring in a particular gene pair in a particular F1 offspring of a treated parent is $26/(318 \times 77) \simeq 10^{-3}$. The probability of it occurring in a particular gene is half this, or 5×10^{-4}. This has been calculated from data concerning only 77 genes but we will assume it applies to all 70000 genes. If so, every haploid set of chromosomes in the F1 offspring of treated parents would have about $5 \times 10^{-4} \times 70000 = 35$ mutant genes and each of the six chromosomes in the haploid set would have about six mutant genes. Every F2 offspring which was homozygous for one of the identifiable mutant genes would probably be homozygous for at least five other mutant genes. If most of the wild-type genes were essential to life, few or no F2 homozygotes would survive. Further experiments were performed and it was deduced from the results that only about 2000 of the 70000 genes are essential to life.

FURTHER READING

GENERAL

Bird, A. F. (1971). *The structure of nematodes.* Academic Press, New York & London.

Lee, D. L. & Atkinson, H. J. (1976). *The physiology of nematodes*, 2nd edn. Macmillan, London.

Nicholas, W. L. (1975). *The biology of free-living nematodes.* Clarendon Press, Oxford.

Wallace, H. R. (1973). *Nematode ecology and plant disease.* Arnold, London.

MUSCLES

Rosenbluth, J. (1967). Obliquely striated muscle. III. Contraction mechanism of *Ascaris* body muscle. *J. Cell Biol.* **34**, 15–33.

Toida, N., Kuriyama, H., Tashiro, N. & Ito, Y. (1975). Obliquely-striated muscle. *Physiol. Rev.* **55**, 700–56.

CUTICLE

Lee, D. L. (1972). The structure of the helminth cuticle. *Adv. Parasitol.* **10**, 347–79.

INTERNAL PRESSURE AND LOCOMOTION

Gray, J. & Lissmann, H. W. (1964). The locomotion of nematodes. *J. exp. Biol.* **41**, 135–54.

Harris, J. E. & Crofton, H. D. (1957). Structure and function in nematodes: internal pressure and cuticular structure in *Ascaris. J. exp. Biol.* **34**, 116–30.

Wallace, H. R. (1971). The movement of nematodes in the external environment. In *Ecology and physiology of parasites*, ed. A. M. Fallis, pp. 201–12. Hilger, London.

FEEDING

Bennet-Clark, H. C. (1976). Mechanics of nematode feeding. In *Organisation of nematodes*, ed. N. A. Croll, pp. 313–42. Academic Press, New York & London.

Doncaster, C. C. & Seymour, M. K. (1974). Passive ingestion in a plant nematode, *Hexatylus viviparus* (Neotylenchidae: Tylenchida). *Nematologica* **20**, 297–307.

Lee, D. L. (1975). Structure and function of the intestinal–cloacal junction of the nematode *Heterakis gallinarum*. *Parasitology* **70**, 389–96.
Roche, M. & Torres, C. M. (1960). A method for *in vitro* study of hookworm activity. *Exp. Parasitol.* **9**, 250–6.

THE INHERITANCE OF STRUCTURE

Brenner, S. (1974). The genetics of *Caenorhabditis elegans*. *Genetics* **77**, 71–94.

13
Snails and some other molluscs

Phylum Mollusca
  Class Monoplacophora (*Neopilina*)
  Class Aplacophora
  Class Polyplacophora (chitons)
  Class Gastropoda
    Subclass Prosobranchia (winkles etc.)
    Subclass Opisthobranchia (sea slugs etc.)
    Subclass Pulmonata (snails and slugs)
  Class Scaphopoda (tusk shells)
(See also chapters 14 and 15)

The molluscs are a large group, and a diverse one. This chapter is mainly about the class Polyplacophora (chitons) and the class Gastropoda (winkles, snails, etc.). These are molluscs which crawl on a large, flat foot. Three other minor classes are mentioned briefly. Chapter 14 is about the clams, which have a shell in two parts, hinged together. Chapter 15 is about squids, octopus and their relatives. The basic similarity in plan of the body which makes it appropriate to include all these animals in a single phylum will become apparent.

The molluscs are the first group of animals described in this book to have the type of body cavity known as a coelom. The distinctive features of coeloms are that they are lined with epithelium (that is, by a continuous sheet of cells) and that they usually house the gonads and excretory organs and have an opening to the exterior. The animals described in the remaining chapters of this book have coeloms of some sort, though they may be small. In annelid worms, echinoderms and vertebrates the main body cavities are coeloms. In most molluscs the coeloms are small and the main body cavity is a blood-filled haemocoel. This cavity, like the pseudocoel of nematode worms, has no epithelial lining but consists simply of spaces in other tissues. The coelom consists merely of the gonads, the kidneys and the pericardial cavity (a space round the heart which is part of the excretory apparatus).

Most molluscs have shells, and this chapter has a section about shells. There is also a section about the mechanisms used by gastropods and chitons when they crawl on their flat feet. Gastropods, chitons and bivalves have planktonic larvae which are discussed in another section. The remaining sections of the chapter deal with various aspects of gastropod physiology and with the special problems of life on shores, on land and in stagnant water.

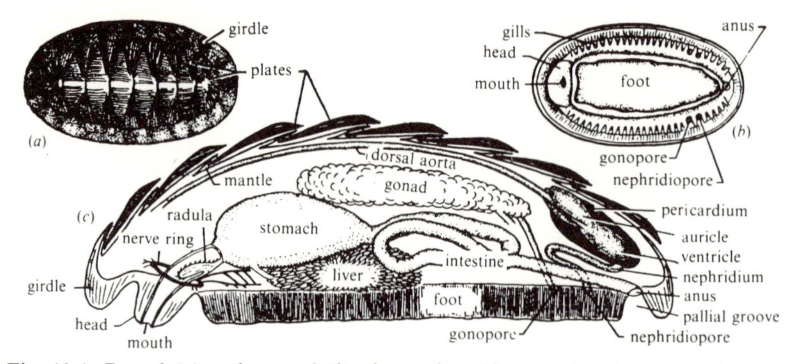

Fig. 13.1. Dorsal (*a*) and ventral (*b*) views of a chiton, and a diagram of its internal structure (*c*). From T. I. Storer & R. L. Usinger (1957). *General zoology*, 3rd edn. McGraw-Hill, New York.

The chitons (Polyplacophora) have peculiar shells. In other respects they are perhaps more like the ancestors of the molluscs than are any of the other common groups. They will serve to introduce the phylum (Fig. 13.1). They are rather flat oval animals, up to 0.3 m long but usually much smaller. They live in the sea where they crawl on the surface of rocks. They have a shell of eight overlapping plates which do not completely cover the body: the girdle which runs round the edge is flexible tissue with spicules of calcium carbonate embedded in it. When attached to rocks they are well protected by the shell and girdle. If dislodged, they roll up like woodlice so that the undersurface is still protected. This undersurface is occupied by the muscular foot. Chitons feed on filamentous algae which they scrape off the surfaces of rocks using an organ called the radula. The mouth is at the anterior end of the body and the anus at the posterior end. The digestive gland (labelled 'liver' in Fig. 13.1) is a diverticulum of the gut where digestive enzymes are secreted and much digestion occurs. There is a ring of nervous tissues round the pharynx from which two main nerves (the pedal cords) run to the foot and two (the pleural cords) to the viscera. Fig. 13.1 shows only the anterior part of the pedal cords. Note the cross-connections, like the rungs of a ladder. There is a line of up to 70 gills on each side of the foot, under the girdle. They are the type of gills called ctenidia, which are described later in this chapter. The heart has three chambers, two auricles and a ventricle. Each auricle receives blood from the gills of its side of the body and passes it to the ventricle, which pumps it through the aorta and other arteries to the tissues. It returns to the heart through the haemocoel, the main body cavity. There is a pair of excretory organs which discharge through openings called nephridiopores on either side of the foot. There is a single gonad.

The structures of gastropods is most easily described by reference to a hypothetical ancestor (Fig. 13.2*a*, *b*, *e*). This animal is presumed to have had a one-piece shell under which was a cavity containing a pair of ctenidia. This

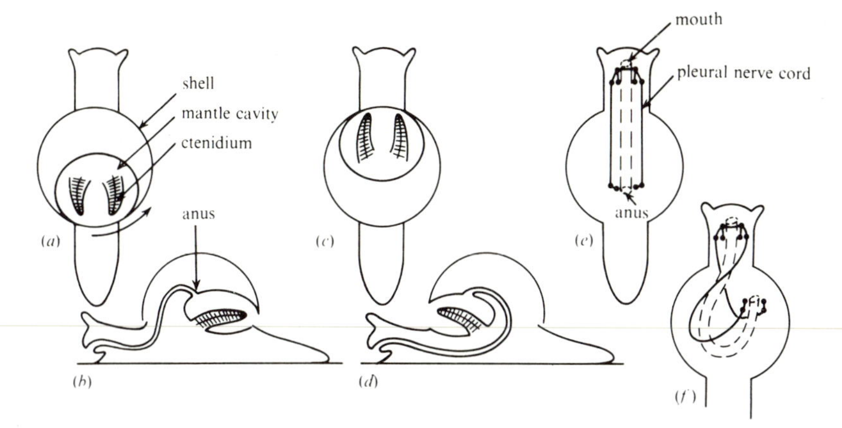

Fig. 13.2. Diagrams of (*a*, *b*, *e*) a hypothetical primitive gastropod and (*c*, *d*, *f*) a member of the subclass Prosobranchia.

is the mantle cavity. It was open posteriorly to the surrounding water, and the anus opened into it. This primitive arrangement is found in gastropod larvae which are described later in the chapter. In the course of evolution the process called torsion occurred. The shell twisted around in the direction indicated by the arrow in Fig. 13.2(*a*) so that the opening of the mantle cavity was anterior, over the head (Fig. 13.2*c*, *d*, *f*). This is the situation in adult members of the subclass Prosobranchia (and also in Pulmonata, which have a mantle cavity but no ctenidia). The ancestral gastropod probably had a pair of pleural nerve cords connecting a ring of ganglia round the mouth to other ganglia near the anus (Fig. 13.2*e*). In Prosobranchia the pleural cords are twisted, due to torsion (Fig. 13.2*f*).

In gastropods of the subclass Opisthobranchia torsion has been reversed. The mantle cavity opens to the side or posteriorly (or is lost) and the pleural cords are untwisted.

Fig. 13.3 shows the anatomy of the snail *Helix*, as an example of the gastropods. It is a member of the subclass Pulmonata, and like other pulmonates has no ctenidia. The mantle cavity has become a lung. The reproductive system is more complicated than in many other gastropods. Prosobranchs have separate sexes but opisthobranchs and pulmonates are hermaphrodite and have a single gonad (the ovo-testis) which produces both eggs and sperm. When *Helix* copulates each partner passes sperm to the other. In many other hermophrodite gastropods sperm passes only one way. One partner acts as male and one as female but the roles may be reversed in later copulations between the same partners.

Fig. 13.4 shows a selection of gastropods. *Haliotis* lives in the sea below low tide level, where it attaches itself so firmly to the rocks that a lever is needed to dislodge it. It eats red algae. The holes in the shell are outlets for the respiratory current (see Fig. 13.18*b*). Species of *Haliotis* can be found in most

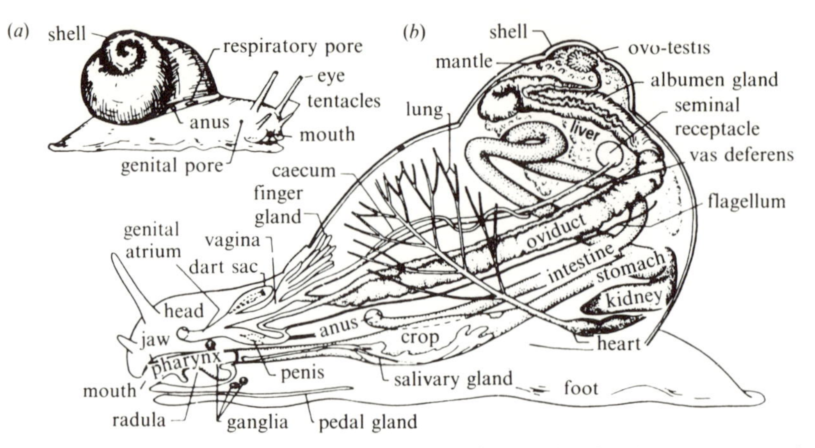

Fig. 13.3. The garden snail, *Helix aspersa*. (*b*) is a diagram showing the arrangement of the internal organs. The label 'lung' points to blood vessels in the wall of the lung. The diameter of the shell of this species is about 35 mm. From T. I. Storer & R. L. Usinger (1957). *General zoology*, 3rd edn. McGraw-Hill, New York.

parts of the world. The whelk *Buccinum undatum* (Fig. 13.4 *b*) is a North Atlantic species, found at and below low tide level. It is a carnivore and scavenger, feeding on crabs, worms and bivalve molluscs and on dead animals. There is a groove (like the spout of a jug) in the shell opening. When the whelk is extended a long tube (the siphon) protrudes from this groove. The respiratory current enters the mantle cavity through the siphon. There is a sense organ in the mantle cavity, and the siphon seems to be used for sniffing prey. The foot bears a hard plate (operculum) which closes the mouth of the shell when the animal retracts into the shell (as in Fig. 13.4 *b*). *Haliotis* and *Buccinum* are prosobranchs and like most other prosobranchs have robust shells, big enough to withdraw into. Most opisthobranchs have reduced shells, or no shell at all (Fig. 13.4 *c*, *d*). *Aplysia* has a reduced shell, far too small to withdraw into and hidden by soft tissues. There are many species, most of them in tropical and subtropical seas: the one which is illustrated (Fig. 13.4 *c*) is from the Mediterranean. They are called sea hares because there are projections like hare's ears on the head. They live near low tide level and feed on seaweeds, which they tend to match in colour. They can swim briefly, by means of the large flaps on either side of the body. There are other opisthobranchs (the pteropods) which swim perpetually. *Aeolidia* (Fig. 13.4 *d*) is an example of the nudibranch sea slugs which have no shell and no ctenidia. It feeds on sea anemones, and the greyish-brown tentacle-like projections on its back make it inconspicuous on certain species of anemone. These projections serve as gills, and they house extensions of the digestive gland. Nematocysts from anemone prey are somehow transported undischarged to the tips of the projections, where they accumulate. It is believed that they protect *Aeolidia* against predators, by discharging when the nudibranch is attacked. *Aeolidia* lives on

Fig. 13.4. A selection of gastropods. (*a*) An abalone, *Haliotis lamellosa* (Prosobranchia, length up to 7 cm). (*b*) A whelk, *Buccinum undatum* (Prosobranchia, up to 12 cm). (*c*) A sea hare, *Aplysia depilans* (Opisthobranchia, up to 25 cm). (*d*) A nudibranch, *Aeolidia papillosa* (Opisthobranchia, up to 8 cm). (*e*) A pond snail, *Lymnaea stagnalis* (Pulmonata, about 5 cm). (*f*) *Planorbis corneus* (Pulmonata, diameter of shell 3 cm). (*a*) to (*d*) are from W. de Haas & F. Knorr (1966). *The young specialist looks at marine life.* Burke, London. (*e*) and (*f*) are from W. Engehardt (1964). *The young specialist looks at pond life.* Burke, London.

European shores at and below low tide level. *Lymnaea* and *Planorbis* (Fig. 13.4*e*, *f*) are snails found in lakes, ponds and ditches. *Lymnaea* is very commonly found on the water weed *Elodea* but does not eat the weed itself: it feeds on filamentous algae and diatoms growing on the weed, which it scrapes off with the radula. *Lymnaea* and *Planorbis* both belong to the subclass Pulmonata which also includes the terrestrial snails (Fig. 13.3) and slugs.

So far, only Polyplacophora and Gastropoda have been described. None of the other classes listed at the beginning of this chapter is at all numerous. The best known of the Monoplacophora is *Neopilina*, dredged from a great depth in the Pacific Ocean. It has a domed shell and a round foot, like a limpet, and has five gills which are not typical ctenidia on each side of the foot. It has six pairs of kidneys which alternate with the five pairs of gills, and eight pairs of muscles running from the shell to the foot. These structures have been claimed as evidence that molluscs are descended from ancestors which were segmented, like annelid worms. The claim is not convincing since more or less regular repetition of parts occurs in many animals which are not regarded as segmented (see for instance Fig. 8.4). The Aplacophora are worm-like animals which have no shell but have a radula, a ctenidium and other characteristics of molluscs. The Scaphopoda have tapering tubular shells and lie buried in sand or mud with the narrow end of the shell protruding.

MOLLUSC SHELLS

Most molluscs have shells into which they can withdraw, so as to be completely enclosed. The main constituent of the shell is calcium carbonate which may be in either of two crystalline forms, calcite and aragonite. The remainder is an organic matrix which consists largely of a protein known as conchiolin and usually makes up less than 5% of the mass of the shell. Bone also consists of salt crystals in an organic matrix but the salt is mainly calcium phosphate and the matrix is mainly collagen and makes up 30% of the mass.

The fine structure of mollusc shells has been studied by various techniques including scanning electron microscopy of broken surfaces. Fig. 13.5 shows some of the arrangements which are found. In each of them blocks or strips of calcium carbonate are separated by thin layers of conchiolin. Much the most common structure is crossed-lamellar (Fig. 13.5*c*). The shell consists of long strips of aragonite laid down in groups. The members of each group are parallel, but different groups lie in different directions. Fig. 13.5(*a*) shows nacre, or mother of pearl. It consists of tiny blocks of aragonite arranged in layers. Many shells have an inner layer of nacre, though they have a different structure through most of their thickness. A few, including the pearly nautilus (*Nautilus*) and the pearl oyster (*Pinctada*), consist mainly or entirely of nacre. Pearls are balls of nacre formed around sand grains and other particles which get into the mantle cavity.

To protect the mollusc, shells must be rather strong. Strips have been cut

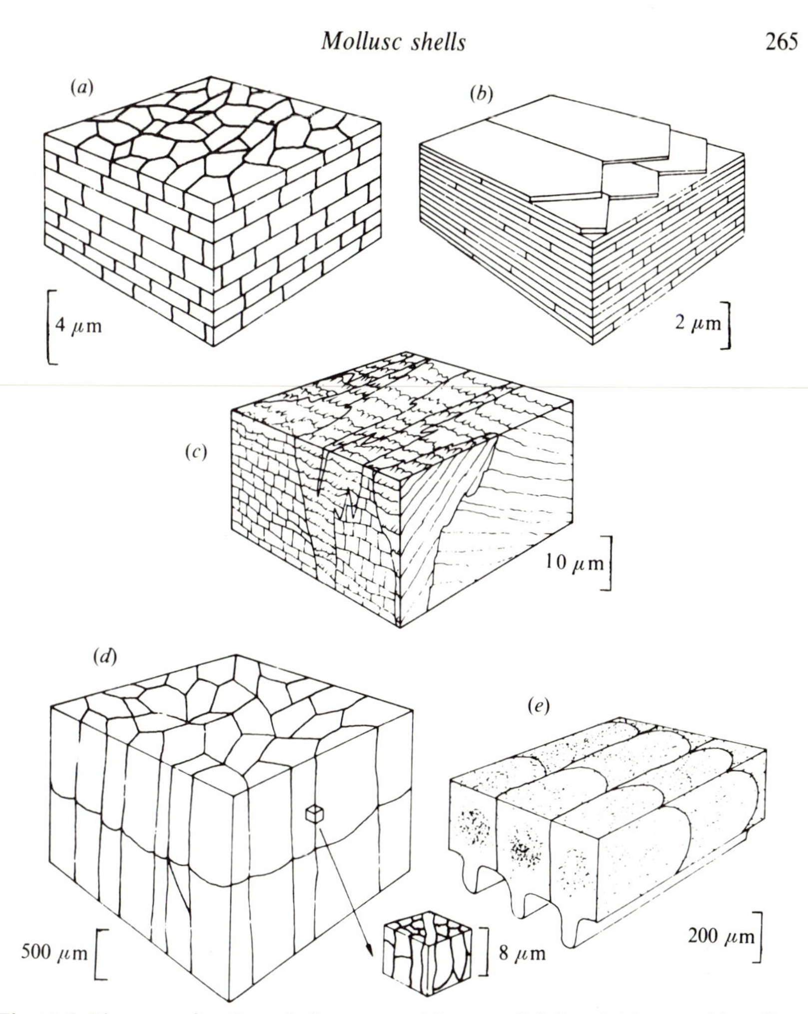

Fig. 13.5. Diagrams of mollusc shell structure. (*a*) nacre; (*b*) foliated; (*c*) crossed-lamellar; (*d*) simple prisms; (*e*) composite prisms. From S. A. Wainwright, W. D. Biggs, J. D. Currey & J. M. Gosline (1976). *Mechanical design in organisms.* Arnold, London.

from the shells of various molluscs and tested in engineers' testing equipment. In most cases the tensile strength lay between 30 and 100 MN m^{-2}, and Young's modulus was about 50 GN m^{-2}. Thus mollusc shell is weaker and stiffer than compact bone, which has a tensile strength of 150–200 MN m^{-2} and Young's modulus around 18 GN m^{-2}. Nacre is generally stronger than crossed-lamellar shell. It is not surprising that shell is stiffer than bone, since it contains a higher proportion of inorganic crystals. It is less obvious why it should be weaker and why nacre should be stronger than crossed-lamellar shell which contains a smaller proportion of organic matter. The next few paragraphs point to a possible reason for they show that the organic matter strengthens shell and bone.

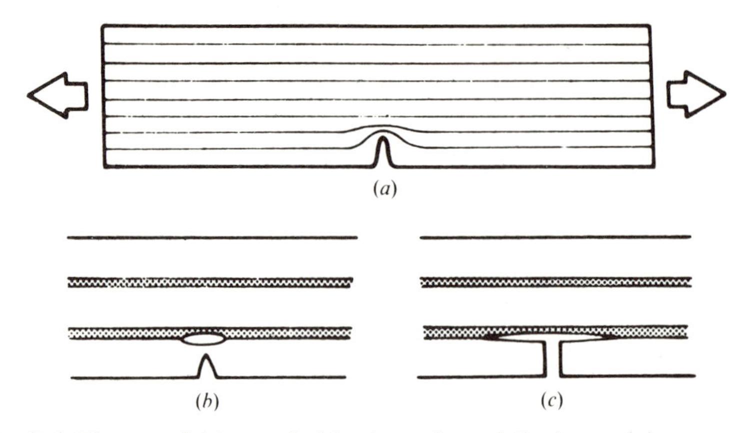

Fig. 13.6. Diagrams of (*a*) a notched bar in tension and (*b*, *c*) a crack in a composite material. From R. McN. Alexander (1975). *Biomechanics*. Chapman & Hall, London.

How does the conchiolin affect the strength of mollusc shell? Breaking a material involves separating two layers of atoms, and the stress required can be calculated. However, the measured strengths of materials are generally far less than theoretical strengths obtained in this way. For instance, the theoretical strength of glass is about 10 GN m^{-2} but ordinary glass breaks at about 200 MN m^{-2}. Fibreglass, which consists of fine glass fibres embedded in a plastic resin, can have tensile strengths up to about 1 GN m^{-2}. Mollusc shell and bone are composite materials like fibreglass: they consist of tiny crystals embedded in an organic matrix. They may owe much of their strength to effects like those which make fibreglass stronger than bulk glass.

Materials tend to break at stresses far below their theoretical strength because stress concentrations develop near irregularities in them. This is illustrated by Fig. 13.6(*a*) which represents a notched bar in tension. The thin lines show the direction of tensile stress and where they are evenly spaced the stress is uniform across the bar. They are diverted round the notch and are close together at the tip of the notch: this indicates that the stress there is high. As the force on the bar is increased the stress at the tip of the notch may pass the tensile stress and a crack may form there, while the stress in the bulk of the bar is still far lower. Once a crack has formed it will tend to spread across the bar because a crack is in effect a very sharp notch: the sharper the notch, the greater the stress concentration. There are always unintentional irregularities in objects made of glass and other brittle materials: they cannot be made so perfect as not to be in danger from stress concentrations.

Cracks may form in fibreglass but cannot spread through it easily. Fig. 13.6(*b*) and (*c*) show why. They represent a bar consisting of strong fibres (shown white) separated by weaker glue (stippled). As a crack approaches, the glue splits so that when the crack reaches the glue it no longer has the sharp tip needed to enable it to continue to the next fibre. A bar in tension has tensile stress acting

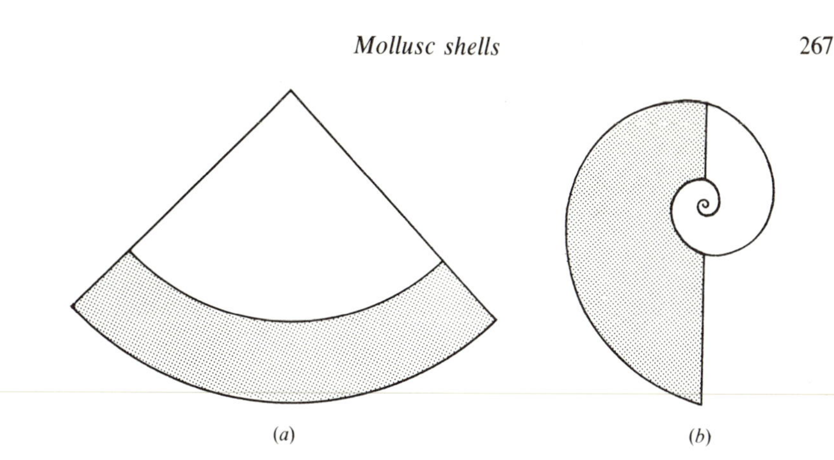

Fig. 13.7. Diagrams of a conical and spiral shell. Addition of the stippled region to the original untinted shell leaves the shape unchanged. From R. McN. Alexander (1971). *Size and shape*. Arnold, London.

along it but where the force is diverted round a notch or crack the stress has a transverse component. It is the transverse component near the tip of the crack in Fig. 13.6(*b*) which makes the glue split. The layers of conchiolin in shells are believed to stop cracks in this way.

Now that we have discussed the strength of mollusc shells we must examine their shape. They are not living material. They can be added to as the mollusc grows but existing parts cannot be altered. The shell of a young snail remains in the adult shell, as the small whorls at the apex.

Most mollusc shells keep the same shape as they grow: the young shell is in effect a scale model of the adult one. This is only possible for a limited range of shapes. If the shell were cylindrical and grew by adding material to its end it would become more slender in its proportions as it grew. If it were a hollow hemisphere it could not be added to without a change of shape. Two shells which can be added to without change of shape are shown in Fig. 13.7. One is a hollow cone like the shells of limpets (*Patella*). The other resembles a *Planorbis* shell (Fig. 13.4*f*) and is in effect a cone coiled into a spiral. The spiral line is a logarithmic spiral: that is, a spiral which increases its radius by the same factor in every turn. The geometry of spiral shells was discussed by Sir D'Arcy Thompson in his classic *Growth and Form*, which was published in 1917. New light has been thrown on it much more recently by Dr David Raup.

Fig. 13.8(*a*) represents a spiral shell. Its centre line is a spiral, with radius r_0 at a point which we will take as a starting point. If we move an angle θ round the spiral we find that the radius r there is given by the equation

$$r = r_0 W^{\theta} \tag{13.1}$$

where W is a constant. This is the equation of a logarithmic spiral. We will assume that the spiral does not lie in a plane but rises helically, like a snail shell (Fig. 13.8*b*). As we move along the spiral through an angle θ we find that the

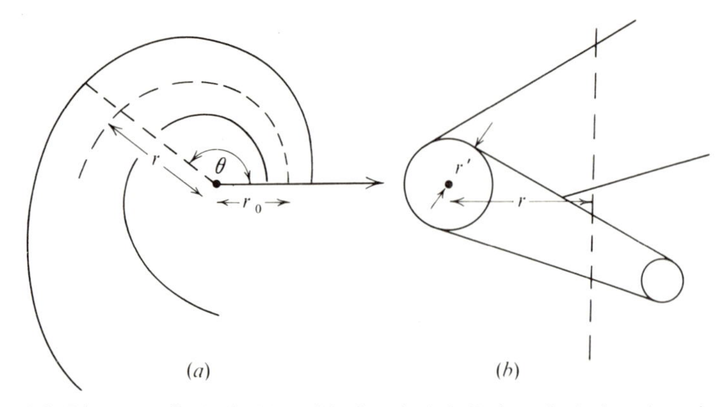

Fig. 13.8. Diagrams of a helical logarithmic spiral shell viewed (*a*) along its axis and (*b*) at right angles to its axis.

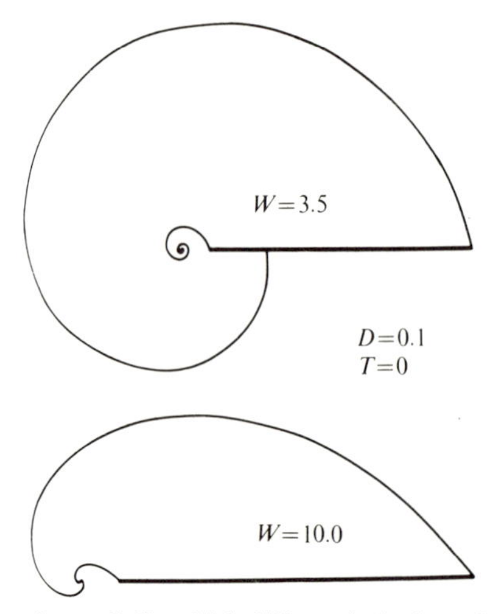

Fig. 13.9. Diagrams of two shells which differ only in the value of the parameter *W*. From D. M. Raup (1966). *J. Palaeontol.* **40**, 1178–90.

spiral moves a distance y along its axis. If growth is not to change the shape of the shell

$$y = T(r - r_0) \qquad (13.2)$$

where T is another constant. For simplicity, the whorls are shown circular in section. At the point where the radius of the spiral is r the radius of the section is r'. To keep the shape of the shell r' grows in proportion to r

$$r' = r(1-D)/(1+D) \qquad (13.3)$$

where D is a third constant. Many very different-looking shells can be described by varying the parameters W, T and D. W indicates the rate at which

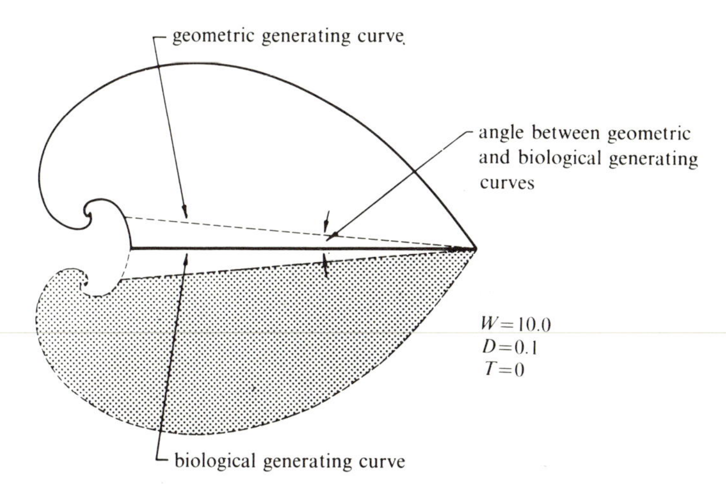

Fig. 13.10. A diagram showing how a bivalve shell can be formed from logarithmic spiral shells, if the mouth of each shell is set at an angle to the radius of the spiral. From D. M. Raup (1966). *J. Palaeontol.* **40**, 1178–90.

the shell expands: the radius increases by a factor W for every radian of revolution. Fig. 13.9 shows that small values of W give tightly coiled shells like snail shells and large values give shells more like the shells of bivalve molluscs: the lower shell resembles a cockle (*Cardium*). T indicates the extent to which the shell rises in a spire. *Lymnaea* (Fig. 13.4*e*) has a large value of T and *Planorbis* (Fig. 13.4*f*) a very small one. D shows how far the inner edge of the shell is from the axis of the spiral. If $D = 0$, $r' = r$ and the inner edges of the whorls reach the axis, as in *Buccinum* (Fig. 13.4*b*), but not in *Planorbis*.

The lower shell in Fig. 13.9 is quite like a cockle shell, but it could not be fitted to another of similar shape to make a bivalve shell because the two points would meet and prevent tight closing. Fig. 13.10 shows how a small modification overcomes this difficulty.

Careful measurements have been made on many mollusc shells and most of them have been found to be described rather accurately by equations (13.1) to (13.3). In some cases, however, W and T change in the course of growth and the adult shell is not the same shape as the young one.

Fig. 13.11 illustrates the range of shape which can be obtained by varying W, T and D and shows the ranges actually found in various groups of animals. Most gastropods have low values of W, but T and D vary over wide ranges. Bivalve molluscs have high values of W and fairly low values of T and D. *Nautilus* and most of the extinct ammonoids (which are also cephalopod molluscs) have plane spiral shells ($T = 0$) with low values of W. Brachiopods have bivalve shells and look superficially like bivalve molluscs, but are very different inside (see chapter 22). They have high values of W and fairly low values of D but unlike bivalve molluscs most of them have $T = 0$.

The gastropod and bivalve types of shell are alternative containers which give

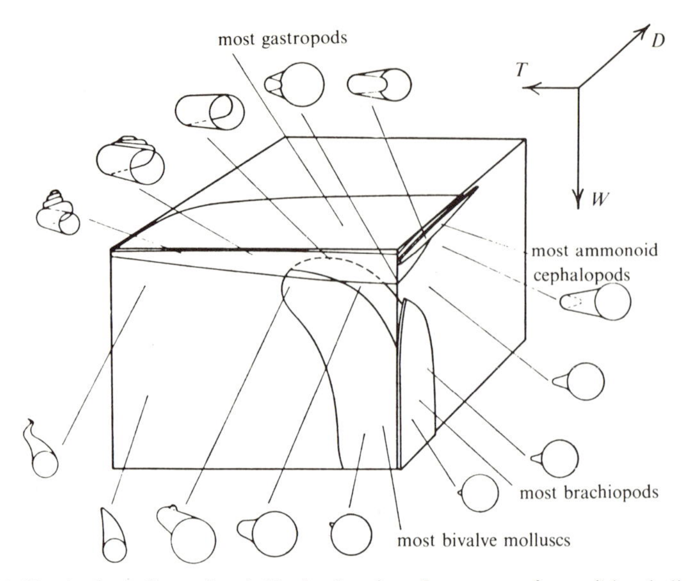

Fig. 13.11. A three-dimensional block showing the range of possible shell forms indicated by equations (13.1) to (13.3), and the ranges actually found in some groups of animals. Some examples of shell shape are shown: they were generated from the equations by means of a computer with a graphic display. From J. D. Currey (1970). *Animal skeletons*. Arnold, London. (After D. M. Raup.)

good protection. A typical gastropod shell has a small opening so that when the snail retracts little of its body is exposed. In some cases the opening can be closed by an operculum. A pair of bivalve shells fit together to enclose the mollusc completely. Some gastropod shells such as those of limpets (*Patella*) and of *Haliotis* have large openings, but these molluscs live on rocks and can protect themselves by holding the shell opening firmly against the surface of the rock.

CRAWLING

Gastropods crawl, but only slowly. A speed of 2.5 mm s^{-1} is fast for a snail (*Helix*) and the maximum speed of the limpet *Patella* is about 1 mm s^{-1}. Fig. 13.12 shows the foot of *Haliotis*, crawling on glass. The stippled areas look darker than the rest, and are raised off the glass. Comparison of successive frames shows that marks on the foot move forward while in a stippled area, and remain stationary while in a white one. In *Haliotis* the light and dark bands move forward over the foot to move the animal forwards but in some other molluscs such as *Patella* the same effect is achieved by bands moving backwards.

Fig. 13.13 shows how forward movement can be produced either by forward-moving or by backward-moving bands. In both cases the front part of the raised portion of the foot is shortening and the rear part elongating, so

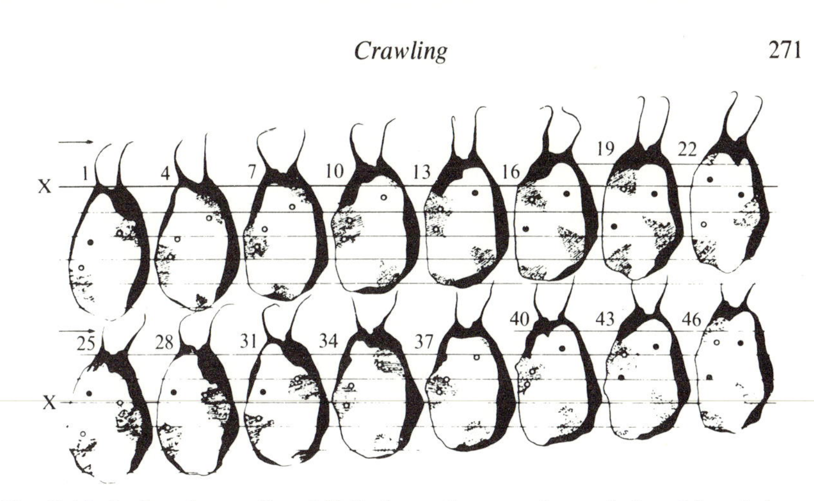

Fig. 13.12. Outlines from a film of *Haliotis* crawling over glass and viewed from below. Three marks had been made on the foot. They are shown as solid circles when stationary and as open circles when moving. From J. Gray (1968). *Animal locomotion*. Weidenfeld & Nicolson, London.

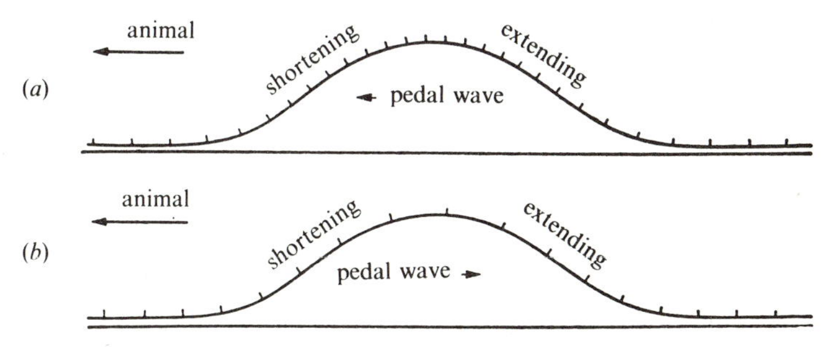

Fig. 13.13. Diagrams showing how forward crawling can be achieved by waves moving either (*a*) forwards or (*b*) backwards along the foot. From H. W. Lissmann (1945). *J. exp. Biol.* **21**, 58–69.

that the body of the mollusc is being moved forward relative to the ground. In (*a*) the raised portion is contracted, so if shortening is occurring at its anterior end the wave of shortening must be travelling forwards. This is what happens in *Haliotis*. In (*b*) the raised portion is elongated and the wave is travelling backwards. This mechanism is the same in principle as the crawling mechanism of nemertean and other worms, shown in Fig. 10.5.

Limpets (*Patella*) use the method shown in Fig 13.13(*b*). The foot is muscular, with 70% of its muscle fibres running vertically from the shell to its sole (most of the other fibres run transversely). There are small, blood-filled cavities among the muscle fibres. Contraction of the vertical muscle fibres lifts the foot and also compresses it dorso-ventrally. If it is prevented from widening by the transverse muscle fibres, it must elongate. Fig. 13.14 shows the sequence of events. Stages 1 to 4 show muscles *a* and *b* contracting, raising

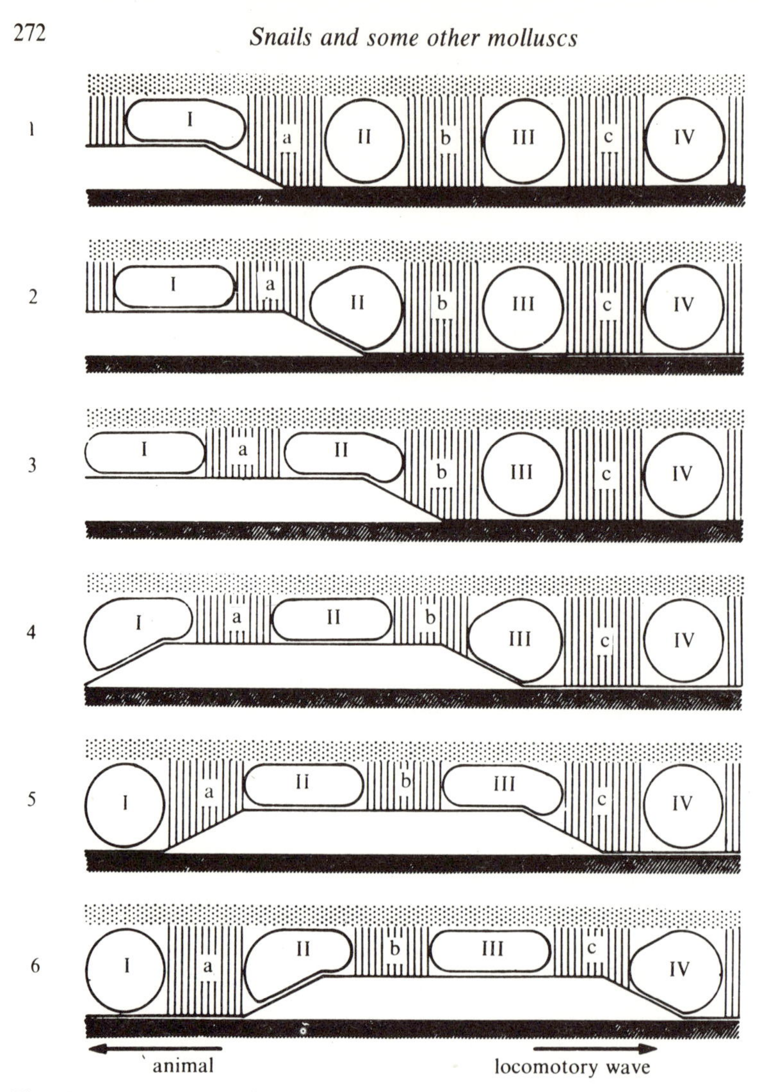

Fig. 13.14. A sequence of diagrammatic vertical sections through the foot of *Patella*, showing the probable mechanism of crawling. *a–c*, groups of vertical muscle fibres; I–IV, blood spaces. From H. D. Jones & E. R. Trueman (1970). *J. exp. Biol.* **52**, 201–16.

the sole and making the space II elongate. At stage 5, *b* is still exerting tension and *c* has started to contract but *a* is relaxed. It is being stretched again, because tension in the other muscles maintains a reduced pressure under the foot. The reduction of pressure can be demonstrated by allowing a limpet to crawl over a hole, connected to a pressure transducer, in a Perspex plate. Whenever the foot over the hole is raised the transducer records a pressure reduction of about 600 N m^{-2}.

Helix crawls in the manner of Fig. 13.13(*a*) and has its muscles differently

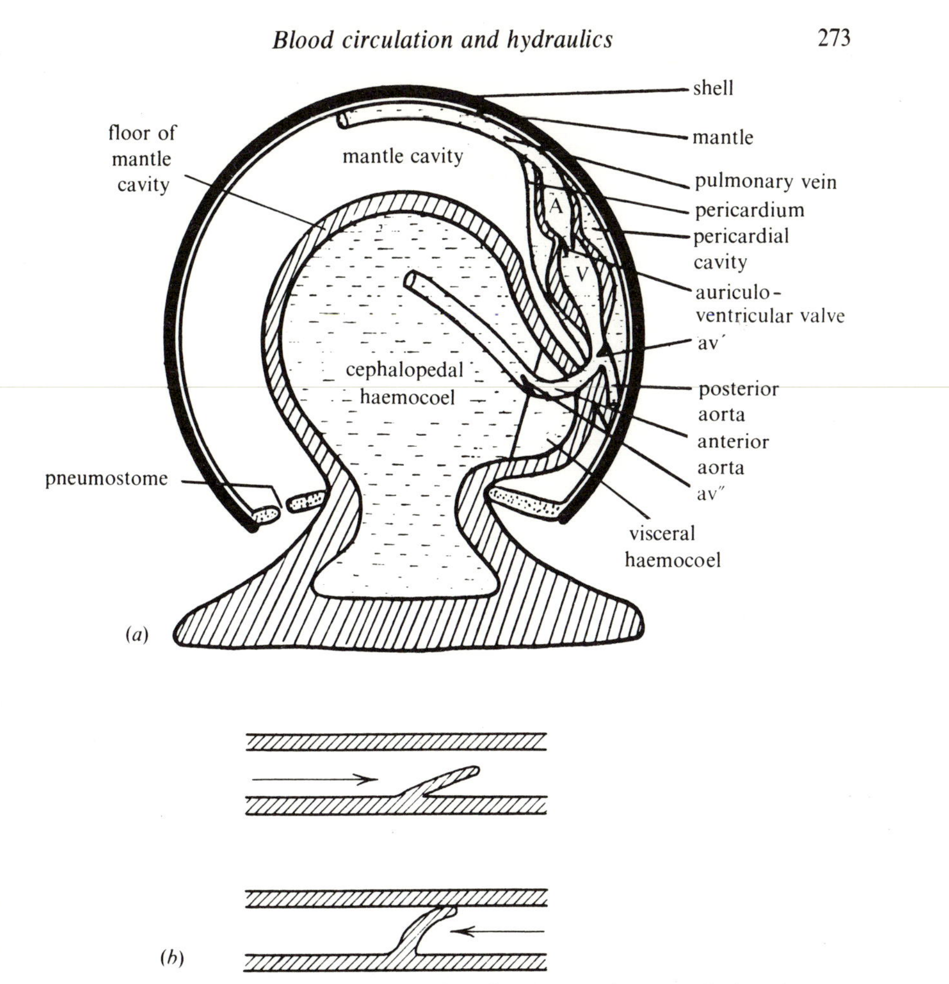

Fig. 13.15. (*a*) A diagrammatic cross-section of *Helix* showing the principal features of its blood system. A, auricle; V, ventricle; av′, av″, aortic valves. From B. A. Sommerville (1973). *J. exp. Biol.* **59**, 275–82. (*b*) Diagrams showing how valves such as the auriculo-ventricular valve, av′ and av″ permit flow in one direction only.

arranged so that parts of the foot which are raised off the ground can be shortened.

BLOOD CIRCULATION AND HYDRAULICS

The circulation of the blood has been studied more thoroughly in the edible snail, *Helix pomatia*, than in any other gastropod. Figs. 13.15 and 13.16 show the principal features of its blood system. The finer details can be seen most clearly in dissection, if coloured latex is injected into the vessels. The heart has two chambers, a thin-walled auricle and a more muscular ventricle. It can be watched beating in the living animal if a transparent window is made in the shell over it, by dropping on concentrated hydrochloric acid. It beats with

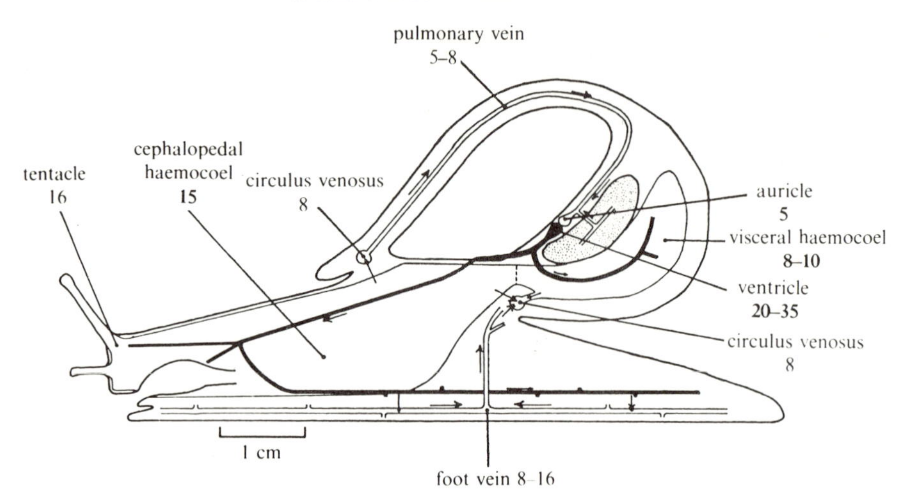

Fig. 13.16. A diagram of blood circulation in *Helix pomatia*. The ventricle and principal arteries are shown black and the kidney is stippled. The arrows show directions of blood flow. The numbers show blood pressure relative to the atmosphere in cm water (1 cm water $\simeq$ 100 Nm^{-2}): pressures shown in bold type were actually measured, and the rest are estimates. From B. Dale (1973). *J. exp. Biol.* **59**, 477–90.

frequencies up to 1 s^{-1}. It will also beat after removal from the animal, provided it is kept in a suitable saline solution. Its beat must be kept going by pacemaker cells, like the cells in the ganglia of jellyfishes (p. 164). Its contraction is muscular and its re-expansion is presumably an elastic recoil. The blood can only flow through it in one direction (from auricle to ventricle) because the valves (Fig. 13.15) would stop reverse flow. These valves are flaps of flexible tissue which open for blood flowing in one direction but close as soon as it starts to move in the opposite direction (Fig. 13.15 *b*).

The blood leaving the heart flows either along the anterior aorta to the head and foot or along the posterior aorta to the digestive gland and other viscera (Fig. 13.16). These arteries divide up into fine branches but there are no capillaries like those of vertebrates. Instead, the fine branches of the arteries open into spaces in the tissues which connect with the foot veins or directly with one or other of the two main divisions of the body cavity. These divisions are the cephalopedal haemocoel in the head and foot and the visceral haemocoel in the shell. From there the blood enters the circulus venosus, a ring of veins around the edge of the mantle cavity. It flows through a network of veins in the lung, and so back to the auricle.

Blood pressure has been measured in various parts of the circulatory system. The snail was held by its shell in a burette clamp. A hypodermic needle connected to a pressure transducer was held in a micromanipulator. The needle was inserted into the heart through a hole in the shell, or into one of the haemocoels, so that the transducer registered the pressure there. The pressure in the ventricle rose to 2000–3500 N m^{-2} above atmospheric each time it contracted and fell to about 500 N m^{-2} each time it relaxed again. Pressures recorded or estimated for various parts of the system are shown in Fig. 13.16. The pressure falls (as it must do) as the blood travels from the ventricle, round the body and back to the auricle again. The pressure in the visceral haemocoel is much lower than in the cephalopedal haemocoel. This is probably due to the first aortic valve (*av′* in Fig. 13.15*a*) which seems likely from its position to close the posterior aorta while blood is flowing out of the heart. The posterior aorta may have to be filled by backflow from the anterior aorta after the ventricle has completed its contraction, though the second aortic valve (*av″*) seems likely to limit the backflow.

For the soft parts of the body to be rigid enough to support the weight of the snail, the body wall must be taut and the fluids inside under pressure. The blood in the haemocoels must thus be kept above atmospheric pressure. The shell and viscera are supported by a narrow stalk which connects them to the head and the foot. They impose a load of 8 g wt (in a typical *Helix*) on a stalk of area 1 cm^2 so a pressure of at least 8 g wt cm^{-2} (800 N m^{-2}) is needed in the cephalopedal haemocoel. The tentacles are kept stiff by the pressure of blood in them and experiments with tentacles cut from narcotized snails show that pressures around 1600 N m^{-2} are needed.

The air in the mantle cavity also has a hydraulic function. When the snail is extended, this air fills much of the cavity of the shell. When it withdraws into the shell the cavity is almost obliterated. It withdraws by contracting muscles which run from the central axis of the shell to the head and foot. This forces air out of the mantle cavity through the pneumostome. To emerge again the snail must pump itself out of the shell. It emerges in a series of jerks with pauses between. During the jerks the pneumostome is open but in the pauses it closes. Fig. 13.15(*a*) shows that if the curved mantle floor contracts it must draw air into the mantle cavity and force the foot out of the shell. It seems that this is what happens. While the floor contracts the pneumostome remains open but while it is relaxing the pneumostome is closed tightly so that air pressure in the mantle cavity prevents the body from moving too far back into the shell.

RESPIRATION

The rates at which gastropods use oxygen vary with size and temperature and between species but are typically around 0.1 cm^3 (g soft tissue)$^{-1}$ h^{-1} or about the same as for flatworms (Fig. 8.6*b*; it seems best to express the rate in terms of soft tissue weight, excluding the shell, in making this comparison). It was shown in chapter 8 that the tissues of an animal cannot be supplied at this rate by oxygen diffusing in from the general body surface if the animal is more than about 1 mm thick or (if cylindrical) 1.5 mm in diameter. Most molluscs are much larger than this and oxygen is distributed round the body in the blood.

Water equilibrated with atmospheric air dissolves about 8 cm^3 oxygen l^{-1} and sea water dissolves about 6 $cm^3 l^{-1}$. If mollusc blood were simply a saline solution which carried oxygen in physical solution, it would carry at most 8 $cm^3 l^{-1}$. To supply oxygen at the rate of 0.1 cm^3 (g tissue)$^{-1}$ h^{-1} needed for metabolism it would have to be pumped through the respiratory organ at *at least* 0.1/8 l $g^{-1} h^{-1}$, or 35 $mm^3 s^{-1}$ for a *Helix* of mass 10 g, excluding the shell. (It would actually have to be pumped faster because it could neither quite reach equilibrium in the lung nor give up quite all its oxygen in the tissues.) It seems unlikely that *Helix* hearts actually pump as fast as this: from experiments with isolated hearts, rates around 20 $mm^3 s^{-1}$ seem more likely. The tissues could not be supplied with oxygen at the required rate unless the heart were larger or beat faster.

The blood is not simply a saline solution, but contains protein. In many gastropods more than 90% of this protein is haemocyanin, which contains copper and can combine with oxygen. *Helix* haemocyanin is a huge molecule, of molecular weight 9 million. It consists of about 180 units of molecular weight about 50000, each containing two copper atoms and capable of combining with one molecule of oxygen. Molluscs and other animals which have haemocyanin in their blood generally have enough to carry about 20 cm^3 oxygen l^{-1}, or 2½–3 times as much as will dissolve in physical solution.

The haemocyanin adds to the colloid osmotic pressure of the blood. Since 1 mole of oxygen occupies 22.4 l, 20 cm^3 oxygen l^{-1} is about 1 mmol l^{-1}. If the haemocyanin molecules were relatively small, each taking up only one molecule of oxygen, their concentration would have to be 1 mmol l^{-1}. The osmotic pressure Π of a solution of molar concentration c is given by the equation

$$\Pi = RTc \tag{13.4}$$

where T is the absolute temperature (about 300 K in this case) and R is the universal gas constant (8.3 J $K^{-1}mol^{-1}$). Hence the osmotic pressure of the haemocyanin would be about 2500 N m^{-2} or 25 cm water. (Note that in using the equation, expressing quantities in SI units, c must be given its value in mol m^{-3}: 1 mmol l^{-1} = 1 mol m^{-3}.) This is similar to the pressure in the contracting ventricle of *Helix* (Fig. 13.16). Since the colloid osmotic pressure of the blood has to be overcome by blood pressure in the kidney, as will be explained in the next section of this chapter, too high a value must be avoided. Aggregation of the haemocyanin into giant molecules containing 180 of the basic units will reduce its contribution to the colloid osmotic pressure by a factor of 180. The total colloid osmotic pressure of *Helix* blood, due to haemocyanin and other proteins, is only 100–300 N m^{-2}.

Haemocyanin is blue when it is oxygenated and colourless when it is not. Blood from the heart of *Helix* is blue and blood from the veins is colourless. It appears that the haemocyanin takes up oxygen in the lung and gives it up in the tissues. It does this because it takes up oxygen where the partial pressure of oxygen is high and releases it where the partial pressure is low. Fig. 13.17 shows that it is about 50% saturated (i.e. it carries about 50% of the maximum

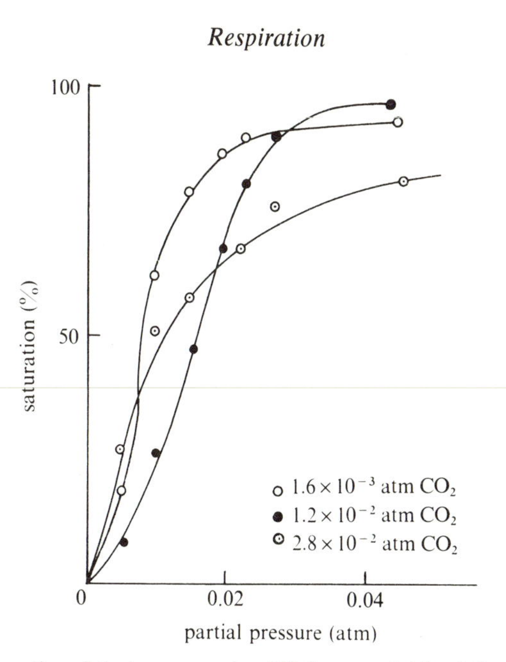

Fig. 13.17. Properties of the haemocyanin of *Helix pomatia* blood. Percentage saturation is plotted against partial pressure of oxygen, for three different partial pressures of carbon dioxide. The measurements were made at 15 °C. Re-drawn from G. L. Spoek, H. Bakker & H. P. Wolvekamp (1964). *Comp. Biochem. Physiol.* **12**, 209–21.

amount of oxygen) when the partial pressure of oxygen is about 0.01 atm. At higher partial pressures it becomes more nearly saturated and below about 0.003 atm it gives up nearly all its oxygen. The partial pressure of oxygen in air is 0.21 atm so the haemocyanin can be expected to become more or less saturated in the lung, but the partial pressure in the tissue must be remarkably low if it loses most of its oxygen there.

Fig. 13.17 shows that the amount of oxygen taken up by the blood depends on the partial pressure of carbon dioxide, as well as oxygen. An increase in the partial pressure of carbon dioxide from 0.0016 to 0.012 atm increases the partial pressure of oxygen needed to saturate the haemocyanin. This is the Bohr effect which is also shown by mammal blood. Further increase in the partial pressure of carbon dioxide to 0.028 atm has a peculiar effect on snail blood: it reduces the partial pressure of oxygen needed for 50% saturation. High partial pressures of carbon dioxide do not affect mammal blood in this way.

The Bohr effect is important in mammals because it aids the release of oxygen from haemoglobin in the tissues, where partial pressures of carbon dioxide tend to be high. It has not been shown to be important in snails.

There are a few gastropods, including *Planorbis* (Fig. 13.4*f*), which have haemoglobin in their blood instead of haemocyanin. Haemoglobin is the red pigment which carries oxygen in vertebrate blood. It is a protein, like haemocyanin, but contains iron, not copper. The haemoglobin of typical

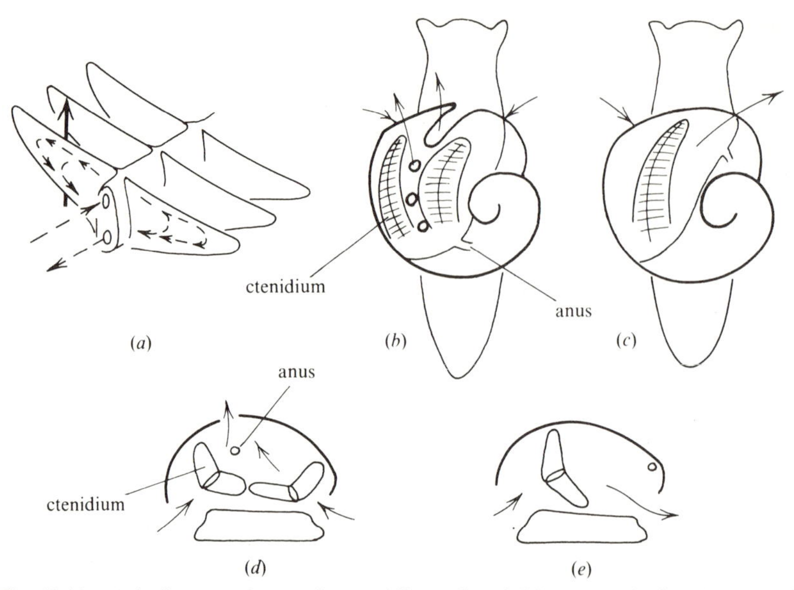

Fig. 13.18. (*a*) A diagram of part of a ctenidium. (*b*), (*d*) Diagrammatic dorsal view and transverse section of a primitive prosobranch. In (*b*) the shell is represented as transparent. (*c*), (*e*) Similar diagrams of a more advanced prosobranch. Broken arrows show the direction of movement of blood and continuous arrows show water movement.

vertebrates has a molecular weight of 68000 and the molecule consists of four units, each capable of carrying a molecule of oxygen. It is enclosed in cells (the red corpuscles). The haemoglobin of *Planorbis* is not in cells but has enormous molecules, each consisting of 90 of the basic units, so it does not add much to the colloid osmotic pressure of the blood.

Prosobranch gastropods have ctenidia (gills) in the mantle cavity. The structure of a ctenidium is shown in Fig. 13.18(*a*). It has a central axis with a row of gill filaments attached on either side. There are two blood vessels in the axis, one carrying blood to the filaments and other carrying it from them to the heart. The blood travels from one vessel to the other through blood spaces in the gill filaments as indicated in the figure by broken lines. Cilia on the flat faces of the filaments drive water through the spaces between the filaments so that the water and the blood cross the filament in opposite directions. This makes it possible for more oxygen to be transferred to the blood than if the water and blood flowed in the same direction because the blood is exposed to the highest concentrations of oxygen in the water just before it leaves the ctenidia.

The more primitive prosobranchs such as *Haliotis* (Fig. 13.4*a*) have two ctenidia which lie more or less horizontally (Fig. 13.18*b*, *d*). Water enters the mantle cavity below them and passes between them to leave through a slit or one or more holes in the shell. These water currents can be shown by releasing drops of a suspension of carmine particles into the water near the mollusc. The anus opens into the dorsal part of the mantle cavity near where the water leaves so faeces are carried away in the current. The kidney opens near it so the urine is carried away without passing over the ctenidia where waste products might diffuse back into the blood. In more advanced prosobranchs such as *Buccinum* (Fig. 13.4 *b*) there is only one ctenidium, and water passes through it from left to right (Fig. 13.18 *c*, *e*). No special slits or holes are needed in the shell. The anus and kidney open near where the water leaves the mantle cavity.

The ctenidium of a *Buccinum* of mass 20 g (excluding the shell) has about 250 filaments of total area 160 cm^2. The external surface area of the animal is very roughly 50 cm^2. Much of this area is normally in contact with a rock or covered by a shell so the whelk could not use it for respiration, but the surface area of the gills is in any case much larger than the external surface area. The same is true of other prosobranchs.

Oxygen diffuses from the water to the blood according to equation (8.1). For fast diffusion the area of the gills should be large and the gradient of partial pressure should be steep (i.e. $-dP/dx$ should be large). The cilia maintain a flow of water through the ctenidia so that the water in contact with the ctenidia is constantly being changed. Only a thin layer of tissue separates the blood in the ctenidia from the water. Thus the distance the oxygen has to diffuse is small and the gradient of partial pressure is correspondingly large.

Some opisthobranchs such as *Aplysia* (Fig. 13.4 *c*) have a single ctenidium, but other such as *Aeolidia* (Fig. 13.4 *d*) do not, and have projections which are not homologous with ctenidia but nevertheless serve as gills. Ctenidia must be rather ineffective respiratory organs in air since air cannot circulate between the gill filaments unless the water between them dries out. Pulmonates have lost their ctenidia but the mantle cavity has become a lung. Its wall has evolved a rich blood supply, and has become ridged so that its surface area is enlarged. The area of the respiratory surface in a large snail (*Helix pomatia*) is about the same as in a *Buccinum* of the same mass (excluding the shell). Only a thin layer of tissues separates the blood from the air. The anus and kidneys do not open into the lung, but on the external surface of the body.

Gastropods which extract oxygen from water need ciliary currents to bring fresh supplies of aerated water to the gills. Simple diffusion suffices in the lungs of pulmonates because the diffusion constant for oxygen diffusing through air is 3×10^5 times the constant for oxygen diffusing through water. The oxygen has to diffuse into the lung through the small pneumostome, across the lung and then through a thin layer of tissue into the blood.

Consider a *Helix* of mass (excluding shell) 10 g. It would use about 1 cm^3 oxygen h^{-1} or 0.3 $mm^3\ s^{-1}$. The respiratory surface of its lung would have an area of about 83 cm^2 (8300 mm^2). The layer of tissue separating the air from the blood would be about 10 μm (0.01 mm) thick. Equation (8.1) can be written

$$J = -AD \cdot \Delta P/x \tag{13.5}$$

where J is the rate of diffusion (volume per unit time), A the area, P the partial pressure difference and x the distance. For diffusion from air to blood in the snail's lung, the diffusion constant D can be taken as 2×10^{-5} mm^2 atm^{-1} s^{-1} so that

$$0.3 = -8300\times2\times10^{-5}\cdot\Delta P/0.01$$
$$\Delta P = -0.02 \text{ atm}$$

Now consider diffusion into the pneumostome, which has a diameter of about 3 mm and an area of 7 mm^2. We need a modified form of the diffusion equation: equation (13.5) is suitable for calculations about diffusion across membranes and also along long slender tubes, but it is not suitable for calculations about tubes which are short, relative to their diameter. When such tubes are considered an allowance must be added to the length to take account of diffusion into the entrance of the tube and away from its exit. If the diameter of the tube is d the modified equation is

$$J = -AD\cdot\Delta P/(x+\tfrac{1}{4}\pi d) \qquad (13.6)$$

The oxygen diffusing in through the pneumostome is diffusing through air so the appropriate value of D is 20 mm^2 atm^{-1} s^{-1}. Since the pneumostome is a pore in a thin membrane the length x is so small that it can be neglected, so we have

$$0.3 = -7\times20\Delta P/\tfrac{1}{4}\pi\times3$$
$$\Delta P = -0.005 \text{ atm}$$

This calculation shows that the difference in partial pressure of oxygen between the atmosphere and the air in the lung is likely to be small (0.005 atm) compared to the difference between the air in the lung and the blood (0.02 atm), provided the pneumostome is kept open. The pneumostome, small though it is, is not a very serious barrier to diffusion. Indeed, *Helix* open and close the pneumostome every 30 s or so and generally keep it open for only 10–20% of the time. The calculation also shows that the area of the lung is ample to support the snail's metabolism, since the estimated partial pressure difference between the air and the blood is only about 10% of the partial pressure of oxygen in atmospheric air.

EXCRETION

Fig 13.19 shows the structure of the excretory organs of *Nerita fulgurans*, a prosobranch gastropod. This particular species has one ctenidium and one kidney: more primitive prosobranchs have two of each. The heart is enclosed in a fluid-filled pericardium which is connected by the renopericardial canal to the kidney, which leads in turn to the mantle cavity. The pericardium is lined by epithelium which covers the whole of its surface including the outer surface of the heart. The epithelium covering the ventricle has not been examined in *Nerita* but in *Viviparus* (another prosobranch) its cells have interlocking processes with slits beween them, which probably serve as ultrafilters like the slits in the flame cells of turbellarians (Fig. 8.11). The muscular wall of the ventricle is spongy and has no inner epithelium, so the peculiar epithelium that

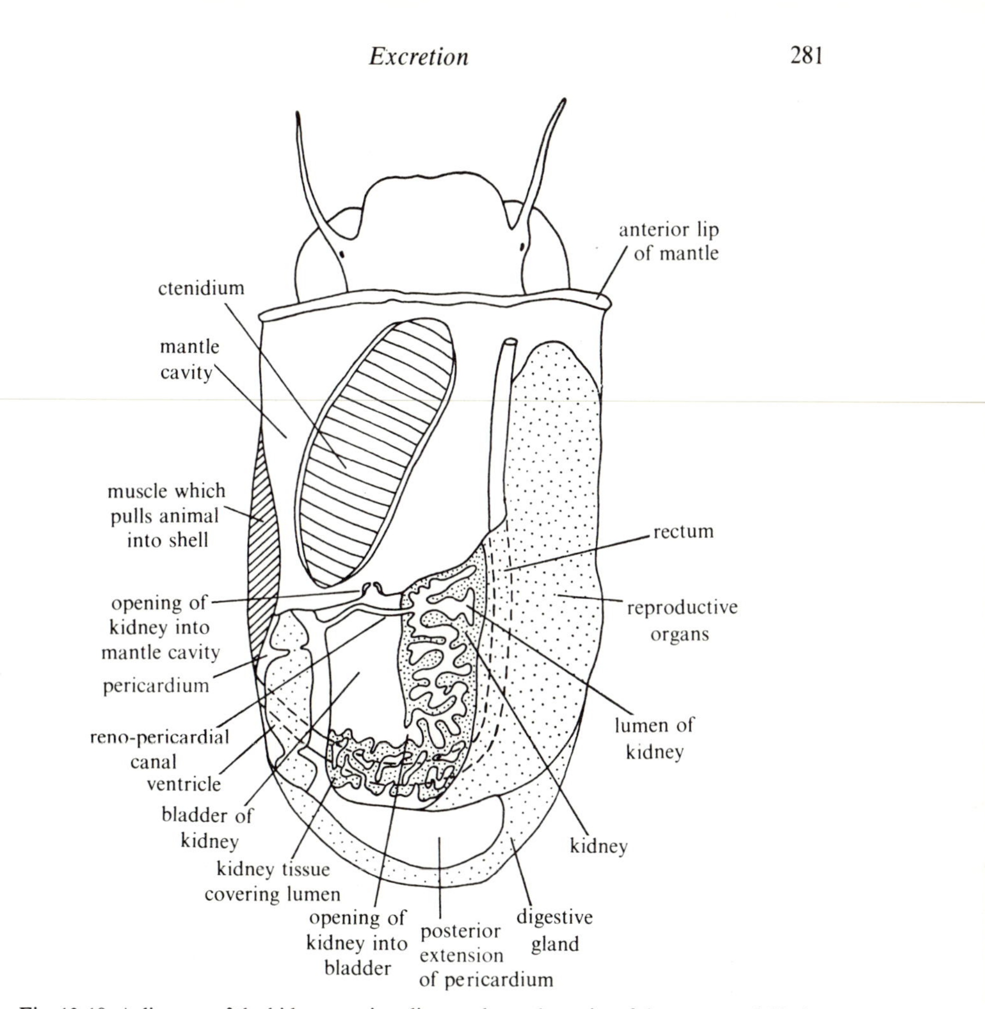

Fig. 13.19. A diagram of the kidney, pericardium and mantle cavity of the gastropod *Nerita fulgurans*. From C. Little (1972). *J. exp. Biol.* **56**, 249–61.

covers the ventricle is the only effective barrier between the blood in the ventricle and the pericardial fluid. When the ventricle contracts the pressure in it rises far above the pressure in the pericardium (Fig. 13.16), and it is believed that ultrafiltration occurs through this epithelium in most molluscs (but it occurs elsewhere in pulmonates). The colloid osmotic pressure of the blood may be too high for ultrafiltration by flame cells to be feasible but it is only a small fraction of the blood pressure (see p. 276).

Samples of the blood, pericardial fluid and urine of several prosobranchs have been analysed. One set of experiments was done on the queen conch, *Strombus gigas*, from the Florida Keys. This species was found convenient because it is so big, up to 30 cm long. Part of the shell, over the pericardium and kidney, was cut away with a miniature circular saw. It was replaced by a sheet of wax which could be lifted whenever samples were to be taken or injections made. Fine glass pipettes were used to take samples of blood and pericardial fluid, which were analysed. It was found that these fluids were almost identical to each other, and to sea water, in ionic composition. However the haemolymph was bright blue (due to haemocyanin) and contained 3.4% organic matter while the pericardial fluid was clear and contained only 0.8% organic matter. This suggests that the pericardial fluid is an ultrafiltrate of the blood. The small quantity of organic matter in it presumably consists of molecules small enough to pass through the filtering epithelium. A further experiment confirmed that fluid was moving from the blood into the pericardium. Radioactive inulin was injected into the blood. Shortly afterwards samples of blood, pericardial fluid and urine (from the kidney) were taken and their radioactivity was measured to find out the concentrations of radioactive inulin. The same concentration was found in all three. Inulin is a polysaccharide which is often used in experiments of kidney physiology because its molecule is small enough to pass through kidney filters and because animals seem unable to metabolize it, to secrete it or to absorb it. The rate at which inulin escaped into the sea water indicated that ultrafiltration was occurring and urine being passed at a rate of 3 cm^3 (kg soft tissue)$^{-1}$ h^{-1} (or 12% of blood volume per day).

Urine from the kidney is brown and contains 2.4% organic matter, three times as much as the pericardial fluid. This suggests that waste products are secreted into it, and experiments with *Haliotis* confirms this. Various substances were injected into the blood and samples were later taken and analysed. In one experiment, phenolsulphonphthalein was injected. Later it was found in much higher concentration in the urine than in either the blood or the pericardial fluid, so it is presumably secreted by the kidney cells into the urine. In other experiments glucose was injected and found in lower concentration in the urine than in the blood or pericardial fluid. Its molecule is small enough to pass through the wall of the ventricle and the animal avoids losing it in the urine by reabsorbing it in the kidney. Another discovery which was made in these experiments was that the left and right kidneys of *Haliotis* (which is one of the primitive prosobranchs with two kidneys) do not have identical functions.

The animal may have to get rid of insoluble particles as well as soluble wastes. This has been investigated by injecting a suspension of thorium dioxide into yet another marine prosobranch. Thorium has a high atomic number so it is very opaque to X-rays. X-rays of the mollusc were taken periodically after the injection. They showed that the thorium dioxide spread round the body within 10 minutes but that after a week most of it had accumulated in the heart and pericardium and was moving into the kidneys. It took 4 to 6 weeks to disappear. Its fate was checked by making microscope sections of tissues of the injected snails. The particles were easily seen in the sections and since thorium is radioactive it was possible to check their identity by autoradiography. Within a few days all the particles had been engulfed by amoebocytes, blood

TABLE 13.1. *Concentrations of ions (in mEquiv l^{-1}) in the blood, muscle cells and urine of a freshwater gastropod* (Viviparus viviparus) *and in the water in which it lives*

	Na^+	K^+	Cl^-
Blood	34.0	1.2	31.0
Muscle cells	13.6	14.6	10.0
Urine	9.0	—	10.0
Water	2.5	0.2	8.0

Data from C. Little (1965). *J. exp. Biol.* **43**, 23–37 and 39–54.

cells which resemble the macrophages of vertebrates (see p. 99). At this stage the amoebocytes are all in the blood, most of them attached to the walls of blood vessels. Later they are found in solid tissues, especially the wall of the heart, and in the pericardium. It seems that they travel to the heart, through its wall into the pericardium and from there to the kidney and out in the urine. Their passage from pericardium to kidney may be helped by the cilia of the wall of the renopericardial canal, which beat towards the kidney. It was also found in this investigation that the total number of amoebocytes increased greatly after the suspension had been injected.

For marine gastropods, the water they live in has almost exactly the same ionic composition as the blood. For freshwater ones, there is a marked difference. This is illustrated by Table 13.1 which gives data for *Viviparus*, a prosobranch found in slow rivers and ditches. The ion concentrations in muscle cells were determined by an experiment using radioactive inulin, just as the concentrations in a protozoan were measured (p. 47). The blood contains much higher concentrations of sodium and chloride than either the external water or the cell contents. The sodium is necessary, if action potentials are to be conducted in the nerves and muscles. It has been shown that *Viviparus* nerves do not conduct action potentials in sodium-free solutions: the reason will be apparent when the mechanism of action potentials has been discussed (chapter 15).

The kidneys have an important role in regulating the ionic composition of the blood. Since the blood has a higher osmotic concentration than fresh water, water will diffuse into the snail and must be excreted if the snail is not to swell. Also, salts will diffuse out and must be replaced. By eating food which contains water and salts and excreting urine containing lower concentrations of salts, the snail can keep both its water content and its salt content constant.

The samples of urine (Table 13.1) were taken by means of a cannula slipped into the kidney opening. Samples of fluid were also taken from the pericardium. The pericardial fluid had the same ionic composition as the blood but the urine was much more dilute. This indicates *either* that water is added to the urine in the kidney *or* that salts are extracted from it. To find out which, radioactive inulin was injected into *Viviparus* and samples were taken after an interval. It was found that the concentrations of inulin in the blood and the pericardial fluid were about the same, but that the concentration in the urine was more than twice as high. Since animals seem unable to secrete or absorb inulin, water must have been absorbed from the urine in the kidney. If some of the water was absorbed from it and it was nevertheless getting more dilute, salts must also have been absorbed.

Like other animals, molluscs have to get rid of the nitrogenous waste products formed by metabolism of proteins. In aquatic gastropods most of this waste is produced as ammonia and no special processes are needed to get rid of it: it simply diffuses out of the blood into the water. It must diffuse most rapidly from the ctenidia where the blood comes closest to the water.

Ammonia is removed very effectively by the water passing over the ctenidia, because it is extremely soluble. To illustrate this, imagine a snail metabolizing only protein. For simplicity, assume that the protein is composed entirely of alanine (an amino acid of moderate molecular weight) and is oxidized according to the equation

$$-NH.CH(CH_3)CO- + 3O_2 = NH_3 + H_2O + 3CO_2$$

One molecule of ammonia will be produced for every three molecules of oxygen which are used. One volume of ammonia will be produced for every three volumes of oxygen. The solubility of ammonia in water is about 140 times the solubility of oxygen. If the water passing over the gills gives up enough oxygen for the partial pressure of the oxygen dissolved in it to fall by n atm, it will receive enough ammonia to build up a partial pressure of only $n/(3 \times 140) = 0.0024\ n$ atm.

Terrestrial pulmonates cannot get rid of ammonia so easily. Ammonia will of course diffuse out of the body, particularly at the lung. The diffusion constants of gases diffusing through gas are inversely proportional to the square roots of their molecular weights so ammonia (molecular weight 17) diffuses $(32/17)^{\frac{1}{2}} = 1.4$ times as fast as oxygen (molecular weight 32). Suppose again that ammonia has to be lost one-third as fast as oxygen is used. Suppose further that gases move between the lung and the outside air entirely by diffusion. Then if the partial pressure of oxygen in the lung is m atm less than in the outside air, the partial pressure of ammonia in the mantle cavity must be $m/(3 \times 1.4) = 0.24\ m$ atm. The partial pressure of oxygen in air is 0.2 atm, and for molluscs in well aerated water n is likely to be about 0.1. The calculation on p. 280 suggests that m could be as low as 0.005 if the pneumostome were perpetually open. Nevertheless, 0.0024 n will be less than 0.24 m. The partial pressure of ammonia in the blood of the aquatic snail would have to be greater than 0.0024 n atm, or the ammonia would not diffuse out at the required rate. The partial pressure in the blood of the terrestrial one would have to be higher, over 0.24 *m* atm. The difference might well be critical, for ammonia is toxic. I have no data on its toxicity to snails but it kills alderfly larvae (*Sialis*) when its partial pressure in their blood reaches 3×10^{-4} atm. Diffusion of ammonia from the respiratory organs is an adequate means of getting rid of nitrogenous waste for aquatic gastropods, but not for terrestrial ones.

Terrestrial pulmonates get rid of most of their nitrogenous waste as purines, compounds containing rings of carbon and nitrogen atoms. These insoluble compounds accumulate in the kidney and are voided only occasionally. Kidneys dissected from *Otala* of tissue weight 5 g commonly contain as much as 0.1 g purines, mainly uric acid. They contain very little ammonia.

The rates at which *Otala* excretes nitrogen as purines, and the water snail *Lymnaea* excretes nitrogen as ammonia, have both been measured. In each case the rate was found to be about 3 mg (kg tissue)$^{-1}$ h^{-1}. Both snails probably used oxygen at rates around 0.1 cm^{-3} (g tissue)$^{-1}$ h^{-1} = 130 mg (kg tissue)$^{-1}$ h^{-1}. If this had been used entirely to oxidize protein, about 20 mg nitrogen (kg tissue)$^{-1}$ h^{-1} would have had to be excreted. (This follows from the equation on p. 284, if the nitrogen is excreted as ammonia. The rate would not be very different for purine excretion.) Since this is much more than the observed rate of excretion, protein was only being used for a small proportion of the snail's metabolism. It would probably account for a much larger proportion of the metabolism in carnivorous species.

In the calculation about ammonia excretion by aquatic and terrestrial gastropods it was assumed that all the metabolism was of protein. This assumption was unrealistic but the conclusion is still valid: if an aquatic and a terrestrial snail use equal proportions of protein in their metabolism and excrete all the nitrogen as ammonia, the terrestrial one is more likely to accumulate a harmful concentration of ammonia in its body.

FEEDING AND DIGESTION

The radula, just inside the mouth, is an important organ in feeding for many molluscs. It is a band of tissue bearing a large number of small teeth, so that it is in effect a flexible rasp. Mature *Helix pomatia* may have as many as 170 rows of teeth in the radula, with 151 teeth in each row, but *Patella* has only 13 teeth in each row (Fig. 13.20). In some molluscs the radular teeth are almost entirely organic in composition, consisting of the polysaccharide chitin and protein. The protein seems to be quinone-tanned. In others, the teeth have hard inorganic crowns. It has been shown by X-ray diffraction that the radular teeth of *Patella* are capped by goethite, $Fe_2O_3 . H_2O$, and that those of chitons are capped by magnetite, Fe_3O_4. Worn teeth are lost at the anterior end of the radula and new ones are secreted at the posterior end, by the radular gland (Fig. 13.21). The rate of growth of the radula of *Helix* has been measured by damaging part of the radular gland, for instance by cautery. The rows of teeth formed by the damaged part are missing or mis-shaped, so if the snail is killed and dissected

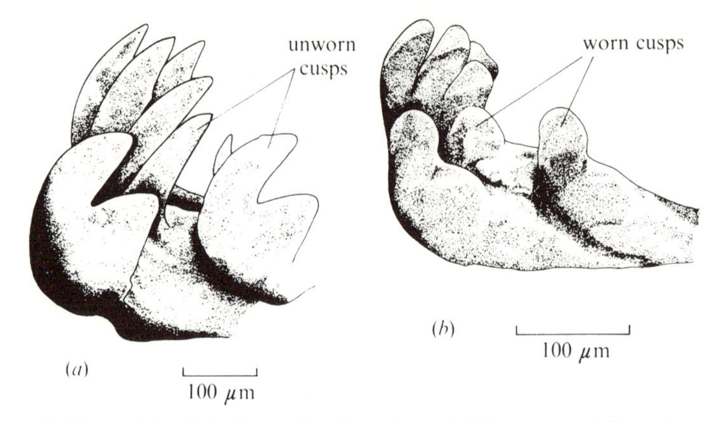

Fig. 13.20. The radula of the limpet *Patella vulgata*. (*a*) Unworn teeth from the posterior end of the radula. (*b*) Worn teeth from the anterior end. Drawn from scanning electron micrographs in N. W. Runham & P. R. Thornton (1967). *J. Zool., Lond.* **153**, 445–52.

some weeks later the part of the radula formed after the damage can be distinguished from the older part. By examining radulae after different intervals it has been shown that new teeth are produced at rates up to 3½ rows per day.

There is a cartilage (the odontophore cartilage) under the radula and a hard jaw made of chitin and protein opposite it. Many muscles are attached to the radula, cartilage and jaws.

Fig. 13.21 shows how the pond snail *Lymnaea* uses its radula. Much of the information was obtained by filming the snail through the glass wall of the aquarium, while it was feeding on algae growing on the glass. The radula was protruded through the mouth (phase 2) and drawn forward, scraping the glass, until it met the jaw. It then returned to the resting position inside the mouth. The complete cycle of movements took 1–2 s. The action of the radula has also been studied by examining the marks left by it after feeding on a thin layer of gelatin spread on glass. The grooves made by individual teeth are quite short, indicating that the radula slides over the odontophore cartilage during rasping, like a belt running over a pulley.

Filamentous algae form a thin layer on rock surfaces, extending down into hollows and even boring a few millimetres into the rock. They are particularly apt to bore in limestone. Some of the gastropods which feed on them, such as winkles (*Littorina*), have wholly organic radular teeth which are too soft to scratch the rock. They can only get the surface layer of algae. Others such as limpets and chitons have mineralized teeth hard enough to rasp away the limestone and get the deeper algae. Grooves scraped by the radular teeth can be found on rocks where chitons live. Fig. 13.20 shows that in spite of their hardness, the radular teeth of limpets get worn.

The radula is also used for other methods of feeding. When *Lymnaea* are

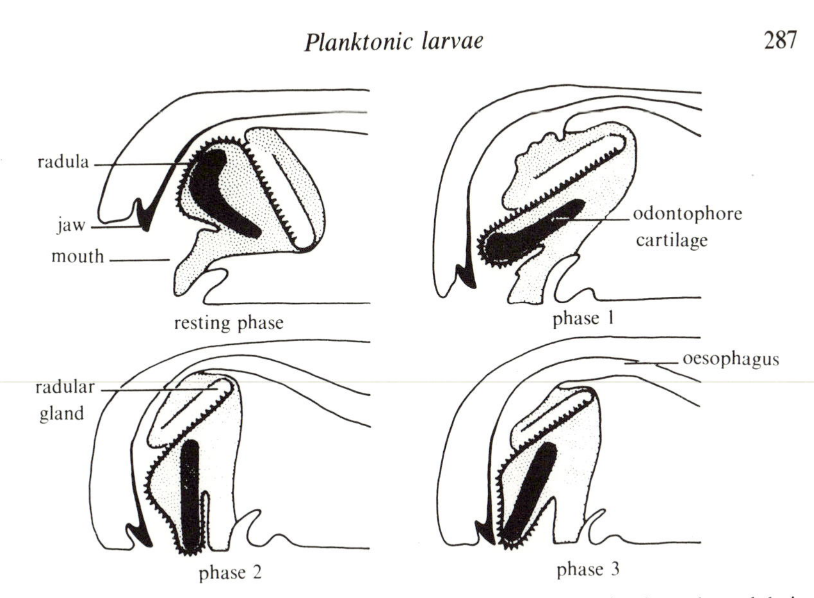

Fig. 13.21. Sagittal sections through the mouth of *Lymnaea* showing how the radula is used in feeding. From N. W. Runham (1975). In *Pulmonates*, vol. 1, ed. V. Fretter & J. Peake, pp. 53–104. Academic Press, New York & London.

given lettuce leaves they do not rasp them but use the radula and jaw to bite pieces off. Slugs (*Agriolimax*) examined after feeding on carrot or lettuce have chunks of these vegetables in the gut: they must have fed by biting rather than rasping. *Aplysia* takes large bites from the leaves of *Ulva* (sea lettuce). The radula is also used in feeding by carnivorous gastropods. The most remarkable of these are some Pacific species of *Conus*, which eat fish as long as themselves. Each radular tooth is hollow like a hypodermic syringe, but has barbs like a harpoon. It is driven into fish which happen to come into contact with the snail, catching them and injecting venom which immobilizes them in a few seconds. They are swallowed whole, with the tooth: a new tooth has to be used to catch the next victim.

A few gastropods such as *Crepidula*, the slipper limpet, are filter feeders. They use essentially the same mechanism of feeding as the typical bivalve molluscs described in chapter 14.

Digestion in gastropods occurs extracellularly in the stomach and intracellularly in the digestive gland. Similar processes occur in bivalves, and are described in more detail in chapter 14.

PLANKTONIC LARVAE

Many marine invertebrates which live on the bottom as adults have larvae which swim for a while in the plankton before settling. They include chitons, most prosobranchs and opisthobranchs, and also some polyclad flatworms, many nemerteans and many of the animals described in later chapters.

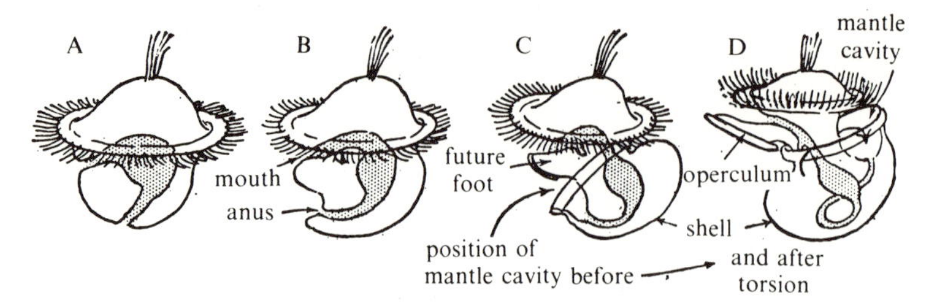

Fig. 13.22. Sketches of four stages in the larval development of the limpet *Patella*. From A. C. Hardy (1956). *The open sea, its natural history*, part 1, *The world of plankton*. Collins, London.

Fig. 13.22 shows larvae of the limpet *Patella*. Stage A is rather like some polyclad and nemertean larvae and very like the early larvae of polychaete worms (Fig. 16.15). By stage C a shell and foot have differentiated but the larva is still planktonic, swimming by means of the ring of cilia. Many gastropods have the ring of cilia extended onto large lobes on either side of the body. At this stage the opening of the mantle cavity is posterior (the tuft of cilia is anterior) but torsion occurs and brings it anterior, as in stage D. In some gastropods torsion is largely completed in a few minutes. Notice that torsion brings the mantle cavity over the head, so that the head can be withdrawn into the larval shell leaving the foot to close the shell opening. This may be why torsion evolved.

The sea is an enormous continuous area of water but only patches of its bottom are generally suitable as habitats for any particular species. If chitons, for instance, produced creeping larvae instead of planktonic ones, each patch of rocky shore would have its own isolated population of chitons. If that population were to die out for any reason, larvae from other populations could not take advantage of the vacant patch. In fact chitons produce trochophore larvae which are distributed by tidal and other currents. (The currents move faster than they can swim so swimming serves only to keep them off the bottom.) This advantage of planktonic larvae has been explored quantitatively, by computer simulation.

Some planktonic larvae, such as those of the worm *Spirorbis*, do not feed, so they must settle and start their adult life quite soon after being released. Most of them settle within 12 hours, which is long enough to take advantage of the distributing effect of a cycle of tides without too much danger of being washed right away from the coast. Other larvae, such as those of oysters (*Ostrea*), feed and grow, and can survive much longer. Oyster larvae in the laboratory settle at an age of 10–14 days. Larval molluscs which feed get their food like *Stentor* (Fig. 3.15) and rotifers, from the currents set up by the ring of cilia.

How long can a larva which does not feed survive? The following calculation is based on oyster larvae for lack of information about gastropods, but the conclusions probably hold for gastropod larvae as well. Newly-hatched oyster larvae use oxygen at a rate of about 6 cm^3 (g dry weight)$^{-1}$ h^{-1} or 1 cm^3 (g wet weight)$^{-1}$ h^{-1}. This is much higher than the metabolic rate of adult molluscs but metabolic rate per unit body weight is generally higher for small animals than for their larger relatives (Fig. 8.6*b*). *Paramecium* is about the same size as a young oyster larva and uses oxygen about as fast. How long can metabolism continue at this rate if the larva does not feed?

Consider the oxidation of a polysaccharide

$$(C_6H_{10}O_5)_n + 6n\,O_2 = 6n\,CO_2 + 5n\,H_2O$$

This equation shows that one mole of polysaccharide (162*n* g) is oxidized by 6*n* moles of oxygen ($6n \times 22.4 = 134n$ l), so 1 l oxygen oxidizes about 1.2 g polysaccharide. Alternatively, it can oxidize about the same mass of protein (as the equation on p. 284 shows) or about 0.5 g fat. If the larva used 6 cm^3 oxygen (g dry weight)$^{-1}$ h^{-1} and did not feed, it would destroy 8–17% of its dry weight in a day and could probably survive 2 or 3 days, considerably longer than non-feeding larvae usually remain in the plankton. Oyster larvae normally feed but young oyster larvae have been kept without food for 2 days and have survived, losing weight at about the calculated rate.

Planktonic larvae must settle in appropriate places if they are to mature and breed successfully. The method of choosing an appropriate site has been studied particularly thoroughly for barnacles and is discussed in chapter 17.

Though planktonic larvae are common in the sea they are rare in fresh water. The freshwater prosobranch *Viviparus* gives birth to creeping young and pulmonates hatch from eggs as creeping young. Planktonic larvae might have a value in distributing a species between patches of suitable habitat in a lake but if they hatched in one pond they could not colonize another, and in a river they would be washed downstream. The swan mussel *Anodonta* (a bivalve mollusc) releases swimming larvae in ponds and slow rivers but these larvae are exceptional. When fish breathe water containing them they cling to their gills and live there for a while as parasites before dropping off and settling. The fish are as likely to carry them upstream as down.

LIFE ON THE SHORE

Many of the molluscs found on shore live only on the lower part of the shore, near and below the level of low water in spring tides. They include prosobranchs such as *Haliotis* (Fig. 13.4*a*) which have holes in their shells and others such as *Aeolidia* (Fig. 13.4*d*) which have no shell at all. Such molluscs could probably not survive long in dry places, because they would lose water rapidly by evaporation. Some gastropods which live higher on the shore are found mainly in damp places, for instance under boulders, and do not generally have to withstand dry conditions. However limpets (*Patella* spp.) are common on bare rock faces between the tide marks, where they are covered and uncovered at every tide and may be left dry for periods of several hours.

The flesh of a limpet contains 86% water and *Patella vulgata* can survive loss of over 50% of this water. It dries out very slowly because it can clamp its shell down firmly on the surfaces of rocks (though not so tightly as to stop respiration). Limpets weighing 7 g allowed to attach to glass and left in a desiccator for 12 hours lost only 2.5% of their flesh weight (3% of their water). This is slow but the desert snails described in the next section lose water much more slowly even than this.

Limpets on south-facing rock faces have to endure heating by the sun but they do not get nearly as hot as the desert snails described in the next section.

LIFE ON LAND

Many pulmonate snails and slugs live on land, most of them in fairly damp places. Slugs spend the day underground or under stones. Snails retire into their shells by day and emerge only at night or after rain, when the relative humidity is high. They are also apt to be inactive, with a very low metabolic rate, for long periods during the cold weather of winter or during summer drought. Such states of dormancy are called hibernation (in winter) or aestivation (in summer). Many snails have the mouth of the shell covered by a sheet of dried mucus, while they are dormant.

Sphincterochila boissieri is a small (4 g) snail with an extraordinary capacity for survival by aestivation. It can be found at midsummer on the surface of the Negev Desert (in Israel) where rain falls only between November and March. It aestivates throughout the drought but when rain falls it emerges and eats mud, from which it digests algae and lichens.

Professor Knut Schmidt-Nielsen and some colleagues studied this snail. They bored tiny holes through the shell and cemented thermocouples into them. They put the snails back on the ground as they had found them, and recorded their temperatures throughout the day. Fig. 13.23 shows the highest temperatures they ever recorded. The ground is far hotter than the air because it is heated by the sun. It is not quite so hot where it is shaded by the snail. The snail is much cooler than the ground because its shell is chalky white and reflects 90–95% of the solar energy which falls on it. The snail retires to the second whorl of the shell so that air in the empty first whorl serves as heat insulation between it and the ground (In one experiment this whorl was filled by injecting water into it, and the snails got hotter.) Fig. 13.23 shows that the tissues of the snail were at only 50 °C, though the ground was at 65 °C. Laboratory tests showed that the snails died if heated to 55 °C for a few hours, though they survived at 50 °C. Most molluscs are killed by much lower temperatures: for instance, *Helix* dies quickly at 45 °C.

The aestivating snails lost weight at an average rate of only 0.7 mg day^{-1}. This was presumably mainly due to loss of water. A snail of mass 4 g (including the shell) initially contains 1400 mg water which is probably enough for several years' drought. So low a rate of water loss is only possible because the

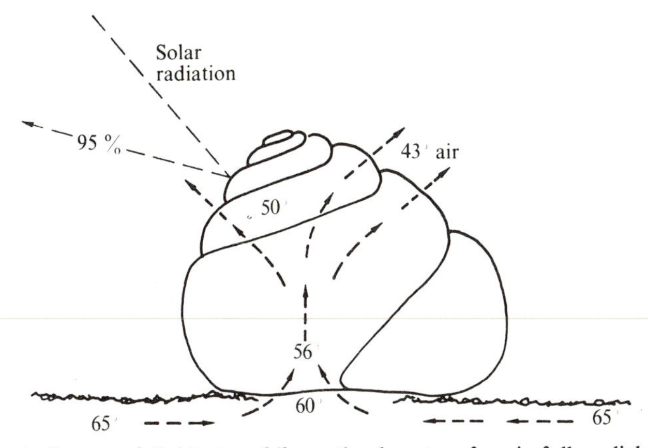

Fig. 13.23. A diagram of *Sphincterochila* on the desert surface in full sunlight showing heat flow (broken arrows) and temperatures. From K. Schmidt-Nielsen, C. R. Taylor & A. Shkolnik (1971). *J. exp. Biol.* **55**, 385–98.

metabolic rate is low: if oxygen could diffuse rapidly into the shell, water could diffuse rapidly out.

The rates at which aestivating snails used oxygen were measured at various temperatures in the laboratory. The rate varied with temperature, but it was estimated that the average rate in the natural habitat, over a complete day and night, was probably about 0.004 cm^3 oxygen (g soft tissue)$^{-1}$ h^{-1}, less than 5% of the rate that might be expected for active snails (see p. 275). The snails contain little fat and probably depend mainly on protein and carbohydrate. In a complete year of drought an aestivating snail would use $0.004 \times 24 \times 365 = 35$ cm^3 oxygen (g tissue)$^{-1}$. This would oxidize only about 40 mg protein and carbohydrate out of the 108 mg protein (g soft tissue)$^{-1}$ and 24 mg carbohydrate (g soft tissue)$^{-1}$ initially contained in the snail. If the metabolic rate were not so low, the available food would be exhausted long before the summer was over.

LIFE IN STAGNANT WATER

There are two main groups of freshwater snails. There are prosobranchs such as *Viviparus* which have ctenidia and depend entirely on dissolved oxygen in the water, and pulmonates such as *Lymnaea* and *Planorbis* which have lungs and use atmospheric oxygen. These pulmonates climb up plants to the surface, open the pneumostome and allow oxygen to diffuse in for a minute or so before submerging again. Dives occasionally last as long as 2 hours. In ponds and ditches choked with water weeds oxygen released by photosynthesis may raise the partial pressure of dissolved oxygen well above the atmospheric value of 0.21 atm and in such conditions the snails get most of their oxygen from the water. They have no gills so the oxygen must diffuse in through the general body surface.

The pulmonate *Biomphalaria* lives in papyrus swamps in Uganda. The partial pressure of dissolved oxygen there is extremely low, for the following reasons. During the day, the water is shaded from the sun by vegetation. At night the water does not radiate heat to a cold sky but to this same vegetation, which is little cooler than the water and radiates back almost as much heat as it receives from the water. Hence the temperature of the water does not fluctuate much. There is little mixing of water by convection currents which would distribute oxygen through the body of the water if the surface water cooled and sank at night. The water is stagnant so there is no turbulence to mix it. The vegetation shields it from disturbances by wind and makes the light so dim that little oxygen can be produced by photosynthesis in the water. These circumstances combine to make the partial pressure of dissolved oxygen extremely low. *Biomphalaria* must depend entirely on atmospheric oxygen.

Even in water containing no dissolved oxygen, *Biomphalaria* only visits the surface once every 25 min. How can a single breath last so long? The volume of its lung is 0.36 cm^3 (g tissue)$^{-1}$. Samples of gas taken from the lungs of snails about to dive contained 17% oxygen (a little less than the 21% in the atmosphere). Samples from snails surfacing after a dive contained only 3% oxygen. Hence the oxygen used during a dive is 14% of $0.36 = 0.05$ cm^3 (g tissue)$^{-1}$. If this is used in 25 min the metabolic rate is 0.12 cm^3 (g tissue)$^{-1}$ h^{-1}. The measurements were made at 25 °C. *Helix* at 15 °C uses about 0.08 cm^3 oxygen (g tissue)$^{-1}$ h^{-1} so the metabolic rate of *Biomphalaria* is about what might be expected, allowing for the difference of temperature. The visits to the surface are frequent enough to enable *Biomphalaria* to maintain a normal metabolic rate.

FURTHER READING

GENERAL

Fretter, V. & Peake, J. (1975–). *Pulmonates.* Academic Press, New York & London.
Morton, J. E. (1967). *Molluscs*, 4th edn. Hutchinson, London.
Purchon, R. D. (1977). *The biology of Mollusca*, 2nd edn. Pergamon, Oxford.
Runham, N. W. & Hunter, R. J. (1970). *Terrestrial slugs.* Hutchinson, London.
Wilbur, K. M. & Yonge, C. M. (1966). *Physiology of Mollusca.* Academic Press, New York & London.
Yonge, C. M. & Thompson, T. E. (1976). *Living marine molluscs.* Collins, London.

MOLLUSC SHELLS

Currey, J. D. & Taylor, J. D. (1974). The mechanical behaviour of some molluscan hard tissues. *J. Zool., Lond.* **173**, 395–406.
Raup, D. M. (1966). Geometric analysis of shell coiling: general problems. *J. Palaeontol.* **40**, 1178–90.

CRAWLING

Jones, H. D. & Trueman, E. R. (1970). Locomotion of the limpet, *Patella vulgata* L. *J. exp. Biol.* **52**, 201–16.
Trueman, E. R. (1975). *The locomotion of soft-bodied animals.* Arnold, London.

BLOOD CIRCULATION AND HYDRAULICS

Dale, B. (1973). Blood pressure and its hydraulic function in *Helix pomatia* L. *J. exp. Biol.* **59**, 477–80.

Sommerville, B. A. (1973). The circulatory physiology of *Helix pomatia* [3 papers]. *J. exp. Biol.* **59**, 275–304.

RESPIRATION

Jones, J. D. (1972). *Comparative physiology of respiration.* Arnold, London.

Spoek, G. L., Bakker, H. & Wolvekamp, H. P. (1964). Experiments on the haemocyanin–oxygen equilibrium of the blood of the edible snail (*Helix pomatia* L). *Comp. Biochem. Physiol.* **12**, 209–21.

Yonge, C. M. (1947). The pallial organs in the aspidobranch Gastropoda and their evaluation throughout the Mollusc. *Phil. Trans. R. Soc.* **232B**, 443–518.

EXCRETION

Boer, H. H. & Sminia, T. (1976). Sieve structure of slit diaphragms of podocytes and pore cells of gastropod molluscs. *Cell Tissue Res.* **170**, 221–9.

Brown, A. C. & Brown, R. J. (1965). The fate of thorium dioxide injected into the pedal sinus of *Bullia* (Gastropoda: Prosobranchiata). *J. exp. Biol.* **42**, 509–19.

Harrison, F. M. (1962). Some excretory processes in the abalone, *Haliotis rufescens. J. exp. Biol.* **39**, 179–92.

Little, C. (1965). Osmotic and ionic regulation in the prosobranch gastropod mollusc, *Viviparus viviparus* Linn. *J. exp. Biol.* **43**, 23–37.

Little, C. (1965). The formation of urine by the prosobranch gastropod mollusc, *Viviparus viviparus* Linn. *J. exp. Biol.* **43**, 39–54.

Little, C. (1967). Ionic regulation in the queen conch, *Strombus gigas* (Gastropoda, Prosobranchia). *J. exp. Biol.* **46**, 459–74.

Sattelle, D. B. (1972). The ionic basis of axonal conduction in the central nervous system of *Viviparus contectus* (Millet) (Gastropoda: Prosobranchia). *J. exp. Biol.* **57**, 41–53.

Speeg, K. V. & Campbell, J. W. (1968). Purine biosynthesis and excretion in *Otala* (= *Helix*) *lactea*: an evaluation of the nitrogen excretory potential. *Comp. Biochem. Physiol.* **26**, 579–95.

FEEDING AND DIGESTION

Kohn, A. J. (1956). Piscivorous gastropods of the genus *Conus. Proc. nat. Acad. Sci. USA* **42**, 168–71.

Runham, N. W. & Thornton, P. R. (1967). Mechanical wear of the gastropod radula: a scanning electron microscope study. *J. Zool., Lond.* **153**, 445–52.

PLANKTONIC LARVAE

Crisp, D. J. (1976). The role of the pelagic larva. In *Perspectives in experimental biology*, vol. 1, ed. P. S. Davies, pp. 145–55. Pergamon, Oxford.

LIFE ON THE SHORE

Davies, P. S. (1969). Physiological ecology of *Patella.* III. Desiccation effects. *J. mar. biol. Assoc. UK* **49**, 291–304.

Davies, P. S. (1970). Physiological ecology of *Patella.* IV. Environmental and limpet body temperatures. *J. mar. biol. Assoc. UK* **50**, 1069–77.

Newell, R. C. (1970). *Biology of intertidal animals.* Logos, London.

LIFE ON LAND

Schmidt-Nielsen, K., Taylor, C. R. & Shkolnik, A. (1971). Desert snails: problems of heat, water and food. *J. exp. Biol.* **55**, 385–98.

LIFE IN STAGNANT WATER

Jones, J. D. (1972). *Comparative physiology of respiration.* Arnold, London.

14
Clams

Phylum Mollusca (cont.)
  Class Bivalvia
    Subclass Protobranchia (*Nucula* etc.)
    Subclass Lamellibranchia (most other genera)

Most bivalve molluscs live in the sea but a few live in fresh water. They are a large group which have great economic importance, partly because some of them are collected or cultivated as human food and partly because they make up a large part of the diet of some commercially important fishes. For instance plaice (*Pleuronectes platessa*) in the Irish Sea feed largely on the bivalves *Cultellus*, *Ensis* and *Abra*.

The most characteristic feature of the class is the bivalve shell. It is in two parts (valves), each of helical logarithmic spiral form with a high value of W (p. 267). These two parts are rigid and consist of calcium carbonate with a small proportion of protein, like gastropod shells. They are connected by a flexible but elastic protein ligament which serves as a hinge. The properties of the hinge ligament and of the adductor muscles which close the shell are described later in the chapter. The two valves lie on the left and right sides of the animal, with the hinge dorsal. Most bivalve molluscs are bilaterally symmetrical and have valves which are mirror images of each other.

Fig. 14.1 compares bivalve molluscs with a chiton. The chiton (Fig. 14.1*a*) is depressed (dorso-ventrally flattened), with a large foot which has a flat sole. Many pairs of small ctenidia are housed in grooves on either side of the foot. Fig. 14.1(*b*) and (*c*) shows two bivalve molluscs. They are compressed (transversely flattened). They have bivalve shells. The foot is relatively small and the grooves on either side of it are represented by a large mantle cavity. There is just one pair of ctenidia, as in gastropods, but the ctenidia are generally very large. They divide the mantle cavity into a ventral inhalant chamber and a dorsal exhalant chamber. Cilia between the gill filaments drive water from the inhalant chamber so that a current is kept flowing into the inhalant chamber, between the gill filaments and out of the exhalant chamber. Most bivalves get their food by filtering small particles such as unicellular algae out of this current of water. The mechanism is quite different from the filter-feeding mechanism of sponges and is described later in the chapter.

Bivalves do not crawl on their feet like chitons and snails but many of them

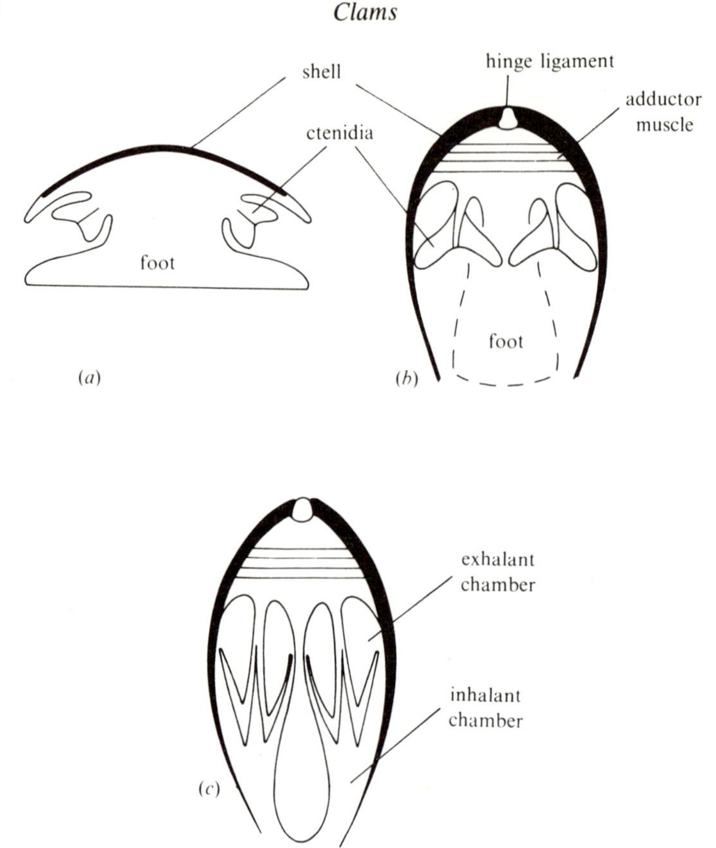

Fig. 14.1. Diagrammatic transverse sections through (*a*) a chiton, (*b*) a primitive bivalve mollusc (a member of the Protobranchia), and (*c*) typical bivalve mollusc (a member of the Lamellibranchia). In (*b*) the foot lies anterior to the ctenidia, in the position indicated by the broken outline.

use the foot for burrowing in sand or mud. The method of burrowing is described later in the chapter.

Fig. 14.1(*b*) and (*c*) shows differences between the two subclasses of Bivalvia. The Protobranchia is a small, primitive group. The foot has a flat sole, although it is not used for crawling. The ctenidia are relatively small and lie posterior to the foot. They are similar in form to the ctenidia of gastropods and chitons. Note that in transverse section each ctenidium forms an inverted V. Most bivalves belong to the other subclass, the Lamellibranchia. Their feet have no flat sole. Their ctenidia are very large, generally almost as long as the animal, and lie on either side of the foot. The gill filaments are greatly elongated and are doubled back so that the ctenidium in section has the shape of a W (see also Fig. 14.9*a*).

Fig. 14.2 shows the structure of a member of the Lamellibranchia in more detail. There is no distinct head but the end which has the mouth must plainly

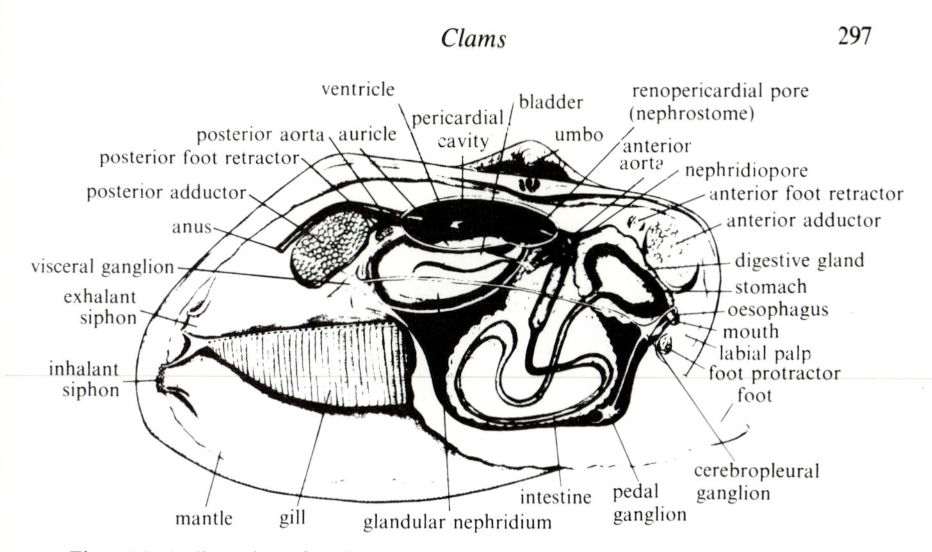

Fig. 14.2. A dissection of a bivalve mollusc. The right valve and ctenidium have been removed and the viscera exposed by further dissection. From P. A. Meglitsch (1972). *Invertebrate zoology*, 2nd edn. Oxford University Press.

be regarded as anterior. The anus is posterior. The mouth opens into the inhalant chamber, and the anus into the exhalant one. There is no radula. There are digestive glands and a stomach which has a complex internal structure (they are described later in the chapter). The nervous system has essentially the same form as in gastropods: a pair of ganglia beside the mouth are connected to ganglia in the foot and viscera by pedal and pleural nerve cords. The blood circulatory system and the kidneys are very like those of gastropods. There are two auricles which receive blood from the ctenidia and pass it to the ventricle. The intestine (curiously) generally runs through the middle of the ventricle. Arteries carry the blood to the tissues but it returns through haemocoels which are the main body cavities. The pedal and visceral haemocoels are connected but the passage between them (Keber's valve) can be closed when required. The heart is enclosed in a pericardium which opens into a pair of kidneys (the kidney shown in Fig. 14.2 is labelled 'glandular nephridium'). The kidneys discharge through nephridiopores into the exhalant chamber. They seem to function in the same way as in gastropods.

Fig. 14.3 shows the shells of two bivalves which do not burrow. *Mytilus* anchors itself to rocks and other solid objects by threads of quinone-tanned collagen, known as byssus threads. Juvenile *Pecten* anchor themselves with byssus threads but the adults do not and are among the very few bivalves which can swim. They do this by repeatedly opening and closing the shell. Flaps of tissues round the edge of the shell enable them to direct the water which squirts out of the shell as it closes, so as to swim either with the hinge leading or with the hinge behind. They can swim only briefly, and on a rather erratic course. Swimming serves as a means of escaping from predators such as starfishes and

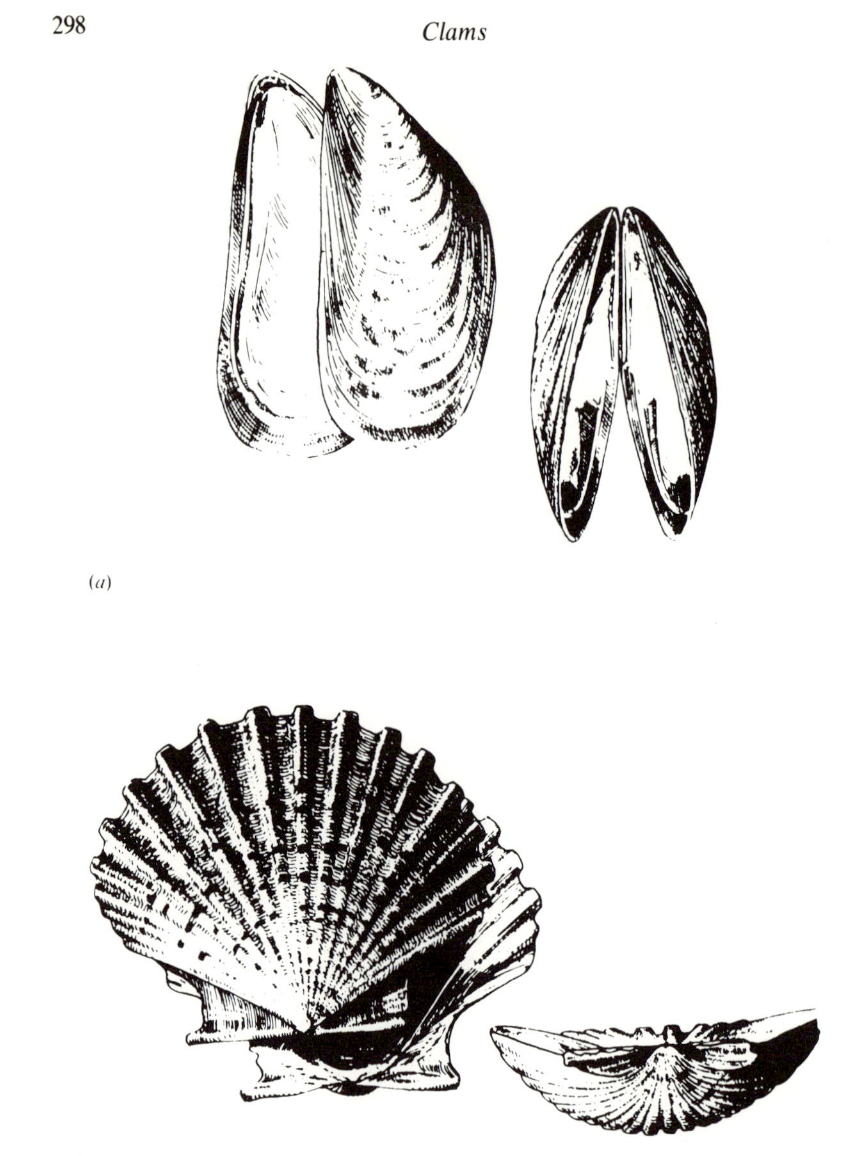

Fig. 14.3. Shells of (*a*) the common mussel, *Mytilus edulis* (length usually not more than 10 cm) and (*b*) a Mediterranean species of scallop, *Pecten jacobaeus* (up to 13 cm). From W. de Haas & F. Knorr (1966). *The young specialist looks at marine life*. Burke, London.

may also be used to move to a new site on the sea bottom. *Mytilus edulis* is very common intertidal species on Atlantic shores and another species of *Mytilus* is abundant on the Pacific coast of N. America. Species of *Pecten* are found offshore on many coasts. *Mytilus edulis* is cultivated as food in Europe and several species of *Pecten* are fished commercially, by dredging.

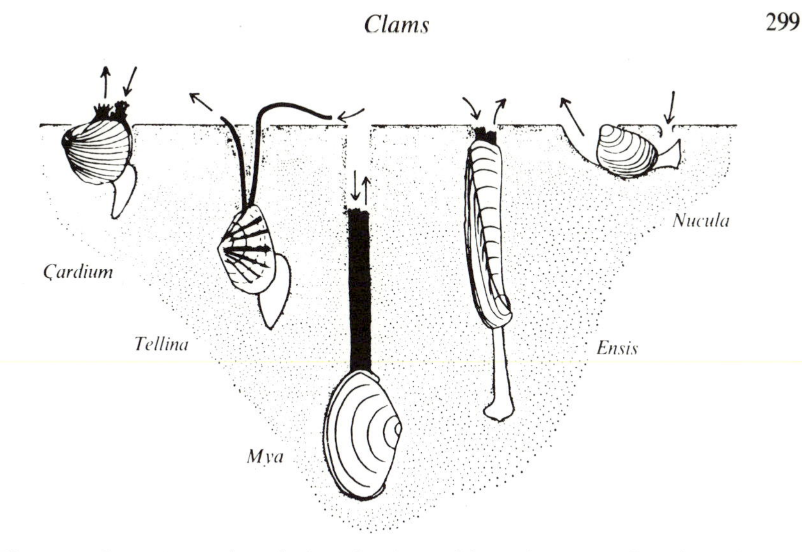

Fig. 14.4. Some burrowing bivalves in the positions they normally adopt. Arrows indicate water currents. *Cardium edule* has valves up to about 5 cm long; *Tellina tenuis* 3 cm; *Mya arenaria* 12 cm; *Ensis ensis* 16 cm and *Nucula nucleus* 1.4 cm. (These are probably the species which the artist intended to represent.) From J. E. Morton (1967). *Molluscs*, 4th edn. Hutchinson, London.

Both *Mytilus* and *Pecten* are filter feeders.

Mytilus and *Pecten* both have logarithmic spiral shells with high values of *W*, but their shells are otherwise very different. *Pecten* is unusual, in having valves which are not mirror images of each other. The right valve is convex (and often used as an ashtray) but the left valve is almost flat.

Fig. 14.4 shows a selection of burrowing bivalves. *Nucula* is a member of the Protobranchia, with a flat sole on its foot. The water current, which is driven by the cilia of the ctenidia, enters the mantle cavity at the anterior end and leaves at the posterior end. In the other genera (which are all Lamellibranchia) it both enters and leaves at the posterior end, through inhalant and exhalant siphons which are tubes formed as extensions of the tissue which lines the shell (the mantle: see Fig. 14.2). Only burrowing bivalves have siphons. All the genera shown in Fig. 14.4 are found in sand or mud, between high and low tide levels or offshore. Some of them bury themselves deeper than others. *Tellina* uses its long inhalent siphon as a vacuum cleaner to suck up detritus (fragments of dead organic matter) from the surface of the mud. *Nucula* has a pair of long tentacles which emerge from the shell and collect food particles. Both *Tellina* and *Nucula* probably also practise filter feeding, which is how the other genera all feed.

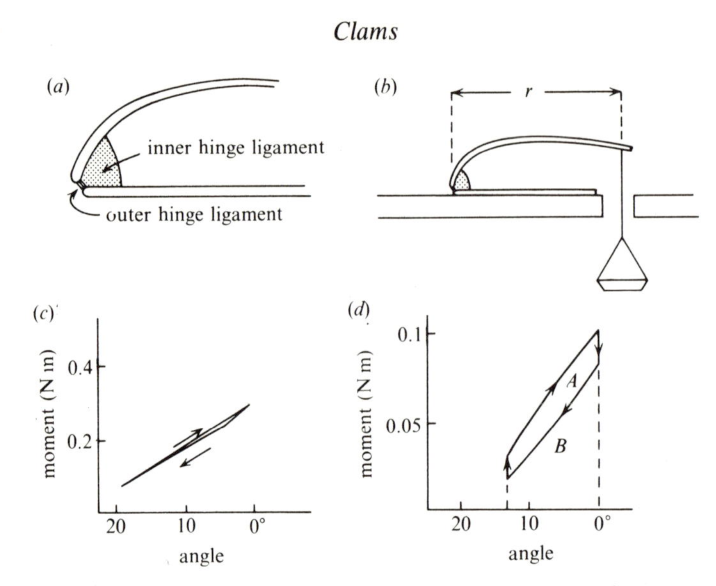

Fig. 14.5. (*a*) A diagrammatic section through the hinge of *Pecten*. (*b*) An experiment on the hinge ligament of *Pecten* (see text for details). (*c*) A graph of the moment acting about the hinge of *Pecten*, against the angle between the valves, determined as shown in (*b*) (but with slightly different apparatus). (*d*) As (*c*), but for *Anodonta cygnea*. (*c*) and (*d*) have been modified from E. R. Trueman (1953). *J. exp. Biol.* **30**, 453–67.

THE HINGE AND ADDUCTOR MUSCLES

Bivalves have either one or two adductor muscles, which connect the two valves of the shell (Figs. 14.1, 14.2). When they shorten, the valves close. Muscles cannot push the shell open from inside, and there is no shell-opening muscle. Instead, the hinge ligament is elastic. It is strained when the valves close, and makes them spring open again when the adductor muscle relaxes. (A different solution to the problem of opening a hinged shell from inside has been evolved by the brachiopods and is described in chapter 22.)

The hinge ligaments of different bivalves differ in structure and mechanical properties. Fig. 14.5(*a*) shows the ligament of *Pecten*, which has two parts. The outer hinge ligament is a narrow strip of flexible but inextensible protein which joins the valves together along the hinge line. The inner hinge ligament is a block of a protein called abductin, which is remarkably like rubber. If you dissect it out and drop it on a hard floor, it will bounce. It is the spring which makes the valves open, when the adductor muscle relaxes.

Fig. 14.5(*b*) shows how the properties of the hinge ligament can be investigated. The adductor muscle has been removed and part of the left (flat) valve has been ground away. A hole has been bored in the right valve so that a weight can be hung from it, and the shell has been placed on a board, over a hole. Each weight added to the pan makes the shell close further, and when weights are removed it opens again. Fig. 14.5(*c*) shows the results of such an experiment. The moment which the weight exerts about the hinge is plotted against the angle between the valves (a weight mg would exert a moment mgr, see Fig. 14.5(*b*)). Young's modulus for abductin can be calculated from the gradient of the graph and the dimensions of the ligament. It is 1–4 MN m^{-2}, or about the same as for soft vulcanized rubbers. Experiments at different temperatures have provided evidence that the elasticity depends on the same molecular mechanism as in rubber and similar cross-linked polymers (see p. 140). Abductin has been hydrolysed to break it down into its component amino acids, and an unusual compound has been found among them. It consists of two tyrosine molecules with a link between them (it is 3,3′-methylene *bis*tyrosine). It probably forms the cross-links. If its tyrosine molecules are incorporated into different peptide chains, it will form a link between the chains.

Some polymeric materials such as mesogloea have marked visco-elastic properties; they continue stretching for some time after a force has been applied to them (Fig. 6.15 *b*). *Pecten* hinge ligament is not like that. When a weight is applied in the experiment shown in Fig. 14.5(*b*) the valves close quickly to their new position, and stay there. When the weight is removed they spring back immediately to almost exactly their original position. Fig. 14.5(*c*) records an experiment in which the moment was gradually increased until the shell was closed, and then decreased again. The graph of moment against angle for increasing moments coincides almost exactly with the graph for decreasing moments.

Most bivalve hinge ligaments are markedly visco-elastic. When a weight is applied, they continue closing slowly for a considerable time. When it is removed they do not spring back immediately to the original position, but approach it slowly. In an experiment in which a series of weights is applied and then removed, the loading and unloading curves do not coincide but form a loop (Fig. 14.5 *d*). The area under a graph of moment against angle represents work done on or by the ligament. In Fig. 14.5(*d*) the area $(A+B)$ represents the work done on it during loading and the area B represents the work done by it in its elastic recoil during unloading. The area A thus represents work done against the viscosity of the ligament, and lost as heat. The fraction of energy returned in the elastic recoil, $B/(A+B)$, is known as the rebound resilience.

When *Pecten* swims it opens and closes its valves 2–3 times per second. If the hinge ligament had marked viscous properties a lot of energy would be wasted as work done against viscosity. The unusual properties of the ligament, with very low viscosity, allow large savings of energy.

The experiment shown in Fig. 14.5 is not absolutely conclusive in showing that *Pecten* ligament has superior properties, because the rebound resilience of a material is a function of the frequency with which it is distorted and allowed to recoil. The valves were closed and allowed to open in the experiment much more slowly than in swimming. However, other experiments have shown that the rebound resilience at the swimming frequency is about 0.91, which is as high as for most rubbers at their optimum frequency.

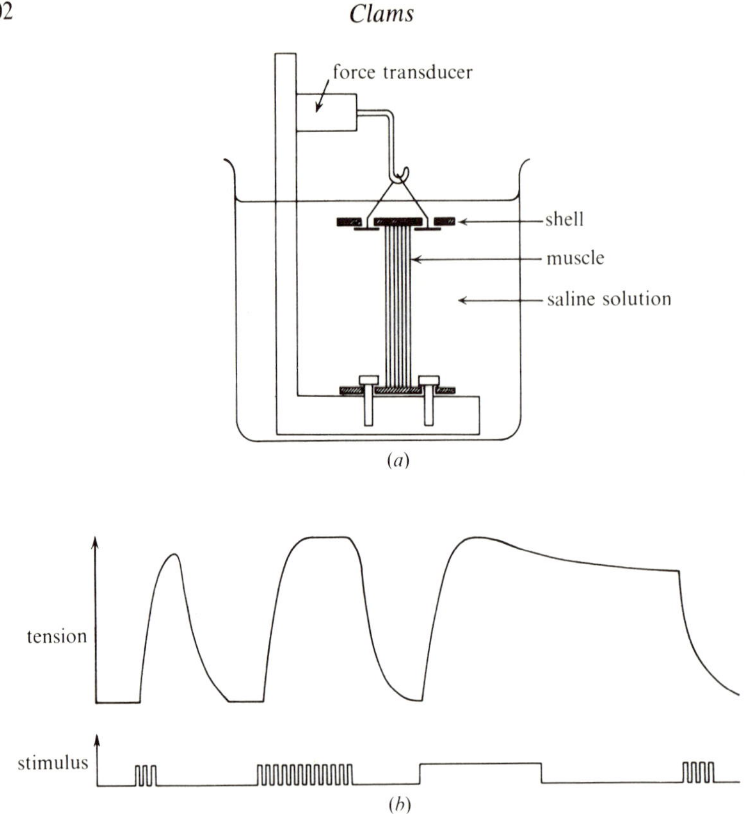

Fig. 14.6. Diagrams showing an experiment with *Crassostrea* opaque adductor muscle, and the results of the experiment.

A disadvantage of the elastic hinge ligament is that bivalves must keep their adductor muscles taut whenever the shell is closed, and even when it is open but not gaping widely (unless it is prevented from gaping, for instance by the walls of a burrow). Bivalves seldom allow the valves to gape widely and sometimes have to hold them tightly closed for long periods. For instance, *Mytilus* remains closed while exposed by the tide. An oyster can remain closed for a month in a moist atmosphere. Keeping a muscle taut uses energy.

This disadvantage is largely overcome by muscles with a remarkable property. They can be locked in the contracted state and will maintain tension for long periods with very little expenditure of energy. This state is called catch, and muscles capable of it are called catch muscles.

The adductor muscles of most bivalves have two parts, a translucent part and an opaque one. The opaque part is the catch muscle. The anterior byssus retractor muscle (ABRM) of *Mytilus* is another catch muscle, and has been used in many physiological experiments. It is a slender muscle which runs from the shell to the base of the bundle of byssus threads.

Fig. 14.6 shows an experiment with the opaque part of the adductor muscle of the oyster *Crassostrea*. A small piece of the muscle was dissected out, together with pieces of shell at either end which were used to attach it to the apparatus. It was fixed between a rigid support and a force transducer and stimulated electrically. Pulses of alternating current or trains of very short DC pulses (for instance, 10 pulses per second) made the muscle develop tension quickly. The tension declined rapidly when the stimulus ended. The time constant for the decay of tension (i.e. the time in which the tension fell to $1/e = 0.37$ of its initial value) was on average about 9 s. If the stimulus was a long one the muscle remained contracted, in a state of tetanus, until it ended, and then relaxed rapidly. Single pulses of direct current lasting several seconds made the muscle develop tension just as quickly, but the tension decayed very slowly when the stimulus ended. The average time constant was 5 minutes, 30 times as long as after an AC stimulus. Thus AC stimulated only a normal tetanus but DC put the muscle into the catch state. Further, an AC stimulus given to a muscle in a state of catch made it relax quickly. The muscle could also be put into catch or released from catch by drugs. Addition of a little acetylcholine to the solution containing the muscle stimulated catch, which persisted after the drug had been washed away but could be released by addition of 5-hydroxytryptamine (5-HT). It is presumed that separate nerves stimulate catch and release and it is suspected that the transmitter substances at the nerve–muscle junctions are acetylcholine and 5-HT, respectively. The mechanism of catch is unknown.

Experiments have been performed on *Mytilus* ABRM, to find out how economical catch is. Oxygen electrodes were used to measure the quantities of oxygen taken up by the muscle from the surrounding solution, during and immediately after periods of tetanus and of catch. Since metabolism using 1 cm^3 oxygen released 20 J, the power requirements for tetanus and catch could be calculated. The results are conveniently expressed as the power required to maintain unit stress in unit mass of muscle: if power is expressed in W, stress in $N\ m^{-2}$ and mass in kg this quantity has units $W\ N^{-1}\ m^2\ kg^{-1}$ or $m^3\ kg^{-1}\ s^{-1}$. It was found to be $3 \times 10^{-6}\ m^3\ kg^{-1}\ s^{-1}$ for ABRM in tetanus, but only $5 \times 10^{-7}\ m^3\ kg^{-1}\ s^{-1}$ in catch. Even the value for tetanus is very low compared to values for vertebrate striated muscle. For instance, frog sartorius muscle uses $8 \times 10^{-4}\ m^3\ kg^{-1}\ s^{-1}$.

The economy of catch can be expressed in another way. When the ABRM is in catch, maintaining a rather high stress of $0.5\ MN\ m^{-2}$, it uses oxygen only 1.5 times as fast as when it is resting.

Compare frog sartorius muscle and *Mytilus* ABRM. The frog muscle is cross-striated and the ABRM obliquely striated, but this difference will not concern us further. The actin filaments have about the same diameter (6 nm) in both but the thick filaments are much thicker in the ABRM (110 nm) than in the frog muscle (11 nm). Those of the ABRM owe their thickness to a core of paramyosin. They are about 20 times as long in the ABRM as in the frog muscle, so an actin filament lying alongside one can be attached to it by 20 times as many cross-bridges. The force which can be exerted on each actin filament should be about 20 times as great as in the frog muscle but there are only about a quarter as many actin filaments as in the frog muscle in unit cross-sectional area. Hence the maximum stress which can be developed should be about five times as great as in the frog muscle. This is confirmed by experiment. Frog sartorius muscle, contracting isometrically, can exert 0.25 MN m^{-2} (a fairly typical value for a vertebrate striated muscle) but the ABRM can exert 1.4 MN m^{-2}. Isometric contraction is contraction at constant length with shortening prevented as in, for instance, the experiment shown in Fig. 14.6.

Long filaments make for slow contraction, as well as large stresses. Since there are only one-twentieth the number of sarcomeres in ABRM as in an equal length of sartorius, each thick filament must move the neighbouring actin filaments 20 times as fast to achieve contraction at the same speed. It is actually found to be about one-twentieth as fast as frog sartorius: it can contract at up to 0.3 lengths s^{-1} and frog sartorius at up to 6 lengths s^{-1}, at 16–18 °C. It is fortuitous that the ratio of speeds agrees so closely with the ratio of filament lengths: vertebrate striated muscles with similar filament lengths have quite a wide range of speeds.

Long filaments also make it possible for tension to be maintained at low energy cost, since fewer cross-bridges need be attached to maintain a given stress in a given mass of muscle. The thick filaments of the ABRM are 20 times as long as those of frog sartorius so the ABRM should be able to maintain tension 20 times as economically as the sartorius. It is in fact 250 times as economical, even in tetanus, which cannot be explained by the length of the filaments alone.

The translucent part of the adductor muscle has thick filaments which are shorter and less thick than in the opaque part but cannot exert such high stresses. In *Crassostrea* the translucent adductor can exert a maximum of 0.5 MN m^{-2} and contract at up to 1.5 lengths s^{-1} while the opaque adductor can exert 1.2 MN m^{-2} (almost as much as the ABRM) but contracts more slowly. *Pinna* opaque adductor cannot contract faster than 0.1 lengths s^{-1}.

Pecten has three parts to its adductor muscle. Two of them correspond to the translucent and opaque parts of the adductors of other bivalves. The third is cross-striated, not obliquely striated, although it has paramyosin in the thick filaments. Its sarcomeres are only about 3 μm long and it can contract at up to 3 lengths s^{-1} (at 14 °C). It is the fastest part of the muscle, and it is the part used in swimming.

When it swims, *Pecten* opens and closes its shell through about 20°. Each time it closes, the swimming muscle shortens by about 0.25 of its length in about 0.4 s, shortening at about 0.6 lengths s^{-1}. This is only 0.2 of the maximum rate of contraction, determined in experiments with isolated muscles.

A muscle can only shorten at its maximum rate when there is no force opposing shortening. The greater the opposing force the more slowly can it

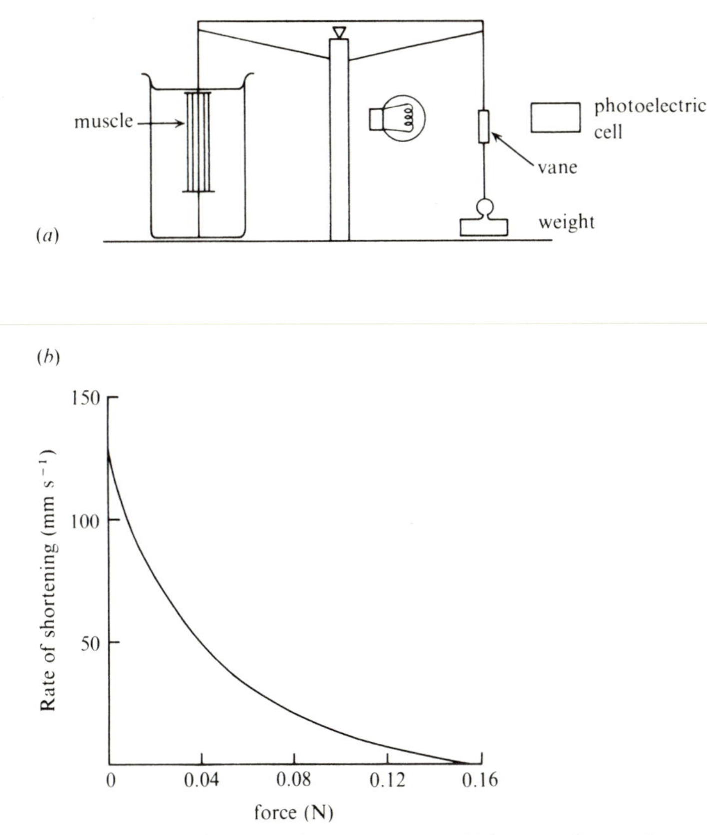

Fig. 14.7. (*a*) Apparatus for measuring the rates at which a muscle can shorten against different forces. (*b*) A graph of rate of shortening against force for a portion of the fast part of the adductor muscle of *Pecten*. From J. Hanson & J. Lowy (1960). In *The structure and function of muscle*, vol. 1, ed. G. H. Bourne. Academic Press, New York & London.

shorten and if the force is too great the muscle will stretch instead of shortening. Fig. 14.7 a shows an experiment to investigate the relationship between force and rate of shortening. When the muscle is stimulated electrically it contracts, lifting the weight attached to the opposite side of the lever. A vane attached to the lever partly interrupts a beam of light aimed at a photoelectric cell. As the lever moves, the illuminated area on the cell changes, so the output of the cell at any instant indicates the position of the lever. This device is used to determine the rate of shortening, when different weights are being lifted.

The results of an experiment with the fast part of the adductor of *Pecten* is shown in Fig. 14.7(*b*). The piece of muscle tested in this particular experiment could exert 0.15 N when contracting isometrically, and could shorten at 130 mm s^{-1} when there was no resisting force. It was estimated that the rate of

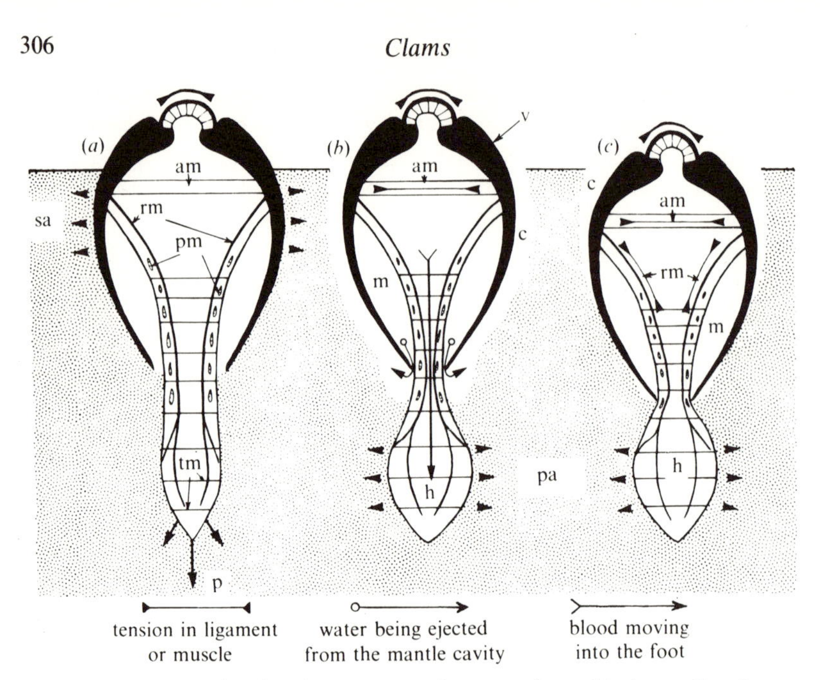

Fig. 14.8. Diagrams showing the sequence of events when a bivalve mollusc burrows. am, adductor muscle; c, sand loosened by water ejected from the mantle cavity; h, haemocoel; m, mantle cavity; p, probing movements of foot; pa, swollen foot anchored in sand; pm, protractor muscle of foot; rm, retractor muscle of foot; sa, open shell anchored in sand; tm, transverse muscles of foot; v, valve of shell. From E. R. Trueman (1968). *Symp. zool. Soc. Lond.* **22**, 167–86.

shortening in swimming is about 0.2 of the maximum rate. At this rate of shortening the muscle can exert about half the force exerted in isometric contraction.

BURROWING

Many bivalve molluscs burrow in sand or mud (Fig. 14.4) and some of them can burrow quite fast. The West Indian surf clam *Donax denticulatus* (which is about 2 cm long) can bury itself completely in about 4 s but most bivalves burrow much more slowly. For instance, a British species of *Donax* takes about 1 minute to bury itself. If bivalves simply spent their lives buried in one place there would be no need to burrow fast, but at least some move about. The most remarkable movements are made by surf clams, which migrate up and down the shore with the tide. They emerge from the sand from time to time and allow the waves to move them up and down the shore before burying themselves again. Other bivales which burrow in sandy shores are liable to be exposed by storms, and to have to burrow again. Bivalves such as *Ensis* climb near the surface at high tide so that their siphons project above the surface, but dig deeper when they are exposed at low tide.

Burrowing can be watched if a glass aquarium is partly filled with sand and a bivalve induced to burrow close to the glass. The sequence of events is shown in Fig. 14.8. So that as many structures as possible can be shown the diagram has been drawn as if the hinge were horizontal; in fact, bivalves normally burrow with the hinge vertical (Fig. 14.4).

Burrowing works like this. The shell is anchored firmly in the sand by allowing the valves to open, while the foot is pushed down deeper (Fig. 14.8*a*). The foot is then anchored by making it swell and the shell is pulled down towards it (Fig. 14.8*c*). Neither shell nor foot can be anchored in this way until the animal has penetrated some distances into the sand, so burrowing starts with the shell lying on the surface and the foot probing to obtain a purchase.

Look at Fig. 14.8 more closely. In (*a*) the adductor muscles are relaxed so that the elasticity of the hinge ligament keeps the valves jammed tightly against the sand. The foot is pushed deeper into the sand by contraction of the transverse muscles which run across its haemocoel. The connection between the pedal and visceral haemocoels (Keber's valve) may be closed at this stage so that contraction of the transverse muscles makes the foot longer and does not merely squeeze blood out of it into the visceral haemocoel. In (*b*) the adductor muscles contract, reducing the volume enclosed by the shell. This drives water out of the mantle cavity round the edge of the shell, and also drives blood from the visceral to the pedal haemocoel. The transverse muscles near the distal end of the foot are relaxed so the foot becomes bulbous, and is anchored firmly in the sand. However, the shell is free to move because the adductor muscles have narrowed it and because the water squirted out of the mantle cavity has loosened the surrounding sand. In (*c*) the retractor muscles contract. These muscles run from the shell to the distal end of the foot so their contraction pulls the shell down into the sand.

Pressures have been measured in burrowing *Ensis*. A hypodermic needle attached to a pressure transducer was inserted between the valves, into the pedal haemocoel. This made it possible to get records of blood pressure during the early stages of burrowing, until the base of the needle came into contact with the surface of the sand and prevented the animal from burrowing deeper. Small fluctuations of pressure up to about 1 kN m^{-2} occurred while the foot was probing the sand (Fig. 14.8*a*). Very much larger pressures, up to about 10 kN m^{-2}, occurred briefly while the valves were being adducted (Fig. 14.8*b*). It was calculated that the stress in the adductor muscles would have to reach 0.5 MN m^{-2} if they alone produced such pressures. *Ensis* has other muscles between the valves which may co-operate with the adductors, in which case the stress required may be considerably lower.

The retractor muscles can exert large forces. This was demonstrated in an experiment in which *Ensis* were allowed to bury the foot in a beaker of sand. While the shell was still clear of the sand it was clamped to a retort stand and the mollusc, trying to pull it down into the sand, lifted the beaker and its contents which weighed 0.4 kg. In further experiments with the foot held by

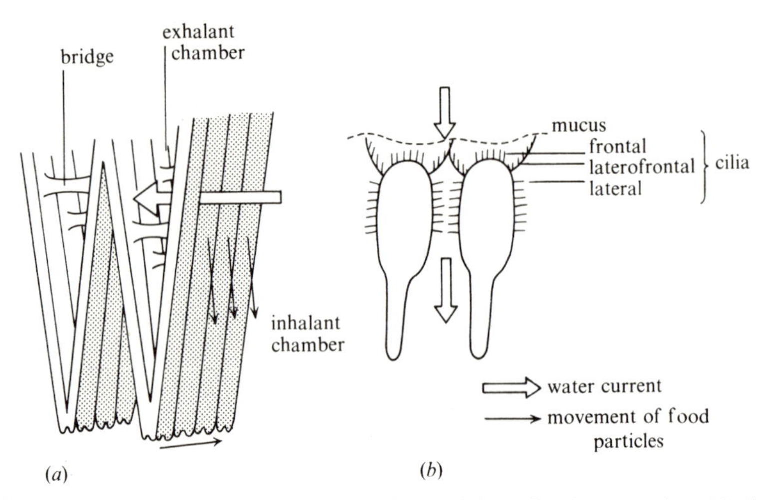

Fig. 14.9. Diagrams showing the structure of a ctenidium of a bivalve such as *Mytilus*. (*a*) shows a few gill filaments and (*b*) is a section through a few filaments.

a cord instead of buried, a 13 cm *Ensis* lifted up to 1.1 kg. It was calculated that stresses up to 0.2 MN m^{-2} must have acted in the retractor muscles.

Some bivalves burrow in substrates which are far less easy to penetrate than sand and mud. Piddocks such as *Zirphaea* bore in rock and the shipworm *Teredo* bores in timber. They use their shells for boring but do not depend on them for protection, and their shells are too small to enclose the whole body. *Zirphaea* bores into limestone, sandstone, shale and stiff clay. It stays in the same burrow throughout its adult life, widening it as it grows. The burrow is bottle-shaped with a relatively narrow neck and the animal can only be extracted by breaking the rock away. Juvenile specimens living in clay can be removed fairly easily without damaging them, and will bore a new burrow if set in a depression in a block of clay. They bore very slowly, at no more than 0.5 mm h^{-1} and the new burrow is not, of course, narrowed at the neck. The anterior ends of the valves have rough, rasp-like surfaces which are used for boring. The process involves a rocking motion of the valves which is only possible because the valves have more freedom of movement than is allowed by the hinges of more typical bivalves.

Rock-boring bivalves gain nothing but protection from the rock but *Teredo* digests and assimilates the wood it bores. It does not depend on the wood alone but also practises filter feeding. Specimens have been analysed and found to contain twice as much organic nitrogen as a quantity of wood equal in volume to the burrow. The best known habitats of *Teredo* are man-made (the piles of piers and the hulls of wooden ships) but it is also found in floating trunks of fallen trees which were, presumably, its original habitat. *Teredo* and related genera do a great deal of damage throughout the world and it has been claimed that this damage exceeds the income from molluscs taken as food.

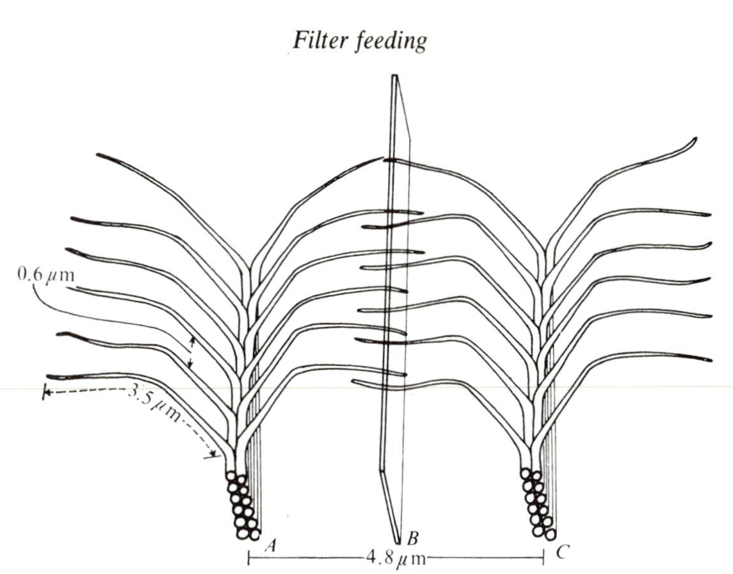

Fig. 14.10. A diagram based on scanning electron micrographs of two laterofrontal cilia on a gill filament of *Mytilus*. From G. Owen (1974). *Proc. R. Soc. B* **187**, 83–91.

FILTER FEEDING

Most bivalves get their food by filter feeding. Protobranchia have relatively small ctenidia (Fig. 14.1 *b*) but Lamellibranchia have greatly enlarged ones (Figs. 14.1 *c*, 14.2). Fig. 14.9 shows the structure of these large ctenidia in more detail. It shows the structure found in lamellibranchs such as *Mytilus*. Bridges of tissue connect the ascending and descending limbs of each filament, but successive gill filaments are held in place, side by side, simply by patches of stiff cilia which interlock in pairs like hairbrushes. Such ctenidia are easily frayed in dissection. Some other lamellibranchs such as *Crassostrea* have bars of tissue connecting successive filaments so that the ctenidium is a more coherent structure.

Water is driven through the ctenidium by the lateral cilia (Fig. 14.9 *b*). The lateral cilia of *Mytilus* have been shown by flash photography to beat in a pattern which is a mirror image of the pattern on *Paramecium* (Fig. 3.4 *a*). (The metachronal waves travel in the opposite direction, relative to the direction of beating, and the cilia also bend in the opposite direction in the recovery stroke. Since both these directions are reversed the cilia do not get in each other's way.) There are also frontal cilia on the edge of the filament which faces the inhalant chamber. They beat ventrally. On either side of them are the large, compound laterofrontal cilia. These beat at about 5 Hz, towards the frontal cilia, and it has been shown by filming them through a microscope that adjacent laterofrontal cilia beat in turn. Each of these compound cilia is a bundle of about 50 cilia of varying length (Fig. 14.10). Each individual cilium is bent near its tip so that the bundle has a feather-like structure. Electron microscope sections show that as far as the bend each cilium has the normal 9+2 structure

(Fig. 2.10*a*) but that beyond the bend the structure is different. It is believed that though the main trunk of the compound cilium beats actively the side branches move passively. The two laterofrontal cilia (*A* and *C*) shown in Fig. 14.10 are not immediate neighbours: the line (*B*) between them indicates the plane of beating of the intervening cilium which is not shown because it is beating out of phase with them. The diagram shows that the side branches of these feather-like cilia are close enough together to form an effective barrier to particles of diameter greater than about 0.6 μm.

Bivalves filter out unicellular algae and other suspended particles from water. There is some doubt about the mechanism. One investigator cut holes in the shells of bivalves, using a dental drill. He cut away the underlying tissue to allow a clear view into the mantle cavity and then cemented glass windows over the holes. He left the animals for 2 weeks to recover from the operation and then watched through the windows while they fed in dilute suspensions of diatoms or detritus. He formed the impression that particles on the ctenidium were trapped in a thin sheet of transparent mucus which covered the whole ctenidium (Fig. 14.9*b*) and was kept moving ventrally by the frontal cilia. This sheet could not be seen; its presence was inferred from the synchronized movements of the particles. It was presumably secreted near the dorsal edges of the ctenidium. Particles were trapped in it as it travelled down the ctenidium. When it reached the ventral edge it formed into a string (with the particles still trapped in it) which was carried anteriorly along a groove in the ctenidium. At the anterior end the string was passed to ciliated flaps of tissue called labial palps (Fig. 14.2). These were held in one of two positions. In one, they passed the mucus and food particles to the mouth. In the other they passed them to the mantle (the tissue lining the shell) to be carried by cilia to the edge of the shell and discarded. Material discarded in this way is known as pseudofaeces to distinguish it from true faeces which have been discarded after passing through the gut.

Other investigators who removed a whole shell valve to see the ctenidium saw no evidence of a mucous sheet: possibly rough handling had stopped normal feeding. They described elaborate mechanisms which seemed to sort particles out on the gills, separating particles to be transported to the mouth from ones to be rejected. These mechanisms seem unlikely to work if a mucous sheet is present. It seems uncertain whether there is a mucous sheet which acts as a filter or whether the water is simply strained between the side branches of the laterofrontal cilia.

Many experiments have been performed to find out how fast bivalves filter water and how fine are the finest particles they can filter out. In some experiments the bivalves were put in a bowl of a suspension which gradually became less concentrated as they passed it through their ctenidia. Samples of the suspension were taken from time to time and their concentrations measured by colorimetry or by means of an electronic particle counter. In other experiments the bivalves have been kept in vessels with the suspension flowing slowly through and the concentrations entering and leaving have been measured. Let the suspension flow through at a rate V (volume per unit time). Let the number of particles per unit volume of suspension be n_{in} in the inflow and n_{out} in the outflow. Then the bivalves must be extracting $(n_{in}-n_{out})\ V$ particles in unit time. They are getting these particles from a suspension of mean concentration $\frac{1}{2}(n_{in}+n_{out})$ so they must be filtering in unit time a volume $2V(n_{in}-n_{out})/(n_{in}+n_{out})$ of suspension. This is the rate at which the suspension must be passing through the ctenidia if the filter is perfect, stopping all the particles. If only a fraction are stopped the suspension must be passing through faster.

Measurements have been made of the rates at which *Mytilus* filters suspensions of diatom colonies (diameter 200 μm) and of colloidal graphite (about 2 μm). Both are filtered effectively. If many of the graphite particles leaked through the filter, filtration rates calculated from experiments with graphite particles should be lower than rates calculated from experiments with diatoms. In fact, they were found to be rather higher (perhaps the big diatom colonies clogged the filter). In other experiments *Mytilus* have been put in a mixed suspension of graphite particles (diameter 1–2 μm) and lobster haemocyanin (diameter 12 nm). The graphite was filtered out but the haemocyanin apparently passed through the filter for its concentration scarcely diminished. It seems that the filter of *Mytilus* is fine enough to stop unicellular algae and bacteria, but not individual protein molecules.

Mytilus has been found to filter water at rates around 250 cm^3 (g soft tissue)$^{-1}$ h^{-1} or (in some experiments) rather faster. Two hundred and fifty cubic centimetres of well-aerated sea water would contain 1.5 cm^3 dissolved oxygen. In some of the experiments the rate of oxygen consumption was measured at the same time as the filtration rate, by using an oxygen electrode to measure the partial pressure of oxygen in the water entering and leaving the experimental chamber. Rates around 0.06 cm^3 oxygen (g soft tissue)$^{-1}$ h^{-1} were found. Thus only about 4% of the dissolved oxygen was being removed from the water as it was filtered. Water was being driven through the ctenidia far faster than would have been necessary if they served solely for respiration.

Neither the rate of filtration nor the rate of consumption of oxygen changed much when the concentration of the suspension was changed, provided the concentration was not raised too high above the range which is usual in coastal water. This makes it possible to calculate how much food is needed in the water, to enable *Mytilus* to grow. One litre of oxygen oxidizes about 1.2 g polysaccharide or protein or about 0.5 g fat. Hence the 0.06 cm^3 oxygen used by 1 g of *Mytilus* tissue in an hour will oxidize 0.03–0.07 mg dry organic matter. If *Mytilus* filters 250 cm^3 water each hour and gets this much food from it, the food content of the water must be at least 0.1–0.3 mg dry organic matter l^{-1}. It must actually be rather higher since analyses of *Mytilus* faeces show that only 70–85% of the food is assimilated.

Mytilus cannot grow unless it assimilates food faster than it uses it in

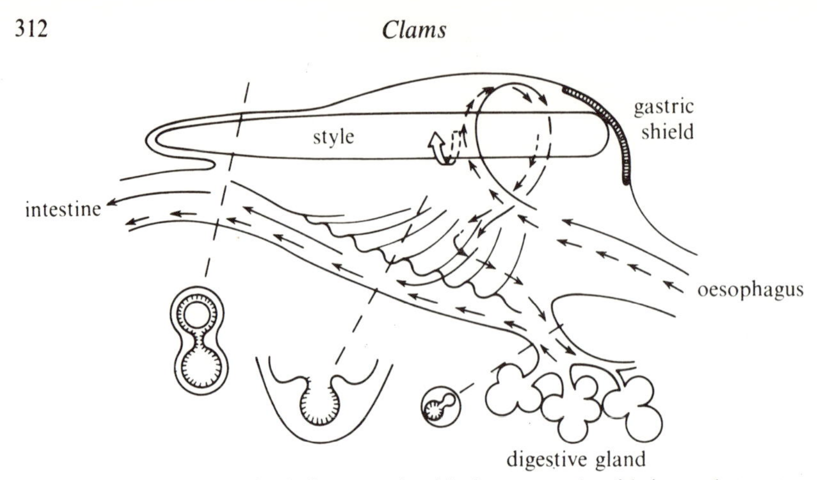

Fig. 14.11. A highly stylized diagram of a bivalve stomach with internal structures revealed by removing the right wall. Three sections cut at right angles to the plane of the paper are also shown. Short and long arrows show movements of particles which do and do not enter the digestive gland, respectively.

metabolism. The experiment just described shows that it cannot be expected to grow in water containing less than 0.1 mg dry organic matter l^{-1}. Concentrations are generally lower than this in mid-ocean, but higher in coastal water. For instance, the phytoplankton concentration in the English Channel generally varies, according to season, between 0.1 and 1.0 mg dry organic matter l^{-1}. It is mainly in coastal waters that bivalves flourish.

DIGESTION

The gut in bivalves has three main parts, the oesophagus, stomach and intestine. The stomach is remarkably complicated (Fig. 14.11) and its workings are not fully understood. Its most peculiar feature is the crystalline style, which is a translucent rod of protein and carbohydrate bound together as mucoprotein. The style is kept revolving at rates of the order of 0.2 revolutions s^{-1} by cilia in the pocket which houses its posterior end. Its anterior end rotates against the gastric shield which is a chitinous plate on the stomach wall. The shield is pierced by the microvilli of the underlying cells.

The style has incorporated in it various enzymes which digest carbohydrates. It is continuously worn away at its anterior end and replaced by secretion at the posterior end. The rate of replacement has been measured by putting bivalves in water containing trypan blue, which stained their styles. They were then returned to clean water and examined at intervals to find out how quickly the blue style was replaced by colourless material. Complete replacement took 4 hours in one species and 24 hours in another. The enzymes of the style, and enzymes secreted by the digestive gland, carry out extracellular digestion in the stomach. As well as releasing enzymes the rotating style stirs the stomach contents.

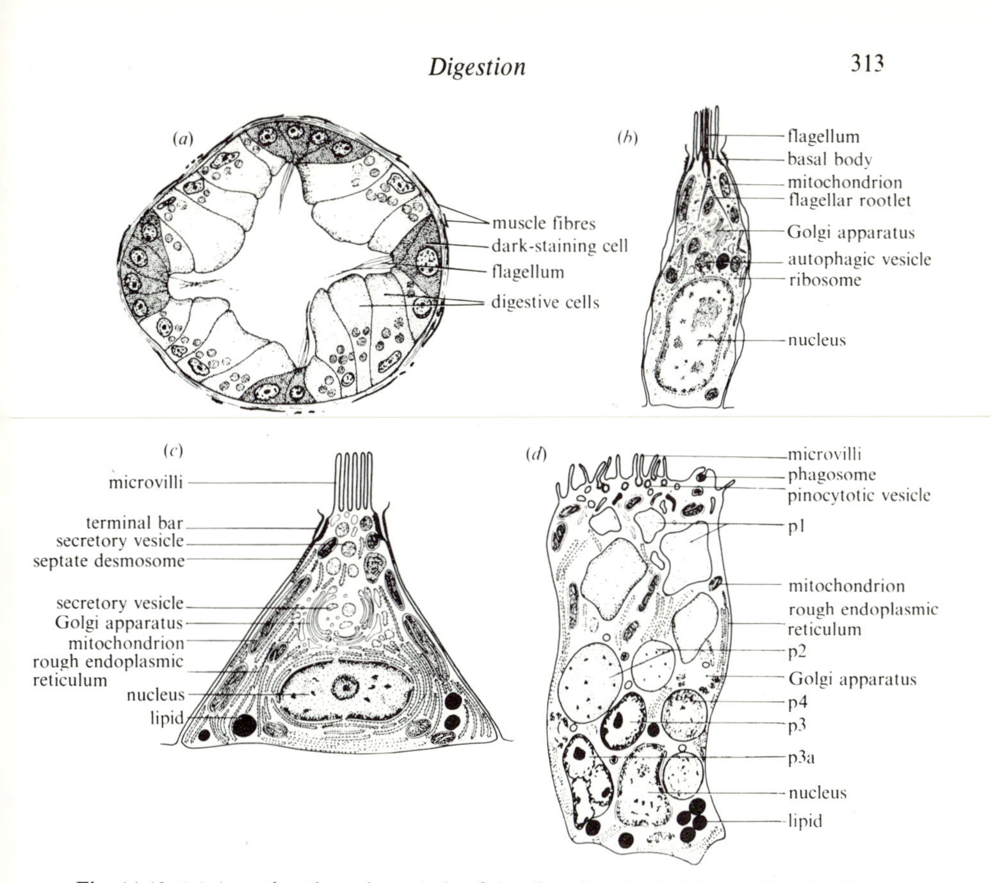

Fig. 14.12. (*a*) A section through a tubule of the digestive gland of the cockle, *Cardium edule*, seen by light microscopy. Tubules like this have diameters of 50–100 μm. (*b*), (*c*), (*d*) Schematic drawings based on electron microscope sections of the three types of cell found in the tubules. (*b*) is an immature cell, (*c*) a secretory cell and (*d*) a digestive cell. *p*1–*p*4, large vesicles. From G. Owen (1970). *Phil. Trans. R. Soc.* **258B**, 245–60.

Parts of the stomach wall are ridged, like a ploughed field. Cilia in the furrows beat along the furrows, towards the intestinal grove in the floor of the stomach. Cilia on the ridges beat at right angles to the ridges. Some particles fall into the furrows and travel direct to the intestinal groove where there are cilia which transport them to the intestine. Others tend to remain on the ridges and are passed by the cilia from ridge to ridge, finally entering the main ducts of the digestive glands. This sorting process has been watched in the opened stomachs of dissected bivalves. Some indication that the sorting is effective in life has been obtained by putting *Mytilus* and other species in a mixed suspension of a unicellular alga (diameter about 30 μm) and alumina particles (mean diameter 18 μm). Both pseudofaeces and faeces were collected and analysed. Pseudofaeces were formed only in concentrated suspensions and when they were formed contained algae and alumina in the same proportions as the suspension. Hence the food entering the mouth also contained algae and alumina in these proportions. The first faeces to be passed were white, consisting mainly of alumina. Later faeces were greener, containing more algae. This shows that the sorting mechanism of the stomach is capable of separating at least some of the alumina from the algae, getting rid of it quickly while the algae are retained for digestion.

The ducts of the digestive glands are partly divided into two channels, one without cilia and the other with cilia beating towards the stomach. The cilia probably drive fluid out of the glands, drawing fluid and particles into the non-ciliated channel. Evidence that particles enter by the non-ciliated channel has been obtained by feeding oysters on algal cultures labelled with radioactive carbon. Radioactivity was detectable in the non-ciliated channels of oysters killed after only 10 minutes, but appeared in the ciliated channels only after 90 minutes.

The ducts of the digestive gland branch, leading to blind-ended tubules. Light microscopy shows that the walls of these tubules contain at least two types of cell which stain differently (Fig. 14.12*a*). Electron microscopy shows that there are at least three types of cell. One type bears flagella and is thought to be immature, capable of replacing one or both of the other types (Fig. 14.12*b*). Another type is pyramidal in shape, packed with rough endoplasmic reticulum (Fig. 14.12*c*). These cells look very like known protein-secreting cells such as the cells which secrete enzymes in the mammalian pancreas. They presumably secrete enzymes involved in extracellular digestion. Vesicles which presumably contain the enzymes seem to form at the Golgi apparatus and can be found between it and the cell surface where they presumably discharge. The remaining cells are the digestive cells, with conspicuous vacuoles containing fragments of food (Fig. 14.12*d*).

To find out how the digestive cells work, cockles (*Cardium*) were fed suspensions of various materials which show up well in electron micrographs. Pigeon blood was used because haemoglobin can be stained to appear conspicuously dark. Graphite particles were used because they look quite different from natural cell contents. After feeding with these materials the cockles were killed and pieces of digestive gland were sectioned for electron microscopy. Blood corpuscles got as far as the main ducts of the glands but apparently broke up there, releasing haemoglobin. Sections were obtained showing haemoglobin being taken into tiny pinocytotic vesicles (Fig. 14.12*d*) which formed between the microvilli of the digestive cells. Other sections showed graphite particles of diameter about 0.3 μm, apparently being taken in by phagocytosis by a phagosome. From the tiny vesicles which took them in both the graphite and the haemoglobin were passed to the much larger vesicles (*p*1–*p*4) where digestion probably occurs. The large vesicles were never seen opening direct to the cell surface. Digestive enzymes are probably brought to the large vesicles by lysosomes formed by the Golgi apparatus. The digestive cells periodically shed chunks of cytoplasm with vesicles in them containing the residues left after digestion. The vesicles labelled *p*3 and *p*3*a* are probably more or less ready for shedding.

FURTHER READING

GENERAL

See the list for chapter 13

THE HINGE AND ADDUCTOR MUSCLES

Alexander, R. McN. (1966). Rubber-like properties of the inner hinge ligament of Pectinidae. *J. exp. Biol.* **44**, 119–30.
Baguet, F. & Gillis, J. M. (1968). Energy cost of tonic contraction in a lamellibranch catch muscle. *J. Physiol., Lond.* **198**, 127–43.
Millman, B. M. (1964). Contraction in the opaque part of the adductor muscle of the oyster (*Crossostrea angulata*). *J. Physiol., Lond.* **173**, 238–62.
Ruegg, J. C. (1968). Contractile mechanisms of smooth muscle. *Symp. Soc. exp. Biol.* **22**, 45–66.
Trueman, E. R. (1953). Observations on certain mechanical properties of the ligament of *Pecten*. *J. exp. Biol.* **30**, 453–67.

BURROWING

Nair, N. B. & Ansell, A. D. (1968). The mechanism of boring in *Zirphaea crispata* (L.) (Bivalvia: Pholadidae). *Proc. R. Soc.* **170B**, 155–73.
Trueman, E. R. (1967). The dynamics of burrowing in *Ensis* (Bivalvia). *Proc. R. Soc.* **166B**, 459–76.
Trueman, E. R. (1968). The burrowing activities of bivalves. *Symp. zool. Soc. Lond.* **22**, 167–86.

FILTER FEEDING

Aiello, A. & Sleigh, M. A. (1972). The metachronal wave of lateral cilia of *Mytilus edulis*. *J. Cell Biol.* **54**, 493–506.
Jørgensen, C. B. (1966). *Biology of suspension feeding*. Pergamon, Oxford.
Jørgensen, C. B. (1975). Comparative physiology of suspension feeding. *Ann. Rev. Physiol.* **37**, 57–79.
MacGinitie, G. E. (1941). On the method of feeding of four pelecypods. *Biol. Bull.* **80**, 18–25.
Owen, G. (1974). Studies on the gill of *Mytilus edulis:* the eu-latero-frontal cilia. *Proc. R. Soc.* **187B**, 83–91.
Widdows, J. & Bayne, B. L. (1971). Temperature acclimation of *Mytilus edulis* with reference to its energy budget. *J. mar. Biol. Assoc. UK* **51**, 827–43.

DIGESTION

Foster-Smith, R. L. (1975). The effect of concentration of suspension and inert material on the assimilation of algae by three bivalves. *J. mar. biol. Assoc. UK* **55**, 411–18.
Owen, G. (1970). The fine structure of the digestive tubules of the marine bivalve *Cardium edule*. *Phil. Trans. R. Soc.* **258B**, 245–60.
Owen, G. (1974). Feeding and digestion in the Bivalvia. *Adv. comp. Physiol. Biochem.* **5**, 1–36.

15

Squids and their relatives

Phylum Mollusca (cont.), Class Cephalopoda
  Subclass Nautiloidea (pearly nautilus)
  Subclass Ammonoidea (ammonites etc., extinct)
  Subclass Coleoidea (octopus, squid, cuttlefish)

The cephalopods are as different from the gastropods and bivalves as these classes are from each other, but are built on the same basic mollusc plan. All cephalopods live in the sea. They can swim by jet propulsion, by squirting water from the mantle cavity. There are sections of this chapter about swimming and buoyancy. Many of them have exceptionally thick neurones to the jet propulsion muscles, which have been used for experiments that would not have been feasible with neurones of normal size. There is a section of the chapter about these experiments, which provided the basis for our understanding of nervous conduction. Cephalopods have large and effective brains and eyes which are also discussed later in the chapter.

Fig. 15.1 shows some cephalopods including the cuttlefish *Sepia*, which will be described in more detail than the rest. *Sepia* is fairly common in European coastal waters. It swims slowly by undulating the fins on either side of its body, and fast by jet propulsion. When kept in an aquarium with a thick layer of sand on the bottom it swims about by night and buries itself by day. It searches for shrimps and other prey by squirting water from the mantle cavity to stir up the sand. It grabs prey with a pair of long tentacles which are normally hidden, folded away in pockets on either side of the mouth.

Figs. 15.2 and 15.3 show the structure of *Sepia* in more detail. There are eight short arms around the mouth, in addition to the long tentacles. The grasping surfaces of the arms and tentacles are covered with small suckers. There is a pair of hard chitinous jaws, like the beak of a parrot, and also a radula. The posterior salivary glands secrete a venom which rapidly disables crustacean prey. There is a large stomach and a digestive gland (hepatopancreas, Fig. 15.2). The digestive gland secretes enzymes which digest food in the stomach and caecum, and little intracellular digestion occurs. The anus opens into the mantle cavity, beside the opening of the ink sac. This is a bag filled with a suspension of particles of melanin (a black pigment) which looks like ink. When a cuttlefish is attacked it discharges ink into the mantle cavity and squirts out the inky water. The cuttlefish moves off by jet propulsion leaving a cloud of inky water which hides it and distracts its attacker.

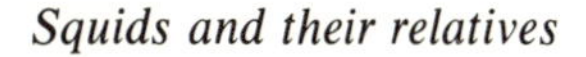

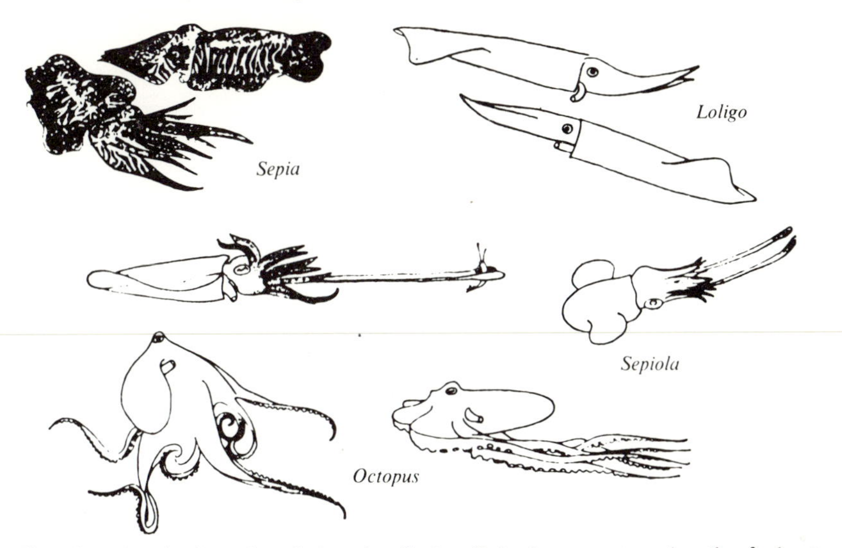

Fig. 15.1. A selection of cephalopods. *Sepia officinalis* grows to a length of about 30 cm (excluding tentacles), *Loligo vulgaris* to 50 cm and *Sepiola rondeletii* to 6 cm. A large *Octopus vulgaris* may have arms 90 cm long. From M. J. Wells (1962). *Brain and behaviour in cephalopods*. Heinemann, London.

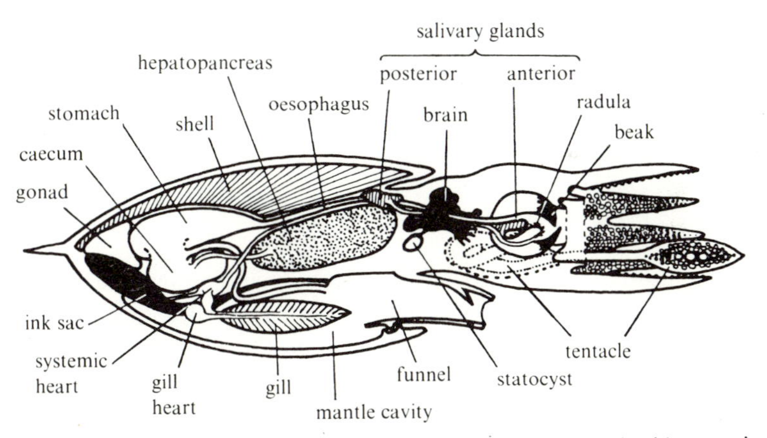

Fig. 15.2. A diagram showing the left half of a specimen of *Sepia*, halved by a sagittal cut. From M. J. Wells (1962). *Brain and behaviour in cephalopods*. Heinemann, London.

The mantle cavity has a thick muscular wall, the mantle. The funnel is the nozzle used in jet propulsion. There is an open slit between it and the mantle but two knobs of the inner face of the mantle fit into sockets in the funnel. In Fig. 15.3 the mantle has been cut along the mid-line and turned back, leaving the funnel intact. There is a single pair of ctenidia (gills) in the mantle cavity. The animal does not depend on cilia to keep water flowing over them. Instead it pumps water through the mantle cavity by gentle rhythmic contractions of the

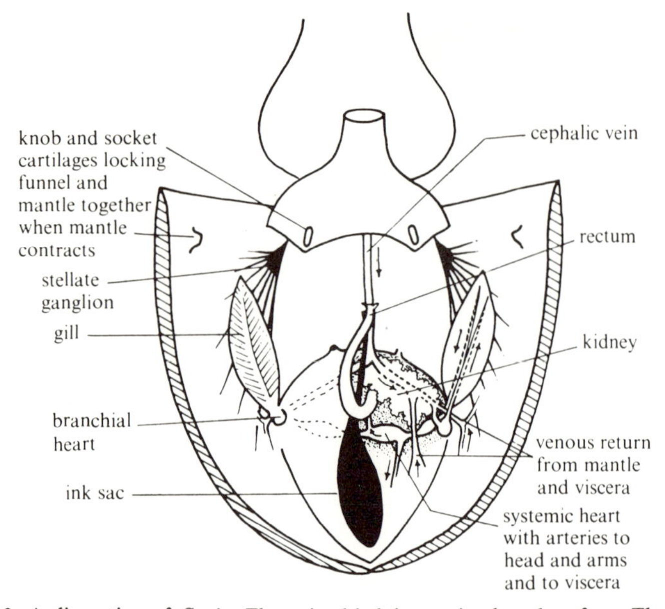

Fig. 15.3. A dissection of *Sepia*. The animal is lying on its dorsal surface. The mantle has been cut along the ventral mid-line and folded back on either side of the animal. Further dissection has exposed part of the heart and the left kidney. From M. J. Wells (1962). *Brain and behaviour in cephalopods*. Heinemann, London.

mantle muscles, similar to the movements of jet propulsion but much more gentle.

The blood of cephalopods contains higher concentrations of haemocyanin than does that of other molluscs. For instance, *Octopus* blood contains about 100 g haemocyanin l^{-1}, enough to carry about 40 cm^3 oxygen l^{-1}. (The oxygen capacity of the blood of most fish is 50–150 $cm^3\ l^{-1}$.) The blood is contained in blood vessels: fine capillaries connect the arteries to the veins, and there are no large haemocoels. Capillaries are not merely gaps in the tissue, but have thin walls formed by endothelial cells. The channels labelled 'longitudinal duct' and 'drainage channel' in Fig. 15.9 are capillaries. The space around the viscera is not a haemocoel but an extension of the pericardial cavity (that is to say, it is a true coelom). Each ctenidium has an auricle which takes blood from it and passes it to the ventricle, which pumps it round the body (The label 'systemic heart' in Fig. 15.3 points to the ventricle). Veins bring the blood back to the ctenidia again. As well as the main (systemic) heart there are two auxiliary (branchial) hearts which boost the pressure of the blood as it enters the ctenidia. The kidneys connect the pericardium to the mantle cavity, as in other molluscs.

Most modern cephalopods have no shell, or only a chitinous rudiment (the pen of squids). *Sepia* has a peculiar shell which consists of calcium carbonate and protein like the shells of gastropods and bivalves but is quite different in

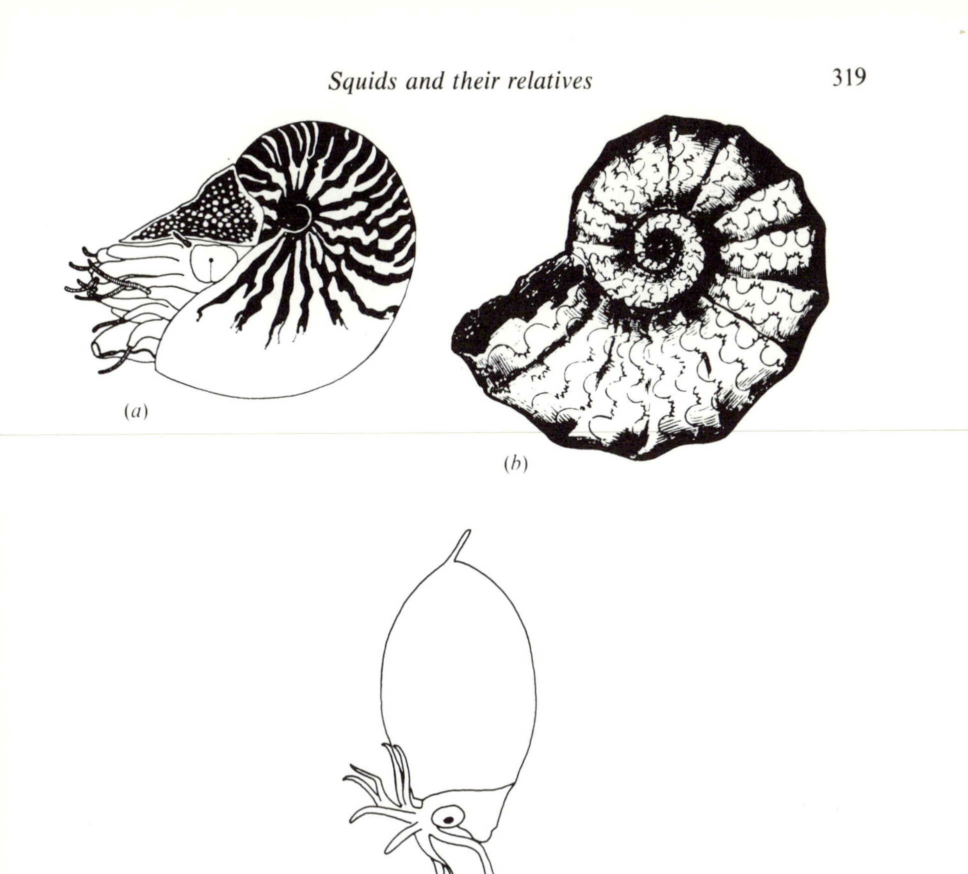

Fig. 15.4. (*a*) *Nautilus*, in normal swimming posture. The diameter of the shell is commonly about 15 cm. From M. J. Wells (1962). *Brain and behaviour in cephalopods*. Heinemann, London. (*b*) *Ceratites nodosus*, an ammonoid fossil from the Trias. Diameter 7 cm. The shell has been removed to show the edges of the folded septa. From H. Woods (1950). *Palaeontology: invertebrate*, 8th edn. Cambridge University Press. (*c*) *Helicocranchia pfefferi*, in the position in which it floats when not swimming actively. Length (excluding tentacles) 5 cm. Drawn from a photograph in E. J. Denton, J. B. Gilpin-Brown & T. I. Shaw (1969). *Proc. R. Soc.* **174B**, 271–9.

structure. It is the cuttlebone which is often given to cage birds to peck. It is a stack of thin-walled chambers largely filled with gas, and serves as a buoyancy organ.

The large brain is a ring of ganglia round the oesophagus. There is a pair of statocysts under the brain which are organs of balance, comparable in function to the otolith organs and semicircular canals of vertebrate ears.

The sexes are separate and fertilization is internal. The male *Sepia* forms a packet of sperm (a spermatophore) and passes it to the female, putting it into a pocket under her mouth. He does this with one of his eight arms, the hectocotylus, which differs in structure from the rest. The spermatophore bursts and the sperm swim into the female's mantle cavity and eventually reach the oviduct and fertilize the eggs. The eggs are quite large, about 8 mm in diameter. The female blows them out of her funnel one by one and sticks them to seaweed or gorgonians. The young hatch out looking like miniature adults.

Figs. 15.1 and 15.4 show some other cephalopods. *Nautilus* (Fig. 15.4*a*) is the only surviving genus of the subclass Nautiloidea, though there are plenty of fossil nautiloids in Palaeozoic rocks. There are several species of *Nautilus*. They live in coastal waters in the S.W. Pacific where they can be caught in traps like lobster pots, set on the bottom at depths around 100 m. They have a beautiful shell of logarithmic spiral form with $T = 0$ (see p. 267). Externally, the shell is shaped like the shell of *Planoris* (Fig. 13.4*f*) but its internal structure is quite different. It is divided by partitions into chambers (Fig. 15.8) which contain enough gas to give the animal about the same density as sea water. The body of the animal occupies the large chamber at the mouth of the shell. This body is much like the body of *Sepia* but it has a very large number of small tentacles and no suckers, and it has four ctenidia.

The subclass Ammonoidea is extinct. Its members are abundant as fossils in Mesozoic rocks (Fig. 15.4*b*). Their spiral shells have diameters up to 1 m. They have chambers like the shell of *Nautilus* but the partitions between them are wrinkled.

The animals shown in Figs. 15.1 and 15.4(*c*) are members of the subclass Coleoidea. Squids of the genus *Loligo* are abundant off European and N. American coasts. *Loligo vulgaris* kept in aquaria swim perpetually, never resting on the bottom. *Sepiola* is a small cuttlefish. *Helicocranchia* is a small oceanic squid caught in deep water, mostly at depths between 100 and 400 m. *Octopus* is a cephalopod which can swim but spends most of its time on the bottom. It is found offshore on both sides of the Atlantic, and in the Pacific. It has no tentacles, but its eight arms are very long. It feeds largely on crabs and lobsters.

The illustrations do not show the largest cephalopod, the giant squid *Architeuthis*. Its trunk and arms are up to 5 m long and the overall length with the tentacles extended may be as much as 15 m.

JET PROPULSION

Cephalopods can propel themselves backwards or forwards, by squirting water from the funnel. The lower *Loligo* in Fig. 15.1 is swimming backwards but the upper one has bent its funnel round to squirt water backwards, and propel itself forwards. *Octopus* swim relatively slowly but squid can swim fast, at speeds comparable to the top speeds of teleost fish. A 20 cm (100 g) *Loligo* has been filmed in an aquarium accelerating from rest to 2.1 m s^{-1} (backwards) by a single contraction of its mantle. The oceanic 'flying squid' *Onycoteuthis* is presumably fast, for it has been known to leap from the water and land on the decks of ships.

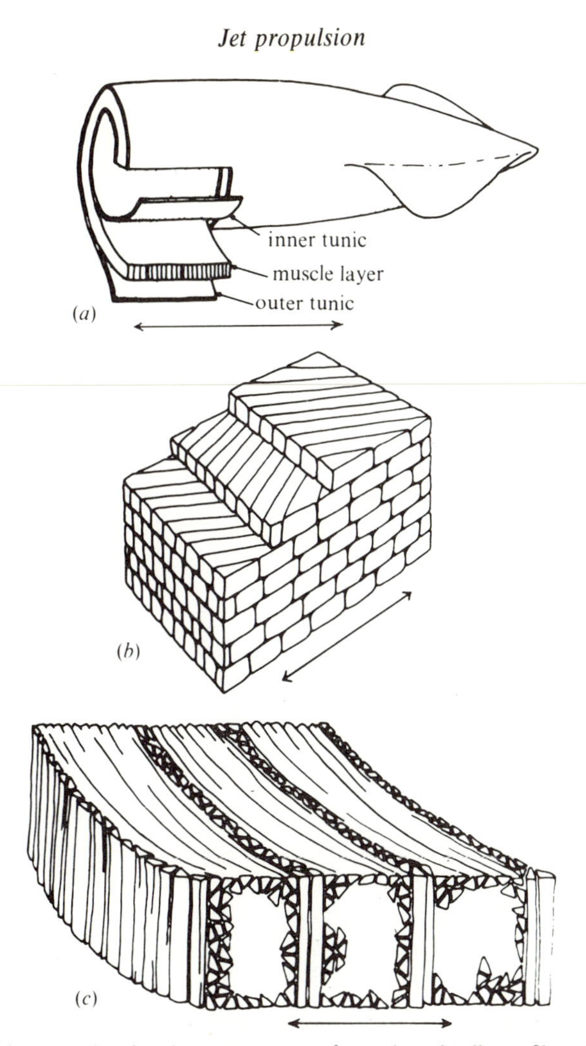

Fig. 15.5. Diagrams showing the arrangement of muscle and collagen fibres in the mantle of the squid *Lolliguncula*. Arrows indicate the longitudinal axis of the animal. (*a*) represents the posterior end of the animal, (*b*) shows part of the outer tunic at much higher magnification and (*c*) shows part of the muscle layer. The muscle layer is 20 times as thick as the outer tunic. From S. A. Wainwright, W. D. Biggs, J. D. Currey & J. M. Gosline (1976). *Mechanical design in organisms*. Arnold, London.

The movements of respiration and jet propulsion are similar but jet propulsion is of course much more vigorous. Water is taken into the mantle cavity when it expands, through the slit between the mantle and the funnel. A valve in the funnel (Fig. 15.2) prevents it from entering through the funnel. However, when the mantle muscles contract the pressure in the mantle cavity keeps the base of the funnel pressed tightly against the mantle, closing the slit. The water has to leave through the funnel. This arrangement ensures that water enters ventral to the ctenidia and passes between the filaments to leave dorsal to them. Water

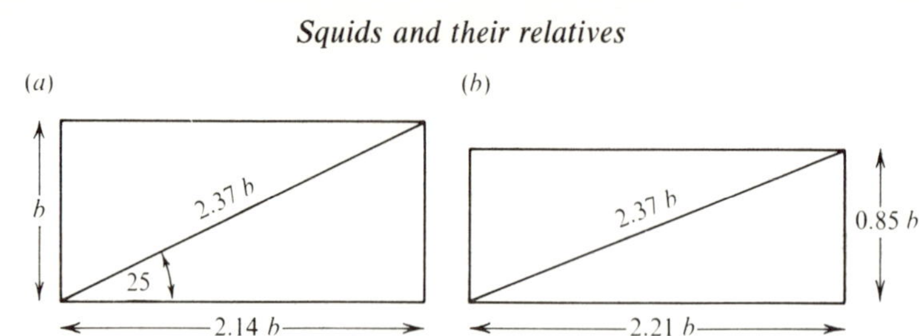

Fig. 15.6. These diagrams are explained in the text.

leaving the funnel of resting *Octopus* has been sampled and analysed and it has been shown that up to 80% of the dissolved oxygen is removed from the water as it passes over the ctenidia.

The muscles responsible for swimming are in the mantle. In *Loligo* they make up 35% of the mass of the body but in *Octopus* much less. (The swimming muscles of teleost fish may be as much as 60% of the mass of the body.) Fig. 15.5 shows how they are arranged. They are sandwiched between two sheets of collagen fibres, the inner and outer tunics. Each tunic has several layers of fibres, with the fibres of successive layers running alternately in left-handed and right-handed helices. This is like the arrangement of *Ascaris* cuticle (Fig. 12.5 *b*), but the fibres run at much smaller angles to the long axis of the body. The angle is about 25° in *Lolliguncula* mantle, but over 55° in *Ascaris* cuticle. The muscle is mainly circular muscle, with fibres running circumferentially round the body, but there are also radial fibres.

When the circular muscle contracts and decreases its diameter, the mantle must either get longer or its wall must get thicker or both so that the volume of muscle remains constant. When the radial muscles contract and make the wall thinner, the mantle must get longer or increase in diameter. The tunics ensure that changes in length are small. Fig. 15.6(*a*) represents a rectangle of width *b*, cut from one of the tunics of an expanded squid mantle. The collagen fibres run at 25° to the long axis of the squid and the rectangle has been cut so that one of these fibres forms its diagonal. The length of this diagonal is $b \operatorname{cosec} 25° = 2.37b$. Fig. 15.6(*b*) represents the same rectangle when the mantle has contracted so as to reduce its diameter to 0.85 of its expanded value (films of *Lolliguncula* swimming show that it contracts about this much). The diagram shows that if the length of the diagonal collagen fibre remains constant the length of the rectangle will increase from $2.14b$ to $2.21b$, by about 3%. A 15% decrease in the diameter of the mantle will be accompanied by only a 3% increase in its length. Since the volume remains constant the thickness of the mantle must increase by 14% ($0.85 \times 1.03 \times 1.14 = 1$). The films of *Lolliguncula* show that when the diameter of the mantle decreases by 15% the length does not change perceptibly.

The fibres of the tunics make the circular and radial muscles antagonistic to each other. Contraction of the circular muscle stretches the radial muscle, and contraction of the radial muscle stretches the circular muscle and enlarges the

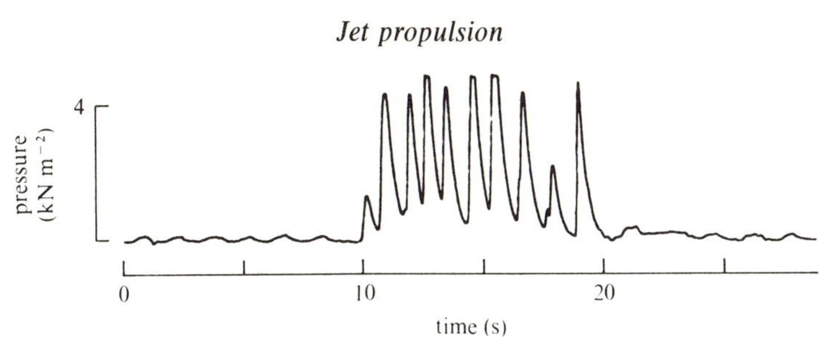

Fig. 15.7. A record of pressure in the mantle cavity of *Loligo* showing small fluctuations during quiet breathing and large ones during jet propulsion. From A. Packard & E. R. Trueman (1974). *J. exp. Biol.* **61**, 411–19.

mantle cavity. The muscles are obliquely striated but have sarcomeres only about 1.6 μm long. Short sarcomeres are to be expected in fast muscles but this is shorter even than the sarcomeres of vertebrate striated muscle.

Fine tubes attached to pressure transducers have been slipped into the mantle cavities of living cephalopods, to obtain records of pressure changes during breathing and swimming. The experiment has only been feasible with tethered specimens so no records have been obtained of free swimming. Fig. 15.7 shows a typical record from *Loligo*. In quiet breathing the pressure rises and falls through about 300 N m^{-2}. When the animal tries to swim the pressure rises much higher. Some of the pressure peaks in Fig. 15.7 have flat tops because the pressure rose beyond the range of the recorder. Other records from the mantle cavity of *Loligo* show pressures up to 30 kN m^{-2} while the circular muscles are contracting.

Consider a 20 cm *Loligo* swimming fast, at 2 m s^{-1}. Its Reynolds number (see p. 41) would be 4×10^5. This is well up in the range of Reynolds numbers in which bodies leave wakes: the pattern of flow around the squid will be as shown in Fig. 2.8(*b*) or possibly Fig. 2.8(*c*). An unstreamlined shape would result in a broad wake, and high drag. Streamlining has no value for Protozoa which swim at Reynolds numbers less than 1, but it is advantageous for squids which swim at much higher Reynolds numbers. Drag at Reynolds numbers well above 1 is not given by equation (2.1) but by

$$\text{Drag} = \tfrac{1}{2}\rho A u^2 C_D \qquad (15.1)$$

where ρ is the density of the fluid, A the frontal area (i.e. the area of a full-scale front view of the body: this is often the cross-sectional area of the thickest part of the body) and u is the velocity. C_D is a quantity called the drag coefficient which depends on the shape of the body and on the Reynolds number. For Reynolds numbers between 10^4 and 10^5 it is about 0.5 for spheres but only 0.05 for well-streamlined bodies, so streamlining can reduce drag dramatically. The best shape is like a torpedo, rounded in front and tapering gently behind. The bodies of squid such as *Loligo* look quite well streamlined.

How can jet propulsion be made most efficient? Consider a squid of mass

m swimming steadily. Each jet accelerates it and it decelerates between jets but its mean velocity is $\bar{u}$ and the mean drag on its body is D. It emits jets with frequency n, taking in each time a mass M of water and ejecting it with velocity U. In unit time it accelerates a mass nM of water to velocity U, giving momentum nMU to water. By Newton's Second Law of Motion force equals rate of change of momentum

$$D = nMU \qquad (15.2)$$

The power required to drive the squid through the water is Du. In addition, power is used accelerating the water: in unit time a mass nM of water is given kinetic energy ½ nMU^2, so this much power is needed. By equation (15.2) it equals ½ DU.

$$\begin{aligned} \text{Efficiency} &= \text{useful power/total power} \\ &= Du/(\tfrac{1}{2}DU + Du) \\ &= 2u/(U + 2u) \qquad (15.3) \end{aligned}$$

The smaller U can be made, the more efficient swimming at a given speed will be. Since nMU must equal the drag (equation 15.2) it is advantageous to jet at high frequency and to eject a large mass of water in each jet.

It seems likely that the swimming of squid is less efficient than the swimming of most teleost fish. Teleosts have large tails which can accelerate large masses of water in unit time, so that relatively low values of U are possible. However, it is difficult to make quantitative comparisons between squid and fish because we have little information about squid swimming performances, and because the undulating movements of fish swimming increase the drag on the body.

BUOYANCY

Cephalopods which have no special buoyancy organs are denser than sea water. They include both *Octopus* which crawls on the bottom and swims only occasionally, and squid such as *Loligo* and *Onycoteuthis* which swim perpetually. Many other cephalopods have approximately the same density as sea water. Some, including *Sepia* and *Nautilus*, owe this to gas-filled spaces in their shells. Others, such as *Helicocranchia*, owe it to spaces filled with ammoniacal solutions of low density. Professor Eric Denton and his colleagues have investigated both types of buoyancy mechanism.

They measured the densities of cephalopods by weighing them in air and water. Consider an animal of weight (in air) W, density ρ_a and volume W/ρ_a. If it is immersed in water of density ρ_w an upthrust $W\rho_w/\rho_a$ acts on it, by Archimedes' Principle, so its weight in water W_w is given by

$$W_w = W - (W\rho_w/\rho_a)$$

and

$$\rho_a = W\rho_w/(W - W_w) \qquad (15.4)$$

If W and W_w are measured, ρ_a can be calculated. It has been shown in this way that the density of *Loligo* is about 1070 kg m^{-3}, which is about the same as the

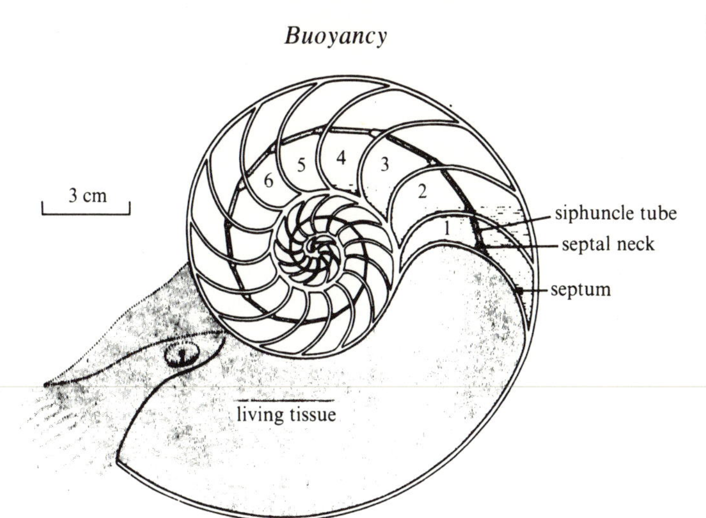

Fig. 15.8. A diagrammatic sagittal section through *Nautilus*. A strand of living tissue, the siphuncle, extends along a tube through all the chambers of the shell. Some of the chambers contain liquid. From E. J. Denton (1974). *Proc. R. Soc.* **185B**, 273–99.

densities of many fish which lack swimbladders. The densities of *Nautilus* and *Sepia* are very close indeed to the density of sea water, about 1026 kg m^{-3}.

Fig. 15.8 is a section through *Nautilus*. The animal occupies the quarter-turn of the shell nearest the opening, and the rest of the shell is divided into compartments. Compartment 1 is the most recently formed and is still full of fluid but the other compartments are filled mainly or entirely with gas. This makes the animal float in the attitude shown. The only contact between the gas spaces and the living tissue is through the wall of a thin tube which runs through the centres of the compartments. This tube contains a strand of living tissue (stippled) called the siphuncle. The walls of the chambers are impermeable shell material but the wall of the tube is porous, as can be shown by a simple experiment. A small piece of blotting paper is rolled up and pushed into the siphincle tube of an empty shell. The adjacent compartment is filled with coloured water which seeps through the wall of the tube and stains the blotting paper.

The walls of *Nautilus* shells are quite thick and they are made of material of density 2700 kg m^{-3}. When the chambers are completely full of gas the density of the shell is about 910 kg m^{-3}. This is not very much less than the density of sea water so quite a large shell is needed to reduce the density of the animal only a little. Consider an animal made up of a volume V_t of tissues of density ρ_t and a volume V_s of shell (or other buoyancy organ) of volume ρ_s. The weights of tissue and shell are $\rho_t V_t$ and $\rho_s V_s$ respectively, so the density ρ_a of the whole animal is given by

$$\rho_a = (\rho_t V_t + \rho_s V_s)/(V_t + V_s)$$

$$V_s/V_t = (\rho_t - \rho_a)/(\rho_a - \rho_s) \qquad (15.5)$$

The soft parts of *Nautilus*, removed from the shell, have density about 1060 kg m^{-3}. Take this as the value of ρ_t, together with $\rho_s = 910$ kg m^{-3} and $\rho_a = 1026$ kg m^{-3}. This gives $V_s/V_t = 0.29$. To give a *Nautilus* the same density as sea water, the volume of the shell must be at least 29% of the volume of the tissues. This is the volume required if the spaces in the shell are completely filled with gas. In fact, an average of 5% of the space in the chambers is filled with liquid, which increases the density of the shell considerably. The volume of the shell is about 60% of the volume of the soft parts. The shell of *Sepia* is much less dense than the shell of *Nautilus* and its volume is only about 10% of the volume of the soft parts.

It might be expected that the pressure in the chambers of a *Nautilus*, caught deep in the sea, might be high. It is not. If a freshly-caught *Nautilus* is held under water while a hole is bored into a chamber no gas bubbles out: rather, water is sucked in. The pressure in the chambers is less than 1 atm and can be calculated from the volume of water sucked in, which can be determined by weighing. In one particular (fairly typical) *Nautilus* the newest chamber was full of liquid, the second chamber contained gas at 0.37 atm and successive older chambers contained gas at successively higher pressures. The sixth chamber (the oldest investigated) contained gas at 0.82 atm. The gas was analysed by mass spectrometry and found to be mainly nitrogen with a little oxygen, argon and water vapour. In the sixth chamber the partial pressures of nitrogen and argon were 0.74 and 0.01 atm.

The partial pressures of nitrogen, oxygen and argon in the atmosphere are 0.78, 0.21 and 0.01 atm. The surface water of the sea is in equilibrium with the atmosphere so these are the partial pressures of the gases dissolved in it. Deeper in the sea the partial pressure of oxygen is less because organisms use it in respiration, but the partial pressures of nitrogen and argon are 0.78 and 0.01 atm at all depths, irrespective of the hydrostatic pressure. They are dissolved in the blood and other tissues of animals at the same partial pressures.

The observations on *Nautilus* indicate that the chambers are not filled by pumping gas into them under pressure; rather, their original liquid contents are drawn out leaving a vacuum into which gases diffuse, until they are in equilibrium with the dissolved gases of the tissues. At equilibrium, the partial pressures of nitrogen and argon would be 0.78 and 0.01 atm. It is only in the older chambers that equilibrium is approached.

Water is probably removed from the chambers by an osmotic mechanism. Samples have been taken of the liquid left in the chambers and their freezing points determined, so that their osmotic concentrations could be calculated. They were all less concentrated than sea water, some as little as 0.4 times as concentrated as sea water. The blood of *Nautilus* has about the same osmotic concentration as sea water. Hence water must tend to move from the chambers to the blood in the siphuncle, through the porous siphuncle tube, until the pressure differences between the blood and the contents of the chamber equals the difference of osmotic pressure.

The osmotic concentration of sea water is about 1 Osmol l^{-1}. Hence by equation (2.4) its osmotic pressure is 24 atm. This is also the osmotic pressure of *Nautilus* blood. Even if the liquid in the chambers were pure water, the difference in osmotic pressure between it and the blood would not exceed 24 atm. Water could not be withdrawn, leaving gas at less than 1 atm, when the animal was at pressures greater than 25 atm. Since the pressure is 1 atm at the surface of the sea and increases by 1 atm for every 10 m descent, the blood should not be able to withdraw water by osmotic means at depths greater than about 240 m. There have been a few records of *Nautilus* being caught at greater depths, but it is not certain whether they are reliable. However there is another cephalopod with gas-filled spaces in its shell which is known to live at depths down to 1200 m. This is *Spirula*. It must extract the liquid from its shell by something other than simple osmotic means. It seems possible that the osmotic concentration of the blood in the siphuncle can be raised well above the osmotic concentration of the rest of the blood, by a countercurrent mechanism.

Whatever the refinement that enables the mechanism to work at great depths, the osmotic pressure of the fluid in the chambers must be reduced. This must presumably be done by the epithelium of the siphuncle which is shown in Fig. 15.9(*a*). The surface which is uppermost in the figure faces the porous wall of the siphuncle tube. It is covered with microvilli. The opposite face of the epithelium is deeply folded and blood capillaries extend out into the folds, close to the microvillous face. The parts of the cells which look striped in Fig. 15.9(*a*) have been shown by electron microscopy to contain numerous mitochondria, which suggest that an active, energy-consuming process occurs there.

Fig. 15.9 (*b*) shows a possible process. The cells may pump salts out into the spaces between them, reducing the cellular concentrations of salts. Water will tend to follow by osmosis. If water and salts diffuse freely across the microvillous surface, the fluid in the chambers will always be in equilibrium with the cell contents.

Nautilus has certainly been caught at 180 m where the pressure is 19 atm, and it probably lives even deeper. It shell must be able to withstand at least 19 atm while the pressure in the chambers is less than 1 atm. It must be strong to avoid being crushed. Its strength has been tested by blocking the siphuncle with epoxy resin, putting the shell into a pressure chamber and increasing the pressure. Shells tested in this way did not shatter until the pressure reached 60–70 atm so that they were amply strong for the pressures at which *Nautilus* is known to live. The shell of *Sepia* is crushed at about 24 atm but *Sepia* does not live deeper than about 150 m where the pressure is 16 atm. About 170 atm is needed to crush the shell of *Spirula*.

Helicocranchia (Fig. 15.4 *c*) is an example of the many cephalopods which achieve about the same density as sea water without a gas-filled shell. It is given buoyancy by the fluid in its coelom which has about the same osmotic

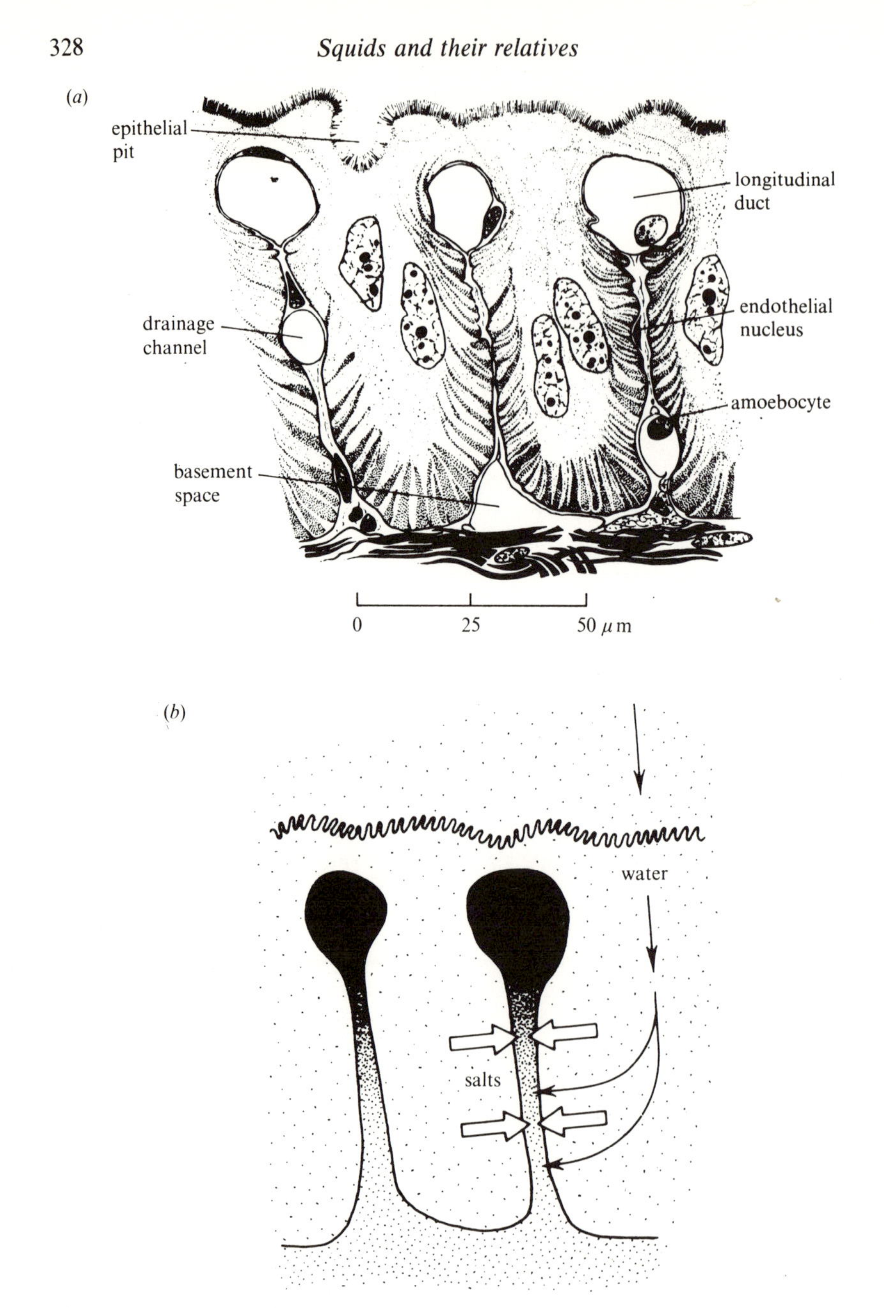

Fig. 15.9. (*a*) A section of the epithelium of the siphuncle of *Nautilus*. From E. J. Denton & J. B. Gilpin-Brown (1973). *Adv. mar. Biol.* **11**, 197–268. (*b*) A diagram of the same epithelium, illustrating a possible mechanism for removal of liquid from the chambers of the shell. Differences in density of stipple represent differences of osmotic concentration.

concentration as sea water but contains far less sodium and an extraordinarily high concentration of ammonium ions. Its composition is almost the same as that of a mixture of one part of sea water with four parts of an ammonium chloride solution of the same osmotic concentration. The density of this fluid is much less than that of sea water, about 1010 kg m^{-3}. The density of the remainder of the body is about 1050 kg m^{-3}. Hence by equation (15.5) the volume of the coelom must be 1.5 times the volume of the rest of the body, to give the whole animal the density of sea water (1026 kg m^{-3}). Its volume is about that, which is why the animal looks bloated.

Helicocranchia is not a unique oddity. Many other species of squid have low-density ammoniacal fluids, either in the coelom or in vacuoles in other tissues. They are seldom collected for they live in mid-ocean, many of them at depths of several hundred metres, but they seem to be common. Sperm whales (*Physeter*) feed on squid and it seems from their stomach contents that ammoniacal squid make up 50–80% of their diet. It has been estimated that there are 1¼ million sperm whales of total mass 11 million tonnes and that they must eat over 100 million tonnes of ammoniacal squid each year. The mass of fish taken annually by the fishing fleets of the world is only 60 million tonnes. Plainly, ammoniacal squid are not rare.

Now consider how a squid like *Loligo* avoids sinking, though considerably denser than the sea water. Fig. 15.10 shows the forces which may act on a *Loligo* of weight W when it swims forwards. Since its density is 1070 kg m^{-3} the Archimedes upthrust is $(1026/1070)W = 0.96W$. The additional upward force of $0.04W$ which is needed for equilibrium is probably provided by the fins, working like aeroplane wings. Since this force acts near the posterior end of the body it must exert a moment about the squid's centre of mass, tending to turn the squid heels over head (clockwise in the diagram). However the thrust exerted by the jet from the funnel acts below the centre of mass, exerting a moment in the opposite direction

The force on the fins has two components, the lift L_f which is needed to prevent the squid from sinking and the drag D_f. Lift acts at right angles to the direction of motion. Drag acts backwards along the direction of motion. Only drag acts on a symmetrical body, moving along its axis of symmetry. Hydrofoils and aerofoils are shaped to give as much lift as possible with as little drag as possible, and they give most lift when tilted at a small angle to the direction of motion like the fins shown in the diagram. No matter how well a hydrofoil is designed lift cannot be obtained without drag and if a squid's fins generate lift amounting to $0.04W$ it is most unlikely that the drag on them will be less than about $0.004W$. This extra drag is the penalty for being denser than the water.

Now suppose that this same squid had a gas-filled shell like that of *Sepia*, which added 10% to its volume and reduced its density to that of sea water. If it was arranged in such a way as to enlarge the body without changing its shape the drag coefficient of the body would be unaltered but the frontal area

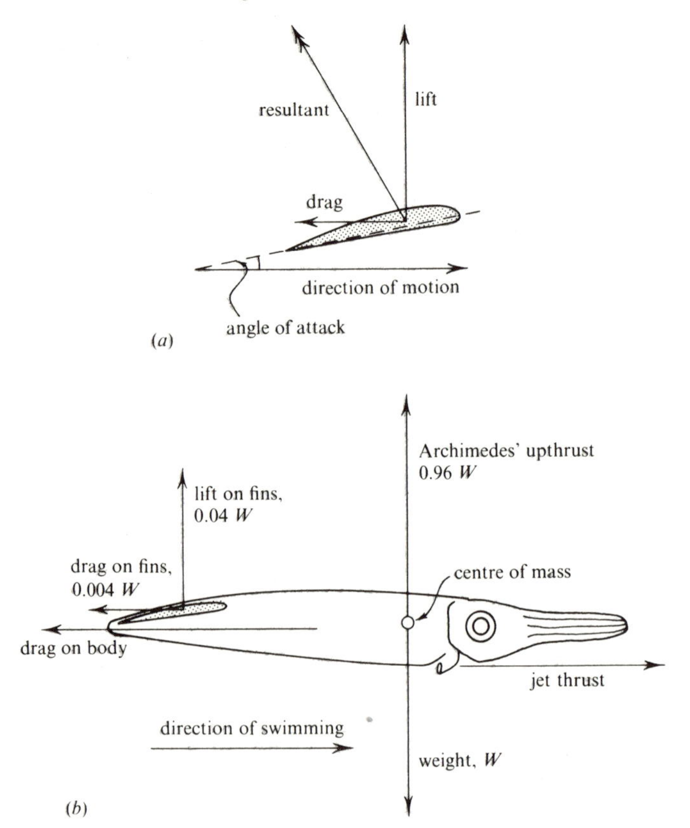

Fig. 15.10. Diagrams showing (*a*) forces on a hydrofoil (seen in section) and (*b*) forces on a squid, swimming forwards by jet propulsion.

and thus the drag would be increased by 7% (For bodies of the same shape area is proportional to $(\text{volume})^{0.67}$ and $1.10^{0.67} = 1.07$.) By equation (15.1) a 7% increase in drag would be an increase of $0.035\rho Au^2C_D$ where ρ is the density of the water, A the frontal area, u the velocity and C_D the drag coefficient.

If our squid has no buoyancy organ, it will suffer an estimated drag of 0.004 W on its fins. If it has a buoyant shell it will be bulkier and so incur drag amounting to $0.035\rho Au^2c_D$. If it swims fast enough (i.e. if u is large) a buoyancy organ will give rise to more extra drag than lifting fins, and will be a disadvantage. We do not know enough to calculate the critical speed but it may well be that *Loligo* and similar squid swim too fast for a buoyancy organ to be useful. Many species of mackerel and tuna (teleost fish of the family Scombridae) lack swimbladders and are denser than sea water. Like *Loligo* they swim perpetually, rather fast.

GIANT AXONS

Nerves radiate from the stellate ganglion (Fig. 15.3) to the muscles of the mantle. Each nerve contains axons of ordinary size and one giant axon. The longest nerves, which go to the posterior end of the mantle, have the thickest giant axons, up to 0.8 mm diameter in *Loligo forbesi*. There are also giant axons between the stellate ganglia and the brain.

If one of the mantle nerves is stimulated electrically, in a freshly-dissected squid, the circular muscles contract in the part of the mantle it serves. The nerve can be dissected so that all the axons except the giant one are cut, or so that the giant one is cut and the rest left intact. In either case, stimulating the nerve will still make the muscles contract. However, much greater forces act when the muscle is stimulated through the giant axons than when it is stimulated through the small ones. Also, the force produced in response to a stimulus to the small fibres depends on the voltage of the stimulus: a small voltage results in a small force and a larger voltage in a larger one. Presumably different voltages excite different numbers of axons. A stimulus to the giant axon evokes the maximum force, provided the voltage is high enough to get any response at all. It is believed that the small axons control the gentle movements of respiration and that the giant ones are responsible for jet propulsion.

The velocity of action potentials in the mantle nerves has been measured, by recording at two points on the nerve the action potentials evoked by a stimulus at a third point. The velocity was calculated from the difference in arrival time of the same action potential at the two recording electrodes. It was found that giant axons of diameter 700 μm conduct at about 20 m s^{-1}, and small axons of diameter 40 μm conduct at 4 m s^{-1}. It seems to be a general rule that thick axons conduct faster than similar thin ones, and there is a theoretical explanation. However, the myelin sheath which invests many vertebrate axons greatly increases conduction speed, making many slender vertebrate axons faster than squid giant axons. Some cat axons of diameter 4 μm conduct at 25 m s^{-1}, and other, thicker vertebrate axons conduct at up to about 100 m s^{-1}.

Cephalopods have not evolved myelin sheaths, so if their axons are to be fast they have to be thick. The giant axons enable cephalopods to escape more quickly by jet propulsion than would otherwise be possible. The furthest part of the mantle of a large *Loligo* is about 30 cm from the stellate ganglion. If there were only slender axons conducting at 4 m s^{-1}, an action potential would spend 75 ms travelling this distance. In a giant axon which conducts at 20 m s^{-1}, it spends only 15 ms. Since the thickest and fastest giant axons serve the most distant parts of the mantle, action potentials arrive more nearly simultaneously all over the mantle than would otherwise be the case.

Squid giant axons were discovered and studied by Professor J. Z. Young but they have since been used in experiments by many other distinguished physiologists. Their size makes them suitable for experiments which would be difficult or even impossible with smaller axons. The theory of nerve conduction

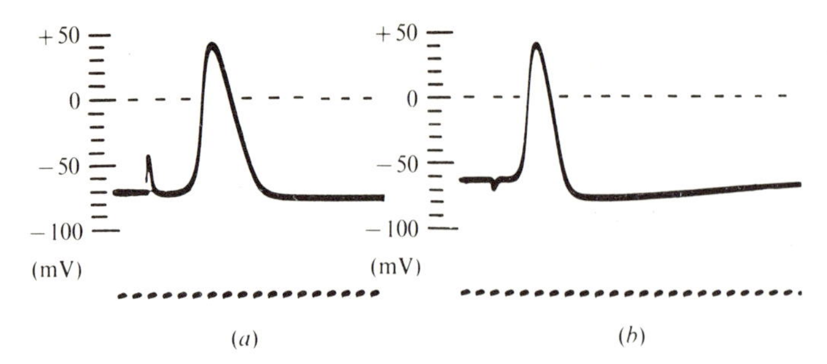

Fig. 15.11. Two action potentials, recorded through electrodes inserted into squid giant axons. (*a*) was recorded from an axon in position in a dissected squid and (*b*) from an isolated axon bathed in sea water. The time marks below the records show a frequency of 2500 s^{-1}. From A. L. Hodgkin (1958). *Proc. R. Soc.* **148B**, 1–37.

which was developed by Sir Alan Hodgkin and Professor A. F. Huxley, and others, was based mainly on experiments with squid giant axons.

An action potential is a brief reversal of membrane potential which travels along the cell. It is possible to record the passage of an action potential with electrodes outside the cell (see for instance Fig. 6.10), but the most revealing records show the potential difference across the cell membrane. To obtain them, an electrode must be inserted into the cell. This was first accomplished with squid giant axons but as techniques have improved it has become possible with very much smaller neurones.

Fig. 15.11(*a*) shows an action potential recorded through an electrode inside a squid giant axon. The resting membrane potential was −70 mv. An electrical stimulus was delivered to a different point on the axon at the time of the small blip on the record (this blip is a stimulus artefact). The membrane potential quickly rose to +40 mV. It fell again within about 2 ms but overshot its original value, falling to −75 mV. It returned slowly to the original value, taking longer than the duration of the section of record which is illustrated. Similar action potentials have been recorded from neurones of various sizes in a great variety of animals.

To get some insight into the mechanism of the action potential we need to know the concentrations of the principal ions, inside and outside the axon. The contents of squid giant axons are easily obtained for analysis, by squeezing them out like toothpaste from a tube. In the squid's body, the fluid outside the axons probably has the same ionic composition as squid blood.

The results of analyses are shown in Table 15.1. The concentrations of ions in squid blood are much the same as in sea water. The concentrations of sodium and chloride in axon contents are much lower, and the concentration of potassium is much higher. The axon also contains high concentrations of isethionate ($CH_2OH.CH_2SO_3^-$) and amino acids. Axon contents are rather different from muscle cell contents, which are more similar in composition to the marine ciliate *Miamiensis* (Table 2.1).

TABLE 15.1. *Concentrations (in mmol (kg water)$^{-1}$) of the principal ions in giant axon contents, muscle cell contents and blood of squid* (Loligo) *and in sea water*

	Na^+	K^+	Cl^-	Isethionate	Amino acids
Axon	50	400	108	270	87
Muscle	31	189	45	—	483
Blood	440	20	560	—	4
Sea water	460	10	540	—	—

Data from A. L. Hodgkin (1958). *Proc. R. Soc.* **148B**, 1–37; R. D. Keynes (1963). *J. Physiol., Lond.* **169**, 690–705; and J. D. Robertson (1965). *J. exp. Biol.* **42**, 153–75.

Consider a squid axon bathed in tissue fluid of the same ionic composition as squid blood. By applying equation (2.3) to the data of Table 15.1 we find that the Nernst potential for potassium is $58 \log_{10} (20/400) = -75$ mV. The Nernst potentials for sodium and chloride are +55 mV and −41 mV, respectively. None of these ions is in equilibrium at the resting potential of −70 mV. Potassium and chloride must be pumped into the cell and sodium must be pumped out. (If chloride were not pumped in its concentration in the axon would fall even lower.) These calculations of Nernst potential depend on the assumption that the activities of the ions inside and outside the axon are in the same ratio as the measured concentrations. This would not be the case if ions were bound in the axon, in non-ionic form. It has been shown that there is no significant binding of potassium or chloride.

The resting membrane potential of squid axon lies between the Nernst potentials for sodium and potassium, but is much nearer the latter. This implies that the cell membrane is much more permeable to potassium than it is to sodium (see p. 64). In the action potential the membrane potential comes quite near the Nernst potential for sodium, suggesting that the membrane has become highly permeable to sodium. It seems that the action potential is very like the response of *Paramecium* to a stimulus at the anterior end (Fig. 3.7), but with sodium playing the part played by calcium in *Paramecium*. However, the membrane potential falls again very much more rapidly than in *Paramecium*, and overshoots. It overshoots to a value very close indeed to the Nernst potential for potassium, suggesting that the permeability of the cell membrane to potassium is abnormally high. The very brief action potential seems to involve a brief increase in sodium permeability, quickly followed by a longer-lasting increase in potassium permeability.

If this interpretation of the action potential is correct sodium must leak into the axon rapidly during the action potential and slowly at other times. It must be pumped slowly out between action potentials. Potassium must leak out

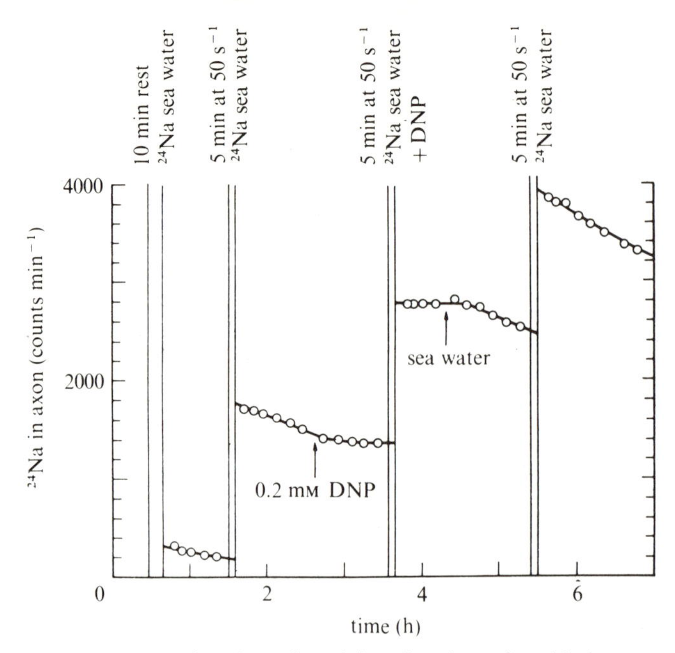

Fig. 15.12. A graph showing the radioactivity of a piece of squid giant axon, plotted against time, during an experiment which is described in the text. From A. L. Hodgkin & R. D. Keynes (1955). *J. Physiol., Lond.* **128**, 28–60.

rapidly during and immediately after the action potential, and slowly at other times. It must be pumped in between action potentials. This has been demonstrated by experiments with giant axons, using radioactive sodium and potassium. The radioactivity of the axons was measured by holding them in flowing sea water over a Geiger counter. Fig. 15.12 shows what happened in one experiment. An axon was put into sea water containing $^{24}Na^+$ for 10 minutes, and was found afterwards to have taken in some radioactivity. Later it was put back into the radioactive sea water and stimulated repeatedly so that many action potentials travelled along it. In 5 minutes of this treatment it took up more radioactivity than in the previous 10 minute period in radioactive sea water. Hence, sodium leaks in faster during an action potential than at other times. During the periods in ordinary sea water, the radioactivity of the axon diminished, showing that sodium was being pumped out of the axon. However this was halted when 2,4-dinitrophenol (DNP) was added to the water. DNP inhibits metabolic processes, probably by preventing ATP synthesis. The sodium pump can also be halted by adding cyanide to the water but an axon poisoned with cyanide will pump out radioactive sodium for a while after ATP has been injected into it. Hence the energy needed to drive the sodium pump is probably supplied to it as ATP.

Many other experiments have been carried out on squid axons. In some of

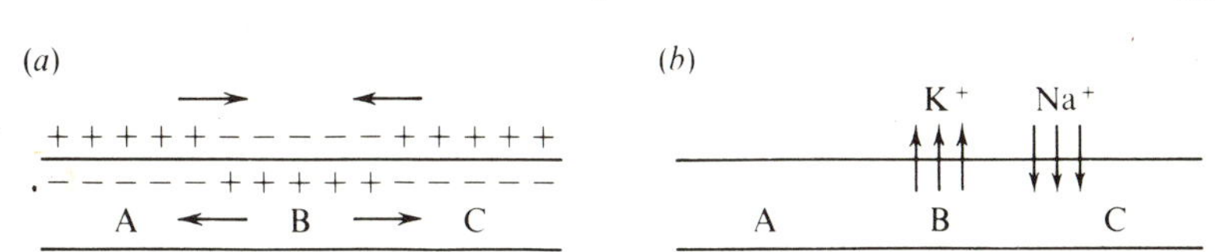

Fig. 15.13. A diagram of an axon showing (*a*) electrical charges and current flow and (*b*) movements of ions through the cell membrane during the passage of an action potential.

the most important, the technique known as voltage clamping was used, to investigate the effects of changes of membrane potential on the permeability of the cell membrane. It was found that an increase from −60 mV (the resting potential in sea water, Fig. 15.11 *b*) to −40 mV caused a brief increase in sodium permeability, lasting about 3 ms. It also caused an increase in potassium permeability which was slower to develop but lasted as long as the membrane potential was kept at −40 mV. Bigger increases of membrane potential caused similar but larger increases in sodium and potassium permeability. It is not yet known why these changes happen, but the mechanism of transmission of action potentials depends on them.

Fig. 15.13 shows an axon at an instant when an action potential is passing point B, travelling towards C. At B the membrane potential is positive but elsewhere it is negative. Current must flow inside the axon from B towards A and C, where the potential is lower. It must flow outside the axon from A and C to B. These currents increase the membrane potential at C, making the membrane more permeable to sodium ions. Sodium ions flow in, bringing positive charges with them and so increasing the membrane potential further and making the membrane still more permeable to sodium ions. Thus the action potential reaches C. Meanwhile potassium ions are leaking out of the axon at B taking their positive charges with them and reducing the membrane potential.

There is a short refractory period after an action potential has passed, during which a second action potential cannot be elicited, or can only be elicited by an exceptionally strong stimulus. The refractory period lasts about as long as the increased potassium permeability and is partly caused by it. Hence A in Fig. 15.13 is just coming out of the refractory state.

EYES

Cephalopods have the largest eyes among the invertebrates. *Nautilus* has very simple eyes without lenses, which work on the principle of the pinhole camera. The other cephalopods have eyes remarkably like the eyes of vertebrates (Fig. 15.14). Light enters the eye through a transparent cornea and is focussed by the lens on the retina, which is a densely-packed layer of light-sensitive (retinula) cells interspersed by supporting cells. The axons attach to the

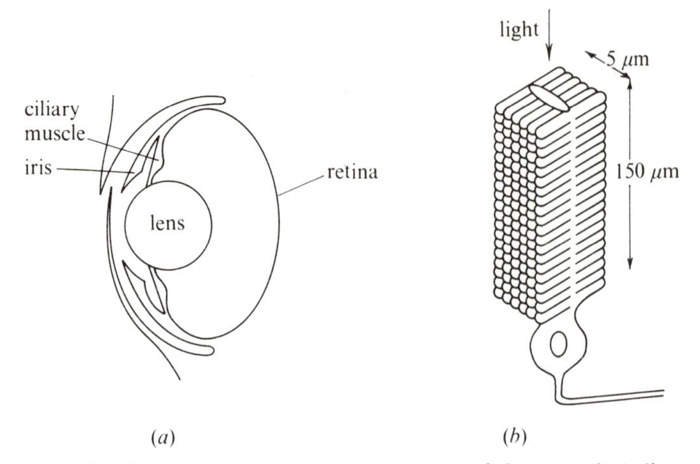

Fig. 15.14. (*a*) A diagrammatic section through the eye of *Octopus*. (*b*) A diagram based on electron micrographs of a retinula cell of *Octopus*.

retinula cells on the outer face of the retina, not on the inner face as in the inverted retina of vertebrates. The other end of each retinula cell bears a long process pointing towards the light, fringed on either side by microvilli (The diagram of a mollusc light receptor in Fig. 7.12, is based on *Helix*, not on cephalopods.) There is an iris which contracts in bright light, making the pupil small. There are ciliary muscles which may be used to move the lens so that near or distant objects can be brought into focus on the retina. There are also muscles outside the eye which turn it in its socket.

Refraction at the cornea is very important in human eyes. Air has a refractive index of 1. Water and the fluids which fill the eye have a refractive index of 1.33. Hence light entering a human eye is strongly converged by refraction at the convex surface of the cornea and quite a weak lens suffices to focus it on the retina. In water there is virtually no refraction at the corneal surface and a much stronger lens is needed. Cephalopods and teleost fish have approximately spherical lenses.

Samples from the cores of cuttlefish lenses (and of teleost ones) have a refractive index of about 1.53, about the same as crown glass. A sphere of this refractive index, immersed in eye fluids of refractive index 1.33, would have a focal length about four times its radius. Images of distant objects would be formed four radii from the centre of the lens and they would be poor, because of spherical aberration. If the lens had the right size and position for light coming through its centre to be focussed on the retina, light coming through the edges would be out of focus (Fig. 15.15 *a*). This fault could be reduced by using a small pupil (Fig. 15.15 *b*) which would reduce the amount of light entering the eye and make the eye less effective in dim light. A better arrangement has been evolved both by cephalopods and by teleosts. Their lenses are not homogeneous but have a refractive index of 1.53 at the centre, and lower

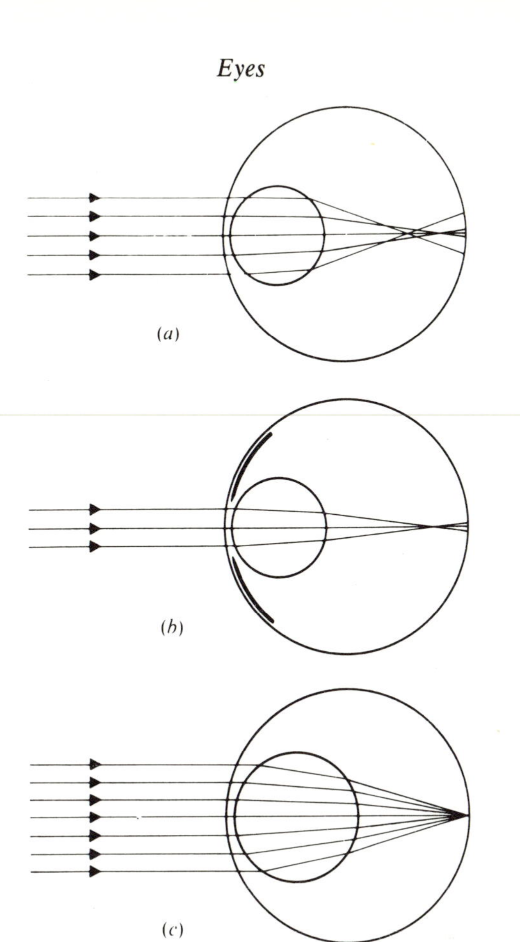

Fig. 15.15. Diagrams of cephalopod (or fish) eyes showing (*a*) the spherical aberration which would occur if the lens were homogeneous and had a wide aperture, (*b*) how this could be reduced by reducing the aperture, and (*c*) how it can be remedied by a non-homogeneous lens. From R. McN. Alexander (1975). *The chordates.* Cambridge University Press.

refractive indices in outer layers. They have focal lengths of only 2.5 radii so a remarkably large lens can be used in a small eye, and they give little or no spherical aberration.

Diffraction makes it impossible for light of wavelength λ passing through an aperture of radius r to form distinct images of objects separated by angles smaller than about $0.6\ \lambda/r$ radians. Consider light entering a cephalopod eye with the pupil wide open so that r is the radius of the lens. Since the focal length of the lens is $2.5r$ the images of objects $0.6\ \lambda/r$ radians apart will be $1.5\ \lambda$ apart on the retina. It would be useful to have retinula cells as close together as this in a perfect eye, so that they could take full advantage of the detail shown in the image.

Blue light has wavelengths around 0.4 μm in air, or 0.3 μm in water. Hence it might be useful to have retinula cells as little as 0.5 μm apart. This would require extraordinarily small cells: even in the fovea of the human eye, the cones are about 2 μm apart. The retinula cells of *Octopus* are about 5 μm apart, and the eye seems to be less acute even than this suggests. In a series of tests *Octopus* seemed unable to distinguish black and white striped boards from plain grey ones if the width of the strips subtended an angle less than 0.3° at the eye (This corresponds to a distance of 50 μm or more on the retina.) Man can distinguish stripes subtending only 0.01°.

BRAINS AND BEHAVIOUR

Cephalopods have the largest brains among the invertebrates, as well as the largest eyes. The brain of an *Octopus* is about equal in size to the brain of a teleost fish of the same body weight. It contains about 170 million cells. The brain of *Octopus* and the behaviour it makes possible have been studied intensively by Professor J. Z. Young and his associates. *Octopus* is more convenient than squid or cuttlefish for experiments on behaviour, because it is easier to keep in aquaria. The experiments have been carried out at the Naples Zoological Laboratory.

An octopus in an aquarium generally spends most of the day at one end of its tank. If a pile of bricks is put in the tank it will sit among them. If a crab is put at the other end of the tank the octopus will leave its end, and pounce on the crab by jet propulsion. It will also attack other moving objects such as a shape cut from sheet plastic, waved on the end of a wire.

Many experiments have been performed with pairs of shapes, for instance a cross and a square. From time to time the cross or the square is dipped into the tank and moved around. Whenever the octopus attacks the square, for instance, it is given food. Whenever it attacks the cross it is given a mild electric shock. If this is repeated often enough the octopus learns to discriminate between the shapes: it will usually attack the square and usually leave the cross alone. Fifty or so trials over a period of about 5 days is enough to train an octopus to make a simple discrimination reasonably reliably. The octopus may make the discrimination successfully several weeks later even if it has not been shown the shapes in the interval.

This method has been used to investigate the ability of octopus to distinguish one shape from another. It has been found that octopus have some unexpected limitations. They distinguished a narrow vertical rectangle from a narrow horizontal rectangle, making the correct choice in 81% of trials. They apparently could not distinguish a rectangle sloping at 45° to the left from one sloping to the right for they made the correct choice only 50% of the time (i.e. they apparently chose at random). They made frequent errors when required to distinguish a disc from a square.

Similar experiments have been carried out to investigate the ability of octopus to distinguish objects by touch. It was necessary to blind the octopus

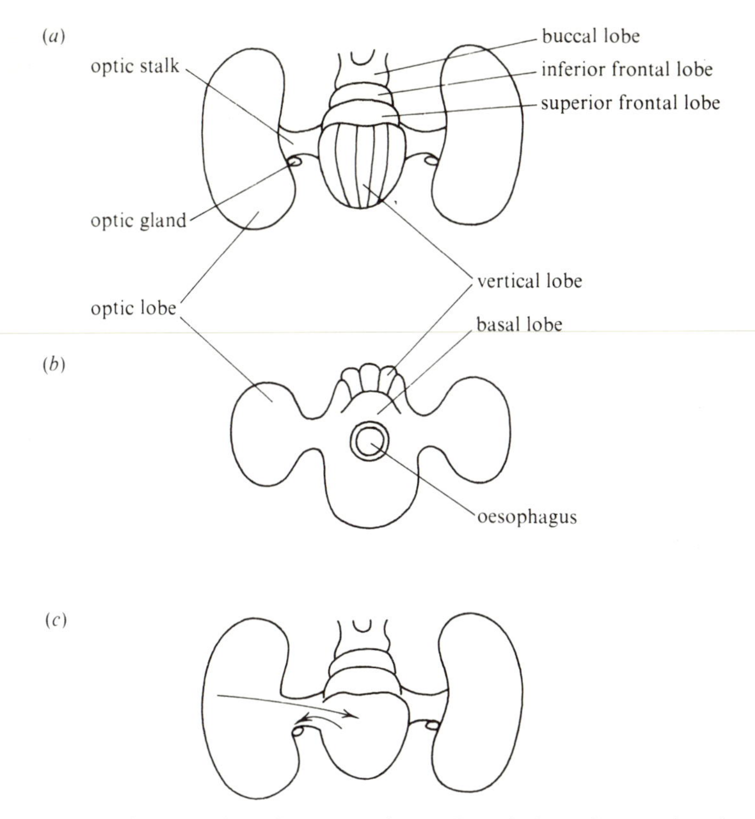

Fig. 15.16. Diagrams of an *Octopus* brain. (*a*) Dorsal view; (*b*) posterior view; (*c*) a dorsal view showing probable inhibitory pathways between the eye and the optic gland.

for these experiments by cutting their optic nerves. An untrained octopus will usually grasp an unfamiliar object and pull it under the body close to the mouth. If the experimental octopus did this with one object of a pair it was rewarded with food but if it did it with the other it was given a mild electric shock. It was found that octopus could discriminate between plastic cylinders with closely spaced grooves and similar cylinders with similar grooves more widely spaced. However, they failed to distinguish a cylinder with lengthwise grooves from one with similarly-spaced circumferential grooves.

These unexpected limitations to the ability of octopus to distinguish objects by sight and by touch are apparently due to deficiencies in the analytical powers of the nervous system, rather than in the sense organs. The octopus can see (or feel) the details but fails to perceive some differences of pattern.

Now look at the structure of the brain (Fig. 15.16). It consists of a ring of ganglia encircling the oesophagus, with very large optic lobes on either side. Attempts have been made to discover the functions of its various parts, by cutting them out. The octopus is anaesthetized, part of its brain is cut out and it is allowed to recover. Only an hour or so later it will be sufficiently recovered to attack prey. Operated animals are tested for deficiencies in behaviour and are then killed so that that their brains can be examined to check the extent of the damage.

Removing the optic lobes blinds the animal but has no other obvious effect on behaviour. (The optic lobes may be sites of visual memory, but visual memory cannot be demonstrated in a blind octopus.) If all the parts of the brain dorsal to the oesophagus are removed the animal can survive for several weeks but lies in a tangled heap and cannot feed. The buccal lobes are needed for feeding and the basal lobes for co-ordinated crawling and swimming. The more dorsal parts of the brain seem to be involved in learning. If the vertical lobe is removed from an octopus trained to distinguish shapes by sight, the effect of the previous training is lost and the animal is very slow to re-learn. This operation also makes animals trained to discriminate by touch less accurate, but it does not abolish touch memory. Complete removal of the subfrontal lobes (which are hidden below the inferior lobes in Fig. 15.16*a*) apparently destroys touch memory and leaves the animal with very little ability to discriminate by touch.

These experiments give some indication of where memory may be stored in the brain, but they tell us nothing about how it is stored. We do not yet know the physical basis of memory, in octopus or any other animal.

A HORMONE AND THE GONADS

Evolution of a blood circulation made possible the evolution of hormones carried by the blood. Hormones are known in gastropods but I have preferred to defer discussion of hormones to this chapter, so that a simple system in *Octopus* can be used to introduce them. The first evidence of its existence was obtained by accident in the experiments on learning which have just been described. It was noticed that young octopus which had been blinded or had had parts of their brains removed, tended to reach sexual maturity precociously. Once this had been noticed it was investigated systematically.

The optic lobes are connected to the main part of the brain by short stalks (optic stalks) which bear glands (optic glands). These glands are much larger in adult octopus than in immature ones. They were found to be enlarged in all experimental animals which had matured precociously. They are believed to secrete a hormone which is released into the blood and stimulates the gonad to develop.

Most of the experiments were performed on immature female *Octopus vulgaris* weighing 0.4 kg or less. At this size their ovaries normally make up less than 0.3% of the weight of the body. Females seldom mature naturally until they weigh at least 1 kg, and mature ovaries may be up to 20% of the body weight. Several operations cause great increases in the weights of the ovaries of immature females, sometimes making them grow within 2 months to 10%

of body weight. The most dramatic effects were obtained after removing the subpedunculate lobe (which lies ventral to the vertical lobe) or cutting the proximal end of the optic stalk. Presumably the optic gland normally receives messages from the subpedunculate lobe which inhibit it from enlarging until the appropriate time. Less rapid enlargement follows cutting of the distal end of the optic stalk, or the optic nerves. This suggests that inhibitory messages start from the eyes. Cutting the distal part of the optic stalk and removing the subpedunculate lobe simultaneously is no more effective than the latter operation alone, which suggests that the inhibitory messages from the eye may not go direct to the optic glands but may travel via the subpedunculate lobes (Fig. 15.16*c*).

Similar operations also make males mature precociously.

Octopus normally live about 2 years, dying after their first and only breeding season. Maturation of the gonads starts in the autumn preceding the spring in which they breed. It may occur in response to shortened day length, which would explain the role of the eyes. *Sepia* kept in aquaria with abnormally short daily periods of light matured precociously, but there seem to have been no similar experiments on *Octopus*.

So far it has been shown that the gonad and the optic glands enlarge in response to the same stimuli, but it has not been shown how the one effect is related to the other. In a series of experiments a blood space behind the eye was opened under anaesthetic and a piece of tissue from another octopus dropped in. Several weeks later the animal was killed and its gonad was examined. In some of the experiments pieces of testis or optic lobe were put into the blood space, and there was no discernible effect on the gonad. In others optic glands were inserted and in about a third of these experiments the gonad enlarged, becoming larger than in any unoperated animals of the same size. This suggests that the optic glands produce a hormone which stimulates enlargement of the gonad. Optic glands from male and female donors were both effective in bringing about maturation both in males and in females, so the hormone is probably the same in both sexes. It made no difference whether the implanted gland was enlarged or not at the time of the operation, presumably because it was free from inhibition in its new site.

It may seem unsatisfactory that only a third of the optic gland implants resulted in enlarged gonads. However, in many cases when no enlargement was observed the implant could not be found at post-mortem: it had presumably slipped out of the wound after the operation, or been destroyed by the host's amoebocytes. In most of the cases in which enlargement did occur the implanted gland was found at post mortem, attached to the wall of the blood space. In several cases Indian ink was injected into the blood system of the host and entered the gland, showing that it had become connected to the host's blood system.

Each of the developing eggs of an octopus is surrounded by several hundred follicle cells which secrete its yolk protein. Eggs with their follicle cells have been taken from maturing octopus and put into a nutritive solution containing a radioactive amino acid ([^{14}C]leucine). The rate at which they took up radioactivity and incorporated it into protein indicated the rate of protein synthesis. It was found to be faster if ground-up optic glands from precociously maturing animals were added to the solution.

FURTHER READING

GENERAL

Nixon, M. & Messenger, J. B. (eds.) (1977). *The biology of cephalopods* (*Symposium of the Zoological Society no. 38.*) Academic Press, New York & London.

Packard, A. (1972). Cephalopods and fish: the limits of convergence. *Biol. Rev.* **47**, 241–307.

Wells, M. J. (1962). *Brain and behaviour in cephalopods.* Heinemann, London.

See also the list for chapter 13

JET PROPULSION

Packard, A. & Trueman, E. R. (1974). Muscular activity of the mantle of *Sepia* and *Loligo* (Cephalopoda) during respiration and jetting and its physiological interpretation. *J. exp. Biol.* **61**, 411–19.

Trueman, E. R. & Packard, A. (1968). Motor performances of some cephalopods. *J. exp. Biol.* **49**, 495–507.

Ward, D. V. (1972). Locomotory function of squid mantle. *J. Zool., Lond.* **167**, 437–49.

BUOYANCY

Denton, E. J. (1974). On buoyancy and the lives of modern and fossil cephalopods. *Proc. R. Soc.* **185B**, 273–99.

Denton, E. J. & Gilpin-Brown, J. B. (1973). Flotation mechanisms in modern and fossil cephalopods. *Adv. mar. Biol.* **11**, 197–268.

GIANT AXONS

Aidley, D. J. (1971). *The physiology of excitable cells.* Cambridge University Press.

Hodgkin, A. L. (1958). Ionic movements and electrical activity in giant nerve fibres. *Proc. R. Soc.* **148B**, 1–37.

EYES

Pumphrey, R. J. (1961). Concerning vision. In *The cell and the organism*, ed. J. A. Ramsay & V. B. Wigglesworth, pp. 193–208. Cambridge University Press.

BRAINS AND BEHAVIOUR

Wells, M. J. (1974) A location for learning. In *Essays on the nervous system*, ed. R. Bellairs & E. G. Gray, pp. 407–70. Clarendon, Oxford.

Young, J. Z. (1964). *A model of the brain.* Clarendon, Oxford.

A HORMONE AND THE GONADS

Wells, M. J. (1976). Hormonal control of reproduction in cephalopods. In *Perspectives in experimental biology*, vol. 1, ed. P. S. Davies, pp. 157–66. Pergamon, Oxford.

16
Annelid worms

Phylum Annelida
  Class Polychaeta (ragworms, lugworms, etc.)
  Class Myzostomaria
  Class Oligochaeta (earthworms etc.)
  Class Hirudinea (leeches)

The bodies of annelids consist of a large number of more or less similar segments arranged in single file. The main body cavities are coeloms and typically there are partitions (septa) between the coeloms of successive segments.

The most familiar annelids are earthworms such as *Lumbricus* (Fig. 16.1). Their segmented structure is immediately apparent. Each of the furrows which encircle the body corresponds to a septum. The mouth is at the anterior end of the body and the anus at the posterior end. The segments of the posterior two-thirds of the body are more or less identical, and one of them is shown in Fig. 16.2. The body wall consists of a thin cuticle of collagen fibres, an epidermis and layers of circular and longitudinal muscle. It is pierced by eight chaetae (two pairs on each side) which are short bristles made partly of chitin. The intestine has an inner epithelium which includes cells that secrete digestive enzymes, and layers of circular and longitudinal muscle. The pleat in the roof of the intestine houses the chloragogen cells which contain oil and glycogen, presumably as a food store.

The space between the body wall and the gut is the coelom. In every segment of the body except the first three and the last the coelom contains a pair of nephridia. These are slender coiled tubes which run from an opening in the coelom of one segment to an external opening through the body wall of the next posterior segment. Their functions and their anatomical relationship to the coelom are discussed later in this chapter. A nerve cord runs along the body ventral to the gut giving off three pairs of nerves in each segment (Fig. 16.3*a*). These nerves send branches to the body wall and also by way of the septa to the gut. There are three main longitudinal blood vessels, one dorsal and two ventral, which send branches in each segment to the body wall and the gut. The chaetae and nephridia and the branches of the nerve cord and of the main blood vessels are repeated regularly, segment after segment, for almost the whole length of the body.

Fig. 16.1(*b*) shows a dissection of the anterior part of *Lumbricus*. Note that

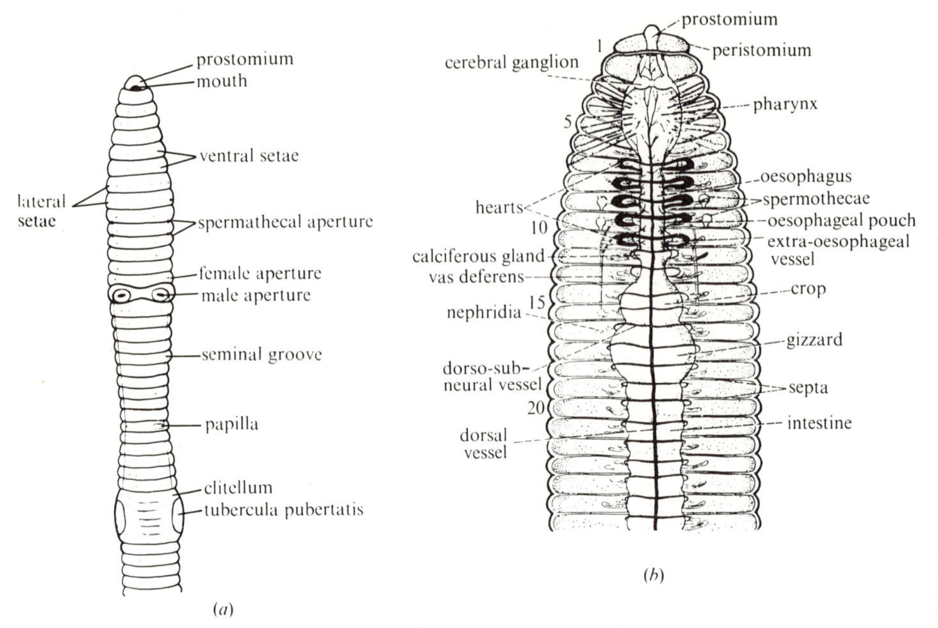

Fig. 16.1. (*a*) The ventral surface of the anterior third of an earthworm, *Lumbricus terrestris*. Members of this species grow to a maximum length of about 30 cm. From C. A. Edwards & J. R. Lofty (1972). *Biology of earthworms*. Chapman & Hall, London. (*b*) A dissection of *Lumbricus*. The body wall has been cut along the dorsal mid-line and pulled aside to show the internal organs. The testis sacs and seminal vesicles have been removed. From A. J. Grove & G. E. Newell (1969). *Animal biology*, 8th edn. University Tutorial Press, London.

anterior to the long intestine, the gut has several distinct parts. The pharynx is muscular and muscles radiating from it can enlarge it to draw food in. The gizzard also has thick muscular walls. Soil is taken into the gut with food and contractions of the gizzard muscles grind the food against the soil particles. Digestive enzymes are secreted by the walls of the pharynx, crop and intestine.

There is a pair of nerve ganglia dorsal to the anterior end of the pharynx (Fig. 16.3 *a*). Connections between them and the ventral nerve cord encircle the pharynx. Nerves from the ganglia and nerve ring run forwards to the snout and inwards to the gut. In the region of the oesophagus, five pairs of hearts connect the dorsal blood vessel to the ventral blood vessel (Fig. 16.3 *b*). The circulation of the blood and its function in respiration are discussed later in the chapter.

Earthworms are hermaphrodite. *Lumbricus* has a pair of ovaries in the coelom of the 13th segment. Ova are released into the coelom and travel along the oviducts which lead from it to the female apertures in the 14th segment. The 10th and 11th segments each have a pair of testes, enclosed in large testis sacs which have been removed from the dissection shown in Fig. 16.1 (*b*). Ducts from the testis sacs convey sperm to the male apertures on the 15th segment.

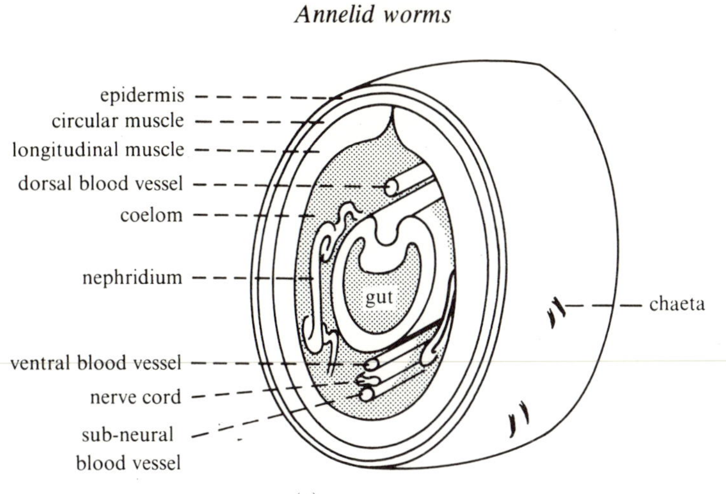

(*a*)

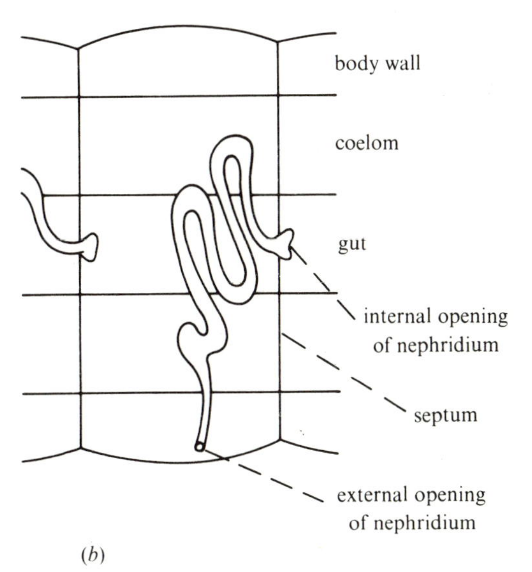

(*b*)

Fig. 16.2. Diagrams of a segment from the intestinal region of *Lumbricus*. (*a*) is an oblique view from behind and to the right and (*b*) is a view from the right side, drawn as if the body wall were transparent.

The 9th and 10th segments each have a pair of spermathecae (seminal receptacles).

When two earthworms copulate they bring the anterior ends of their bodies together, pointing in opposite directions. The spermathecal apertures of each worm are in contact with the clitellum of the other (the clitellum is a region

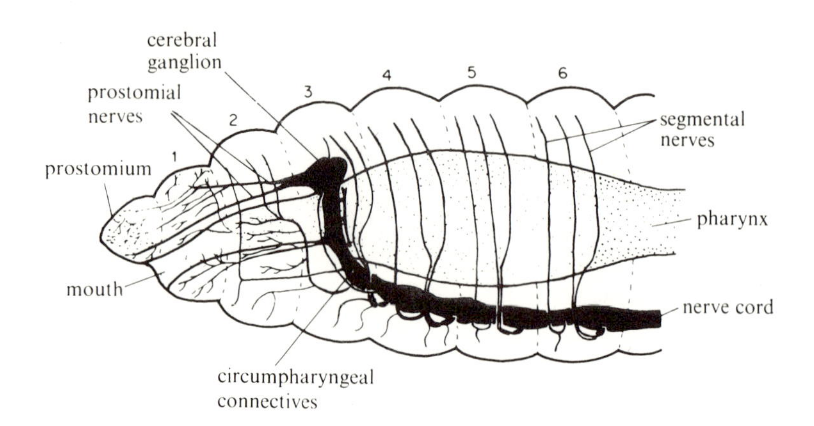

(*b*)

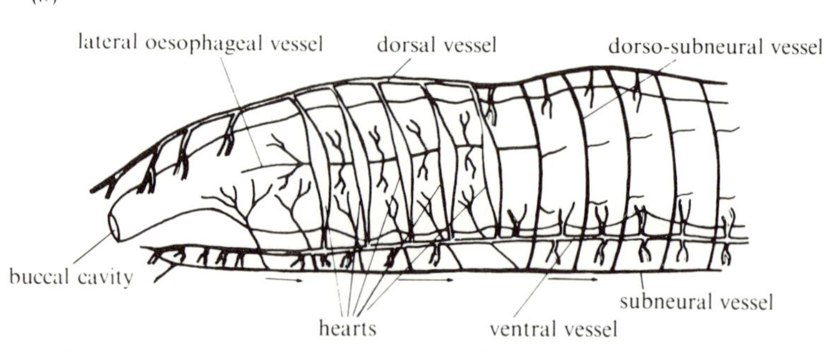

Fig. 16.3. (*a*) A lateral view of the nervous system of the first six segments of *Lumbricus*. (*b*) A lateral view of the anterior blood vessels of *Lumbricus*. From C. A. Edwards & J. R. Lofty (1972). *Biology of earthworms*. Chapman & Hall, London.

of thickened body wall, Fig. 16.1 *a*). The worms secrete mucus which binds them together. Sperm released from the male apertures of each worm moves along the seminal grooves (Fig. 16.1 *a*) to the clitellum, where it enters the spermathecae of the partner. There are muscles which deepen the grooves, forming depressions which travel posteriorly along them conveying the sperm. After copulation the worms separate. The clitellum of each secretes a short tube which is moved forward along the worm. Eggs and sperm are discharged into it as it passes the female aperture and spermathecal apertures. The ends of the tube close as it comes off the anterior end of the worm, forming a cocoon from which young worms eventually hatch. Several fertile cocoons can be produced after a single copulation.

Earthworms live in burrows in the soil. *Lumbricus terrestris* feeds on fallen leaves which it collects from the surface and pulls into its burrow. Other earthworms eat leaves or other plant materials, and dung. Particles of soil are often swallowed with food, and the gut may seem to be filled mainly with soil. The mechanism of burrowing and the effects of earthworms on the soil are discussed later in the chapter.

Lumbricus and the other earthworms are members of the class Oligochaeta. The enchytraeids are much smaller members of the class, which also live in soil. Adults of different species are 1–50 mm long. There are also aquatic oligochaetes such as *Tubifex* which lives in mud in the bottoms of rivers and streams and is sold as a food for aquarium fish. All these oligochaetes are fairly similar to each other, except in size.

The class Polychaeta is much more diverse. Its members are nearly all marine. Most of them have a pair of lateral projections called parapodia on every segment. The chaetae project from the parapodia and are much more numerous than in oligochaetes. Nearly all species have separate male and female individuals, and gonads develop in many segments. There is generally no copulation: gametes are shed into the water. The young do not hatch as miniature worms but as trochophore larvae very like the trochophores of gastropod and bivalve molluscs (Fig. 13.22).

The ragworm *Nereis diversicolor* (Fig. 16.4 *a*) is one of the species with large parapodia. There is an account later in the chapter of these parapodia and of the ways in which they are used in locomotion. *Nereis* is common on muddy shores where it occupies a semi-permanent burrow. Anglers dig it up for use as bait. It is often particularly common in estuaries where sea water and fresh water mix and it may have to withstand large fluctuations of salinity. Its ability to do so is discussed later in the chapter. It feeds on dead animals and pieces of algae. The pharynx can be turned inside out so that it projects as a proboscis in front of the mouth, as in Fig. 16.4(*a*). It bears a pair of sharp jaws which are hidden inside it when it is retracted but appear when it is protruded.

The lugworm *Arenicola* is another common polychaete which anglers dig up to use as bait (Fig. 16.4 *b*). Its parapodia are no more than low ridges so it looks much like an earthworm, but it has 13 pairs of gills on its dorsal surface (these are the tufts which are shown in the figure). It lives between the tide marks in muddy sand. It makes a J-shaped burrow (Fig. 16.5) and may occupy the same burrow for several months. It can be observed in a burrow if it is induced to burrow in a narrow layer of sand between two vertical glass plates. Its head faces the blind end of the burrow, where it swallows sand. This leaves a cavity in the sand which is filled by sand collapsing into it from above. Because of this there is a depression in the surface of the sand, over the blind end. The collapsing sand carries with it detritus from the surface, so the sand which the worm eats contains a fair proportion of organic matter. This is the worm's food. Every 45 min or so the worm crawls backwards to the open end of its burrow and defecates. The faeces are mainly sand and form a little mound around the

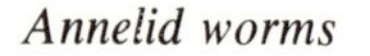

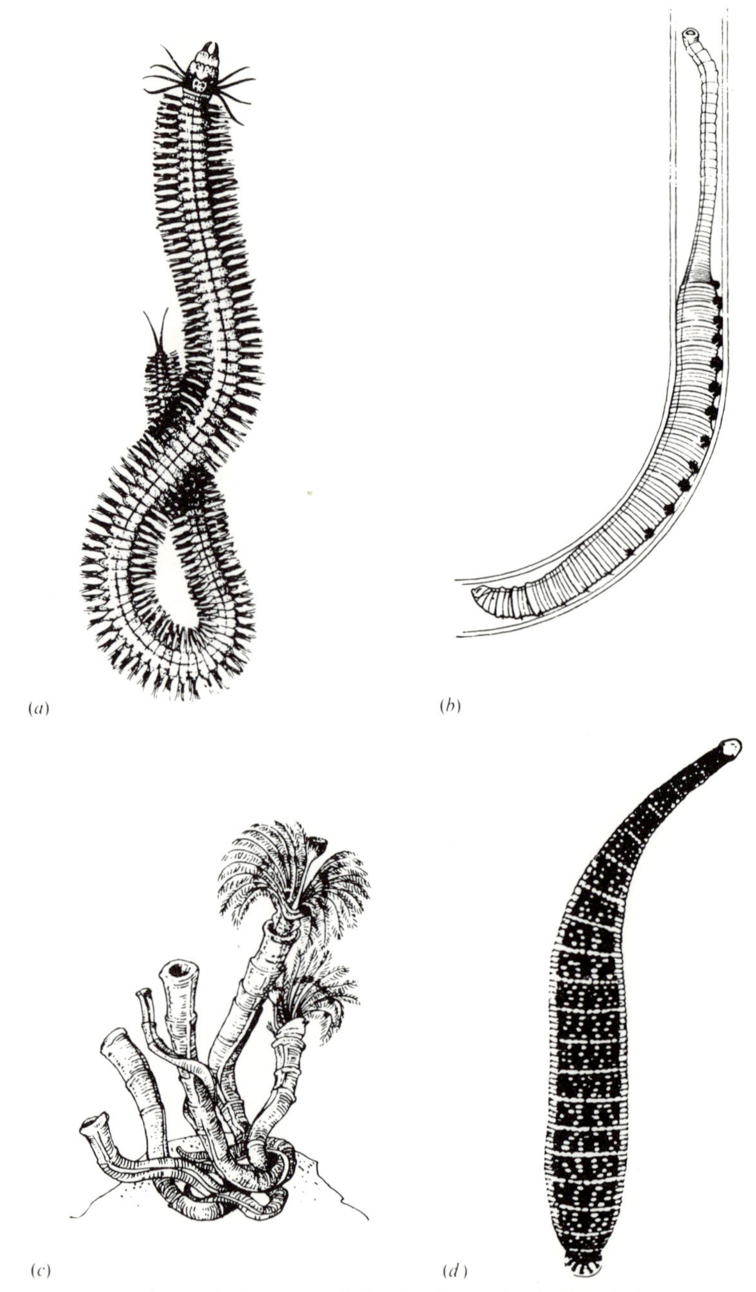

Fig. 16.4. (*a*) to (*c*) three polychaetes, and (*d*) a leech. (*a*) *Nereis diversicolor* (up to about 10 cm long); (*b*) *Arenicola marina* (25 cm) in its burrow, head downwards; (*c*) *Serpula vermicularis* (5 cm); and (*d*) *Erpobdella octoculata* (6 cm). (*a*) and (*c*) from W. de Haas & F. Knorr (1966). *The young specialist looks at marine life.* Burke, London. (*b*) from G. P. Wells (1950). *J. mar. biol. Assoc. UK* **29**, 1–44. (*d*) from W. Engelhardt (1964). *The young specialist looks at pond life.* Burke, London.

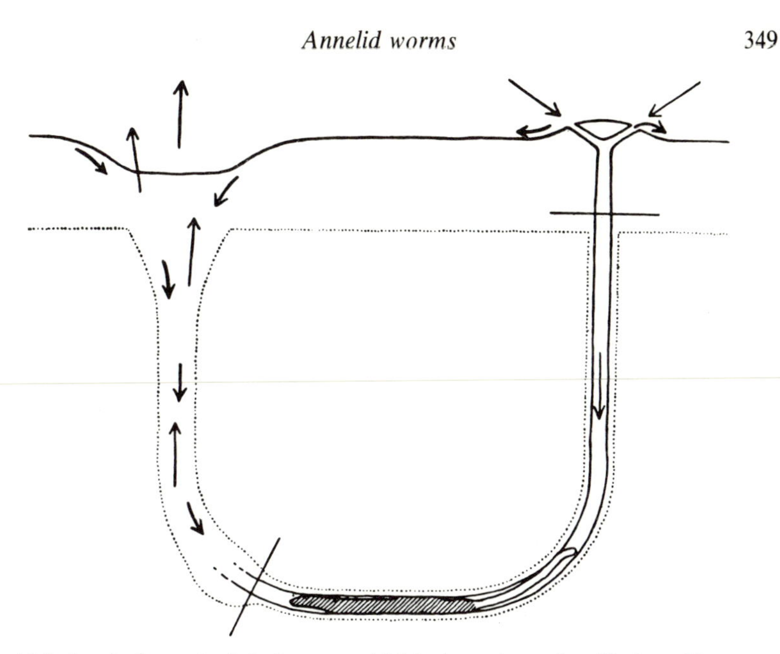

Fig. 16.5. *Arenicola marina* in its burrow, which is shown in section. The long thin arrows show movements of water driven by peristaltic movements of the worm. The short thick arrows show movements of sand. The dotted line shows the boundary between yellow sand (at the surface) and black sand (deeper). From G. P. Wells (1950). *Symp. Soc. exp. Biol.* **4**, 127–42.

open end of the burrow. In laboratory conditions lugworms irrigate their burrows for a few minutes after each defecation, by peristaltic movements. Water is drawn in at the open end of the burrow and escapes at the closed end by percolating through the sand. This brings a fresh supply of dissolved oxygen to the gills. In nature, water currents must tend to draw water through the burrow, just as they draw water through sponges (p. 109: note that the open end of the burrow is on a mound). This may provide adequate irrigation for respiration and peristaltic irrigation may have a different main function, for instance to loosen the sand at the blind end of the burrow.

Terebella (Fig. 16.6) is another polychaete with small parapodia. It often makes its burrow directly under a stone. It has branched gills at its anterior end and also long writhing tentacles which it uses to collect detritus from the surface of the sand. There are cilia on one side only of the tentacles, beating towards the mouth. Muscles inside the tentacles (Fig. 16.6*f*) can curl up their edges so that the cilia lie in a groove (Fig. 16.6*b*) or flatten them out (Fig. 16.6*c*). The tentacles flatten and extend over the sand, apparently crawling on their cilia just as flatworms do. When a particle of food is encountered the tentacle forms a groove for it (Fig. 16.6*b*). The particle is moved along the tentacle to the mouth by the squeezing action of the edges of the groove, or it is gripped in the groove and the tentacle retracts to pass it to the mouth.

Serpula (Fig. 16.4*c*) does not burrow. It secretes a protective tube of calcium

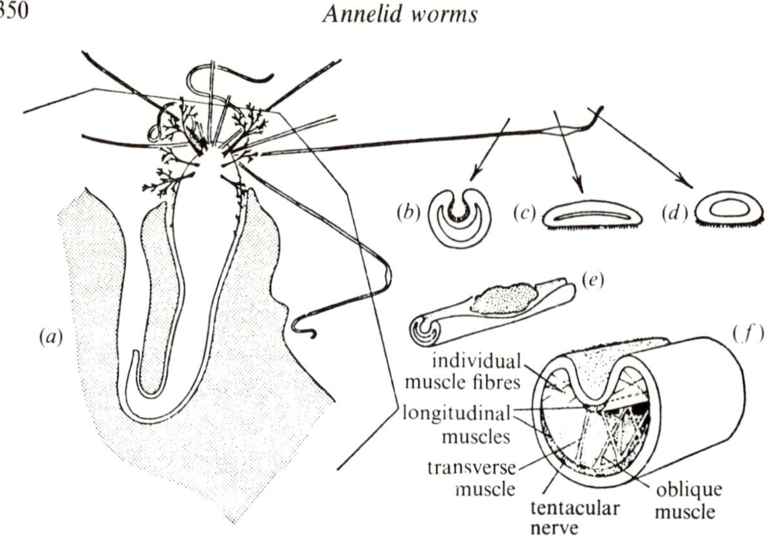

Fig. 16.6. (*a*) *Terebella lapidaria* in its burrow. The length of the worm, excluding tentacles, is about 5 cm. (*b*) to (*d*) Sections of a tentacle. (*e*) and (*f*) Diagrams of pieces of tentacle. There is a particle of food on the tentacle in (*e*). From R. P. Dales (1955). *J. mar. biol. Assoc. UK* **34**, 55–79.

carbonate (with a small proportion of organic matter) on stones, shells and seaweeds. It has stiff feathery tentacles with cilia which drive a current of water upwards between them. This serves for filter feeding. The tentacles can be withdrawn into the tube, and a modified tentacle serves as a stopper to close the mouth of the tube.

Erpobdella (Fig. 16.4*d*) is a leech, a member of the class Hirudinea. Most leeches are dorso-ventrally flattened. They have a sucker at the posterior end of the body and often another round the mouth. The coelom has largely disappeared: the space it occupies in other annelids has been invaded by botryoidal tissue. Leeches are hermaphrodite. Most of them live in fresh water. *Erpobdella* is common in ponds where it feeds on insect larvae and other invertebrates. Some other leeches attach themselves to fish and suck their blood.

The class Myzostomaria consists of a few parasites of echinoderms.

LOCOMOTION

The parapodia of polychaetes such as *Nereis* have a great many muscles arranged in a complicated way and can be moved rather like legs (Fig. 16.7). They are hollow, with the coelom extending into them. Muscles between their front and rear faces keep them flat, preventing them from being inflated like balloons when the muscles of the body wall increase the pressure in the coelom.

Fig. 16.7(*a*) and (*b*) shows parapodial oblique muscles (not labelled) running across the longitudinal muscles to the hinge lines at the base of the

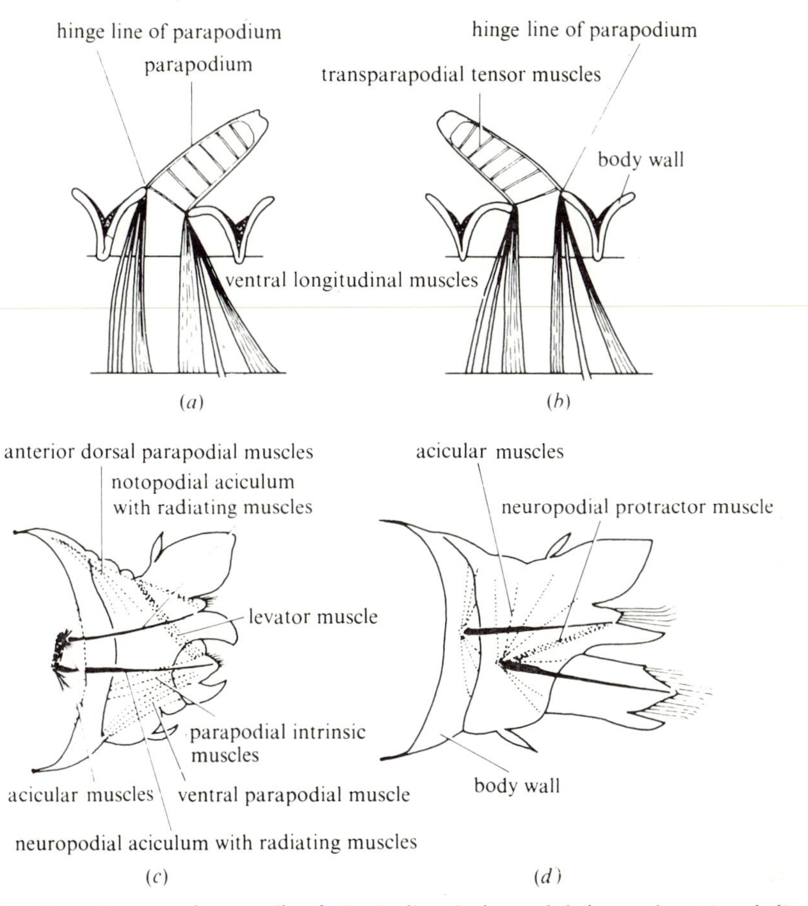

Fig. 16.7. Diagrams of parapodia of *Nereis diversicolor*, and their muscles. (*a*) and (*b*) are horizontal sections. (*c*) and (*d*) are face views of a contracted parapodium and an extended one, with the positions of contracting muscles indicated by dotted lines. Acicula are shown black. From C. Mettam (1974). *J. Zool., Lond.* **153**, 245–75.

parapodium. When the anterior oblique muscles contract the parapodium swings forward and when the posterior ones contract it swings back.

The parapodium can also be partly withdrawn into the body or extended from it. Fig. 16.7(*c*) shows a parapodium withdrawn, and the dorsal and ventral muscles which probably withdraw it. The levator muscle helps them and also lifts the parapodium off the ground. Fig. 16.7(*d*) shows a parapodium extended. It is probably extended largely by the pressure which acts in the coelom when the muscles of the body wall contract.

Each parapodium has two bundles of chaetae, each with a thick central chaeta known as the aciculum. Muscles run from the acicula to the body wall. They may help to move the parapodium as a whole. They are also used to protrude

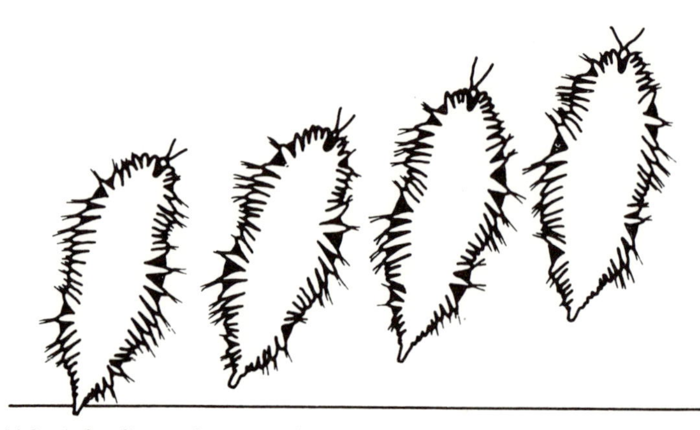

Fig. 16.8. *Aphrodite aculeata* crawling on glass, seen from below. The outlines were traced from photographs taken at intervals of 1 s. This species grows to a maximum length of about 18 cm. From C. Mettam (1971). *J. Zool., Lond.* **163**, 489–514.

the chaetae from the parapodium. Fig. 16.7(*c*) shows the chaetae retracted into the parapodium and Fig. 16.7(*d*) shows them protruded.

The sea mouse *Aphrodite* is an exceptionally stout polychaete which crawls by movements of its parapodia alone (Fig. 16.8). Each parapodium swings forward and back, like a leg. In the back stroke it is extended with its chaetae protruded, and rests on the ground. In the forward stroke it is contracted with the chetae retracted, and lifted off the ground. Thus the parapodia push the worm along. They do not lift the ventral surface of the worm off the ground, but simply slide it along. They move in strict order. The left and right parapodia of a segment move alternately. Each parapodium moves slightly after the parapodium on the same side of the next segment behind. Thus waves of parapodial movement travel forward along the worm, alternating on the left and right sides.

Long thin polychaetes such as *Nereis* crawl slowly like this, but when they crawl fast they undulate the body as well as moving the parapodia (Fig. 16.9). The waves of bending travel forward along the body at the same rate as the waves of parapodial movement. The same two parapodia are marked by arrows on each outline in Fig. 16.9. Consider the one nearer the left of the page. In (*a*) it has just been set down. It is reaching well forward, partly because it has been swung forward relative to the trunk by its oblique muscles, and partly because of the way the trunk is bent. In (*b*) and (*c*) the parapodium is being swung backwards, by its oblique muscles and by the changing bend of the trunk. Its tip is stationary on the ground so the worm is moved forwards. The distance moved forward betwen (*a*) and (*c*) is longer than it would be if the worm did not undulate. In (*d*) the parapodium is off the ground, being brought forward for the next step.

Earthworms have no parapodia and crawl quite differently, in the same way

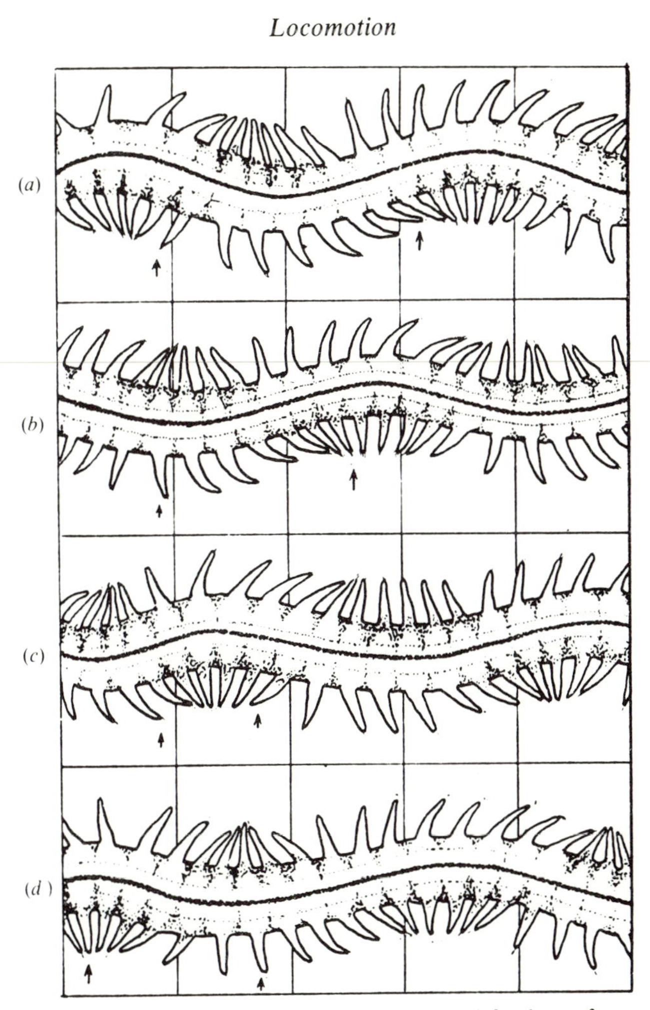

Fig. 16.9. *Nereis diversicolor* crawling fast towards the left, drawn from a cine film. Arrows show successive positions of two parapodia. The lines on the floor of the aquarium are 1 cm apart. From R. B. Clark (1964). *Dynamics in metazoan evolution.* Clarendon, Oxford (after a film by Sir James Gray).

as nemerteans such as *Lineus* (Figs. 10.3, 10.5). Each segment is made alternately short and fat (by contraction of its longitudinal muscles) and long and slender (by contraction of its circular muscles). Peristaltic waves travel backwards along the body. Fat segments take most of the weight of the worm and tend to remain stationary, so the worm moves forwards. The chaetae also help to anchor the fat segments. They are not at right angles to the body but point somewhat posteriorly, as can be felt by running a finger forwards and back along a worm. Hence they allow their segments to slide forward freely but tend to prevent them from sliding back. Also, the chaetae are protruded when the segment is fat and retracted when it is thin. The crawling action also serves

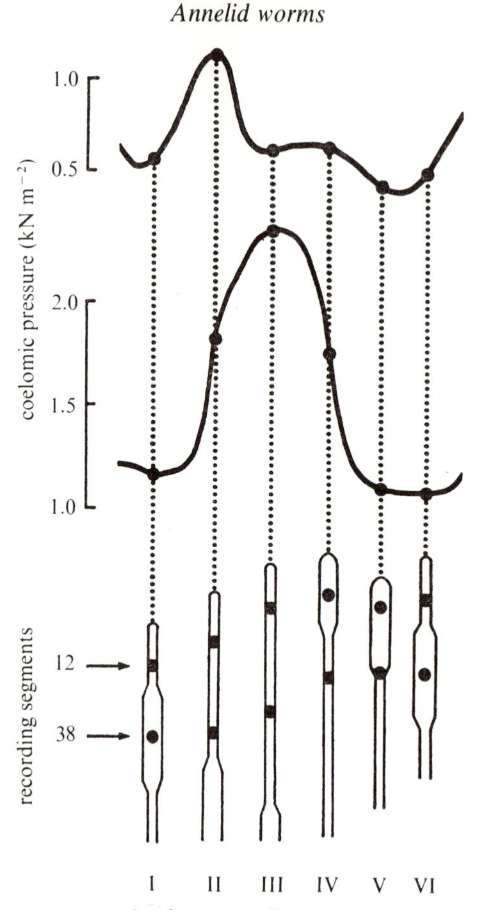

Fig. 16.10. Simultaneous records of pressure from segments 12 (upper record) and 38 (lower record) of *Lumbricus* crawling over damp earth. Positions of the worm are shown below. The worm advanced 2–3 cm between positions I and VI. This movement probably occupied about 5 s. From M. K. Seymour (1969). *J. exp. Biol.* **51**, 47–58.

for burrowing, as can be seen when an earthworm burrows alongside the wall of a glass tank filled with soil. The fat segments are anchored by being jammed against the wall of the burrow.

Pressures have been recorded from the coelom of *Lumbricus*, through fine steel tubes slipped into the coelom and connected by flexible tubing to pressure transducers. Fig. 16.10 shows records made simultaneously from two segments of a crawling worm. The pressures are quite small, and are greatest in each segment when it is slender. The pressure is not the same all along the worm. At position III for instance the pressure is only 0.5 kN m^{-2} in segment 12 but 2.5 kN m^{-2} in segment 38. Since the septa are thin they cannot maintain much pressure difference between successive segments, but quite substantial pressure differences can build up along a series of segments.

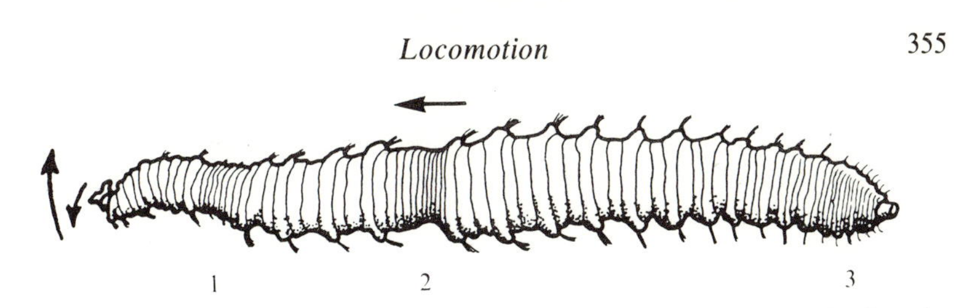

Fig. 16.11. A drawing made from photographs of *Polyphysia crassa* burrowing into the surface of mud. Worms of this species are typically about 8 cm long. From H. Y. Elder (1973). *J. exp. Biol.* **58**, 637–55.

Records have also been obtained of pressures in the coeloms of worms burrowing in loose earth. They show that in burrowing, in contrast to crawling, the pressure is greatest when the segment is fat. While the anterior segments are slender and moving forward, they probably slip into existing crevices in the soil. This requires little force so the pressure in the coelom can be quite low. When the longitudinal muscles contract the segments get fatter, enlarging the crevice, and it is for this that high pressures are required. It is difficult to get representative records of pressures during burrowing because the ground gets in the way of the tube connecting the worm to the transducer. Only records of burrowing in loose earth have been obtained and the pressures they show are no more than about 2.5 kN m^{-2}. Pressures up to 7.5 kN m^{-2} have been recorded from *Lumbricus* squirming in air and it is believed that much higher pressures may act in burrowing.

Let the circular muscles occupy a fraction C of the cross-sectional area of an earthworm and let the longitudinal muscles occupy a fraction L. Let these muscles contract exerting stresses σ_C and σ_L, respectively, and setting up an excess pressure P in the coelom. Plainly

$$P = L\sigma_L \tag{16.1}$$

Let the radius of the worm be r and let the thickness of the layer of circular muscle be x. The stress σ_C will set up a circumferential tension $x\sigma_C$, which is enough to balance an excess pressure $x\sigma_C/r$ (see p. 140).

$$P = x\sigma_C/r$$

However, the area occupied by the circular muscle in a transverse section of the worm is about $2\pi rx$ so

$$C \simeq 2\pi rx/\pi r^2 = 2x/r$$

and

$$P \simeq \tfrac{1}{2}C\sigma_C \tag{16.2}$$

In typical *Lumbricus* $C = 0.11$ and $L = 0.28$. Hence five times as much stress is needed in the circular muscles as in the longitudinal ones, to set up a given excess pressure. When the excess pressure was 7.5 kN m^{-2} in a squirming worm, the stresses in the circular and longitudinal muscles must have been 140 kN m^{-2} and 27 kN m^{-2}, respectively.

When a worm is in a burrow contracting its longitudinal muscles there is no need for tension in the circular muscles: indeed, it is better for them to be slack so that the force on the wall of the burrow is as large as possible. We know that the circular muscles can exert stresses up to 140 kN m^{-2}. If the longitudinal muscles can do the same they can set up pressure up to 39 kN m^{-2} (from equation 16.1) when the body wall is pressed against the wall of a burrow. Such a pressure may be developed in burrowing in hard earth. If it acted all along the worm, all the segments would have to be pressed against the wall of the burrow because the circular muscles could not withstand it. However, the septa make it possible for different segments to have different pressures in them. Part of a worm may develop a very high pressure by contracting its longitudinal muscles while another part may be at much lower pressure, so that the circular muscles can make its segments long and thin.

Polyphysia (Fig. 16.11) is a polychaete which has no septa, except between a few segments at the ends of its body. This enables it to crawl and burrow by a technique which is impossible for earthworms. An earthworm segment must get thicker when it shortens and thinner when it elongates. *Polyphysia* segments can get shorter and thinner simultaneously, displacing coelomic fluid into other segments (region 2 in Fig. 16.11). *Polyphysia* crawls by sending waves of simultaneous shortening and constriction forward along its body. This moves it forward in just the same way as the forward-moving waves in the foot of *Haliotis* move it forward (Figs. 13.12, 13.13*a*). *Polyphysia* lives in soft flocculent mud below low tide level. It burrows in the mud moving forward in this way while excavating the burrow by side-to-side movements of its head. *Arenicola* and many other polychaetes are also without septa for much of the length of their bodies.

Some annelids can swim. Leeches swim like nemerteans and *Turbatrix* (Fig. 12.8*a*), by sending waves of dorso-ventral bending posteriorly along the body. *Nereis* and other polychaetes with large parapodia swim by side-to-side undulation and the waves travel forwards: their swimming movements are very like fast crawling (Fig. 16.9) but the waves have greater amplitude and wavelength. These worms are propelled in the same direction as the waves partly because the parapodia have an effect like the flimmer filaments of some flagella (p. 42) and partly because of the movements of the parapodia relative to the trunk. The parapodia are more expanded for the backstroke than the forward stroke so their movements tend to drive the worm forwards. Also, they are clumped together (on the inside of a bend) in the forward stroke, which reduces the backward force they exert then on the worm.

Nereis diversicolor swims only occasionally, and not very well. Some other species of *Nereis* metamorphose when sexually mature, developing much larger parapodia along much of the length of the body. This enables them to swim faster, and they swim to the surface to spawn.

Many annelids including earthworms respond to strong stimuli by shortening very rapidly indeed. The longitudinal muscles all along the body shorten more

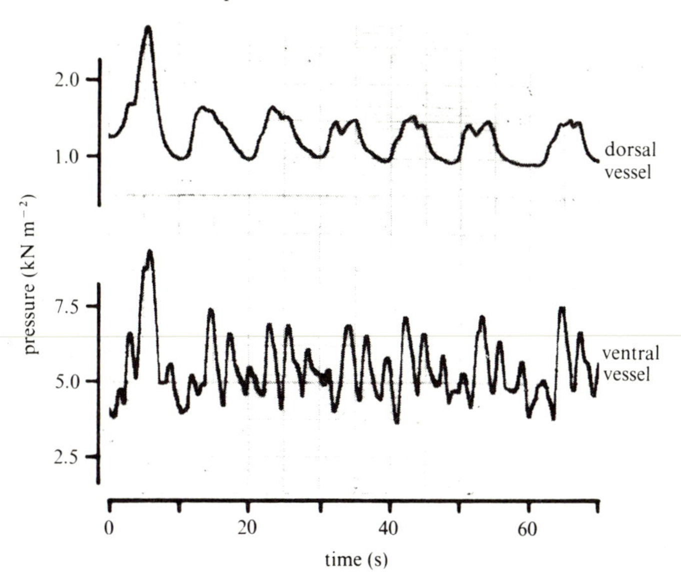

Fig. 16.12. Simultaneous records of pressure in the dorsal and ventral blood vessels of *Glossoscolex*. From K. Johansen & A. W. Martin (1965). *J. exp. Biol.* **43**, 333–47.

or less simultaneously. This is made possible by giant fibres in the nerve cord which transmit action potentials very fast. The largest of the giant fibres in *Lumbricus* is only 90–160 μm in diameter but it has a thin myelin sheath and conducts at 20–45 m s^{-1}, much faster than squid fibres of similar diameter. The giant fibre response enables worms to disappear very rapidly into their burrows, when they are attacked.

RESPIRATION AND BLOOD CIRCULATION

Earthworms and many other annelids have haemoglobin in their blood. Most have it as large polymeric molecules in solution but some polychaetes have it in corpuscles. Some polychaetes have a different iron compound (chlorocruorin) instead of haemoglobin, and some have no respiratory pigment at all. The blood is circulated by peristalsis of blood vessels supplemented in some cases by pulsation of hearts. Valves allow it to flow round the body in only one direction, anteriorly in the dorsal vessel and posteriorly in the ventral one.

The circulation of the blood has been studied must thoroughly in the giant Brazilian earthworm *Glossoscolex*. This worm grows to lengths over 1 m and masses over 0.5 kg. In some experiments a suspension of material opaque to X-rays was injected into the blood system. X-radiographs taken subsequently showed the main blood vessels clearly. Peristaltic waves could be seen moving forward along the dorsal vessel at rates around 12 segments per second.

In other experiments thin plastic tubes connected to pressure transducers were inserted into the dorsal and ventral blood vessels, so that blood pressure could be recorded. Fig. 16.12 is a typical record. It shows much higher pressures in the ventral vessel than in the dorsal vessel, due to the hearts. It also shows that the peaks of pressure in the ventral vessel (due to the heart beat) were much more frequent than the peaks in the dorsal vessel (due to peristalsis). The hearts usually beat in synchrony, and seem to have been doing so when this record was made.

Earthworms have no gills but rely for their supply of oxygen on diffusion through the whole outer surface of the body. If there were no circulatory system to distribute oxygen round the body they could hardly grow larger than about 1.5 mm in diameter (p. 185). Even with a circulatory system there must be a maximum feasible diameter for worms without gills. A short fat worm (shaped like a sausage, for instance) would have a smaller surface area than a long slender worm of the same mass and might not be able to get enough oxygen.

Consider a cylindrical worm of length l and diameter d which uses oxygen at a rate m per unit volume of tissue. Its volume is $\pi ld^2/4$ so it uses oxygen at a rate $\pi ld^2m/4$. Its surface area is about πld. Blood is brought by capillaries to a distance s below the surface, all over the body, so diffusion must occur through a layer of tissue of thickness s. The partial pressure of oxygen is P_o outside the worm and P_b in the blood. By equation (8.1):

$$\pi ld^2m/4 = -\pi ldD(P_b - P_o)/s$$

$$d = (4D/sm)(P_o - P_b) \qquad (16.3)$$

D is the diffusion constant for oxygen diffusing through tissue and is probably about 2×10^{-5} mm^2 atm^{-1} s^{-1}.

The metabolic rates of various (tropical) earthworms are shown in Fig. 8.6. The largest had a mass of about 18 g and used about 0.06 cm^3 oxygen g^{-1} h^{-1}. Since their densities must have been about 1 g cm^{-3} this is about 0.06 cm^3 oxygen cm^{-3} h^{-1}, or 1.7×10^{-5} s^{-1}. This value of m will be used. The cuticle and epidermis of a typical earthworm are together about 50 μm thick but the most superficial blood vessels from loops within the epidermis and 30 μm is probably a realistic estimate of s. This is much more than the distance between air and blood in a snail's lung, but the lung is a delicate structure in a protected position while the worm's epidermis must be able to withstand abrasion. If the worm is in air $P_o = 0.21$ atm. P_b must be appreciably greater than zero so it seems clear that $(P_o - P_b)$ must be less than 0.2 atm. Putting these values in equation (16.3) we find that d cannot be more than about 30 mm. An earthworm more than about 30 mm in diameter would not be feasible unless it had a lower metabolic rate than has been assumed or its blood came nearer the surface of the body. The thickest earthworms such as *Rhinodrilus fafner* of S. America have diameters around 25 mm.

Though earthworms have no gills, many polychaetes do. They include *Arenicola* (Fig. 16.4*b*) and *Terebella* (Fig. 16.6). Earthworms generally live in soil with plenty of air spaces thrcugh which oxygen can diffuse rapidly. Oxygen diffuses much more slowly through water, so the burrows of polychaetes must be irrigated if the partial pressure of oxygen in them is to be kept reasonably high. Many polychaetes including *Arenicola* irrigate their burrows by peristalis. However, irrigation is impossible for intertidal species at low tide. Samples of water taken from *Arenicola* burrows at low tide contained oxygen at partial pressures less than 0.02 atm. This may be why *Arenicola* needs gills.

NEPHRIDIA AND COELOMS

Earthworms in damp soil face the same osmotic problems as animals living in fresh water. Professor Arthur Ramsay took tiny samples of fluid from nephridia of *Lumbricus* and determined their freezing points. He showed that worms with coelomic fluid of concentration 170 mOsmol l^{-1} produced urine of only about 30 mOsmol l^{-1}. Fluid from near the coelomic opening had the same osmotic concentration as the coelomic fluid, but the concentration fell as the fluid travelled along the nephridium. Thus the nephridia seem to work in the same way as the kidney of the gastropod *Viviparus* (p. 283). High pressures developed in the coelom in burrowing would squeeze coelomic fluid out through the nephridia, were there not sphincter muscles around their external openings. Excess pressure in the coelom must help to drive fluid through the nephridia when the sphincters are open, but the pressure in resting worms is low and there are cilia in the nephridia which can drive fluid along them.

Professor Edwin Goodrich studied the nephridia and coeloms of annelids, looking for evidence of how they evolved. He made a telling comparison between polychaetes and nemertean worms. Nemerteans (Fig. 16.13 *a*) have a row of gonads down each side of the body. Each gonad is a clump of germ cells enclosed in epithelium, and when it is ripe it develops a genital pore which allows its products to escape into the surrounding water. Nemerteans have protonephridia. In some, all the flame cells of each side of the body are connected together by their ducts, as in turbellarians (Fig. 8.4 *c*). In others there are clusters of flame cells at intervals along each side of the body, each cluster conected to a separate external aperture.

Polychaetes such as *Nereis* have gonads in most of their segments but others such as *Arenicola* have gonads in only a few. The gonads are clusters of germ cells in the coelom, with no enclosing epithelium. Their products detach as they develop and float free in the coelom. In some polychaetes, such as *Nereis*, the body wall ruptures to release them. In others a coelomoduct develops as they ripen, connecting the coelom to the exterior and allowing the gametes to escape. Goodrich and others argued that coeloms must have evolved from gonads like those of nemerteans, by enlargement of the cavity without corresponding enlargement of the cluster of germ cells. The epithelium which lines the coelom (the peritoneum) is homologous to the epithelium which encloses the nemertean gonad. Notice how the peritonea of the left and right

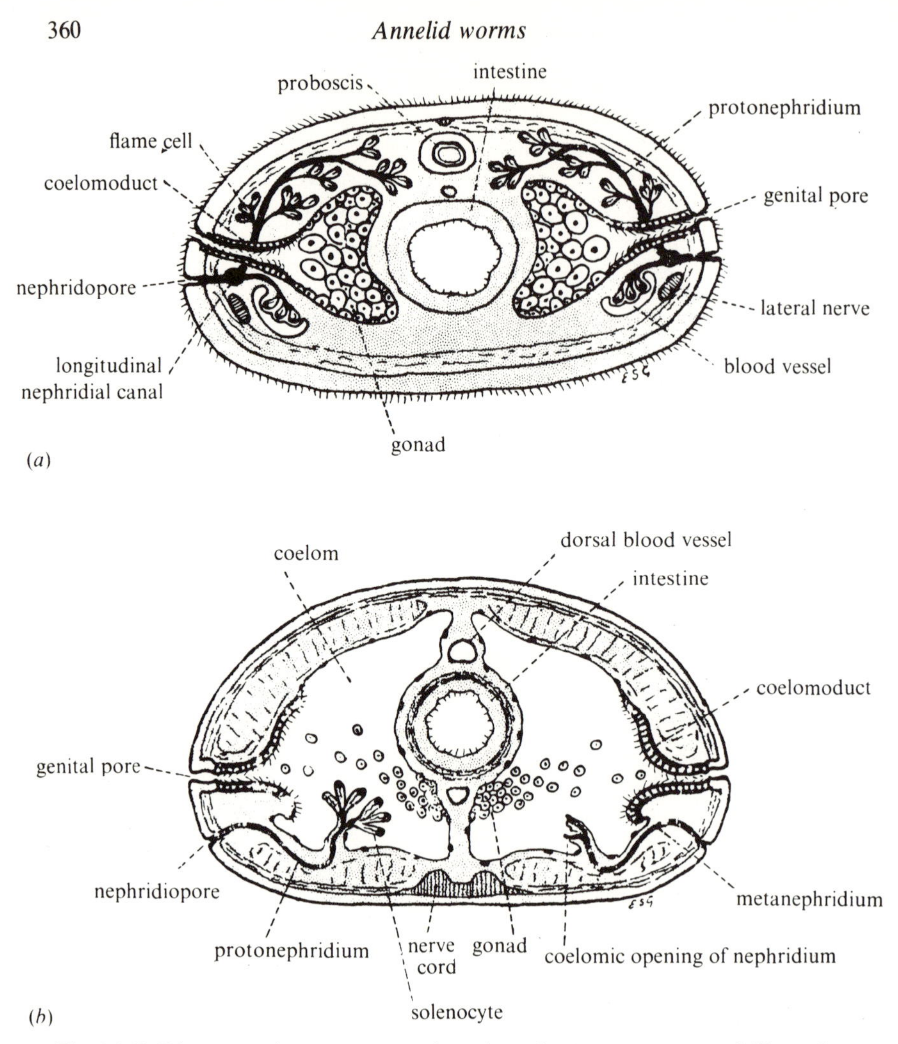

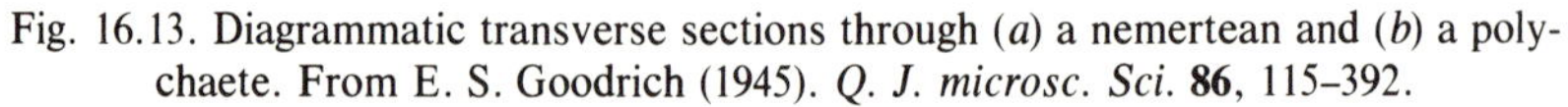
Fig. 16.13. Diagrammatic transverse sections through (*a*) a nemertean and (*b*) a polychaete. From E. S. Goodrich (1945). *Q. J. microsc. Sci.* **86**, 115–392.

coeloms lie back-to-back above and below the gut, forming mesenteries (Fig. 16.13*b*). The septa between segments are also formed by two layers of peritoneum, lying back-to-back (Fig. 16.14).

Trochophore larvae have protonephridia, and so do some adult polychaetes. These resemble the protonephridia of flatworms and nemerteans but have solenocytes (with a single flagellum) instead of flame cells (with a bunch of flagella). Other polychaetes have metanephridia, tubes open at both ends like the nephridia of earthworms. Some polychaetes have entirely separate coelomoducts and metanephridia (Figs. 16.13*b* and 16.14, right-hand side). The

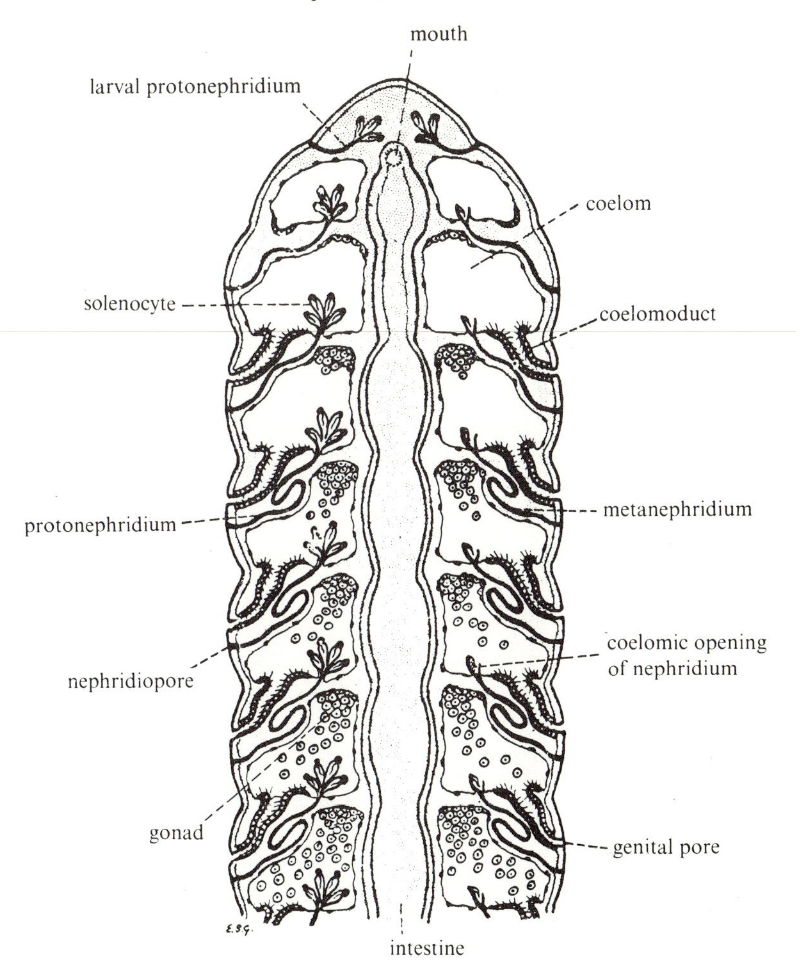

Fig. 16.14. Diagrammatic horizontal section of the anterior end of a polychaete. From E. S. Goodrich (1946). *Q. J. microsc. Sci.* **86**, 115–392.

coelomoduct develops from the peritoneum but the nephridium develops quite separately, from a cell lying between two segments. No known polychaete has entirely separate coelomoducts and protonephridia, as shown on the left sides of the diagrams, though this is believed to have been the primitive condition. Many polychaetes have coelomoducts and nephridia combined, either as a forked structure in which both lead to the same external aperture or a single tube which serves both functions. In such cases the embryology of the combined structure provides evidence of how it probably evolved. The kidneys of molluscs function like nephridia but probably evolved from coelomoducts.

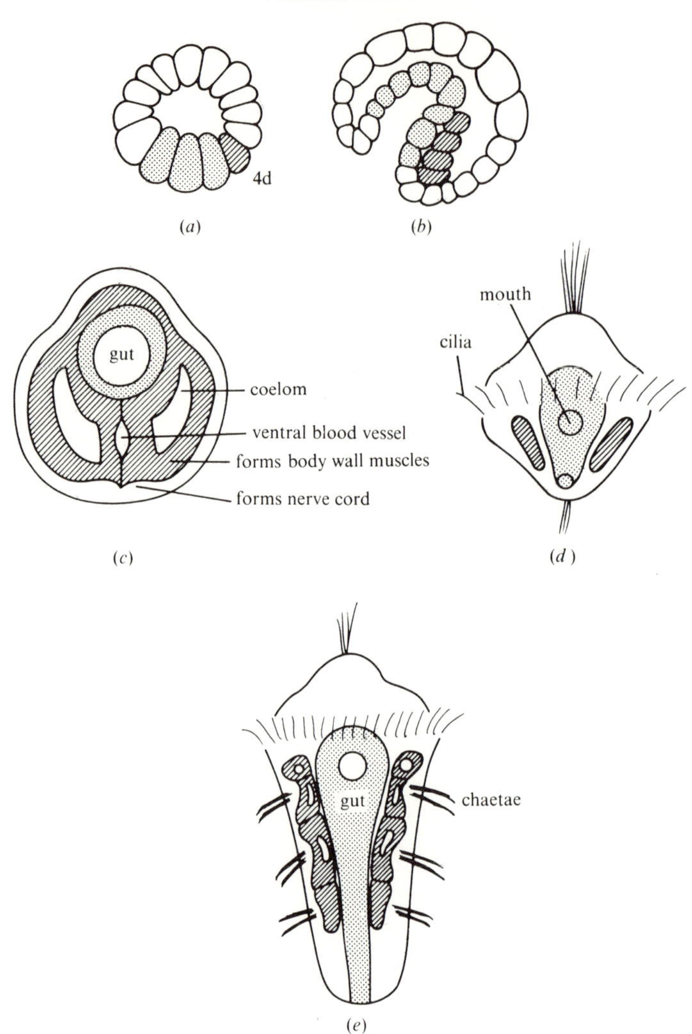

Fig. 16.15. Diagrams illustrating the development of typical polychaetes. (*a*), (*b*) and (*c*) Sections of a blastula, a gastrula and a later larva. (*d*) and (*e*) A trochophore and a later larva. Mesoderm (derived from cell 4d) is hatched; endoderm (derived from cells 4a–c and 4A–D) is stippled.

DEVELOPMENT

Annelids and molluscs (excluding cephalopods) develop from the zygote by spiral cleavage, like polyclad flatworms (Fig. 8.14). However, cells 4a–c and 4A–D survive and form the lining of the gut. Fig. 16.15 shows the course of development of typical polychaetes. Division initially produces a hollow ball of cells, the blastula (Fig. 16.15*a*). The cells at one end of the ball are the products of division of cells 1a–d. Those at the other end come from 4A–D, and have the products of 4a–d next to them. This end invaginates (folds in) to form the next embryonic stage, the gastrula (Fig. 16.15 *b*). The products of 4a–c and 4A–D are now in the appropriate position to form the lining of the gut. The products of 4d, however, have slipped into the cavity of the blastula and formed two bands of cells, one on each side of the gut. The gastrula develops to a trochophore, very like the trochophores of molluscs (Fig. 16.15 *d*; compare Fig. 13.22). The single opening of the gut of the gastrula becomes the mouth, and an anus forms at this stage.

It is convenient to call the cells derived from 4a–c and 4A–D the endoderm, those derived from 4d the mesoderm, and the rest the ectoderm. In general, the ectoderm covers the external surface of the body, the endoderm lines the gut and the mesoderm lies between. However, there are exceptions. For instance, the anterior part of the gut is formed separately from the rest, by invagination of ectoderm, at the time the mouth forms. It is therefore lined by ectoderm.

The mesoderm bands grow and form cavities which become the coeloms of successive segments (Fig. 16.15 *e*). The most anterior segments are formed first and more segments are added at the posterior end so that the most posterior segment is always the newest one. This is quite different from the situation in tapeworms where the newest proglottid is always the one next to the scolex. The larva sinks and settles on the bottom as it becomes a young worm.

The mesoderm forms the peritoneum and gonads. It also forms the muscles of the body wall and the walls of the principal blood vessels, as indicated in Fig. 16.15(*c*). However, the nerve cord is formed from ectoderm which sinks into the body so as to be enclosed by the body wall. The nephridia also develop from ectoderm.

LIFE IN ESTUARIES

For animals living in estuaries, the salt concentration of the water around them may fluctuate widely. It will fall as the tide falls and rise again as the tide rises. Its value at a given state of the tide may fluctuate seasonally. *Nereis diversicolor* is common in estuaries and is sometimes most abundant where the salt concentration is most variable. It can survive in fresh water, provided the calcium concentration is not too low. *N. limnicola* is very similar worm which is found in estuaries and even in fresh water on the Pacific coast of N. America.

Many experiments have been carried out on the ability of *Nereis* to survive changes of salt concentration. In one, *N. diversicolor* were kept in mixtures of different proportions of sea water and distilled water. After 6 or more days samples of their coelomic fluid were taken, for measurements of osmotic concentration. Fig. 16.16(*a*) shows that worms kept in undiluted sea water (1 Osmol l^{-1}) or 50% sea water (0.5 Osmol l^{-1}) had coelomic fluid only slightly more concentrated than the water. Those kept in more dilute mixtures (5–300 mOsmol l^{-1}) had coelomic fluid much more concentrated than the water.

If *Nereis* are wiped dry, droplets of urine can be watched forming at the

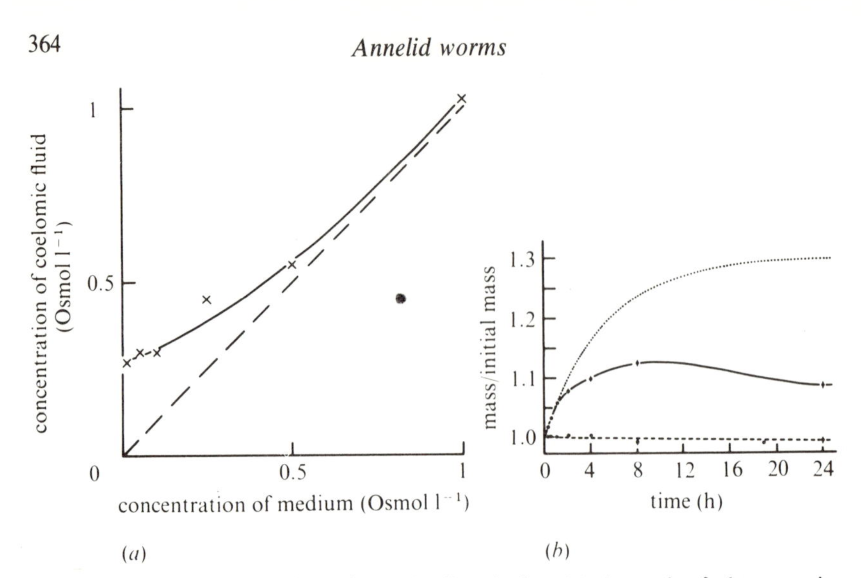

Fig. 16.16. Osmotic regulation of *Nereis diversicolor.* (*a*) A graph of the osmotic concentration of the coelomic fluid against that of the water in which the worms had been kept. Data from C. R. Fletcher (1974). *Comp. Biochem. Physiol. A* **47**, 1199–214. (*b*) A graph of body mass (expressed as a multiple of initial mass) against time after transfer from 70% sea water to (continuous line) 50% sea water or (broken line) another dish of 70% sea water. The dotted line is explained in the text. After C. R. Fletcher (1974). *Comp. Biochem. Physiol. A* **47**, 1221–34.

nephridial apertures. These droplets have been collected in fine pipettes, and their osmotic concentration determined by measuring their freezing point. It was found that urine from worms kept in very dilute sea water is only 50–60% as concentrated as the coelomic fluid. The worms were apparently maintaining the osmotic concentration of their body fluids in the same way as *Lumbricus*, by secreting a dilute urine.

At concentrations between 50 and 100% sea water *Nereis* coelomic fluid is only slightly more concentrated than the water and the worms do not face any severe osmotic problem. However, a problem arises when the osmotic concentration of the medium is changed. Suppose for instance that a worm is moved from 70% sea water to 50% sea water. Immediately after transfer the worm's body fluids are quite a lot more concentrated than the water; water tends to diffuse in, diluting the body fluids and making the worm swell. If the surface of the worm were a semipermeable membrane no solutes would diffuse out and equilibrium would not be attained until the quantity of water in the worm was $70/50 = 1.4$ times its initial value. Since the worm would have contained initially about 80% water, its final volume would be about 1.32 times its initial volume. It would swell gradually to this volume as indicated by the dotted line in Fig. 16.16(*b*). What happened when the experiment was tried is shown by the continuous line. The worms stopped swelling after 8 hours and after 24 hours were only 1.09 times their initial volume.

Some of the worms were dissolved in potassium hydroxide solution and analysed for chloride. It was found that the total quantity of chloride in the body fell by about 25% in the 24 hours following transfer to 50% sea water. It was concluded from this and other observations that the worms must have limited their swelling by passing urine of about the same osmotic concentration as the coelomic fluid. This got rid of some of the water which diffused in and also got rid of some salt, reducing the amount of swelling needed to restore equilibrium.

EARTHWORMS AND THE SOIL

Pastures, woods and orchards in Europe often have 50–120 g earthworms per square metre. An average pasture will support cattle at a density of about 0.5 tonne ha^{-1} or 50 g m^{-2}, so many pastures support a greater mass of earthworms than of cattle. Enchytraeid worms are also common in soil, particularly in moorland soils where up to 50 g m^{-2} have been found.

Animals in woodland soil get most of their energy from the leaves that fall from the trees. Earthworms feed directly on fallen leaves. Enchytraeids feed largely on fungi and bacteria which have themselves fed on dead leaves. A deciduous wood in England had a mean earthworm population of 120 g m^{-2}. The mean mass of individual worms was 2.7 g and it was estimated from laboratory measurements of metabolic rate at various temperatures that their mean metabolic rate, over a whole year, would have been 0.02 cm^3 oxygen g^{-1} h^{-1}. This amounts to 20 l oxygen m^{-2} yr^{-1}. A coniferous wood in Wales had a mean enchytraeid population of 11 g m^{-2}. The mean body mass of these worms was only 0.1 mg so their metabolic rate could be expected to be high (Fig. 8.6). It was calculated from laboratory measurements that their mean metabolic rate over a year would have been 0.3 cm^3 oxygen g^{-1} h^{-1}. This amounts to 30 l oxygen m^{-2} yr^{-1}. It was estimated that the rate of fall of leaves was about 300 g dry weight m^{-2} yr^{-1} in each wood. Complete metabolism of the leaves would use about 300 l oxygen m^{-2} yr^{-1}, so earthworm metabolism accounted for about 7% of the litter in the deciduous wood and enchytraeid metabolism for about 10% of the litter in the coniferous one. There must have been some of the other group of worms in each wood so oligochaete respiration probably accounted for 10–15% of the energy in the litter in each case. Much of the rest must have been used by bacteria, fungi and nematodes.

Though earthworms metabolized less than 10% of the energy content of the litter in the deciduous wood they probably broke a much larger proportion of the leaves into fragments and passed them through their guts. The importance of earthworms in breaking up litter was illustrated when 2.5 cm discs of oak and beech leaves were buried out of doors, in nylon bags of various mesh. They were examined periodically and the area lost from each disc was recorded. Discs in 7 mm mesh bags (big enough to admit earthworms) disappeared 2–3 times as fast as discs in 0.5 mm mesh bags (which excluded earthworms).

Earthworms in pastures feed largely on dung. Though cattle are much larger than earthworms they have a metabolic rate (per unit body mass) around 10 times as high. About 30% of the energy content of the grass they eat remains in the dung. Hence a given mass of cattle probably produces ample dung to support an equal mass of earthworms.

FURTHER READING

GENERAL

Dales, R. P. (1967). *Annelids*, 2nd edn. Hutchinson, London.

Edwards, C. A. & Lofty, J. R. (1972). *Biology of earthworms*. Chapman & Hall, London.

LOCOMOTION

Clark, R. B. & Tritton, D. J. (1970). Swimming mechanisms in nereidiform polychaetes. *J. Zool., Lond.* **161**, 257–71.

Elder, H. Y. (1973). Direct peristaltic progression and the functional significance of the dermal connective tissue during burrowing in the polychaete *Polyphysia crassa* (Oersted). *J. exp. Biol.* **58**, 637–55.

Mettam, C. (1967). Segmental musculature and parapodial movement of *Nereis diversicolor* and *Nephthys hombergi* (Annelida: Polychaeta). *J. Zool., Lond.* **153**, 245–75.

Seymour, M. K. (1969). Locomotion and coelomic pressure in *Lumbricus terrestris* L. *J. exp. Biol.* **51**, 47–58.

RESPIRATION AND BLOOD CIRCULATION

Johansen, K. & Martin, A. W. (1965). Circulation in a giant earthworm, *Glossoscolex giganteus*. I. *J. exp. Biol.* **43**, 333–48.

NEPHRIDIA AND COELOMS

Goodrich, E. S. (1946). The study of nephridia and genital ducts since 1895. *Q. J. microsc. Sci.* **86**, 115–392.

Ramsay, J. A. (1949). The osmotic relations of the earthworm. *J. exp. Biol.* **26**, 46–56.

Ramsay, J. A. (1949). The site of formation of hypotonic urine in the nephridium of *Lumbricus*. *J. exp. Biol.* **26**, 65–75.

DEVELOPMENT

Anderson, D. T. (1973). *Embryology and phylogeny in annelids and arthropods*. Pergamon, Oxford.

LIFE IN ESTUARIES

Fletcher, C. R. (1974). Volume regulation in *Nereis diversicolor* [3 papers]. *Comp. Biochem. Physiol.* **47A**, 1199–234.

Smith, R. I. (1970). Hypo-osmotic urine in *Nereis diversicolor*. *J. exp. Biol.* **53**, 101–8.

EARTHWORMS AND THE SOIL

Burges, A. & Raw, F. (1967). *Soil biology*. Academic Press, New York & London.

17
Crustaceans

Phylum Arthropoda, Class Crustacea
  Subclass Cephalocarida
  Subclass Branchiopoda (water fleas etc.)
  Subclass Mystacocarida
  Subclass Ostracoda
  Subclass Copepoda
  Subclass Branchiura
  Subclass Cirripedia (barnacles)
  Subclass Malacostraca (crabs, shrimps, woodlice, etc.)
(For other classes see chapters 18 to 21)

This is the first of four chapters about the largest phylum of all, the Arthropoda. The members of this phylum have a jointed, exterior skeleton like a suit of armour (an exoskeleton). They have segmented bodies, typically with a pair of limbs on each segment. The main body cavity is a haemocoel.

By far the majority of the arthropods are insects, which are described in the next two chapters. This chapter introduces the phylum and deals particularly with the class Crustacea, which has about 25000 known species. The class Osteichthyes (bony fish) has about the same number of species and the only classes of animals which have more are the Gastropoda and Insecta.

Large sections of this chapter deal with the exoskeleton and the muscles which move it, and with legs. Other sections deal with sense organs and reflexes because research on crabs and other large crustaceans has provided excellent examples of the ways in which nervous systems work. There are also sections about planktonic Crustacea and about mechanisms used by Crustacea for filter feeding.

It has long been traditional to describe the crayfish *Astacus fluviatilis* as a typical crustacean, to introduce the class. It is big enough to be easily dissected and it is more typical of the class than the equally large crabs. It is a European species found in streams, especially in limestone districts. It spends much of the day under stones or in burrows in the bank but emerges at night to feed on water snails, insect larvae, etc.

The body of *Astacus* has two main parts, an anterior cephalothorax which is roofed by a rigid piece of exoskeleton (the carapace) and a posterior flexible abdomen. The most conspicuous limbs are the pair which bear the chelae (pincers) and the four pairs of walking legs. The chelae are used to seize and

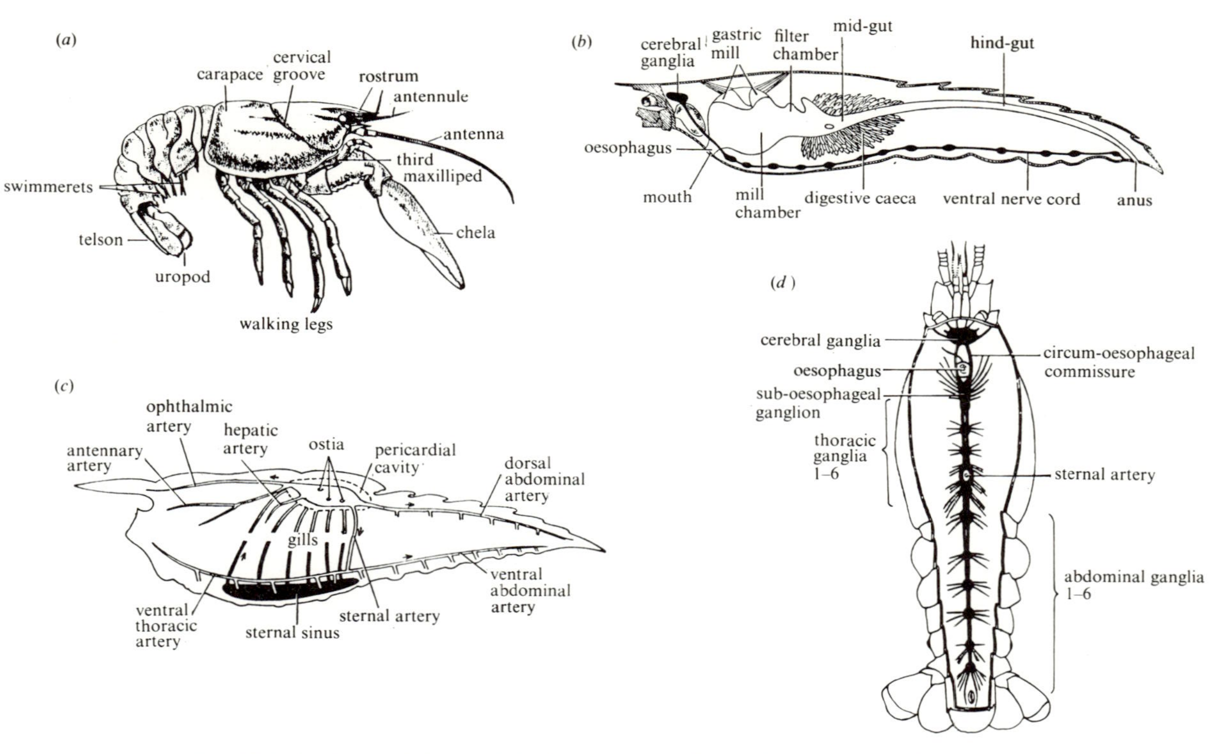

Fig. 17.1. Anatomy of the crayfish *Astacus fluviatilis*. (*a*) A lateral view of the intact animal. (*b*), (*c*) and (*d*) Diagrams of dissections showing the gut, the blood circulation and the nervous system, respectively. Length up to 15 cm. From G. Chapman & W. B. Barker (1966). *Zoology for intermediate students*. Longman, London.

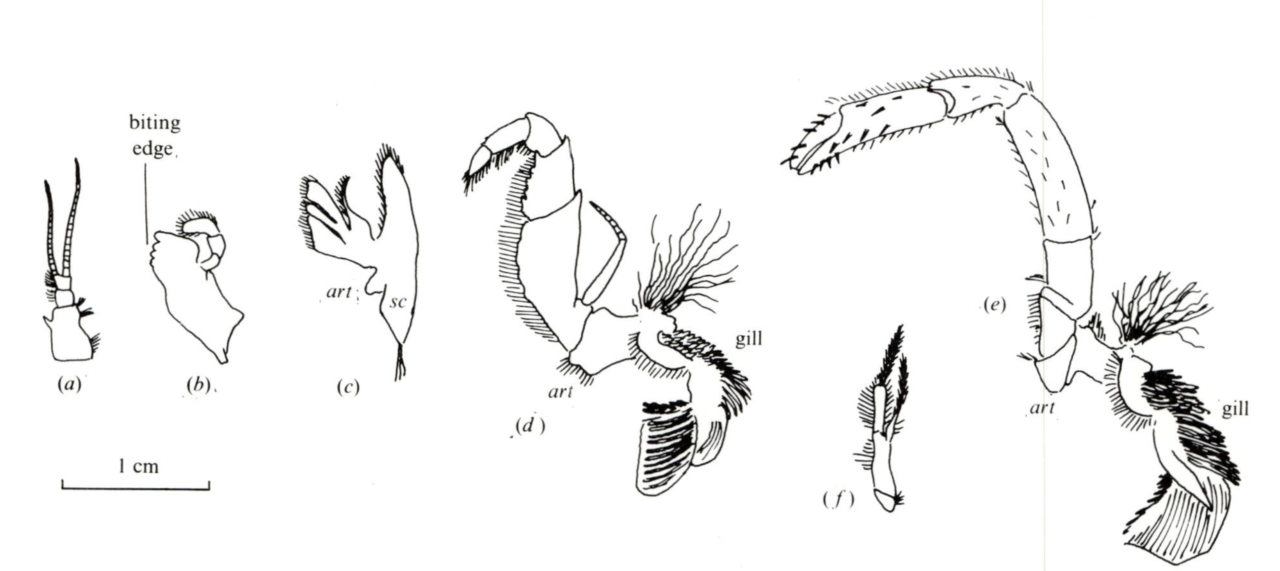

Fig. 17.2. Sketches of crayfish limbs. Each sketch represents a left limb viewed from its posterior face. (*a*) Antennule; (*b*) mandible; (*c*) second maxilla; (*d*) third maxilliped; (*e*) walking leg; (*f*) swimmeret. *art*, point of articulation with the trunk; *sc*, scaphognathite. After T. H. Huxley (1880). *The crayfish*. Paul, London.

TABLE 17.1. *The segments, limbs and nerve ganglia of the crayfish*

Segment	Limbs	Ganglion[a]
	Cephalothorax	
1	None	X
2	Antennules	(cerebral)
3	Antennae	
4	Mandibles	
5	First maxillae	X
6	Second maxillae	(sub-oesophageal)
7	First maxillipeds	
8	Second maxillipeds	
9	Third maxillipeds	X
10	Chelae	X
11	First walking legs	X
12	Second walking legs	X
13	Third walking legs	X
14	Fourth walking legs	X
	Abdomen	
15	First swimmerets	X
16	Second swimmerets	X
17	Third swimmerets	X
18	Fourth swimmerets	X
19	Fifth swimmerets	X
20	Uropods	
21	None	

[a] Each X indicates one ganglion or pair of ganglia.

crush food, and to pass it to the mouth. These and the other limbs are listed in Table 17.1 and some of the limbs are illustrated in Fig. 17.2. The antennules and antennae are limbs modified for sensory functions. The antennules are branched and have chemo-sensory organs along one of the two branches. In addition the bases of the antennules house the statocysts, organs of balance which are described later in the chapter. The antennae are used as organs of touch and are often waved about in front of the animal but are also often carried trailing posteriorly. The next six pairs of limbs are arranged around the mouth. The mandibles are biting jaws and the maxillipeds are used to hold and manipulate food. The first maxillae have no apparent function but the second maxillae are important in respiration; their movements pump water over the gills through the paddle-like action of processes called scaphognathites. The gills are projections from the bases of the limbs of segments 8 to 14. The swimmerets on the abdomen move forward and back in seemingly futile fashion as the crayfish walks. The pair of uropods form a fan which can be spread or folded. *Astacus* normally moves about by walking on its walking legs but it can swim rapidly backwards by repeatedly bending the abdomen ventrally with the uropods spread.

Table 17.1 shows that every segment except the first and last bears a pair of limbs or modified limbs. The first segment seems not to be distinguishable in adult animals but can be found in embryos. The situation is rather like that of polychaete worms, in which every segment bears a pair of parapodia. However, the parapodia are all (typically) more or less identical whereas the limbs of crayfish are not.

Fig. 17.2 shows that many of the limbs are two-branched. It is likely that the ancestors of Crustacea had two-branched limbs, all more or less alike, all along the body. If so the mandibles, chelae and walking legs have each lost a branch in the course of evolution. The blades of the mandibles are enlarged limb bases.

Fig. 17.1(*b*) shows the arrangement of the gut. Only the mid-gut is endodermal; the mill chamber and filter chamber and the hind-gut are formed from ectoderm which tucks in at the mouth and anus as the embryo develops. These ectodermal parts of the gut are lined with cuticle, similar to the cuticle of the outer surface of the body. Most of the gut cuticle is flexible, like the flexible membranes between the stiff plates of the exoskeleton. However, there are thick stiff plates in the cuticle of the mill chamber. There are muscles to move these plates and they form the gastric mill which breaks up chunks of food into smaller pieces. The food cannot move on to the mid-gut until it is finely ground because it is strained through fringes of setae which project from the cuticle of the filter chamber.

A pair of digestive glands open into the mid-gut. Each consists of a large number of finger-like caeca. They secrete most or all of the digestive enzymes, but enzymes seep forward into the mill chamber and digestion starts there. Digestion seems to be entirely extracellular and the products of digestion are absorbed through the walls of the mid-gut and digestive glands. There are no cilia in the gut and the food is moved along the gut entirely by peristalsis. The gut muscles are striated, not unstriated like the gut muscles of vertebrates.

Fig. 17.1 (*b*) and (*d*) show the central nervous system. There is a ventral nerve cord (divided for most of the length of the body into two strands) with ganglia which link the strands together. The cerebral ganglia lie anterior to the mouth and the two strands of the nerve cord pass on either side of the mouth. The central nervous systems of annelids are very like this (Fig. 16.3 *a*). Primitive arthropods presumably had a ganglion in each segment, but the crayfish has separate ganglia only for segments 9 to 19. The cerebral ganglia are connected by nerves to the antennules and antennae, and also to the eyes which are compound eyes like those of insects (see chapter 18). The sub-oesophageal ganglion provides the nerves for the closely-spaced mouthparts (except the third maxillipeds).

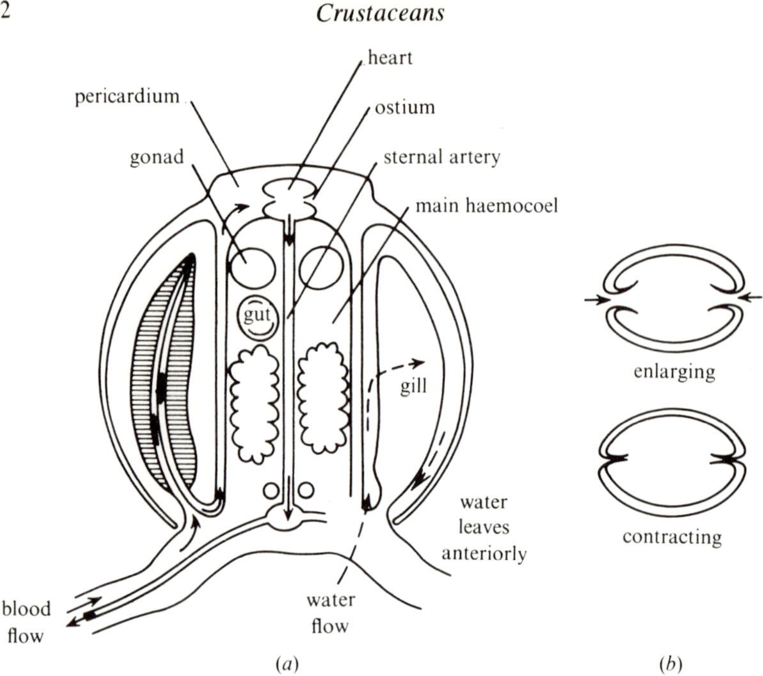

Fig. 17.3. (*a*) A diagrammatic transverse section through the cephalothorax of a crayfish. Short arrows indicate blood flow and long broken arrows indicate the flow of water over the gills. (*b*) Diagrammatic transverse sections of the heart showing how the valves of the ostia open when the heart is enlarging but close when it is contracting.

The gills are hidden under the carapace (Fig. 17.3). Each consists of numerous fine filaments, projecting from a central stalk. The total area of the gills of a 40 g crab (*Carcinus maenas*) is about 300 cm^2, about the same as would be expected for a gastropod of the same mass (see p. 279). The gills are covered by a very thin layer of cuticle. The path of water over the gills has been investigated by watching the movements of dyes or milk, released into the water near the animal. Water enters the gill cavity through the gaps between successive pairs of limbs in segments 9 to 14, and leaves by a pair of exhalant openings on either side of the mouth. The movement of dyed water within the gill chamber has been observed in crabs which had had part of the carapace replaced by a window of transparent plastic. Fig. 17.3 shows the direction of flow. The water is apparently propelled by the movements of the scaphognathites which are just inside the exhalent openings. The scaphognathites beat 1 to 5 times per second, in the crab *Carcinus*, and set up a pressure difference across the gills of 2–6 mm water (20–60 $N\ m^{-2}$; this has been measured by means of pressure transducers). The use of scaphognathites instead of cilia to drive water over the gills is a striking difference from gastropods and bivalves.

Figs. 17.1(*c*) and 17.3 show the arrangement of the blood system and the directions of flow of blood. The directions were discovered by watching the movement of injected dyes. There is a dorsal heart enclosed in a blood-filled pericardium. Three pairs of openings, the ostia, admit blood from the pericardium into the heart. They have valves which prevent flow in the reverse direction. When the heart contracts it drives blood out through the arteries. When it expands (presumably by elastic recoil) it draws blood in through the ostia. The frequency of the heart beat is 0.5–1 s^{-1}. Peak pressure in the heart of a lobster (*Homarus*) during contraction was about 1800 N m^{-2} when the lobster was resting and 3600 N m^{-2} when it was active: similar pressures occur in the hearts of snails (p. 274). There are arteries to all parts of the body including the legs. The sternal artery pierces the ventral nerve cord (Fig. 17.1 *d*). At the tissues the blood escapes from fine branches of the arteries into the haemocoel. It returns to the pericardium by way of the gills. This circuit is rather similar to the circuit of the blood in gastropods (Fig. 13.16).

The blood contains a haemocyanin, which enables it to transport (in the crab *Cancer*) 34 cm^3 oxygen l^{-1}. The haemocyanin is present as large polymeric molecules dissolved in the blood, but the molecules of crustacean haemocyanin are not nearly as big as those of snail haemocyanin (p. 276).

The main body cavity is a haemocoel and there is hardly any trace of coeloms. However, there is a pair of excretory organs (the green glands), each of which consists of a rudimentary coelom and coelomoduct; they can be identified as coeloms because they develop from mesoderm. The green glands lie in the haemocoel anterior to the mouth and their ducts open to the exterior through the pores in the basal segments of the antennae. They are supplied with blood by the antennary artery (Fig. 17.1 *c*). Ultrafiltration occurs from the blood into the rudimentary coelom and urine is excreted through the duct. The urine of freshwater crayfish may be even more dilute than that of the freshwater gastropod *Viviparus* (Table 13.1), though the crayfish has much more concentrated blood. Crayfish with 184 mEquiv l^{-1} chloride in their blood excreted urine with only 3.4 mEquiv l^{-1} chloride.

The sexes are separate and the reproductive organs are simple. The gonads lie in the main haemocoel, dorsal to the gut (Fig. 17.3). The female has a pair of ovaries with oviducts which lead to openings near the bases of the second walking legs. The male has a pair of testes with vasa deferentia leading to openings near the bases of the fourth walking legs. Fertilization is external. The eggs adhere to the females' swimmerets and the young which hatch from them at first hold on to her swimmerets by their chelae. The newly-hatched young are more or less similar in form to the adults, though much smaller.

Most Crustacea have larval stages quite unlike the adults. In many cases the earliest larval stage is a nauplius with three pairs of limbs corresponding to the antennules, antennae and mandibles of the adult (Fig. 17.4).

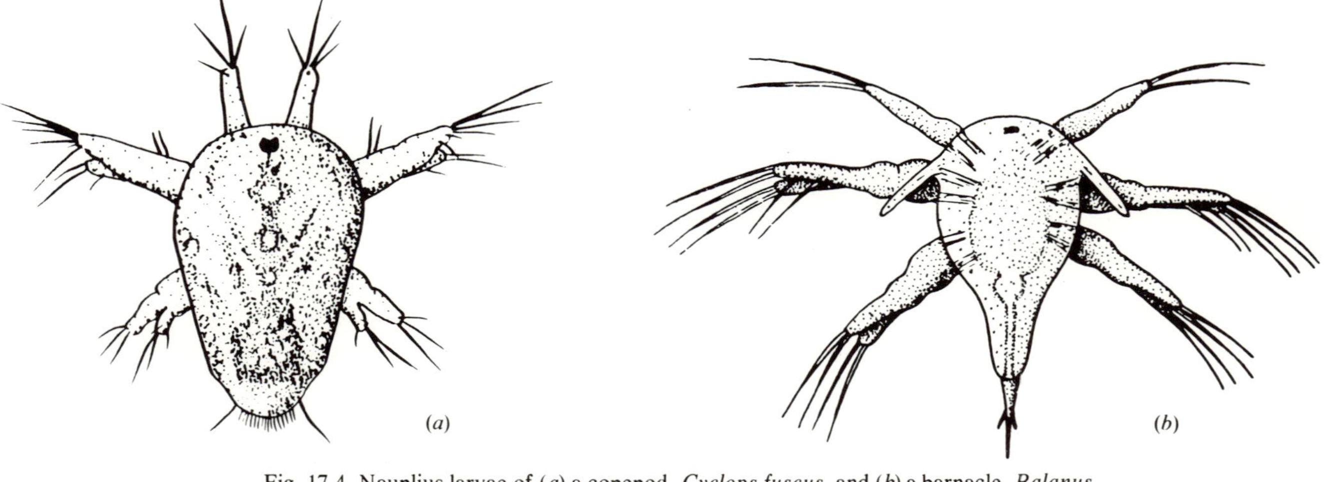

Fig. 17.4. Nauplius larvae of (*a*) a copepod, *Cyclops fuscus*, and (*b*) a barnacle, *Balanus* (length about 0.3 mm). From J. Green (1961). *A biology of Crustacea*. Witherby, London.

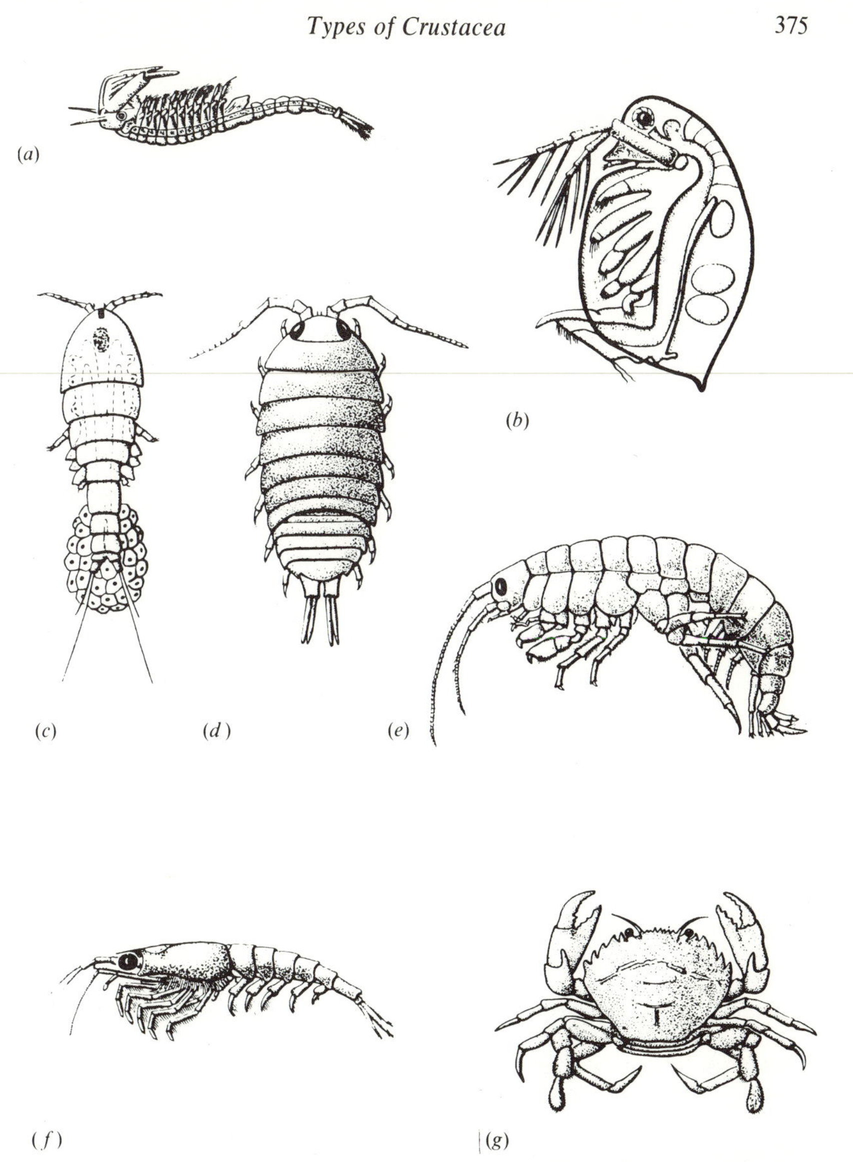

Fig. 17.5. A selection of Crustacea. (*a*) *Cheirocephalus* (Branchiopoda: length about 12 mm); (*b*) *Daphnia* (Branchiopoda: 1–4 mm); (*c*) *Tigriopus* (Copepoda); (*d*) *Ligia* (Malacostraca: 25 mm); (*e*) *Gammarus* (Malacostraca: 15 mm); (*f*) *Thysanoessa* (Malacostraca: 30 mm) and (*g*) *Portunus* (Malacostraca: carapace 10 cm wide). From A. P. M. Lockwood (1968). *Aspects of the physiology of Crustacea*. Oliver & Boyd, Edinburgh.

TYPES OF CRUSTACEA

Figs. 17.5 and 17.6 give an impression of the variety of Crustacea. *Cheirocephalus* and *Daphnia* both belong to the subclass Branchiopoda. Both are filter feeders which produce feeding currents and filter the water by means of broad paddle-like limbs fringed with fine setae. *Cheirocephalus* is occasionally found

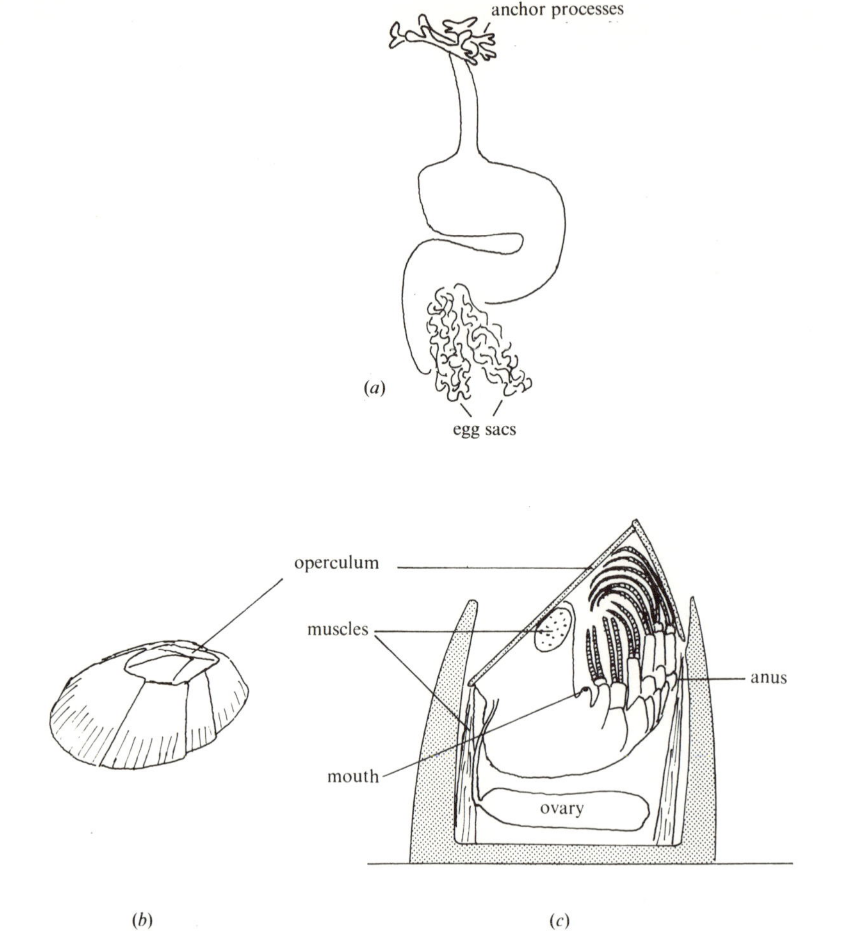

Fig. 17.6. (*a*) An adult female *Lernaeocera branchialis* (length about 3 cm). After T. & A. Scott (1913). *The British parasitic Copepoda.* Ray Society, London. (*b*) An acorn barnacle, *Balanus balanoides* (diameter about 10 mm). (*c*) A diagram of *Balanus* cut in half to show its internal structure.

in temporary rainwater pools in S. England. It has a relatively long body and generally swims ventral side up. *Daphnia* and similar branchiopods are known as water fleas. They are extremely common in ponds. They have relatively short, stocky bodies and the paddle-shaped limbs are enclosed in a carapace. They swim by movements of their antennae. In many species of water flea, as in many species of rotifer (p. 232), reproduction is usually by parthenogenesis and males occur only occasionally. Rotifers and the smaller water fleas are similar in size and live largely in the same habitats.

Most Branchiopoda live in freshwater but Copepoda are abundant both in fresh water and in the sea. They are among the most plentiful and important constituents of the marine zooplankton. They are the principal food of the herring (*Clupea harengus*) which is one of the most important fish landed commercially in Europe. The copepod illustrated in Fig. 17.5(*c*) is a female, carrying eggs. Many species have a pair of egg sacs but this species has just one. Copepods swim by movements of their antennae and other limbs. They feed mainly by filter feeding, on phytoplankton. As well as free-living copepods there are many parasites. Adult female *Lernaeocera* (Fig. 17.6*a*) live on the gills of cod (*Gadus morhua*) and related teleosts. Their heads are buried in the tissues of their hosts and they feed on blood. It is by no means obvious from their appearance that they are crustaceans, and the pair of egg sacs are the only recognizable copepod features. However, the earliest larval stage is a typical naupilus and there are later larval stages which look quite like typical copepods. Much of the larval life of both males and females is spent attached as parasites to the gills of the flounder (*Platichthys flesus*). Both sexes eventually detach and become free-swimming, fertilization occurs, and the female settles on the gills of cod and assumes the adult form.

Balanus (Fig. 17.6*b*, *c*) and the other barnacles form the subclass Cirripedia. Their earliest larvae are nauplii (Fig. 17.4*b*) and there are several planktonic larval stages, but the adults live attached to rocks, floating wood and ships. *Balanus balanoides* lives on intertidal rocks and is the commonest European species. Large numbers are often found packed so closely together that the underlying rock is completely hidden. There are six pairs of two-branched limbs. Fig. 17.6(*c*) shows them and also shows the positions of the mouth and anus. The animal is attached to the rock by the dorsal surface of its head. It is enclosed in a ring of fixed calcareous plates with a lid (operculum) of movable plates. The lid opens and closes, rather like a double door. The animal can retract into the shell and close the operculum for protection (for instance when it is exposed at low tide). It can open the operculum and extend the limbs for filter feeding. There is a later section of this chapter which describes the process of filter feeding, and another which describes how barnacle larvae select the sites where they settle and metamorphose to the adult form.

There are parasitic barnacles as well as free-living ones. One of the most remarkable is *Sacculina*, a parasite of crabs. As an adult it is a bag of tissue protruding from the crab's abdomen with root-like processes permeating the whole of the crab's body. However its earliest larvae are nauplii with the pair of horns characteristic of barnacle nauplii (see Fig. 17.4*b*) and its later larvae also resemble those of typical barnacles.

Astacus is a member of the subclass Malacostraca, and so are *Ligia*, *Gammarus*, *Thysanoessa* and *Portunus* (Fig. 17.5*d–g*). This subclass includes the majority of species of Crustacea. All its members have the same number of segments (Table 17.1) but they vary a lot in appearance. *Ligia* is one of the isopods, a group of depressed (dorso-ventrally flattened) Malacostraca which also includes the terrestrial woodlice. *Ligia* is found under stones and weed on the upper part of the shore and feeds on pieces which it bites off seaweeds.

Gammarus is one of the amphipods, a group of compressed (flattened from side to side) Malacostraca. The genus includes fresh, brackish and seawater species which live on the bottom and feed on detritus. *Thysanoessa* is an example of the group of planktonic crustaceans known as krill, which are very plentiful in the oceans. They feed both on smaller zooplankton and on phytoplankton and are themselves the principal food of the whalebone whales (i.e. of the whales of the suborder Mysticeti, which includes most of the large whales). *Portunus* is an example of the crabs, which are very similar in structure to the crayfish but have a very broad carapace and a very small abdomen which is folded forward. Crabs eat a wide variety of foods including molluscs, whose shells they break with their chelae. *Portunus* lives on shores and, unlike most other crabs, can swim.

THE EXOSKELETON

Crustaceans are enclosed in jointed skeletons, like mediaeval knights in armour. We will examine the structure and properties of the skeletal material before considering the structure and working of the skeleton as a whole.

The skeleton consists of hardened parts of the non-living cuticle which covers the whole body. It has three main constituents, calcium salts, protein and the polysaccharide chitin. The salts are mainly calcium carbonate but they include a little phosphate, and make up anything from 0 to 90% of the dry mass of the cuticle. The highest proportions are found in the hardest skeletal parts, for instance at the tips of crabs' legs. The rest of the dry mass of the cuticle is about 70% chitin, 30% protein, and much of the protein seems to be bound to the chitin by covalent bonds. The skeleton is stiffened and strengthened by quinone tanning (p. 143) as well as by incorporation of calcium salts.

Electron microscope sections of arthropod cuticle often show strange, curving lines (Fig. 17.7*a*). The lines are chitin fibrils and their diameter in crustacean cuticle is typically about 10 nm. They look as if they run in hooped paths but careful study of insect cuticle has shown that this is an illusion. The actual arrangement is like a complex version of plywood. Successive layers in a sheet of plywood have the grain running north–south and east–west. Insect cuticle is laid down in layers with all the chitin fibrils in a given layer running in the same direction, but the directions are not at right angles in successive layers; rather, they differ only by a small angle. Fig. 17.7(*b*) is a vertical section through a series of layers in which the direction changes from parallel to the page (at the top), to at right angles to the page (half way down), to parallel to the page again (at the bottom). Fig. 17.7(*c*) shows an oblique section through the same set of layers. At a lower magnification it could easily be mistaken for an array of hooped fibrils. Cuticle is commonly made up of many sets of layers with the fibril direction changing through 180° in each set.

Fig. 17.7(*b*) and (*c*) are drawn as though the fibril direction changed 22½° between each layer and the next. Arthropod cuticle generally has much smaller

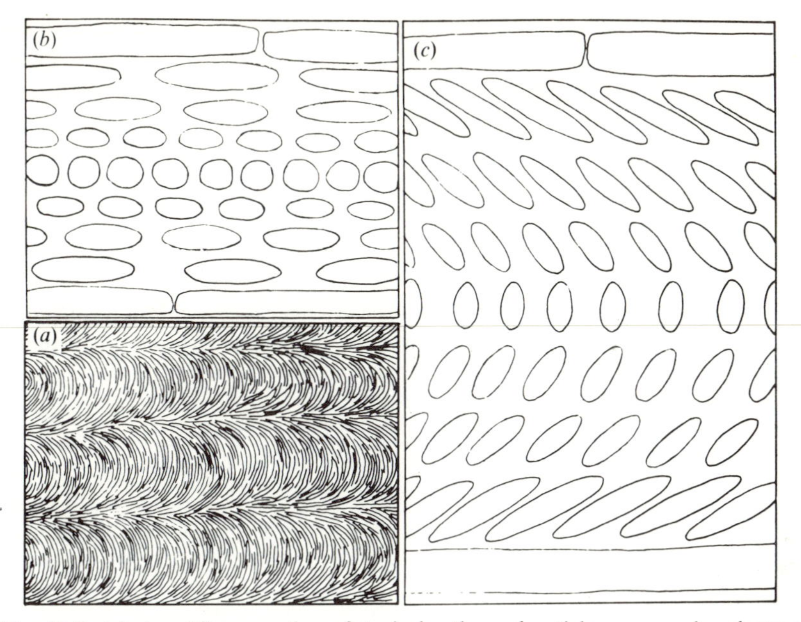

Fig. 17.7. (*a*) An oblique section of typical arthropod cuticle, as seen by electron microscopy. (*b*) A diagrammatic vertical section and (*c*) a diagrammatic oblique section of the same cuticle, on a larger scale than (*a*). These diagrams are explained in the text. From S. A. Wainwright, W. D. Biggs, J. D. Currey & J. M. Gosline (1976). *Mechanical design in organisms.* Arnold, London.

changes of angle between successive layers: changes as small as 1° have been observed.

It seems fairly clear that insect cuticle commonly has the structure shown in Fig. 17.7, but there is controversy about the structure of crustacean cuticle. Scanning electron micrographs of broken crayfish cuticle seem to show curved fibrils running between the layers.

The structure of arthropod cuticle is not uniform throughout its thickness (Fig. 17.8). There is a thin outer waxy layer known as the epicuticle. The main thickness of the cuticle consists of an exocuticle in which the protein is generally tanned and an inner endocuticle in which it is not. Crustacea generally have calcium salts in the exocuticle and the outer parts of the endocuticle.

Pieces of crab carapace have been stretched and broken on engineers' testing equipment. They had lower strength than many mollusc shells but they were also less dense so the value of strength/mass was about the same as for typical mollusc shells (Table 17.2). They were less stiff (had a lower Young's modulus) than many mollusc shells, as might be expected since they contained a smaller proportion of calcium salts.

Wood splits easily along the grain but is harder to break across the grain.

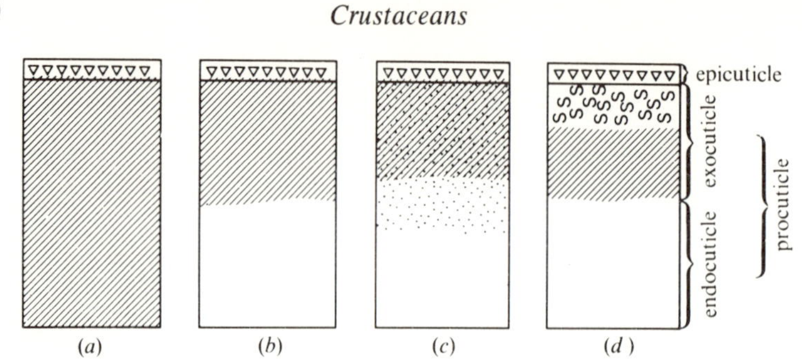

Fig. 17.8. Diagrammatic vertical sections through the cuticle of (*a*) some insects; (*b*) most insects, spiders, mites and centipedes; (*c*) crustaceans; and (*d*) some scorpions. Triangles indicate waxes; SSS, proteins cross-linked by sulphur; hatching, quinone-tanned proteins; stipple, calcium salts. From S. A. Wainwright, W. D. Biggs, J. D. Currey & J. M. Gosline (1976). *Mechanical design in organisms.* Arnold, London.

It is much stiffer along the grain than across the grain: Young's modulus for Douglas fir timber is about 10 times as high for tension along the grain as for tension across the grain. Plywood with grain running at right angles in successive layers is almost equally strong and stiff for tension in any direction in its own plane. The arrangement of fibres in arthropod cuticle is probably even better because the smallness of the angle between successive layers makes the layers less likely to come apart under stress.

The cuticle of a crustacean covers the whole external surface of the body. Movement is possible because there are flexible regions between the stiff ones. These arthrodial membranes are continuous with the stiff cuticle and have the same basic structure but most of their protein is not tanned and they are not impregnated with calcium salts. Fig. 17.9(*a*) shows, very diagrammatically, the exoskeleton of a crayfish. The cephalothorax is covered by an unjointed carapace but the abdomen has a separate dorsal and ventral plate of stiff cuticle for each segment, with strips of arthrodial membrane between. The plates overlap like tiles on a roof (Fig. 17.9*b*) so that the relatively weak arthrodial membranes are not exposed. The antennae are covered by rings of stiff cuticle joined by rings of arthrodial membrane which make them flexible. The legs are covered by tubes of stiff cuticle with arthrodial membrane at the joints. The stiff cuticle of the more distal segment involved in a joint forms a pair of knobs which fit into sockets in the more proximal segment, so that the joint is a hinge and can rotate only about an axis through the two knobs (Fig. 17.9*c*, *d*: the proximal end of a limb is the end attached to the trunk and the distal end is the free end). Successive joints along the leg have their hinge axes at right angles to each other, making a wide range of leg positions possible.

Vertebrates and arthropods can move in broadly similar ways. They can bend their trunks, walk on their legs, and so on. The exoskeleton of arthropods allows the same sorts of movements as the endoskeleton of vertebrates. What are the advantages of each type of skeleton?

TABLE 17.2. *Mechanical properties of some skeletal materials*

	Tensile strength ($MN\ m^{-2}$)	Young's modulus ($GN\ m^{-2}$)	Density ($kg\ m^{-3}$)
Mollusc shells	30–100	50	2700
Crab carapace	32	13	1900
Locust leg cuticle	95	9.5	1200
Locust apodeme	600	19	
Compact bone	150–200	18	2000

Data from S. A. Wainwright, W. D. Biggs, J. D. Currey & J. M. Gosline (1976). *Mechanical design in organisms.* Arnold, London.

Obviously an exoskeleton protects internal organs, but it is also rather easily damaged because there is no soft tissue between it and the body surface to cushion a blow. To understand what this implies we need to consider the mechanics of breaking things.

If a piece of material is stretched until it breaks, a certain amount of work is done stretching it against the elastic restoring force. It cannot be broken unless this amount of work is done. If a cup is dropped on the floor it will not break unless it is dropped from high enough to have enough kinetic energy at the moment of impact to provide the work needed to break it. It is less likely to break if the floor is carpeted because some of the energy is used deforming the carpet, leaving less to break the cup. Similarly, soft tissues protect endoskeletons from blows. This was demonstrated in a series of experiments with metatarsal bones from the feet of rabbits. In the intact rabbit, the metatarsals are covered by furry skin. They were broken by a blow from a heavy pendulum and it was found that 37% more energy was needed to break them when they were covered with skin than when they were not. Bones deeper in the body are even better protected. The exoskeletons of arthropods do not have this sort of protection.

Imagine a series of animals each with a different mass m. Let them all have exoskeletons of the same material and let the exoskeleton be the same proportion of body mass in each case. Then the energy required to fracture the exoskeleton will be km, where k has the same value for all the animals. Now let each animal collide with a rigid wall while travelling with velocity u. The exoskeleton will break if the kinetic energy $\frac{1}{2}mu^2$ is greater than km, i.e. if $u > \sqrt{(2k)}$. Whatever the size of the animal, fracture can be expected at the same impact velocity. An exoskeleton will thus be dangerously fragile for any animal which moves fast, but there are no really fast arthropods. On the other hand, many vertebrates, which have endoskeletons, can move at great speed.

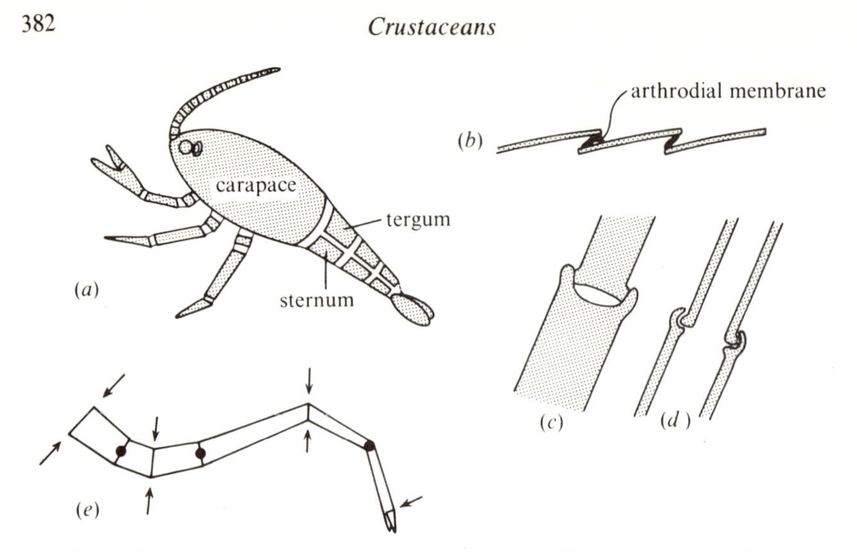

Fig. 17.9. Diagrams of the exoskeleton of a crayfish. (*a*) The complete animal, grossly simplified with the number of parts reduced; (*b*) a longitudinal section through a few of the dorsal plates of the abdomen; (*c*) a sketch of and (*d*) a section through a leg joint; (*e*) a sketch of a leg with arrows representing axes of joints lying in the plane of the paper and dots representing axes at right angles to the paper. Stiff cuticle is stippled in (*a*) to (*d*).

This simple argument will mislead readers if they take it too literally, but the principle it presents is sound.

As well as protecting internal organs, exoskeletons have another advantage. Consider a rod which has to withstand forces tending to bend it. It is neither as strong nor as stiff as a hollow tube of the same length, made from the same mass of the same material. This is why tubes are used for making scaffolding and the frames of bicycles. A stout, thin-walled tube is better than a more slender thick-walled tube of the same length and mass, provided its wall is not so thin that it is apt to kink like a drinking straw. The exoskeletons of crustacean legs are thin-walled tubes, stout enough to accommodate the muscles and other soft tissues within them. Endoskeletons made of the same mass of the same material would be neither as stiff nor as strong in bending.

The skeletons of arthropods, like the shells of molluscs, cannot be re-shaped once they have been formed. Molluscs grow by adding to the existing shell (p. 267). Arthropods grow by shedding the skeleton and forming a new, larger one. The process is similar to the moulting of nematodes (Fig. 12.6; but nematode cuticle can grow after the final moult). The inner layers of the old cuticle are partially broken down so that the cuticle is detached from the epidermis. The breakdown products are resorbed: the crab *Carcinus* recovers about 80% of the organic matter of its old cuticle and 20% of the calcium salts in this way. New, soft cuticle begins to form inside the old.

The animal often retires to a relatively safe place before the old cuticle splits.

Carcinus, for instance, hides among seaweed. The osmotic concentration of the blood rises (in the crab *Pachygrapsus*) from around 0.7 Osmol l^{-1} between moults to 1.4 Osmol l^{-1} just before moulting. The crab swells, presumably by osmotic uptake of water (the osmotic concentration of sea water is about 1.0 Osmol l^{-1}). The old cuticle splits and the animal withdraws from it. The water taken up increases the volume of the crab by about 40% (in *Pachygrapsus*) and stretches the new cuticle. It is only after this that the new cuticle is hardened by quinone-tanning and deposition of salts.

Some crustaceans which live on shores and in estuaries are exposed to large changes of salt concentration. We have already seen how the worm *Nereis diversicolor* responds to such changes: when the concentration is reduced it may swell by 9% or more of its initial volume and yet survive (Fig. 16.16*b*). The exoskeletons of crustaceans do not allow so much swelling. The crab *Porcellana platycheles* can survive concentrations from 50% sea water to 150% sea water. When moved to 60% sea water it swells by only 4% and returns to its initial volume within 24 h. However if the ducts of the green glands are blocked with wax it swells by about 10% and dies after about 3 hours. Few other species of crab survive changes of salt concentration as well as *P. platycheles*, but the Chinese mitten crab *Eriocheir sinensis* spends part of its life in fresh water and part in the sea.

HOW MUSCLES WORK JOINTS

Crabs have movable eyes mounted on stalks. Their eye movements have been studied in great detail by Professor Adrian Horridge and others and provide convenient examples of the ways in which muscles work joints. The eye itself is shown black in Fig. 17.10. It is on the end of a cup-shaped piece of rigid cuticle which fits over the tubular eyestalk. The cup is attached to the stalk by a ring of arthrodial membrane which allows a wide variety of movements: the eyecup can be pointed up or down and to one side or the other, and it can be withdrawn over the stalk like a telescope being shortened. The eyestalk is in a hollow under the carapace so withdrawal pulls the eye into a protected position.

Some of the muscles responsible for these movements are shown in Fig. 17.10. Muscle 18 runs between the eyestalk and the main body skeleton and moves the eyestalk. The others run from the eyestalk to the eyecup and move the cup in various directions.

When a man watches a train go by his eyes fix on one carriage and follow it for a while, then flick back to fix on another. Crabs watch moving objects in rather similar fashion. Fig. 17.11 shows apparatus for studying the phenomenon. The crab is fixed to the centre of a turntable covered by a striped drum. Either the turntable or the drum can be rotated and the crab watches the moving stripes. Small holes are bored in the eyecup and fine wire electrodes are pushed through into the muscles to record the electrical activity

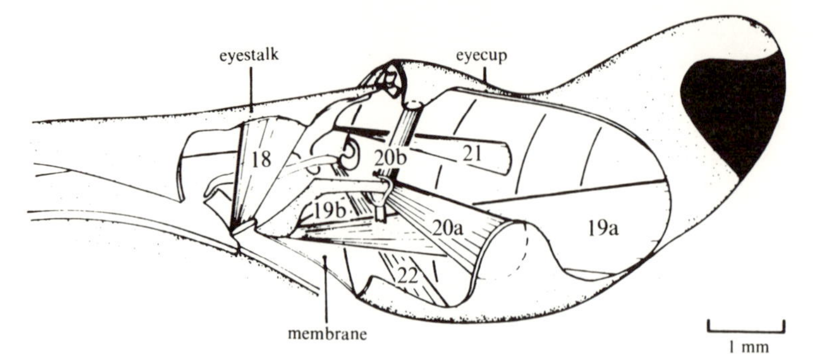

Fig. 17.10. A lateral view of a dissection of the right eyecup of the crab *Carcinus*. Numbers refer to muscles, discussed in the text. From M. Burrows & G. A. Horridge (1968). *J. exp. Biol.* **49**, 285–97.

which occurs when the muscles are active. The crab is killed and dissected after the experiment to make sure that the electrodes are lodged in the intended muscles.

Consider muscle 20a, which attaches to the lateral side of the eyecup and must tend to turn the eye laterally. Fig. 17.12 shows records made from it in apparatus similar to that of Fig. 17.11, but with an additional device for recording the direction in which the eyecup is pointing. In (*a*) the drum was turning laterally, relative to the eye from which the record was made. Muscle 20a is active while the eye is turning slowly, following the drum, but not during the quick flick back. In (*b*) the drum is rotating the other way and the muscle is active only during the quick flick. Muscle 20a is apparently active (as one would expect) whenever the eye is turning laterally. Further experiments showed that muscles 19b and 21 are active when the eye is turning medially. Muscle 22 is used with 20a when the eye is being turned laterally but it is also used when the eye is being turned down. Muscle 20b is used when the eye is turned up. There are three more eyecup muscles which are not illustrated: there are more than enough muscles to turn the eye in any desired direction.

Muscle 19a is the biggest of the eyecup muscles but it is not used when the eye is turned from side to side or up and down. It serves to withdraw the eye, but is assisted by all the muscles shown in Fig. 17.10: electrical activity can be recorded from all of them during withdrawal. The eye is presumably extended again when they relax by the pressure in the haemocoel.

The muscles shown in Fig. 17.10 are small, with few fibres. Muscle 20a, for instance has about 25 fibres. In contrast, consider the very much larger muscles which work the crab's chelae (Fig. 17.13 *a*). The lower claw of each chela is fixed. The upper claw is movable, with knob-and-socket joints on either side at O forming a hinge joint. Muscles attach to it through apodemes, which are plates formed by ingrowth of the cuticle. The plates themselves are stiff but their attachments to the movable claw at P and Q are flexible. Large

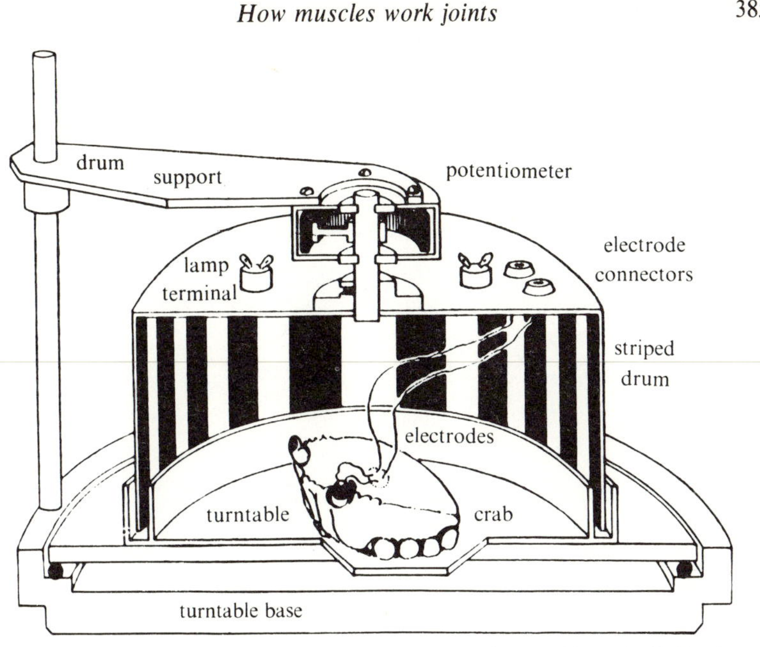

Fig. 17.11. Apparatus for studying the eye movements of crabs. From D. C. Sandeman & A. Okajima (1973). *J. exp. Biol.* **58**, 197–212.

numbers of muscle fibres run obliquely from the exoskeleton to the apodemes (Fig. 17.13 *b*). The muscle which attaches to the upper apodeme opens the claw and the (much larger) muscle of the lower apodeme closes the claw.

The muscles of the crab chela are pennate: that is, their fibres attach obliquely to an apodeme. Many other muscles are parallel-fibred, with all their fibres parallel to each other and to the direction in which the muscle pulls (Fig. 17.13 *c*). What are the merits of the two arrangements? Suppose the crab had parallel-fibred claw muscles. If they filled the same space in the chela their fibres would be about twice as long as in the actual crab (compare Fig. 17.13 *c* with *b*). However there would only be room for about half as many of these long fibres, so that they could only exert half as much force. Suppose the fibres of the pennate muscle exert a total force F. They run at 30° to the apodeme (this angle changes a little as the claw opens and closes) so the component of force along the apodeme is $F \cos 30° = 0.87\,F$. The parallel-fibred muscle with half as many fibres could exert only $0.5\,F$. The pennate arrangement almost doubles the force, in this particular case. It is important for the force to be large as crabs use their chelae for breaking open the shells of the molluscs which they eat.

The fibres must not be too short or the range of movement of the chela will be restricted. If the claw is to be capable of opening and closing through 1 radian (57°) the muscle that closes it must be capable of lengthening and shortening by an amount equal to OQ. It is unlikely to be able to do this unless the

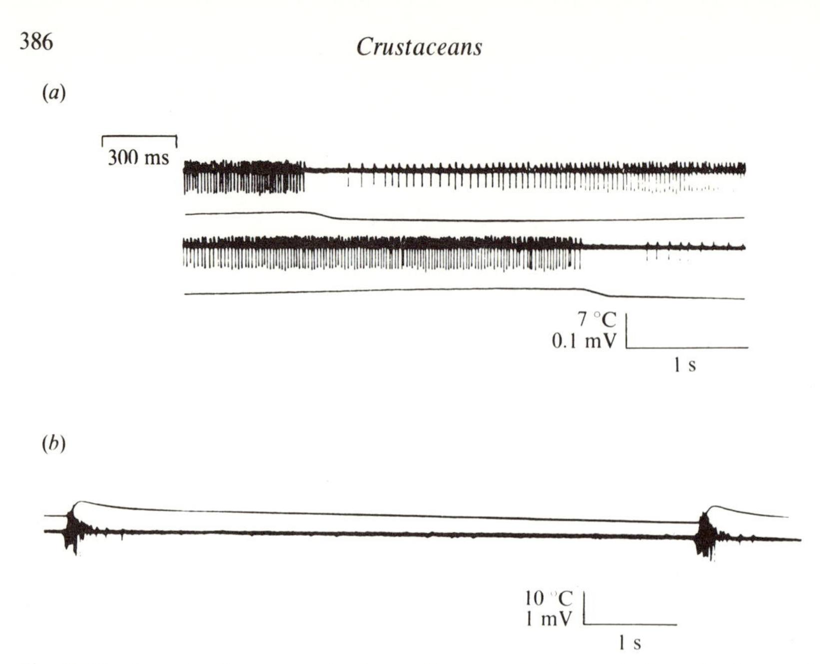

Fig. 17.12. Records of movements of a crab's eye and of electrical activity in muscle 20a of Fig. 17.10. The lines with spikes on them are the records of electrical activity. The other lines show angular displacement of the eye. The crab is watching stripes moving away from the midline in (*a*) and towards it in (*b*). From M. Burrows & G. A. Horridge (1968). *J. exp. Biol.* **49**, 223–50.

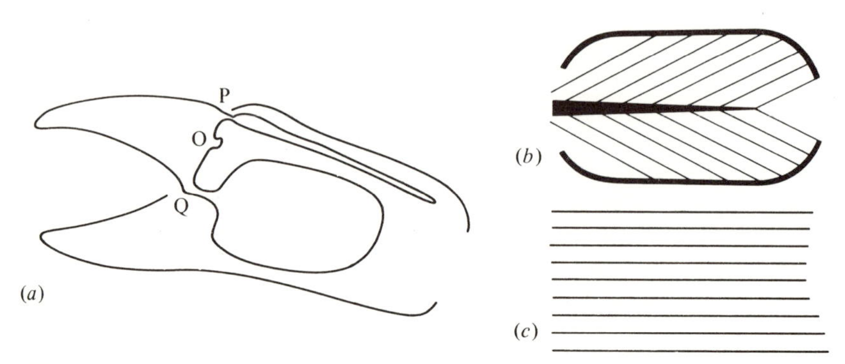

Fig. 17.13. (*a*) The exoskeleton and apodemes of a crab's chela. (*b*) A horizontal section through the chela showing the pennate arrangement of the muscle fibres. (*c*) A parallel-fibred muscle of the same dimensions as the pennate muscle shown in (*b*).

extended length of the fibres is about 2 OQ, or more (see p. 244). The principal leg muscles of crabs are also pennate.

CONTROL OF MUSCLES

Arthropod muscles are all striated. There seem to be no obliquely striated muscles like those of nematodes, molluscs and annelids, nor any unstriated

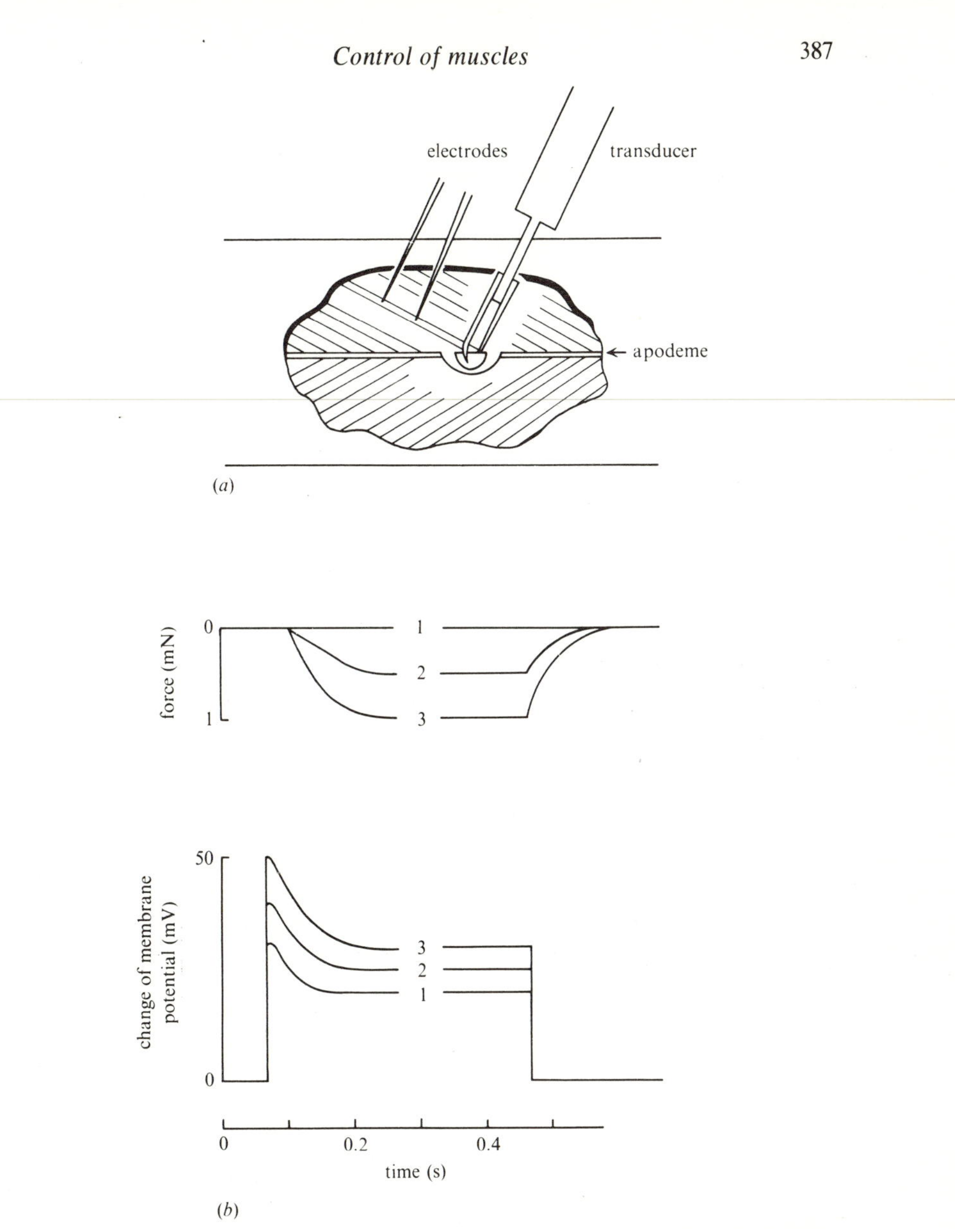

Fig. 17.14. (*a*) Apparatus for measuring the force exerted by a single fibre of a crab leg muscle, in response to changes of membrane potential. (*b*) Records of (above) force and (below) change of membrane potential, from the experiment shown in (*a*). Several records, distinguished by numbers, have been superimposed. After H. L. Atwood, G. Hoyle & T. Smyth (1965). *J. Physiol.* **180**, 449–82.

fibres like those of vertebrate guts and blood vessels. Muscle fibres with different properties are often mixed in the same muscle so experiments with intact muscles are apt to give confusing results. Fig. 17.1 (*a*) shows a technique which has been used to study the properties of individual fibres. A crab leg

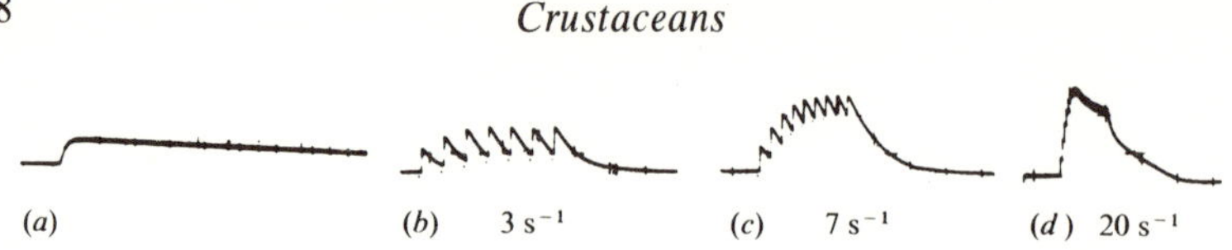

Fig. 17.15. Records of membrane potential in a muscle fibre in a leg muscle of a crab (*Chionecetes*). The motor axon was stimulated by (*a*) a single shock and (*b–d*) shocks at the frequencies shown. Potential is shown on the same scale in each case but the time scales vary. From H. L. Atwood (1965). *Comp. Biochem. Physiol.* **16**, 409–26.

has been cut open to expose a pennate muscle. A small piece of the apodeme has been cut free from the rest and all but one of the attached muscle fibres cut away from it. This piece of apodeme is held in a tiny clamp attached to a force transducer which registers the force when the fibre is stimulated to contract. Two electrodes have been struck into the fibre; current is passed through one to alter the membrane potential of the fibre and the other is used to measure the resulting changes in membrane potential. The fibre only contracts when its membrane potential is increased (i.e. made less negative than its resting value of about −70 mV).

Fig. 17.14(*b*) shows results from experiments with this apparatus. Records from a series of tests are superimposed: the stimulus was stronger in successive tests. A stimulus which raised the membrane potential only slightly caused no contraction. However when the membrane potential was raised by more than about 20 mV the fibre developed tension. The further beyond this threshold the membrane potential was taken, the greater the force registered by the transducer.

Some of the muscle fibres developed tension faster than others, at the same membrane potential. A few transmitted action potentials when the membrane potential passed a threshold. However, the responses shown in Fig. 17.14(*b*) are typical of a great many arthropod muscles. Properties like this make possible a system for controlling muscle force which is quite unlike the system used in mammals.

In mammals each skeletal muscle has a large number of motor axons, each serving a different group of muscle fibres. Each muscle fibre is served by just one axon. If a small force is required, action potentials are sent to the muscle along only a few of the axons and only a few of the muscle fibres are activated. If a large force is required action potentials are sent along all the axons and all the fibres contract.

In arthropods each muscle receives only a few motor axons. Often there is only one motor axon which serves all the muscle fibres, and when there are several they may all serve all the muscle fibres. Fig. 17.15 shows the results of an experiment which indicates how action potentials in a single motor axon can cause different intensities of contraction. The nerve to a crab leg muscle was stimulated by a series of short electric shocks, each of which made an action potential travel along the single motor axon. These set up excitatory post-synaptic potentials (EPSPs) in the muscle fibres. When they came at low frequencies

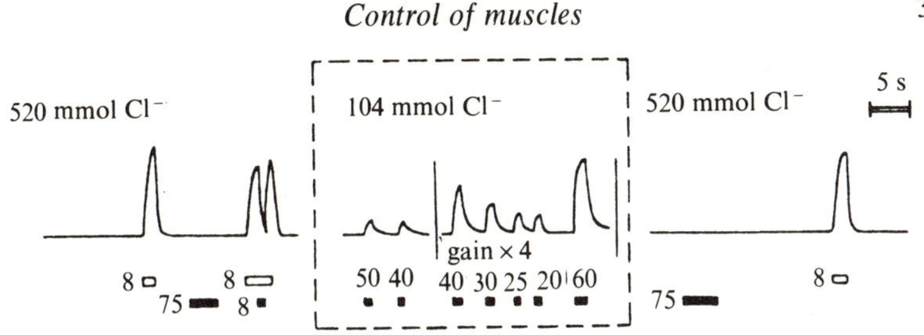

Fig. 17.16. Records of force exerted by a crab leg muscle. The motor axon was stimulated during the periods indicated by white bars and the inhibitory axon during those indicated by black bars. The numbers indicate frequency of stimulation (s^{-1}). The part of the record framed by broken lines was obtained while the muscle was in a solution of low chloride concentration, as explained in the text. From E. Florey & W. Rathmayer (1972). *Pflügers Arch.* **336**, 359–62.

each EPSP had died away by the time the next occurred. At higher frequencies each arrived before the previous one had died away and the membrane potential rose by a staircase effect. The higher the frequency of stimulation (within limits) the higher the membrane potential rose and the greater the force the muscle must have exerted. The crab can control the force by adjusting the frequency of action potentials in the motor axons. In muscles with several motor axons control may be more subtle, with action potentials in different axons producing different sized EPSPs in the same muscle fibre.

As well as motor axons, there are inhibitory axons serving most arthropod muscles. Fig. 17.16 illustrates their effect on a crab leg muscle. Ignore for the present the part of the record enclosed by broken line. Stimulation of the motor axon alone makes the muscle develop tension. Stimulation of the inhibitory axon alone (even at a very high frequency) has no apparent effect. However if the inhibitory axon is stimulated (even at quite low frequency) while the motor axon is being stimulated, the muscle relaxes its tension.

The synapses between motor nerves and muscles work as follows. Action potentials in the axon cause the release of a transmitter substance at the synapse. This substance has an effect on the muscle cell membrane; it increases the permeability of the membrane to sodium ions so the membrane potential rises towards the Nernst potential for sodium (see p. 64). This rise in membrane potential is the EPSP.

Action potentials in an inhibitory axon cause release of a different transmitter substance at the inhibitory synapse. The effect of this substance is to increase the permeability of the muscle cell membrane to chloride, moving the membrane potential towards the Nernst potential for chloride. If this potential is lower (more negative) than −50 mV or so (the threshold for tension development) action potentials in the inhibitory axon will tend to eliminate tension. The concentration of chloride ions in crab muscle cytoplasm is about 54 mEquiv l^{-1} and in crab blood 520 mEquiv l^{-1}, so the Nernst potential for chloride should be about $58 \log_{10} (54/520) = -56$ mV. At the beginning and end of the experiment shown in Fig. 17.16 the muscle was in a saline solution of the same ionic composition as crab blood. In the middle of the experiment (where the record is enclosed by a broken line) it was put in a solution of the same osmotic concentration, but with most of the chloride replaced by propionate. This solution contained only 104 mEquiv l^{-1} chloride so the Nernst potential for chloride of the muscle in it was about $58 \log_{10} (54/104) = -17$ mV, well above the threshold for tension. While the muscle was in this solution stimulation of the inhibitory action made it develop tension.

There is evidence that the transmitter substance at crustacean motor nerve–muscle synapses is glutamic acid ($HOOC.CH_2.CH_2.CHNH_2.COOH$) and at inhibitory ones γ-aminobutyric acid ($HOOC.CH_2.CH_2.CH_2NH_2$). The evidence includes the results of experiments in which minute quantities of the substances were injected into synapses.

WALKING

Nearly all arthropods stand with their feet well out on either side of the body as shown in Fig. 17.17(*a*). This makes them very stable. Suppose a wind or water current acts towards the right of the diagram, exerting a force F on the body. It will tend to bowl the arthropod over, but the weight W will tend to prevent this. The animal will be bowled over when the clockwise moment exerted by F about the downwind feet exceeds the anticlockwise moment exerted by W, i.e. when

$$Fy > Wx \tag{17.1}$$

The force needed to bowl the animal over is Wx/y, so the greater the value of x/y the more stable the animal will be.

A stance with the legs well apart is particularly important for Crustacea which live on shores. The density of a typical crab is about 1150 kg m^{-3} so when it is submerged in sea water (density 1026 kg m^{-3}) the effective weight W available to stabilize it is only 11% of its weight in air (see equation 15.4). Waves and tides are liable to set up quite rapid water movements which may exert large forces tending to bowl the animal over.

Terrestrial arthropods such as insects stand with their feet wide apart but most mammals do not. For instance x/y is typically about 2 for a blowfly and 0.2 for a dog. However, F is the proportional to the area of the animal; for animals of the same shape but different size it will be proportional to (body weight)$^{0.67}$. At the same wind speed F/W will thus be larger for a small animal than a large one. Small animals are more liable to be blown over than large ones unless they stand as arthropods do with their legs well apart. (A wind is slower close to the ground, but not sufficiently so to invalidate the argument.)

Fig. 17.17(*b*) shows *Ligia* walking. Notice how similar the action is to the crawling of *Aphrodite* (Fig. 16.8). An important difference is that *Ligia* runs with only its feet on the ground; its belly is well clear.

Not all crustaceans move their legs in as regular a rhythm as *Ligia*. The

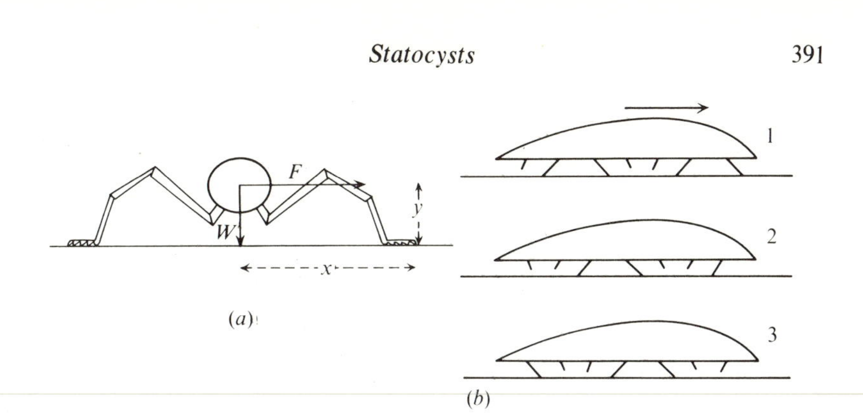

Fig. 17.17. (*a*) A diagrammatic front view of a standing arthropod. Further explanation is given in the text. (*b*) Diagrams of *Ligia* walking, showing three successive positions. Only the right legs are shown.

lobster *Homarus* has eight walking legs, like the crayfish. It moves the legs of each pair alternately, but the order of movement of successive pairs is variable. Crabs run sideways ('crabwise'). Most run at unremarkable speeds but ghost crabs (*Ocypode*) only 2–3 cm across the carapace can attain 2 m s^{-1}. In doing this they make about 17 strides per second, each stride 12 cm long.

STATOCYSTS

Sensory 'hairs' are common on arthropods. They consist of a stiff hair-like projection from the cuticle, hinged at the base, with sensory cells which detect any bending of the hinge. The example shown in Fig. 17.18 is from a statocyst but there are very similar sensory hairs on the external surfaces of arthropods. Three sensory cells are attached to an apodeme-like invagination of the cuticle at the base of the hair. Each cell is attached to this invagination through a process containing a ring of nine double microtubules, which has obviously evolved from a cilium. Cilia modified to various extents occur in the statocysts of medusae (p. 166) and many other sense organs. When the hair in Fig. 17.18 is bent to the left the cilia are stretched but when it is bent to the right their tension is relaxed. These changes of tension change the frequency of action potential in the nerves to the hairs.

It is possible to record action potentials in the axons of the sensory hair cells, in the statocyst nerve of a dissected lobster. Fig. 17.19 shows records from an experiment in which this was done. A single sensory hair was moved and action potentials in one of its sensory cells were recorded from a fine bundle of axons separated from the statocyst nerve. (The electrodes were extracellular and the uniformity of height of the action potentials indicates that all were in the same axon.) Initially the frequency of action potentials was very low. It increased as the hair has bent posteriorly but decreased again slightly when the angle of bending became very large. It decreased to the initial level when the hair was allowed to return to the vertical but was not the same at 35° during

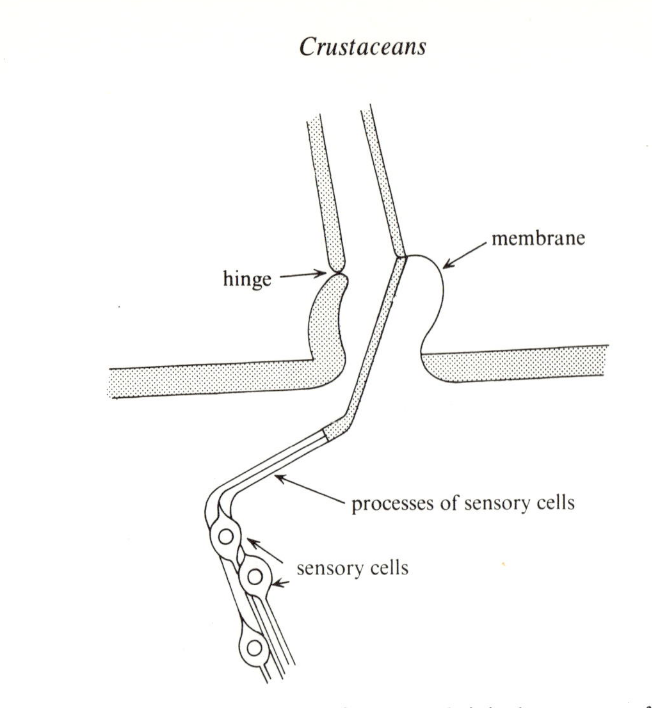

Fig. 17.18. A section through the base of a sensory hair in the statocyst of the crayfish *Astacus*. After H. Schone & R. A. Steinbrecht (1968). *Nature, Lond.* **220**, 184–6.

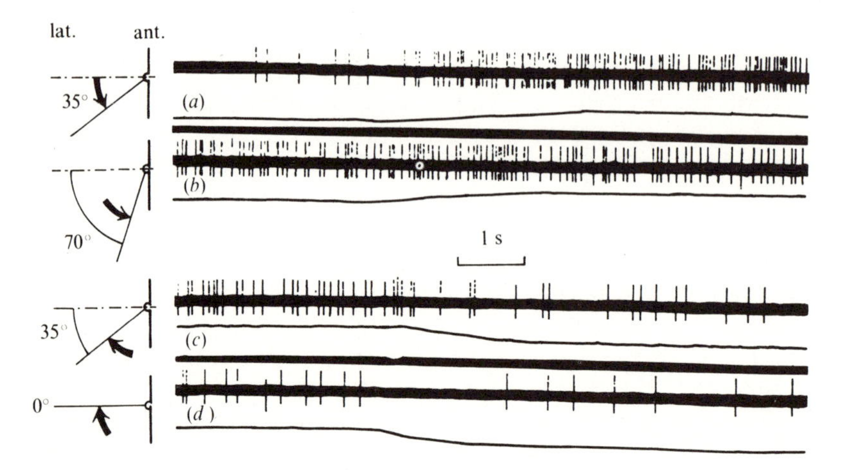

Fig. 17.19. Action potentials from a single sensory cell of a sensory hair in the statocyst of a lobster, *Homarus americanus*. The hair was initially vertical but was bent by means of a glass needle mounted in a micromanipulator. During each of the records *A*, *B*, *C* and *D* it was bent through 35°, as indicated on the left. Several seconds elapsed between each record and the next. The line below the record of action potentials in each case shows the position of the hair. From M. J. Cohen (1960). *Proc. R. Soc.* **152 B**, 30–49.

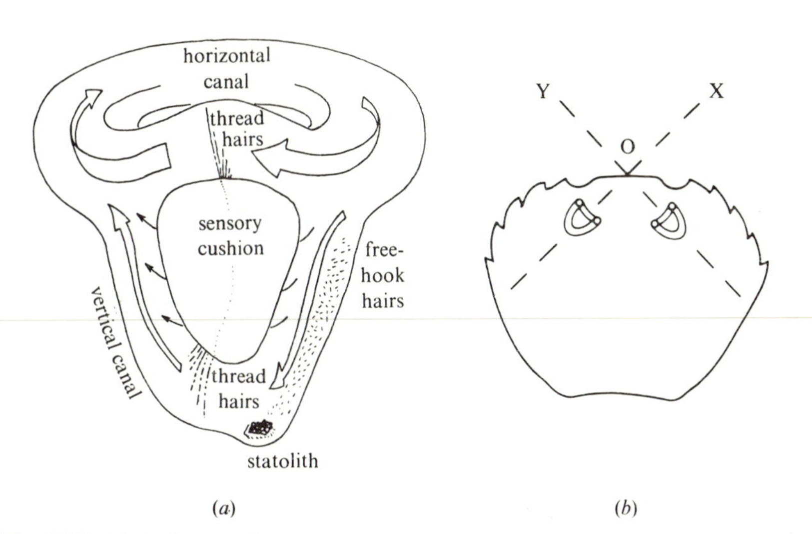

Fig. 17.20. (*a*) A diagram showing the structure of a crab statocyst. The arrows show how the water in it moves when the crab is rotated anticlockwise about a vertical axis. From D. C. Sandeman & A. Okajima (1972). *J. exp. Biol.* **57**, 187–204. (*b*) A diagrammatic plan view of a crab showing the orientation of the statocysts.

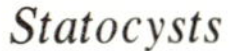

the return, as it had been at 35° during bending. Bending anteriorly decreased the frequency. The relationship between the angle of the hair and the frequency of action potentials is not a simple one, and it is different for different types of sensory hair.

Fig. 17.20(*a*) shows the structure of a statocyst. It is a water-filled cavity open to the sea. Its wall is uncalcified cuticle. There is a horizontal ring (the horizontal canal) and a vertical chamber. A bulge called the sensory cushion almost obliterates the middle of the vertical chamber so that it is, effectively, another ring-shaped canal. At the ventral-most point of the vertical canal is the statolith, which is a clump of sand grains, cemented together, resting on very short sensory hairs. There are other, longer sensory hairs elsewhere in the canals.

The statolith and its sensory hairs form a gravity receptor, like the statocysts of medusae (p. 166). This was demonstrated by an experiment with prawns (*Palaemon*) which was only possible because the statolith is lost and has to be replaced at each moult, when the cuticle lining the statocyst is shed. Iron filings were substituted for sand in the aquarium and the prawns used them when they moulted to replace their statoliths. Therefore they could be made to turn upside-down by holding a magnet over the aquarium so that the direction of gravity seemed to be reversed.

The other sensory hairs detect rotation, like the semicircular canals of vertebrates. When the crab turns, the water in the canals tends to lag behind, because of its inertia. It therefore moves relative to the sensory hairs, tending to bend them. The movement can be seen by dissecting a statocyst out of a crab, injecting it with a drop of dye, and turning it. It has been shown that even very gentle movements of the water alter the frequencies of action potentials in the axons of the sensory cells.

Rotation about a vertical axis causes fluid movement in the horizontal canals (and also in the vertical canals, Fig. 17.20 *a*). The vertical canals are set at 45° to the transverse plane. The left one is set at right angles to the axis OX (Fig. 17.20 *b*) so rotation about OX must set its fluid in motion. Similarly rotation about OY moves the fluid in the right vertical canal. Any rotation can be described as a combination of rotations about OX, OY and the vertical axis OZ, so if the crab can detect the direction and speed of rotation in each canal it can analyse any rotation it may experience.

REFLEXES

When a crab is poked anywhere near an eye, it withdraws the eye to safety. This is a simple reflex due to axons from sense organs on the body surface being connected to a motor neurone which activates muscle 19a (Fig. 17.10). The motor neurone has an exceptionally large axon (up to 50 μm diameter in *Carcinus*) which can be seen under a microscope in intact, unstained crab cerebral ganglia (Fig. 17.21). The cell body and nucleus have not been located.

The eye can be stimulated to withdraw by touch, or by electrical stimulation of the nerve from the antenna or carapace. When this is done action potentials can be recorded in the nerve to the eye, simply by setting this nerve on an electrode. The potentials recorded in this way are small (as action potentials recorded extracellularly always are) but they are bigger than any of the other action potentials recorded from the nerve and so can be presumed to be in the thickest axon. Confirmation that they are in the thickest axon was obtained by using as an electrode a very fine glass capillary filled with potassium ferrocyanide. The nerve to the eye was pierced with this electrode until it was apparent from the record that the tip of the electrode was actually in the motor axon of the withdrawal reflex. Ferrocyanide was then released from the electrode tip. The animal was killed and the brain was treated with ferric chloride to precipitate the ferrocyanide as Prussian blue. It was examined under a microscope and the blue precipitate was found in the thick axon shown in Fig. 17.21 (*a*).

There are probably a very large number of sensory axons which synapse with the big motor neurone and can fire the reflex. The reflex only works if several are stimulated at about the same time. Fig. 17.21 (*b*) shows results of an experiment in which the nerve to the carapace was progressively divided into finer and finer strands, which were stimulated electrically. Recordings were made from an electrode in the nerve to the eye. Some of these records are shown: the big action potentials are from the motor neurone of the withdrawal reflex. The top record was obtained when half the nerve to the carapace was

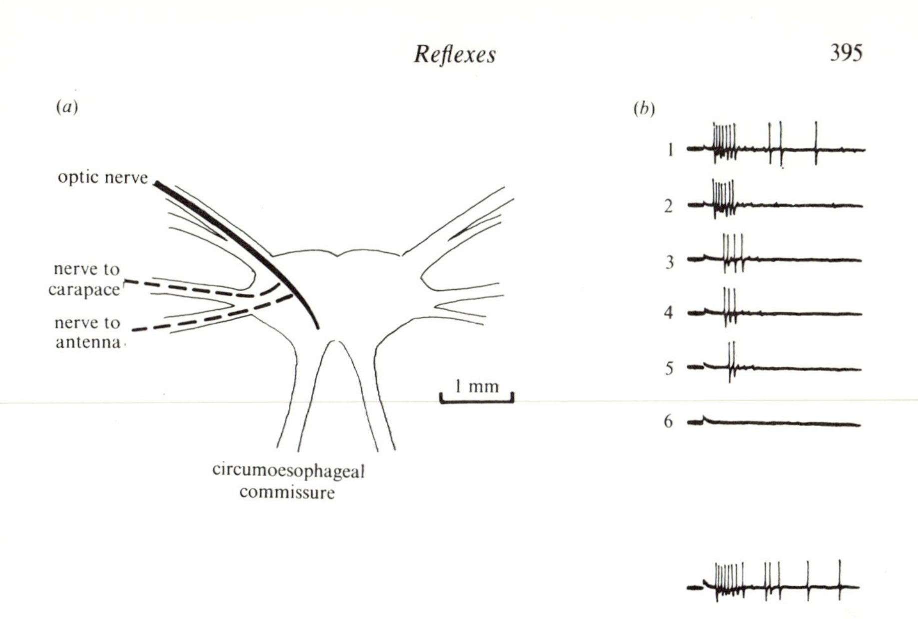

Fig. 17.21. (*a*) A sketch of the cerebral ganglia of a crab showing the large motor axon described in the text and (by broken lines) the paths of sensory axons which synapse with it. (*b*) Records of action potentials in the axon shown in (*a*) in an experiment which is described in the text. From D. C. Sandeman (1969). *J. exp. Biol.* **50**, 87–98.

stimulated. This half was split into two quarters and one of them was stimulated to obtain the second record. For the third to sixth records, successively finer strands were stimulated. Finally, the strands were gathered together to re-form the initial half nerve, and the record at the bottom was obtained. The records show that the finer the strand that is stimulated (and so the fewer the sensory axons that are stimulated) the less the response in the motor axon. No action potentials followed stimulation of the finest strand.

It is believed that the eye withdrawal reflex is a very simple one, in which the sensory neurones synapse directly with the motor neurone. Many other reflexes are more complicated. When the striped drum in Fig. 17.11 is turned, the eyes alternately follow the stripes and flick back. These movements must be controlled by a reflex involving the sensory cells of the eyes. When the drum is stationary but the turntable is turned similar eye movements occur, even in darkness, so there must be another reflex involving (presumably) the sensory hairs of the statocysts. This reflex can be elicited by water movements in the statocyst canals, produced by injecting water into a canal through a pipette. These eye-movement reflexes must require very much more complicated connections between neurones, than the eye-withdrawal reflex. It has not so far been possible to identify the neurones involved or to establish how they are connected.

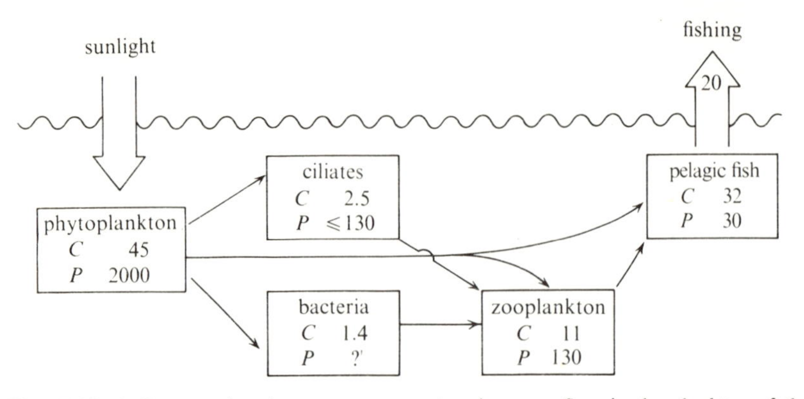

Fig. 17.22. A diagram showing energy content and energy flow in the plankton of the western Mediterranean. Numbers labelled *C* represent energy content expressed in kilojoules per square metre of the sea's surface. Those labelled *P* represent annual production in the same units. Data from a diagram by Margalef reproduced by P. Bougis (1976). *Marine plankton ecology*. North-Holland, Amsterdam.

MARINE ZOOPLANKTON

There is a great variety of planktonic invertebrates in the sea. They include foraminiferan and radiolarian Protozoa, medusae, siphonophores, krill, copepods and the larvae of many invertebrates. However, the copepods are generally predominant both in numbers and in ecological importance. It therefore seems appropriate to discuss the zooplankton in general terms in this chapter.

The zooplankton has of course been mentioned in earlier chapters. Its effects on phytoplankton populations were discussed in chapter 2. The daily vertical migrations of planktonic siphonophores were discussed in chapter 7. Many planktonic crustaceans make similar vertical migrations, so that they are nearer the surface at night than during the day. Some euphausiaceans (krill) move up and down daily through several hundred metres, like the siphonophores of the deep scattering layers. Copepods make much smaller vertical migrations.

The food of copepods can be identified by examining their gut contents and the pellets of faeces which they produce. Many copepods, including the most common members of the marine plankton, are filter feeding herbivores. They feed on diatoms and flagellates down to 3–4 μm long. Some other copepods are predators which take animal prey, including smaller copepods.

The role of the zooplankton in the ecology of the sea is illustrated by Fig. 17.22, which is based on observations made in the Western Mediterranean. The estimate of phytoplankton production is about half the maximum value for unpolluted water given on p. 33. It is primary production, deriving its energy directly from the sun by photosynthesis. The zooplankton production is secondary production which uses as its raw materials the foodstuffs produced by photosynthesis by the phytoplankton. It has been estimated from zooplankton catches together with laboratory measurements of growth rates.

Metabolic rates per unit body mass are generally lower for large organisms than for small ones (Fig. 8.6*b*). Similarly growth rates per unit mass are generally lower for large organisms: a large organism cannot double its mass as fast as a small one. The zooplankton are larger than most of the phytoplankton and do not grow so fast. There is about one quarter as much zooplankton as phytoplankton, but there is far less than a quarter as much zooplankton production as phytoplankton production. Pelagic fish such as sardines feed mainly on zooplankton. They are much larger than the zooplankton and grow correspondingly more slowly so that though there is several times as much fish as zooplankton, fish production is only a small fraction of zooplankton production. A large proportion of it is removed by fishing, but even so the energy content of the fish catch is only about 1% of the phytoplankton production.

SETTLEMENT OF BARNACLE LARVAE

Many of the Crustacea in the plankton are larvae of species which spend their adult life on the shore and sea floor. If they are to mature and flourish as adults they must settle on a suitable patch of sea floor. Crabs and other crustaceans which walk about on the sea floor as adults have some margin for error. Barnacles which settle on a particular spot and stay there must find a good spot at the outset. The planktonic larvae of molluscs and polychaetes must similarly find suitable places to settle, but discussion of the problem has been deferred to this chapter because Professor Dennis Crisp has made a particularly enlightening study of settlement by barnacle larvae.

The barnacle *Balanus balanoides* flourishes only on rock and similar surfaces between the tidemarks. A larva may make contact with the bottom in all sorts of unsuitable places, on sand or mud, at extreme high tide level, or offshore. The most reliable sign that a surface is suitable is that there are barnacles living there already.

In their experiments, Professor Crisp and a colleague offered barnacle larvae small rectangles of slate to settle on. Hollows were ground in the slates to make them more attractive. Slates treated in different ways were laid out round the perimeter of a large trough of sea water. Barnacle larvae were released into the sea water. The trough was rotated slowly so that factors such as uneven lighting should not make the larvae congregate on one side. After a few hours the slates were removed and the numbers of young barnacles which had settled on each were counted.

Before the experiments, all the slates were thoroughly clean. Some were then soaked for a few hours before use in an extract of barnacles, made by chopping and grinding adult barnacles and removing the solid debris by filtration. In a series of experiments larvae were given the choice of equal numbers of clean slates and ones treated with barnacle extract. One hundred settled on the clean slates and 2197 on the treated ones. Some substances present in adult barnacles apparently encourages settlement.

Further experiments showed that extracts of the mollusc *Mytilus* had little or no effect on settlement but that extracts of crabs (*Carcinus*) and cockroaches (*Blaberus*) were very effective. The substance which encourages settlement is apparently present in arthropods in general, not just in barnacles. However, it is not identical in all arthropods: the barnacles *Balanus* and *Elminius* are each more sensitive to extracts of their own species than to extracts of the other. This has a possible advantage, for *Elminius* flourishes in more sheltered situations than *Balanus*.

Extracts of whole crab or of crab carapace favour settlement but extracts of crab viscera, blood or muscle do not. The active substance must be present in the cuticle. It was shown that its molecule was too large to pass through a cellulose filter with pores of diameter 5 nm. It is probably a cuticle protein.

Some other planktonic larvae find suitable sites for settlement by other means. For instance, *Spirorbis borealis* is a polychaete which spends its adult life like *Serpula* (Fig. 16.4*c*) in a permeability fixed tube, which it builds on a frond of the seaweed *Fucus*. The larvae seems to be able to identify *Fucus* by chemical means.

FILTER FEEDING

Many small Crustacea get their food by filter feeding. They include Branchiopoda such as *Cheirocephalus* and *Daphnia*, most planktonic Copepoda, the Cirripedia (barnacles) and Malacostraca such as *Thysanoessa*. The mechanisms are quite different from those used by bivalve molluscs: the necessary water currents are driven by limb movements and the food particles are strained out by setae.

The mechanism of filter feeding is most easily studied in barnacles, since they do not move about as they feed. Fig. 17.23 shows the barnacle *Balanus* making a cycle of normal feeding movements. This barnacle has six pairs of limbs, three short and three long, all fringed by setae. At the beginning of the cycle (frame 1) the barnacle has its operculum closed with all the limbs hidden below it. The operculum opens and the three pairs of long limbs emerge (frame 9). They unroll until they are fully extended and inclined posteriorly (frame 25). The unrolling is probably caused by pressure in the haemocoel, for there are no muscles in the limbs themselves which could have this action. Next the limbs move rapidly forward, rolling up again, and the operculum closes over them (frame 39).

In their forward beat the long limbs are moving fast with their setae extended. Any particles in their path are apt to be caught, and are passed to the mouth and swallowed. Only fairly large particles can be caught in this way because the setae are fairly widely spaced. The limb movements drive a current of water slowly over the barnacle so that even in still water the limbs do not filter the same water in successive beats.

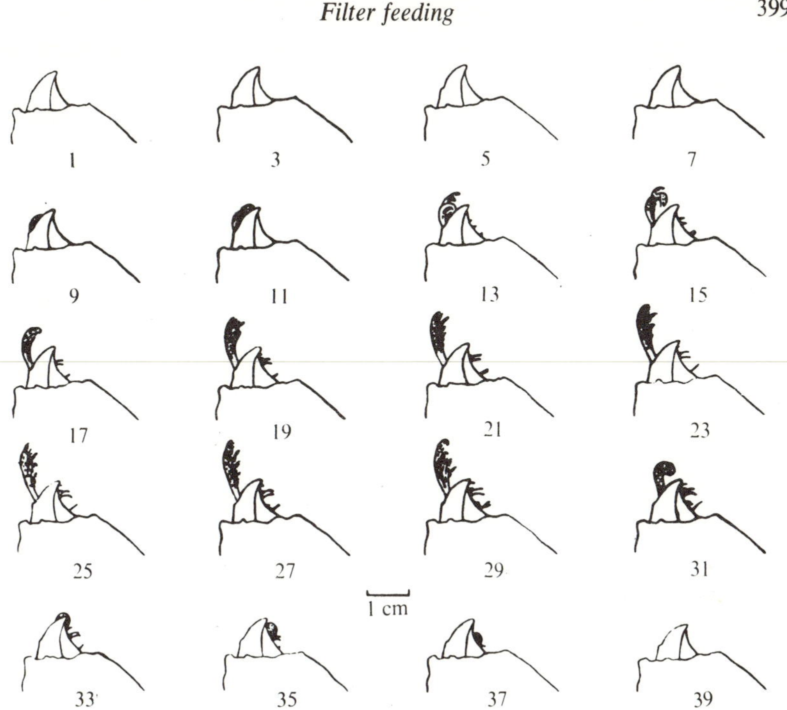

Fig. 17.23. Feeding movements of *Balanus balanus*. These outlines were traced from alternate frames of a cine film taken at 16 frames per second. From D. J. Crisp & A. J. Southward (1961). *Phil. Trans. R. Soc.* **243B**, 271–307.

The current driven by the limbs can be observed by releasing drops of milk into the water near a barnacle. This technique shows that as well as the main flow over the barnacle there is some flow through the opercular cavity. Opening the operculum enlarges the cavity and draws water in anteriorly. Closing it drives water out posteriorly. As the water moves through the opercular cavity it passes between the setae of the three pairs of short limbs. These setae are spaced only 1–2 μm apart so that they can filter out very small particles such as bacteria and flagellate protozoa. In each beat a relatively large volume of water is coarsely strained by the long limbs and a smaller volume is more finely strained by the short limbs.

The rate of filtration has been measured for several species of barnacles, in the same way as for bivalve molluscs (p. 310). One series of experiments used suspensions of crustacean nauplii, 0.3 mm or more long. *Balanus balanoides* filtered these suspensions at rates of about 10 cm^3 h^{-1} per barnacle, or about 100 cm^3 (g soft tissue)$^{-1}$ h^{-1}. This is rather less than the usual filtration rate of the bivalve *Mytilus* (250 cm^3 h^{-1}, see p. 311). The natural foods of *Balanus* range from flagellate Protozoa to small Crustacea.

FURTHER READING

GENERAL

Green, J. (1961). *A biology of Crustacea.* Witherby, London.

Hughes, G. M., Knights, B. & Scammel, C. A. (1969). The distribution of P_{O_2} and hydrostatic pressure changes within the branchial chambers in relation to gill ventilation of the shore crab *Carcinus maenas* L. *J. exp. Biol.* **51**, 203–20.

Lockwood, A. P. M. (1968). *Aspects of the physiology of Crustacea.* Oliver & Boyd, Edinburgh.

Waterman, T. H. (1960). *The physiology of Crustacea.* Academic Press, New York & London.

THE EXOSKELETON

Currey, J. D. (1967). The failure of exoskeletons and endoskeletons. *J. Morph.* **123**, 1–16.

Dalingwater, J. E. (1975). SEM observations on the cuticles of some decapod crustaceans. *Zool. J. Linn. Soc.* **56**, 327–30.

Davenport, J. (1972). Volume changes shown by some littoral anomuran Crustacea. *J. mar. biol. Assoc. UK* **52**, 863–78.

Joffe, I., Hepburn, H. R., Nelson, K. J. & Green, N. (1975). Mechanical properties of a crustacean exoskeleton. *Comp. Biochem. Physiol.* **50A**, 545–9.

Welinder, B. S. (1974–5). The crustacean cuticle: I to III [3 papers]. *Comp. Biochem. Physiol.* **47A**, 779–87; **51B**, 409–16; **52A**, 659–63.

HOW MUSCLES WORK JOINTS

Burrows, M. & Horridge, G. A. (1968). The action of the eyecup muscles of the crab *Carcinus* during optokinetic movements. *J. exp. Biol.* **49**, 223–50.

CONTROL OF MUSCLES

Atwood, H. L. (1972). Crustacean muscle. In *The structure and function of muscle*, 2nd edn, vol. 1, ed. G. H. Bourne, pp. 421–89. Academic Press, New York & London.

Florey, E. (1975). The integrative capacity of chemical transmission at arthropod neuromuscular synapses. In *Simple nervous systems*, ed. P. N. R. Usherwood & D. R. Newth, pp. 323–41. Arnold, London.

WALKING

Alexander, C. G. (1972). Locomotion in the isopod crustacean *Ligia oceanica* (Linn.). *Comp. Biochem. Physiol.* **42A**, 1039–47.

Burrows, M. & Hoyle, G. (1973). The mechanism of rapid running in the ghost crab *Ocypode ceratophthalma*. *J. exp. Biol.* **58**, 327–49.

Macmillan, D. L. (1975). A physiological analysis of walking in the American lobster (*Homarus americanus*). *Phil. Trans. R. Soc.* **270B**, 1–59.

Manton, S. M. (1953). Locomotory habits and the evolution of the larger arthropodan groups. *Symp. Soc. exp. Biol.* **7**, 339–76.

STATOCYSTS

Cohen, M. J. (1960). The response patterns of single receptors in the crustacean statocyst. *Proc. R. Soc.* **152B**, 30–49.

Sandeman, D. C. & Okajima, A. (1972–3). Statocyst-induced eye movements in the crab *Scylla serrata*: I and II [2 papers]. *J. exp. Biol.* **57**, 187–204; **58**, 197–212.
Schone, H. & Steinbrecht, R. A. (1968). Fine structure of statocyst receptors of *Astacus fluviatilis*. *Nature, Lond.* **220**, 184–6.

REFLEXES

Sandeman, D. C. (1969). The synaptic link between the sensory and motoneurones in the eye-withdrawal reflex of the crab. *J. exp. Biol.* **50**, 87–98.
Sandeman, D. C. (1969). The site of synaptic activity and impulse initiation in an identified motoneurone in the crab brain. *J. exp. Biol.* **50**, 771–874.

MARINE ZOOPLANKTON

Bougis, P. (1976). *Marine plankton ecology*. North-Holland, Amsterdam.

SETTLEMENT OF BARNACLE LARVAE

Crisp, D. J. & Meadows, P. S. (1962). The chemical basis of gregariousness in cirripedes. *Proc. R. Soc.* **156B**, 500–20.
Crisp, D. J. & Meadows, P. S. (1963). Adsorbed layers: the stimulus to settlement in barnacles. *Proc. R. Soc.* **158B**, 364–87.

FILTER FEEDING

Crisp, D. J. & Southward, A. J. (1961). Different types of cirral activity of barnacles. *Phil Trans. R. Soc.* **243B**, 271–307.

18
Insects in general

Phylum Arthropoda (cont.), Class Insecta

Two chapters of this book are allotted to insects. About three-quarters of all known species of animal are insects, so two chapters out of 26 may seem a meagre allocation. However, the insects are all fairly similar to each other, which is why they are all included in the same class. There seems to me to be no more variety within the insects than within other large classes such as the Gastropoda and Crustacea.

Because the class Insecta is so large and important these chapters have been arranged differently from the chapters on other groups. This chapter consists of accounts of the structure of typical insects, and of some of the major topics in general insect physiology. Chapter 19 gives brief descriptions of some of the larger orders of insects, and accounts of some aspects of their biology.

Fig. 18.1 shows the external appearance of a locust. The long hind legs are a specialization for jumping but in other respects the locust is a reasonably typical insect. Its body is enclosed in an exoskeleton constructed on the same principles as the exoskeletons of Crustacea. It consists of stiff plates or tubes joined by flexible arthrodial membranes. Except in the head, each segment has a dorsal and a ventral plate. The legs are enclosed in skeletal tubes jointed together in the same way as in crabs (Fig. 17.9). However, the exoskeleton is not impregnated with calcium salts: it consists simply of chitin and tanned protein with a waxy outer layer.

The body has three regions: head, thorax and abdomen. Embryology has shown that the head develops from six segments which correspond to the first six segments of Crustacea (Table 17.1), but the adult head bears only four pairs of modified limbs. These are the antennae (segment 2) and three pairs of mouthparts (segments 4 to 6). The head also bears a pair of compound eyes and a few simple eyes (ocelli). A later section of the chapter is concerned with the optics of compound eyes. The thorax has three segments, labelled I to III. Each of the three bears a pair of walking legs and the second and third each bear a pair of wings. The abdomen has 11 segments and no limbs except for the small cerci of the last segment.

When they walk, most insects move their legs in groups of three (the first and third legs of one side and the second leg of the other). This ensures that there are always three feet on the ground, forming a stable tripod.

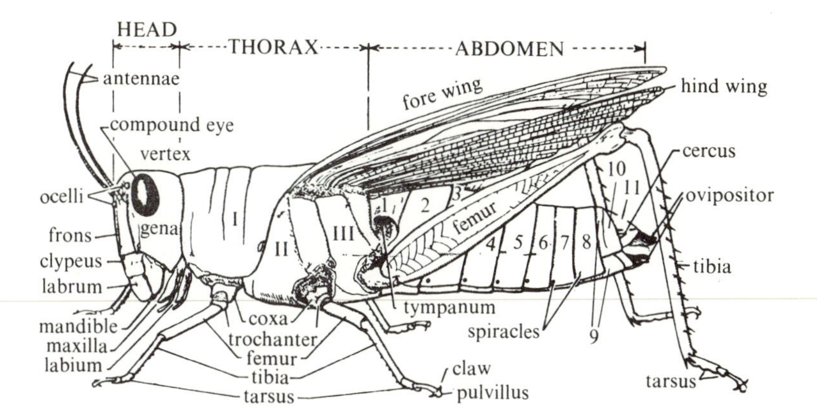

Fig. 18.1. A diagram showing the main external features of a locust. From T. I. Storer & R. L. Usinger (1957). *General zoology*. McGraw-Hill, New York.

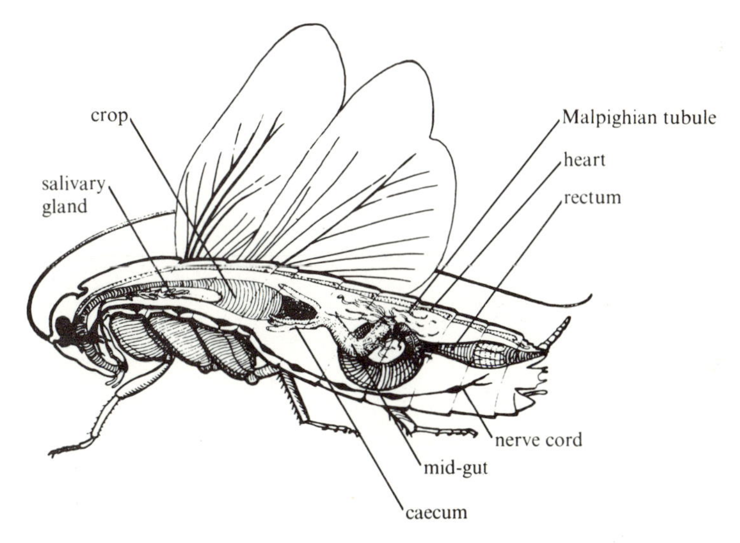

Fig. 18.2. The internal anatomy of a cockroach. After V. B. Wigglesworth (1964). *The life of insects*. Weidenfeld & Nicolson, London.

Fig. 18.2 shows the internal arrangement of a cockroach, another reasonably typical insect. (Locusts and cockroaches are very similar in internal structure.) Notice the similarity to crayfish (Fig. 17.1). The main body cavity is a haemocoel. There is a nerve cord ventral to the gut and a heart dorsal to it. The nerve cord has a ganglion in each segment of the thorax and abdomen. In the head it encircles the oesophagus: there is a brain dorsal to the oesophagus and a sub-oesophageal ganglion ventral to it. Many insects have

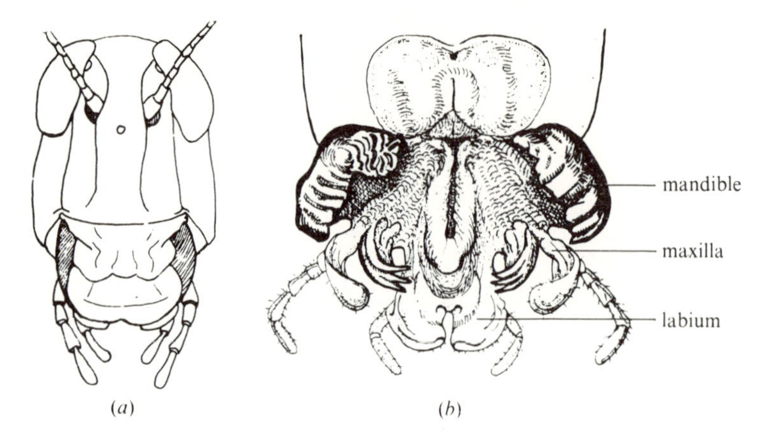

Fig. 18.3. Mouthparts of a grasshopper. (*a*) An anterior view of the head. (*b*) The labrum has been folded dorsally and the mouthparts separated to expose the mouth. After V. B. Wigglesworth (1964). *The life of insects.* Weidenfeld & Nicolson, London.

a short heart like the crayfish but the cockroach has a heart almost as long as the thorax and abdomen, with a pair of ostia in each segment. There is only one artery, which runs forward from the heart and opens into the haemocoel of the head.

We will examine the mouthparts (Fig. 18.3) before the gut. The most anterior mouthparts are a pair of biting mandibles. These are simple blades, with no distal joints like those of crustacean mandibles (Fig. 17.2). Next are a pair of complicated maxillae which are used for manipulating food. Finally there is the labium which is a pair of second maxillae fused together to form a lower lip. There is a pair of large salivary glands (Fig. 18.2) which discharge saliva through an opening in the labium. The saliva contains an enzyme which digests starch, and is spread on the food before it is ingested. Cockroaches feed on crumbs of food and other waste matter left around by man: the natural food is probably dead animals. Grasshoppers and locusts, with very similar guts and mouthparts, eat vegetation.

The anterior part of the gut is a large crop where food is stored and some initial digestion takes place. The food in the crop is mixed with saliva and also with other digestive enzymes which get squeezed forward from the mid-gut. The crop is followed by the gizzard which is muscular and has its lining of cuticle thickened to form teeth. Food is broken up in the gizzard, by the action of the muscle and teeth. The mid-gut is the only part of the gut which is not lined by cuticle, but even here the epithelium is protected from abrasion by a loose chitinous membrane. Enzymes secreted by the mid-gut epithelium diffuse freely through this membrane into the lumen, and products of digestion diffuse out through the membrane to be absorbed by the epithelium. In the rectum, water and inorganic ions are absorbed from the gut contents before they are passed as faeces.

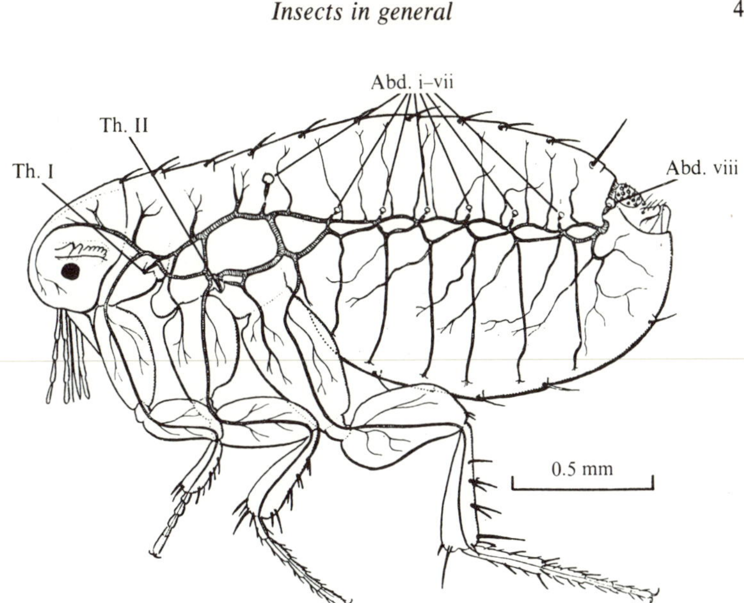

Fig. 18.4. The tracheal system of a flea (*Xenopsylla*). Th I, II, thoracic spiracles; Abd. i–viii, abdominal spiracles. From V. B. Wigglesworth (1935). *Proc. R. Soc.* **118B**, 397–419.

At either end of the mid-gut, blind-ended tubes are attached. The relatively thick caeca at the anterior end are simply extensions of the mid-gut surface: their epithelium secretes enzymes and absorbs products of digestion. The much thinner Malpighian tubules are excretory organs. Their function and the function of the rectum are discussed later in the chapter, in the section on water balance.

Some quite conspicuous organs are omitted from Fig. 18.2. They include the fat body, a diffuse gland dorsal to the gut. In function it parallels the mammalian liver. Globules of fat and granules of glycogen and protein are stored in its cells. It is the major site of intermediary metabolism. For instance, it converts glucose absorbed from the gut to the disaccharide trehalose which is the principle sugar in the blood and one of the principal fuels for flight.

Fig. 18.2 also omits the muscles, the respiratory system and the gonads. Among the muscles, the ones which flap the wings are particularly large. They are described later in the chapter. The respiratory system is a branching system of air-filled tubes, quite different from anything encountered in earlier chapters. The tubes are called tracheae, and open to the atmosphere through holes called spiracles in the sides of the thorax and abdomen. Fig. 18.1 shows the spiracles of a grasshopper and Fig. 18.4 shows the arrangement of the tracheal system of a flea. The tracheae permeate all the tissues of the body. They are lined by cuticle which is shed when the insect moults. This cuticle is strengthened by thickenings, either rings or helices, which prevent the tracheae from collapsing while leaving them flexible. The finest branches of the tracheae, called tracheoles, have diameters of the order of 1 μm.

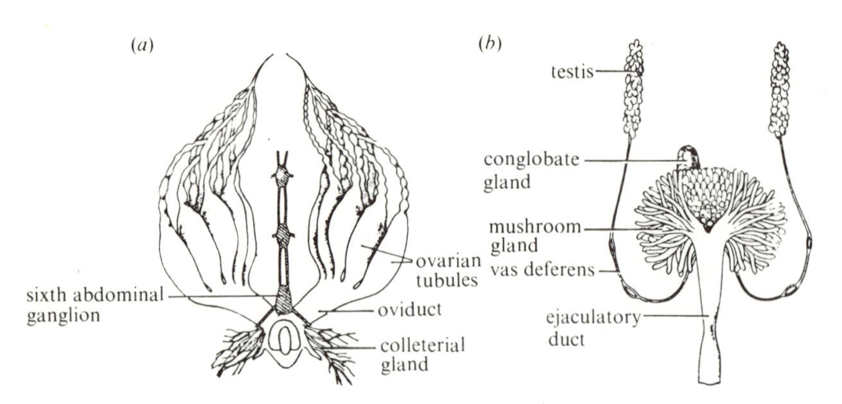

Fig. 18.5. Female (*a*) and male (*b*) reproductive organs of the cockroach (*Periplaneta*). From G. Chapman & W. B. Barker (1966). *Zoology for intermediate students.* Longman, London.

Oxygen diffuses from the atmosphere through the spiracles and tracheae to the tracheoles, and from them to the tissues. In many insects parts of the tracheae are enlarged to form thin-walled collapsible sacs, and diffusion is supplemented by pumping. The capacity of the tracheal system to supply oxygen to the tissues and particularly to the flight muscles is discussed later in the chapter.

The spiracles of most insects are fitted with valves which can be opened and closed by muscles, as required to control airflow or restrict water loss.

Male insects have a pair of testes and females a pair of ovaries (Fig. 18.5). They lie in the haemocoel of the abdomen and ducts lead from them to a single external opening near the posterior end of the abdomen. The reproductive organs include accessory glands as well as the gonads themselves.

Cockroaches lay their eggs in batches of 12 or more enclosed in an egg capsule. The capsule is soft and white when first formed, but later becomes hard and dark and eventually almost black. It is secreted by the female accessory glands (the collaterial glands, Fig. 18.5). The left gland is the larger. Its contents are white and can be kept indefinitely without change of colour, but they darken when mixed with the (clear) fluid from the right gland, or with compounds such as 3,4-dihydroxybenzoic acid

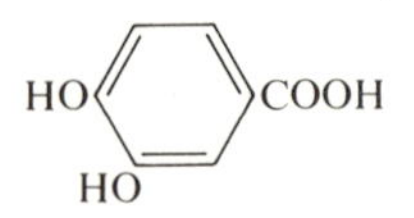

Analysis has shown that the glands contain the following substances:

left	*right*
protein	glucosidase
glucoside of 3,4-dihydroxybenzoic acid	
phenol oxidase	

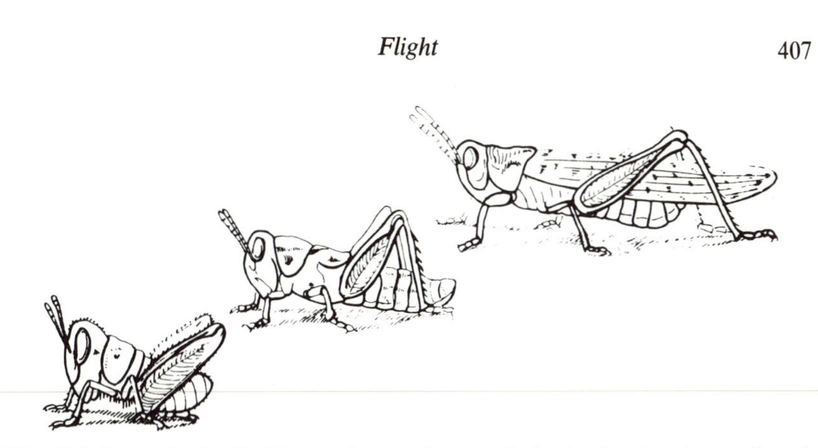

Fig. 18.6. Stages in the life history of a grasshopper. Left, the first larval stage (length 3 mm); centre, an intermediate larval stage (12 mm); right, the adult (30 mm). From V. B. Wigglesworth (1964). *The life of insects.* Weidenfeld & Nicolson, London.

When the contents of the two glands are mixed the glucosidase breaks down the glucoside into its components, glucose and 3,4-dihydroxybenzoic acid. The acid is oxidized by the phenol oxidase to a quinone which tans the protein as explained in the account of gorgonin on p. 142. This was the first case of quinone tanning in nature to be recognized and explained. Insect cuticle consists of tanned protein and chitin but cockroach egg capsules consist of tanned protein alone.

Though cockroaches lay their eggs in batches, many other insects lay their eggs singly.

Insects hatch from their eggs as larvae or nymphs. They moult several times, growing between moults. At each moult their structure changes to a greater or lesser extent until they eventually reach the adult stage: thereafter, they do not moult. In some insects such as locusts (Fig. 18.6) the changes from stage to stage are relatively small, in which case the young stages are called nymphs. Each nymphal stage is a little different from its predecessor in shape and only the adult has fully developed wings and reproductive organs. In many other insects there is a profound metamorphosis: the adult looks entirely different from the larvae. These insects include the butterflies and moths (Lepidoptera) and the flies (Diptera). A caterpillar looks quite unlike the adult butterfly and a maggot looks quite unlike the adult fly. In such extreme cases an inactive non-feeding stage (the pupa) intervenes between the final larval stage and the adult. During the pupal stage the whole insect is remodelled.

FLIGHT

Insects are the only invertebrates which can fly. Their wings develop as outgrowths of the body wall and so consist of two cuticle-covered membranes, back to back. Most of the area of the wing is very thin but it is stiffened by

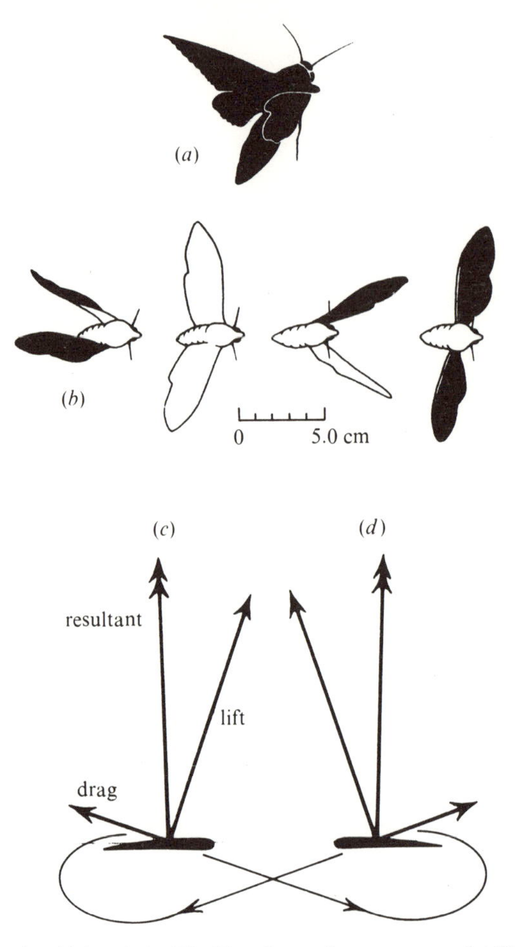

Fig. 18.7. Hovering flight of the Florida tobacco hornworm moth, *Manduca sexta*. (*a*) was drawn from a flash photograph. (*b*) was traced from a cine film and shows the moth from above; successive outlines show it at intervals of 12 ms. The ventral surfaces of the wings are shown black. (*c*) and (*d*) show a section through the wings and the forces which act on them in their forward and backward strokes, respectively. (*a*) and (*b*) are from T. Weis-Fogh (1973). *J. exp. Biol.* **59**, 169–230.

a network of tubular veins. The cavities of the veins are connected to the haemocoel and are filled with blood. They also contain tracheae and sometimes nerves. Even the veins are quite thin, too thin to make a flat wing sufficiently stiff. Insect wings are lightly pleated, with pleats running from the base of the wing towards the tip. This stiffens them, just as a sheet of paper can be stiffened by pleating.

Many insects hover like helicopters, keeping themselves stationary in mid

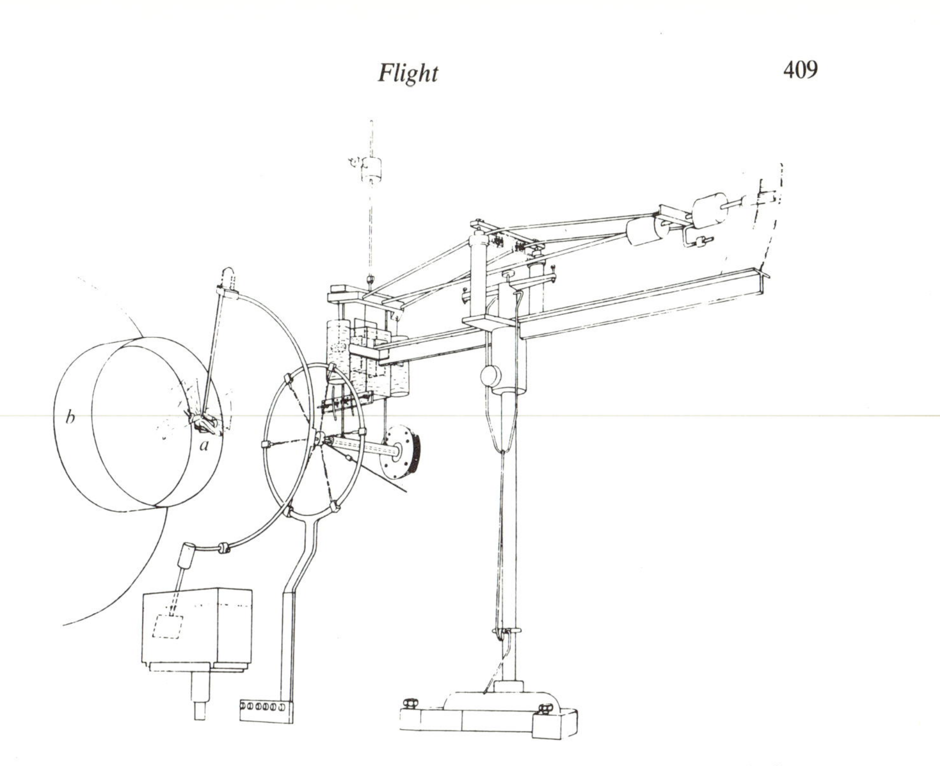

Fig. 18.8. A locust flying in Weis-Fogh's apparatus. The locust *a* is suspended in front of the nozzle *b* of a wind tunnel. From T. Weis-Fogh (1956). *Phil. Trans. R. Soc.* **239B**, 459–510.

air by beating their wings. Moths such as *Manduca* (Fig. 18.7) hover in front of flowers at night, extending their long tongues into them to feed on the nectar. Fig. 18.7(*a*) shows a hovering *Manduca* in side view and Fig. 18.7(*b*) shows one from above. The film (*b*) was taken in a laboratory, of a moth taking sugar solution from an artificial flower. The camera had to be run at a high framing rate because *Manduca* beats its wings 28 times per second.

The pictures show that when *Manduca* hovers, it beats its wings forward and back in a horizontal plane. The wings have their dorsal side uppermost in the forward stroke but turn upside-down so that their ventral surfaces (shown black in (*b*)) are uppermost in the backward stroke. Fig. 18.7(*c*) shows how this keeps the insect airborne. The wings act as aerofoils so lift and drag forces act on them (see p. 329). Both in the forward stroke and in the backward one they have an angle of attack such that the resultant of lift and drag acts upwards. This effect could be achieved without turning the wings upside-down, but turning them over ensures that the anterior edge of the fore wing is the leading edge for the backward stroke as well as for the forward one. This edge is stiffer than the remainder of the wings because it has closely-spaced veins and rather deep pleating. If the leading edge were not stiff the wing would flutter uncontrollably. This technique of hovering is similar in principle to the hovering of helicopters. However, a helicopter rotor keeps turning in the same direction while insect wings move alternately forward and back.

When a helicopter is hovering the rotor needs only to produce a vertical force. When the helicopter flies forward drag acts on the fuselage so the force produced by the rotor must have a forward component. This is obtained by tilting the rotor forward. Similarly insects change from hovering to forward flight by altering the plane of beating of their wings. It is even more difficult to film insects in free flight than to film them hovering, and many studies of insect flight depend on making the air move while the insect is stationary. Fig. 18.18 shows how this was done in apparatus designed and used by Professor Torkel Weis-Fogh. The insect is stuck by wax to the end of a wire which is pivoted at the top. The insect is, in effect, the bob of a pendulum. Its feet cannot reach any solid support. In such circumstances many insects beat their wings as if they were flying. The insect hangs in a jet of air blown from the nozzle of a wind tunnel. The tunnel is designed so that the velocity of the air is as nearly uniform as possible, across the whole width of the nozzle, and so that there is as little turbulence as possible.

In free flight at constant velocity the drag on the body would be precisely balanced by the forward thrust produced by the wings, so that the net horizontal force would be zero. This is imitated in the experiment by making the movements of the pendulum control the speed of the electric fan which drives the air through the wind tunnel. If the insect is blown backwards the speed of the fan is automatically reduced, but if the insect pulls the pendulum forward the speed is increased. An electromagnet compensates for the drag on the wire supporting the insect. Thus the speed of the jet is adjusted so that the drag on the insect matches the thrust produced by its wings.

In free flight the weight of the insect would be exactly balanced by the upward forces produced by the wings, so the net vertical force would be zero. This condition cannot be obtained automatically in the apparatus but the pendulum hangs from a balance which measures the net vertical force. When it registers zero the forces on the insect are the same as if it were free, flying forward in still air at the speed of the wind tunnel jet.

Fig. 18.9 is based on films of blowflies taken in this way. Fig. 18.9(*a*) shows that the wings move forward and down and then backward and up relative to the fly's body. At the same time the air is moving backwards past the fly, equivalent to the fly moving forward through the air. Hence the path of a wing tip relative to the air is sinuous, as shown in Fig. 18.9(*b*). The wings twist so that the anterior edge leads in the upstroke as well as in the downstroke. They probably produce an upward and forward force, like a tilted helicopter rotor, in both strokes. Forces probably act briefly in other directions while the wings are twisting at the top and bottom of their strokes.

Movements like this are typical of the forward flight of small insects which beat their wings at high frequencies. Large insects such as locusts beat their wings at lower frequencies, and their movements in forward flight are more like those of birds.

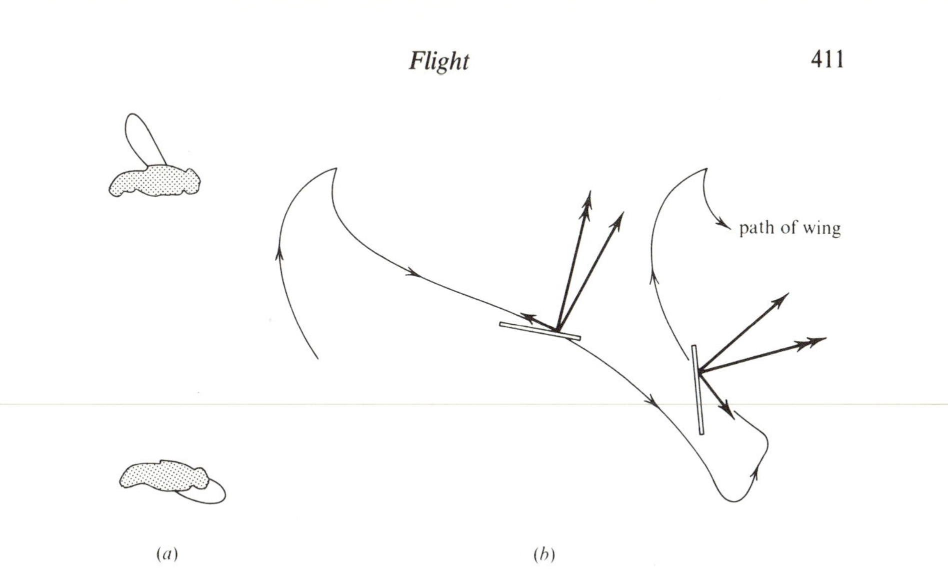

Fig. 18.9. Simulated forward flight of a blowfly (*Phormia*) in a wind tunnel. The fly was making 120 cycles of wing movements per second and the speed, to which the tunnel adjusted automatically, was 2.75 m s^{-1}. It was filmed by a high-speed cine camera. (*a*) shows the positions of the wings at the top of the upstroke and the bottom of the downstroke. (*b*) shows the path of the wing tips through the air, with sections through a wing in the middle of the downstroke and the middle of the upstroke. Based on illustrations in W. Nachtigall (1966). *Z. vergl. Physiol.* **52**, 155–211.

Aeroplane wings and helicopter blades have relatively thick, streamlined cross-sections, as in Fig. 15.10(*a*). Insect wings are thin pleated membranes. A magnified insect wing would perform very badly indeed as an aeroplane wing or a helicopter rotor. However, these aerofoils operate at much higher Reynolds numbers than insect wings. Tests with models have shown that at the Reynolds numbers at which insect wings operate, pleated membranes are just as good as smooth streamlined aerofoils. They have the advantage that they can be made lighter: the importance of lightness will soon become apparent.

How much power is needed for flight? The usual way of measuring the power consumption of an animal which depends on aerobic metabolism is to measure the rate at which it uses oxygen. Professor Weis-Fogh found it more convenient to measure the rate of heat production in his studies of the metabolism of flying insects. Muscle has rather low efficiency. Only 25% or less of the metabolic energy it uses is converted to mechanical work and the rest is lost as heat. Thus the heat released gives a good indication of the metabolic energy used.

The flight muscles are in the thorax. When an insect flies the thorax heats up until the rate of loss of heat from it equals the rate of production of heat by the muscles. Professor Weis-Fogh made a small hole in the thorax of an insect and inserted a thermistor. He then flew the insect in the apparatus shown in Fig. 18.8 and used the thermistor to record the temperature in the thorax. The thoraxes of locusts (*Schistocerca*), dragonflies (*Aeschna*) and hornets (*Vespa*) warmed up until they were 7–11 K above room temperature. He found out what rates of heat production this represented by experiments with dead insects, fitted with thermistors and with small electric heating elements in their thoraxes. Each dead insect was put in the wind tunnel with the jet blowing at the appropriate speed and the current through the heating element was adjusted until the thermistor registered the same temperatures as in the experiment with the live insect. The rate of heat production by the element was calculated by multiplying the current by the voltage. The rate of heat production by the flying insect must have been the same and the metabolic rate was estimated by adding a fraction to it to represent the mechanical work done on the air. Metabolic rates of 80 W kg^{-1} were found for flying locusts and 120 W kg^{-1} for flying dragonflies and hovering hornets. Since metabolism using 1 cm^3 oxygen releases 20 J, these metabolic rates correspond to oxygen consumptions of 14–22 cm^3 g^{-1} h^{-1}. Similar metabolic rates are achieved by flying birds and bats but not, so far as is known, by any other animals (see Fig. 8.6). These metabolic rates are particularly remarkable as they are almost entirely due to the metabolism of the wing muscles which represent only 20–35% of the body mass. The metabolic rates of the wing muscles must be around 100 cm^3 oxygen g^{-1} h^{-1}. No other tissue in any animal is known to use oxygen so fast.

Mechanical power is used, in insect flight, in two main ways. First there is the aerodynamic power needed to overcome the drag on the wings. This can be calculated by applying propeller theory, just as the power needed to drive a helicopter rotor can be calculated. For a hovering hornet, for instance, it is 21 W kg^{-1}. Secondly, there is the inertial power needed to accelerate the wings at the beginning of each stroke. A hornet has wings 24 mm long which beat through an angle of 2.1 radians. Hence the wing tips travel $24 \times 2.1 = 50$ mm in each downstroke and 50 mm in each upstroke, a total of 0.1 m in each cycle of wing movements. The wing frequency is 100 s^{-1} so the average speed of the wing tips must be $0.1 \times 100 = 10$ m s^{-1}, and their peak speed must be higher. Even light wings, moving so fast, have appreciable kinetic energy. The energy must be given to them at the beginning of each stroke and is lost at the end. It must be supplied 200 times per second, for 100 downstrokes and 100 upstrokes. It has been calculated that this inertial power requirement for a hornet is 48 W kg^{-1}, more than double the aerodynamic power requirement. It would be even higher if the wings were not so light.

Here we have a paradox. The aerodynamic power of 21 W kg^{-1} and the inertial power of 48 W kg^{-1} add up to a total power requirement of 69 W kg^{-1}. The metabolic rate, which is the rate of utilization of the chemical energy of food, is 120 W kg^{-1}. Hence the efficiency of conversion of chemical energy to mechanical work would seem to be $69/120 = 0.58$. No muscle is known to be as efficient as this, or even to approach this efficiency.

The paradox was resolved by Professor Weis-Fogh who pointed out that energy could be saved by elastic structures in the thorax. If you fix one end of a hacksaw blade in a vice and twang the other, the blade will vibrate at its natural frequency. In the middle of each vibration it moves fastest and has maximum kinetic energy. At the extremes of the vibration it is momentarily stationary: it has no kinetic energy but it is bent and so has elastic strain energy. As the blade vibrates energy is converted back and forth, from kinetic energy to elastic strain energy and vice versa. A little energy is lost (as heat) at each conversion and the vibration gradually dies down, but the whole kinetic energy does not have to be supplied afresh each time. Similarly insect wings have elastic structures attached to them which take up kinetic energy as elastic strain energy at the end of each stroke, and restore it in an elastic recoil for the next stroke. Provided the insect beats its wings at the natural frequency, the mechanical power which the muscles must supply is little more than the aerodynamic power.

Part of the elasticity is in the muscles, which are described in the following section. Part is in the cuticle of the thorax, particularly in parts made of a protein called resilin which was discovered by Weis-Fogh. Resilin is an amorphous cross-linked protein with rubber-like properties very similar to those of the abductin of scallop hinge ligament (p. 300). Indeed its properties are even more remarkable for the rebound resilience of locust resilin, at the wingbeat frequency, is no less than 0.97.

The resilin is arranged in different ways in different insects. In dragonflies the apodemes of some of the wing muscles are made of resilin. In locusts the wings are attached to the thorax by flexible elastic hinges made of resilin.

WING MUSCLES

When the insects evolved wings, muscles serving other functions in the thorax were brought into use as wing muscles. This seems apparent from the arrangement of the wing muscles, which are shown in Fig. 18.10. Muscles 1, 2, 3 and 4 serve to move the legs as well as the wings. Muscle 5 is a dorsal longitudinal muscle comparable to the dorsal longitudinal muscles of wingless segments. How do they work?

Each segment of the thorax is enclosed in two main plates of cuticle, a dorsal one and a ventral one. A wing is attached to the dorsal plate at *X* (Fig. 18.10*a*) and to the ventral plate at *Y*. If the dorsal plate, and thus *X*, is pulled down, the wing must rise. If the dorsal plate rises, the wing must swing down again. Fig. 18.10(*a*) shows two muscles (1 and 2) which pull the dorsal plate down and raise the wing. Fig. 18.10(*b*) shows two muscles (3 and 4) which depress the wing and raise the dorsal plate. The longitudinal muscles (5, Fig. 18.10*c*) also raise the dorsal plate and depress the wings: they raise the dorsal plate by making it arch upwards.

Muscles 1 and 3 tend to make the leg swing posteriorly and muscles 2 and 4 tend to swing it anteriorly. 1 and 2 acting together can exert balanced forces on the leg so that they raise the wing without moving the leg. Similarly 3 and 4 can depress the wing without moving the leg. However, 1 and 3 acting

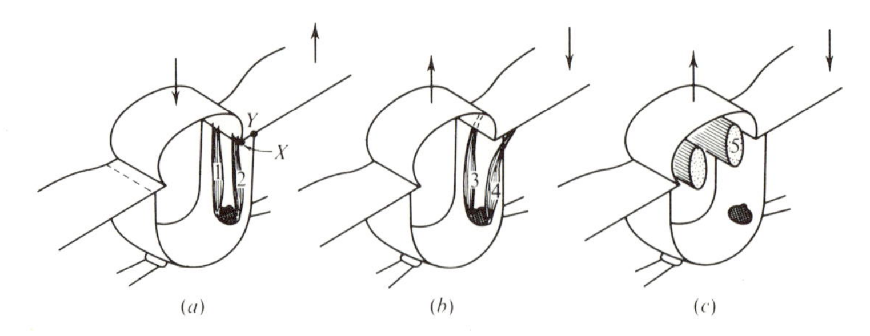

Fig. 18.10. Diagrams showing the wing muscles in one of the segments of the thorax of a grasshopper. (*a*) shows muscles which raise the wings, while (*b*) and (*c*) show muscles which depress them.

together, or 2 and 4 acting together, can move the leg without moving the wing. It has been demonstrated by electrical recording from their motor axons that these four muscles act both as wing muscles and as leg muscles.

Actual wing mechanisms are more complicated than Fig. 18.10 indicates. Many insects are able to lay the wings parallel to the abdomen when they are resting, and to spread them for flight. Flies have a mechanism which twists the wing automatically to give it the appropriate angles of attack for the upstroke and the downstroke (see Fig. 18.9*b*).

The wing muscles of locusts, dragonflies and some other insects are fairly conventional striated muscles. They can deliver a very high sustained power output and to make this possible contain a lot of mitochondria (30% by volume of locust wing muscle). In other respects they are not very different from the striated muscles of other arthropods and of vertebrates. A burst of action potentials must arrive in the motor axon to initiate every contraction. All insects with this type of muscle beat their wings at fairly low frequencies.

The wing muscles of flies (Diptera), wasps and their relatives (Hymenoptera) and many other insects have different and very remarkable properties. They do not need action potentials to initiate every contraction. In one experiment, action potentials were recorded from the wing muscle of a tethered blowfly. The insect was beating its wings at a frequency of 120 s^{-1} but action potentials were occurring at a frequency of only 3 s^{-1}. Occasional action potentials suffice to keep the muscles oscillating at the much higher frequency. Muscles which have this oscillatory property are known as fibrillar flight muscles. Without this property the highest wing beat frequencies might not be attainable. However, the highest wing beat frequency reliably recorded for an intact insect seems to be only 600 s^{-1} (for the mosquito, *Aedes*) which is lower than the maximum frequency of action potentials recorded from the electric organs of American knife fishes (*Gymnotoidei*).

Any structure which has mass and elastic stiffness has a natural frequency of vibration. The example of a hacksaw blade clamped in a vice has already

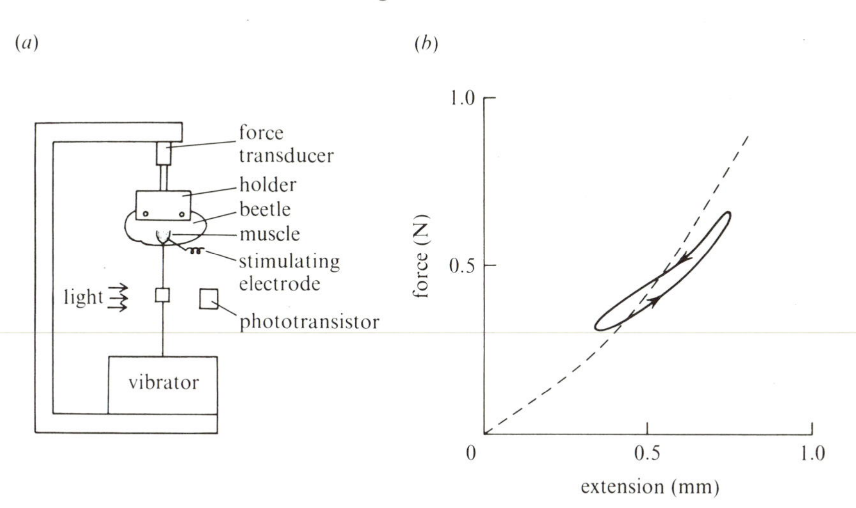

Fig. 18.11. (*a*) Apparatus for investigating the properties of fibrillar flight muscles. From R. McN. Alexander (1968). *Animal mechanics*. Sidgwick & Jackson, London. (*b*) The result of an experiment on *Oryctes*. It is explained in the text. Redrawn from K. E. Machin & J. W. S. Pringle (1959). *Proc. R. Soc.* **151B**, 204–25.

been cited. The greater the mass the lower the frequency (the natural frequency of the hacksaw blade can be reduced by attaching a lump of clay to its free end). The greater the stiffness, the higher the frequency. Active fibrillar muscle apparently oscillates automatically at the natural frequency of the system of which it is part, irrespective of the frequency of the action potentials it receives. An indication of this can be obtained by amputating the wings of a fly. This removes mass from the system without affecting its stiffness and the wing stumps vibrate at much higher frequency than the original wingbeat frequency.

Professor John Pringle and his colleagues have studied the properties of fibrillar flight muscle. They naturally found it more convenient to experiment with reasonably large muscles working at reasonably low frequencies than with smaller muscles working at higher frequencies. They therefore used some of the largest insects, the Indian rhinoceros beetle *Oryctes* and the giant tropical water bug *Lethocerus*. Species of *Lethocerus* grow up to 11 cm long. Though these insects have quite low wing beat frequencies, their wing muscles are fibrillar.

Fig. 18.11(*a*) shows how some of the experiments were done. The insect is firmly fixed in a holder attached to a force transducer. One of the wing muscles is dissected out and attached by a wire to a moving-coil vibrator. A vane on the wire partly blocks a beam of light aimed at a phototransistor, so that when the vane moves the amount of light reaching the phototransistor changes. The electrical outputs of the force transducer and the phototransistor record the force exerted by the muscle and its changes of length, respectively.

Fig. 18.11 (*b*) shows the results of experiments with the apparatus. The muscle was stimulated electrically throughout. The broken line shows what happened when the vibrator was used to stretch the muscle without letting it oscillate. The muscle behaved as an elastic material, exerting progressively larger forces as it was stretched. The continuous line shows what happened when the vibrator was connected to a circuit which made it behave like a mechanical system with mass, stiffness and viscosity. The muscle oscillated. It exerted more force as it shortened than as it stretched so the graph of force against length is an anticlockwise loop. The area of the loop represents the work done by the muscle, in each cycle of its oscillation, against the simulated viscosity of the vibrator. This is the work which in the flying insect would supply the aerodynamic power. Contrast this *anticlockwise* loop with the *clockwise* loops in Fig. 14.5. The areas of the latter do not of course represent work done by the mollusc hinge ligament, but work done on it. The frequency of oscillation of the wing muscle increased when the simulated stiffness of the circuit was increased and decreased when the simulated mass was increased: the resonant frequency of a passive system would have changed in the same way.

Similar oscillatory contractions are obtained when the experiment is performed with isolated fibres, treated with glycerol to break down the cell membrane (see p. 44) and bathed in a suitable saline solution containing ATP. The oscillation is plainly a property of the muscle alone, not of the nervous system.

RESPIRATION

Insect wing muscles use oxygen faster than any other known tissue, and need a very efficient system to supply them with oxygen. The supply comes through the tracheae from the spiracles.

The tracheal supply to the wing muscles differs in detail between groups of insects. The arrangement found in dragonflies is shown as an example in Fig. 18.12. There are two pairs of spiracles on the thorax and 10 pairs on the abdomen. Each muscle has an axial trachea running through its centre, connected at its ventral end to one or other of the first three spiracles and at its dorsal end to air sacs which are considerably larger than the diagram indicates. The air sacs in the thorax are also connected to the abdominal tracheae. Flight movements make the thorax enlarge and contract at each wing beat, so that air is drawn into the air sacs and blown out of them again. Some of this air may flow from and to the abdomen but much of it must flow through the axial tracheae of the wing muscles. As the thorax expands, air must be drawn from the first three pairs of spiracles through the muscles to the air sacs. As it contracts, air must be blown out by the same route. Thus the air in the axial tracheae is continually changed.

The details have been worked out more fully for locusts, which have a slightly more complicated tracheal system than dragonflies. It is difficult to estimate the sizes of the air sacs in a dissection but a technique has been devised for filling them with coloured jelly and making the rest of the locust

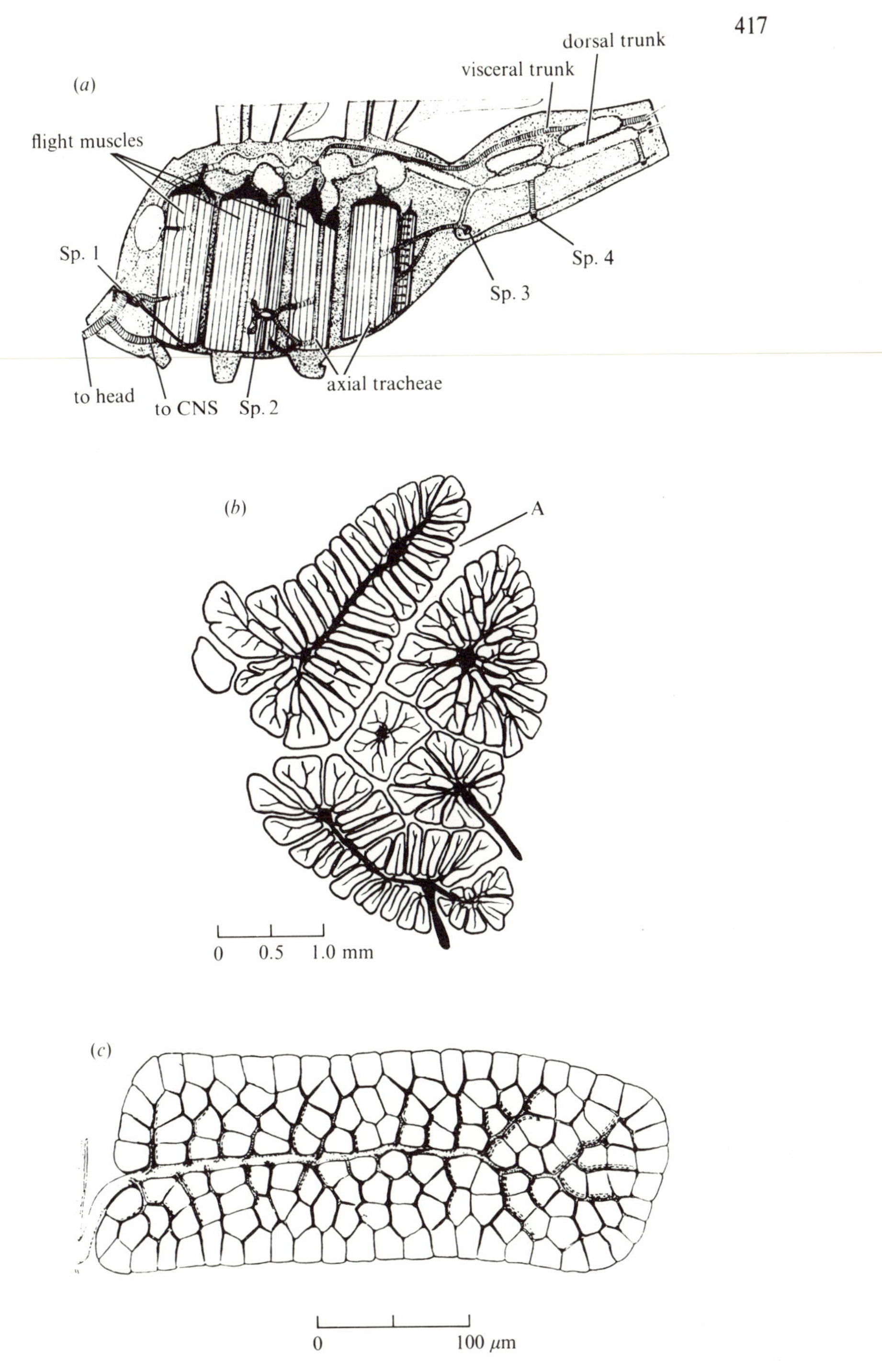

Fig. 18.12. The tracheal supply to the wing muscles of dragonflies. (*a*) A diagram showing the tracheae and air sacs in the thorax and anterior abdomen of a dragonfly such as *Aeschna*. Sp., spiracle. From P. L. Miller (1962). *J. exp. Biol.* **39**, 513–35. (*b*) A horizontal section through the hind-wing muscles of *Aeschna*. Tracheae are drawn black. (*c*) Lobe A from (*b*), greatly enlarged. (*b*) and (*c*) from T. Weis-Fogh (1964). *J. exp. Biol.* **41**, 229–56.

transparent. This makes possible measurements which show that the thoracic air sacs occupy 100–150 mm^3 in a 2 g locust. The roof of the thorax rises in the downstroke of the wings and falls in the upstroke (see p. 413). It rises and falls through about 0.6 mm which is enough to enlarge and contract the thorax by about 25 mm^3. This volume may seem small but since the wings beat at a frequency of 17 s^{-1} the total ventilation volume is $17 \times 25 = 425$ mm^3 s^{-1} or 1500 cm^3 h^{-1}.

Some of this airflow bypasses the muscles and so is ineffective in supplying oxygen to them. The effective rate of ventilation has been estimated in another way. Locusts were fitted with a device which took small samples of air from the second thoracic spiracle, while they were flying in a wind tunnel. The samples were analysed and found to contain, on average, 13% oxygen. Dry air contains 21% oxygen but moist air at the temperature of the flying locusts' thoraxes contains only 20%, so the oxygen used by the locusts amounted to 7% of the air. A 2 g locust uses about 28 cm^3 oxygen h^{-1} in flight and if this is 7% of the volume of the air pumped through the muscles that volume must be 400 cm^3 h^{-1}. This is only about a quarter of the volume calculated from the total volume change of the thorax, but it is still ample.

The pumping action renews the air in the axial tracheae, but oxygen can only move from there to the muscle fibres by diffusion. Sections through the muscles show lobes, each of them a bundle of muscle fibres (Fig. 18.12*b*, *c*). Each lobe is served by radial branches from the axial trachea. The branches subdivide into fine tracheoles which pass between the muscle fibres. Though the fibres are only 20 μm in diameter, each has tracheoles between itself and its neighbour. The arrangement allows fast diffusion between the axial trachea and the muscle fibres, since oxygen diffuses much faster in air than in water or tissues (see the discussion of pulmonate snails on p. 279).

Since movement of oxygen from the axial trachea depends on diffusion there must be a maximum practical radius for insect wing muscles. In the muscles shown in Fig. 18.12, no muscle fibre is more than about 1 mm from an axial trachea. How near the limit is this?

The question can be answered in the same way as the question of the maximum feasible thickness for a flatworm (p. 83). The diffusion constant for oxygen diffusing in air is 20 mm^2 atm^{-1} s^{-1}. A section through a muscle lobe such as the one shown in Fig. 18.12(*c*), cut at right angles to the direction of diffusion, would show that the air passages made up about 1% of the area. Hence we can estimate that the diffusion constant D for oxygen diffusing through this muscle is 1% of the value for diffusion through air, or 0.2 mm^2 atm^{-1} s^{-1}. The metabolic rates m of insect wing muscles in flight are around 0.03 mm^3 oxygen (mm^3 muscle)$^{-1}$ s^{-1}. The partial pressure P_s of oxygen in the axial trachea is 0.13 atm. Applying equation (8.2) to determine the maximum diffusion distance we find

$$s \leqslant (2\,DP_s/m)$$
$$\leqslant (2 \times 0.2 \times 0.13/0.03) = 1.7 \text{ mm}$$

The diameter of the muscle is 2*s* plus the diameter of its axial trachea, so this calculation indicates that insect wing muscles of diameter 3.5 mm are feasible. The vast majority of insects have wing muscles more slender than this. Species of the giant water bug *Lethocerus* have wing muscles of diameter up to 5 mm, but these muscles have a peculiar arrangement of tracheae which probably ensures that air is pumped through the radial tracheae as well as the axial ones.

Oxygen diffuses radially outwards from the axial tracheae of insect wing muscles, but the fuels for flight have to diffuse in from the blood in the haemocoel. The principal fuels are the sugar trehalose, and fatty acids. The lobed structure of the muscles ensures that the fuels do not have to diffuse far. In the muscles shown in Fig. 18.12, no fibre is more than about 0.2 mm from the nearest blood. When the muscles contract they squeeze blood out of the clefts between their lobes. When they extend they draw fresh blood into the clefts. Nevertheless, high fuel concentrations are needed in the blood. The blood of locusts and bees contains about 2% sugars (mainly trehalose). In contrast, the sugar content of human blood does not normally rise above 0.2%.

Even when they are resting on the ground the larger species of insects pump air through their tracheae. Locusts do this mainly by expanding and contracting the abdomen. They open their anterior and posterior spiracles in turn so that air enters through the anterior ones and leaves through the posterior ones. The rate at which air is pumped through the system is (appropriately) far less than in flight.

WATER BALANCE

Terrestrial snails and slugs are generally active only at night or in damp weather, when the relative humidity is high. During the day snails retire into their shells and slugs into holes, and the whole of a hot dry season may be passed in a dormant state (p. 290). Thus snails and slugs avoid losing too much water by evaporation. Many insects, however, are active in very hot dry conditions, even in deserts. They need much more effective protection against water loss than snails.

Evaporation is a particularly severe problem for small terrestrial animals, simply because they are small. Compare two animals of the same shape, one twice as long as the other. The larger one has four times the surface area of the smaller so it is likely to lose water by evaporation about four times as fast. However, it has eight times the volume of the smaller animal and so starts off with eight times as much water. It will take twice as long to dry up.

Water loss from the spiracles is inevitable. Consider an insect in dry air at 40 °C (for instance, a desert beetle). In some period of time it pumps a litre of air through its tracheae. This air enters the spiracles dry but it inevitably leaves them saturated with water vapour, containing 60 mg water. The air enters containing 210 cm^3 oxygen and leaves containing considerably less. It is unlikely that more than about a third of the oxygen will be removed for

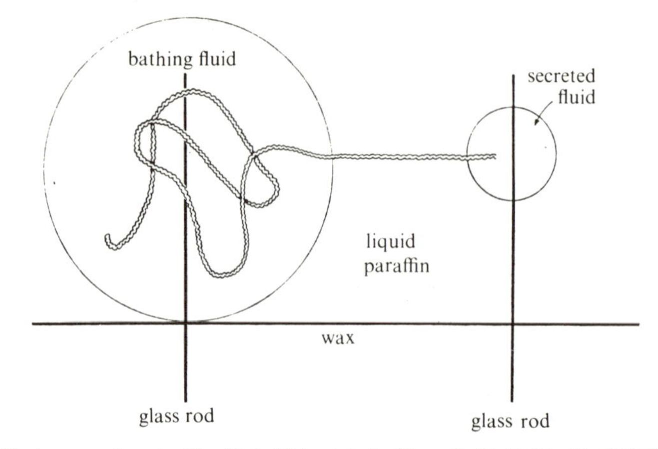

Fig. 18.13. An experiment with a Malpighian tubule. From S. H. P. Maddrell (1971). *Phil. Trans.* **262B**, 197–207.

respiration, so that the 60 mg of water is lost for the uptake of 70 cm^3 oxygen. Thus 0.9 mg water must be lost for every cubic centimetre of oxygen used. More would be lost if the spiracles were left perpetually open.

Insects can be made to open their spiracles by mixing carbon dioxide with the air. For instance, the bug *Rhodnius* keeps its spiracles perpetually open in an atmosphere containing 5 % carbon dioxide. In an experiment, a *Rhodnius* was kept starved in dry air for 6 days. It was weighed every day. For the first two days and the last three it was kept in normal (though dry) air. On each of these days it lost less than 2 mg, mainly by evaporation of water. On the third day its spiracles were kept open by means of carbon dioxide and it lost 11 mg.

The spiracles are not the only route for water loss, and not necessarily the most important. *Eleodus* is a desert beetle which at temperatures around 40 °C uses 0.4 cm^3 oxygen g^{-1} h^{-1}. If it opened its spiracles only when necessary to admit or expel air and used one third of the oxygen from this air it would inevitably lose through the spiracles 0.36 mg water g^{-1} h^{-1} or 0.036% of its weight per hour. It actually loses a total of 0.2% of its weight per hour, mostly through the cuticle which is not perfectly waterproof. It would lose water faster if the cuticle did not have its outer layer of wax.

Water is also lost with faeces and urine. Insects excrete most of their waste nitrogen as uric acid, which can be excreted with far less water than would be needed to get rid of ammonia (p. 284). Terrestrial snails similarly excrete uric acid and other purines, and gain the same advantage.

The Malpighian tubules and the rectum co-operate in the process of excretion. Many experiments have been done with single Malpighian tubules, removed from the insect. The basic technique was invented by Professor Arthur Ramsay and is illustrated in Fig. 18.13. Everything in the picture is submerged in liquid paraffin. Most of the length of the tubule is in a drop of fluid which represents the blood of the intact insect: this fluid may be an imitation of insect blood or it may be some other solution devised for the experiment. The open end of the tubule (the end which was connected to the gut in the intact insect) is pulled out from this drop of bathing fluid into the paraffin. The fluid secreted by the tubule emerges from this end and can be collected for analysis.

The secreted fluid generally has about the same osmotic concentration as the bathing fluid, but contains much less sodium and much more potassium. The rate of secretion is very much reduced if the bathing fluid contains no potassium. These observations have led to the theory that the main active process is secretion of potassium taken from the bathing fluid and secreted into the tubule lumen. The lumen will become positively charged and the potential difference will tend to drive anions from the blood to the lumen; if potassium is moved, anions will follow. The movement of potassium and anions will increase the osmotic concentration of the fluid in the lumen so water will be drawn into the lumen with them.

Measurements with electrodes confirm that the lumen is positively charged relative to the bathing fluid: the potential is usually about 30 mV. Thus the potassium moving into the lumen is moving against both a concentration difference and a potential difference. The secretion of potassium must be an active, energy-consuming process.

The walls of the Malpighian tubules are only one cell thick. Fig. 18.14(*a*) shows one of the cells. Notice the long clefts on the side facing the blood, and the microvilli on the lumen side. It is thought likely that the active processes of secretion occur across the cell membrane of the clefts and microvilli so that gradients of concentration build up as shown in Fig. 18.14(*b*). The concentration of ions is low at the point marked *b* because ions are being pumped from there across the cell membrane, and ions to replace them have to diffuse down the narrow cleft. The concentration is high at *c* because ions are being secreted into the cell there and can only move towards *d* by diffusion. There are therefore large differences of osmotic concentration between *a* and *c* and between *b* and *d*. Water will be drawn across the cell membrane quite rapidly by osmosis, although the ionic concentrations in the blood and in the centre of the cell may be almost identical. Gradients of concentration probably build up along the microvilli, as well as along the clefts. These gradients resemble the gradients believed to develop in the clefts in the siphuncle of *Nautilus* (Fig. 15.9).

Electron probe X-ray microanalysis is being used by a group of zoologists at Cambridge to measure the concentrations of ions in different parts of Malpighian tubule cells. The results should confirm or deny the theory represented by Fig. 18.14(*b*). The initial results show that the concentrations of sodium and potassium in the centres of the cells are very similar to the concentrations in the secreted fluid. This is as might be expected: it seems to be a general rule that cells contain much more potassium than sodium.

Uric acid is secreted from the blood into the Malphigian tubules. It may

(*a*)

(*b*)

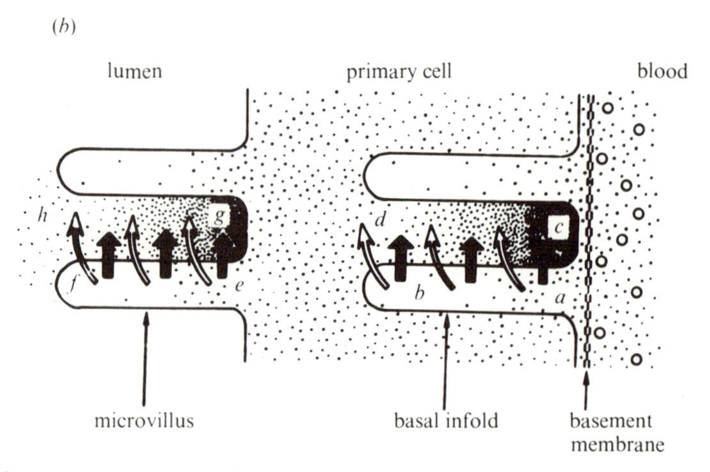

Fig. 18.14. (*a*) An electron microscope section of a cell in the wall of a cockroach Malpighian tubule. The surface on the right is the outer surface of the tubule and the surface on the left faces the lumen. From J. L. Oschmann & M. J. Berridge (1971). *Fed. Proc.* **30**, 49–56. (*b*) A diagram of the same cell. Black arrows indicate active transport of salts and open arrows indicate passive flow of water. From M. J. Berridge & J. L. Oschmann (1969). *Tissue and Cell* **1**, 247–72.

remain in solution if the urine is flowing very copiously (for instance in *Rhodnius* just after a meal of blood) but it generally precipitates, making the urine cloudy.

The fluid produced by the Malpighian tubules has roughly the same osmotic concentration as the blood. It is secreted rapidly, at rates up to 50% of the total blood volume per hour. If it were passed as urine the insect would lose water

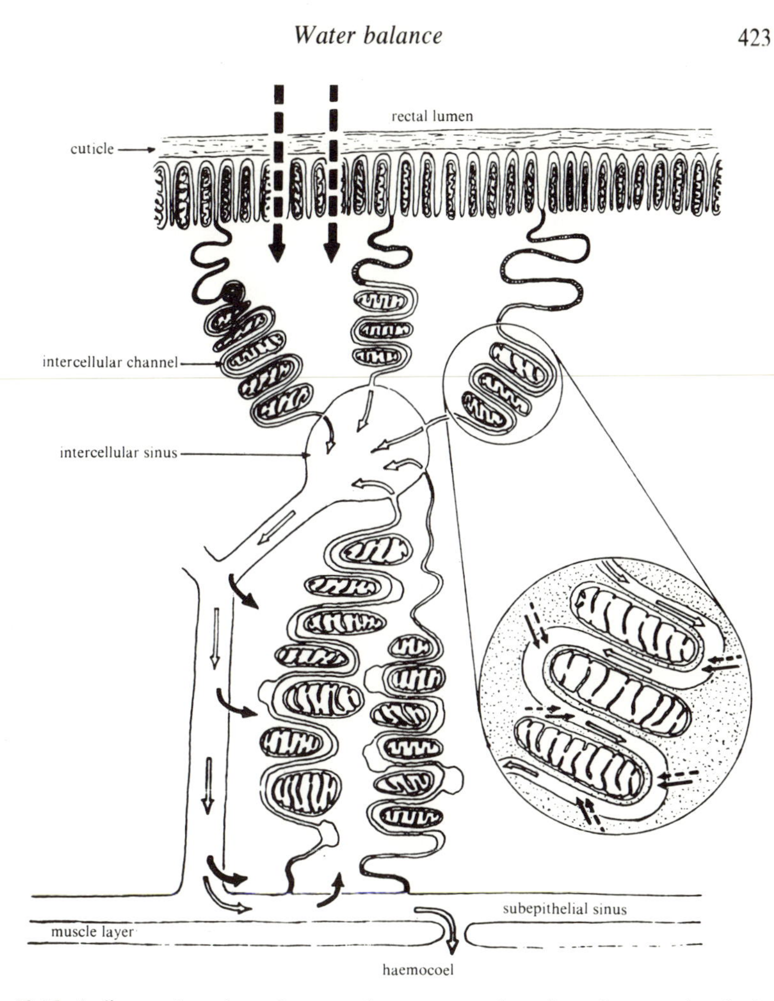

Fig. 18.15. A diagram based on electron microscope sections through a rectal pad of a cockroach (*Periplaneta*). Solid black arrows represent active transport of salts, broken arrows represent passive movement of water across membranes and hollow arrows represent flow of fluid along intercellular spaces. From J. Noble-Nesbitt (1973). In *Comparative physiology*, ed. L. Bolis, K. Schmidt-Nielsen & S. H. P. Maddrell, pp. 333–51. North-Holland, Amsterdam.

very rapidly indeed. However, water is removed from the urine, and from the faeces, in the rectum. The concentration of the fluid finally excreted depends on the needs of the insect at the time. Cockroaches excrete fluid at osmotic concentrations up to 1 Osmol l^{-1} (more than twice the concentration of the blood). Some other insects, which are better adapted to life in dry conditions, excrete solid pellets of faeces and uric acid. They include mealworms (larvae of the beetle *Tenebrio*) which live in grain stores.

Cockroaches and some other insects have distinct patches on the rectal wall where resorption of water occurs. A section through one of these rectal pads is shown in Fig. 18.15. There are narrow clefts between adjacent cells, closed at both ends by desmosomes. There are many mitochondria in the cytoplasm alongside the clefts, suggesting that an active process occurs there. Though closed at their ends and clefts connect to broader intercellular sinuses which discharge into the blood.

These sinuses are large enough, in cockroaches, for it to be possible to thrust micropipettes into them and take samples of their fluid. Cockroaches were kept without water for a few days and then anaesthetized and opened. The rectum was filled by the experimenters with a solution which became more concentrated as the cockroach extracted water from it. Samples of fluid were taken from the rectal lumen and from the intercellular sinuses. It was found that however high the osmotic concentration in the lumen became, the fluid in the sinuses was always more concentrated. This suggests that cockroaches use an osmotic mechanism to remove water from the lumen. They probably build up high osmotic concentrations in the wall of the rectum by salt secretion, so that water is drawn osmotically from the lumen.

Some insects including mealworms can take up water vapour from a damp atmosphere. The most remarkable example known is the firebrat *Thermobia*, which lives in bakeries and kitchens. Firebrats kept without food or water gain weight in air of relative humidities down to 45%, presumably by uptake of water. They do not gain weight if the anus is blocked by a ligature. This suggests that uptake is by way of the rectum and that *Thermobia* extracts water vapour from the air by the same mechanism as is used to dry the faeces. However, it seems almost inconceivable that an osmotic mechanism in an animal should be capable of removing water from air of relative humidities down to 45%. Saturated solutions of sodium and potassium chloride are in equilibrium with air at relative humidities of 75% and 85%, respectively, so they cannot remove water from drier air. Some other much more soluble solute would be needed to make a solution which would take up water down to 45% relative humidity. *Thermobia* may take up water by a non-osmotic mechanism, perhaps active transport of water.

COMPOUND EYES

The eyes of insects, like those of most Crustacea, are compound, composed of a large number of ommatidia. The very large eyes of some dragonflies have as many as 28000 ommatidia.

Fig. 18.16 shows the structure of the compound eye of an insect. Each of the many ommatidia points in a slightly different direction. Fig. 18.17 shows the structure of one of them. The lens is transparent cuticle and is shed and replaced, like the rest of the cuticle, at every moult. The so-called crystalline cone behind it is not in fact crystalline: it is a transparent, extracellular jelly. The eight long retinula cells are the sensory cells. Microvilli project inwards

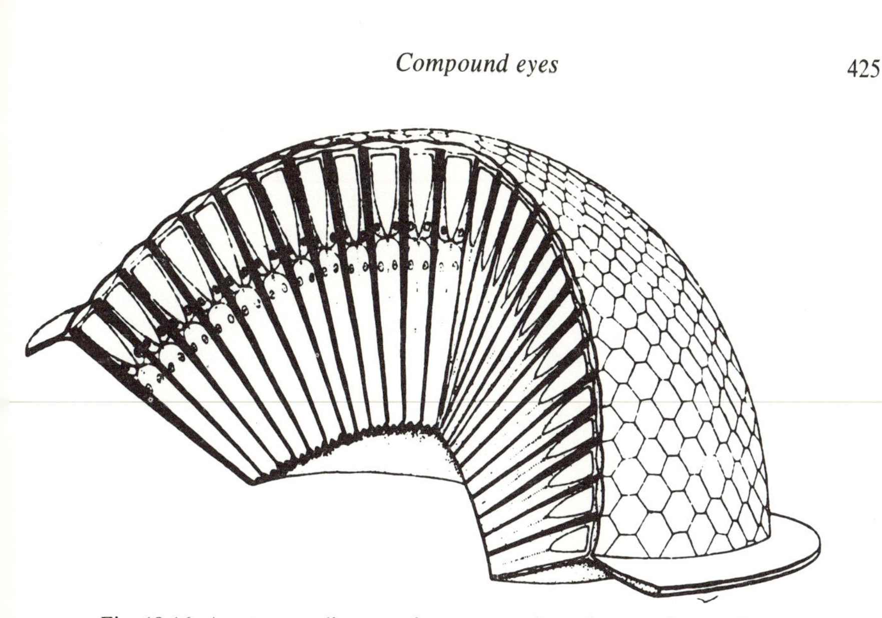

Fig. 18.16. A cut-away diagram of a compound eye from an insect. From V. B. Wigglesworth (1964). *The life of insects*. Weidenfeld & Nicolson, London.

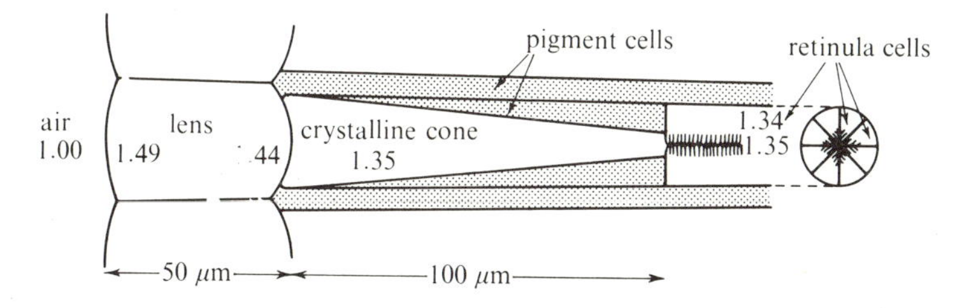

Fig. 18.17. Diagrammatic sections through a single ommatidium in the eye of a bee. The numbers (other than those indicating dimensions) are refractive indices. Based on data in F. G. Varela & W. Wiitanen (1970). *J. gen. Physiol.* **55**, 336–58.

from them towards the central axis of the ommatidium (compare this arrangement with that of the microvilli on the *Octopus* retinula cell in Fig. 15.14 *b*). Pigment cells between the ommatidia prevent light from straying from one ommatidia to the next.

The refractive indices shown in Fig. 18.17 were measured in microscope sections of bee eyes, by means of interference microscopy. The lens has a refractive index ranging from 1.49 (almost as high as for crown glass) near the outer surface to 1.44 near the crystalline cone. The transparent structures deeper in the eye all have refractive indices little different from water (1.33). The lens is convex on both faces. It has a higher refractive index than either the air outside it or the crystalline cone inside. It therefore acts as a converging lens: light passing through it is brought to a focus in the crystalline cone. It can be calculated from the refractive indices and the curvature of the faces of the lens that distant objects will be focussed about two-thirds of the way down the crystalline cone (not, as might be expected, on the ends of the retinula cells).

The outer surface of the eye of a bee is part of a sphere of radius 1.2 mm (1200 μm). The diameters of the lenses are 32 μm, so each ommatidium is set at an angle $32/1200 = 0.027$ radians (1.5°) to the next. The eye is presumably unable to distinguish objects separated by angles less than this: a bee viewing a page 30 cm away would be unable to distinguish spots less than 8 mm apart. This makes it, apparently, much inferior to the octopus eye (p. 335) which can distinguish stripes only 0.3° apart. Could the resolution of the bee eye be improved by giving it more ommatidia, of smaller diameter?

The resolution is limited by diffraction, which makes it impossible for light of wavelength λ passing through an aperture of radius r to form separate images of objects less than about $0.6\ \lambda/r$ radians apart. The radius of the bee lens is about 16 μm. Consider light of wavelength 0.5 μm passing through it. The minimum angle for separate images is $0.6 \times 0.5/16 = 0.019$ radians, only a little less than the angle between ommatidia. Reducing the diameter of the ommatidia would increase the angle of diffraction and very little improvement of resolution would be possible.

The light-sensitive molecules are believed to lie in the cell membranes of the microvilli of the retinula cells. In an ideal ommatidium, light entering the retinula cells would travel along the core of microvilli until it was absorbed by these molecules. This tends to happen because the refractive index of the core is slightly greater than that of the surrounding cytoplasm. Light cannot escape from a material of high refractive index into one of lower refractive index if it strikes the boundary at too shallow an angle: instead, it is reflected back into the material of higher refractive index. This is the principle of fibre optics.

In the eyes of bees and many other insects which are active by day, pigment cells separate adjacent ommatidia. Light which passes through the lens of one ommatidium cannot reach the retinula cells of another. Light which enters obliquely is absorbed by the pigment in the pigment cells, and is no use for vision. This arrangement works badly in dim light because little light reaches the retinula cells. Moths and other insects which fly at night have a different type of eye with mobile pigment. In bright light adjacent ommatidia are largely screened from each other by pigment. In dim light the pigment becomes concentrated between the lenses, allowing light to pass freely from one ommatidium to another, at the level of the crystalline cones. This allows far more light to reach the retinula cells and so makes the eye more sensitive to dim light. However, it probably also makes the eye unable to distinguish fine detail, for the light is probably not at all well focussed.

If four of the retinula cells in an ommatidium have their microvilli oriented east–west, the other four have their oriented north–south. It has been shown by experiments with crayfish that the microvilli absorb twice as much polarized light if the electric vector of the light is parallel to their long axes, than if it

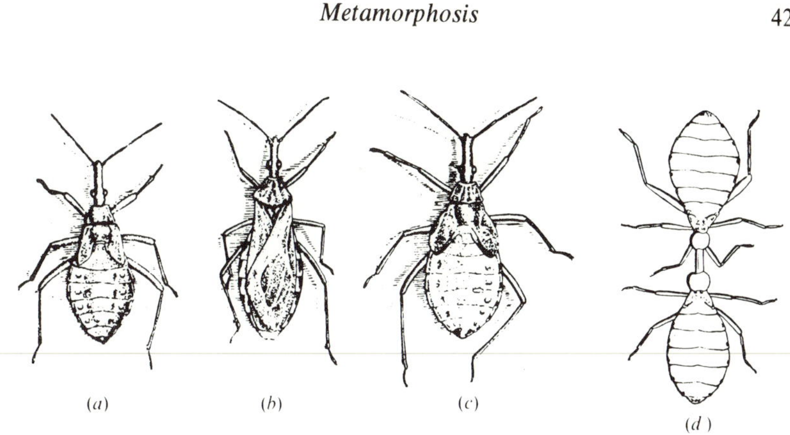

Fig. 18.18. *Rhodnius prolixus*. (*a*) The final (fifth) larval stage. (*b*) An adult. (*c*) A sixth larval stage which does not occur naturally but was produced in experiments described in the text. (*d*) Two decapitated fourth-stage larvae joined by a glass tube. From V. B. Wigglesworth (1974). *Insect physiology*, 7th edn. Chapman & Hall, London.

is at right angles to them. (This is a consequence of the sensitive molecules being parallel to the surfaces of the microvilli.) Light polarized in different planes is absorbed preferentially by different retinula cells within the same ommatidium. This makes it possible for arthropod eyes to distinguish the plane of polarization of light, which the human eye cannot do. Bees use this facility in navigation, as is explained in the next chapter.

METAMORPHOSIS

In the course of its life an insect moults several times, and each time it is transformed. The transformations may all be relatively slight, as in the locust (Fig. 18.6), or there may be dramatic changes from larva to pupa and from pupa to adult. How is moulting initiated, and how are the transformations controlled so that the right change occurs at the right stage in the life history?

The research of Sir Vincent Wigglesworth has gone a long way towards answering these questions. Its success depended on a wise choice of experimental animal. This animal was *Rhodnius*, a South American blood-sucking bug (Fig. 18.18). One of its advantages is that it survives drastic surgery very well. Another is a peculiar convenient life history. It passes through five larval stages. During each it takes a single gigantic meal of blood, amounting to several times its body mass. Ten to twenty days later (the interval differs between larval stages) it moults and is transformed to the next stage. The experimenter can induce moulting at will (by feeding the bug), and once the meal has been given he can predict accurately when moulting will occur.

Many of the experiments used fourth-stage larvae, which normally moult 14 days after a meal (at 26 °C). If one of these larvae is decapitated more than 4 days after a meal, and the neck is sealed with wax to prevent loss of blood, it will survive and moult at the expected time. It will probably wear the old cuticle like an overcoat instead of shedding it completely, but in other respects the moult is normal. However, if it is decapitated less than 4 days after a meal it fails to moult. It seems that the head gives the signal for moulting about 4 days after a meal and that once the signal has been given the moult will occur even without the head.

A simple experiment revealed the nature of the signal. It used two *Rhodnius* which will be called A and B. A was fed, and a week later B was fed. The day after B's meal both bugs were decapitated and joined neck-to-neck by a glass capillary tube (Fig. 18.18*d*). If the decapitated bugs had been kept separate A would have moulted and B would not. Joined together, both moulted. It seems that the signal to moult must have been transmitted from A to B through the glass tube: it must have been transmitted in the blood. The signal must be a hormone secreted in the head and distributed round the body by the blood.

The hormone is called the activation hormone. It is secreted by modified neurones in the dorsal part of the brain. Fourth-stage *Rhodnius* decapitated less than 4 days after a meal would not normally moult, but they can be made to moult by implanting this part of the brain into the abdomen. Other parts of the brain are ineffective. However, even the part which secretes the activation hormone is ineffective if the abdomen is separated from the thorax by a tight ligature. It seems that the activation hormone does not act directly. It activates a gland in the thorax which secretes a second hormone, called ecdysone. This has been demonstrated by experiments in which glands taken from the thorax were implanted into isolated abdomens. Ecdysone, like many vertebrate hormones, is a steroid.

The activation hormone and ecdysone give the signal for moulting but a third hormone is needed to decide whether the moult is to produce another larval stage or an adult. This is the juvenile hormone, so called because its presence prevents metamorphosis to the adult form. It is a lipid. Wigglesworth demonstrated its action by two types of experiment. In one he made first-stage larvae metamorphose to tiny precocious adults. In the other, he made fifth-stage larvae (which ought to have metamorphosed directly to adults) produce giant sixth- and even seventh-stage larvae (Fig. 18.18*c*).

The miniature adults were produced by joining first-stage larvae, decapitated the day after a meal, to moulting fifth-stage larvae. The join was made by a tube, as in Fig. 18.18(*d*), but the head of the larger larva was left almost intact. The miniature adults produced in this way were not perfect, but they were much more like adults than normal second-stage larvae.

Giant larvae were produced by joining fifth-stage larvae, decapitated the day after a meal, to fourth-stage larvae fed a week earlier. Alternatively, glands from earlier larvae were implanted into fifth-stage larvae. The gland in question is the corpus allatum, which lies immediately posterior to the brain. It secretes juvenile hormone until the fifth larval stage is reached, and then ceases.

An effect of juvenile hormone is illustrated by the experiment shown in Fig.

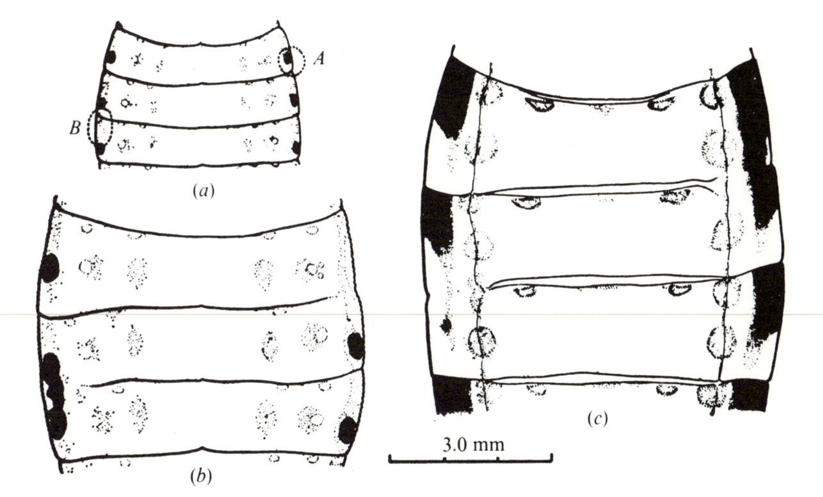

Fig. 18.19. Dorsal views of part of the abdomen of a *Rhodnius*. The third-stage larva (*a*) was burned at *A* and *B*. It developed through the fourth stage to the fifth stage (*b*) and finally became adult (*c*). From V. B. Wigglesworth (1970). *Insect hormones*. Oliver & Boyd, Edinburgh.

18.19. Larval *Rhodnius* have black spots near the posterior corners of their abdominal segments. Third-stage larvae (Fig. 18.19*a*) were given burns which destroyed a spot and the underlying epidermis (at *A*) or the epidermis between spots (at *B*). The wounds were repaired by an influx of neighbouring epidermal cells, and later larvae had an altered pattern of spots (Fig. 18.19*b*). Burn *A* resulted in the absence of a spot because the cuticle where the spot should have been was now underlain by non-spot epidermis. Burn *B* resulted in two spots being joined together where their epidermis had filled the gap left by the wound. However, larval spot epidermis is capable of producing pale cuticle and non-spot epidermis of producing black cuticle, as appears when the adult stage is reached (Fig. 18.19*c*). Normal adults have spots at the anterior corners of their abdominal segments, not the posterior corners, so the effects of burns *A* and *B* are reversed when the larva becomes adult. It seems that each epidermal cell has two sets of instructions, one specifying the type of cuticle it should produce in the presence of juvenile hormone and the other specifying the type to be produced in the absence of the hormone.

Wigglesworth used *Rhodnius* for his classic experiments on the activation hormone, ecdysome and juvenile hormone, but experiments on other insects have shown that their moulting and metamorphosis are controlled in similar fashion.

FURTHER READING

GENERAL

Chapman, R. F. (1971). *The insects, structure and function*, 2nd edn. English Universities Press, London.
Imms, A. D. (1951). *A general textbook of entomology*, 8th edn. Methuen, London.
Rockstein, M. (ed.) (1973–4). *Physiology of Insecta*, 2nd edn. Academic Press, New York & London.
Snodgrass, R. E. (1935). *Principles of insect morphology*. McGraw-Hill, New York.
Wigglesworth, V. B. (1964). *The life of insects*. Weidenfeld & Nicolson, London.
Wigglesworth, V. B. (1974). *Insect physiology*, 7th edn. Chapman & Hall, London.

FLIGHT

Alexander, R. McN. (1977). Flight. In *Mechanics and energetics of animal locomotion*, R. McN. Alexander & G. Goldspink, pp. 249–78. Chapman & Hall, London.
Rainey, R. C. (ed.) (1975). *Insect flight. Royal Entomological Society Symposium 7*. Blackwell, Oxford.
Weis-Fogh, T. (1973). Quick estimates of flight fitness in hovering animals, including novel mechanisms for lift production. *J. exp. Biol.* **59**, 169–230.

WING MUSCLES

Wilson, D. M. (1962). Bifunctional muscles in the thorax of grasshoppers. *J. exp. Biol.* **39**, 669–77.
Usherwood, P. N. R. (ed.) (1975). *Insect muscle*. Academic Press, New York & London.

RESPIRATION

Weis-Fogh, T. (1964). Diffusion in insect wing muscle, the most active tissue known. *J. exp. Biol.* **41**, 229–56.
Weis-Fogh, T. (1967). Respiration and tracheal ventilation in locusts and other flying insects. *J. exp. Biol.* **47**, 561–87.

WATER BALANCE

Ahearn, G. A. (1970). The control of water loss in desert tenebrionid beetles. *J. exp. Biol.* **53**, 573–96.
Gupta, B. L. (1976). Water movement in cells and tissues. In *Perspectives in experimental biology*, vol. 1, ed. P. Spencer-Davies, pp. 25–42. Pergamon, Oxford.
Maddrell, S. H. P. (1971). Fluid secretion by the Malpighian tubules of insects. *Phil. Trans. R. Soc.* **262B**, 197–207.
Noble-Nesbitt, J. (1973). Rectal uptake of water in insects. In *Comparative physiology*, ed. L. Bolis, K. Schmidt-Nielsen & S. H. P. Maddrell, pp. 333–51. North-Holland, Amsterdam.

COMPOUND EYES

Horridge, C. A. (1974). (ed.). *The compound eye and vision of insects*. Clarendon, Oxford.
Horridge, G. A. (1977). The compound eye of insects. *Sci. Am.* **237**(1), 108–20.

METAMORPHOSIS

Wigglesworth, V. B. (1970). *Insect hormones*. Oliver & Boyd, Edinburgh.

19

A review of the insects

Phylum Arthropoda (cont.), Class Insecta
- Order Thysanura (silverfish)
- Order Collembola (springtails)
- Order Odonata (dragonflies)
- Order Ephemeroptera (mayflies)
- Order Plecoptera (stoneflies)
- Order Dictyoptera (cockroaches)
- Order Isoptera (termites)
- Order Orthoptera (locusts and crickets)
- Order Thysanoptera (thrips)
- Order Hemiptera (bugs)
- Order Trichoptera (caddis-flies)
- Order Lepidoptera (moths and butterflies)
- Order Diptera (flies)
- Order Siphonaptera (fleas)
- Order Hymenoptera (ants, bees and wasps)
- Order Coleoptera (beetles)
- and other orders

This chapter is a brief review of the main orders of insects. It draws attention to some of the characteristic features of the orders and discusses a few of their interesting peculiarities.

THYSANURA (SILVERFISH)

The Thysanura are the most primitive of all insects. They have no wings (Fig. 19.1 *a*). The larvae look like the adults and have similar habits, and moulting does not cease when the adult stage is reached. There are small leg-like appendages on the abdomen, reminiscent of the swimmerets of the crayfish (Fig. 17.1). There are also three long processes at the posterior end of the abdomen, two cerci and a median 'tail'.

Silverfish (*Lepisma*) are thysanurans often found in kitchens. They feed there on scraps of food, and also on paper. The firebrat (*Thermobia*, Fig. 19.1 *a*) used to be common in bakeries, when they were kept less clean than is usual now. Its remarkable ability to absorb water vapour from air was described on p. 424. *Petrobius* is a thysanuran which lives on shores near high tide level. At high tide it retreats into cracks in rocks so that though it may be below water level, it is enclosed in a pocket of air. It eats algae and detritus.

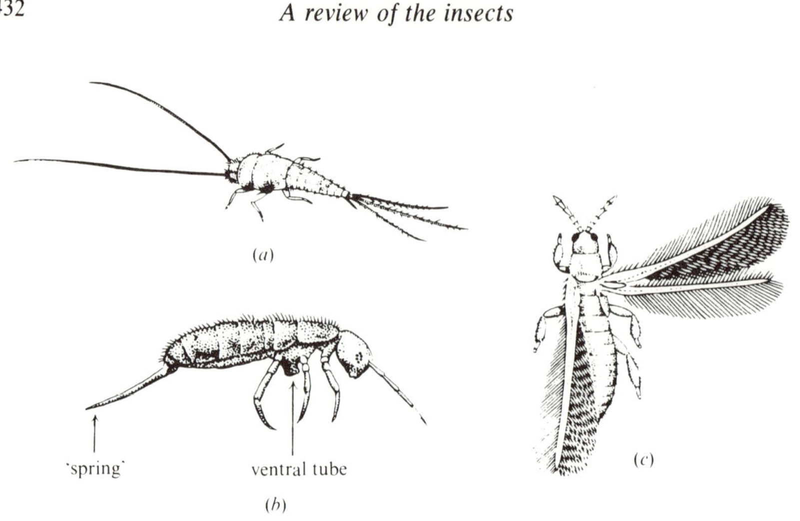

Fig. 19.1. (*a*) The firebrat, *Thermobia domestica* (order Thysanura). Length about 1.5 cm. (*b*) A springtail (order Collembola). (*c*) A thrips (order Thysanoptera). From M. Chinery (1976). *A field guide to the insects of Britain and northern Europe*, 2nd edn. Collins, London.

COLLEMBOLA (SPRINGTAILS)

The springtails are another order of wingless insects, quite different in appearance from the silverfish (Fig. 19.1 *b*). They are probably not very closely related to the other insects. They are small, most of them 1–3 mm long. They have only six segments in the abdomen instead of the 10 to 12 found in other insects. There are two long projections of the abdomen, like two prongs of a fork. These are normally folded forward under the abdomen and held in place by a catch but they can be swung ventrally and posteriorly very suddenly (to the position shown in Fig. 19.1 *b*) to make the animal jump. The ventral tube, shown in Fig. 19.1 (*b*), is characteristic of the order, but its function is unknown.

Springtails are extremely common in soil and leaf litter.

ODONATA (DRAGONFLIES)

The dragonflies are winged insects with long, slender abdomens. They have large eyes and are active by day, feeding on smaller insects which they catch in flight. They hover in a most unusual way which has not been fully explained. They keep the body horizontal and beat the wings up and down (Fig. 19.2 *a*). Most insects have the body more vertical when they hover, and beat their wings horizontally (Fig. 18.7). Fig. 19.2 (*a*) also shows that the wings beat out of phase with each other.

Dragonflies lay their eggs in fresh water. The nymphs (Fig. 19.2 *b*) stay in the water until they are ready to moult to the adult stage (this takes 2 years in the case of *Aeschna*). Like the adults, they are carnivorous. The labium is

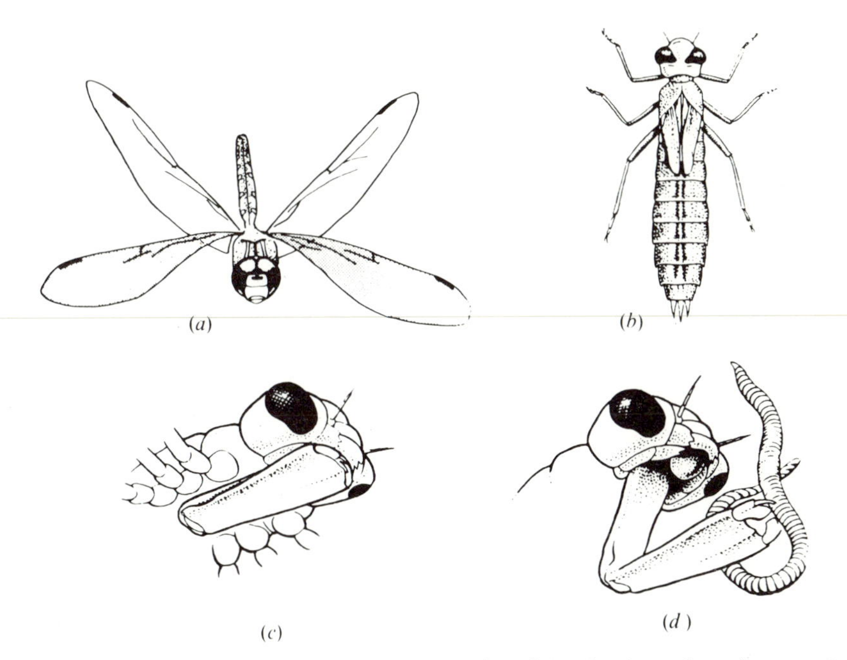

Fig. 19.2. Dragonflies (order Odonata). (*a*) Front view of *Aeschna juncea* hovering, traced from a film. From R. A. Norberg (1975). In *Swimming and flying in nature*, vol. 2, ed. T. Y.-T. Wu, C. J. Brokaw & C. Brennen, pp. 763–81. Plenum, New York. (*b*), (*c*) and (*d*) Nymph of a dragonfly such as *Aeschna* (overall length about 4 cm). The large drawings of the head show the labium (*c*) folded and (*d*) extended to catch a worm. From M. Chinery (1976). *A field guide to the insects of Britain and northern Europe*, 2nd edn. Collins, London.

very long and is normally held folded under the head, but it can be shot out rapidly to seize prey between two hooks at its end (Fig. 19.2*c*, *d*).

Though the nymphs live in water, they have tracheae like terrestrial insects. *Aeschna* and related dragonfly nymphs have gills in the rectum. These gills are leaf-like structures, projecting inwards from the wall of the rectum. They have very thin cuticle, and they have tracheoles looping through them. There are muscles which alternately enlarge the rectum, drawing water in through the anus, and contract the rectum, squirting the water out again. Thus the water bathing the gills is continually renewed. Oxygen dissolved in it must diffuse out of solution into the tracheoles in the gills, and carbon dioxide must diffuse in the opposite direction. The water is normally squirted fairly gently from the rectum but it is sometimes squirted violently, driving the nymph forward by jet propulsion. Nymphs swim in this way when they are disturbed.

The nymphs of another group of dragonflies have gills which project from the posterior end of the abdomen.

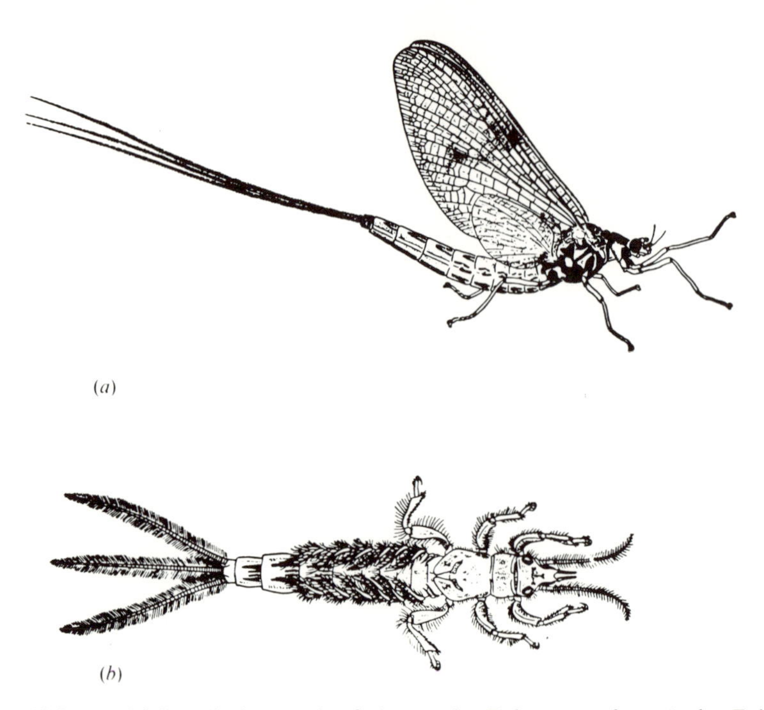

Fig. 19.3. (*a*) Adult and (*b*) nymph of the mayfly *Ephemera vulgata* (order Ephemeroptera). Each is about 2 cm long, excluding 'tails'. From W. Engelhardt & H. Merxmuller (1964). *The young specialist looks at pond life*. Burke, London.

EPHEMEROPTERA (MAYFLIES) AND PLECOPTERA (STONEFLIES)

Mayflies (Fig. 19.3) have quite a long life as nymphs in streams and ponds but their adult, aerial life is very short indeed; it is often less than a day. The nymphs feed on dead plant material and the adults do not feed at all. The feather-like structures along the sides of the abdomen of the nymph in Fig. 19.3 (*b*) are gills.

Mayflies have three 'tails', both as nymphs and as adults (Fig. 19.3). The hind wings of the adults are much smaller than the fore wings. Stoneflies are rather similar insects with similar life-histories but they have only two 'tails' and the hind wings are generally larger than the fore wings.

DICTYOPTERA (COCKROACHES)

A cockroach has already been introduced as an example of a typical insect (Fig. 18.2). Cockroaches are nocturnal, mainly tropical, insects which have become established in temperate regions in kitchens, warehouses and bakeries. The commoner domestic species rarely fly. The praying mantises belong to the same order.

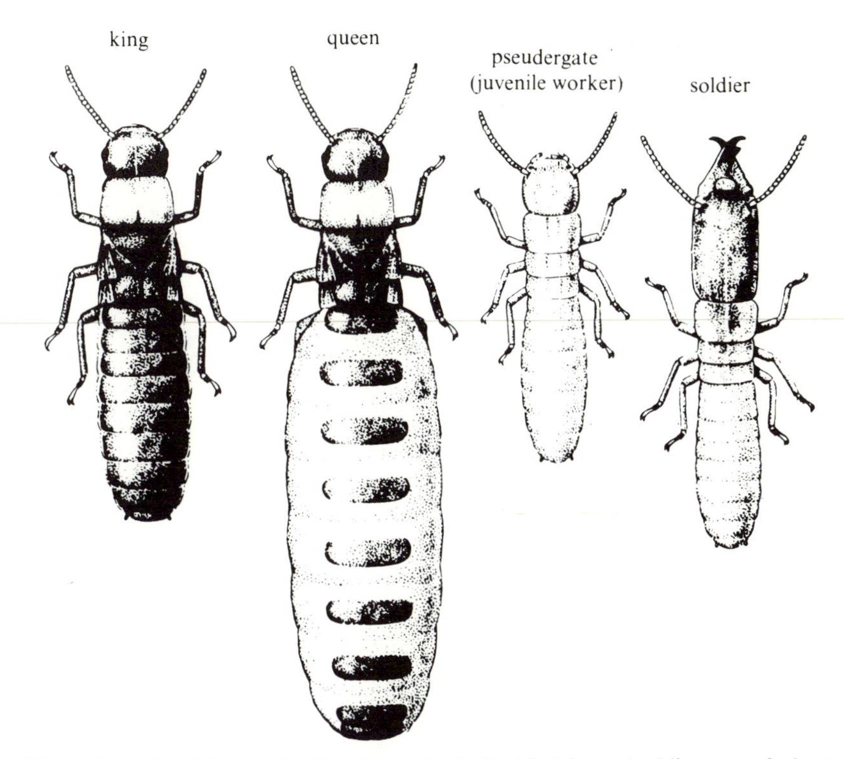

Fig. 19.4. Castes of the termite *Kalotermes flavicollis*. The king and soldier are each about 1 cm long. From M. Chinery (1976). *A field guide to the insects of Britain and northern Europe*, 2nd edn. Collins, London.

ISOPTERA (TERMITES)

The termites are in many ways similar to the cockroaches but they live in large, highly organized colonies. Most of them live in the tropics but there are two European species. One of them is *Kalotermes flavicollis*, which infects dead and diseased vines and trees. Its colonies live in the wood, gradually eating it away. *Kalotermes* is not a serious pest but other termites which live in wood do very serious damage to buildings in the tropics. The damage may be very severe before it is noticed: a painted board may look sound although most of the timber has been eaten away, leaving little more than a skin of paint.

A typical colony of *Kalotermes* consists of 500–1000 individuals, only two of them sexually mature. These two are the king (a male) and the queen (a female). They are shown in Fig. 19.4. Notice that the queen's abdomen is greatly swollen, leaving wide strips of arthrodial membrane between the hard plates. It contains very large ovaries.

Unless the original king or queen has been lost, all the other members of the colony are offspring of the king and queen. Most of them are nymphs but about

3% (or more in small colonies) are adult but sterile soldiers (Fig. 19.4). The king and queen stay together and undertake all the reproduction in the colony. The soldiers face intruding animals and snap their big mandibles at them, defending the colony. All other work is done by advanced nymphs (including pseudergates, Fig. 19.4). These feed on the wood and dispense food from mouth and anus to king, queen, soldiers and younger nymphs. They secrete no enzyme capable of digesting cellulose but have in their guts flagellate protozoa which digest cellulose anaerobically. This produces mainly acetic acid which is absorbed and used by the termite. The termites depend on the flagellates in just the same way as cattle depend on the ciliates and bacteria in the rumen (p. 81). Young termites become infected with flagellates by feeding from the anus of older termites. The extent to which food passes from individual to individual was demonstrated by an experiment in which termites removed from a colony were fed 1 or 2 days on filterpaper impregnated with radioactive phosphate. These individuals, numbering 9% of the colony, were returned to the colony. Within 20 hours, 70% of the members of the colony were radioactive.

Pseudergates also remove the eggs as the queen lays them and take them to another part of the colony where they are reared.

From time to time some of the nymphs in the colony moult to become winged adults which leave through tunnels bored to the surface of the wood by the pseudergates. They fly off and form pairs which shed their wings and mate. Each successful pair becomes king and queen of a new colony composed entirely of their own offspring. The stumps of the shed wings of the king and queen can be seen in Fig. 19.4.

A pseudergate may moult at intervals without growing or changing its form, remaining a pseudergate until death. Alternatively it may in appropriate circumstances, become a winged adult (and eventually a king or queen), or a wingless reproductive adult to replace a dead king or queen, or a soldier. How are the alternatives controlled?

When the king and queen are removed from a colony, even if only for 24 hours, replacements start to develop and reach maturity 4–7 days later. Several are produced but all but two are killed and eaten by the other members of the colony. If a colony is divided by a fine wire gauze screen with the king and queen both on the same side, replacement adults appear on the other side but are promptly eaten. However, a double wire gauze screen allows the orphaned half of the colony to produce its replacement reproductives. It is believed that the information that king and queen are present is conveyed by antennae through the single screen.

If a king or queen is fixed in a partition between two halves of a colony, so that one half has access only to its anterior end and the other to its posterior end, replacement reproductives develop in the anterior half only. This is so even if its abdomen is varnished, but if its anus is blocked replacements appear in the posterior half as well. It seems that some substance released from the anus by the king and queen and eaten by other members of the colony inhibits the

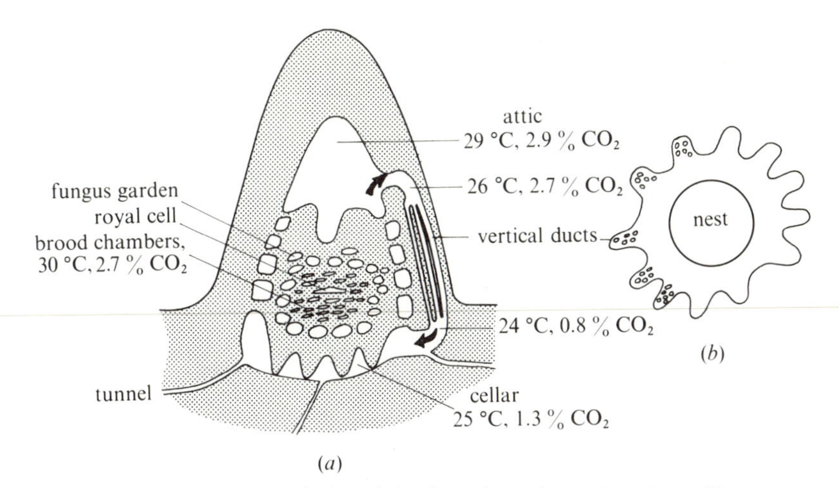

Fig. 19.5. Diagrammatic vertical and horizontal sections through a *Macrotermes natalensis* mound. Temperatures and carbon dioxide concentrations are as measured by Dr Martin Luscher (1961, *Sci. Am.* **205** (1), 138–45). Arrows show how air must circulate by convection.

development of replacements. The inhibition is passed from member to member of the colony by anal feeding: it can be transmitted from one half of a divided colony to the other by pseudergates fixed in the partition, with their heads towards the king and queen. There must be separate inhibitory substances for the two sexes. They are examples of the class of substances known as pheromones which transmit information between members of a species. There is a passage later in this chapter about pheromones produced by moths.

Kalotermes is a rather primitive termite. *Macrotermes natalensis* is a more advanced one. It lives in Africa in colonies of up to at least 2 million members. Though the colonies are so large they still have only one king and one queen. The queen may be as much as 14 cm long, with a relatively tiny head and thorax appended to a huge abdomen. The colony does not depend on juvenile pseudergates to do the routine work: there is an adult worker caste as well as the soldier and reproductive castes. The colony does not live in wood but builds a mound of earth with a nest inside. The mound may be 3 m high, or even more. Termite mounds are very conspicuous features of the East African landscape.

Fig. 19.5 shows the structure of a *Macrotermes* mound. There is a strong outer wall of earth with vertical ridges on its outer surface. Various tunnels run through the wall, especially groups of vertical ducts in the ridges. There are underground tunnels radiating some tens of metres from the mound. In the centre of the mound is the nest, a delicate structure built of clay mixed with saliva. It is divided into interconnecting chambers. There is a royal cell for the king and queen. Around it are brood chambers where the eggs are taken to hatch and where the young nymphs are cared for. Around these are fungus gardens containing spongy 'combs' built of faeces, on which grows a fungus (*Termitomyces*) not found outside termite nests.

Macrotermes has been kept alive for 18 months in a laboratory, feeding on rotten wood without the fungus. The fungus is plainly not essential to its survival. However it has been shown by staining fungus gardens and looking for dye in the guts of the termites, that the combs are systematically eaten and rebuilt. The termites feed largely on dead branches around the mound, travelling to them through the radiating tunnels. The wood is largely cellulose, with 20–30% lignin. Most of the cellulose is digested by bacteria in the gut (not by flagellates in this case). The faeces, which are used to build the combs, contain a little cellulose and a high proportion of lignin. Apparently the fungus breaks down the lignin and so makes the rest of the cellulose accessible for digestion by the termites.

The climate in *Macrotermes* nests has been investigated by Dr Martin Luscher. He bored holes into them with a hollow bit so that he could insert thermometers and suck out gas samples for analysis. The air in the nests was very damp, with relative humidity 96–100% (termites die quickly in dry air). It was warmer than the air outside and its temperature fluctuated much less between day and night. The high temperature must have been due to the metabolism of the termites and fungus.

The vertical ducts in the ridges connect the air space ('attic') above the nest to the 'cellar' below it (Fig. 19.5). The high temperature in the nest must set up convection currents up through the nest and down through the vertical ducts. This brings the air near the surface of the mound where oxygen can diffuse into it and carbon dioxide can diffuse out through the porous earth wall. Fig. 19.5 shows that the carbon dioxide content of the air falls from nearly 3% in the attic to 1% in the cellar. The 2 million termites in a fairly large colony would produce about 10 l of carbon dioxide, and use 10 l of oxygen, every hour.

How can the termites and other social insects have evolved? Only a tiny proportion of the members of a species reproduce. There must be genes which prevent the rest from becoming reproductive adults. How could genes which prevented most of their possessors from breeding appear in an increasing proportion of the population from generation to generation and so become established? While the social habit was evolving, termites prevented by the new genes from breeding would have to compete against termites which bred freely.

It is tempting to suggest that division of labour is efficient and good for the species, and therefore evolved; that some individuals gave up the ability to reproduce, for the greater good of the species. This is nonsense. Suppose an insect appeared, in a non-social species, with a new gene which made it devote its life to the care of its fellows, without reproducing. There might as a result be more insects in the next generation, but none of them would possess the new gene.

An acceptable theory for the origin of social insects has been devised by Dr

William Hamilton. A diploid animal (such as a termite) gets half its genes from each parent. A gene in either parent has a 50% chance of appearing in its offspring. A termite which does not become a king or queen has no offspring, but devotes its time to rearing its brothers and sisters. Two siblings each get 50% of their genes from the king and 50% from the queen. They will on average share 50% of their genes (25% from each parent). Hence a termite's siblings are as similar to it as its offspring would be. A termite does as much to propagate its genes by rearing siblings as it would by rearing an equal number of its own offspring. If rearing siblings is more efficient it will be favoured by natural selection. Notice that this argument depends on the non-reproductive members of the colony all being offspring of the same parents.

Many members of the order Hymenoptera are social. They are discussed later in this chapter.

ORTHOPTERA (LOCUSTS AND CRICKETS)

The members of this order eat plants. Many species of locusts have been serious pests in the past and the desert locust *Schistocerca gregaria* still does dreadful damage to crops in N. Africa and the Middle East.

Locusts have two phases, solitary and gregarious, which are morphologically distinct. Solitary locusts do little damage but in crowded conditions their nymphs develop into gregarious adults which travel in swarms. In favourable conditions the swarms may become enormous, and destroy nearly all the vegetation in their path. In a swarm there may be more than 100 adult locusts per square metre of ground, each able to eat 1.3 g fresh grass daily.

Fig. 18.6 shows a grasshopper. Like other Orthoptera it has long hind legs and uses them for jumping. Desert locusts can jump distances up to 0.8 m but crickets jump less well and are as likely to run away when disturbed, as to jump.

Many Orthoptera produce sounds. Grasshoppers (Acrididae) are not very noisy but mole crickets (*Gryllotalpa vineae*) are. The sound intensity 1 m from the burrow of a mole cricket is 90 dB (1 mW m^{-2}), about the same as the intensity 15 m from a heavy truck or pneumatic drill. Mole crickets can be heard at a distance of 600 m.

The mechanism of sound production is like running a finger along the teeth of a comb: there is a scraper which runs along a toothed file. Grasshoppers have the file on the inner face of the hind leg and the scraper on a wing. Crickets and mole crickets have files on the undersurfaces of the forewings, and the scrapers are veins which project upwards from the fore wings (Fig. 19.6*a*). The sound is made with the wings raised a little from the surface of the body, overlapping right over left (Fig. 19.6*b*). They are vibrated so that the left scraper runs along the right file. It slides easily on the opening stroke because of the way the teeth slope, but the teeth catch on the closing stroke and make both wings vibrate (Fig. 19.6*c*). The region called the harp vibrates particularly strongly.

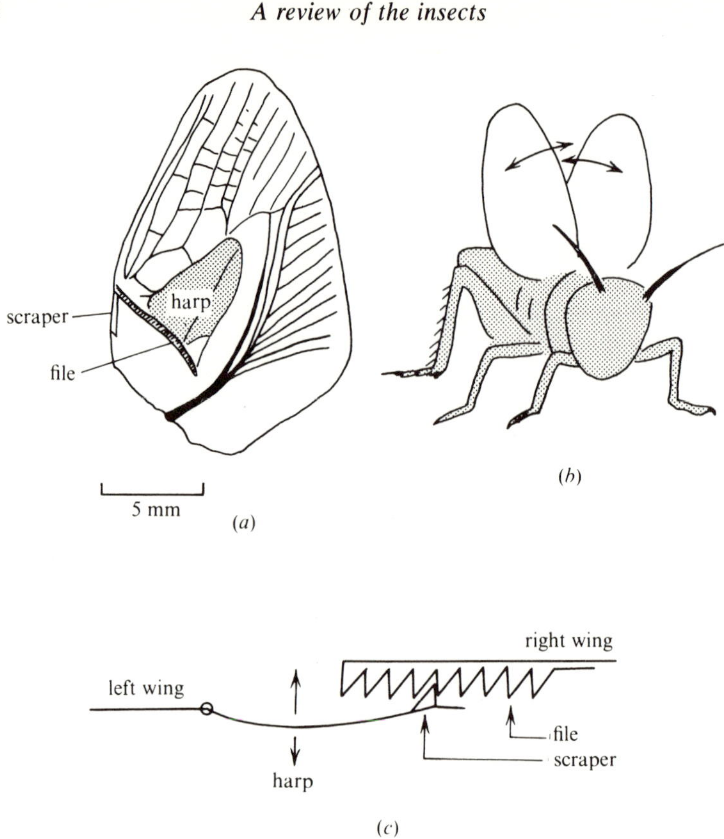

Fig. 19.6. The mechanism of sound production of crickets and the mole cricket. (*a*) Ventral view of the right fore wing of *Gryllotalpa*, re-drawn from H. C. Bennet-Clark (1970). *J. exp. Biol.* **52**, 619–52. (*b*) A sketch of a cricket showing how the fore wings are moved. (*c*) A diagrammatic section through the wings.

In each closing stroke the scraper of *Gryllotalpa vineae* runs over about 30 teeth in about 8.5 ms. Since each tooth contact produces a vibration, this produces a pulse of sound of frequency $30/(8.5\times10^{-3}) = 3500\ s^{-1}$ (Fig. 19.7*b*). The harp has a resonant frequency (like a drumskin) which is also $3500\ s^{-1}$, so the action of the file and scraper makes it vibrate strongly. The resonant frequency has been demonstrated in experiments with cricket (*Gryllus*) wings. Fine cork powder was spread over the wing and musical tones were played through a loudspeaker. At the resonant frequency the powder outside the harp was still but the powder on the harp was vigorously agitated.

The harp acts as a loudspeaker. Loudspeakers are very inefficient if they are small compared to the wavelength of the sound. Ideally the diameter should be at least one third of the wavelength, that is about 3 cm for sound of frequency $3500\ s^{-1}$ in air. The harp of *Gryllotalpa* is only about 5 mm long. In

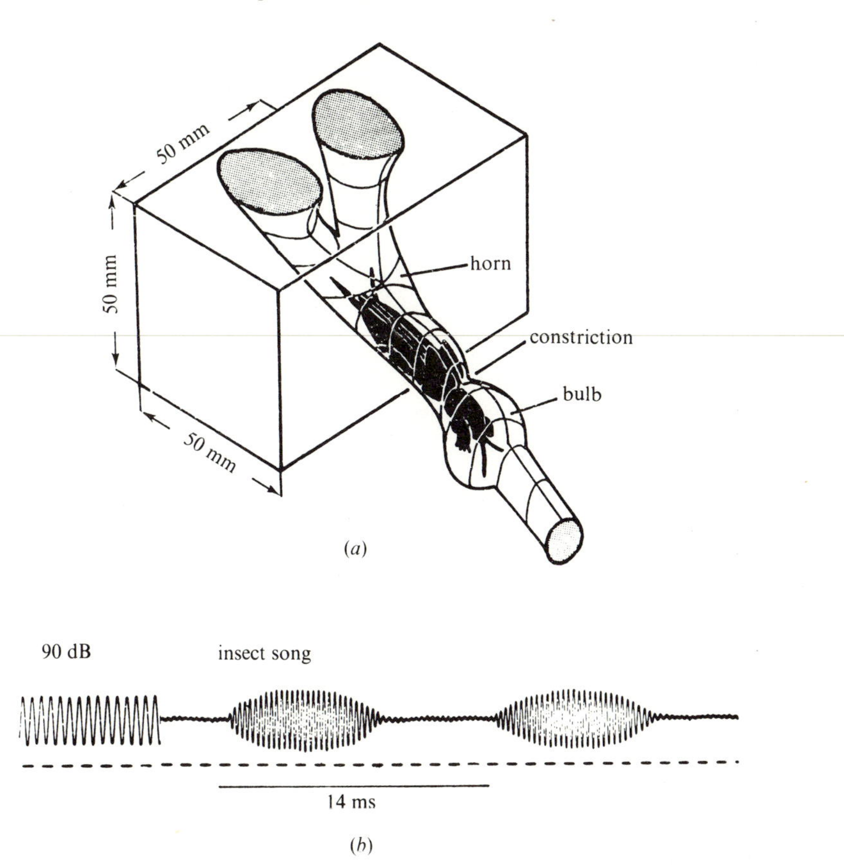

Fig. 19.7. (*a*) *Gryllotalpa vineae* in its burrow. The form of the burrow was discovered by pouring in plaster of Paris, allowing it to set and digging it up. (*b*) An oscilloscope record of the song of *Gryllotalpa vineae*, with a 90 dB calibration signal on the left. From H. C. Bennet-Clark (1970). *J. exp. Biol.* **52**, 619–52.

crickets efficiency is improved by the harp being surrounded by the rest of the wing: a small loudspeaker is more efficient if it is set in a baffle. *Gryllotalpa* has a much more effective arrangement. It uses its short (mole-like) fore legs to dig a remarkable burrow (Fig. 19.7*a*). The burrow has two openings, of diameter about one-third the wavelength of the sound it produces, which are the wide ends of tapering horns. When it is singing, the insect occupies a constriction where the burrow fits quite closely round it. The horizontal channel (on the right of the diagram) connects to a system of tunnels in which the insect searches for food. The horns of the burrow increase the efficiency of sound production in the same way as did the horns of old-fashioned gramophones. (Modern record players do not need horns because their amplifiers make high efficiency unnecessary.) Interference between the sound waves emitted by the two horns concentrates the sound in the insect's median plane, so that relatively little sound energy is directed laterally.

Orthoptera have organs on their legs or abdomens which function as ears, enabling them to hear the sounds they produce.

Only male crickets and mole crickets sing. The song attracts females for mating.

THYSANOPTERA (THRIPS)

Thrips are minute insects, most of them less than 3 mm long. They are most easily found in flowers. They have peculiar mouthparts which are used to pierce plant cells and suck out the contents.

Some thrips have no wings but others have the type of wing shown in Fig. 19.1(*c*), and fly quite well. Some tiny wasps and beetles have similar wings.

HEMIPTERA (BUGS)

The Hemiptera are a large order of insects with sucking mouthparts. Some suck the blood of vertebrates or insects: for instance *Rhodnius* (Fig. 18.18) sucks the blood of mammals. Others with very similar mouthparts suck the juices of plants. They include the aphids, the greenfly and blackfly which are so troublesome to gardeners and farmers.

Fig. 19.8 shows the mouthparts of an aphid. The maxillae and mandibles are long slender stylets. The labium is equally long but stouter, and has a groove which houses the stylets. Only the stylets pierce the plant: the labium folds as they are driven in (Fig. 19.9). The two maxillary stylets interlock so as to form two tubes, the salivary duct and the food canal (Fig. 19.8*b*). The salivary glands (Fig. 19.9) secrete saliva containing an enzyme (pectinase) which probably attacks the middle lamella between adjacent plant cells. Saliva is pumped down the salivary duct to the tips of the stylets and helps to clear a path for them between the cells of the plant. Most aphids insert their stylets into the phloem, which contains sap under turgor pressure. This pressure drives the sap up the food canal, into the aphid's gut. When the stylets of a feeding aphid were cut but left embedded in the plant, sap exuded from them for several days. However, aphids are capable of sucking when necessary. The muscles of the pharynx (Fig. 19.9) can enlarge it, sucking in food.

Aphids can feed continuously for long periods but blood-sucking bugs may succeed best if they can feed quickly before being noticed and dislodged by their victims. Fifth instar larvae of *Rhodnius* of initial mass 0.05 g can drink 0.3 g blood in 15 minutes. They are seldom noticed by their victims, and probably owe this in part to the fineness of their stylets. Since the stylets are fine the food canal is necessarily also fine: its diameter at the tip is about 8 μm. To suck blood fast through so fine a tube requires a large pressure difference, and it has been calculated that absolute negative pressures must be developed

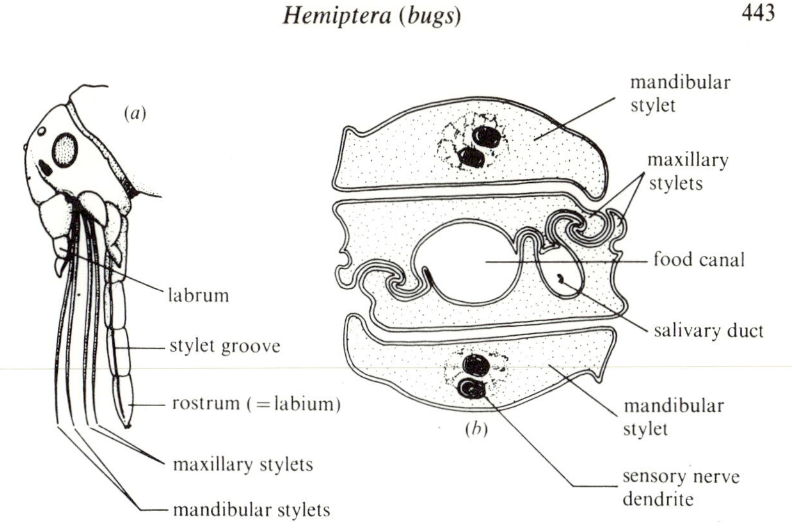

Fig. 19.8. (*a*) The mouthparts of an aphid, separated for clarity, and (*b*) a section through the stylets. From R. Blackman (1974). *Aphids*. Ginn, London.

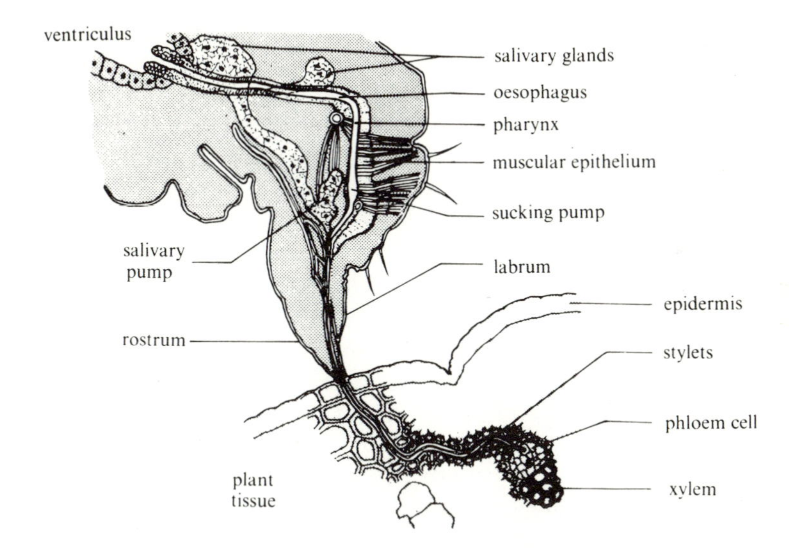

Fig. 19.9. A section through the head of an aphid feeding on a plant. From R. Blackman (1974). *Aphids*. Ginn, London.

in the pharynx. Absolute negative pressures cannot be attained in ordinary water pumps, because they make the water boil, but they also occur in the xylem of trees.

Typical aphid populations consist solely of females during summer. They reproduce parthenogenetically and viviparously: no fertilization is required, and the young are born as nymphs, not laid as eggs. Males appear in autumn and sexual reproduction occurs then, producing eggs which do not hatch until the spring. This life cycle is rather like that of rotifers of the class Monogononta (p. 232). As in the rotifers, parthogenesis makes very rapid multiplication possible. Multiplication is particularly fast since the females of many species contain embryos at birth and start giving birth themselves only 10 days after being born. If the mean generation time is 14 days and each female produces 50 offspring (these are reasonably typical figures) the population is potentially capable of multiplying by a factor of 50 in 14 days or 2500 in 28 days, a very high rate for such large and advanced organisms. Fortunately predation and other causes of mortality prevent the potential from being realized. The offspring of parthenogenetic females are wingless if they are born in uncrowded conditions, and feed on the same plant as the mother. In crowded conditions winged offspring appear and fly off to find other plants.

Their feeding method has made bugs important vectors of disease. *Rhodnius* transmits the trypanosome which causes Chagas' disease, as described on p. 96. Sixty-five per cent of known plant viruses are transmitted by aphids. Some, such as cucumber mosaic virus, are transmitted rather haphazardly; virus particles remaining on the stylets after feeding on an infected plant may infect another plant if the aphid moves to it. Other viruses, such as potato leaf-roll virus, circulate and in some cases probably multiply in the aphid's blood, reach the saliva and are injected with the saliva when the aphid feeds again.

Many of the bugs live on or in water. Pond-skaters such as *Gerris* walk on the surface of water, supported by surface tension. The foot depresses the surface (Fig. 19.10 *a*, *b*) so that the surface tension T exerts an upward force on it. This works particularly well since the contact angle α is obtuse, allowing the surface tension to act vertically when the foot is pressed well down into the water. The contact angle between water and the waxy surface of insect cuticle is about 105°.

Walking on water is only feasible for small animals. The surface tension of an air–water surface is 70 mN m^{-1}. A large *Gerris* might have a mass of 50 mg and so need a force of 0.5 mN to support it. It could be supported by surface tension acting along edges of total length $0.5/70 \text{ m} = 7 \text{ mm}$. The part of each foot which rested on the water (Fig. 19.10 *a*) would be about 1 mm long so six feet each with two long edges supply edges of total length 12 mm, which is ample. Now consider an animal of the size of a sheep, one million times as heavy as the pond-skater. To stand on water it would need feet of total edge length 7 million mm: that is, 7 km. No large animal could conceivably evolve feet big enough to support it by surface tension on water.

Water boatmen such as *Corixa* are bugs which swim in water below the surface. They row themselves along by means of their hind legs, which are fringed with long setae. The details of the swimming action seem not to have been studied. The water beetle *Acilius* (one of the Coleoptera) also swims by means of fringed legs. Its setae are hinged at their bases so as to spread out in

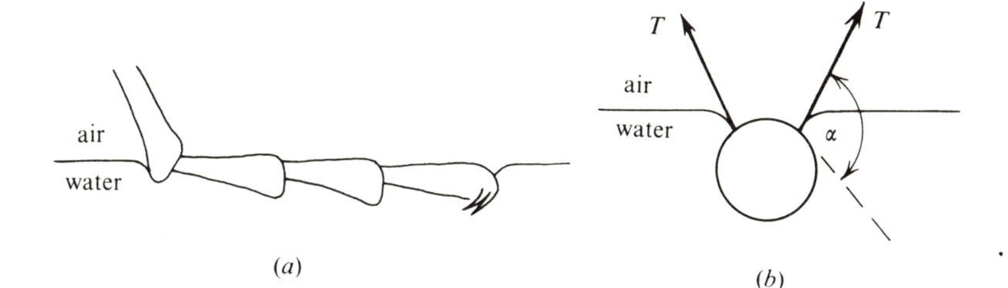

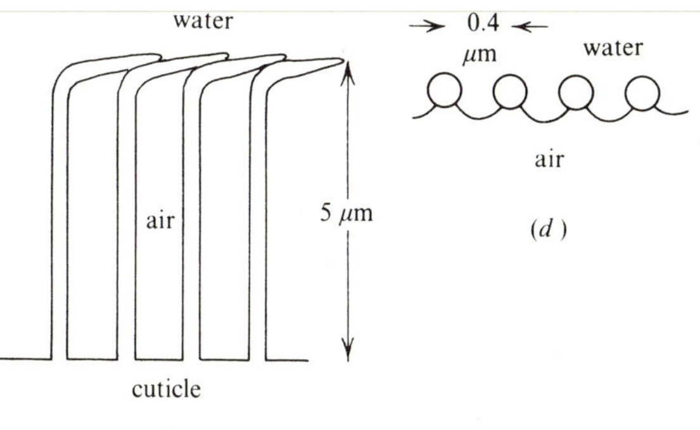

Fig. 19.10. (*a*), (*b*) Diagrams showing in side view and in section the foot of an insect standing on the surface of water. T is a force exerted on a leg by surface tension; α is the contact angle. (*c*) A diagrammatic section through the plastron of *Aphelocheirus*. (*d*) A diagrammatic section on a larger scale through the hooked-over tips of the hairs of the plastron.

the power stroke when they function as the blade of an oar, and trail behind in the forward recovery stroke.

Water boatmen have no gills and have to visit the surface to breathe air. They carry air with them when they dive, most if it under their wings. This air covers the spiracles so oxygen can diffuse from it into the tracheae.

A bubble of air lasts longer than might be thought because, as its oxygen is used up, dissolved oxygen from the water tends to diffuse into it. Air is about 80% nitrogen and 20% oxygen so well-aerated water contains dissolved nitrogen at a partial pressure of 0.8 atm and dissolved oxygen at 0.2 atm. A bubble of air freshly taken by a water boatman contains nitrogen and oxygen at these partial pressures (it is assumed that the insect remains near the surface so that the total pressure in its gas bubble is not much more than 1 atm). As the insect uses oxygen from its bubble the partial pressure of oxygen in the bubble will fall below 0.2 atm, and oxygen will diffuse in from the water. Carbon dioxide will be released into the bubble and will diffuse out into the water. Carbon dioxide has a very much higher diffusion constant in water than oxygen does, so the partial pressure of carbon dioxide in the bubble will rise far less than the partial pressure of oxygen falls. The total pressure in the bubble must remain 1 atm (or slightly more, because of hydrostatic pressure and surface tension) so if the partial pressure of oxygen falls below 0.2 atm, that of nitrogen must rise above 0.8 atm. Nitrogen will diffuse out of the bubble which will get gradually smaller and eventually disappear.

In spite of this, the bubbles last a remarkably long time. *Notonecta* (another water bug which breathes in the same way) survived 7 hours in water saturated with air although prevented from reaching the surface. With a bubble of oxygen in water saturated with oxygen they survived only 35 minutes, for the following reason. Since there was no nitrogen in the bubble the partial pressure of oxygen in it never fell below the partial pressure of dissolved oxygen in the water, so no additional oxygen diffused in. In water saturated with nitrogen they survived only 5 minutes.

The bug *Aphelocheirus* has a much more perfect mechanism of underwater respiration. This is the plastron, a coating of tiny hairs over most of the surface of the body. This coating traps a layer of air which communicates with the air in the tracheae. The layer is very thin since the hairs are only 5 μm high, but it does not diffuse away like *Corixa*'s bubble.

Fig. 19.10(*c*) shows some of the hairs. Notice that they have a right-angle bend at the top. Fig. 19.10(*d*) shows how surface tension prevents water from penetrating between the hairs, even when the pressure of the water is greater than the pressure in the layer of air. (The same principle makes gaberdine raincoats waterproof, although they are porous.) The hairs of the plastron are about 0.4 μm apart (Fig. 19.10*d*) so the total length of hooked-over tops is 2500 mm per square millimetre of plastron, or 2.5×10^6 m m^{-2}. The total length of edge on which surface tension can act is double this. Since the surface tension is 0.07 N m^{-1} the plastron is capable of resisting a pressure of $0.07\times2\times2.5\times10^6$ N m^{-2}, or 3.5 atm, provided the hairs are stiff enough not to buckle under the pressure.

Aphelocheirus under water has a characteristic sheen, due to the plastron. When high pressures are applied to it the sheen fades at 4 atm. It is believed that the hairs keep the thickness of the plastron more or less constant until this pressure is reached: the air is kept at 1 atm while the external pressure rises to 4 atm, making a pressure difference of 3 atm. The fading at this pressure is probably due to the hairs collapsing.

The advantage of the plastron is that since the hairs keep the volume constant, removal of oxygen by respiration does not increase the partial pressure of nitrogen and there is no tendency for nitrogen to diffuse out. The gas in the plastron is held permanently and never needs renewing.

Aphelocheirus lives in the River Volga at depths down to at least 7 m. The pressure at this depth is 1.7 atm, but the partial pressure of dissolved gases in the water is 1 atm or a little less since the gases come from the atmosphere. The air in a plastron can only last at this depth if it is kept at a pressure 0.7 atm or more below the pressure of the surrounding water. The plastron is well able to maintain such a pressure difference.

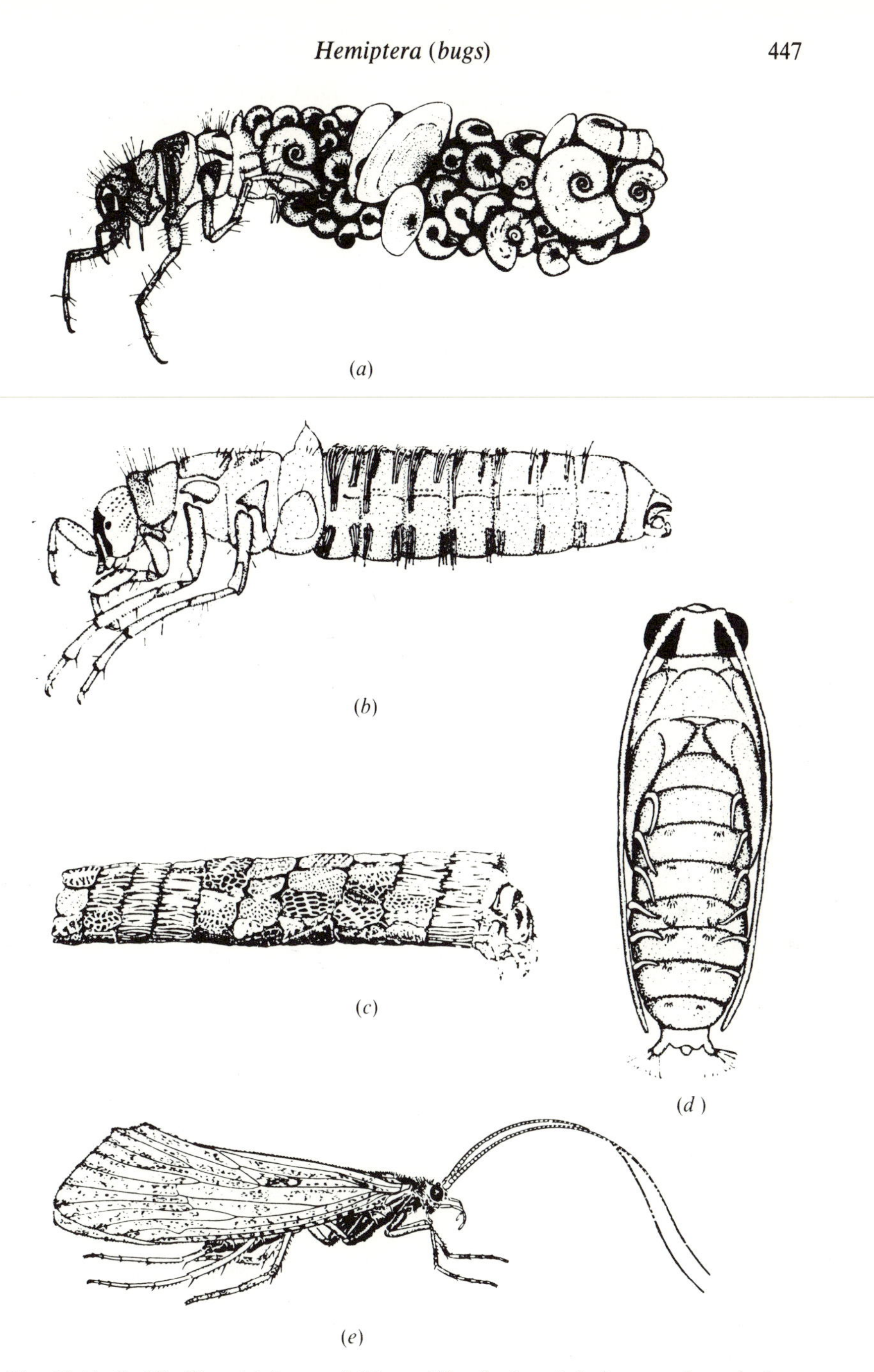

Fig. 19.11. Caddis flies. (*a*) Larva of *Limnephilus flavicornis* in its case. Length about 25 mm. (*b*) Larva of *Limnephilus* sp. removed from its case (20 mm). (*c*) Larva of *Phryganea* sp. in its case (50 mm). (*d*) Pupa of *Plectrocnemia conspersa* (20 mm). (*e*) Adult *Phryganea* sp. (20 mm). (*a*), (*b*) and (*d*) from N. E. Hickin (1952). *Caddis*. Methuen, London. (*c*) and (*e*) from W. Engelhardt & H. Merxmuller (1964). *The young specialist looks at pond life*, Burke, London.

Cicadas are terrestrial bugs which sing. Only males sing and their song attracts females, like the songs of crickets. However, the mechanism of sound production is quite unlike the mechanism in crickets. Membranes on either side of the abdomen are made to vibrate by the direct action of fibrillar muscles. The songs of different species have distinctive rhythms.

TRICHOPTERA (CADDIS FLIES)

Adult caddis flies are insects of unremarkable appearance which are generally found near ponds and streams and seldom fly unless disturbed (Fig. 19.11). Their wings have a coating of fine hairs. Their larvae, known as caddis worms, nearly all live in water. Most of them build protective cases of leaf-fragments, sand or other material bound together with silk. The cases are usually tapering tubes, open at both ends. The larva increases the mean diameter of the case as it grows by adding to the wide end and removing material from the narrow one.

The orders of insects described so far have nymphs more or less like the adults but without wings (see Figs. 18.6 and 18.18). The Trichoptera and the orders which follow have larvae very different from the adult. A more or less inactive pupal stage, in which the structure of the insect is extensively reorganized, intervenes between larva and adult (Fig. 19.11 *d*).

LEPIDOPTERA (MOTHS AND BUTTERFLIES)

Adult Lepidoptera have wings and bodies covered by small scales, which have apparently evolved from hairs like those of caddis flies. Hairs, scales and a range of intermediate structures can be found on the same wing. The colours of butterflies and moths, which are often brilliant, are in the scales. Much of this section is about colour patterns and their functions.

The butterflies are a group of Lepidoptera which are active by day. Most of the other Lepidoptera (the moths) are active at night. Butterflies have unusually large wings for their body mass and beat them at unusually low frequencies. For instance, a cabbage white butterfly (*Pieris brassicae*) of mass 0.14 g had wings 34 mm long and beat them at a frequency of 11 s^{-1}, whereas a cerambycid beetle of the same mass had wings of length 16 mm and beat them at 80 s^{-1}.

Adult Lepidoptera feed mainly on nectar, which is sucked through a long tube formed by the two maxillae. This tube is 15 cm long in the convolvulus hawk moth (*Agrius convolvuli*, wing span about 10 cm) and even longer in some tropical moths. Such long tubes can reach the nectaries of very long, tubular flowers. When not in use the tube is coiled in a tight spiral, ventral to the head. There are muscles which unroll it for use but it rolls up again afterwards by elastic recoil of a strip of resilin.

Nectar is a sugar solution containing hardly any protein or amino acids. Experiments have shown that adult Lepidoptera cannot assimilate proteins mixed with their food. The proteins they need for producing eggs are carried over from the larval stages.

Larval Lepidoptera, known as caterpillars, are quite unlike the adults. They have the three pairs of thoracic legs which are usual in insects and a number of leg-like outgrowths on the long abdomen. They have biting mandibles, and nearly all of them feed on flowering plants. Most of them eat leaves but some eat roots and the caterpillar of the goat moth (*Cossus*) lives in tree trunks, eating the wood. Clothes moth caterpillars (*Tineola*) eat animal material, especially hair and feathers which are both made of the protein keratin. Their natural habitat is probably the nests of birds and small mammals but they cause man considerable annoyance and expense by eating woollen and silk clothes and carpets. Each species of Lepidoptera lays its eggs on the plant or other food appropriate to its caterpillars. In many cases, only one species of plant is used.

The pupa (chrysalis) is often enclosed in a cocoon of silk secreted by modified salivary glands. The cocoons of the cultivated silk moth (*Bombyx mori*) are wound from a single strand of silk which may be more than a kilometre long.

The colours of Lepidoptera (and of other insects) are produced in various ways. Many of them are the colours of pigments. For instance, the yellow of the brimstone butterfly (*Gonepteryx rhamni*) and the orange of the orange-tip (*Anthocharis cardamines*) are pterines, and the reds and browns of many other butterflies are probably ommochromes. Some other insect colours do not involve pigments. The coloured parts of the body are built from materials which in bulk would be colourless, and the colours are due to interference or other optical phenomena.

South American butterflies of the genus *Morpho* have extraordinarily vivid blues on their wings. These blues are produced by interference of light from structures which consist, in effect, of alternating layers of cuticle and air. The principle is the same as the one which makes an oil film floating on water look coloured.

Fig. 19.12(*a*) shows a stack of alternating layers of cuticle and air. Some of the light striking the top of the stack is reflected from there, some passes through the top layer of cuticle and is reflected from the lower surface of the layer, and some penetrates deeper before it is reflected. At every cuticle–air interface the refractive index changes and reflection is apt to occur. Consider the various rays which are reflected into the eye. Rays α and γ come from the same source but are reflected from the upper surfaces of successive layers of cuticle. Ray γ travels $2(a+b)$ further than ray α on its way to the eye. Light of wavelength $2(a+b)$ will arrive in phase in the two rays but light of other wavelengths will arrive more or less out of phase and be partially destroyed by interference. Rays β and δ are reflected from the lower surfaces of successive layers of cuticle and light of wavelength $2(a+b)$ will arrive in phase in them. Light of this wavelength in ray β might be expected to arrive out of phase with light of the same wavelength in ray α since the difference in distance

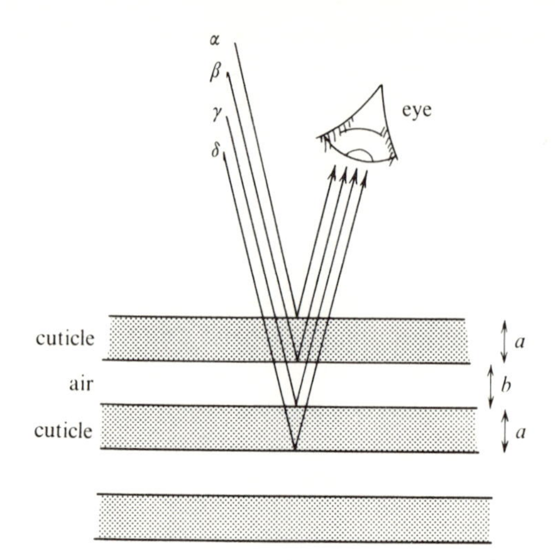

(a)

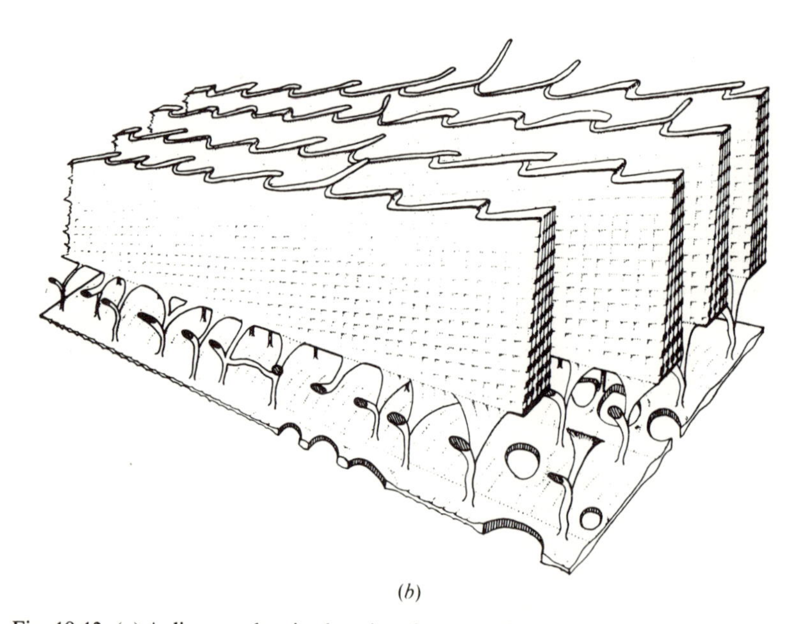

(b)

Fig. 19.12. (a) A diagram showing how interference colours are produced by alternating layers of materials of different refractive index. It is explained in the text. (b) The surface of a scale of the butterfly *Morpho*, from T. F. Anderson & A. G. Richards (1942). *J. appl. Phys.* **13**, 748–58.

for this pair of rays is only $2a$. However the phase gets reversed on reflection from a material of higher refractive index, but not from one of lower refractive index: α and γ have their phase reversed on reflection, but β and δ do not. Light of a particular wavelength will arrive at the eye in phase in all four rays, if a and b are each one-quarter of a wavelength.

It is necessary to explain more precisely what this means. Consider, for instance, blue light of wavelength 400 nm. If it is to be reflected without loss by interference, the air layers should each be 100 nm thick. However the refractive index of cuticle is about 1.5 so the wavelength of the light in cuticle is $400/1.5 = 267$ nm and the thickness of the cuticle layers should be a quarter of this, 67 nm.

Fig. 19.12(*b*) shows the layered structure which produces the interference colours of *Morpho*. The blue scales on the wings have closely-spaced vanes standing vertically on them. Thicker bars run horizontally along the vanes. Since the bars of adjacent vanes are side by side, almost touching, the effect is almost the same as if there were alternate layers of cuticle and air, parallel to the scale surface.

The structure shown in Fig. 19.12(*b*) cannot be seen by light microscopy since the distance between adjacent bars is too small a fraction of the wavelength of light. It was first revealed by electron microscopy, in the very early days of the technique. The first account of it was published in 1942.

For optimum reflection of blue light the bars should all be 67 nm deep and the spaces between them 100 nm deep. In fact the bars vary in depth and the total depth of (bar + space) seemed in the early electron microscope preparations to be only 130 nm. However, it was shown that the structure shrunk badly when the electron beam was turned on.

The colours of Lepidoptera serve various functions. In some cases they are effective as camouflage. The peppered moth (*Biston betularia*), in its original form, is white with black speckling on the wings and body. It often rests on tree trunks and is very inconspicuous on a background of lichen. The Industrial Revolution which started in England in the late eighteenth century had a marked effect on trees near industrial areas. Smoke pollution made lichens much less plentiful, leaving bare bark which was blackened by soot. In 1850 a black form of the peppered moth was caught for the first time, in Manchester. It is far less conspicuous than the white form on bare, sooty bark and has become extremely common. It has become the predominant form in the industrial areas of Britain, and to the east of them where pollution is carried by prevailing winds. It remains rare in the rural areas in the west and north of Britain, in Cornwall, N.W. Wales and the Scottish Highlands. It has been shown by crossing the black and white forms that the black form is controlled by a single dominant gene.

Many species of moth, and some of other insects, have evolved black forms since the Industrial Revolution. These forms are called industrial melanics. Their evolution is perhaps more spectacular than any other evolutionary change

Fig. 19.13. A caterpillar of the eyed hawk moth (*Smerinthus ocellatus*) on a twig of willow (*Salix* sp., one of its natural food plants). The caterpillar on the left is in its normal resting position, with the ventral surface uppermost. From H. B. Cott (1975). *Looking at animals: a zoologist in Africa.* Collins, London.

witnessed by man. (Their only obvious rivals are some cases of the evolution of pesticide resistance.)

The efficacy of the camouflage of the two forms of peppered moth has been tested in field experiments. Equal numbers of marked specimens of the two forms were released in a rural wood in Dorset, where the natural population was entirely or almost entirely white. Flycatchers (*Muscicapa striata*) and other birds were seen feeding on them. Later, traps were set which attracted the moths by strong lights or by pheromones (see p. 455). Fourteen per cent of the white moths were recaptured but only 5% of the black ones. Predators had presumably found the black moths more easily than the white ones. The experiment was repeated in a wood near Birmingham where the natural population was predominantly black. There 28% of the black specimens were recaptured, but only 13% of the white ones.

There are many examples of camouflage among Lepidoptera (and other animals), some of them involving much more detailed resemblance than the resemblance of the peppered moth to bark. Look at Fig. 19.13. The caterpillar is green with oblique yellow lines, matching the green leaves with their oblique yellow veins. Not only are the colours right but the caterpillar on the left (in its natural position) looks flat like a leaf rather than cylindrical. This is because its uppermost (ventral) surface is dark and its lower (dorsal) one is lighter, counteracting the effect of the shadow thrown by light falling on it from above. When the caterpillar is turned the other way up (on the right) it is far more conspicuous. Many animals are countershaded in this way but since most of

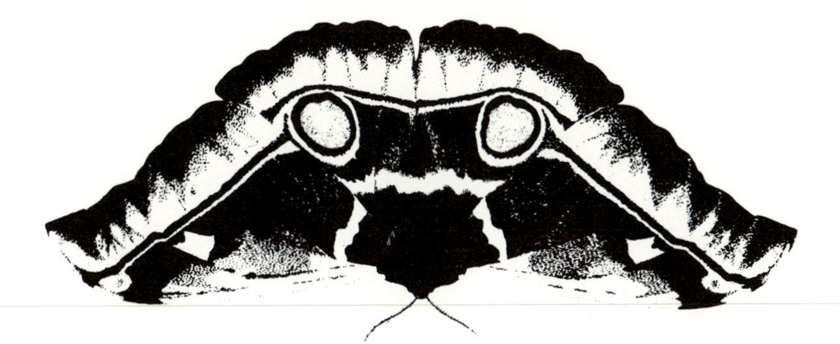

Fig. 19.14. An East African moth, *Bunaea alcinoë*, displaying its eye spots. Wing span about 14 cm. From H. B. Cott (1975). *Looking at animals: a zoologist in Africa.* Collins, London.

them habitually keep the dorsal surface uppermost, most of them are darker on the dorsal than on the ventral surface.

Many moths and butterflies have patterns on their wings which resemble the eyes of mammals. An example from Africa is shown in Fig. 19.14. When the wings are folded the eye spots are hidden by the fore wings. When they are spread the moth looks disconcertingly like the face of a predator such as the genet (*Genetta*). When a moth like this is found by a bird which threatens to eat it, it spreads its wings. The bird suddenly finds itself confronted by the appearance of a dangerous predator, and may well take flight. The effect of such displays on birds has been tested. A small horizontal glass screen was arranged so that patterns could be projected onto it from below. It was put in a cage with a bird. A mealworm (*Tenebrio* larva) was placed on the screen and when the bird approached to eat it, a pair of eyes or other patterns were made to appear suddenly on either side of it. It was found that feeding was discouraged to some extent, by each of the patterns that were tried. However, eye-like patterns were more effective than circles which in turn were more effective than crosses. This was true when the experiments were done with hand-reared birds which had never encountered a vertebrate predator. The birds were chaffinches (*Fringilla coelebs*), yellowhammers (*Emberiza citrinella*) and great tits (*Parus major*).

Warning coloration is another common phenomenon in Lepidoptera and other insects. Many animals are protected against predators by having a sting (like wasps, *Vespula* etc.), or an unpleasant taste, or by some other device. These devices would not be very effective if predators did not learn to recognize and avoid the protected animals: it is far better to be avoided than to be seized and spat out again, probably mutilated. Protected animals are often conspicuously coloured in bold patterns. For instance wasps (order Hymenoptera) have black and yellow striped abdomens. Birds stung by wasps learn not to attack them

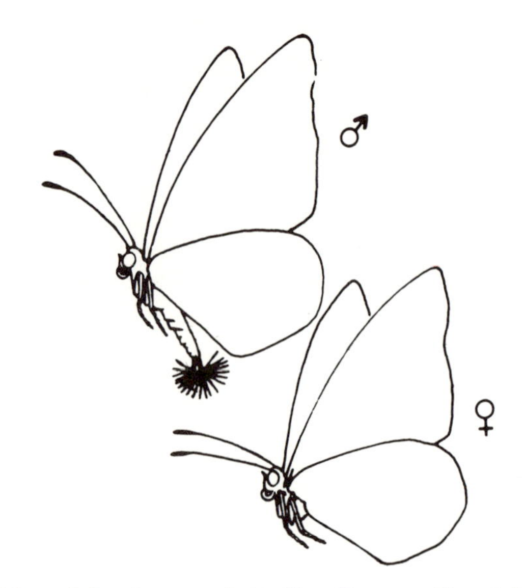

Fig. 19.15. Male and female queen butterflies (*Danaus gilippus*) in courtship flight. From L. P. Brower, J. v. Z. Brower & F. B. Cranston (1965). *Zoologica, NY* **50**, 1–39.

and it has been shown that they remember not to attack them for at least several months. The caterpillars of the cinnabar moth (*Callimorpha jacobaeae*) are striped black and orange. It has been shown that birds find their taste very unpleasant.

Wasps and cinnabar moth caterpillars both have encircling stripes, of black and either yellow or orange. The resemblance may benefit both. It has been shown experimentally that birds which have encountered wasps are less likely to attack cinnabar moth caterpillars, when they first meet them, than birds which have not. Similarities of appearance between noxious species which are believed to give this sort of advantage are described as Müllerian mimicry.

There are also many cases of apparently harmless and palatable insects resembling ones which are protected by stings or unpleasant taste. The moth *Sesia apiformis* has transparent, almost scale-free wings and a black and yellow striped abdomen, like a wasp. Many hover-flies (family Syrphidae, order Diptera) also have black and yellow striped abdomens. The superficial resemblance to wasps probably makes predators avoid these apparently harmless mimics. This type of mimicry is known as Batesian mimicry.

Pheromones are chemicals used for communication between members of a species. Certain pheromones play major roles in the reproduction of Lepidoptera. Fig. 19.15 shows how the male queen butterfly induces the female to copulate. He flies after her, and as he overtakes her he spreads two tufts of hairs which scatter a dust over her. The active ingredient in the dust is an alkaloidal ketone. It induces the female to alight and submit to copulation.

Removal of the tufts drastically reduces mating success. Males of many other Lepidoptera dispense aphrodisiac pheromones in similar fashion.

Many female moths emit volatile pheromones which attract males from a distance. These sex attractants bring the sexes together in the same way as do the songs of male crickets and cicadas (p. 448). Different species use different compounds just as different cicadas use different rhythms, but most of the compounds seem to be unsaturated 12- or 14-carbon acetates or alcohols. Some of them have been used in attempts to control moths of which the caterpillars are destructive pests. In some cases traps have been baited with synthetic sex attractant, to capture male moths. A different approach has been tried with the pink bollworm (*Pectinophora gossypiella*) which is the most important pest of cotton crops in many countries. Small quantities of a synthetic compound which acts as a sex attractant (though not quite identical with the natural pheromone) were placed, closely spaced, all over the fields. The odour was everywhere, and the males had difficulty locating the females. At the end of the season the number of caterpillars per boll of cotton in the treated fields was less than 10% of the number in untreated ones.

DIPTERA (FLIES)

The Diptera include the house fly (*Musca domestica*), the mosquitoes and many other familiar insects. They have only one pair of wings, the anterior pair. The posterior pair is represented only by a pair of small club-shaped structures, the halteres. These look useless, but flies cannot fly without them. A fly with its halteres cut off can still beat its wings at the normal frequency and through the normal angle, but it loses control in the air. It cannot fly a straight course and it is liable to fall on its back. It can fly again if a thread is attached to its abdomen so as to trail behind in flight. The thread tends to stabilize flight, keeping the fly flying in a constant direction, just as the feathers at the rear end of an arrow stabilize its flight.

It thus seems that the halteres have an important function in the control of flight. They vibrate at the same frequency as the wings. They act like gyroscopes which keep reversing their direction of rotation. If the insect deviates from a straight path torques must act on the halteres, like the torques which act when you try to turn the axis of a spinning gyroscope to a new direction. The halteres are far too small to have any appreciable direct stabilizing effect on the direction of flight (like the effect of the gyroscopic stabilizers used in ships) but they have sense organs in them which can detect gyroscopic torques. It has been shown that they can do this, by electrical recording from their nerves. The halteres apparently serve as rotation detectors, similar in function to the semicircular canals of vertebrate ears though quite different in mechanism. They control the reflexes which make stable flight possible, just as the semicircular canals and otoliths of vertebrates control the reflexes of balance.

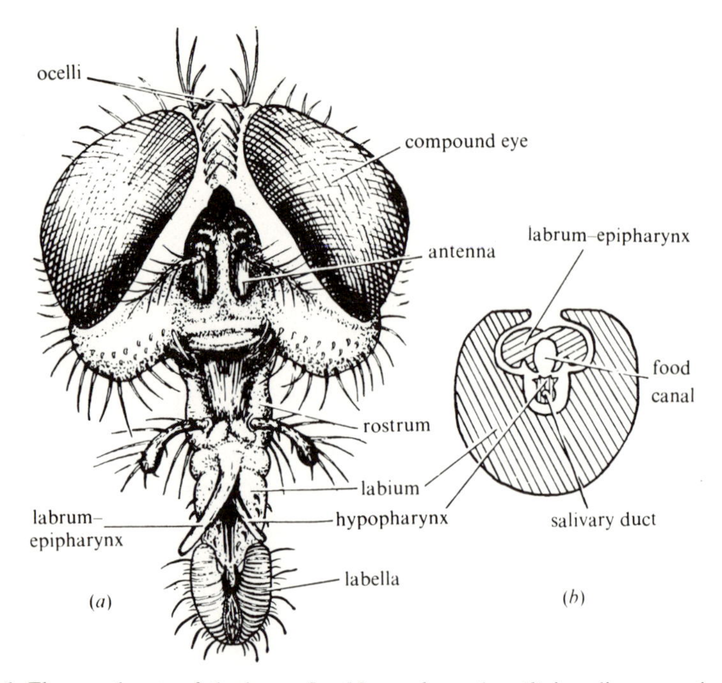

Fig. 19.16. The mouthparts of the house fly, *Musca domestica.* (*b*) is a diagrammatic section. From V. A. Little (1963). *General and applied entomology*, 2nd edn. Harper & Row, New York.

Neither haltere can detect rotations about its own axis of vibration but since the axes of the left and right halteres are not parallel to each other, the two between them can detect rotation about any axis. They can detect roll, pitch and yaw in any combination.

The thoraxes of Diptera have curious elastic properties. If a fly such as the blow-fly *Sarcophaga* is anaesthetized with carbon tetrachloride its wings behave like an electric light switch. A switch will stay put in the 'on' position or in the 'off' position but not in intermediate positions. Similarly the fly's wings will rest fully up or fully down but not in any other position. The mechanism which is responsible is called the click mechanism. It does not work in dead flies because it depends on tension in a pair of small muscles.

Diptera have mouthparts adapted for drinking liquid food. Fig. 19.16 shows the mouthparts of a house fly. The end of the labium is a grooved pad, the labella. The grooves converge on the food canal, which is formed by the labrum (upper lip) and hypopharynx (a projection from the floor of the mouth). Flies feed on a great variety of liquid foods, or foods made liquid by spreading saliva on them. The labella is adapted for mopping these liquids up. Mosquitoes have piercing mouthparts superficially very like the mouthparts of aphids (Fig. 19.8), but close examination shows that they are constructed on the same pattern as those of the fly. The food canal is enclosed by the labrum dorsally and the hypopharynx ventrally, not by the pair of maxillá as in aphids.

Mosquitoes feed mainly on nectar and other sugar solutions which they suck up this food canal. Some species feed only on these foods and must draw on reserves for the proteins needed for egg production. However the females of many species need a meal of blood before producing each batch of eggs. *Aedes aegypti* is one of these species. It can be fed conveniently in the laboratory by providing pads of cotton saturated with nutrient solutions. In an experiment, females kept on pads soaked in diluted honey laid no eggs but remained alive and vigorous for several months. Others given the same honey solution with skim milk added laid on average one egg per day each, and ones given the honey solution with haemolysed beef blood laid five eggs each per day.

Mosquitoes and other flies which suck human blood are important vectors of disease. *Aedes aegypti* transmits the virus which causes yellow fever. Various other mosquitoes transmit the protozoa which causes malaria (p. 92) and the nematodes which cause elephantiasis (p. 238). Sand-flies (*Phlebotomus*) and tsetse flies (*Glossina*) transmit the Protozoa which cause leishmaniasis and sleeping sickness (p. 96).

The fruit fly *Drosophila melanogaster* has played an outstandingly important part in genetics. It is 3 mm long or less and is relatively easy to rear in large numbers in bottles. Only 9 days is needed for each generation (at 25 °C). There are only four pairs of chromosomes. There are many viable mutant forms with peculiarities of eye colour, wing structure, etc., which are easy to recognize under a low-power microscope.

The usefulness of *Drosophila* first became apparent betwen 1910 and 1915, when T. H. Morgan and A. H. Sturtevant carried out the experiments which showed for the first time how genes are arranged in line along the length of chromosomes. The phenonemon of crossing-over enabled them to discover the order in which the genes are arranged along each chromosome. By 1915 their maps showed the relative positions of 50 genes, and many hundreds of genes have been mapped since. These very important experiments are explained in textbooks of genetics.

Twenty years later another advantage of *Drosophila* was exploited. In common with other Diptera, *Drosophila* has giant chromosomes in its larval salivary glands. The DNA chains are no longer in these than in ordinary chromosomes but they have many replicates: each giant chromosome is a bundle of identical DNA chains arranged side by side, and is accordingly thick enough for easy observation under a microscope. The giant chromosomes can be stained by a technique which does not colour them uniformly, but in a complex pattern of bands which makes it possible to distinguish one part of the chromosome from another even after mutations which disturb the order of the parts. A combination of genetic experiments with examination of giant chromosomes made it possible to determine not merely the order of the genes along the chromosomes, but their actual positions.

Drosophila has been used in a great many other investigations which have

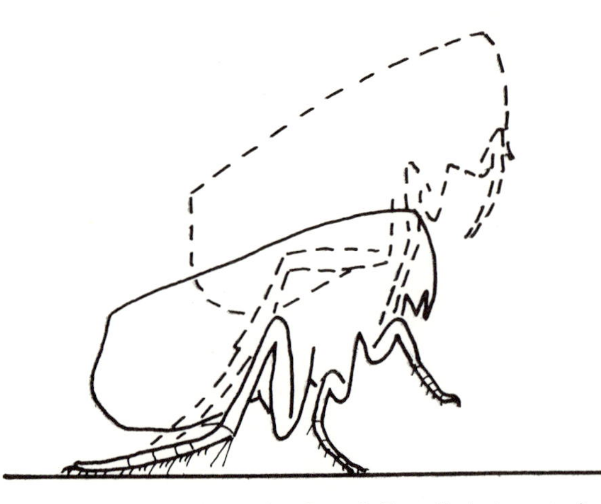

Fig. 19.17. Superimposed outlines of a flea (*Spilopsyllus*) about to jump and (broken outline) taking off. The flea is 1.5 mm long. From drawings based on high-speed cine films by H. C. Bennet-Clark & E. C. A. Lucey (1967). *J. exp. Biol.* **47**, 59–76.

added enormously to our knowledge of genetics. A particularly fascinating recent investigation concerned a group of genes which change appendages drastically. One of them makes flies develop tiny legs on their heads in place of antennae. Others transform the halteres into wings. The normal forms of these genes seem to have an important function in development: they subject a group of cells in the appropriate part of the body to the influence of the various genes which specify the details of an antenna, leg, wing or haltere.

SIPHONAPTERA (FLEAS)

Adult fleas are ectoparasites on the skin of birds and mammals, and drink blood through piercing mouthparts. They have no wings (Fig. 18.4). Their larvae are worm-like and there is a pupal stage.

Fleas drop off their host after feeding and have to find a new host to feed again. They respond to the heat radiated by a warm-blooded animal by jumping, and with luck a jump will land them among the hair or feathers of a potential host. They jump by extending their large hind legs (Fig. 19.17).

The legs have to extend very rapidly. Consider the rabbit flea *Spilopsyllus*, which is only 1.5 mm long but can jump to a height of 35 mm. Cine films show that it leaves the ground at a velocity of 1.0 m s^{-1} (This would be enough to take it to a height of 50 mm if it jumped precisely vertically and were not slowed down by air resistance.) When it extends its legs, its body moves forward only about 0.4 mm before the feet leave the ground (Fig. 19.17). Hence it has to accelerate from rest to 1.0 m s^{-1} in this tiny distance. Its mean velocity over this distance must be about half the final velocity so it must extend its legs and travel the 0.4 mm in about 0.8 ms. This is about the time taken for a single downstroke of the wings of a mosquito, but only oscillating fibrillar muscles can contract so fast. No known muscle can complete an isolated contraction in so short a time.

The flea's jump is made possible by a catapult mechanism. When a boy uses a catapult he stretches the rubber relatively slowly, storing elastic strain energy. This energy is released much more quickly, in the elastic recoil. The elastic elements of the flea's catapult are blocks of the elastic protein resilin, which is so important in the flight of other insects (p. 413). There is one block in each hind leg which is distorted by a large muscle, storing energy. The leg does not move until a smaller muscle releases a trigger mechanism. This allows the block to make its elastic recoil and extend the leg very suddenly.

HYMENOPTERA (ANTS, BEES AND WASPS)

Hymenoptera have the first segment of the abdomen joined to the thorax in such a way as to seem part of the thorax. Most of them have a pronounced waist in the second abdominal segment, which enhances the effect. Wasps and ants have biting mouthparts quite like those of grasshoppers (Fig. 18.3) but bees such as *Apis* have only small mandibles, and a long tongue-like extension of the labium. This is deeply grooved so as to be almost tubular, and nectar is drunk through it.

Ants are social insects, and so are many species of wasps and bees. Their colonies resemble termite colonies in many respects. Each generally includes just one reproducing female, the queen. Only a small proportion of the members of the colony become capable of reproduction: most members are sterile workers or (in ants) soldiers. Two differences from termite colonies are that there is no king in the colony (the queen is fertilized before founding the colony, and stores the sperm she receives) and that all the workers and soldiers are female.

The Hymenoptera cannot be divided into a primitive non-social group and an advanced social one. Rather, many of the taxonomic groups include both social and non-social species. Zoologists who have studied them closely believe that social habits have evolved at least eleven times within the order. They seem to have evolved only once among other insects, in the termites. This suggests that the Hymenoptera have some special characteristic which makes them particularly apt to evolve social habits.

The characteristic in question seems to be their mechanism of sex determination. The great majority of metazoan animals have both sexes diploid, but one sex has two X chromosomes and the other an X and a Y. In Hymenoptera the males are haploid and the females diploid. Ova are haploid. Ova which are left unfertilized remain haploid and develop into males but ones which are fertilized receive a second haploid set of chromosomes from the spermatozoon, and become diploid females.

Since males are haploid, all the spermatozoa produced by an individual male carry the same genes. Two sisters share all the genes they get from their father but on average only half the genes they get from their diploid mother. A gene which is present in a female has a 75% chance of appearing in a sister but only (as in termites) a 50% chance of appearing in a son or daughter. Female Hymenoptera make more of their genes appear in succeeding generations by rearing reproductive sisters than they would by rearing an equal number of their own offspring.

Rearing brothers is less effective. A gene present in a female has only a 25% chance of appearing in her brother, because the only genes he shares with her are half of the ones she got from her mother. A female hymenopteran which devotes its life to rearing brothers and sisters will only do better (from a genetic point of view) than it could by rearing an equal number of sons and daughters, if it devotes more effort to rearing sisters than brothers. Some brothers (or other males) must of course be reared, and the scarcer males are the better the chance each one has of breeding. It can be shown that the best strategy for a worker with the option of rearing sisters carrying 75% of her genes or brothers carrying 25% of them, is to devote three times as much effort to rearing sisters as to rearing brothers. Intact colonies of many species of ant have been dug up, weighed and counted. It has been found that the weight of the young reproductive females being reared in the colony is typically about three times the weight of the young males.

A gene in a male hymenopteran has a 100% chance of appearing in a particular daughter but only a 50% chance of appearing in a brother or sister, so a male gets more genetic advantage from daughters than from rearing an equal number of brothers or sisters (since males develop from unfertilized eggs, males have no sons). He will do better as a father than he would as a worker.

It follows that females are particularly prone to evolve into workers, but males are most unlikely to do so. This is presumably why all the workers are female. The males are all capable of reproduction and play no part in the care of the young.

Worker ants have no wings and find food by running over the ground. In some species a worker which has found a piece of food too big to bring back single handed lays a scent trail back to the nest. Other workers follow the trail to find the food.

Honey bees (*Apis mellifera*) fly to find the nectar and pollen on which they feed. Plainly, they cannot lay scent trails through the air. Nevertheless those that find good sources of food apparently inform their fellows where the food is. It may be several hours before any bees at all find a dab of honey left on cardboard out of doors. Once it has been found by one bee, many more soon arrive.

The method used to transmit the information was discovered by Professor Karl von Frisch. He showed that bees which had found good sources of food presented samples to their fellows when they returned to the hive, and

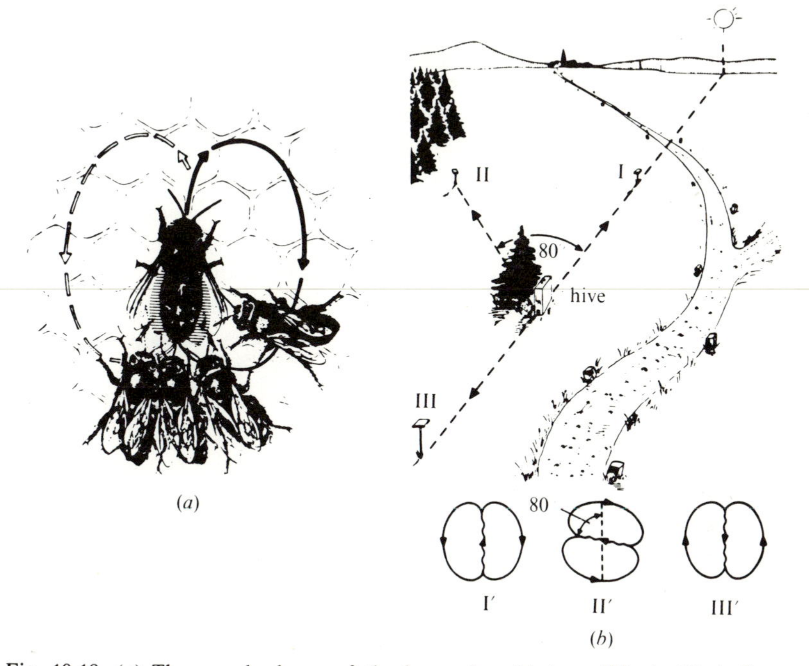

Fig. 19.18. (*a*) The waggle dance of the honey bee (*Apis mellifera*). (*b*) A diagram showing how direction is indicated when the dance is performed on a vertical surface. I′, II′ and III′ are dances at the hive indicating food at I, II and III. From K. von Frisch (1967). *The dance language and orientation of bees.* Harvard University Press, Cambridge, Mass.

performed symbolic dances. The samples informed the others of the scent of the food source and the dances told them where it was.

If the food is very near the hive the dancer simply runs round in circles, reversing the direction of circling from time to time. If it is more than about 100 m away the dance shown in Fig. 19.18 is performed. The dancer makes a short straight run (up the page in Fig. 19.18 *a*) and then returns in a semicircle to its starting point. It repeats this several times, returning in semicircles alternately to the left and to the right. During the straight part of the run it wags its abdomen rapidly from side to side and buzzes. The further the food is, the longer is the straight run: it lasts about 0.5 s when the food is 500 m from the hive, 2 s at 2000 m and 4 s at 4500 m. This has been demonstrated by placing dishes of sugar solution at different distances from the hive. Bees marked with paint so as to be recognizable individually were observed feeding at the dishes and subsequently dancing at the hive.

Further experiments tested the accuracy of the information transmitted by the dance. A dish of scented sugar solution was placed at a spot which a few marked bees had been trained to visit. Cards with the same scent but no food were placed in line with the hive and dish, some between them and some beyond the dish. An observer waited beside each card. When the marked bees found the dish they returned to the hive, gave samples of the scented solution to other bees and danced. These other bees set out, and the number which visited each card was recorded. If the food was 300 m from the hive, few bees visited cards less than 200 m from the hive or more than 400 m from it. If it was 2000 m away, few visited cards less than 1700 m or more than 2300 m from the hive. It seems that the dance conveys distances accurate to within a few hundred metres.

Other experiments with cards set out in an arc of a circle, centred on the hive, showed that the dance also conveys information about direction: few bees visited cards more than 15° to either side of the direction of the food. More elaborate experiments have been performed to confirm that it really is the dance which conveys this information.

Fig. 19.18(*b*) shows how direction is indicated. The dance is normally performed in the hive, on the vertical surface of a comb. If the direction of the food is towards the sun the straight (tail-wagging) run is made vertically upwards. If it is at an angle θ to the sun the run is made at an angle θ to the vertical (Fig. 19.18*b*). The direction relative to the sun can be indicated correctly even when the whole area is in the shadow of a mountain so that the sun cannot be seen from the hive, from the food or from any intermediate point. There is good evidence that bees estimate the direction of the hidden sun by observing the plane of polarization of the light from a patch of sky. A suggestion was made on p. 426 about a possible mechanism for detecting the plane of polarization.

The waggle dance is remarkably informative. In recent experiments designed to test direction-finding more precisely than von Frisch did, the standard deviation of the angle at which bees flew in response to a dance was about 4°, so few bees were more than 8° off course (in a normal distribution, 96% of items fall within two standard deviations of the mean). The dance can apparently show which 16° sector includes the required direction, so it can specify $360/16 = 22$ distinct directions. In other tests the standard deviation of the distance flown averaged about 60 m, so few bees were more than 120 m wrong and distance was being effectively specified in 240 m bands. Since the distance may be up to 12 km, $12000/240 = 50$ distinct distances can be specified. Thus the dance can specify $22 \times 50 = 1100$ different combinations of direction and distance. It can probably convey even more information than this since rich sources of food provoke particularly vigorous dances.

Information is measured in 'bits'. One bit is the information given by answering 'yes' or 'no' to a question which permits no other answer, so n bits suffice to distinguish between 2^n alternatives. Since $1100 \simeq 2^{10}$, the waggle dance can convey at least 10 bits of information. This is more than any other animal except man is known to be able to convey by any form of language. It is far more than can apparently be conveyed by the calls of birds. About 15 (nearly 2^4) distinct calls have been recognized for chaffinches (*Fringilla coelebs*) and each of several other species which have been studied closely, so a call can apparently convey only 4 bits of information.

COLEOPTERA (BEETLES)

The Coleoptera is the largest order in the animal kingdom. About 0.3 million species have been described, out of 1–1.5 million species of animal. The fore wings of beetles are thick and stiff and are not flapped in flight. They are called elytra (singular: elytron) and protect the delicate hind wings when the beetle is not flying. The hind wings are generally longer than the elytra but are folded at rest so as to be completely covered. Beetles have biting mouthparts both as larvae and as adults. Most of them eat plant material but most ladybirds (Coccinellidae) eat aphids and other insects. Water beetles such as *Dytiscus* are also carnivorous.

The beetles include many destructive pests. The boll weevil (*Anthonomus grandis*) is a notorious pest of cotton. Both the larvae and the adults feed on the flower buds and developing fruit. Wireworms are larvae of click beetles (*Elateridae*). They live underground and damage crops by feeding on roots. The larvae of woodworm (*Anobium punctatum*) and death watch beetle (*Xestobium rufovillosum*) bore in dead wood and ruin furniture and the structural timbers of buildings. Mealworms (*Tenebrio* spp.) and flour beetles (*Tribolium* spp.) infest food stores. However, some of the predaceous beetles are useful in agriculture, feeding on pests. A ladybird (*Rodolia cardinalis*) was used in the late nineteenth century in the first successful case of biological control of a pest. The cottony-cushion scale (*Icerya purchasi*, Hemiptera) had been imported accidentally into California from Australia. It had multiplied enormously and had become a serious threat to the citrus fruit industry. It is not a pest in Australia and it was found that it is preyed on there by *Rodolia*. *Rodolia* was introduced into California and the epidemic was checked.

FURTHER READING

GENERAL

See the list for chapter 18

ISOPTERA

Hamilton, W. D. (1964). The genetical evolution of social behaviour [2 papers]. *J. theoret. Biol.* **7**, 1–52.

Howse, P. E. (1970). *Termites: a study in social behaviour.* Hutchinson, London.

Krishna, K. & Weesner, F. M. (ed.) (1969–70). *Biology of termites*, 2 vols. Academic Press, New York & London.

Luscher, M. (1961). Social control of polymorphism in termites. *Symp. R. entomol. Soc.* **1**, 57–67.

Luscher, M. (1961). Air-conditioned termite nests. *Sci. Am.* **205**(1), 138–45.

Wilson, E. O. (1972). *The insect societies.* Harvard University Press, Cambridge, Mass.

ORTHOPTERA

Bennet-Clark, H. C. (1970). The mechanism and efficiency of sound production in mole crickets. *J. exp. Biol.* **52**, 619–52.

Chapman, R. F. (1976). *A biology of locusts.* Arnold, London.

Michelsen, A. & Nocke, H. (1974). Biophysical aspects of sound communication in insects. *Adv. Insect Physiol.* **10**, 247–96.

HEMIPTERA

Blackman, R. (1974). *Aphids.* Ginn, London.

Dixon, A. F. G. (1973). *Biology of aphids.* Arnold, London.

Hinton, H. E. (1976). Plastron respiration in bugs and beetles. *J. Insect Physiol.* **22**, 1529–50.

Mittler, T. E. (1957–8). Studies on the feeding and nutrition of *Tuberolachnus salignus* (Gmelin) (Homoptera, Aphididae) [2 papers]. *J. exp. Biol.* **34**, 334–41; **35**, 74–84.

Thorpe, W. H. (1950). Plastron respiration in aquatic insects. *Biol. Rev.* **25**, 344–90.

LEPIDOPTERA

Anderson, T. F. & Richards, A. G. (1942). An electron microscope study of some structural colours of insects. *J. appl. Phys.* **13**, 748–58.

Birch, M. C. (ed.) (1974). *Pheromones.* North-Holland, Amsterdam.

Blest, A. D. (1957). The function of eyespot patterns in the Lepidoptera. *Behaviour* **11**, 209–55.

Hinton, H. E. (1976). Recent work on the physical colours of insect cuticle. In *The insect integument*, ed. H. R. Hepburn, pp. 475–96. Elsevier, Amsterdam.

Kettlewell, B. (1973). *The evolution of melanism: the study of a recurring necessity.* Clarendon, Oxford.

Sheppard, P. M. (1975). *Natural selection and heredity*, 4th edn. Hutchinson, London.

DIPTERA

Gillett, J. D. (1971). *Mosquitoes.* Weidenfeld & Nicolson, London.

Lea, A. O., Dimond, J. B. & DeLong, D. M. (1956). Role of diet in egg development by mosquitoes (*Aedes aegypti*). *Science* **123**, 890–1.

Oldroyd, H. (1964). *The natural history of flies.* Weidenfeld & Nicolson, London.

Pringle, J. W. S. (1948). The gyroscopic mechanism of the halteres of Diptera. *Phil. Trans. R. Soc.* **233B**, 347–84.

Shorrocks, B. (1972). *Drosophila.* Ginn, London.

SIPHONAPTERA

Bennet-Clark, H. C. & Lucey, E. C. A. (1967). The jump of the flea; a study of the energetics and a model of the mechanism. *J. exp. Biol.* **47**, 59–76.

HYMENOPTERA

Free, J. B. (1977). *The social organization of honeybees.* Arnold, London.

Gould, J. L. (1975). Honey bee recruitment: the dance-language controversy. *Science* **189**, 685–93.

Sudd, J. H. (1967). *An introduction to the behaviour of ants.* Arnold, London.

Trivers, R. L. & Hare, H. (1976). Haplodiploidy and the evolution of the social insects. *Science* **191**, 249–63.

von Frisch, K. (1967). *The dance language and orientation of bees.* Harvard University Press, Cambridge, Mass.

COLEOPTERA

Evans, G. (1975). *The life of beetles.* Allen & Unwin, London.

20

Spiders and horseshoe crabs

Phylum Arthropoda (cont.)
- Class Merostomata (horseshoe crabs)
- Class Arachnida
  - Order Acari (ticks and mites)
  - Order Scorpiones (scorpions)
  - Order Araneida (spiders)
  - and other orders

Fig. 20.1 shows a common British spider of a genus also found in N. America. It belongs to the group of spiders which build orb webs to catch insects, and part of this chapter is about the webs.

Like other spiders, *Araneus* has its body divided, by a constricted waist, into an anterior prosoma and a posterior opisthosoma. The prosoma bears all the legs and corresponds roughly to the head and thorax together of an insect. The opisthosoma corresponds to the abdomen. There are four pairs of walking legs. Immediately anterior to them are a pair of pedipalps which look like legs but are smaller. The maxilliary lobes are projections from the bases of the pedipalps which lie on either side of the mouth and are used for manipulating food. The pedipalps of males have cavities at their tips which are used in mating, as will be described. The most anterior appendages of all are the chelicerae which have long fangs with poison glands opening at their tips (Fig. 20.2*a*). They are used to inject poison into prey.

Two of the principal joints in the legs of spiders have muscles to bend them but none to extend them. Instead, the legs are extended hydraulically. The muscles of the body wall of the prosoma contract, driving blood into the legs and straightening the joints. The pressures involved have been recorded by means of a transducer. They are generally low, but rise to brief peaks of up to 60 kN m^{-2} (0.6 atm), when the spider moves vigorously.

Fig. 20.3 shows the internal organs of a spider. The main body cavity is a haemocoel as in other arthropods and there is the usual dorsal heart with ostia. The brain encircles the oesophagus and there is another big ganglion posterior to it in the prosoma. This is the whole of the central nervous system. The gut has many diverticula both in the prosoma and in the opisthosoma where the branching diverticula are known as the digestive gland. There are branched Malphigian tubules which are believed to function like the Malpighian tubules of insects. There are in addition a pair of coxal glands which open

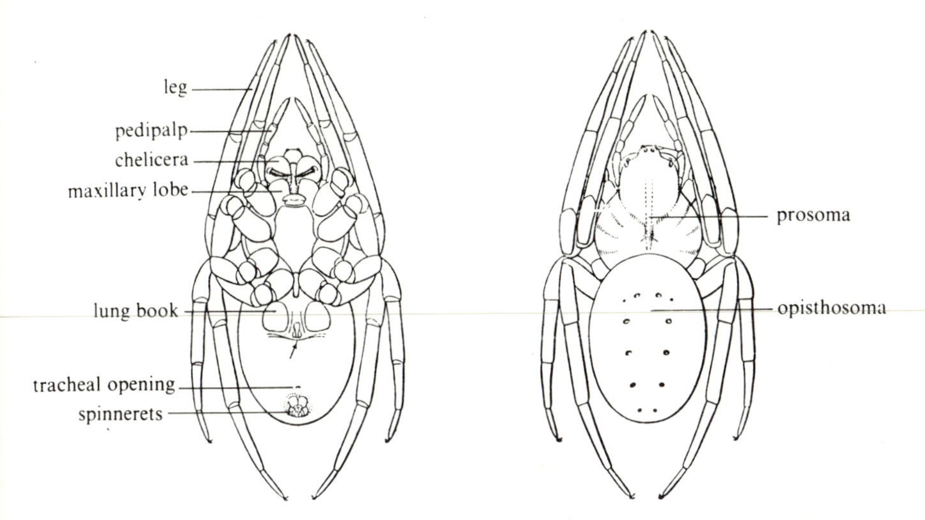

Fig. 20.1. Ventral and dorsal views of the garden spider, *Araneus diadematus*. Length (excluding legs) about 12 mm. After W. S. Bristowe (1971). *The world of spiders*, 2nd edn. Collins, London.

at the bases of the first pair of legs. They resemble the green glands of crustaceans and are presumably excretory organs.

As well as two types of excretory organ there are two types of respiratory organ. There are tracheae which open through a ventral spiracle near the posterior end of the opisthosoma. They are branching air-filled tubes like the tracheae of insects. Only their distal ends are shown in Fig. 20.3. There is also a pair of lung books which open more anteriorly on the ventral surface of the opisthosoma. They are cavities with leaves like those of a book extending into them. The leaves are hollow, filled with blood which circulates through them. Pillars cross their cavities so that blood pressure does not inflate them and obliterate the air spaces between them (Fig. 20.2*b*). The cuticle of the leaves is very thin, so oxygen can diffuse rapidly from the air to the blood inside.

Spiders have complicated equipment for producing silk for webs and other structures. There are many silk glands which open through separate nozzles on the spinnerets at the posterior end of the body. The next section of this chapter is about silk and webs.

The reproductive organs of *Araneus* are quite simple and both sexes have a genital opening on the ventral surface of the opisthosoma. The male makes a little mat of silk and deposits on it a drop of sperm from which he fills the cavities in his pedipalps. He then goes in search of a female. When he finds one he approaches her cautiously and if she submits to his advances he inserts his pedipalps one at a time into the openings which lead to her spermathecae (Fig. 20.3), filling the spermathecae with sperm which is stored in them until it is needed.

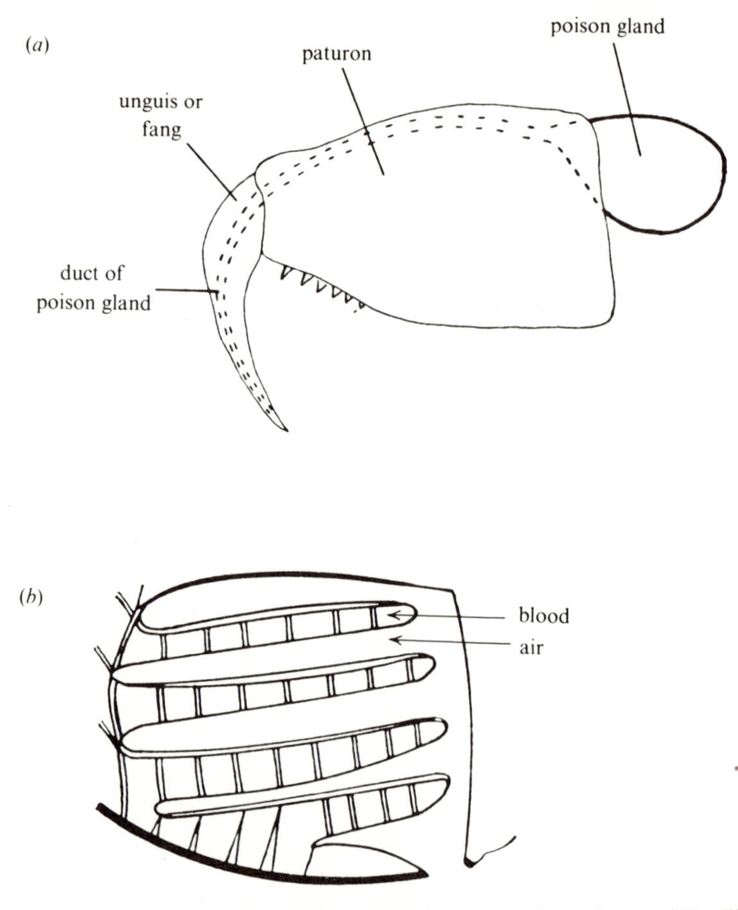

Fig. 20.2. (*a*) A spider's chelicera and its poison gland. From K. R. Snow (1970). *The arachnids: an introduction.* Routledge & Kegan Paul, London. (*b*) A diagrammatic section through a lung book. Only four leaves are shown but the lung books of most spiders have many more, up to 150. From T. Savory (1964). *Arachnida.* Academic Press, New York & London.

In the autumn the female lays a batch of several hundred eggs which she covers and fastens to a solid surface with a layer of silk. She dies shortly afterwards but the eggs do not hatch until spring. Young spiders are similar in shape to adults. They moult several times as they grow and become adult in their second summer.

Fig. 20.4 shows two other members of the Arachnida. *Ixodes* (Fig. 20.4 *a*, *b*) is one of the ticks, relatively large parasitic members of the Acari. It is an ectoparasite of sheep and cattle, and sucks their blood. The chelicerae bear sharp blades which cut an incision in the skin of the host. The toothed

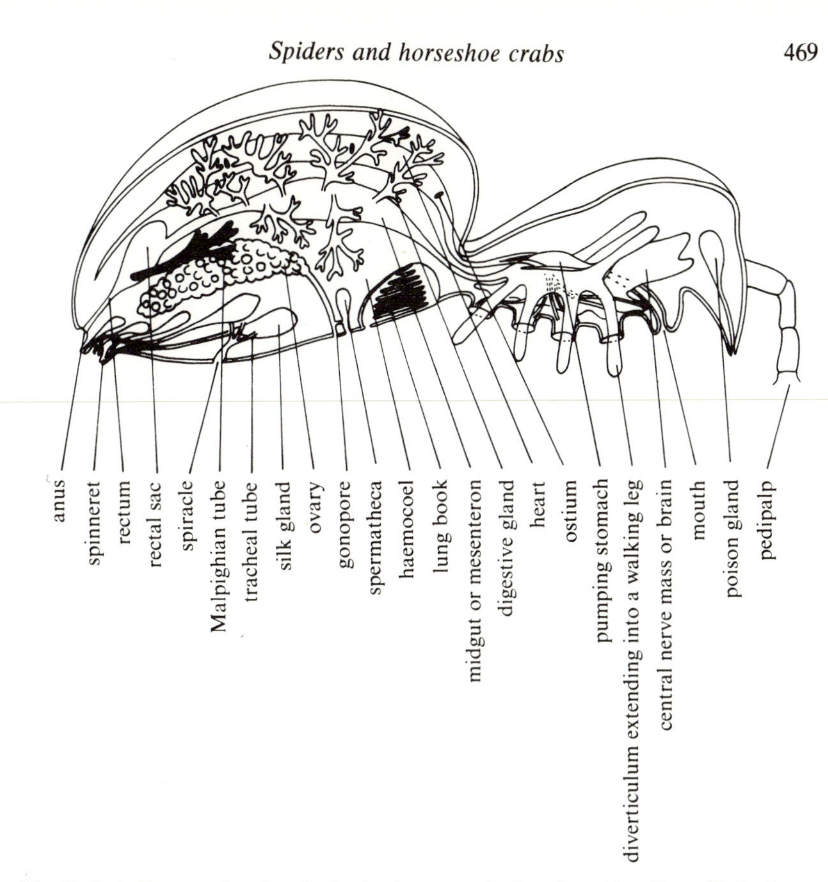

Fig. 20.3. A diagram showing the internal organs of a female spider. From K. R. Snow (1970). *The arachnids: an introduction.* Routledge & Kegan Paul, London.

hypostome is pushed into the wound and anchors the tick while it feeds. The tick remains attached for several days taking an enormous meal of blood. Though it lives for 3 years it feeds only once in each of its two larval stages, and once as an adult.

Ixodes has pedipalps on either side of the chelicerae, and four pairs of walking legs. There is no division between prosoma and opisthosoma.

The smaller members of the Acari are called mites. Some of them are parasitic but many are not. Mites are extremely common in soil.

Buthus (Fig. 20.4 *c*, *d*) is a scorpion, a member of the order Scorpiones. It is common in the Mediterranean countries. The big pincers which look like the chelae of crabs are pedipalps. The chelicerae are a much smaller pair of pincers. There are four pairs of walking legs attached to the prosoma. The opisthosoma is segmented. Its anterior end is as broad as the prosoma but it tapers to a narrow tail which has at its end the dreaded sting, a spine with a pair of poison glands opening near its tip. Poison is squeezed from the glands

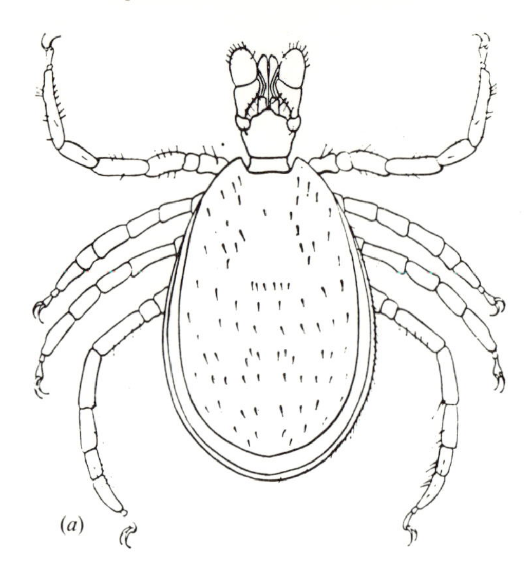

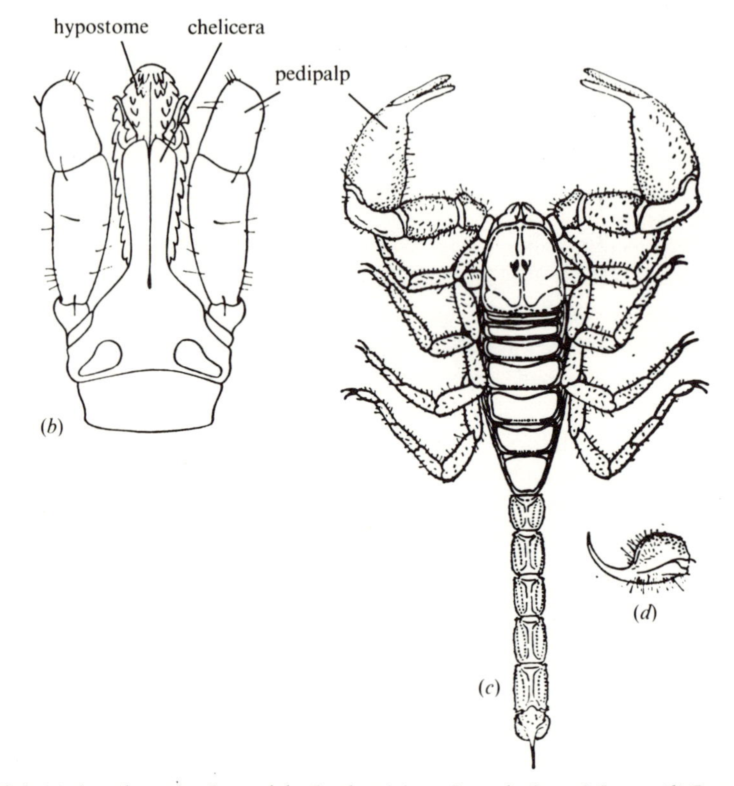

Fig. 20.4. (*a*) A male castor bean tick, *Ixodes ricinus*. Length about 2.5 mm. (*b*) Dorsal view of the snout of *Ixodes ricinus*. (*c*) A scorpion, *Buthus occitanus*. (*d*) A lateral view of the sting of *Buthus*. From T. Savory (1964). *Arachnida*. Academic Press, New York & London.

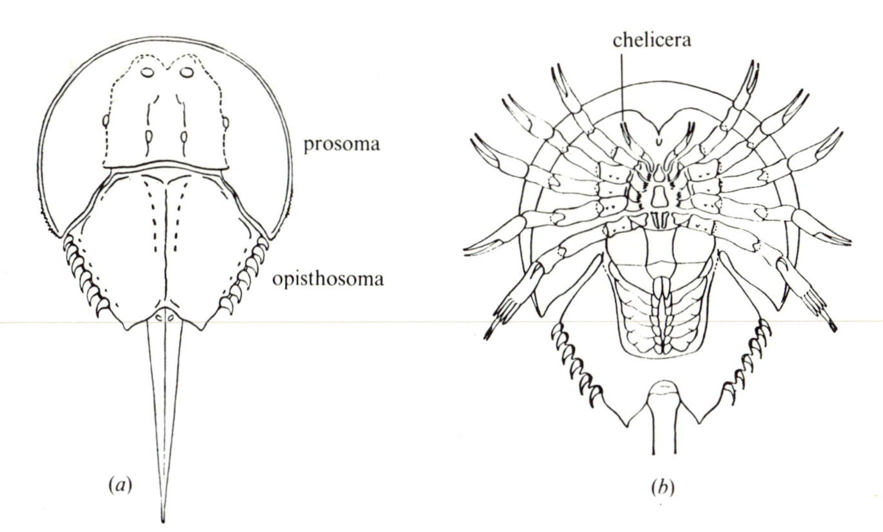

Fig. 20.5. Dorsal (*a*) and ventral (*b*) views of the horseshoe crab *Limulus polyphemus*. The limbs are normally kept bent so that they are not visible in a dorsal view. Width of carapace up to about 38 cm. From T. Savory (1964). *Arachnida*. Academic Press, New York & London.

as the sting strikes. The sting of *Buthus* causes fever in man, and considerable pain. It has been known to kill children, and some other scorpions are even more dangerous.

Scorpions are predators. They grab an insect or spider with their pedipalps, tear through the body wall with their chelicerae and suck out the juices.

There are only five living species in the class Merostomata. It used to be customary to include them in the Arachnida, which they resemble in many ways. One of them is *Limulus polyphemus* (Fig. 20.5) which lives in shallow water on the Atlantic coast of N. America. It crawls on the bottom, feeding on worms and molluscs.

Limulus has a prosoma and an opisthosoma (joined by a hinge) and in addition a spine at the posterior end. It has no mandibles. The first pair of appendages are short, chelate chelicerae. There are five pairs of walking legs: the first pair correspond to the pedipalps of Arachnida but are very like the others. The bases of the legs lie alongside the mouth and bear processes which are used for breaking up food. Six pairs of flaps underneath the opisthosoma are modified limbs with gills on their posterior surfaces. Movements of the flaps pump water over the gills, in through the slots between the prosoma and opisthosoma and out posteriorly.

There are four eyes on the prosoma, a pair of simple eyes near the anterior edge and a pair of compound ones more laterally. The compound eyes are discussed later in the chapter.

Fossils so like *Limulus* that they are put in the same genus are found in

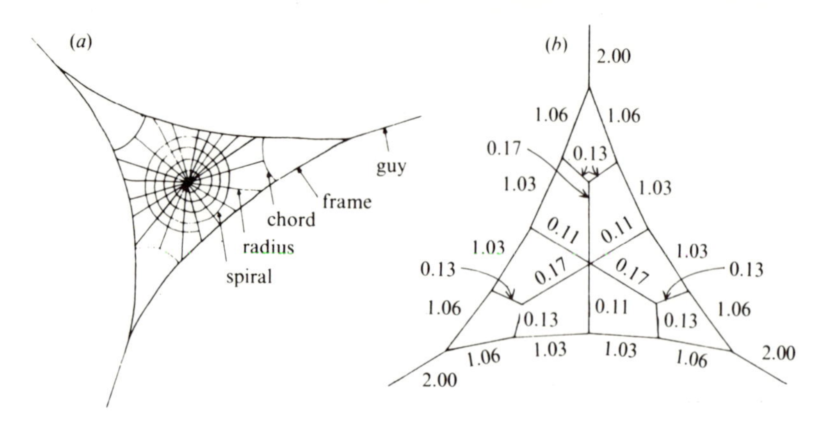

Fig. 20.6. (*a*) A web of *Araneus*. A real web would be less regular than this diagram and would have more turns in the spiral. Diameter about 35 cm. (*b*) A simplified diagram of the web showing the force which must act in each thread when a tensile force of 2 units is exerted on one guy. From S. A. Wainwright, W. D. Biggs, J. D. Currey & J. M. Gosline (1976). *Mechanical design in organisms*. Arnold, London.

Triassic rocks (about 200 million years old). Other very similar fossils are found in Devonian rocks (almost 400 million years old).

SILK AND WEBS

Different spiders make different kinds of web. Among them, the orb webs of *Araneus* (Fig. 20.6) are particularly well known. They are made of two distinct kinds of silk, secreted by different glands. The guys, chords and radii are made of frame silk, which is not sticky. The spiral is made of viscid silk coated with a glue-like substance which forms droplets so that the spiral looks like a string of tiny glass beads.

The web of *Araneus* is a net for catching insects. It is build in a vertical plane so that flying insects are apt to collide with it and be trapped by the glue on the spiral. When this happens the spider runs up to the prey, injects it with poison from the chelicerae, wraps it in silk and carries it off to eat.

To appreciate the design of the web we need to know the mechanical properties of the two types of silk. Fig. 20.7 shows apparatus which has been used to measure them. The silk is glued to the yoke and to the force transducer. A motor turns the capstan, tilting the beam and stretching the silk. The force on the silk is registered by the force transducer. The movement of the beam, and so the extension of the silk, is registered by the displacement transducer (an LVDT). Results from the tests are shown in Table 20.1. Both frame silk and viscid silk are remarkably strong, about as strong as cellulose fibres from flax and 10 times as strong as tendon collagen. Though they are about equal in strength, viscid silk is by far the more extensible: it stretches to about three

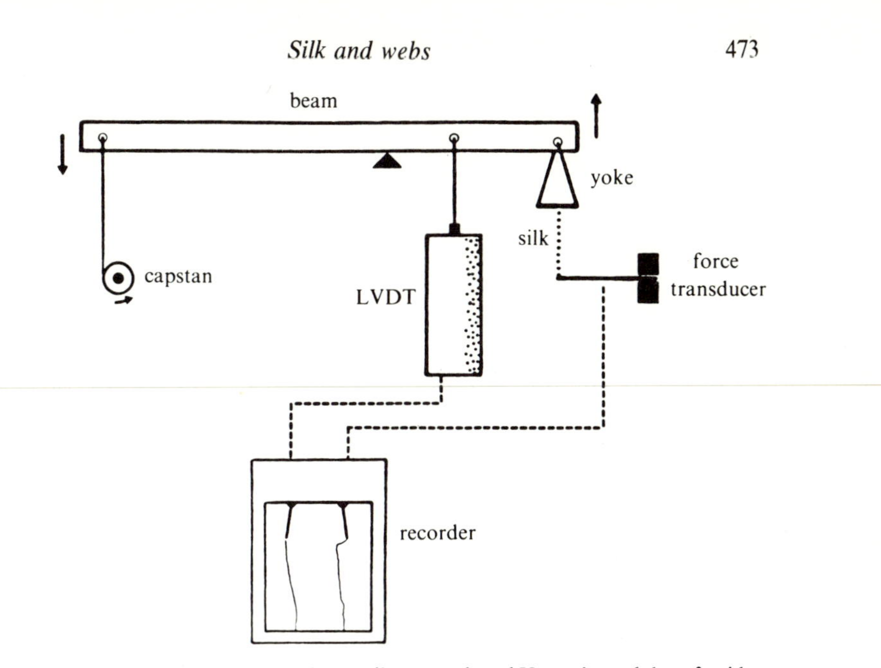

Fig. 20.7. Apparatus used to measure the tensile strength and Young's modulus of spider silk. LVDT, linear variable differential transformer. From M. Denny (1976). *J. exp. Biol.* **65**, 483–506.

TABLE 20.1. *Composition and properties of silks*

	Bombyx mori (silkworm)	*Araneus sericatus* (spider)	
	Cocoon silk	Frame silk	Viscid silk
Tensile strength (GN m^{-2})	0.6	1	1
Strain at break	0.2	0.25	2.0
Young's modulus (GN m^{-2})	7	4	0.6
Composition (% of amino acid residues):			
Glycine	45	32	44
Alanine	29	36	8
Serine	12	6	3

Data from S. A. Wainwright, W. D. Biggs, J. D. Currey & J. M. Gosline (1976). *Mechanical design in organisms.* Arnold, London.

times its initial length (i.e. to a strain of 2) before breaking but frame silk stretches to only 1.25 times its initial length (a strain of 0.25). Silkworm silk has properties like frame silk.

The ability of viscid silk to stretch so much suggests that it is a more or less amorphous cross-linked polymer, like abductin (p. 300) and resilin (p. 413). The more restricted extensibility of frame silk and silkworm silk suggests fibrous structure with parts of the molecules lined up in crystalline array (Fig. 6.16*c*). These suggestions are supported by observations in the polarizing microscope: frame silk and silkworm silk are birefringent even when slack but viscid silk shows little or no birefringence. The partly crystalline nature of frame silk and silkworm silk has also been demonstrated by X-ray diffraction.

The silk in the silk glands is a liquid, soluble in water. No crystalline structure can be demonstrated (in samples from silkworms) by X-ray diffraction. When the liquid passes through the nozzles on the spinnerets it immediately becomes a solid fibre. The mechanism that works the change seems to be this. When a liquid flows along a tube it flows faster near the centre than near the walls. If a long molecule lies at right angles to the flow, one of its ends will tend to travel faster than the other until it is aligned with the flow. Thus the molecules are arranged parallel to each other and crystallization is apt to occur. The silk molecules crystallize and form a fibre as they pass through the slender nozzles. A similar effect has been obtained artificially by rubbing liquid (silkworm) silk between a rotating plate and a stationary one. This aligned the molecules which crystallized to form insoluble fibrous clots.

Silkworm silk and frame silk contain very large proportions of the amino acids glycine and alanine, and some serine (Table 20.1). It seems likely that parts of the molecules consist of glycine and alanine or serine arranged alternately and that these parts crystallize readily. (The molecules of alanine and serine are very similar in size.) *Araneus* viscid silk contains far less alanine and little serine, and may have its amino acids arranged mainly in non-repeating patterns or in orders unfavourable to crystallization. Presumably some crystallization occurs, enough to cross-link the silk and make it solid.

A spider does not squirt silk out but sticks it to an external object and pulls. Silk can be pulled from the larger nozzles on the spinnerets of an active spider, simply by touching the tip of the nozzle with a needle and then drawing the needle away. This cannot be done with an anaesthetized spider because relaxation of its muscles allows the pressure in the opisthosoma to fall, and the liquid silk withdraws from the nozzle. A spider descending on a thread lets its weight pull the silk from the nozzle and may stop its descent by reducing the pressure in its opisthosoma. Spiders also draw silk from the nozzles by pulling with their hind legs, grasping the silk thread between their claws.

A taut web of inextensible thread would be easily broken because small forces at right angles to the threads would set up large tensile forces in them (this is why a guitarist can snap a steel string so easily). A web which is extensible enough to sag under a weight or balloon out in the wind will be harder to break because the threads are pulled more nearly parallel to the applied force. The extensibility of frame silk is near optimal for resisting static loads. The extra extensibility of viscid silk is useful when the web suffers dynamic loads, for instance when an insect collides with it.

Insects are likely to hit the web at full flying speed, typically 3–4 m s^{-1}. The force required to stop an insect is its kinetic energy divided by the deceleration distance, so the more the web stretches the smaller the force that is needed. The viscid silk stretches more than frame silk could do, and so enables the web to catch larger insects than if frame silk had been used throughout. Also, once an insect is caught there is nothing it can push hard enough on in its efforts to free itself: the viscid silk stretches so much, for such small forces, that the insect's efforts are frustrated.

A web would be ineffective if insects bounced off it like gymnasts off a trampoline. The glue on the spiral helps to prevent this, and so does the low rebound resilience of the silk. When either type of silk is stretched and allowed to recoil, only about one third of the energy is returned in the recoil.

Some of the threads in the web need to be stronger than others. Fig. 20.6(*b*) shows the forces which would act in each thread of a simplified web when a tensile force of 2 units was exerted on one of the guys. The forces were calculated by a standard engineering method. The largest forces act on the guys, rather smaller ones on the frame threads and much smaller ones on the radii. Appropriately different thicknesses of thread are used, consisting of different numbers of strands. All the frame silk is produced in strands of diameter about 2 μm but each guy typically consists of eight to ten strands, each frame thread of six to eight strands and each radius only two strands. There would be no advantage in making the radii as strong as the guys: it would use more silk without increasing the overall strength of the web.

There is also no advantage in making the web so strong that it might catch large insects, which might hurt the spider before it could overcome them. The ability of webs to catch large insects has been assessed by holding the webs horizontally and dropping fragments of wood onto them. An object dropped from 0.85 m reaches a velocity of 4.1 m s^{-1}, a fairly typical insect flying speed. A 25 mg piece of wood dropped from this height broke *Araneus sericatus* webs in about 50% of trials and was caught in about 50%. This crude test suggests that the webs are capable of catching insects up to about 25 mg: they could catch 12 mg house flies (*Musca*) but not 100 mg honey bees (*Apis*). This seems to be true. The mass of adult *Araneus sericatus* is 100–150 mg.

Araneus diadematus builds a new web daily. The quantity of silk required is quite small (about 0.1 mg) and the spider recovers most of it by eating the old web.

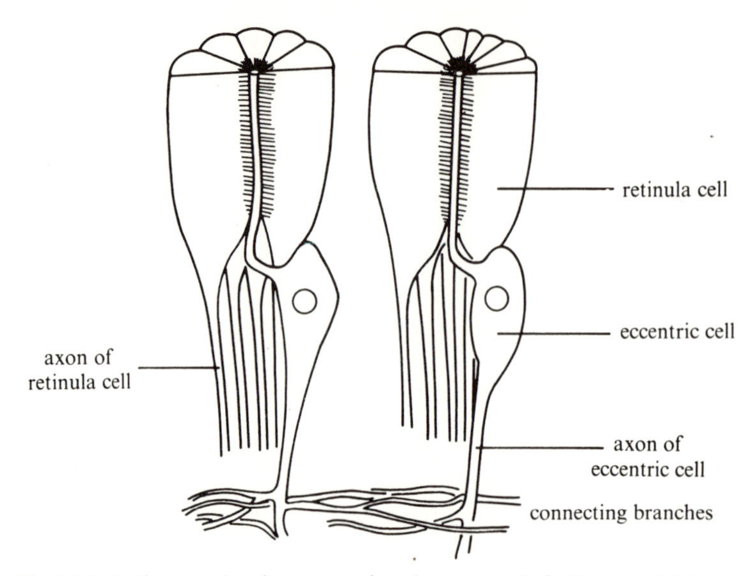

Fig. 20.8. A diagram showing connections between cells in the eye of *Limulus*.

THE EYES OF HORSESHOE CRABS

The lateral (compound) eyes of *Limulus* were used by Dr Haldan Hartline for a series of experiments which revealed a preliminary stage of processing of visual information for analysis in the brain. The principle he discovered he later found to operate also in the eyes of vertebrates.

A peculiarity of structure made *Limulus* eyes particularly suitable for the investigation. The retinula (sensory) cells of the compound eyes of other arthropods send axons direct to the brain and all processing of information occurs there. In *Limulus* each ommatidium has one or two eccentric cells (Fig. 20.8). It is probably the retinula cells which are directly stimulated by light, but the information is transmitted to the brain by the eccentric cells. Each eccentric cell is in contact with the retinula cells of its own ommatidium, and also has branches which synapse with neighbouring eccentric cells.

The axons of the eccentric cells seem to be the only ones in the optic nerve which carry action potentials. It is not too difficult to fray the nerve until a single axon has been separated from the rest and can be picked up on electrodes so that recordings can be made from it.

Fig. 20.9(*a*) shows action potentials recorded from an eccentric cell, in response to different intensities of light on its ommatidium. Increasing the light intensity by a factor of 10000 increased the frequency of action potentials by a factor less than 20. There is only a limited range of frequencies which can be used. Frequencies higher than about 100 s^{-1} are beyond the capability of the axon, and frequencies below about 1 s^{-1} are relatively little used because the animal would have to wait so long before it knew what the intensity was.

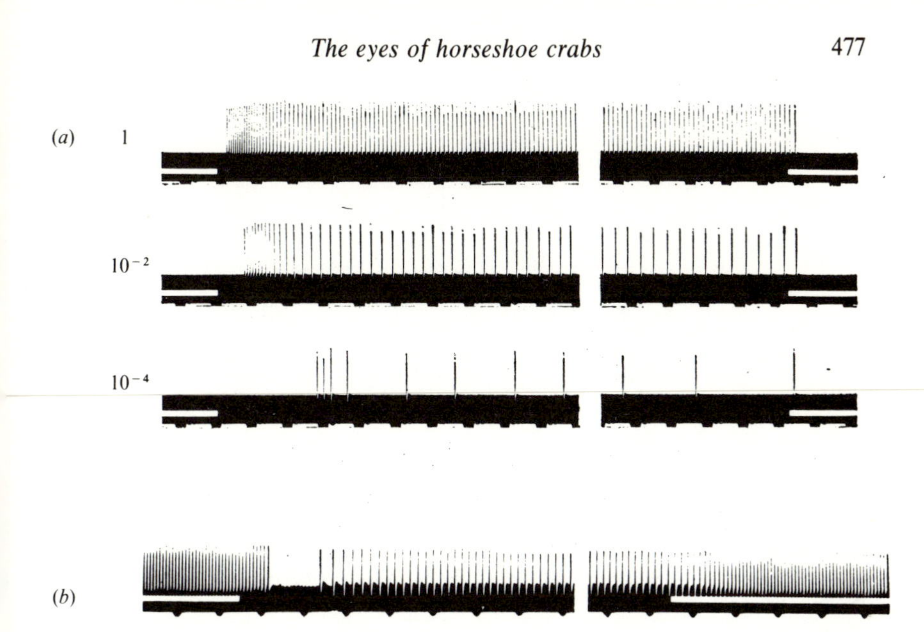

Fig. 20.9. Action potentials recorded from the optic nerve of *Limulus*. (*a*) Records from a single axon when the eye was illuminated at three different intensities. The highest intensity (top) was 10000 times the lowest one (bottom). Blackening of the white bands shows when the light was switched on. (*b*) A record from another axon, showing inhibition by neighbouring ommatidia. The ommatidium from which the recording was made was illuminated at constant intensity but a ring of surrounding ommatidia was illuminated only during the period marked by blackening of the white band. Intervals of 0.2 s are marked under all the records. From H. K. Hartline (1969). *Science* **164**, 270–8.

If the eye is to differentiate between intensities over a wide range, small differences in intensity will give very small differences of frequency.

However, contrast is enhanced at boundaries between lighter and darker parts of the visual field by the process called lateral inhibition. Each eccentric cell is influenced by its own ommatidium and also by neighbouring ones. Action potentials in neighbouring eccentric cells inhibit it. The effect is illustrated in Fig. 20.9(*b*). The frequency of action potentials in the eccentric cell fell sharply when neighbouring ommatidia were illuminated although the intensity on its own ommatidium was unchanged.

The practical effect of lateral inhibition is shown in Fig. 20.10(*a*), which shows frequencies of action potentials in eccentric cells on either side of a sharp boundary between bright and dim illumination. The ommatidia at (i) and (ii) are equally brightly lit but eccentric cells at (ii) receive less inhibition because the ommatidia to the right of them are dimly lit. The ommatidia at (iii) and (iv) are equally dimly lit but the eccentric cells at (iii) receive more inhibition because they have brightly-lit ommatidia to their left. Thus lateral inhibition enhances the contrast between (ii) and (iii).

Fig. 20.10(*a*) refers to a spatial contrast between light and darkness. Fig.

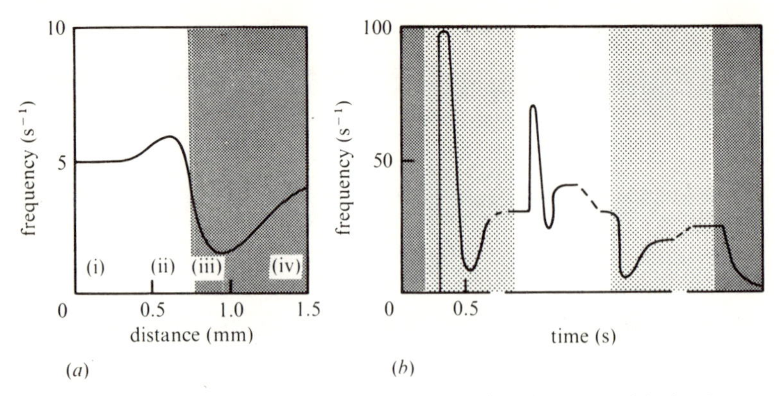

Fig. 20.10. Diagrammatic graphs of frequency of action potentials in the axons of eccentric cells in the eye of *Limulus*. (*a*) Frequencies in different eccentric cells, along a transect across the eye. (*b*) Frequency in a single cell, plotted against time. Stipple is used to indicate intensity of illumination: the lighter the stipple, the brighter the light. Based on data in H. K. Hartline (1969). *Science* **164**, 270–8.

20.10(*b*) refers to a temporal contrast, to a change of brightness over the whole eye. When brightness increases, the frequency of action potentials rises initially to a level higher than the one at which it eventually settles. This is largely due to the phenomenon of adaptation which is common to a great many sense organs. However the effect is enhanced and complicated by lateral inhibition, which occurs with a small delay. An increase of brightness results in a burst of action potentials at a high frequency which falls sharply as the inhibitory effect of high frequencies in neighbouring ommatidia is felt.

FURTHER READING

GENERAL

Barthel, K. W. (1974). *Limulus*, a living fossil. *Naturwissenshaften* **61**, 428–33.
Bristowe, W. S. (1971). *The world of spiders*, 2nd edn. Collins, London.
Savory, T. (1964). *Arachnida*. Academic Press, New York & London.

SILK AND WEBS

Denny, M. (1976). The physical properties of spiders' silk and their role in the design of orb-webs. *J. exp. Biol.* **65**, 483–506.
Wilson, R. S. (1969). Control of drag-line spinning in certain spiders. *Am. Zool.* **9**, 103–11.

THE EYES OF HORSESHOE CRABS

Hartline, H. K. (1969). Visual receptors and retinal interaction. *Science*, **164**, 270–8.

21
Other arthropods

Phylum Arthropoda
  Class Trilobita
  Class Onychophora
  Class Merostomata (see chapter 20)
  Class Arachnida (see chapter 20)
  Class Crustacea (see chapter 17)
  Class Insecta (see chapters 18 and 19)
  Class Chilopoda (centipedes)
  Class Diplopoda (millipedes)
  and other classes

Above is a list of the main classes of the phylum Arthropoda. Four of them have been discussed in previous chapters. This chapter deals with the remaining four.

The Trilobita are a primitive class which lived in the Palaeozoic era, about 300–600 million years ago. They were particularly common in the early part of the era, and at least 10000 species are known. Most of the fossils which have been found are 2–10 cm long but a few are longer, up to 70 cm. They are found in sedimentary rocks laid down in the sea, most of them in rocks with a fine texture which indicates that they were originally mud rather than sand. The trilobites presumably crawled about on muddy parts of the sea bottom.

Fig. 21.1 shows a typical trilobite. There were plates of cuticle on the dorsal surface but any cuticle on the ventral surface must have been thin, for it has not been preserved. There was a single plate on the head and one covering several posterior segments but for most of the length of the body each segment had its own plate, which overlapped the plates of neighbouring segments and was presumably joined to them by flexible arthrodial membrane. It seems likely that trilobites rolled up like woodlice (isopods, p. 377) to protect the ventral surface when necessary, for many of the fossils are rolled.

The structure of the limbs can be seen in some exceptionally well preserved specimens. There is a pair of distinct antennae but all the rest of the limbs are alike, with two branches. The ventral branches are leg-like and must have been used for crawling. The dorsal ones are fringed by filaments which presumably served as gills. Four of the pairs of two-branched limbs are on the head.

The most primitive living arthropods are the 100–200 species which make up the class Onychophora (Fig. 21.2 *a*). They live in damp habitats on land, in the

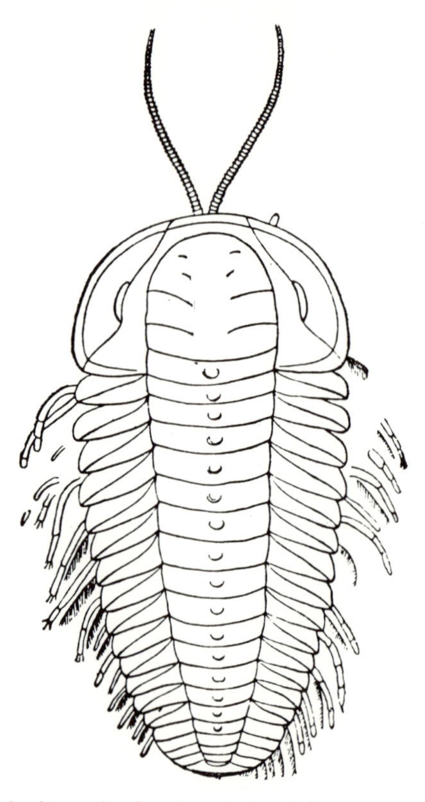

Fig. 21.1. *Triarthrus becki*, a trilobite. Length (excluding antennae) about 2 cm. From A. Sedgwick (1909). *A student's textbook of zoology*, vol. 3. Swan, Sonnenschein & Co., London.

tropics and subtropics. They spend the day in crevices in the soil or in rotting logs, or under stones, and emerge at night to search for prey.

The Onychophora have cuticle which is tanned in its outer layers, like insect cuticle. However it is very thin and has tiny furrows so that it can be stretched concertina-fashion. Consequently the cuticle is soft and flexible, more like the cuticle of a caterpillar than that of most adult insects. It allows the animal to deform its body so that it can slip through extraordinarily fine cracks in logs to the safety of cavities within.

The first three pairs of appendages are a pair of antennae, a pair of cutting jaws and a pair of knobs called oral papillae. The rest of the limbs are walking legs, all of them alike. The jaws have stiff cuticle, and there is a pair of stiff cuticular claws on the end of each walking leg. The walking legs are otherwise flexible so they need no joints, and they are kept extended only by pressure in the body cavity. They are superficially rather like the parapodia of polychaetes but the arrangement of muscles inside them is more like the

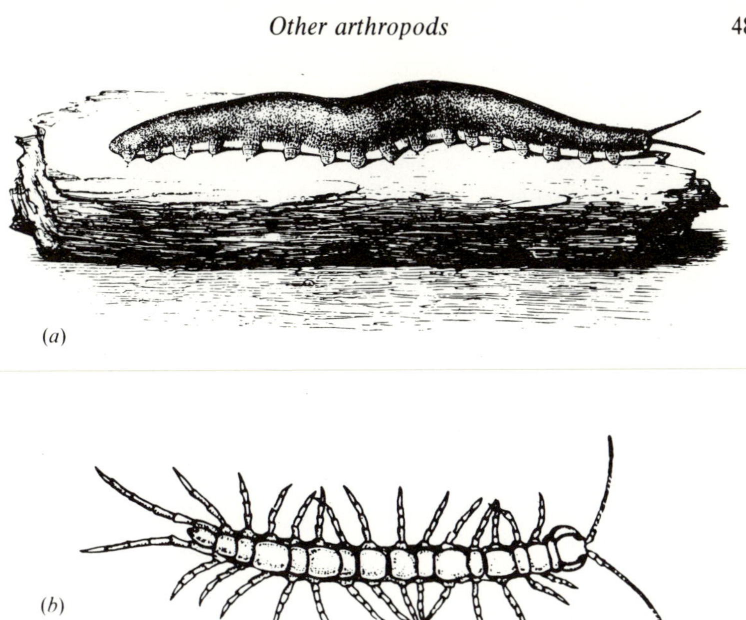

(c)

Fig. 21.2. (*a*) *Peripatus capensis*. Length 8 cm. From A. Sedgwick (1895). In *The Cambridge natural history*, ed. S. F. Harmer & A. E. Shipley. Macmillan, London. (*b*) *Lithobius* and (*c*) *Iulus*. From J. G. Blower (1972). In *Textbook of zoology. Invertebrates*, ed. A. J. Marshall & W. D. Williams. Macmillan, London.

arrangement in other arthropod limbs than in parapodia (see Fig. 16.7). They have nothing comparable to the acicula of polychaetes.

There is a pair of huge glands which discharge through pores in the oral papillae. Their secrection can be squirted out and rapidly congeals to form sticky threads. It is used for defence.

The main body cavity is a haemocoel. There are nephridia which open at the bases of the legs. Though so much more numerous than the green glands of Crustacea (p. 373) they have the same basic structure. There is a dorsal heart with ostia, there are tracheae which open through numerous spiracles and there is a pair of ventral nerve cords. In all these respects, Onychophora resemble other arthropods.

It is generally believed that the arthropods evolved from segmented soft-bodied ancestors which had all the segments more or less alike along most of the length of the body. These ancestors would have been very like annelid worms. In the course of evolution different parts of the body became modified to serve different functions and the segmentation became less monotonous. This happened particularly in the head where segments coalesced and their appendages were modified as antennae or mouthparts, each pair different from the next. The Trilobita and the Onychophora both seem to show early stages in this process. However, the two classes are in many respects very different. It is possible that the trilobites may be close to the ancestry of a line which led to the arachnids and horseshoe crabs, and that the Onychophora resemble early ancestors on a different line which led to the insects and myriapods. The crustaceans would have evolved along a third line. There is a fair amount of evidence for this possibility.

The centipedes (class Chilopoda) and the millipedes (class Diplopoda) have sometimes been put together in a single class Myriapoda. *Lithobius* (Fig. 21.2 *b*) is a genus of centipedes which is common in Europe. By day they generally hide under leaf litter or in any suitable crevice. At night they hunt for small worms, slugs and arthropods.

Centipedes have antennae, mandibles and two pairs of maxillae. *Lithobius* has only 15 pairs of legs but other centipedes have up to 177 pairs. The first trunk segment bears a pair of sharp-pointed pincers with venom glands opening at their tips. They are used to grasp prey and to immobilize it by injecting venom. *Lithobius* has long and short terga (dorsal skeletal plates) on alternate segments. Centipedes have tracheae and a pair of Malpighian tubules.

Iulus (Fig. 21.2 *c*) is a millipede. It burrows in soil and feeds as earthworms do on dead and decaying vegetation. Its cuticle is calcified like crab cuticle and protects the animal well when it rolls up in a spiral, which it does when molested. The head bears antennae, mandibles and peculiar posterior mouthparts. The trunk bears walking legs and at first sight there seem to be two pairs per segment (except on the first few segments). This is because a single ring of cuticle encloses each pair of segments. Millipedes have peculiar tracheae and a pair of Malpighian tubules.

WALKING

Many crustaceans move their legs, when they walk, in rather irregular sequence (see p. 390). Centipedes and millipedes move theirs in beautifully regular metachronal rhythms which have been studied by cinematography. Many centipedes walk in the manner of *Scolopendra* (Fig. 21.3 *a–d*). Metachronal waves travel posteriorly along the body so that when two adjacent feet are on the ground the more anterior one is at a later stage in the step, and the two legs converge. The phase difference between successive pairs of legs depends on the speed. At the low speed shown in Fig. 21.3 (*a*) each leg is 1/5 cycle out of phase with the next so that the first walking leg moves in phase with leg 6, leg 2 with leg 7, and so on. At the high speed shown in Fig.

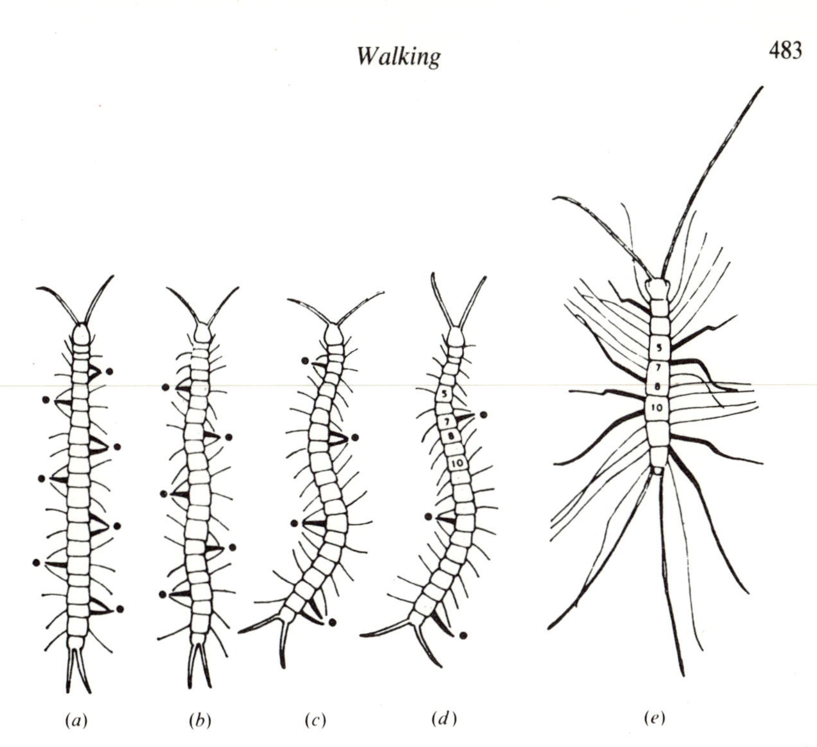

Fig. 21.3. Outlines traced from photographs of the centipedes *Scolopendra* (*a–d*) and *Scutigera* (*e*) running. Legs in contact with the ground are indicated by thick lines. (*a*) to (*d*) show running at progressively higher speeds. From S. M. Manton (1965). *J. Linn. Soc.* (*Zool.*) **45**, 251–484.

21.3(*d*) each leg is 1/13 cycle out of phase with the next so that leg 1 moves in phase with leg 14. The number of feet on the ground at any instant also changes with speed. At low speed, half the feet are on the ground at any instant (and each foot is on the ground for half the time). At high speed only 0.15 of the feet are on the ground at any instant (and each foot is on the ground for 0.15 of the time).

Suppose a centipede has legs of length l. It will probably swing each leg through about 60° and so advance a distance about equal to l while a particular foot is on the ground. If each foot is on the ground for a fraction β of the time, the centipede will advance l/β in each complete cycle of leg movements. To run at velocity u it must make $u\beta/l$ cycles of leg movements per unit time (i.e. each leg must step with frequency $u\beta/l$). Very high stepping frequencies would be needed at high speeds if β were not reduced as u increased. Even with the reductions of β which occur, very high frequencies are used. For instance, *Cryptops* is a centipede which resembles *Scolopendra* in its proportions. A specimen 38 mm long could run at 0.26 m s^{-1} and in doing so used a stepping frequency of 25 s^{-1}. This is similar to the wing beat frequencies of large insects (the moth *Manduca*, 28 s^{-1}; the locust *Schistocerca*, 17 s^{-1}).

Scutigera (Fig. 21.3*e*) is the fastest known centipede. A specimen only 22 mm long ran at 0.42 m s^{-1}, but because the legs are so long this involved a stepping frequency of only 13 s^{-1} (when l is large, $u\beta/l$ is relatively small for given values of u and β).

Adjacent legs of *Scolopendra* converge while they are on the ground. If *Scutigera* ran in the same way its long legs would cross over each other while on the ground and would be apt to get in each other's way. It moves differently, as Fig. 21.3(*e*) shows. The metachronal waves run forwards along the body so that adjacent legs diverge while on the ground. A foot which is off the ground may have to be lifted over one which is on the ground but legs are not crossed while both are on the ground.

The leg actions of *Scutigera* and *Ligia* (Fig. 17.17*b*) and the parapodial action of crawling polychaetes (Fig. 16.8) all involve forward-moving metachronal waves.

Most centipedes run on the surface or through existing crevices. Millipedes often burrow by forcing their way through soil. To enable millipedes to exert the large forces required, they have very large numbers of closely spaced legs. They use high values of β, especially at low speeds, so that up to 0.8 of the legs are on the ground at any instant, pushing simultaneously. The metachronal waves run forwards. The left and right legs of a pair move together, not alternately as in centipedes.

Even at their top speeds, millipedes seldom have fewer than 0.4 of their feet on the ground. They cannot run fast and do not need to. They do not need speed to catch their (vegetable) food and they do not rely on speed to escape enemies; they simply curl up.

FURTHER READING

Blower, J. G. (ed.) (1974). *Myriapoda* (*Symposium of the Zoological Society of London* 32). Academic Press, New York & London.

Manton, S. M. (1977). *The Arthropoda: habits, functional morphology and evolution.* Oxford University Press.

22

Bryozoans and brachiopods

Phylum Phoronida
Phylum Bryozoa
  Class Phylactolaemata
  Class Stenolaemata
  Class Gymnolaemata
Phylum Brachiopoda
  Class Inarticulata
  Class Articulata

This chapter is about three phyla which resemble each other in various ways, most notably in having a characteristic crown of tentacles called a lophophore. The Phoronida is a tiny phylum of about 15 species. The Bryozoa (sometimes called Ectoprocta) is a moderately large phylum of about 4000 species which has until recently attracted surprisingly few research workers. Living Brachiopoda are relatively scarce (there are only about 300 species) but fossil Brachiopoda are enormously abundant in marine sedimentary rocks of the Palaeozoic period. Brachiopods seem to have been the dominant bottom-living animals in many Palaeozoic marine communities.

All known Phoronida and Brachiopoda, and most Bryozoa, live in the sea. There are a few freshwater Bryozoa.

PHORONIDA

Phoronis hippocrepia (Fig. 22.1) is a European member of the phylum Phoronida. It lives on rocky shores, often in clusters. It occupies a chitin tube which usually has sand grains adhering to it. Fig. 22.1(*a*) shows animals with their lophophores spread and others withdrawn into their tubes. They withdraw on the slightest disturbance.

The gut is U-shaped so the mouth and anus are both (conveniently) at the open end of the tube. The mouth is surrounded by the bases of the tentacles of the lophophore, but the anus is not. The tentacles are not arranged in a simple ring, but around the perimeter of a crescent. Some species have the horns of the crescent spirally coiled. The tentacles immediately between the mouth and the anus are the youngest and smallest: this is where new tentacles are added as the animal grows. The tentacles have cilia on them which set up currents as indicated by arrows in Fig. 22.1(*b*). These currents serve for filter feeding.

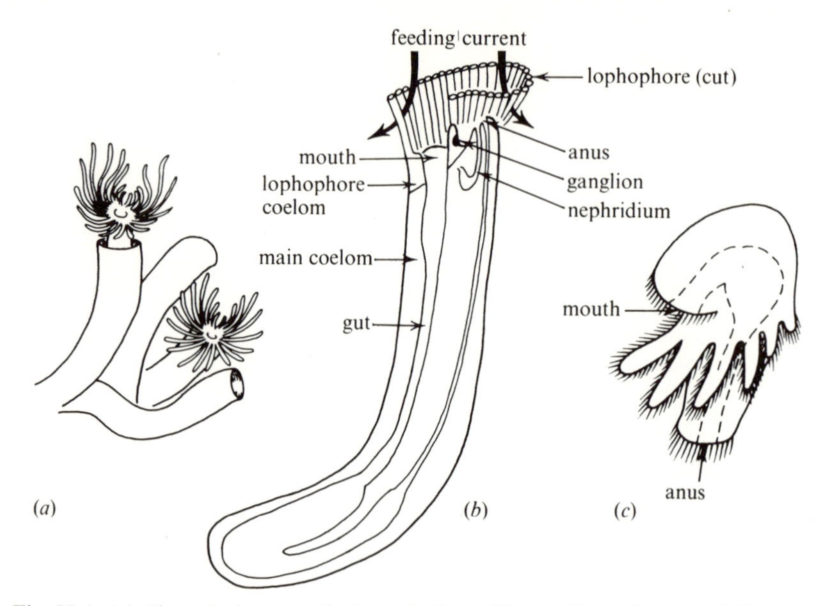

Fig. 22.1. (*a*) *Phoronis hippocrepia*. Length about 10 mm. (*b*) A diagram of *Phoronis*, cut in half longitudinally. The arrows show the direction of the feeding current. (*c*) The larva of *Phoronis*. Length about 0.5 mm.

There is a coelom which is divided by a septum into a small lophophore coelom and a large main cavity. The lophophore coelom extends into the tentacles. A pair of metanephridia lead from the main coelomic cavity to openings on either side of the anus. Under the epidermis there is a nerve net which is thickened to form a ganglion between the mouth and the anus. A giant nerve fibre runs posteriorly along the body from this thickening. An intact animal withdraws rapidly into its tube when disturbed but this response is lost if the giant fibre is cut near the lophophore. There is a simple system of blood vessels in which blood is circulated by peristalsis. There are corpuscles in the blood, containing haemoglobin.

Phoronis is hermaphrodite, with simple gonads in the coelom. The eggs and sperm are shed into the coelom where fertilization occurs, and the fertilized eggs escape through the nephridia. They adhere to the base of the lophophore for a while before hatching to release ciliated larvae (Fig. 22.1 *c*) which live in the plankton for a few weeks and then settle and metamorphose.

BRYOZOA

The Bryozoa are colonial. Several examples are shown in Figs. 22.2 to 22.5. *Cristatella* (Fig. 22.2) is a member of the class Phylactolaemata and lives, like the other members of the class, in fresh water. It lives in lakes and clear ponds and is often found on the undersides of waterlily (*Nymphaea*) leaves.

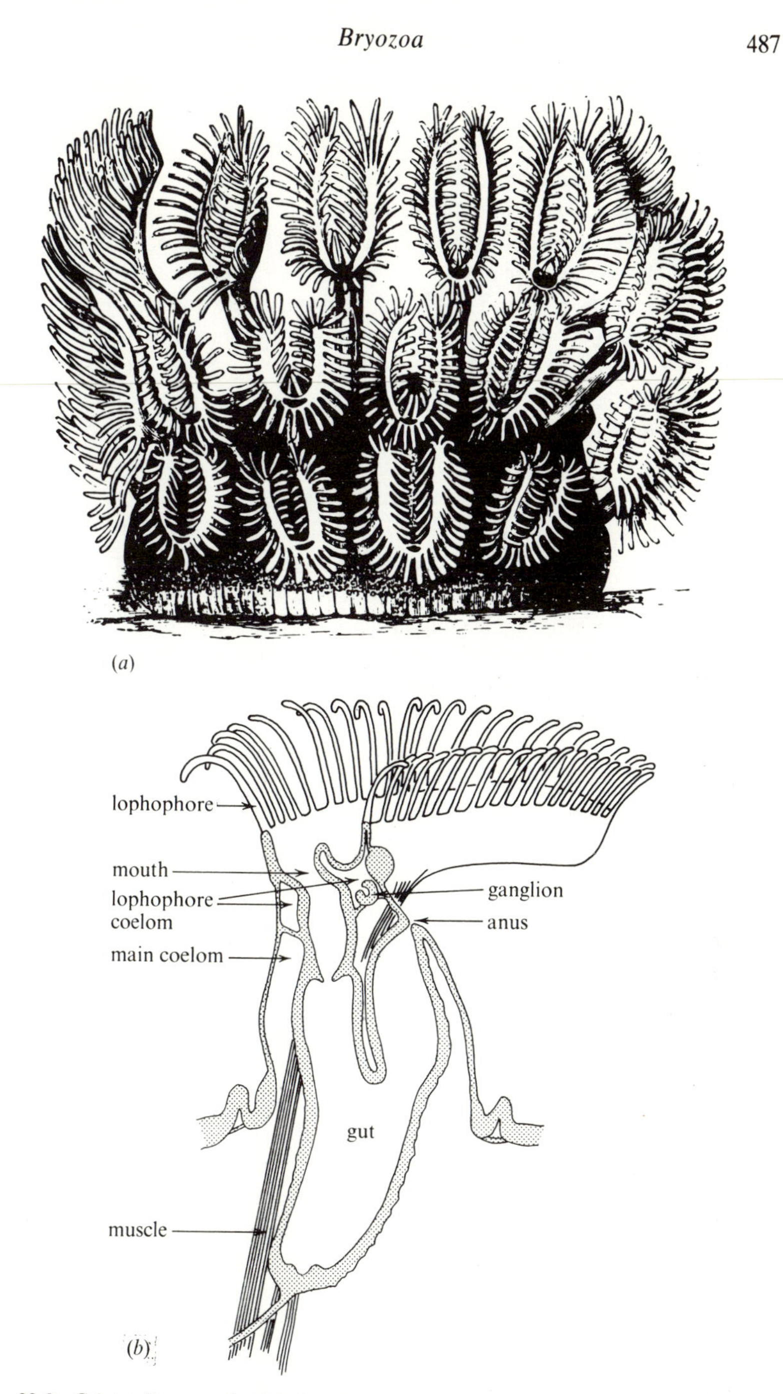

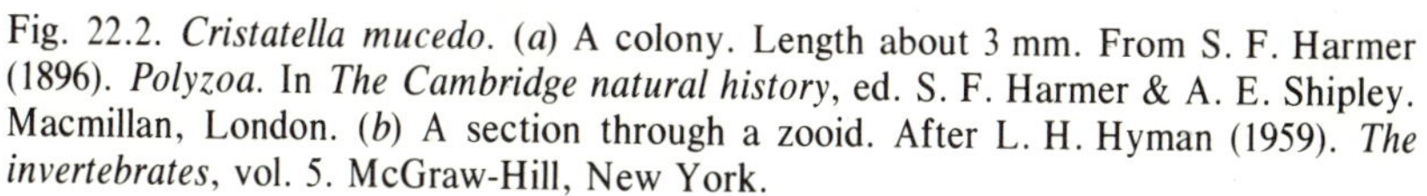
Fig. 22.2. *Cristatella mucedo.* (*a*) A colony. Length about 3 mm. From S. F. Harmer (1896). *Polyzoa.* In *The Cambridge natural history*, ed. S. F. Harmer & A. E. Shipley. Macmillan, London. (*b*) A section through a zooid. After L. H. Hyman (1959). *The invertebrates*, vol. 5. McGraw-Hill, New York.

A colony of *Cristatella* looks rather like a feathery slug. Like slugs, and unlike most Bryozoa, it can crawl. Its speed is unimpressive (about 10–15 mm per day) but it is interesting to find a colony which can behave in an integrated way as if it were a single individual. There are of course other examples, notably the swimming colonies of siphonophores (p. 158).

Each individual member of the colony (zooid) has a lophophore which looks extremely like the lophophore of *Phoronis*. Their cilia set up feeding currents in the same way as in *Phoronis*. Fig. 22.2(*a*) shows the lophophores projecting above the surface of the colony but they are withdrawn into the colony when disturbed by touch or vibration. Fig. 22.2(*b*) shows the structure of a zooid. The U-shaped gut, the lophophore and the two cavities of the coelom are arranged in the same way as in *Phoronis*. There is a ganglion in almost the same position as in *Phoronis* but there is no circulatory system and no nephridia. The main coeloms of the individuals interconnect.

Cristatella is hermaphrodite with simple gonads in the coelom. Embryos develop in the coelom and are released as young colonies. Asexual reproduction also occurs. The colony may split in two. It also forms statoblasts, which are asexually-formed buds enclosed in protective chitinous shells. The statoblasts are generally not released until the beginning of winter when the colony dies and disintegrates. They remain dormant through the winter and develop into new colonies in the spring.

Electra (Fig. 22.3) is a member of the Gymnolaemata and lives in the sea like nearly all the other members of the class. It is found at all depths from the shore down to about 100 m. It forms a thin layer encrusting seaweeds, stones and mollusc shells. Each zooid has a box-like exoskeleton with a hinged lid. The boxes are rigid, made of calcium carbonate, protein and a polysaccharide related to chitin. The lophophore is ring-shaped, not crescent-shaped, but in other respects the individuals are very like those of *Cristatella*.

There is a flexible body wall under the perforated roof of each rigid box. Extensor muscles crossing the coelom pull this part of the body wall down into the box, forcing the lophophore out through the lidded opening. Other (retractor) muscles have the opposite effect, pulling the lophophore down into the box.

Electra is hermaphrodite. It releases eggs through the intertentacular organ (Fig. 22.3 *b*) and sperm through pores in the tips of the tentacles. Fertilization occurs in the sea and there is a ciliated larva like the one shown in Fig. 22.3 *c*. The larva settles and metamorphoses to the adult form. New zooids are produced by budding, so that a single larva gives rise to a whole colony.

Fig. 22.4 shows another member of the Gymnolaemata, also found on rocky shores. *Bugula* differs from *Electra* in three striking ways. First, the colony resembles a little tree instead of encrusting the rock. Secondly, the embryos develop in special chambers (ovicells) and receive nourishment from the colony. Thirdly, there are two types of zooid, ordinary ones and avicularia. Avicularia are shaped like crabs' chelae, with a movable claw. There are small muscles

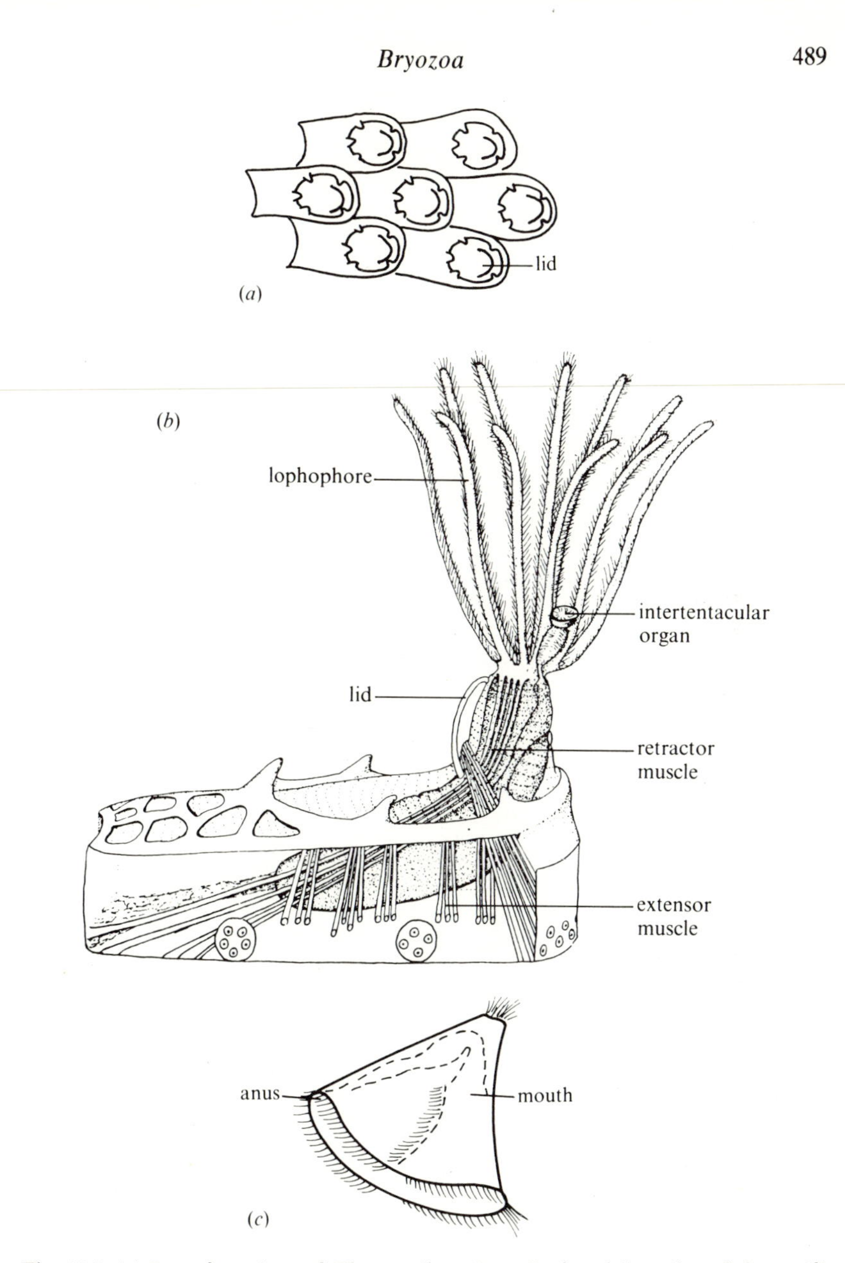

Fig. 22.3. (*a*) Part of a colony of *Electra pilosa*. Length of each box about 0.6 mm. (*b*) A zooid of the same species with the lophophore extended. From R. B. Clark (1964). *Dynamics in metazoan evolution*. Clarendon, Oxford. (*c*) A larva of *Membranipora*. Length about 0.8 mm. Based on a drawing in J. S. Ryland (1970). *Bryozoans*. Hutchinson, London.

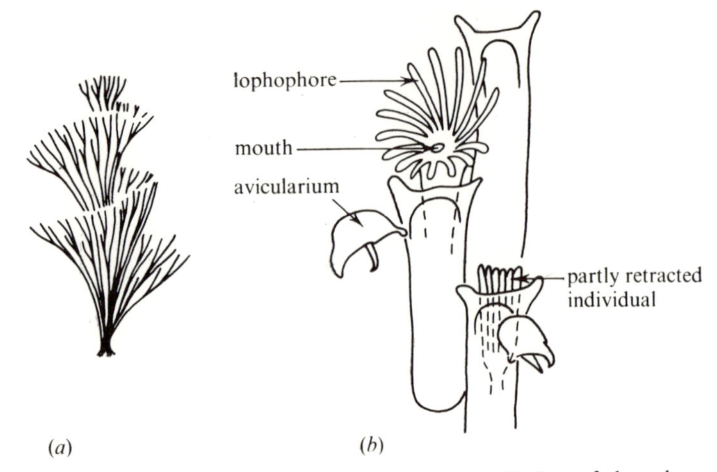

Fig. 22.4. *Bugula turbinata*. (*a*) A colony (height about 2 cm). (*b*) Part of the colony, enlarged. Based on S. F. Harmer (1896). Polyzoa. In *The Cambridge natural history*, ed. S. F. Harmer & A. E. Shipley. Macmillan, London.

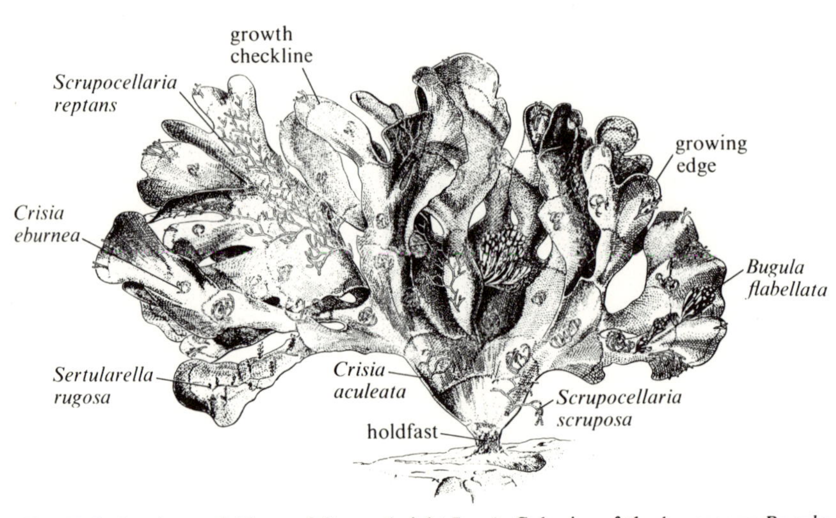

Fig. 22.5. A colony of *Flustra foliacea* (height 7 cm). Colonies of the bryozoans *Bugula flabellata*, *Crisia aculeata* and *eburnea* and *Scrupocellaria reptans* and *scruposa* are attached to it, as also are colonies of the hydrozoan *Sertularella rugosa*. From A. R. D. Stebbing (1971). *J. mar. Biol. Assoc. UK* **51**, 283–300.

to open the claw, and a big one to close it. There is a tuft of bristles which are apparently sensory, for if they are touched the avicularium snaps shut, often catching the object which touched the bristles. Objects of up to about 50 μm diameter can be caught and held firmly. The avicularia are about 0.5 mm apart and they nod continually (though slowly), so small animals which crawl on the

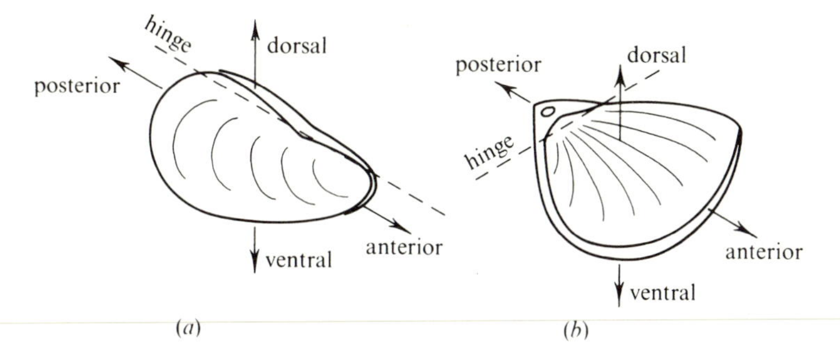

Fig. 22.6. Diagrams of (*a*) a bivalve mollusc shell and (*b*) a brachiopod shell.

colony are very likely to be caught. The avicularia do not feed, but they give the colony some protection.

Colonies of *Bugula* at Wood's Hole, Massachusetts, are infested by amphipod crustaceans (*Corophium* and *Jassa*). These build tubes among the branches of the colony, inhibiting its reproduction. However, the amphipods are apt to be caught by the avicularia whenever they emerge from their tubes. Their antennae and limbs are seized and they have to spend a lot of time wriggling free. An observer watching a particular colony found that this occupied more than half their time. The avicularia do not succeed in evicting the amphipods, but they restrict their damaging activities.

Flustra (Fig. 22.5) is yet another example of the Gymnolaemata. It forms large, rather stiff colonies which could be mistaken for seaweed, and are found attached to stones off European shores. Other Bryozoa in turn often attach themselves to it: the colony shown in the illustration has five species of Bryozoa and one of Hydrozoa attached to it.

BRACHIOPODA

Brachiopods live in bivalve shells and look very different from phoronids and bryozoans. Most laymen would identify them as bivalve molluscs but there is a marked difference from the molluscs in the symmetry of the shell and of the animal inside (Fig. 22.6). In bivalve molluscs, each valve is typically asymmetrical but one is a mirror image of the other (*Pecten*, Fig. 14.3, is an exception). In brachiopods the valves are not mirror images but each is symmetrical about a plane perpendicular to the axis of the hinge. The soft parts are also more or less symmetrical about this plane. Bivalve molluscs have a left valve and a right one but brachiopods have a dorsal valve and a ventral one. Confusingly brachiopods often rest with the valve which is conventionally called ventral uppermost.

Brachiopods, like bivalve molluscs, have logarithmic spiral shells with high

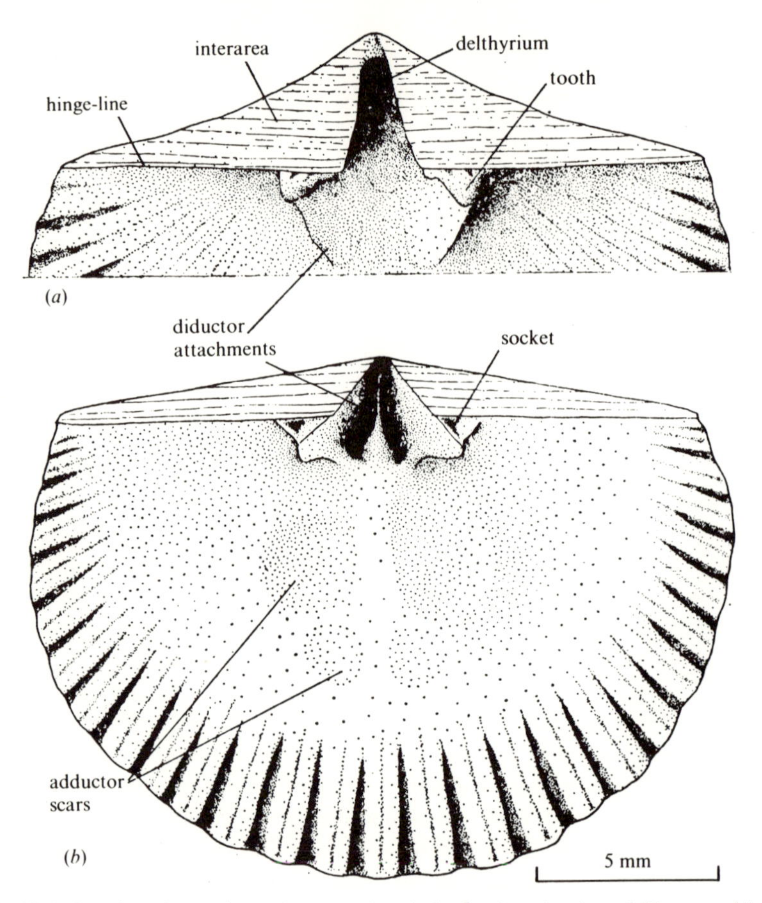

Fig. 22.7. Interior views of (*a*) the ventral and (*b*) the dorsal valve of *Hesperorthis*, a fossil brachiopod of the Ordovician period. From M. J. S. Rudwick (1970). *Living and fossil brachiopods*. Hutchinson, London.

values of W (p. 267; Figs. 13.9 and 13.11). Unlike most bivalve molluscs they have $T = 0$ (and must do, to achieve their symmetry).

The brachiopods illustrated in Figs. 22.7 to 22.9 belong to the class Articulata. Their shells are mainly calcium carbonate with a very small proportion of protein. The valves are joined together by a hinge joint; two teeth on the ventral valve move in sockets on the dorsal one (Fig. 22.7). There is no abductin such as bivalve molluscs have (p. 300), to make the valves spring open when the adductor muscles relax. Instead there are diductor muscles which run posterior to the axis of the hinge, so that when they shorten the valves open (Fig. 22.8). The adductor muscles seem to have 'quick' and 'catch' portions, like those of bivalve molluscs.

Most Articulata live attached by a short stalk to rock, a loose stone or a shell.

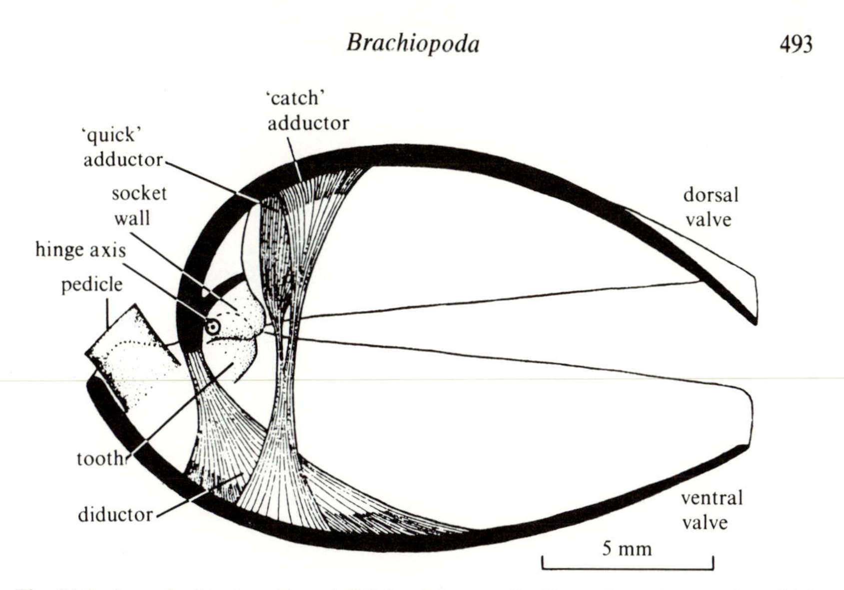

Fig. 22.8. An articulate brachiopod (*Waltonia*) cut sagittally to show the muscles which open and close the shell. From M. J. S. Rudwick (1970). *Living and fossil brachiopods.* Hutchinson, London.

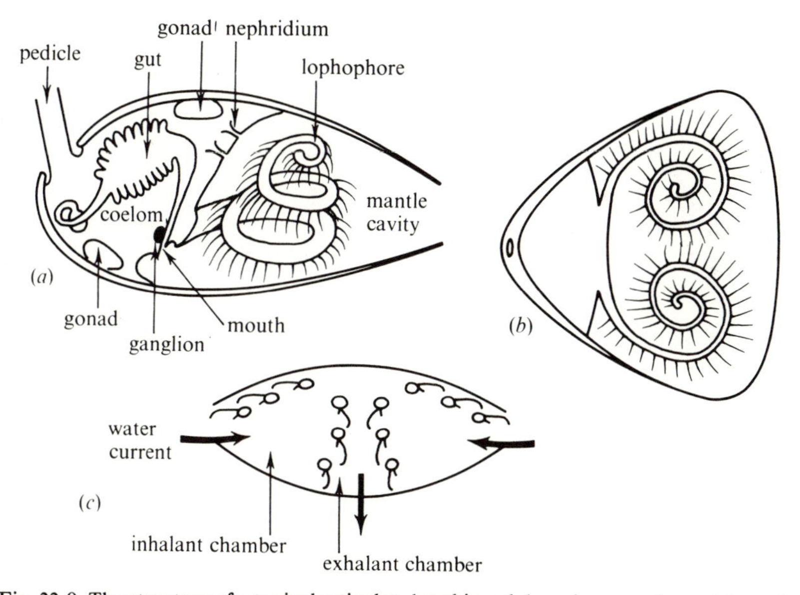

Fig. 22.9. The structure of a typical articulate brachiopod, based on members of the order Rhynchonellida. (*a*) represents a specimen cut in half sagittally; (*b*) a specimen with the dorsal valve removed; and (*c*) a transverse section.

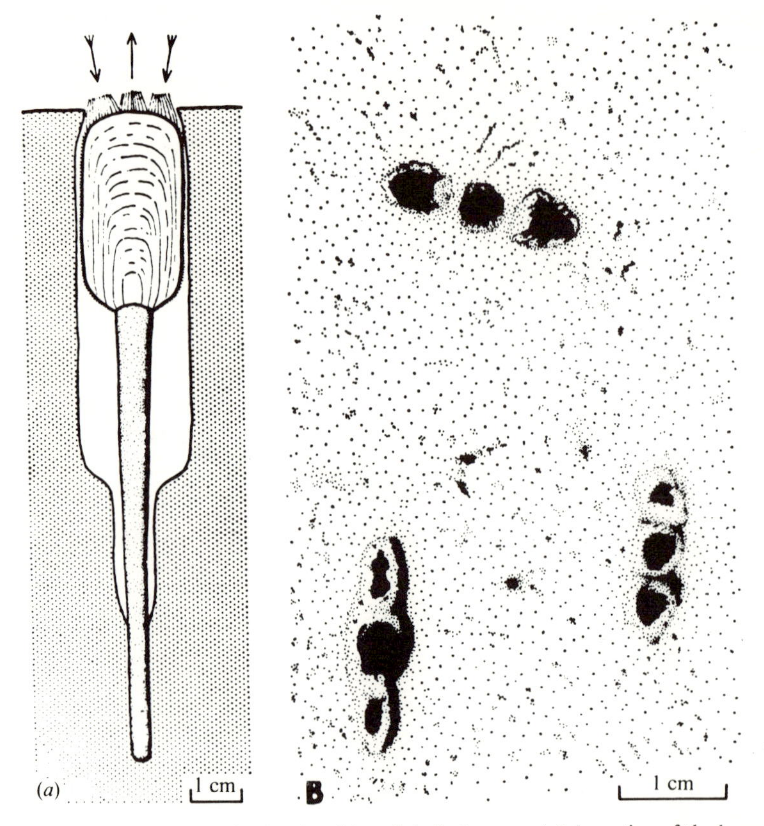

Fig. 22.10. *Lingula*, an inarticulate brachiopod, in its burrow. (*a*) A section of the burrow with the animal in position, with arrows showing the directions of the feeding current. (*b*) Surface view. From M. J. S. Rudwick (1970). *Living and fossil brachiopods.* Hutchinson, London.

The stalk is called the pedicle and consists of connective tissue covered by epithelium. The animal has muscles which enable it to swivel on its stalk but the stalk itself contains no muscles.

Fig. 22.9 shows how the lophophore and viscera are arranged, in a typical articulate brachiopod. Much of the space between the valves is occupied by a cavity comparable to the mantle cavity of molluscs. This contains the lophophore, which is a simple ring of tentacles encircling the mouth in young brachiopods but extends along coiled supports in most adults. Different shapes of coil are formed in different groups of brachiopods: in the example which is illustrated the lophophore is broken so that it is no longer a complete loop. The tentacles divide the mantle cavity into two lateral inhalant chambers and a median exhalant one. Their cilia draw water in between the valves on either side and drive it out anteriorly. The mouth is on the inhalant side of the row of tentacles so the direction of flow is the same as in phoronids and bryozoans.

Diatoms and other small planktonic organisms are filtered from the water and transported by cilia to the mouth.

Articulate brachiopods have no anus. Diverticula from the gut form a digestive gland where most of the digestion occurs, as in molluscs. There are one or two pairs of metanephridia. There is a blood circulatory system which has not been studied in detail. There is a simple nervous system radiating from a ganglion near the mouth. The coelom is undivided.

The sexes are separate in most species. There are two pairs of gonads in the coelom, and their products escape through the nephridia. Ciliated larvae develop from the fertilized eggs.

Lingula (Fig. 22.10) is a member of the class Inarticulata, but in some ways a very unusual one. There are about a dozen species in the Pacific and Indian Ocean. Inarticulate brachiopods have no hinge joining their valves. Their pedicles are muscular. They have an anus. Most of them (including *Lingula*) have a shell of protein, chitin and calcium phosphate. *Lingula* is unusual in that it lives buried in mud. It is also remarkable for having changed very little in 450 million years of evolution. Fossils so similar to the modern species that they are put in the same genus are found in Ordovician rocks.

FURTHER READING

Kaufman, K. W. (1971). The form and functions of the avicularia of *Bugula* (Phylum Ectoprocta). *Postilla* **151**, 1–26.

Rudwick, M. J. S. (1970). *Living and fossil brachiopods.* Hutchinson, London.

Ryland, J. S. (1970). *Bryozoans.* Hutchinson, London.

Ryland, J. S. (1976). Physiology and ecology of marine bryozoans. *Adv. mar. Biol.* **14**, 285–443.

23
Acorn worms

Phylum Hemichordata
  Class Enteropneusta (acorn worms)
  Class Pterobranchia

This short chapter introduces a small group of animals which have figured predominently in discussions of the evolution of the invertebrates. They show features of resemblance to the lophophorate phyla (chapter 22), the echinoderms (chapter 24) and the chordates (chapter 25).

The members of the clas Enteropneusta are worm-shaped animals. Our example, *Saccoglossus ruber* (Fig. 23.1), is found buried in sand on the lower parts of beaches in Wales. It occupies an irregular U-shaped burrow and defecates little mounds of sand, like lugworm casts (p. 347) at the rear opening of the burrow.

The body of *Saccoglossus* has three parts: a proboscis, a short collar and the long trunk. The proboscis is the most muscular part, and is used for burrowing. It moves through the sand by peristalsis, pulling the rest of the body behind it.

There is a mouth at the anterior end of the collar and an anus at the extreme posterior end of the body. There are also many pores on each side of the anterior part of the trunk, connecting the gut cavity to the outside. These pores are believed to be homologous with the gill slits of fishes. They are U-shaped internally, but oval externally. The gut is generally full of sand. The epidermis of the proboscis has gland cells which secrete mucus and cilia which beat towards the mouth. Sand grains and detritus are trapped in the mucus, transported to the mouth and swallowed. The detritus and any organic matter adhering to the sand grains are digested.

Cilia in the gill slits beat outwards, drawing water in at the mouth and driving it out through the gill pores. If the collar is a reasonably close fit in the burrow this will tend to draw water in at one end of the burrow and drive it out at the other, ensuring that the partial pressure of dissolved oxygen in the water does not fall too low. It was suggested on p. 349 that the peristaltic irrigation movements of the lugworm may have the same function. It is widely believed that the gill slits function as gills (i.e. that a large proportion of the oxygen needed for metabolism diffuses into the body through them) and this may be the case. However, their area is not particularly large and their

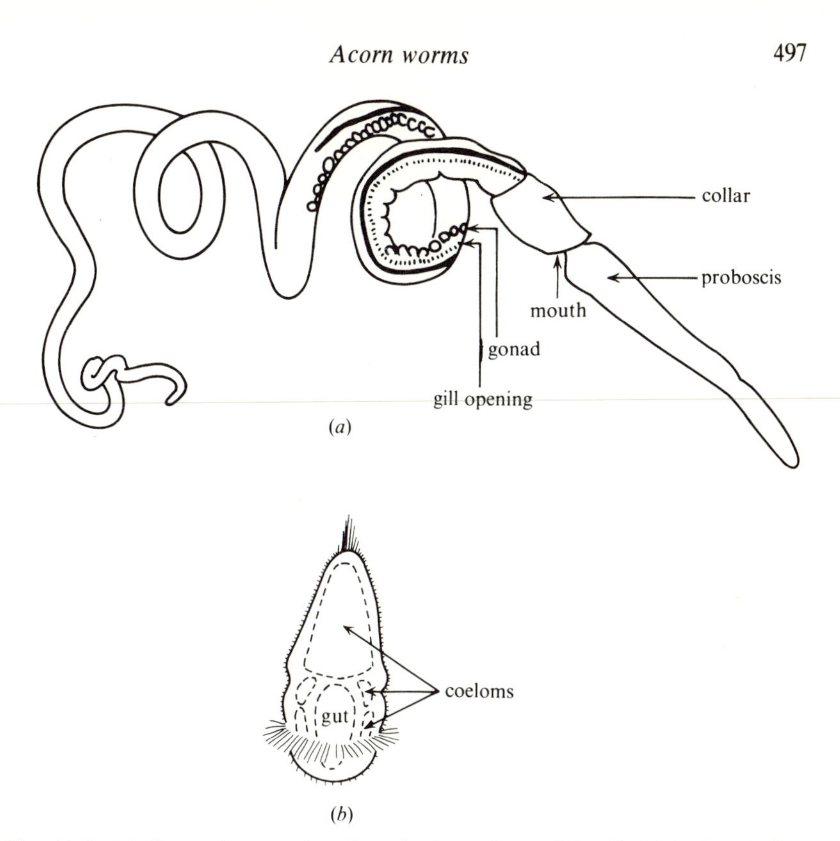

Fig. 23.1. (*a*) *Saccoglossus ruber*. Length about 5 cm. After F. W. R. Brambell & H. A. Cole (1939). *Proc. zool. Soc. Lond.* **109B**, 211–36. (*b*) Larva of *Saccoglossus horsti*. Length 0.3 mm. After C. Burdon-Jones (1952). *Phil. Trans. R. Soc.* **236B**, 553–89.

epithelium is not particularly thin, so diffusion through the general body surface may be more important.

There is a simple blood system. There is no heart but the main blood vesels are contractile. They are a dorsal longitudinal vessel in which the blood flows anteriorly, and a ventral longitudinal vessel in which it flows posteriorly. The nervous system is extremely simple. There is a network of nerve cells among the bases of the epidermal cells, as in Cnidaria (Fig. 6.2). This network is consolidated to form two main nerve tracts, a dorsal one running the whole length of the body and a ventral one in the trunk. A third tract forms a ring encircling the anterior end of the trunk and connecting the dorsal and ventral tracts.

Each of the three parts of the body has its own coelom or pair of coeloms. The proboscis is mainly filled with muscle but has a central coelomic cavity. This cavity is incompletely divided into left and right halves which suggests that it may have evolved by coalescence of a pair of coeloms. It is connected to the exterior at the posterior end of its left half by the proboscis pore. There

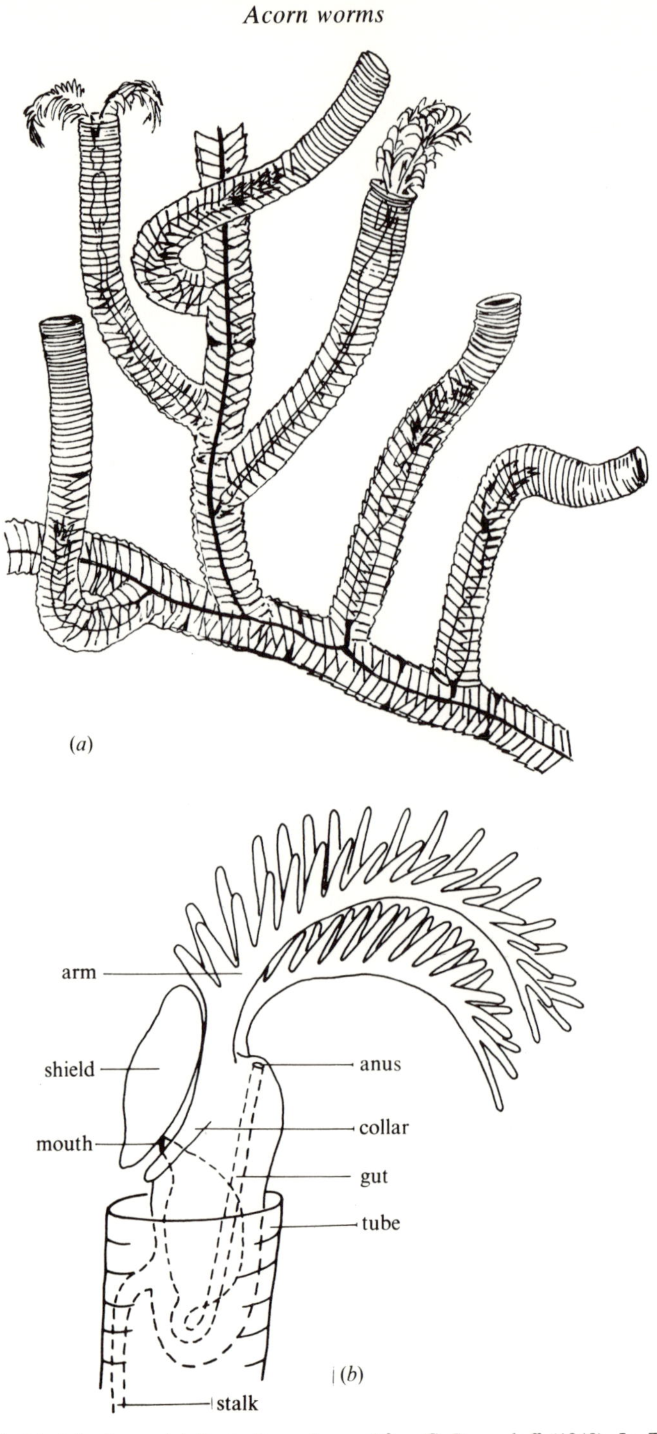

Fig. 23.2. *Rhabdopleura*. (*a*) Part of a colony. After C. Dawydoff (1948). In *Traité de zoologie*, ed. P. P. Grassé. Masson, Paris. (*b*) An individual. Length (excluding stalk) about 1 mm.

is a pair of coeloms in the collar, each with a duct leading to an opening in the first gill slit. There is also a pair of coeloms in the trunk, separated by an incomplete partition.

The gonads are numerous swellings on the trunk, each with a pore direct to the exterior. There is a ciliated larval stage with coeloms as in the adult (Fig. 23.1 *b*).

There are only three genera in the class Pterobranchia. One of them is *Rhabdopleura* (Fig. 23.2) which is dredged mainly from depths between 100 and 300 m in the Atlantic. It is a colonial animal which inhabits branching, translucent tubes attached to stones or shells. The individual members of the colony are connected together by a strand of tissue, like the individuals in a hydroid colony (Fig. 7.3). The colony is founded by a sexually produced larva which metamorphoses and then generates other individuals by budding.

The individuals have U-shaped guts so that mouth and anus are close together. They have a pair of arms, each fringed by two rows of tentacles. The tentacles do not encircle the mouth but the arrangement is nevertheless strikingly like a lophophore. It functions like a lophophore: cilia on the tentacles set up a filter-feeding current. Diatoms, radiolarians and crustacean larvae are found in the gut. Colonies have been kept alive in aquaria. The individuals came to the mouths of their tubes and spread their arms to feed, but they withdrew into the tubes when disturbed.

The shield (Fig. 23.2) has glands which secrete the tube. It is plainly homologous with the proboscis of *Saccoglossus*. Adjacent to it is a narrow collar which bears the arms. There is a single coelom in the shield (with two pores), a pair of coeloms (with pores) in the collar and a pair in the trunk. The collar coeloms extend up the arms. There are no gill slits.

Cephalodiscus, another member of the Pterobranchia, has several arms on each side of the collar, and a pair of gill slits.

FURTHER READING

Barrington, E. J. W. (1965). *The biology of Hemichordata and Protochordata.* Oliver & Boyd, Edinburgh.

24
Starfish and sea urchins

Phylum Echinodermata
  Class Crinoidea (sea lilies and feather stars)
  Class Asteroidea (starfish)
  Class Ophiuroidea (brittle stars)
  Class Echinoidea (sea urchins)
  Class Holothuroidea (sea cucumbers)
  (and several extinct classes)

The echinoderms all live in the sea. A few live in dilute sea water, for instance in parts of the Baltic Sea which contain 0.8% salts instead of the usual 3.5%, but none lives in fresh water.

The starfish *Asterias* (Fig. 24.1 *b*) will be used to introduce the phylum. The British species *A. rubens* and the American *A. forbesi* are very similar. They live on the bottom of the sea at depths down to about 200 m. They eat molluscs and crabs. They are able to open the shells of living bivalve molluscs forcibly: the method is described later in the chapter.

Asterias has a mouth in the centre of its lower surface and an anus near the centre of the upper one. It has five arms, and almost perfect five-fold symmetry. The symmetry is, however, spoilt by a single porous plate, the madreporite, and some internal structures connected to it. The madreporite is on the upper surface, with the anus, and they cannot both be placed centrally. The madreporite is conspicuous and is shown in Fig. 24.1(*b*) but the anus is inconspicuous.

The lower surface bears about 1200 tube feet on which the animal crawls. Some of them are visible in Fig. 24.1(*b*), and they are described later in the chapter. A starfish may crawl with any one of its five arms leading and the probability that a particular arm will be leading on a particular occasion is not very different from the probability that any other arm will be leading. Starfish crawl less often with two arms leading, side by side.

Since there is no strongly preferred direction of crawling no arm can be called anterior. The terms anterior and posterior cannot be used appropriately in describing adult starfish. Similarly the terms dorsal, ventral, left and right, which are designed for describing bilaterally symmetrical animals, are inappropriate. The surface which bears the mouth is called the oral surface and the opposite one is called the aboral surface, as in describing Cnidaria (Fig. 8.1). 'Oral' and 'aboral' are preferrable to 'lower' and 'upper' because not all echinoderms keep the aboral surface uppermost.

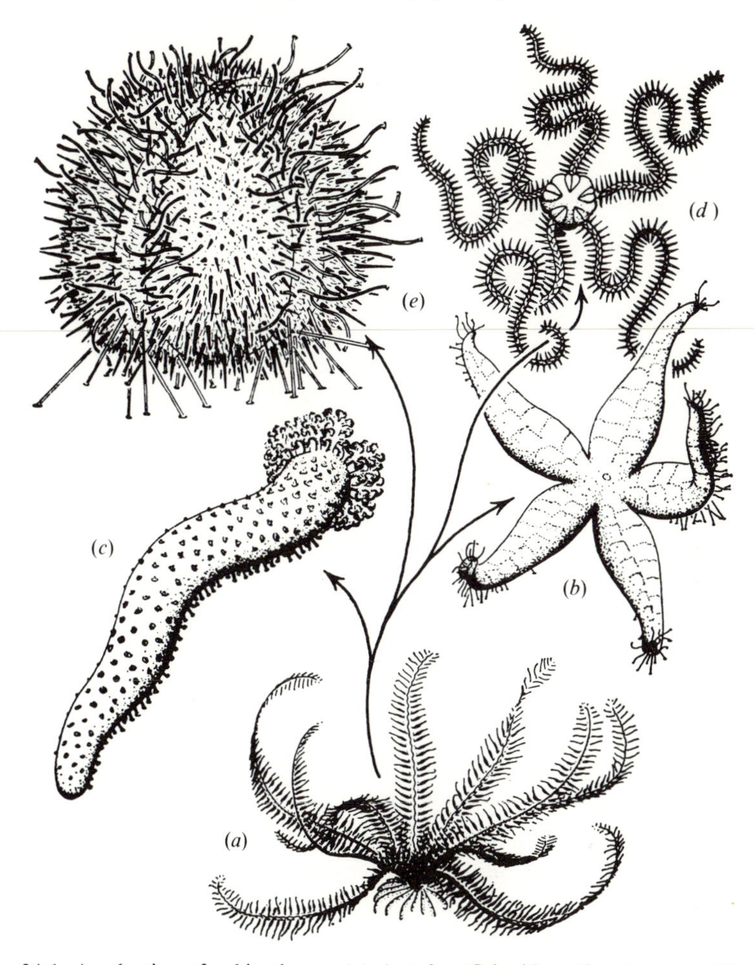

Fig. 24.1. A selection of echinoderms. (*a*) *Antedon* (Crinoidea; diameter up to 20 cm); (*b*) *Asterias* (Asteroidea; 30 cm); (*c*) *Holothuria* (Holothuroidea: length 20 cm); (*d*) *Ophiothrix* (Ophiuroidea: diameter 15 cm); (*e*) *Echinus* (Echinoidea: up to 16 cm). From D. Nichols (1969). *Echinoderms*, 4th edn. Hutchinson, London.

Scattered over the surface of the body are small spines and pedicellariae (Fig. 24.2*d*). The latter resemble tiny crab pincers. They grab, kill and discard small animals (for instance, small polychaetes) which touch them. Their main function may be to prevent the larvae of sessile animals such as barnacles and ectoprocts from settling on the starfish. They are rather similar to the avicularia of ectoprocts (Fig. 22.4).

Fig. 24.2(*a*) shows how the main internal organs are arranged. The gut has a central stomach and five pairs of caeca, one pair in each arm. There are ten gonads, two in each arm, with separate openings on the aboral surface. There is a water vascular system, a peculiarity of the phylum. Its main vessels are a circum-oral water ring encircling the mouth, a radial water canal along each

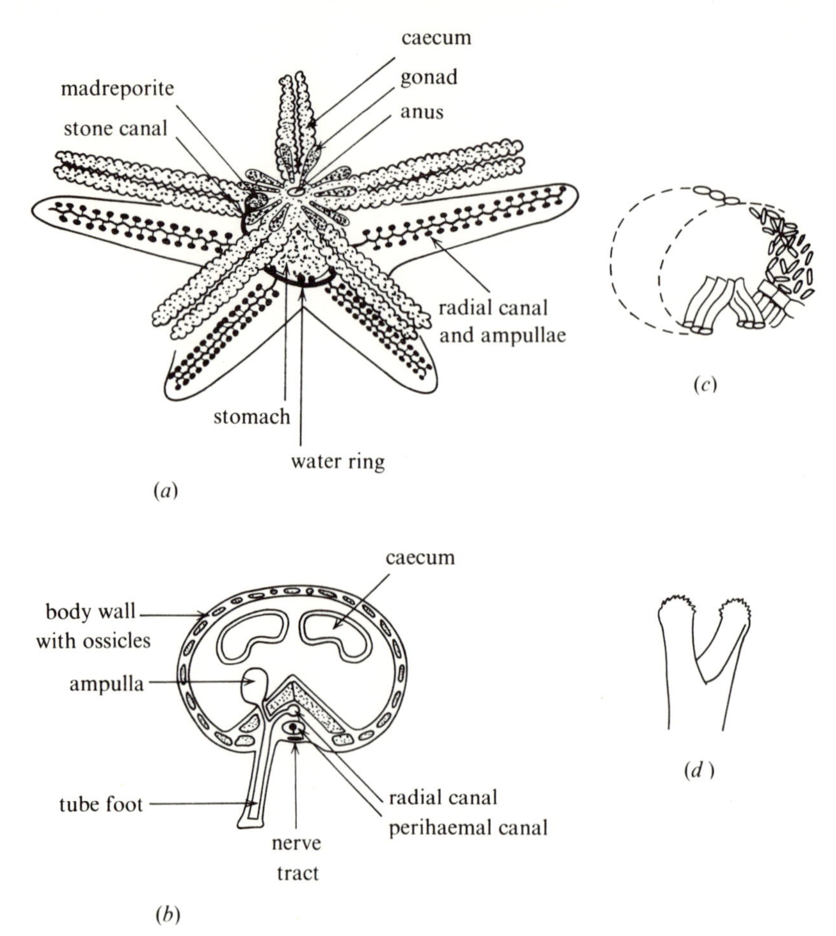

Fig. 24.2. Diagrams illustrating the structure of *Asterias rubens*. (*a*) The internal organs. The gut is lightly stippled, the gonads more heavily stippled and the water vascular system shown black. (*b*) A section through an arm; (*c*) the skeleton of a short piece of arm; (*d*) a pedicellaria.

arm and a stone canal (so called because it has hard spicules in its wall) leading to the madreporite.

The radial water canals have branches to the tube feet, which lie on either side of them (Fig. 24.2*b*). Each tube foot has its own ampulla, a thin-walled sac which lies in the main body cavity.

There are perihaemal canals and haemal strands running largely parallel to the water vascular canals. They are not shown in Fig. 24.2(*a*) but Fig. 24.2(*b*) shows a perihaemal canal running below (oral to) the radial water canal, enclosing a haemal strand. There is a haemal ring oral to the water ring, a haemal strand parallel to the stone canal and other haemal strands to the caeca and

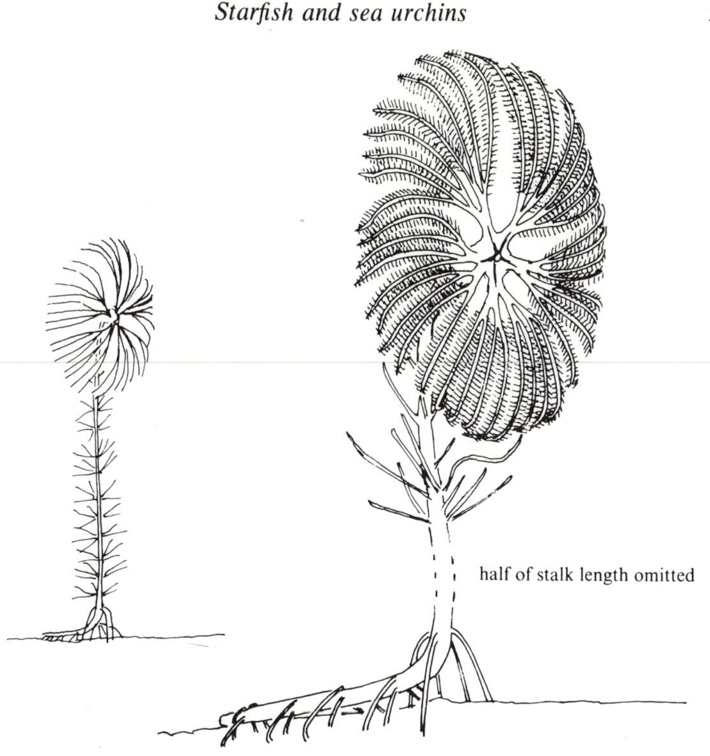

Fig. 24.3. *Cenocrinus asterias* (Crinoidea: height up to almost 1 m).

gonads. Most of the haemal strands are enclosed in perihaemal canals. The stone canal and the haemal strand which runs parallel to it are both enclosed by a perihaemal canal known as the axial sinus. The haemal strands consist of spongy tissue, and some observations on sea urchins suggest that they are concerned in the phagocytosis and destruction of any micro-organisms which may get into the animal.

There is a net of nerve cells under the epidermis, all over the body. It is thickened into more definite nerve tracts which run along the oral surface of each arm and form a ring round the mouth.

The body wall has ossicles of calcium carbonate embedded in it. Their fine structure will be described later. Their arrangement in the arms is shown in Fig. 24.2(*c*). There are close-fitting ossicles in the oral wall and an irregular lattice in the aboral one. The connections from the tube feet to their ampullae run between the ambulacral ossicles, which are the largest ossicles. Even where the ossicles fit closely they are joined flexibly, so the muscles of the body wall can bend the arms orally, aborally or to either side.

The main body cavity is a coelom, and so are the cavities of the water vascular and perihaemal systems, as will become apparent when the development of echinoderms is described later in the chapter. All these cavities are filled with fluid which differs little from sea water. The fluid in the main body cavity has the same osmotic concentration as sea water but very careful measurements have shown that the fluid in the tube feet is 0.02 Osmol l^{-1} more concentrated, owing to higher concentrations of potassium and chloride ions.

The most primitive of living echinoderms are the sea lilies, which are members of the class Crinoidea. They have stalks, and their arms radiate from the top of the stalk like the petals of a flower (Fig. 24.3). The arms often branch repeatedly and all the branches are fringed with pinnules so the disc of arms forms in effect a circular net. Any particle of food which falls on this net or is carried through it by a current is likely to be intercepted. Food is probably caught in mucus secreted by the tube feet and carried by cilia to the mouth, but this has not been actually observed. Most sea lilies live at depths of 200 m or more so observation is difficult.

Fig. 24.3 is based on photographs taken from a submersible vehicle at a depth of about 300 m off Jamaica. Photographs taken where there were currents show the disc of arms spread vertically with the mouth facing downstream. Others taken where there was no current show the arms spread horizontally with the mouth facing upwards. The distal end of the stalk is embedded in the sediment. The cirri which project from it help to root it firmly.

Sea lilies were extremely common in Palaeozoic seas. The ossicles of their stalks are very abundant fossils, especially in limestones of the Carboniferous period.

Most modern species of the class Crinoidea are not sea lilies but feather stars like *Antedon* (Fig. 24.1 *a*). They have no stalks but can cling to rocks and other objects by means of a group of cirri on the aboral surface. They can also swim by beating their arms up and down. *Antedon* lives offshore in relatively shallow water and feeds in the manner described as probable for sea lilies.

Ophiothrix (Fig. 24.2 *d*) is one of the brittle stars, the members of the class Ophiuroidea. It has no anus, and there are no gut caeca in its slender arms. It lives on gravel off European shores, often in extraordinarily dense patches of over 1000 animals per square metre. Scuba divers find that in gentle currents it rests on two or three arms, extending the others vertically to catch food. In stronger currents, in rough weather or when the tide is flowing fastest, it links arms with its neighbours. Individuals detached from their neighbours are apt to be swept along by the current but the patch as a whole stays put. Experiments with radioactive phytoplankton have shown that *Ophiothrix* is capable of filtering out unicellular algae.

Echinus (Fig. 24.2 *e*) and the other sea urchins belong to the class Echinoidea. *Echinus* is almost spherical, and is enclosed in a rigid test made of closely-fitting ossicles. The test is not external like the exoskeletons of arthropods and the shells of molluscs, for the ossicles are embedded in the body wall and are covered externally by epidermis. There are much longer spines than *Asterias* has. They are attached to the outer surface of the test by ball-and-socket joints and have muscles which can point them in any direction.

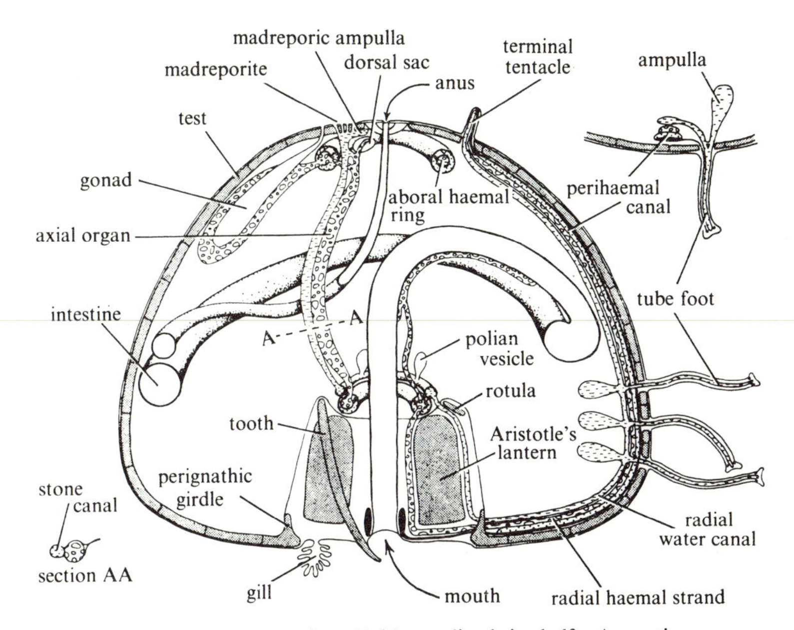

Fig. 24.4. A diagram representing *Echinus* sliced in half. A section across an ambulacrum is shown on the right and one through the stone canal and associated haemal strand (the axial organ) on the left. The water vascular, haemal and perihaemal systems are distinguished by distinctive textures. From D. Nichols (1969). *Echinoderms*, 4th edn. Hutchinson, London.

There are no arms but there are five double rows of tube feet running from the mouth at the centre of the lower surface to the anus at the centre of the upper one. Each double row of tube feet, with its associated structures, is called an ambulacrum. The water canals are inside the test and are connected to the tube feet through holes in the test.

Fig. 24.4 shows the internal organs of *Echinus*. Most of them are arranged as in *Asterias*, with distortions due to the difference between the star shape and the sphere. However the gut is quite long and coiled, without caeca. There is also a remarkable structure around the mouth, called Aristotle's lantern. It is a framework of ossicles supporting five long teeth and has muscles to move it. There is a ring of so-called gills round the mouth but most of the oxygen which the animal uses probably diffuses in through the tube feet.

Echinus lives just offshore on gravel or rocks but is sometimes left exposed at low tide. It eats the 'fur' of attached algae which coats the rocks, scraping it off with the five teeth.

Echinus has almost perfect five-fold symmetry but some other Echinoidea do not. One of them is the heart urchin *Echinocardium* (Fig. 24.5). It is by no means spherical but has the shape of the hearts printed on playing cards. This shape is bilaterally symmetrical. The anus is at the point of the heart, between

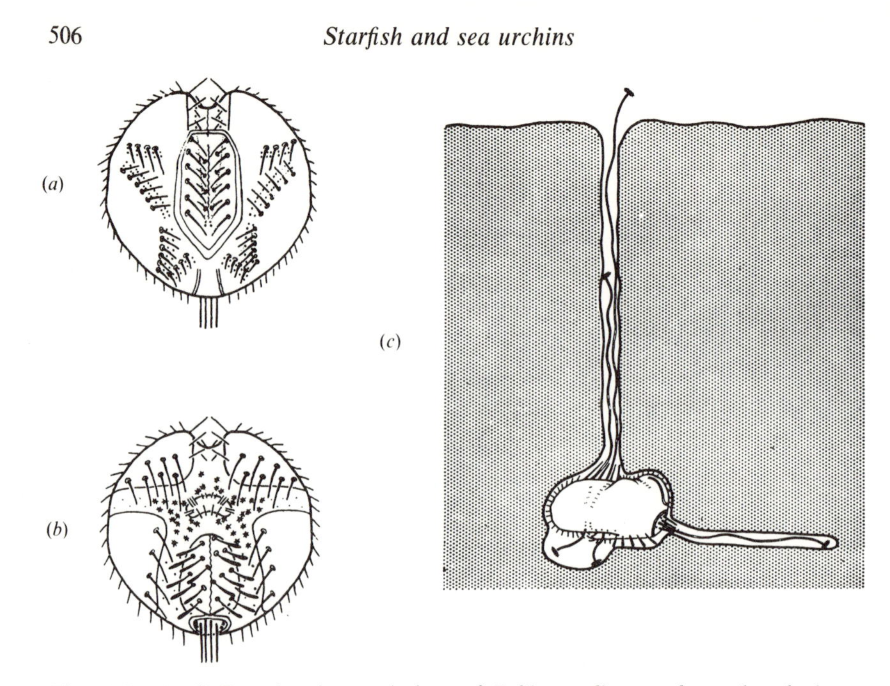

Fig. 24.5. (*a*), (*b*) Dorsal and ventral views of *Echinocardium cordatum* (length about 4 cm). (*c*) A section through the burrow of *Echinocardium* showing the animal in its normal position. Only a few of the tube feet are illustrated. From D. Nichols (1959). *Phil. Trans. R. Soc.* **242B**, 347–437.

two ambulacra, and its position may be regarded as posterior. The oral surface can be considered ventral. The mouth is near its centre but points anteriorly. There is no Aristotle's lantern.

Echinocardium lives close offshore, buried in sand. Fig. 24.5(*c*) is based on observations made through the bottom of a glass aquarium full of sand covered with sea water. The long tube feet near the mouth pick up particles of sand and pass them to the mouth. The film of organic matter on them is digested off and they are passed out through the anus. Thus sand is shifted from in front of the animal to behind it. The animal moves slowly forward at up to 8 cm h^{-1}, levering itself along by means of the spines on the oral surface. Very long dorsal tube feet maintain a vertical shaft to the surface of the sand. This shaft has to be abandoned and replaced by a new one from time to time, as the animal moves forward. Another, posterior, group of tube feet maintains a horizontal tunnel behind the animal.

The epithelium of the whole animal is covered by cilia. Their direction of beating has been investigated by experiments with suspensions of carmine particles. It seems that they draw water down the vertical shaft and drive it away along the horizontal one, whence it percolates into the sand. This water supplies the oxygen which the animal needs.

The sea cucumber *Holothuria* (Fig. 24.1*c*) is an example of the Holo-

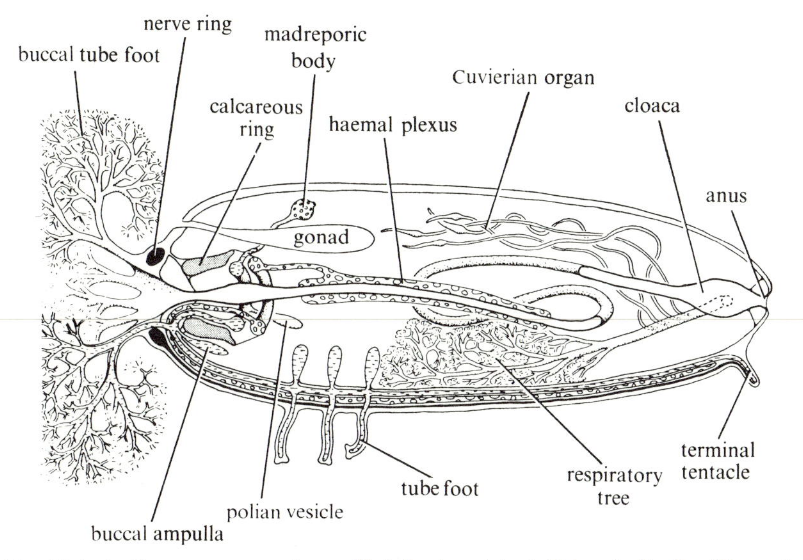

Fig. 24.6. A diagram representing a *Holothuria* cut in half longitudinally. The water vascular and haemal systems are distinguished by the same textures as in Fig. 24.4. From D. Nichols (1969). *Echinoderms*, 4th edn. Hutchinson, London.

thuroidea. Like other members of the class it is bilaterally symmetrical but this symmetry has evolved differently from that of *Echinocardium. Echinocardium* is an echinoid which always moves with a particular ambulacrum leading. That ambulacrum has become anterior and the mouth is ventral. *Holothuria* is like an echinoid which has lain on its side and taken to crawling on the ambulacra of that side. The mouth is anterior and the anus posterior. Three ambulacra are ventral and have tube feet with suckers which are used for locomotion. The other two ambulacra are dorso-lateral. Their tube feet are scattered, not arranged in parallel rows, and have no suckers.

There is a circle of 20 branched tentacles around the mouth which can be withdrawn into the body when the animal is disturbed. They are highly modified tube feet. Each has its own ampulla and is connected to one of the radial water canals (Fig. 24.6). There is no external madreporite but the water vascular system opens into the main body cavity through pores in a madreporic body. There is little trace of a perihaemal system. The body wall is tough but flexible, for instead of large ossicles it has only small, scattered spicules embedded in it.

There is only one gonad, and there is a pair of branched organs in the body cavity called respiratory trees. These are departures from five-fold symmetry but they fit into the scheme of bilateral symmetry. Movements of the cloaca pump water in and out of the respiratory trees. It has been shown by experiment that the animal gets 60% of its oxygen from this water.

Each respiratory tree has a bundle of Cuvierian organs associated with it.

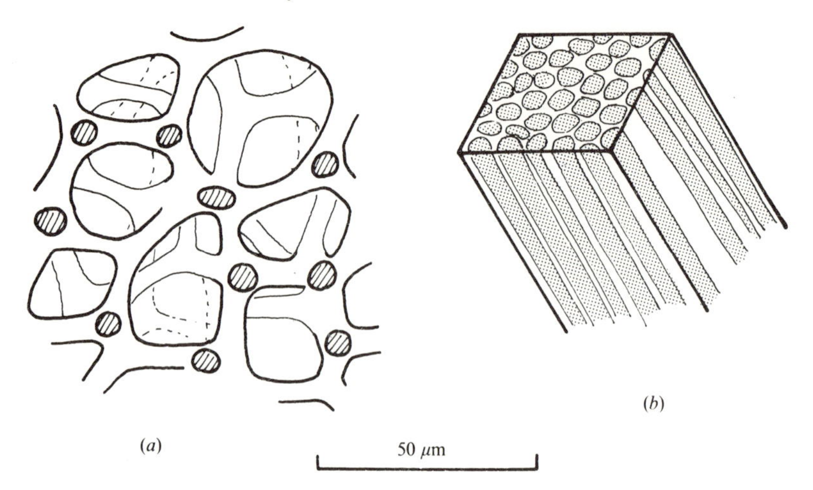

Fig. 24.7. Sketches based on scanning electron micrographs showing the structure of (*a*) an echinoderm ossicle and (*b*) the core of a sea urchin tooth.

These are long tubules filled with collagen and polysaccharide. If the animal is molested the contents of the Cuvierian organs swell, splitting the tubules and bursting through the wall of the cloaca to escape as a mass of long sticky threads which entangle the molester.

Holothuria crawls along on the bottom of the sea, using its tentacles to sweep up the particles of detritus which it eats.

SKELETON

The ossicles of echinoderms are a mixture of calcium and magnesium carbonate, typically 90% calcium and 10% magnesium. They are not solid blocks but three-dimensional networks (Fig. 24.7*a*). The spaces between the bars account for, typically, about 50% of the volume, and are filled in life by soft tissue. Adjacent ossicles are connected by collagen fibres laced through the holes, and tendons are attached to ossicles in the same way.

Each ossicle seems to be a single crystal, like the spicules of calcareous sponges (p. 113). Ossicles have been broken and examined by scanning electron microscopy. Micrographs of the broken surfaces look just like micrographs of broken surfaces of calcite crystals. They look quite unlike the much rougher surfaces which are obtained by breaking mollusc shells, which consist of tiny crystals separated by protein.

The strength of mollusc shell depends on its composite structure (p. 266). Cracks which would spread right through a single solid crystal are halted by the layers of protein between the crystals. The same result is apparently achieved in echinoderm ossicles by the network structure. A crack in one bar will probably spread across the bar but it is unlikely to cross the gap to the next bar. The ossicles are reasonably strong.

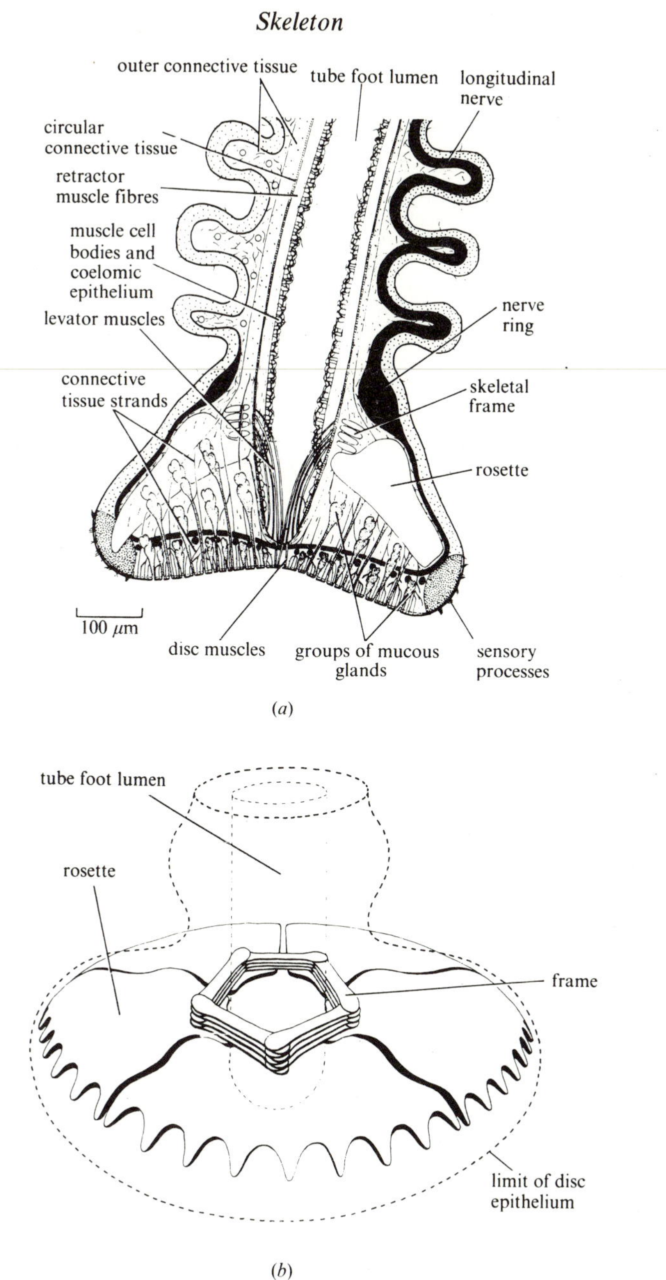

Fig. 24.8. A tube foot of *Echinus*. (*a*) shows the distal part in section and (*b*) shows the skeleton of the distal part. From D. Nichols (1961). *Q. J. microsc. Sci.* **102**, 157–80.

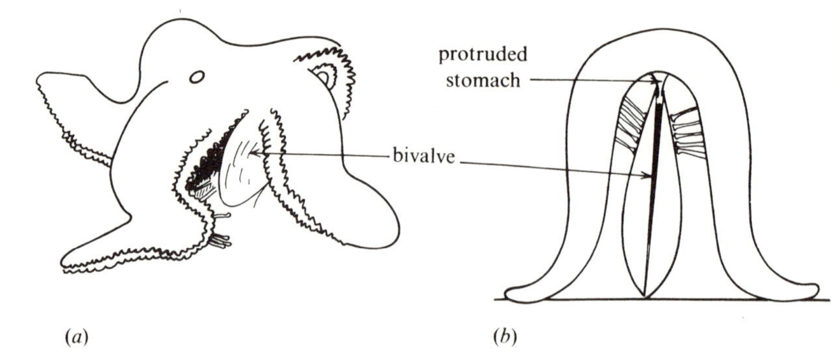

Fig. 24.9. (*a*) A sketch and (*b*) a diagrammatic section of *Asterias rubens* feeding on a bivalve mollusc. (*a*) is based on a photograph by R. Buchsbaum (1948). *Animals without backbones*, 2nd edn. University of Chicago Press.

The teeth of sea urchins have a different structure which makes them more resistant to abrasion (Fig. 24.7*b*). They have very fine rods of calcite running lengthwise along them, occupying about 55% of their volume. Each rod is a single crystal. The rest of the volume is also calcium carbonate, apparently in amorphous form. Only the core of the tooth has this structure; the rest has the same structure as ordinary ossicles and wears down quickly so that the hard core stands proud at the tip of the tooth.

TUBE FEET AND LOCOMOTION

The five classes of echinoderm have different types of tube foot, but a tube foot of *Echinus* will serve as an example (Figs. 24.4, 24.8). Two holes through the test connect it to its ampulla. It is lined with cilia which drive fluid down one of its sides and up the other so there is continuous flow of water from the ampulla through one of the holes in the test, down the tube foot, up again and back through the other hole into the ampulla. This flow is probably important in respiration: oxygen diffuses from the sea into the tube foot, is carried to the ampulla by the flow and diffuses from there into the main body cavity.

There are muscles in the wall of the tube foot and its ampulla. There is a valve between the ampulla and the radial water canal which must close when the pressure in the ampulla rises. Hence contraction of the muscles of the ampulla drives water into the tube foot, and vice versa. The tube foot has longitudinal muscles but no circular ones. Its diameter is limited by a layer of connective tissue with circularly-arranged fibres, but it can extend considerably in length. Contraction of its muscles shortens it and contraction of the ampulla lengthens it, without altering its diameter much. The foot can be pointed in any direction, presumably by differential contraction of its muscles.

The distal end of the tube foot has a sucker which is stiffened by small ossicles (the rosette and frame, Fig. 24.8*b*). Notice that these ossicles are arranged in rings of five which suggests that the mechanism which gives the animal its five-fold symmetry is operating here on a smaller scale. Crinoids and ophiuroids have no suckers on their tube feet.

Tube feet with suckers can be used for pulling as well as for pushing. Muscles running to the centre of the sucker (Fig. 24.8*a*) may help to attach the sucker firmly, by raising its centre. The tube feet of a starfish have to pull particularly hard when they are used to open the shells of bivalve molluscs. Fig. 24.9 shows how this is done. The starfish holds the bivalve with the hinge away from its mouth, and pulls with the tube feet to open the shell slightly. It turns its stomach inside out and slips it through the opening, and starts digesting the tissues of the mollusc. The fluid products of digestion travel into the starfish along ciliated channels in the wall of its stomach.

To insert its stomach the starfish needs only to open the shell slightly. (It has been claimed that 0.1 mm is enough.) Even so small an opening requires a large force for the adductor muscles of bivalve molluscs are large and capable of exerting high stresses (p. 304). *Asterias forbesi* has been given mussels with the adductor muscles cut and replaced by springs which held the valves closed. They could open these shells enough to insert the stomach even when the spring was adjusted so that a force of 30 N was needed.

The pull which a single tube foot can exert has been measured. A starfish was allowed to attach a single tube foot to a fragment of glass suspended from a thread. The force on the thread needed to detach the glass was typically about 0.3 N. The average area of the suckers was about 2 mm^2 so the pressure reduction under the sucker was about $0.3/2 \times 10^{-6} = 1.5 \times 10^5$ N m^{-2}, or 1.5 atm. Absolute negative pressures must occur under suckers as they do in the pharynx of *Rhodnius* (p. 442). The pull must be exerted by the longitudinal muscles of the tube foot which have a cross-sectional areas of about 1 mm^2, so the stress they must exert is about 3×10^5 N m^{-2}. This is similar to the maximum stresses exerted by many other muscles, contracting isometrically (p. 304).

A large *Asterias* has about 1200 tube feet. If each can exert 0.3 N they are theoretically capable of exerting a total of 360 N, or 180 N on each half of a bivalve shell. This is far more than is needed to open any mollusc a starfish might eat but it is also probably far more than the muscles of the arms could withstand. The arms have to be held stiff by the muscles of the body wall while the tube feet pull.

When a starfish crawls, all its tube feet step in the appropriate direction. However, they do not step in phase with each other, nor is there any perceptible pattern to their movements like the metachronal waves of cilia (Fig. 3.4) or of the legs of myriapods (Fig. 21.3). This can be seen most conveniently when a starfish is set upside-down on a glass tube (Fig. 24.11*a*). So placed, it cannot turn over, but its tube feet continue stepping. Each tube foot moves as shown in Fig. 24.10(*a*)–(*d*), very much as human legs move in walking.

The forces exerted by individual tube feet have been measured by the method

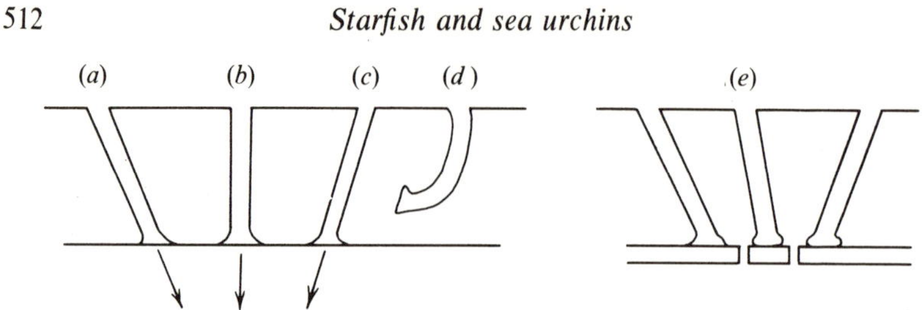

Fig. 24.10. (*a*)–(*d*) Successive positions of a starfish tube foot during stepping. The arrows represent forces exerted on the ground. (*e*) Apparatus for recording the force exerted on the ground.

indicated in Fig. 24.10(*e*). The starfish is walking over a platform which has a small loose section, just big enough to accommodate one tube foot. This loose section is mounted on springs so that a downward force can push it down slightly and a forward or backward force can push it forward or backward slightly. These movements are recorded and the forces can be calculated from them. The details of the apparatus are not shown because the experiment was carried out in 1952 using simple equipment which, though adequate, would not be used today.

The arrows in Fig. 24.10(*a*)–(*d*) show the direction of the force on the ground at each stage in the step. The tube foot always pushes along its own long axis. Similarly, when man walks the force on the ground is more or less in line with the leg, throughout the step.

The forces exerted in crawling are small, especially when the animal is crawling under water with most of its weight supported by buoyancy. Even so, the fluid in the tube feet must be under pressure. Pressures up to about 4000 N m^{-2} have been measured in tube feet by a direct but rather unsatisfactory method. Such pressures in the tube feet would force water slowly out through their walls if it were not for the small difference of osmotic concentration between the fluid inside them and sea water (p. 504). The difference is 0.02 Osmol l^{-1}, which corresponds to an osmotic pressure difference of 50000 N m^{-2}, far more than is needed to balance the highest pressure which has been measured. Perhaps it is needed to draw water in by osmosis fast enough to compensate for leaks in the system.

It has often been suggested that water is drawn in through the madreporite to replenish the water vascular system. The previous paragraph shows this is unnecessary and experiments with dyes have failed to demonstrate uptake through the madreporite. The madreporite has no known function.

Though the tube feet do not show any orderly metachronal rhythm, some co-ordinating system is needed to keep them stepping in the same direction. Information about this has been obtained by experiments with inverted starfish (Fig. 24.11 *a*). Single arms cut off with no nerve ring left attached step towards their bases (Fig. 24.11 *b*) but ones which include a piece of nerve ring step towards their tips (Fig. 24.11 *c*). In a starfish which has had its nerve ring cut

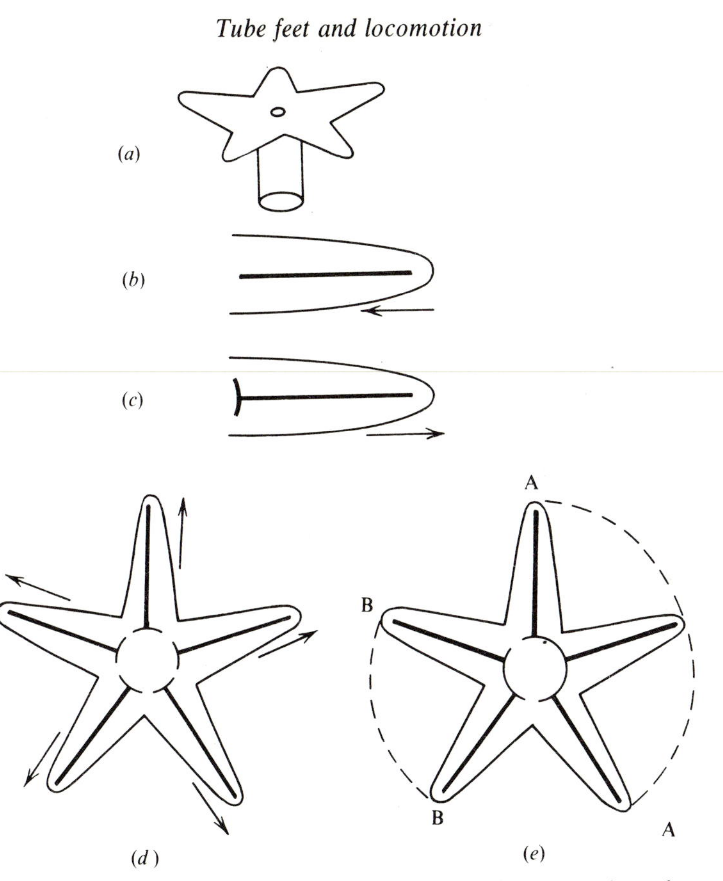

Fig. 24.11. (*a*) A starfish set upside-down on a glass tube, for observations of stepping. (*b*)–(*e*) Experiments which are described in the text. Arrows indicate directions of stepping.

between all the arms, each arm steps towards its tip (Fig. 24.11 *d*). In one which has had its nerve ring cut in two places, separating two arms from the other three, the two groups of arms work independently (Fig. 24.11 *e*). At any instant the group of three arms is always stepping in a common direction but this direction changes from time to time and may be anywhere in the arc AA. The other two arms also step in a common direction, but in the arc BB.

These experiments have been interpreted as indicating that there is a nerve centre in the nerve ring at the base of each arm. Each centre acting alone would drive stepping in the direction of its arm. When the nerve ring is intact one centre dominates the others, making the whole animal step in the direction of its arm. Dominance shifts from time to time to another centre, and the whole animal changes direction. However, the supposed centres have not been found in anatomical studies, and may not exist.

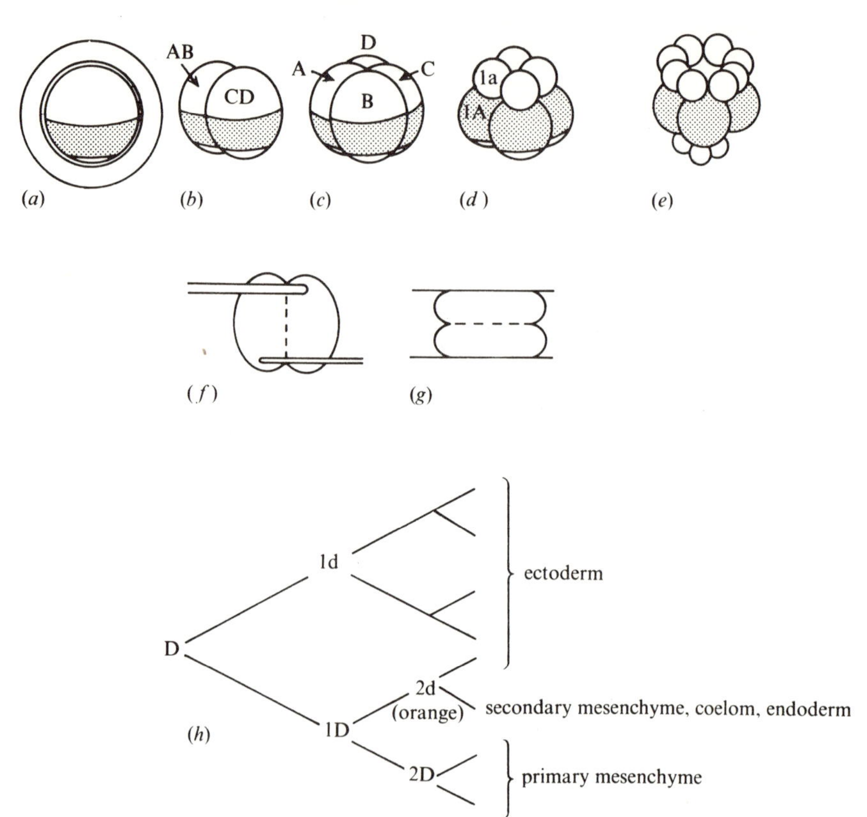

Fig. 24.12. (*a*) A zygote of *Paracentrotus lividus* and (*b*)–(*e*) early stages in its development. (*f*), (*g*) Experiments which are described in the text. (*h*) A diagram showing the fate of the products of division of one of the four cells shown in (*c*).

DEVELOPMENT

The eggs and larvae of sea urchins have played an extremely important part in cell biology. They have proved convenient for research for various reasons. It is reasonably easy to achieve artificial fertilization in the laboratory by mixing ripe eggs and sperm. Development is straightforward with none of the complications which occur in the development of large yolky eggs. It is also rapid: the zygote becomes a mature larva in only 48 h. The larvae are transparent and have only 1000–2000 cells. *Paracentrotus lividus* has been found particularly convenient for study because one end of the egg is distinguished from the other by a band of orange pigment (Fig. 24.12 *a*).

The zygote is enclosed in two layers of membrane, an outer fertilization membrane and an inner hyaline membrane. They are both present at all the stages shown in Fig. 24.12 but are only shown in Fig. 24.12(*a*). The first two divisions of the zygote occur symmetrically with respect to the orange band so the four cells they produce look identical (Fig. 24.12*c*). They seem in fact to be identical. They can be separated by dissection with fine glass needles and each will develop into a small but otherwise normal larva.

The next division separates four pale cells at one end of the embryo from four mainly orange ones at the other end (Fig. 24.12*d*). Neither the top four cells nor the bottom four are capable, without the others, of producing a normal embryo. Top halves of embryos fail to develop a gut and bottom halves fail to develop the usual complement of cilia. Fig. 24.12(*e*) shows the next stage of division and Fig. 24.12(*h*) shows the eventual fate of the products of the next few divisions. This has been discovered partly by studying the imperfect larvae which develop from early embryos after some of the cells have been removed, and partly by experiments in which parts of an egg or embryo were coloured with dyes and the position of the colour observed at a later stage in development.

Compare Fig. 24.12 with Fig. 8.14, which shows early stages in the development of a polyclad flatworm. Notice first a difference at the eight-cell stage. The sea urchin embryo has each upper cell immediately on top of the corresponding lower cell. The manner of division which gives this effect is called radial cleavage. In the polyclad the top four cells alternate with the bottom four as a result of spiral cleavage. Now look at Fig. 24.12(*h*). It applies equally to any of the four cells A, B, C and D. Any one of the four has parts capable of producing all the tissues of the larva (and can indeed develop into a complete larva). In contrast, Fig. 8.14(*b*) applies only to cell D, for cell 4d gives rise to the gastrodermis and parenchyma while 4a, 4b and 4c vanish. The early development of molluscs and of nemertean and polychaete worms proceeds in essentially the same way as in polyclads. (For a minor difference see p. 362.)

The first division of the sea urchin zygote has often been studied as an example of cell division, in the hope of gaining insight into the process of cell division in general. It is of course easier to observe and interpret than the division of a cell within a tissue. The undivided cell is kept spherical by tension in its membrane, just as surface tension tends to keep a raindrop spherical. When the cell divides additional tension is needed round the 'waist' to constrict it. Electron microscope sections show a girdle of microfilaments here, just under the cell membrane. This presumably provides the constricting force. The force has been measured by sticking fine glass needles through a dividing zygote and observing how much they are bent by the constriction (Fig. 24.12*f*). It has also been calculated from observations of the shapes which dividing zygotes adopt when they are squeezed (Fig. 24.12*g*). It is found that the force is about 6×10^{-8} N, corresponding to a stress in the girdle of microfilaments of 0.1 MN m^{-2}. This is comparable to the stresses of about 0.3 MN m^{-2} which most muscles can develop.

Later stages in development have been studied by time-lapse cinematography of embryos trapped in the meshes of a nylon net. When the film is projected at normal speed it shows the process of development greatly speeded up. The

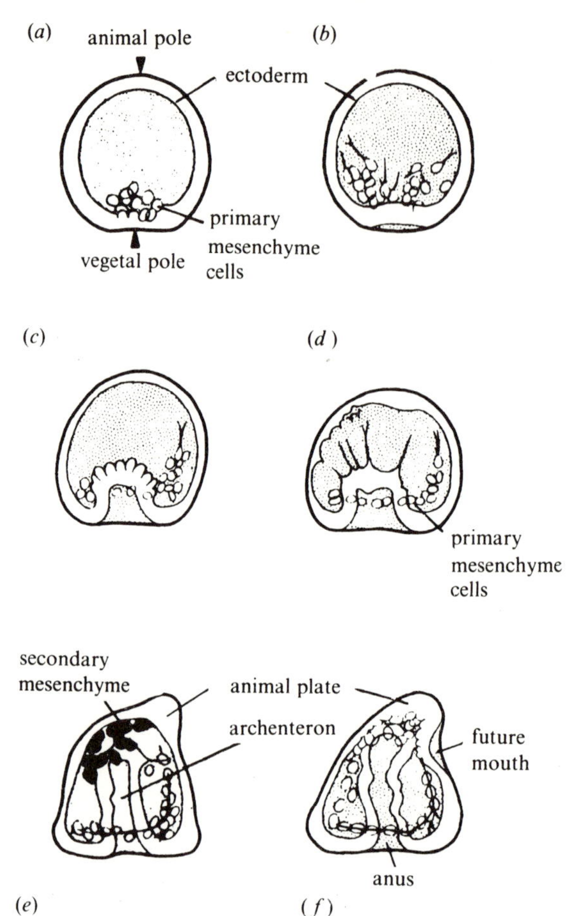

Fig. 24.13. Sections of embryos of *Psammechinus miliaris*, showing stages of development from blastula to gastrula. From T. Gustafson & M. Toneby (1971). *Am. Sci.* **59**, 452.

net is needed to keep the same embryo constantly in the field of the microscope.

The most informative observations have been made on the sea urchin *Psammechinus miliaris*. The zygote divides about 10 times at intervals of about an hour producing about $2^{10} = 1024$ cells. The cells adhere to the hyaline membrane and their planes of division are at right angles to the membrane, so the divisions produce a single layer of cells adhering to the inside of the membrane, which becomes greatly stretched. Thus a hollow blastula is produced (Fig. 24.13 *a*).

The blastula sheds the fertilization membrane and starts swimming by means of cilia which protrude through the hyaline membrane. There is little division

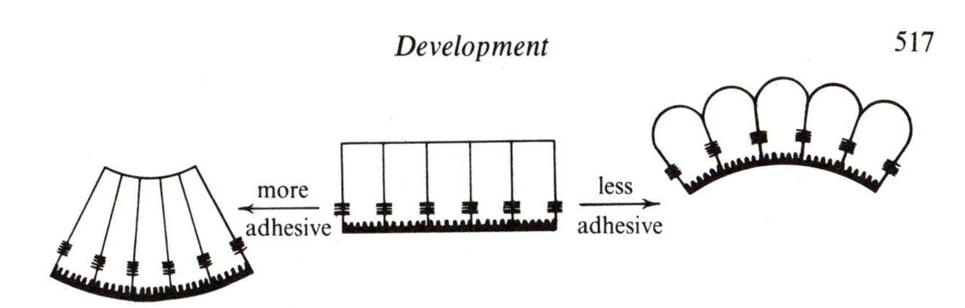

Fig. 24.14. Diagrams of part of a blastula, showing a few cells and part of the hyaline membrane. This diagram is explained further in the text.

in the next 36 hours or so but there is a lot of cell movement, producing a larva which is quite complex in structure but has only about 2000 cells (about twice as many as the rotifer *Epiphanes*, p. 230).

The cells of the blastula have processes which protrude into the hyaline membrane, which is presumably why they adhere to the membrane so firmly. They are also attached to their neighbours by desmosomes, close under the membrane. If cells like this become more adhesive they will tend to become taller and thinner so as to have larger areas in contact with each other. If the area of the membrane is more or less fixed this will make it more convex, as shown in Fig. 24.14. Conversely if the cells become less adhesive they will tend to make the membrane concave.

A dimple appears at the vegetal pole of the blastula (i.e. at the end which is mainly orange in *Paracentrotus*), with a ring around it where the cells are thicker (Fig. 24.13*a*, *b*). This may be due to a group of cells becoming less adhesive (producing the dimple) while the cells around them become more adhesive (producing the thick ring). The cells at the centre of the dimple become so much less adhesive that they become detached from the layer of cells. They are the primary mesenchyme cells which will secrete the larval skeleton.

These cells send out fine pseudopodia which seem stiff, and wave around like bristles. The pseudopodia attach at their tips to other cells and then contract, pulling their own cells behind them. Since the pseudopodia are particularly apt to attach to the very adhesive cells of the thick ring this process tends to arrange the primary mesenchyme cells in a ring around the dimple (Fig. 24.13 *c*). This is where they start to secrete the skeleton.

At this stage another group of cells on the dimple is extending pseudopodia. Each cell extends several pseudopodia, in different directions. These are the secondary mesenchyme cells. Unlike the primary mesenchyme cells they do not detach from their original positions so when their pseudopodia contract the dimple is pulled deep into the cavity of the blastula in the process of gastrulation.

Electron microscope sections of secondary mesenchyme cells show that their pseudopodia have microfilaments and microtubules in them. There is a discussion on p. 36 of the roles of the microfilaments which are present in

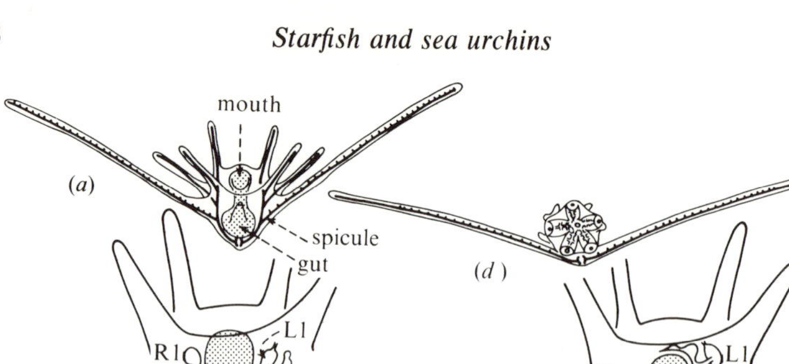

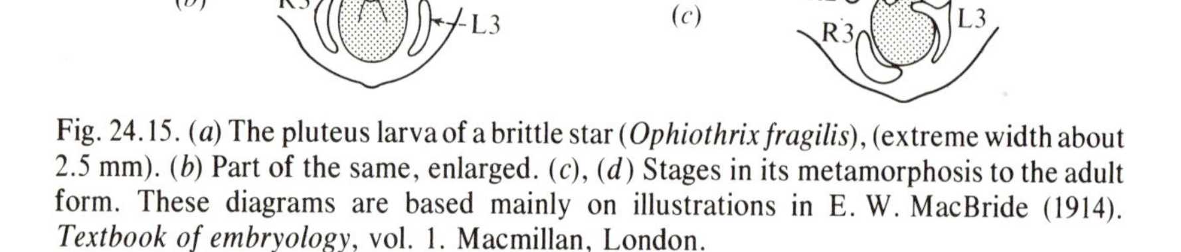

Fig. 24.15. (*a*) The pluteus larva of a brittle star (*Ophiothrix fragilis*), (extreme width about 2.5 mm). (*b*) Part of the same, enlarged. (*c*), (*d*) Stages in its metamorphosis to the adult form. These diagrams are based mainly on illustrations in E. W. MacBride (1914). *Textbook of embryology*, vol. 1. Macmillan, London.

Amoeba and of the microtubules in the pseudopodia and axopodia of some other protozoans. It is likely that the mechanisms of cell movement are the same in echinoderm embryos as they are in protozoans.

Thus the early stages of sea urchin development seem to depend on quite simple processes involving changes in the adhesiveness of cells and the extension and contraction of pseudopodia. The same processes are probably equally important in the development of other metazoan embryos.

The coelom is formed as a cavity which splits off from the blind end of the gut of the gastrula. A mouth is formed (the original opening of the gastrula becomes the anus). The primary mesenchyme secretes three-pointed spicules (rather like sponge spicules) which push the body wall out into long processes. If skeleton growth is inhibited by sodium fluoride the processes fail to develop, which suggests that they are produced by the stretching action of the growing spicules.

Fig. 24.15(*a*), (*b*) show the mature larva of a brittle star. It has long processes stiffened by spicules, like the larvae of sea urchins. There are bands of cilia along the processes. Larvae like this are called pluteus larvae and are characteristic of Echinoidea and Ophiuroidea. The larvae of other classes of echinoderm also have bands of cilia but they lack the processes.

The coelom divides to form (typically) three coelomic cavities on each side of the body. It will be convenient to call these cavities L1, L2, L3 (on the left, numbering from the anterior end) and R1, R2, R3 (on the right). All six cavities are present in the brittle star pluteus shown in Fig. 24.15(*b*), but the larvae of some echinoderms lack R1 and R2. L2 remains connected to L1 and L1 develops an opening to the exterior. This arrangement of coeloms is very like the arrangement found in hemichordates, which have a partly divided proboscis coelom (L1+R1) with an opening to the exterior on the left, a pair of collar coeloms (L2 and R2) and a pair of trunk coeloms (L3 and R3). Young starfish larvae are strikingly similar to the larva of *Saccoglossus*.

The change from the larva to the adult echinoderm is a drastic metamorphosis. The details differ between the classes. Fig. 24.15 shows how it happens in brittle stars. L2 curls round the mouth to form a ring canal from which sprout five radial water canals: it becomes the water vascular system. The connection from it to L1 becomes the stone canal and the external opening of L1 becomes the madreporite. L1 itself becomes the axial sinus but most of the rest of the perihaemal system develops from L3. L3 and R3 form the main body cavity. R1 vanishes and R2 is reduced to a rudiment. These and other changes produce a tiny brittle star in the centre of the pluteus (Fig. 24.15*d*). The long processes of the pluteus fall off leaving a little brittle star which has almost perfect five-fold symmetry though it has developed from a bilaterally symmetrical larva.

FURTHER READING

GENERAL

Binyon, J. (1972). *Physiology of echinoderms*. Pergamon, Oxford.

Bonham, K. & Held, E. A. (1963). Ecological observations on the sea cucumbers *Holothuria atra* and *H. leucospilota* at Rongelap Atoll, Marshall Islands. *Pacific Sci.* **17**, 305–14.

Macurda, D. B. & Meyer, D. L. (1974). Feeding posture of modern stalked crinoids. *Nature, Lond.* **247**, 394–6.

Millot, N. (ed.) (1967). *Echinoderm biology* (*Symposium of the Zoological Society of London* 20). Academic Press, New York & London.

Nichols, D. (1959). Changes in the chalk heart-urchin *Micraster* interpreted in relation to living forms. *Phil. Trans. R. Soc.* **242B**, 347–437.

Nichols, D. (1969). *Echinoderms*, 4th edn. Hutchinson, London.

Warner, G. F. (1971). On the ecology of a dense bed of the brittle-star *Ophiothrix fragilis*. *J. mar. biol. Assoc. UK* **51**, 267–82.

SKELETON

Brear, K. & Currey, J. D. (1976). Structure of a sea urchin tooth. *J. Mater. Sci.* **11**, 1977–8.

Currey, J. D. (1975). A comparison of the strength of echinoderm spines and mollusc shells. *J. mar. biol. Assoc. UK* **55**, 415–24.

Eylers, J. P. (1976). Aspects of skeletal mechanics of the starfish *Asterias forbesi*. *J. Morph.* **149**, 353–68.

Nichols, D. & Currey, J. D. (1968). The secretion, structure and strength of echinoderm calcite. In *Cell structure and its interpretation*, ed. S. M. McGee-Russell & K. F. A. Ross, pp. 251–61. Arnold, London.

TUBE FEET AND LOCOMOTION

Kerkut, G. A. (1953). The forces exerted by the tube feet of starfish during locomotion. *J. exp. Biol.* **30**, 575–83.

Nichols, D. (1961). A comparative histological study of the tube feet of two regular echinoids. *Q. J. microsc. Sci.* **102**, 157–80.

Prusch, R. D. & Whoriskey, F. (1976). Maintenance of fluid volume in the starfish water vascular system. *Nature, Lond.* **262**, 577–8.

Smith, J. E. (1950). Some observations on the nervous mechanisms underlying the behaviour of starfishes. *Symp. Soc. exp. Biol.* **4**, 196–220.

DEVELOPMENT

Gustafson, T. & Wolpert, L. (1967). Cellular movement and contact in sea urchin metamorphosis. *Biol. Rev.* **42**, 442–98.

Hörstadius, S. (1973). *Experimental embryology of echinoderms*. Clarendon, Oxford.

Tilney, L. G. & Gibbins, J. R. (1969). Microtubules and filaments in the filopodia of the secondary mesenchyme cells of *Arbacia punctulata* and *Echinarachnius parma*. *J. Cell Sci.* **5**, 195–210.

Yoneda, M. & Dan, K. (1972). Tension at the surface of the dividing sea-urchin egg. *J. exp. Biol.* **57**, 575–88.

25

Sea squirts and amphioxus

Phylum Chordata, Subphylum Urochordata
- Class Ascidiacea (sea squirts)
- Class Larvacea
- Class Thaliacea (salps)

Subphylum Cephalochordata (amphioxus)
Subphylum Vertebrata (vertebrates)

The vertebrates are not given a phylum to themselves but are put in the phylum Chordata with the sea squirts and their relatives and the amphioxus. This chapter is about the animals which are included in the chordates but not in the vertebrates.

The sea squirt *Ciona* (Fig. 25.1 *a*) will serve to introduce the subphylum Urochordata. It is a member of the class Ascidiacea. It lives attached to rocks on European shores. It is a translucent sack-shaped animal with no very obvious external features except two openings called siphons at the upper end. It would be hard to imagine an animal which looked less like a vertebrate but the larva has some striking similarities to vertebrates, as will be shown.

Fig. 25.2 shows the main features of the internal structure of the adult. There is a protective outer tunic which consists of protein and polysaccharide (including cellulose) with very few cells embedded in it. Much of the space inside is occupied by the pharynx and by the space called the atrium. The wall between the pharynx and the atrium is perforated by so many holes that it is in effect a fine network. The inhalant siphon is the mouth, leading into the pharynx, and the exhalant siphon opens into the atrium. Water is drawn through the inhalant siphon into the pharynx, passing through a ring of tentacles which keep large particles out. It passes through the perforated pharynx wall into the atrium and is driven out through the exhalant siphon. Food is filtered from it. It has been found by examining gut contents that diatoms and other unicellular algae are the main foods.

Feeding works in very much the same way as in bivalve molluscs (p. 309). The bars of the network which forms the wall of the pharynx bear two sets of cilia, frontal cilia which project into the pharynx and lateral cilia which project across the perforations. Careful observation of sea squirts with transparent tunics reveals that the inner face of the wall of the pharynx is covered by a thin sheet of mucus, as shown in Fig. 25.2(*c*). Notice how similar this diagram is to Fig. 14.9(*b*) which represents a section of part of a ctenidium

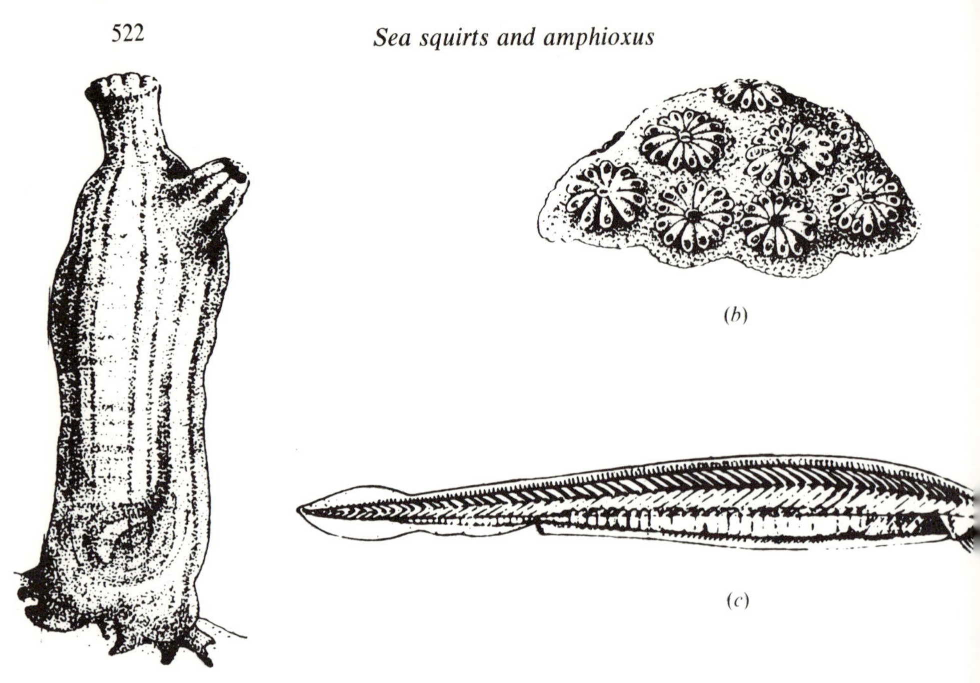

Fig. 25.1. (*a*) *Ciona intestinalis*. Height up to 15 cm. (*b*) A colony of *Botryllus schlosseri*. Diameter of flower-like groups of individuals about 7 mm. (*c*) *Branchiostoma lanceolatum*. Length up to 6 cm. From W. de Haas & F. Knorr (1966). *The young specialist looks at marine life*. Burke, London.

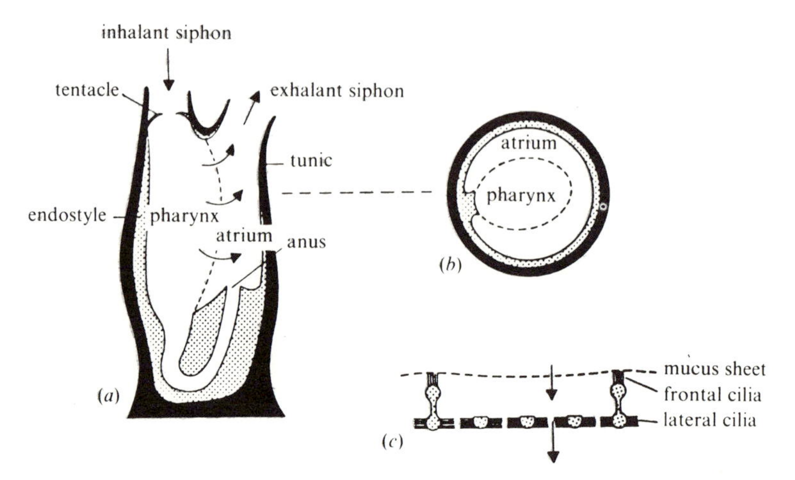

Fig. 25.2. Diagrammatic sections of *Ciona*. (*a*) Longitudinal section, (*b*) transverse section, and (*c*) a greatly magnified section through part of the wall of the pharynx. Arrows indicate flow of water. From R. McN. Alexander (1975). *The chordates*. Cambridge University Press.

of a bivalve mollusc. The lateral cilia drive the water through and the frontal cilia move the mucus sheet.

At least most of the mucus is secreted by gland cells in the endostyle (Fig. 25.2*a*) which runs along the ventral edge of the pharynx. It is by no means obvious from the figure that this edge is ventral, but comparison of the larva with the adult shows that it is. Food particles in the water passing through the mucus sheet are trapped by the mucus. The frontal cilia gradually move the sheet with the trapped particles towards the dorsal edge of the pharynx, where other cilia drive them posteriorly to the digestive part of the gut. Faeces are expelled from the anus into the atrium and are carried away in the exhalant current.

Experiments like the ones described on p. 311 have shown that colloidal graphite particles only 1–2 μm in diameter are efficiently retained by the mucus filter, but that haemoglobin molecules (diameter 3 nm) pass freely through. The perforations in the pharynx wall are about 50 μm wide.

The rest of the anatomy of *Ciona* is very simple. There is a single ganglion in the body wall between the siphons. There are no special respiratory organs: the large surface area of the pharynx must be ample to enable the animal to take up the oxygen it needs from the feeding current. There seem to be no special organs of excretion or of osmotic or ionic regulation: ammonia formed as a waste product of protein metabolism simply diffuses out of the body.

There is a system of blood vessels and a heart which is simply a U-shaped tube. Muscular constrictions travel along the heart, driving the blood through. For a few minutes all the constrictions travel in the same direction and the blood flows one way round the circulation. Then constrictions start travelling in the opposite direction and flow is reversed for a few minutes. The rate of flow has been calculated from the size of the heart and the frequency of the beat. It seems that the blood makes about 20 circuits in one direction before reversing.

The perforations in the wall of the pharynx are gill openings, like the gill openings of hemichordates. They do not open direct to the exterior but are enclosed by the atrium so that the very delicate wall of the pharynx is protected by the tunic, much as the delicate gills of teleost fish are protected by the bony operculum. These gills openings are the only obvious point of resemblance between adult sea squirts and vertebrates. However, adult sea squirts develop from swimming larvae which look like tiny tadpoles and resemble vertebrates in many ways (Fig. 25.3). These larvae do not feed and they swim for only a few hours before settling and starting their metamorphosis.

The tail of the larva contains a notochord. This is a row of 40–42 cylindrical cells with large vacuoles, enclosed in a sheath of connective tissue fibres. On either side of the notochord are a few rows of muscle cells. When the larva swims the muscles on the left and right contract alternately, bending the tail from side to side. The notochord presumably functions in the same way as in vertebrates, as a flexible structure of constant length. It cannot shorten

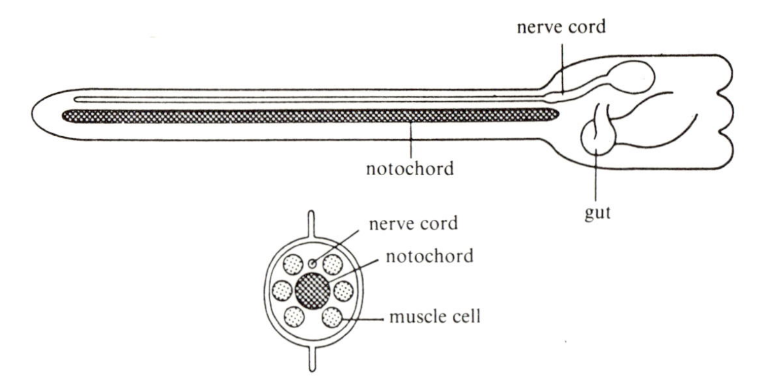

Fig. 25.3. Diagrammatic lateral view of the larva of a typical sea squirt, and a transverse section through its tail. Length about 1 mm. From R. McN. Alexander (1975). *The chordates.* Cambridge University Press.

because its sheath prevents it from swelling. It if were not there the muscles on both sides of the tail could contract simultaneously, shortening the tail. Because it is there, shortening of the muscles of one side lengthens those of the other: it makes the muscles on either side of the tail antagonistic to one another. Compare this design with that of a nematode worm (p. 248). The whole body of the nematode, enclosed in the body wall, serves the mechanical function which the notochord serves in the larval sea squirt.

Dorsal to the notochord the larva has a tubular nerve cord, similar to the tubular dorsal nerve cords of vertebrates. The anterior end of the nerve cord is swollen and contains simple sense organs, a statocyst and an organ which seems from its structure to be a light detector. However, this swelling is not at all like a vertebrate brain. The part of the nerve cord which lies in the trunk is relatively stout and seems to be nervous tissue. The part in the tail is slim and its cells do not look like nerve cells. It has no obvious function.

Fig. 25.1 (*b*) shows *Botryllus*, a colonial member of the class Ascidiacea which grows on stones and seaweed on European shores. The colony is gelatinous with orange or blue flower-like patterns on it. Each 'petal' of a 'flower' is an individual animal with its own inhalant siphon. The hole in the centre of the 'flower' is an exhalant siphon shared by all the members of the group. The colony is founded by a single larva which (after metamorphosis) produces the other individuals by budding.

Doliolum (Fig. 25.4) is a salp, a member of the class Thaliacea. It is a planktonic urochordate with a transparent barrel-shaped body. It has an inhalant siphon at one end and an exhalant one at the other, and its filter-feeding current drives it along by jet propulsion. The pharynx fills most of the anterior (inhalant) half of the body and the atrium fills most of the posterior half. The wall which separates them is not a fine mesh but has relatively few large openings. Effective pumping of water in *Ciona* depends on the gill openings

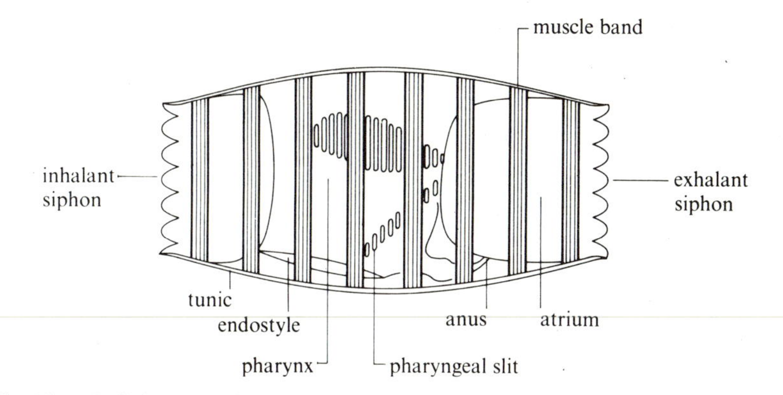
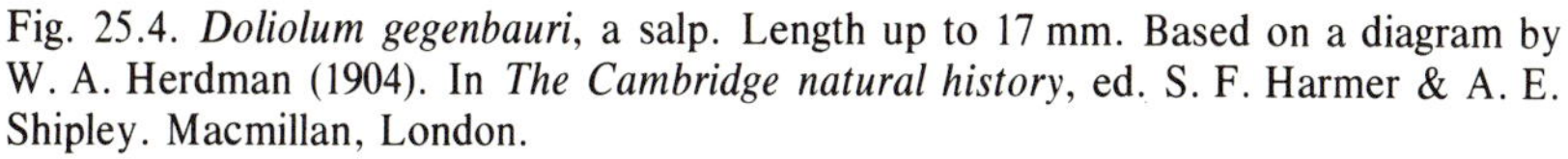

Fig. 25.4. *Doliolum gegenbauri*, a salp. Length up to 17 mm. Based on a diagram by W. A. Herdman (1904). In *The Cambridge natural history*, ed. S. F. Harmer & A. E. Shipley. Macmillan, London.

being small enough for the cilia to span them but *Doliolum* does not use cilia to drive its feeding current. The body is encircled by hoops of muscle which contract rhythmically, drawing water in and driving it out again. Muscles around the siphons close the exhalant siphon while the body is enlarging and close the inhalant one while it is contracting, so the water is always drawn in through the inhalant siphon and driven out through the exhalant one.

The water is filtered through a cone-shaped sheet of mucus. Mucus is added to the cone around the rim, and the point of the cone is drawn into the digestive part of the gut by cilia.

Thaliacea have no tadpole-like larvae, but develop direct to the adult form. Larvacea, in contrast, remain tadpole-like throughout their lives. They feed and develop gonads but retain their tails. *Oikopleura* (Fig. 25.5) is a member of this class and is quite common in marine plankton. It lives in an extraordinary gelatinous 'house', so delicate and transparent as to be extremely difficult to study. The house is secreted by the animal and is believed to be homologous with the tunic of sea squirts. Undulation of the tail drives water through the house, in through a coarse grid and out through a very much finer one which has a large area. This current drives the house slowly along.

The fine grid is so fine that it stops particles down to less than 1 μm in diameter. The water flows through it but the particles accumulate as a concentrated suspension near the animal's mouth (Fig. 25.5). The animal has a pair of gill openings and no atrium. Cilia on the gill openings draw the concentrated suspension in through the mouth. It is filtered through a sheet of mucus: the particles are retained by the mucus but the water passes out through the gill openings. Since the suspension is already concentrated the rate at which it has to be filtered in the pharynx is small. The two gill openings, each small enough for the cilia to reach to the middle of the opening, are sufficient.

The only members of the subphylum Cephalochordata are the various

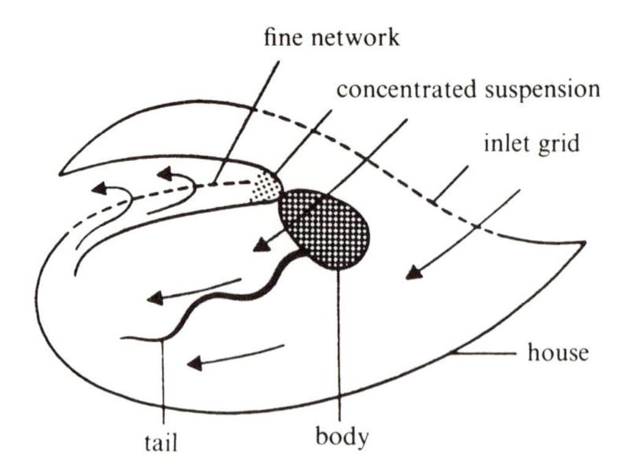

Fig. 25.5. A simplified section of *Oikopleura* in its house. Length of house about 10 mm. Arrows indicate water movement. From R. McN. Alexander (1975). *The chordates*. Cambridge University Press.

species of *Branchiostoma* (Fig. 25.1 *c*) and *Asymmetron*. *Branchiostoma* was long called *Amphioxus*, but *Branchiostoma* is the correct name, according to the rules of zoological nomenclature, because it was used in the original description. The name amphioxus (without an initial capital, and not italicized) is often used as the common English name. Different species of *Branchiostoma* are found in many parts of the world in sand on the sea bottom, completely buried if the sand is coarse but with the head projecting if it is finer.

Fig. 25.6 shows the main features of the anatomy of *Branchiostoma*. There is a notochord and a hollow nerve cord. On either side are longitudinal muscles, divided into segmental blocks like the swimming muscles of fish. They enable the animal to make the eel-like movements which it uses both for swimming and for burrowing.

There is a large pharynx with long gill slits which open into an atrium, which in turn has an opening to the exterior about half way along the body. The gill slits are narrow enough for the lateral cilia to reach half way across them. The cilia drive a filter-feeding current in at the mouth, through a sheet of mucus and out through the gill slits and atrium. The mucus is secreted by a ventral endostyle, as in sea squirts. It is moved dorsally by frontal cilia, and eventually posteriorly to the digestive part of the gut. The caecum (Fig. 25.6) is an outgrowth of the intestine on the right side of the pharynx only, and seems to function rather like the digestive gland of molluscs (p. 314). Cilia beat anteriorly on its ceiling and posteriorly on its floor. Food particles are taken into it and digested, partly by extracellular enzymes and partly intracellularly in food vacuoles.

Feeding is possible even when the animal is completely buried in reasonably coarse sand, because the water can permeate between the sand grains.

Amphioxus resemble fishes in many ways but their filter feeding system is exceedingly like that of sea squirts. The atrial opening is ventral instead of dorsal

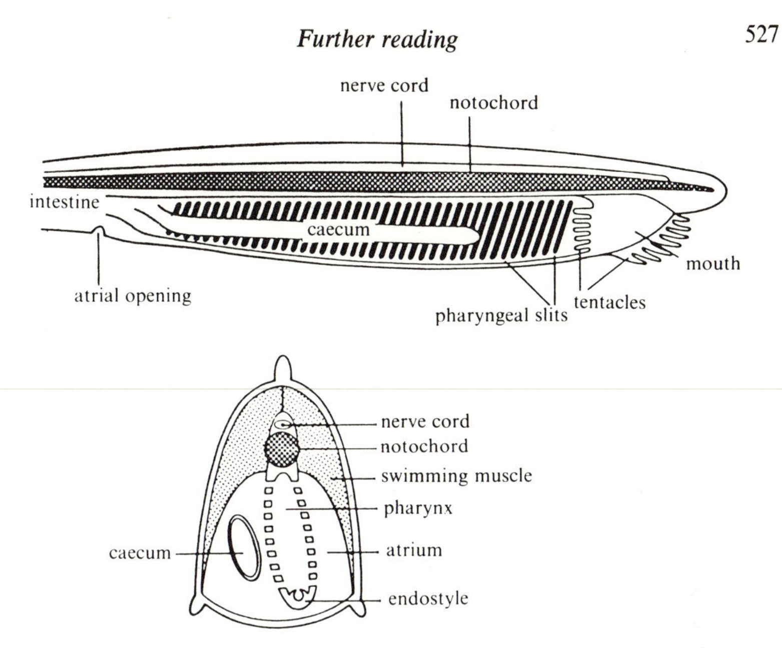

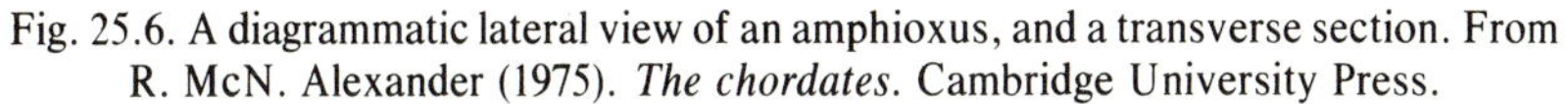

Fig. 25.6. A diagrammatic lateral view of an amphioxus, and a transverse section. From R. McN. Alexander (1975). *The chordates*. Cambridge University Press.

and there are long gill slits instead of the small squarer gill openings of *Ciona*, but otherwise the feeding systems are very alike. Unlike fishes, amphioxus have nothing like a vertebrate brain, no skull, no vertebrae and no kidneys. They have about 100 pairs of protonephridia with solenocytes, like the protonephridia of some annelid worms (p. 360).

FURTHER READING

Alldredge, A. (1976). Appendicularians. *Sci. Am.* **235**(1), 94–102.

Barrington, E. J. W. (1965). *The biology of Hemichordata and Protochordata*. Oliver & Boyd, Edinburgh.

Goodbody, I. (1974). The physiology of ascidians. *Adv. mar. Biol.* **12**, 1–149.

Millar, R. H. (1971). The biology of ascidians. *Adv. mar. Biol.* **9**, 1–100.

26
Evolution of the invertebrates

Many books have a lot of pages about how the various phyla of invertebrates have evolved from some ancestral animal. The evidence on which these discussions are based is so slight and any conclusions to insecure that it seems best to keep this discussion short.

Direct evidence of the course of evolution can only come from fossils, but the fossils which have been found so far tell us nothing about the origin of the invertebrate phyla, except that it happened long ago. Very few animal fossils indeed have been found from rocks earlier than the beginning of the Palaeozoic era, about 570 million years ago. The only ones that have been found are impressions of soft-bodied animals which are tantalizingly difficult to interpret but seem to include medusae and annelid worms. Suddenly, from the beginning of the Palaeozoic, there are quite abundant fossils of the hard parts of animals representing nearly all the major phyla. Some of the early ones are shown in Fig. 26.1. The gastropod mollusc shell (*a*) is remarkably like the shells of some modern limpets though the soft parts of the animal may of course have been very different. The brachiopod (*c*) is also very like modern representatives of its phylum. The trilobite (*b*) is quite unlike any modern animal but is clearly a member of the phylum Arthropoda. The echinoderm (*d*) is fairly similar to modern crinoids. Shells of foraminiferans and bivalve molluscs, spicules of sponges and skeletons of corals are also found in very early Palaeozoic rocks. The earliest fossils of vertebrates (small fragments of fish) are only a little later. The major phyla must have originated well before the Palaeozoic, in an earlier period for which there is hardly any fossil evidence.

Fossils tell us a good deal about the subsequent evolution of some of the phyla, but for hints as to how the phyla originated we have to look at modern animals.

It seems almost self-evident that unicellular animals must have evolved before multicellular ones, so the earliest animals were presumably Protozoa. The idea that they may have been Sarcomastigophora is attractive, for this is the subphylum in which the animal and plant kingdoms overlap. In any case the peculiar nuclei of the Ciliophora make them seem unlikely ancestors for the other animals, and the parasitic protozoans could hardly have evolved until there were other animals to parasitize.

There are two groups of theories about the evolution of multicellular animals

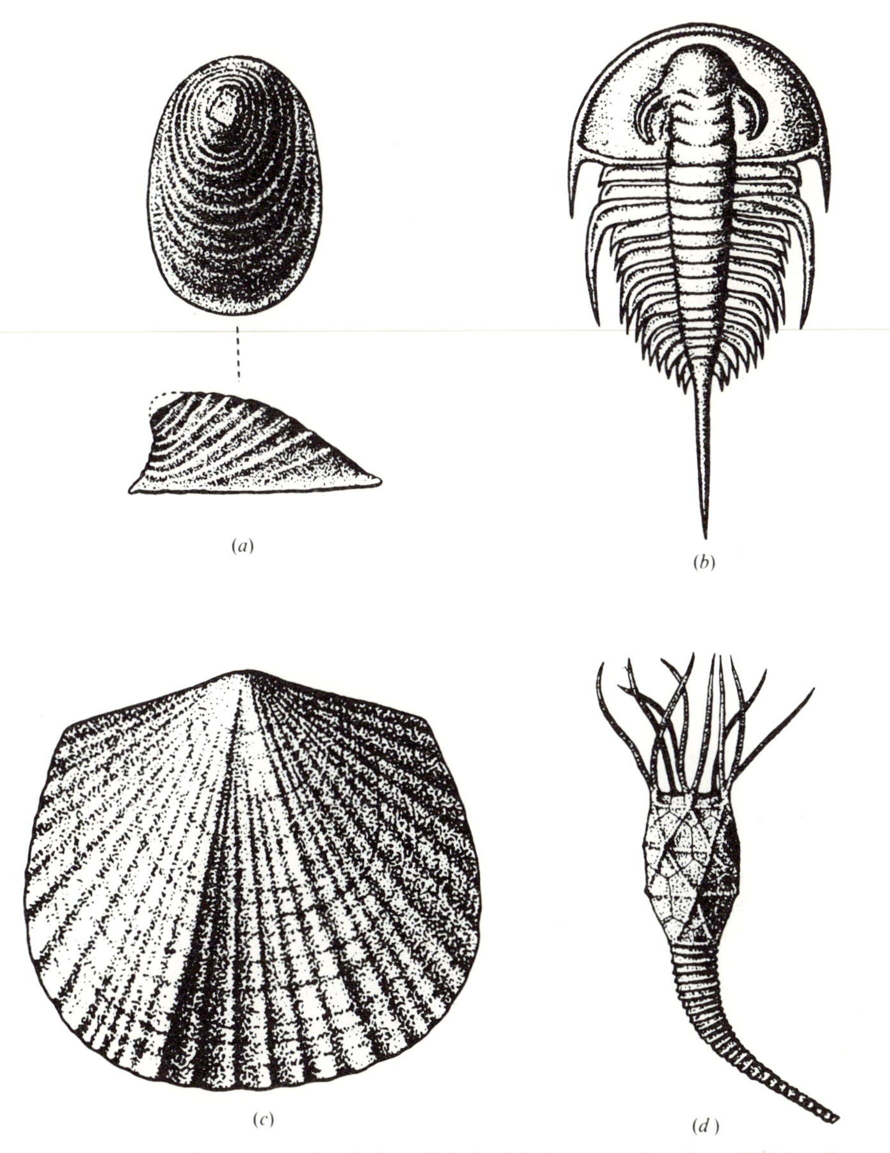

Fig. 26.1. Early invertebrate fossils from Britain. (*a*) A gastropod mollusc (*Helcionella subrugosa*, length 15 mm); (*b*) a trilobite (*Olenellus lapworthi*, length 30 mm); (*c*) a brachiopod (*Orusia lenticularis* diameter 10 mm); and (*d*) an echinoderm (*Macrocystella mariae*, length 50 mm). (*a*), (*b*) and (*c*) are from the Cambrian period (about 500–570 million years ago) and (*d*) from the early part of the Ordovician period (just under 500 million years ago). From British Museum (Natural History) (1964). *British Palaeozoic fossils*. British Museum (Natural History), London.

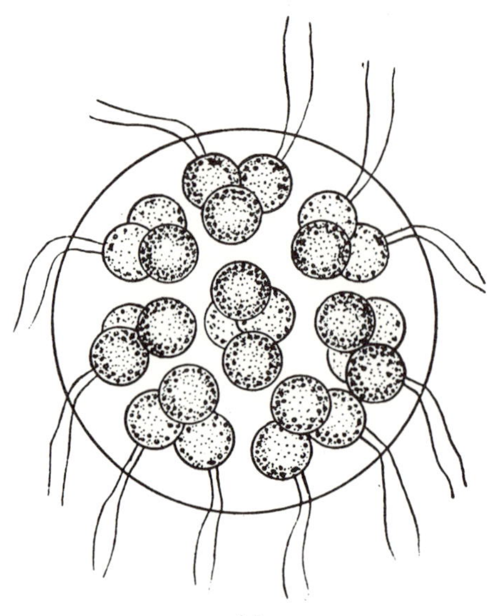

(*a*)

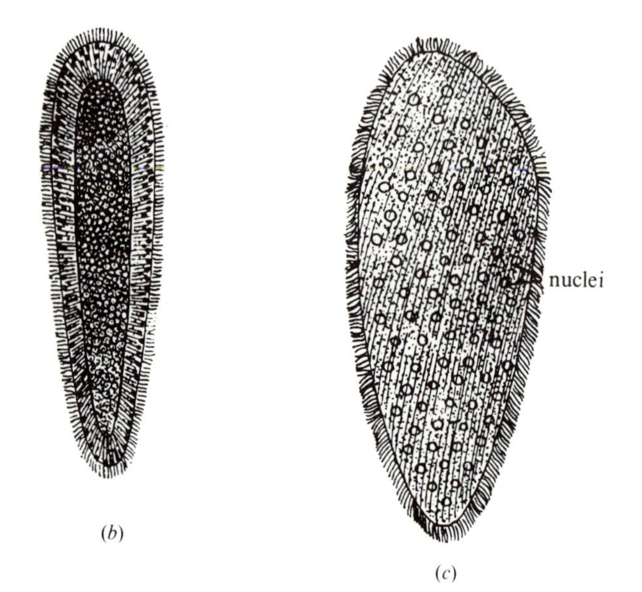

(*b*)

(*c*)

Fig. 26.2. (*a*) *Eudorina* sp., diameter about 0.1 mm. (*b*) The planula larva of a hydroid, length probably about 1 mm. (*c*) *Opalina* sp., length about 0.4 mm. From L. H. Hyman (1940). *The invertebrates*, vol. 1. McGraw-Hill, New York.

from unicellular ones. One group sees a protozoan colony as the intermediate form. *Eudorina* (Fig. 26.2 *a*) is a hollow ball of 32 cells, each very like an entire *Chlamydomonas* (Fig. 2.1 *a*). It is obviously a colony rather than a multicellular individual. *Volvox* is similar but larger, with up to 20000 cells and a diameter of 1–2 mm. Its constituent cells are not all alike (only some of them are capable of reproduction) but it is nevertheless regarded as a colony of unicellular individuals. *Eudorina* and *Volvox* are plants containing chlorophyll, but it is easy to imagine similar colonies of animal cells without chlorophyll. Such a colony might have evolved to become a simple, hollow multicellular animal, an ancestral sponge or coelenterate. This suggestion leaves open the question as to how the single layer of cells which encloses the central cavity of *Eudorina* and similar colonies could have become the double layer which forms the body wall of sponges and coelenterates. The idea that sponges might have evolved in this way seems particularly attractive because the behaviour of the cells of a sponge does not seem highly integrated: a sponge is more like a colony and less like an individual than other multicellular animals. Also, the choanocytes of sponges are very like choanoflagellate protozoans (Fig. 5.2). Some choanoflagellates form colonies (Fig. 5.2 *d*), but these colonies are not at all like those of *Eudorina*.

The other group of theories suggests that multicellular animals evolved from non-colonial protozoans with multiple nuclei. *Opalina* (Fig. 26.2 *c*) is one example of such a protozoan, and the giant amoeba *Pelomyxa* is another. If the cytoplasm of *Opalina* divided so that each nucleus was in a separate cell, the result would be a multicellular animal. It would be a solid lump of cells covered by cilia, not a hollow ball like *Eudorina*. The planula larvae of Cnidaria (Fig. 26.2 *c*) and the flatworms of the class Acoela are small, solid and covered by cilia, and could have evolved from an ancestor like this. It is not at all clear whether the Cnidaria or the flatworms came first. On the one hand it can be argued that the Cnidaria are simpler (with only two layers of cells), that they must have evolved from a planula-like ancestor and that all more complex animals must have evolved from them. On the other hand it is possible to argue that the only planula-like adult animals are the Acoela and that the other multicellular animals including the Cnidaria must have evolved from them.

Fig. 26.3 shows some similarities between phyla which have been interpreted as hints about the course of evolution. It is assumed that multicellular animals evolved from protozoans and that the animals which have coeloms evolved from the ones which do not (all the most complex animals belong to phyla in which coeloms occur).

The similarity between choanoflagellates and the choanocytes of sponges has already been noted. There is also a striking resemblance between the nematocysts of Cnidaria (Fig. 6.19) and the polar filaments of the protozoans of the subphylum Cnidospora (Fig. 4.6), but these two groups of animals are otherwise so dissimilar that they seem unlikely to be closely related. They may have evolved similar structures independently.

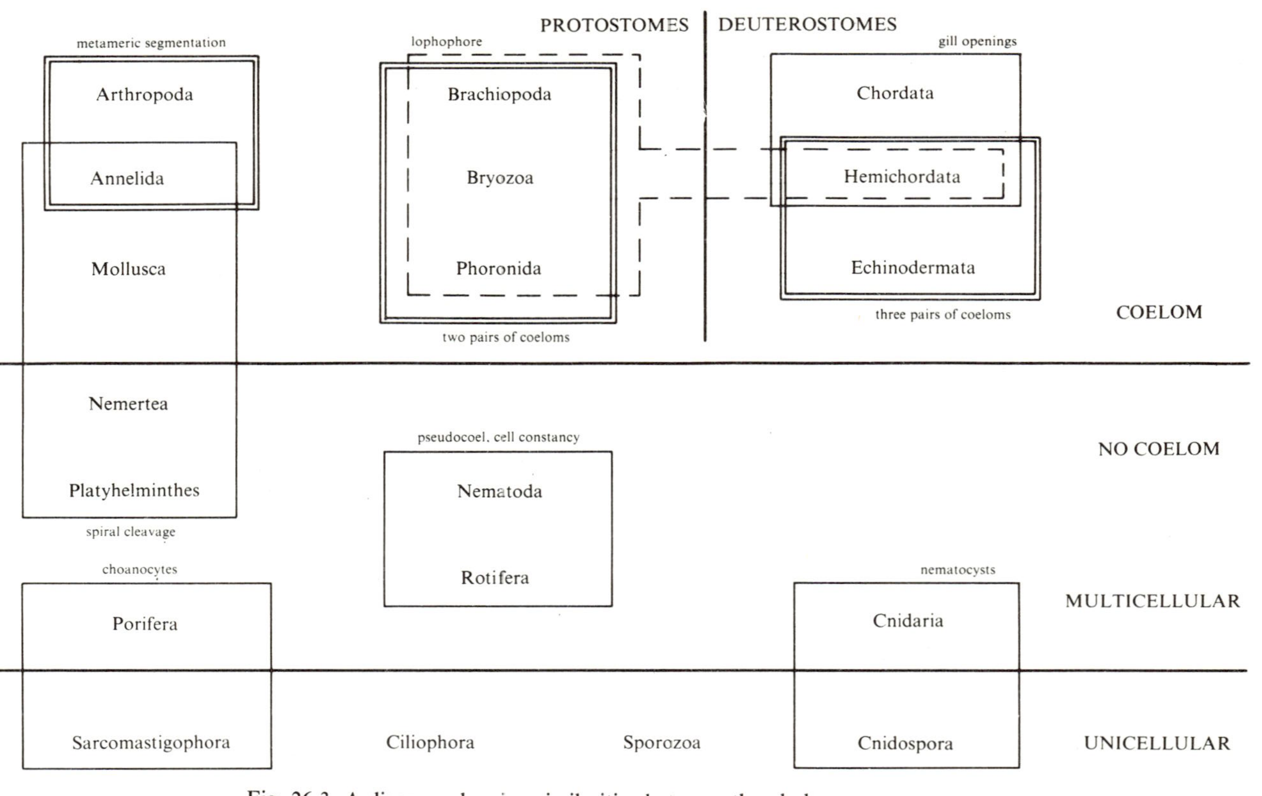

Fig. 26.3. A diagram showing similarities between the phyla.

The nematodes and rotifers are both pseudocoelomate (that is, they have body cavities without an epithelial lining, and no true coelom). The number and arrangement of cells is the same in all normal members of a given species, in many of the organ systems of nematodes and in the whole body of rotifers. There are also other, less striking, similarities. Some zoologists put the nematodes and rotifers, with a few other small groups of animals, in a single phylum Aschelminthes.

Zygotes of polyclad platyhelminths, nemerteans, molluscs and annelids divide (with some exceptions) by spiral cleavage (Fig. 8.14). The exceptions are large zygotes such as those of cephalopod molluscs, in which only part of the cytoplasm cleaves. Spiral cleavage proceeds in essentially the same way in all cases, and cell 4d forms the mesoderm. Arthropods lay large eggs which cleave incompletely but they resemble annelids in being segmented and in the general arrangement of their internal organs. The sensory structures in the eyes of all these phyla are normally derived from microvilli, not cilia (Fig. 7.13).

The similarities between Phoronida, Bryozoa and Brachiopoda were stressed in chapter 22. The members of these phyla have lophophores, and two coelom segments. The Hemichordata have three coelom segments but some of them have an array of tentacles rather like a lophophore (Fig. 23.2 *b*). The larvae of Echinodermata also have three coelom segments and many early echinoderm larvae are extremely like hemichordate larvae. The zygotes of Bryozoa, Hemichordata and Echinodermata divide by radial cleavage, and so do those of amphioxus and ascidians (phylum Chordata). The Chordata and Hemichordata have gill openings. The sensory structures in the eyes of Chordata and Echinodermata are modified cilia (Fig. 7.13).

The coelomate animals can be divided into three groups:

(i) Mollusca, Annelida, Arthropoda

(ii) Phoronida, Bryozoa, Brachiopoda

(iii) Hemichordata, Echinodermata, Chordata

The evidence so far tends to link groups (ii) and (iii), but there is other evidence which links groups (i) and (ii). The process of gastrulation produces a sack-like embryo with a single opening into the cavity which will become the gut (Figs. 16.15 *b*, 24.13 *e*). In groups (i) and (ii) this opening becomes the mouth while a new opening, formed later, becomes the anus. In group (iii) the converse is true: the original opening becomes the anus. The members of groups (i) and (ii) are often referred to collectively as the protostomes, and the members of group (iii) as the deuterostomes.

Though we know so much about the structure and habits of invertebrates and about how their bodies work, our attempts to work out how the phyla evolved are little better than guesses.

FURTHER READING

Clark, R. B. (1964). *Dynamics in metazoan evolution. The origin of the coelom and segments.* Clarendon, Oxford.

Index

Page numbers in italics refer to illustrations.